Student & Parent

One-Stop Internet Resources

T5-CQD-442

Log on to

www.mathmatters2.com

Online Study Tools

- Extra Examples
- Self-Check Quizzes
- Concepts in Motion
- Personal Tutor
- Chapter Assessment Practice
- Standardized Test Practice

Online Resources

- Math*Works* Careers
- Links to Chapter Themes
- Multilingual Glossary

GLENCOE MATHEMATICS

MathMatters 2

An Integrated Program

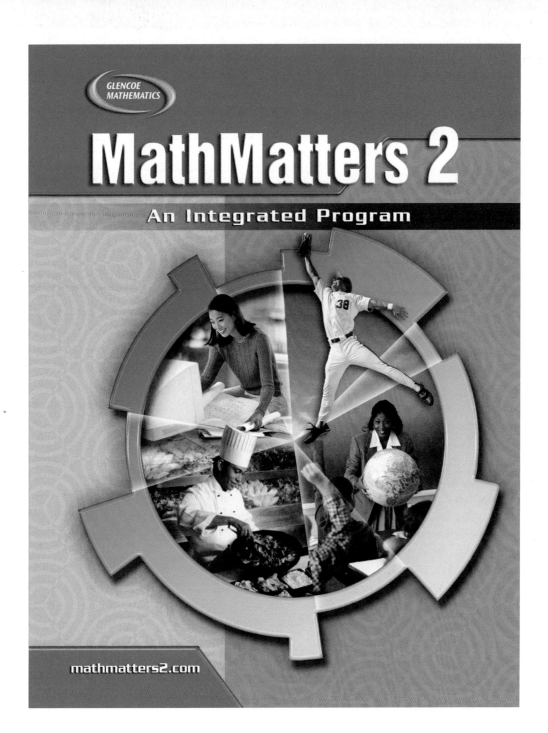

mathmatters2.com

Lynch
Olmstead
De Forest-Davis

Glencoe

New York, New York
Columbus, Ohio
Chicago, Illinois
Peoria, Illinois
Woodland Hills, California

Copyright © 2009 by the McGraw-Hill Companies, Inc. All rights reserved. Printed in the United States of America. Except as permitted under the United States Copyright Act, no part of this publication may be reproduced or distributed in any form or by any means, or stored in a database or retrieval system, without prior written permission of the publisher.

Send all inquiries to:
Glencoe/McGraw-Hill
8787 Orion Place
Columbus, OH 43240-4027

ISBN: 978-0-07-880571-4   *MathMatters 2 Student Edition*
MHID: 0-07-880571-6

1   2   3   4   5   6   7   8   9   10   058/111   13   12   11   10   09   08

Contents in Brief

Chicha Lynch currently teaches Honors Advanced Algebra II at Marin Catholic High School in Kentfield, California. She is a graduate of the University of Florida. She was a state finalist in 1988 for the Presidential Award for Excellence in mathematics teaching. Currently, Ms. Lynch is a participating member of the National Council of Teachers of Mathematics as well as a long-time member of California Math Council North.

Eugene Olmstead is a mathematics teacher at Elmira Free Academy in Elmira, New York. He earned his B.S. in Mathematics at State University College at Geneseo in New York. In addition to teaching high school, Mr. Olmstead is an instructor for T^3, Teachers Teaching with Technology, and has participated in writing several of the T^3 Institutes. In 1991 and 1992, Mr. Olmstead was selected as a state finalist for the Presidential Award for Excellence in mathematics teaching.

Kenneth De Forest-Davis is the mathematics department chairperson at Beloit Memorial High School in Beloit, Wisconsin. Mr. De Forest-Davis earned his B.A. in Mathematics and Computer Science from Beloit College in Wisconsin. He later completed his M.A.T from Beloit College, where he received the Von-Eschen-Steele Excellence in Teaching Award.

Reviewers and Consultants

These educators reviewed every chapter and gave suggestions for improving the effectiveness of the mathematics instruction.

Tamara L. Amundsen
Teacher
Windsor Forest High School
Savannah, Georgia

Kyle A. Anderson
Mathematics Teacher
Waiakea High School
Milo, Hawaii

Murney Bell
Mathematics and Science
    Teacher
Anchor Bay High School
New Baltimore, Michigan

Fay Bonacorsi
High School Math Teacher
Lafayette High School
Brooklyn, New York

Boon C. Boonyapat
Mathematics Department
    Chairman
Henry W. Grady High School
Atlanta, Georgia

Peggy A. Bosworth
Retired Math Teacher
Plymouth-Canton High School
Canton, Michigan

Sandra C. Burke
Mathematics Teacher
Page High School
Page, Arizona

Jill Conrad
Math Teacher
Crete Public Schools
Crete, Nebraska

Nancy S. Cross
Math Educator
Merritt High School
Merritt Island, Florida

Mary G. Evangelista
Chairperson, Mathematics
    Department
Grove High School
Garden City, Georgia

Timothy J. Farrell
Teacher of Mathematics and
    Physical Science
Perth Amboy Adult School
Perth Amboy, New Jersey

Greg A. Faulhaber
Mathematics and Computer
    Science Teacher
Winton Woods High School
Cincinnati, Ohio

Leisa Findley
Math Teacher
Carson High School
Carson City, Nevada

Linda K. Fiscus
Mathematics Teacher
New Oxford High School
New Oxford, Pennsylvania

Louise M. Foster
Teacher and Mathematics
    Department Chairperson
Frederick Douglass High School
Altanta, Georgia

Darleen L. Gearhart
Mathematics Curriculum
    Specialist
Newark Public Schools
Newark, New Jersey

Faye Gunn
Teacher
Douglass High School
Atlanta, Georgia

Dave Harris
Math Department Head
Cedar Falls High School
Cedar Falls, Iowa

Barbara Heinrich
Teacher
Wauconda High School
Wauconda, Illinois

Margie Hill
District Coordinating Teacher
    Mathematics, K-12
Blue Valley School District
    USD229
Overland Park, Kansas

Suzanne E. Hills
Mathematics Teacher
Halifax Area High School
Halifax, Pennsylvania

Robert J. Holman
Mathematics Department
St. John's Jesuit High School
Toledo, Ohio

Eric Howe
Applied Math Graduate Student
Air Force Institute of Technology
Dayton, Ohio

Daniel R. Hudson
Mathematics Teacher
Northwest Local School District
Cincinnati, Ohio

Susan Hunt
Math Teacher
Del Norte High School
Albuquerque, New Mexico

Todd J. Jorgenson
Secondary Mathematics
  Instructor
Brookings High School
Brookings, South Dakota

Susan H. Kohnowich
Math Teacher
Hartford High School
White River Junction, Vermont

Mercedes Kriese
Chairperson, Mathematics
  Department
Neenah High School
Neenah, Wisconsin

Kathrine Lauer
Mathematics Teacher
Decatur High School
Federal Way, Washington

Laurene Lee
Mathematics Instructor
Hood River Valley High School
Hood River, Oregon

Randall P. Lieberman
Math Teacher
Lafayette High School
Brooklyn, New York

Scott Louis
Mathematics Teacher
Elder High School
Cincinnati, Ohio

Dan Lufkin
Mathematics Instructor
Foothill High School
Pleasanton, California

Gary W. Lundquist
Teacher
Macomb Community College
Warren, Michigan

Evelyn A. McDaniel
Mathematics Teacher
Natrona County High School
Casper, Wyoming

Lin McMullin
Educational Consultant
Ballston Spa, New York

Margaret H. Morris
Mathematics Instructor
Saratoga Springs Senior High
  School
Saratoga Springs, New York

Tom Muchlinski
Mathematics Resource Teacher
Wayzata Public Schools
Plymouth, Minnesota

Andy Murr
Mathematics
Wasilla High School
Wasilla, Alaska

Janice R. Oliva
Mathematics Teacher
Maury High School
Norfolk, VA

Fernando Rendon
Mathematics Teacher
Tucson High Magnet
  School/Tucson Unified
  School District #1
Tucson, Arizona

Candace Resmini
Mathematics Teacher
Belfast Area High School
Belfast, Maine

Kathleen A. Rooney
Chairperson, Mathematics
  Department
Yorktown High School
Arlington, Virginia

Mark D. Rubio
Mathematics Teacher
Hoover High School—GUSD
Glendale, California

Tony Santilli
Chairperson, Mathematics
  Department
Godwin Heights High School
Wyoming, Michigan

Michael Schlomer
Mathematics Department Chair
Elder High School
Cincinnati, Ohio

Jane E. Swanson
Math Teacher
Warren Township High School
Gurnee, Illinois

Martha Taylor
Teacher
Jesuit College Preparatory School
Dallas, Texas

Cheryl A. Turner
Chairperson, Mathematics
  Department
LaQuinta High School
LaQuinta, California

Linda Wadman
Instructor, Mathematics
Cut Bank High School
Cut Bank, Montana

George K. Wells
Coordinator of Mathematics
Mt. Mansfield Union
  High School
Jericho, Vermont

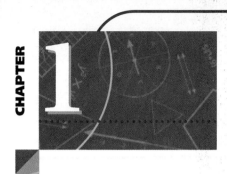

CHAPTER 1

Theme: Market Research

Math*Works* Career
Product Tester 15
Market Researcher 33

Applications
advertising 35
business 34, 40, 41
economics 23
education 12, 28
entertainment 7, 18, 37
fitness 17
food service 8, 36
manufacturing 7
market research 19, 23, 36
medicine 21
music 11, 36
political science 9
recreation 8, 41
retail 8, 17, 26, 41
sports 19, 31
weather 36

Standardized Test Practice
Multiple Choice 46
Short Response/Grid In 47
Extended Response 47

Sample and Display Data

Page 2

Foundations of Algebra

Theme: Population

Math*Works* Career
Concession Stand Operator 61
Commercial Aircraft
Designer 81

Standardized Test Practice
Multiple Choice 98
Short Response/Grid In 99
Extended Response 99

Page 49

Equations and Inequalities

Theme: Physics

Math*Works* Career
Mechanical Engineer 113
Automobile Designer 131

Applications
art 122
business 124, 126, 135
communications 132
engineering 139
entertainment 115
finance 111, 120
fitness 115, 134
geography 129
government 127
health 134
hobbies 135
machinery 123
manufacturing 124
mechanics 119, 139
packaging 111
part-time job 115
physics 107, 110, 114, 119, 125,
129, 136, 138
retail 125
safety 107, 135, 138
sports 119, 128
transportation 120
travel 106

Standardized Test Practice
Multiple Choice 144
Short Response/Grid In 145
Extended Response 145

Page 138

CHAPTER 4

Theme: Games

Math*Works* Career
Baseball Player 157
Board Game Designer 177

Applications
cooking 173
entertainment 174, 181
food service 159
games 161, 163, 164, 171, 175, 180
government 170
health 170
landscaping 180
market research 150, 151
medicine 155
music 160, 174
part-time job 154
retail 152, 159
safety 180
sports 155, 160, 171, 174, 180
travel 174

Standardized Test Practice
Multiple Choice 186
Short Response/Grid In 187
Extended Response 187

Page 172

Logic and Geometry

Theme: Navigation

Math*Works* Career
Cattle Rancher 201
Ship Captain 221

Applications
art 195, 215, 229
civil engineering 218
construction 194, 204, 217
engineering 214
entertainment 232
food service 228
health 199
hobbies 224
industry 233
interior design 209
market research 227
music 205
nature 224
navigation 194, 199, 205, 209, 219, 229, 233
photography 194
physics 198
recreation 214, 229
retail 219
safety 205, 224
travel 208

Standardized Test Practice
Multiple Choice 238
Short Response/Grid In 239
Extended Response 239

Page 188

Graphing Functions

Theme: Business

Math*Works* Career
Music Store Owner 253
Restaurateur 273

Applications
agriculture 270, 278
business 260, 261, 267, 275, 285
business travel 251
community service 247
economics 275
engineering 284
finance 278
geography 244
industry 284
market research 245, 261
music 282, 284
part-time job 278
physics 269, 270, 272, 276, 277, 278, 283, 285
recreation 264
retail 257
science 257
space 278
sports 270
travel 284

Standardized Test Practice
Multiple Choice 290
Short Response/Grid In 291
Extended Response 291

Page 273

Coordinate Graphing and Transformations

Theme: Architecture

Math*Works* Career

Campus Facilities Manager 305
Architect 323

Applications

architecture 299, 303, 309, 312, 318
art 301, 309, 313, 321
entertainment 318
gardening 319
hobbies 321
industry 311
landscaping 303
machinery 307
music 298
nature 320
navigation 302
photography 299, 317, 318
weather 312

Standardized Test Practice

Multiple Choice 328
Short Response/Grid In 329
Extended Response 329

Page 309

Systems of Equations and Inequalities

Theme: Sports

Math*Works* Career
Runner 343
Coach 361

Applications
business 358
construction 345, 351, 357
economics 341
entertainment 359, 362
finance 365
landscaping 350
movies 356
recreation 346
retail 364
safety 341
sports 337, 340, 347, 356, 357, 362
transportation 344, 359, 360
travel 337, 360

Standardized Test Practice
Multiple Choice 370
Short Response/Grid In 371
Extended Response 371

Page 330

CHAPTER 9

Theme: **Geography**

Math*Works* Career
Truck Driver 385
Air Traffic Controller 403

Applications
finance 393, 399, 406
geography 379, 382, 389, 393, 399, 401, 406, 411
interior design 389
landscaping 388
manufacturing 383
money 401
part-time job 379, 392
photography 383, 399
physics 401, 406
recreation 398
retail 388
sports 382, 392
travel 379, 391, 392, 401, 411

Standardized Test Practice
Multiple Choice 416
Short Response/Grid In 417
Extended Response 417

Polynomials

Page 406

Three-Dimensional Geometry

Theme: History

Math*Works* Career
Urban Planner 431
Exhibit Designer 451

Applications
architecture 428, 437, 439, 443
astronomy 457
construction 462
Earth science 455
engineering 447
fitness 456
food service 435
geography 433
history 425, 428, 435, 438, 445, 449, 455, 459
hobbies 455
horticulture 458
interior design 438, 449
landscaping 463
machinery 424, 429, 455
packaging 435, 454, 463
recreation 424, 444, 453
retail 427
safety 449
sports 458

Standardized Test Practice
Multiple Choice 468
Short Response/Grid In 469
Extended Response 469

Page 433

Right Triangle Trigonometry

Theme: Photography

Math*Works* Career
Aerial Photographer 483
Camera Designer 503

Applications

advertising 480
archaeology 496
architecture 476, 505
art 477
astronomy 501
business 509
construction 496
Earth science 481
engineering 495
entertainment 506
fitness 486, 508
geography 509
health 478
history 484
hobbies 487
monuments 481
nature 478, 506
navigation 491
photography 477, 481, 487,
490, 497, 501, 506
recreation 480, 485, 496
safety 500
sports 475, 486, 506
travel 491, 500, 501

Standardized Test Practice
Multiple Choice 514
Short Response/Grid In 515
Extended Response 516

Page 470

Logic and Sets

Student Handbook

Theme: Music

Math*Works* Career
Symphony Orchestra
    Conductor 529
Research Professor of Music
    History 547

Applications
advertising 534, 545
biology 525, 540
cooking 526
food service 523, 540
geography 534
history 522
hobbies 524
horticulture 543
landscaping 550
machinery 526
music 522, 527, 535, 544, 551
physics 531
retail 544
safety 550
travel 540, 549
weather 533

Standardized Test Practice
Multiple Choice 556
Short Response/Grid In 557
Extended Response 557

Page 527

How to Use Your MathMatters Book

Welcome to *MathMatters*! This textbook is different from other mathematics books you have used because *MathMatters* combines mathematics topics and themes into an integrated program. The following are recurring features you will find in your textbook.

Chapter Opener This introduction relates the content of the chapter and a theme. It also presents a question that will be answered as part of the ongoing chapter investigation.

Are You Ready? The topics presented on these two pages are skills that you will need to understand in order to be successful in the chapter.

Build Understanding The section presents the key points of the lesson through examples and completed solutions.

Try These Exercises Completing these exercises in class are an excellent way for you to determine if you understood the key points in the lesson.

Practice Exercises These exercises provide an excellent way to practice and apply the concepts and skills you learned in the lesson.

Extended Practice Exercises Critical thinking, advanced connections, and chapter investigations highlight this section.

Mixed Review Exercises Practicing what you have learned in previous lessons helps you prepare for tests at the end of the year.

MathWorks This feature connects a career to the theme of the chapter.

Problem Solving Skills Each chapter focuses on one problem-solving skill to help you become a better problem solver.

Look for these icons that identify special types of exercises.

 CHAPTER INVESTIGATION Alerts you to the on-going search to answer the investigation question in the chapter opener.

 TECHNOLOGY Notifies you that the use of a scientific or graphing calculator or spreadsheet software is needed to complete the exercise.

 WRITING MATH Identifies where you need to explain, describe, and summarize your thinking in writing.

 ERROR ANALYSIS Allows you to review the work of others or your own work to check for possible errors.

 MANIPULATIVES Shows places where the use of a manipulative can help you complete the exercise.

Sample and Display Data

THEME: Market Research

Market research involves using polls, surveys, and statistical data to study the characteristics and actions of a market. A market is the specific audience of a product or service. The purchaser of a product or service may be different than the user of that product or service. For this reason, market research strategies vary depending on the type of audience being targeted.

Data must be displayed in a clear and concise way for the reader to easily interpret and draw conclusions. In this chapter you will learn to present data in various formats.

- Marketers cannot make claims about a product or service based on polls and surveys alone, so they hire product testers. **Product testers** (page 15) provide proof of a claim as required by the Federal Trade Commission.

- **Market researchers** (page 33) gather data and then decide a format in which to advertise a product or service that will reach a targeted market. Visual displays of data are critical to relaying messages.

Math
Online

mathmatters2.com/chapter_theme

Participation in Leisure Activities by Gender and Age

	U.S. Adult population (million)	See movies	Watch sports	Visit amusement park	Exercise	Play sports	Perform charity work	Make home improvements and repairs	Access the Internet
Total	209.1	66%	41%	57%	75%	45%	43%	66%	52%
Gender									
Male	101.0	66%	49%	58%	75%	56%	40%	71%	53%
Female	108.1	65%	34%	57%	77%	35%	46%	61%	51%
Age									
18 to 24	27.1	88%	51%	76%	85%	67%	35%	57%	64%
25 to 34	39.9	79%	51%	70%	82%	63%	41%	63%	63%
35 to 44	45.1	73%	46%	68%	79%	52%	50%	76%	61%
45 to 54	37.7	65%	42%	53%	77%	40%	46%	75%	60%
55 to 64	24.2	46%	33%	40%	69%	19%	44%	71%	42%
65 to 74	18.4	38%	21%	29%	65%	23%	40%	55%	13%
Over 75	16.6	28%	16%	18%	56%	13%	40%	44%	9%

Data Activity: Participation in Leisure Activities

Use the table for Questions 1–4.

1. If you were marketing athletic shoes, to which two age groups would you gear your advertising? Explain your reasoning.

2. About how many people over the age of 75 participated in accessing the Internet?

3. Is the percentage of people 45 to 54 years old who attend movies greater than or less than the percentage that participate in home improvements and repairs?

4. Draw a bar graph to represent the population 35 to 44 years old who participate in leisure activities.

CHAPTER INVESTIGATION

Part of any successful advertising campaign is knowing characteristics of the audience. When conducting a survey, you need to know if the person is a buyer or user of the product. Then learn about the likes and dislikes that influence a purchase.

Working Together

Suppose the school cafeteria manager wants your group to do market research and recommend how she can advertise the cafeteria products and services to students. Conduct a survey of the school population regarding the selection of lunch available in the cafeteria. Make visual representations of your findings and summarize your recommendations for the cafeteria manager. Use the Chapter Investigation icons to check your group's progress.

The skills on these two pages are ones you have already learned. Use the examples to refresh your memory and complete the exercises. For additional practice on these and more prerequisite skills, see pages 576-584.

READING GRAPHS

In this chapter it will be useful to be able to interpret graphs such as circle graphs, pictographs, bar graphs and line graphs.

Refer to the circle graph for Exercises 1–4.

1. How many chose red as their favorite color?

2. What percentage chose blue as their favorite color?

3. How many chose black as their favorite color?

4. What percentage chose purple as their favorite color?

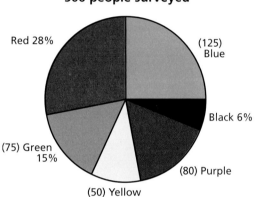

Favorite Colors
500 people surveyed

Red 28%
(125) Blue
Black 6%
(80) Purple
(50) Yellow
(75) Green 15%

Refer to the pictograph for Exercises 5–8.

5. What is the value of half a football?

6. How many more touchdowns were scored by the Bengals than by the Jets this season?

7. How many touchdowns did the Broncos and Rams score in all?

8. How many more touchdowns did the Steelers score than the Broncos?

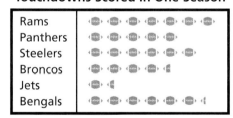

Touchdowns Scored in One Season

| Rams |
| Panthers |
| Steelers |
| Broncos |
| Jets |
| Bengals |

= 4 touchdowns

Refer to the bar graph for Exercises 9–12.

9. Which of these cities has the greatest projected population in 2015?

10. Which of these cities has a projected population of 23 million for 2015?

11. What is the projected population of Bombay and Sao Paulo altogether?

12. How many more people are projected to be living in Mexico City than in Los Angeles?

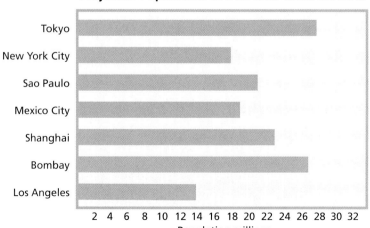

Projected Population of Selected Cities for 2015

Tokyo
New York City
Sao Paulo
Mexico City
Shanghai
Bombay
Los Angeles

2 4 6 8 10 12 14 16 18 20 22 24 26 28 30 32
Population millions

Refer to the line graph for Exercises 13–16.

Monthly Normal Temperature for Pittsburgh, PA

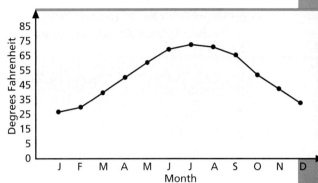

13. Is the normal temperature for Pittsburgh greater in March or November?

14. For what month is the normal temperature in Pittsburgh the highest?

15. What is the difference between the normal temperature in August and the normal temperature in December?

16. Which month has the lowest normal temperature?

ROUNDING

Example   Round 467,376 to the nearest thousand.
The digit in the thousands place is 7. The number to the right of 7 is less than 5, so don't change 7. The rounded number is 467,000.

Round each number to the place of the underlined digit.

17. 438,160	18. 718,760	19. 483.0967	20. 1.987052
21. 953,187	22. 185,438	23. 27.84827	24. 18.69328
25. 761,825	26. 521,618	27. 51.8535	28. 0.83715

ESTIMATION

Front-end estimation is a way to estimate an answer or check an answer.

Examples   Estimate the sum.   Estimate the difference.

$$\begin{array}{r} 6849 \\ +8164 \end{array} \qquad\qquad \begin{array}{r} 8953 \\ -4729 \end{array}$$

Mentally round each number to the place to the far left. Then add or subtract.

$$\begin{array}{rcr} 6849 & \longrightarrow & 7000 \\ +8164 & \longrightarrow & +8000 \\ \hline & & 15{,}000 \end{array} \qquad\qquad \begin{array}{rcr} 8953 & \longrightarrow & 9000 \\ -4729 & \longrightarrow & -5000 \\ \hline & & 4000 \end{array}$$

Estimate each sum or difference.

29. 4618 +3904	30. 7183 −2468	31. 5679 +1538	32. 2871 +5062
33. 3894 +2156	34. 9297 −5564	35. 7428 −2986	36. 8310 +1938
37. 6498 −3781	38. 1368 +5632	39. 8488 −6541	40. 5277 −1409

1-1 Surveys and Sampling Methods

Goals
- Identify sampling methods.
- Recognize biased surveys.

Applications   Manufacturing, Research, Marketing, Politics

Work in groups of three to five students.

1. Brainstorm a list of at least five ways to find out which recording groups or artists are most popular among teens.

2. Review the list and discuss any methods that may not be feasible due to time, money or people limitations.

3. As a group, select the four best methods discussed. Share them with the class.

◣ BUILD UNDERSTANDING

One way to make a decision about an entire group, or **population**, is to collect data from members of that population. To do this, you need to conduct a survey or poll. Tools used to collect data include questionnaires, interviews and records of events.

Since surveying all members of a population can be costly and time-consuming, a more efficient way is to survey a representative part, or **sample**, of the population.

The first step in collecting data from a sample of a population is to choose an appropriate method of sampling. The following table explains four different methods of sampling.

Method	Explanation	Example
Random sampling	Each member of the population has an equal chance of being selected.	The names of all students in a school are placed in a box. Fifty names are drawn and only those students are surveyed.
Cluster sampling	Members of the population are randomly selected from particular parts of the population and surveyed in clusters.	A number of classrooms in a school are selected at random. All students who have classes in those rooms during the first period of the day are surveyed.
Convenience sampling	Members of a population are selected because they are readily available, and all are surveyed.	All students in the school bookstore before the school opens are surveyed.
Systematic sampling	Members of a population that have been organized in some way are selected according to a pattern.	As students in your school pass through the cafeteria line during one day, every tenth person is surveyed.

Example 1

ENTERTAINMENT A concert organizer wants to identify the type of music most popular with adults in her city. Listed below are three ideas for collecting the data. Which type of sampling method is represented by each?

a. Ask the first 30 adults who arrive for a concert at Symphony Hall one evening.

b. Ask all adults who shop in randomly selected music stores in the city.

c. Ask all adults living in the city whose phone numbers end with the digit 3.

Solution

a. convenience sampling

b. cluster sampling

c. random sampling

Sometimes survey findings are **biased**, or not truly representative of the entire population. A biased survey can be a result of the sampling method used. For example, if the only adults surveyed about their favorite type of music are those attending a country music concert, the survey findings are likely to indicate that country music is the type of music most popular with adults.

Survey findings may also be biased because some people do not respond to surveys, and some do not tell the truth on surveys. For example, suppose your teacher gives all students in your class a questionnaire to fill out about the time spent on homework each night. Some students may not fill out the questionnaire. Others may say they spend more time on homework than they actually do.

> **Check Understanding**
>
> Give reasons why you think Example 1, parts b and c, may or may not return biased results.

Example 2   **Personal Tutor at mathmatters2.com**

MANUFACTURING A cereal manufacturer wants to identify the most popular type of breakfast food. The manufacturer decides to include a questionnaire in every tenth box of cereal that it packages.

a. What method of sampling does this represent?

b. Explain whether or not the results are likely to be biased.

Solution

a. This method is systematic sampling. The members of the population have been selected according to a pattern. Every tenth box of cereal contains a questionnaire.

b. The survey results are likely to be biased. The only people who will be polled are those who eat one particular type of breakfast food, cereal.

Some people may not return the questionnaire, and others may not fill it out accurately.

FOOD SERVICE The owners of a restaurant are planning a new menu and want to know the type of sandwich customers like best. What sampling method is represented by each possibility that the owners are considering?

1. Ask every fifth customer who arrives at the restaurant during one day.

2. Ask all customers seated at the table closest to the cashier between 11:30 A.M. and 2:30 P.M. on one day.

3. Ask all customers served during one day by a waiter chosen at random.

4. Randomly select numbers beforehand, and ask customers with those numbers on their receipts.

To identify the time most parents set as a curfew for their teenagers on Friday nights, all teenagers attending the 9:00 P.M. show at one movie theater are surveyed.

5. What method of sampling does this represent?

6. Are the results of this survey likely to be biased? Explain.

7. Explain how the survey can be altered so that it represents systematic sampling.

8. Explain how the survey can be altered so that it represents cluster sampling.

Name the sampling method represented. Then give one reason why the results from each method can be biased.

RECREATION You conduct a survey to find out the favorite sport among teenagers in your town.

9. Ask every eighth teenager who enters a stadium for one particular game.

10. Ask eight teenagers whose names are drawn from all students in your town.

11. Ask the first eight teenagers who arrive at a party.

12. Randomly select a team from eight teams and poll the members of the team.

RETAIL A local radio station conducts a survey to determine the most popular brand of pizza in the area.

13. Have people call into the radio station with their responses.

14. Ask people who exit the Pizza Palace.

15. Call ten people selected randomly from a local phone book.

16. Ask every tenth person entering the supermarket.

17. Describe a sampling procedure that you believe will provide the most accurate results for the survey in Exercises 9–12. Explain your choice.

18. Describe a sampling procedure that you believe will provide the most accurate results for the survey in Exercises 13–16. Explain your choice.

19. DATA FILE Refer to the data on teen attendance at various events on page 562. How do you think the data were collected?

20. TALK ABOUT IT Jamie says that a random sampling is always the best. Do you agree with Jamie? Explain your answer.

POLITICAL SCIENCE A town council is considering a law that requires citizens to recycle all glass and plastic containers. The town holds a public meeting to discuss the proposed law. Before the meeting begins, a council member asks the 50 citizens who arrive first if they are in favor of the recycling law.

21. What sampling method is the town council using?

22. How might the results of their survey be biased?

23. How can the sampling method be changed to produce a convenience sample that is definitely biased in favor of the proposed law?

24. How can the sampling method be changed to produce a convenience sample that is definitely biased against the proposed law?

25. WRITING MATH Set up a table that lists the four types of sampling with advantages and disadvantages of each type.

EXTENDED PRACTICE EXERCISES

26. How can the sampling method used in Exercises 21–24 be changed to produce a cluster sample that is less biased?

27. How can the sampling method used in Exercises 21–24 be changed to produce a random sample?

28. CRITICAL THINKING Would the results of the random survey in Exercise 27 likely be biased?

29. CHAPTER INVESTIGATION Choose a sampling method for your survey. Determine the method of delivery and how to eliminate polling the same students more than once. Write a summary of your plan. Write your survey questions and execute your plan.

MIXED REVIEW EXERCISES

Add. (Basic math skills)

30. 4839 +2763	**31.** 2096 +3887	**32.** 460.76 +915.8	**33.** 118.37 +537.86
34. 8763 +9031	**35.** 7318 +4555	**36.** 62.904 +56.72	**37.** 1048.3 + 392.144

Subtract. (Basic math skills)

38. 7618 −4296	**39.** 6100 −4629	**40.** 5394 −1625	**41.** 8461 −2996
42. 8000 −3917	**43.** 6432 −5981	**44.** 3056 −2967	**45.** 5306 −3198

1-2 Measures of Central Tendency and Range

Goals
- Calculate the mean, median, and mode of data.
- Find the range of a set of data.

Applications Statistics, Sports, Measurement, Education

Use the table for Questions 1–3.

Quiz	1	2	3	4	5	6	7	8	9	10
Ana	9	7	10	9	8	8	9	10	2	9
Stefan	9	8	10	10	7	7	8	9	8	8

1. Which score did Ana receive most often? Which score did Stefan receive most often?

2. Whose average score do you think is higher? Why?

3. Verify your prediction in Question 2 by dividing the total of the quiz scores by 10 for both Ana and Stefan.

■ BUILD UNDERSTANDING

The **mean**, or *arithmetic average*, is the sum of the values in the data set divided by the number of data. The mean is most appropriate to use when there are no extreme values in the data. The mean may or may not be an actual number in the set.

The **median** is the middle value of the data when the data are arranged in numerical order. The median is the most appropriate measure to use when there are extreme values in the data. If the number of data is odd, the median is a number in the set. If the number of data is even, the median is the average of the two middle numbers. It may not be an actual number in the set.

The **mode** is the number that occurs most often in a set of data. A set of data may have one mode, more than one mode, or no mode. Use the mode to describe the most characteristic value of the data. The mode is an actual number in the set.

The mean, median, and mode are **measures of central tendency** because they represent the average value of a data set. Statisticians report the measures that explain the data most appropriately.

Technology Note

Most scientific and graphing calculators have a statistics menu for computing measures of central tendency. Many use the symbol $\bar{x}$ for the mean.

Example 1

Find the mean, median, and mode.

| 4 | 12 | 21 | 33 | 9 | 4 | 78 |

Solution

Mean: $(4 + 12 + 21 + 33 + 9 + 4 + 78) \div 7 = 23$ Divide by 7 since there are 7 items in the set.

Median: 4 4 9 [12] 21 33 78   Write the data in numerical order from least to greatest.

Mode: The number 4 appears twice.

The mean is 23. The median is 12. The mode is 4.

Another number useful in describing a set of data is the range. The **range** is the difference between the greatest and least values in a set.

Example 2 Personal Tutor at mathmatters2.com

MUSIC South Central High School's band and flag corp scored the following number of points during its performances this past season.

$$27 \quad 32 \quad 6 \quad 24 \quad 29 \quad 30 \quad 8 \quad 26 \quad 30 \quad 32$$

Find each of the following. Use a calculator for parts a–b.

a. mean   b. median   c. mode   d. range

e. Which measure of central tendency is the best indicator of the typical number of points scored per game?

Solution

a. Enter the data into list L1, as shown. Return to the main menu by pressing [2nd] [Quit]. Then press [2nd] [List] [▶] [▶] 3 [2nd] [List] 1 [ )] [ENTER] to find that the mean is 24.4.

b. Repeat the same process as in part a to find the median, however, choose 4 in the menu for median. The median score is 28.

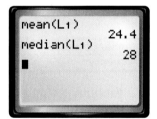

c. The data has two modes, 30 and 32, since both occur twice.

d. Subtract the lowest value, 6, from the highest value, 32. The range $(32 - 6)$ is 26 points.

e. The best indicator of points scored per game is the median score since it is not affected by the extreme values 6 and 8.

Check Understanding

Describe how to solve Example 2, parts a and b, without a calculator.

Example 3

EDUCATION Joanie has scores of 78, 99, 78, 85, and 95 on math tests this grading period. She needs an average of at least 88 to receive a B in the class. She has one more test. What is the minimum score she can get to earn a B for the grading period if all the tests are of equal value?

Solution

Find the total of Joanie's test scores. $78 + 99 + 78 + 85 + 95 = 435$

If her desired average is 88 and there are 6 tests, find the total points she needs by multiplying 88 and 6. $88 \cdot 6 = 528$

Find the score she must earn on the last test by subtracting the points already earned from the required points to get a B. $528 - 435 = 93$

Joanie must score at least 93 on the last test to earn a B for the grading period.

◥ TRY THESE EXERCISES

Find the mean, median, mode, and range for each set of data. Round answers to the nearest hundredth.

1. Davon biked the following number of miles on each day of his trip.

    12.3   12.8   12.4   18.4   27.1   14.9   17.5   12.7   11.4   13.5

2. Juanita spent the following on restaurant meals last month.

    $4.32 $5.16 $3.59 $6.18 $15.87 $8.81 $7.49 $10.00 $2.88

3. In Exercise 1, which measure of central tendency is the best indicator of the typical number of miles Davon biked each day?

4. In Exercise 2, which measure of central tendency is the best indicator of the typical amount of money Juanita spent on lunch?

5. **WRITING MATH** Explain how removing the $15.87 meal from the data in Exercise 2 affects the mean, median, mode, and range. By how much is each affected? Round answers to the nearest cent.

◥ PRACTICE EXERCISES • For Extra Practice, see page 585.

Find each of the following.

Number of Glasses of Water Consumed Daily

Number of glasses	3	4	5	6	7
Frequency	12	18	9	15	6

6. mean   7. median   8. any modes   9. range

10. Which measure of central tendency best indicates the amount of water the average person drinks daily?

11. **WRITING MATH** If your teacher were to allow you to choose the measure of central tendency that is used to determine your grade, which measure would you choose? Why?

EDUCATION Mrs. Jones returned to college to finish her degree. She must have an average of 85% for her employer to reimburse 85% of the tuition fees. She has scored 84, 75, 87, 96 and 75 on her exams. All exam scores are given in percents.

12. Will she receive an 85% if her professor uses the median to determine her grade? Explain.

13. If her professor uses the mean, what score must she receive on the last exam to raise her average to 85%? Assume all exams have equal value.

14. If Mrs. Jones can earn a 90%, her employer will reimburse 90% of her tuition fees. If her professor uses the mean, can Mrs. Jones score enough points to receive a 90% average on six exams? If not, what is her highest possible average? Round to the nearest tenth.

15. **DATA FILE** Refer to the data on sleep times of animals on page 561. Explain whether you think "average" refers to the mean, median, or mode.

Find the unknown number in each data set so that the mean of the set is the value given.

16. 6, 14, 18, 22, 11, ■
    mean = 15

17. 22, 22, 45, 98, ■
    mean = 40

18. 105, 168, 197, ■
    mean = 194

Determine whether each statement is *sometimes*, *always*, or *never* true. If the answer is *sometimes* or *never*, create sample data that supports your answer.

19. If you change one number in a set of data, the mean of the data will change.

20. If you change one number in a set of data, the median of the data will change.

21. If you change one number in a set of data, the mode of the data will change.

22. If you change one number in a set of data, the range of the data will change.

23. YOU MAKE THE CALL Meiying received a 70 on the first test and a 90 on the second test, for a test average of 80. On the third test, she received a 100. She reasoned that her average is 90. Is she correct? Explain.

24. DATA FILE Refer to the data about the coldest U.S. temperatures recorded on page 577. Find the mean, median, mode, and range of the record coldest temperatures.

▧ EXTENDED PRACTICE EXERCISES

To give greater emphasis to the grades students make as a semester progresses, Ricardo's teacher uses a *weighted mean* to assign grades. The first test counts as one grade. The second test counts as two grades. The third test counts as three grades and the fourth test as four grades. Due to the weighting, it is as if there is a total of 10 grades.

25. Ricardo's test scores are 60, 70, 80 and 90, in that order. Find his weighted total.

26. What number must you divide by to find Ricardo's weighted mean? What is his weighted mean?

27. Lamesha is in the same math class as Ricardo. Her test scores are 90, 80, 70 and 60 in that order. Find her weighted mean.

28. CRITICAL THINKING When does a weighted mean have the same value as the arithmetic mean of a set of data?

▧ MIXED REVIEW EXERCISES

Multiply. (Prerequisite skills)

29. 372
  × 58

30. 976
  × 42

31. 522
  × 77

32. 861
  × 19

33. 187
  × 65

34. 318
  × 65

35. 855.6
  × 17.52

36. 412.13
  × 80.09

Divide. Round answers to the nearest thousandth. (Prerequisite skills)

37. 9286 ÷ 52

38. 8104 ÷ 96

39. 7328 ÷ 60

40. 1053 ÷ 64

41. 5000 ÷ 43

42. 6218 ÷ 24

43. 382.4 ÷ 9.685

44. 50.562 ÷ 19.7

Review and Practice Your Skills

A pretzel manufacturer wants to identify the most popular type of snack food. Name the sampling method represented by each of the following.

1. Place a questionnaire in every sixth package of pretzels produced.

2. Place a questionnaire in every package of pretzels sold in a certain state.

3. Give a questionnaire to all of the people who tour the plant where the pretzels are processed and packaged.

4. Dial phone numbers around the country and ask to speak with someone over age 18.

A major city wants to build a new sports stadium. The decision on where to build the stadium is controversial. A survey will be given. Name the sampling method represented by each of the following.

5. Choose a city building and survey everyone entering on a particular day.

6. A survey is given to every fifth commuter going over the bridge into downtown.

7. Give a survey to everyone at the baseball game.

8. Which sampling method above would have the most bias. Why?

Find the mean, median, mode, and range for each set of data. Round answers to the nearest tenth.

9. Yearly incomes
$22,000 $26,000 $27,000 $29,000 $22,000 $35,000 $30,000

10. Heights of students
57 in.   49 in.   57 in.   60 in.   48 in.   58 in.   55 in.   57 in.

11. Price of a hotel room
$65 $71 $89 $105 $155 $80 $105

12. Pupils in a class
22   23   21   24   23   19   23   20   24

13. Airfare prices
$275 $325 $410 $260 $305

14. Weight of athletes
199 lb   189 lb   170 lb   192 lb   202 lb   192 lb   160 lb   189 lb

Find the unknown number in each data set so that the mean of the set is the given value.

15. 75, 87, 90, 99, 72, ■
mean = 80

16. 11, 8, 6, 7, 9, 12, ■
mean = 9

17. 220, 250, 325, ■
mean = 255

PRACTICE ◣ LESSON 1-1–LESSON 1-2

A community wants to know if the residents would support an increase in property taxes to pay for a community swimming pool. (Lesson 1-1)

18. How can a random sample of 50 residents be selected?

19. How can a clustered sampling of 50 residents be selected?

20. How can a convenience sampling of 50 residents be selected?

21. How can a systematic sampling of 50 residents be identified?

Find the mean, median, mode, and range for each set of data. Round answers to the nearest tenth. (Lesson 1-2)

22. Test scores
| 88 | 92 | 81 | 88 | 95 | 70 |

23. Miles run each day
| 6 mi | 8 mi | 7 mi | 6 mi | 5 mi | 5.5 mi | 6.4 mi | 9 mi | 6 mi | 7 mi |

24. Create a set of data with six numbers whose range is 70. (Lesson 1-2)

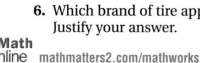

MathWorks Career – Product Tester
Workplace Knowhow

A product tester evaluates claims that a manufacturer makes about its products. He or she must decide what population or sample of the population to survey, what method to use, and concisely report the survey results. The table lists the stopping distances on wet pavement for the same car using two different brands of tires.

| Brand X | 36 ft | 34 ft | 37 ft | 40 ft | 32 ft | 38 ft |
| Brand Y | 33 ft | 42 ft | 35 ft | 34 ft | 38 ft | 34 ft |

1. Calculate the mean stopping distance for each tire brand.

2. Find the mode for Brand X. Find the mode for Brand Y.

3. Determine the median stopping distance for the two brands of tires.

4. Calculate the range of stopping distances for Brands X and Y.

5. Which measure of central tendency most accurately represents the stopping distance data for each brand of tire?

6. Which brand of tire appears to have a shorter stopping distance? Justify your answer.

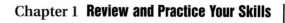

Histograms and Stem-and-Leaf Plots

Goals
■ Use and create histograms to solve problems.
■ Use and create stem-and-leaf plots to solve problems.

Applications Education, Statistics, Inventory, Sports, Entertainment

EDUCATION For each math test, Mr. Lyons makes a frequency table of the class results. Marissa's test paper for one test is missing, but Mr. Lyons knows that he graded it and recorded it in the table.

1. If he uses a table like Table 1, can he identify the missing score? Explain.

2. If he uses a table like Table 2, can he identify the missing score? Explain.

Test Scores (max. score = 20)	Frequency
0 - 5	II
6 - 10	II
11 - 15	ℍℍ ℍℍ III
16 - 20	ℍℍ ℍℍ I

Table 1

Test Scores (max. score = 20)	Frequency
3	I
4	I
6	I
8	II
11	III
12	ℍℍ
13	I
14	IIII
16	III
17	II
18	IIII
19	I
20	I

Table 2

◤ BUILD UNDERSTANDING

It is difficult to see a pattern or trend in data that is not organized. The manner in which the data is organized affects its usefulness. A **frequency table** records the number of times a response occurs but does not offer a visual display.

Frequencies can be shown in a bar graph called a histogram. A **histogram** differs from other bar graphs in that no space is between the bars and the bars usually represent numbers grouped by intervals.

Most graphing utilities can display a histogram. With technology you do not have to make a frequency table first. The calculator does the organizing for you. The intervals are determined by the *x*-scale in the window setting.

Example 1

TECHNOLOGY Use a graphing utility to display the data below in a histogram. For each bar, name the interval and its frequency.

12	15	22	36	45	10	51	12	20	42	16	33
14	23	40	37	15	54	16	22	18	47	50	11

Solution

Enter the data into a list, L1. Choose *histogram* in the statistics plot menu. Set the viewing window as follows.

Xmin = 10   Xmax = 55   Xscl = 5
Ymin = 0   Ymax = 10   Yscl = 1

Intervals and frequencies: 10 to 15: 5   15 to 20: 5   20 to 25: 4
25 to 30: 0   30 to 35: 1   35 to 40: 2
40 to 45: 2   45 to 50: 2   50 to 55: 3

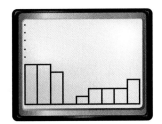

Example 2

 Personal Tutor at mathmatters2.com

FITNESS A gym teacher tested the number of sit-ups students in two classes could do in 1 min. The results are shown.

a. Make a histogram of the data. Title the histogram.

b. How many students were able to do 25–29 sit-ups in 1 min?

c. How many students were unable to do 10 sit-ups in 1 min?

d. Between which two consecutive intervals does the greatest increase in frequency occur? What is the increase?

Sit-Ups Done in 1 Minute

Number of sit-ups	Frequency
0 - 4	8
5 - 9	12
10 - 14	15
15 - 19	6
20 - 24	18
25 - 29	10

Solution

a. Use the same intervals as those in the frequency table on the horizontal axis. Label the vertical axis with a scale that includes the frequency numbers from the table.

b. Ten students were able to do 25–29 sit-ups in 1 min.

c. Add the students who did 0–4 sit-ups and 5–9 sit-ups. So 8 + 12, or 20, students were unable to do 10 sit-ups in 1 min.

d. The greatest increase is between intervals 15–19 and 20–24. These frequencies are 6 and 18. So the increase is 18 − 6 = 12.

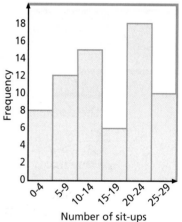

Another method for organizing and displaying data is a **stem-and-leaf plot**. One is shown in Example 3. Each number of the data set is represented by a *leaf* and a *stem*. A leaf is the digit(s) in the place farthest to the right in the number. A stem is the digit(s) that remain when the leaf is dropped. A vertical line separates the stems and leaves.

Outliers are values much greater or less than most of the other values. **Clusters** are isolated groups of values. **Gaps** are large spaces between values. A stem-and-leaf plot helps you notice outliers, clusters and gaps. The plot must include a key.

Example 3

RETAIL The stem-and-leaf plot gives the number of shirts sold daily for 2 wk at Nims department store. Find the following:

a. outliers, clusters, and gaps

b. median   c. mode   d. range

Number of Shirts Sold Daily For 2 Weeks at Nims

4	3
7	2 3 4 5 5 5 9 9
8	0 0 3 4
10	2

4|3 represents 43 shirts.

Solution

a. Since 43 is much less than the other data and 102 is much greater than the other data, both are possible outliers. Clusters of data are in the low to middle 70s and around 79 and 80. The greatest gaps occur between the outliers and the rest of the data and between 75 and 79.

b. The median is the mean of 75 and 79. The median is 77 shirts.

c. Since 75 appears most often, the mode is 75 shirts.

d. The range is the difference between 102 and 43. The range is 59 shirts.

Example 4

 Online Personal Tutor at mathmatters2.com

Display the data in a stem-and-leaf plot. Include a key that explains the plot.

a. Miles ran each day: 3.5, 4.6, 3.9, 4.1, 2.7, 3.4

b. Price of stereos: $121, $129, $125, $117, $124, $119, $110, $117, $120

Solution

a. Use 2, 3 and 4 for the stems.

```
2 | 7
3 | 4  5  9
4 | 1  6
```

2|7 represents 2.7 mi.

b. Use 11 and 12 for the stems.

```
11 | 0  7  7  9
12 | 0  1  4  5  9
```

11|0 represents $110.

TRY THESE EXERCISES

Use the data shown for Exercises 1–5.

1. Use a frequency table to make a histogram of the data.

2. How many families own two to three pets?

3. How many families own more than three pets?

4. To the nearest percent, what percent of families own no pets?

5. Name the median, mode, and range of the data.

Number of Pets Per Family

1	2	3	1	0	2	1	0
1	0	1	4	1	2	0	0
0	1	1	2	2	5	1	0

ENTERTAINMENT Tickets for a rock concert were sold each day as follows: 149, 253, 366, 169, 297, 421, 183, 256, 303, 427, 189, 229, 367, 520, 147, 168, 253, 146, 182, 305, 412, 277

6. Make a stem-and-leaf plot.

7. Name any outliers, clusters, or gaps.

8. Find the median.   9. Find the mode.

10. Find the range.

Make a stem-and-leaf plot of each set of data.

11. Miles biked each day: 12.8, 11.6, 14.8, 15.0, 15.1, 11.6, 13.0, 16.8, 11.6, 10.5

12. Age of Stella's grandchildren: 4, 9, 12, 16, 23, 14, 7, 1, 16, 14, 19, 22, 20, 13, 5, 29, 6, 2, 1, 17

PRACTICE EXERCISES • For Extra Practice, see page 586.

Use the histogram for Exercises 13–16.

13. Which interval contains the most evergreen seedlings?

14. Which intervals contain an equal number of trees?

15. Which intervals contain 95% of the data?

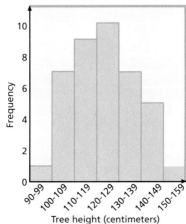

Height of Evergreens in Reforestation Project

16. WRITING MATH Explain how the histogram on page 18 changes if intervals of 20, instead of 10, are used.

SPORTS The data are the percents of free throws Alan made during the basketball games he played.

55	59	56	46	43
79	53	60	39	57
86	45	66	72	46
24	64	47	43	41

17. Organize the data into a stem-and-leaf plot.

18. Make a histogram of the data in intervals of 10.

19. In how many games did Alan make at least 60% of his free throws?

20. What is the median percent of scoring free throws Alan made per game?

21. Is the median the best indicator of Alan's successful free throws during the season? If yes, explain. If not, name the best indicator and explain.

22. WRITING MATH When given data organized in a stem-and-leaf plot, explain the process of identifying the mean, median, and mode.

23. Measure the lengths of your fingers in centimeters. Make a stem-and-leaf plot.

24. DATA FILE Refer to the data about how often earthquakes occur on page 564. Make a histogram of the data.

25. MARKET RESEARCH A civil engineer is studying traffic patterns. She counts the number of cars that make it through one rush-hour green-light cycle. Organize her data into a frequency table, and then make a histogram.

15 16 10 8 8 14 9 7 6 9 10 11 14 10 7 8 9 11 14 10

◣ EXTENDED PRACTICE EXERCISES

A value must meet certain criteria to be named an *outlier*. Follow this process:

    a. Locate the median of a set of data.
    b. Find the median of the lower half of the data. Name it Q_1.
    c. Find the median of the upper half of the data. Name it Q_3.
    d. Subtract Q_1 from Q_3. Name this value IQR since it is the interquartile range.
    e. An **outlier** is any data in the set that is less than $Q_1 - 1.5$ (IQR) or that is greater than $Q_3 + 1.5$ (IQR).

26. Refer to Example 3. Use steps a through e to determine if the possible outliers are indeed outliers.

27. Refer to Exercise 7. Use steps a through e to determine if the possible outliers are indeed outliers.

28. CRITICAL THINKING Describe why histograms are not used to identify mean, median, and mode.

◣ MIXED REVIEW EXERCISES

Find the mean of each set of data. Round to the nearest tenth. (Lesson 1-2)

29. Cost of dinner: $12, $16, $19, $27, $32, $15, $17, $18, $21, $13, $14, $12

30. Miles driven: 37, 42, 47, 38, 40, 29, 34, 36, 44, 45, 43, 46

31. DATA FILE Refer to the data on the ways people get to work on page 575. If there are 657,688 people in Austin, Texas, how many commute to work in a carpool? (Prerequisite Skill)

1-4 Scatter Plots and Lines of Best Fit

Goals
- Use scatter plots to solve problems.
- Use a graphing utility to determine a line of best fit.

Applications Retail, Education, Sports, Statistics, Insurance

Work as a class to answer Questions 1–5.

1. Record the number of miles each student lives from school and the number of minutes it takes to get from home to school.

2. Display these data in a graph using one axis for distance and the other for time. Plot a point on the graph to represent each student.

3. Find the mean of the distances. Find the mean of the times.

4. Plot the point (*a*, *b*) where *a* is the mean of the distances and *b* is the mean of the times.

5. What patterns do you see in your graph? Explain.

◥ BUILD UNDERSTANDING

This type of visual display for data is a **scatter plot,** which shows the relationship of two sets of data. The data are grouped as ordered pairs and graphed as points on a grid. There can be more than one point for any number on either axis.

Example 1

Use the scatter plot.

a. How many people were at the pool on the day the high temperature was 91°F?

b. Find the mode of the high temperatures.

c. Find the range of the daily attendance.

d. Find the median of the daily attendance.

Daily Swimming Pool Attendance for July

y-axis: Number of people (0 to 500)
x-axis: Highest daily temperature (°F) (80 to 96)

Solution

a. Locate 91 on the horizontal axis. Move up to the point, then left to the vertical axis. The day the temperature reached 91°F, 425 people were at the pool.

b. Find the temperature with the greatest number of points above it. The mode of the high temperatures is 86°F.

c. Find the difference between the greatest attendance (500) and the lowest attendance (25). 500 − 25 = 475. The range of the daily attendance is 475.

d. Since there are 31 days in July, count to the sixteenth point on the graph, starting with the least number of people in attendance. The median of the daily attendance is 350.

To create a scatter plot, you must choose a scale for each axis and arrange the data into ordered pairs. To display a scatter plot on a graphing utility, enter the data into lists. The scales are determined by the setting for the viewing window.

COncepts
in MOtion
Interactive Lab
mathmatters2.com

Example 2

MEDICINE The list below shows the age and weight of 15 children who visit Dr. Warren's office during one week for their well-child checkup.

Use a graphing calculator to display a scatter plot of the data.

1 year old, 20 lb	12 year old, 108 lb	5 year old, 60 lb	10 year old, 120 lb	2 year old, 38 lb
3 year old, 35 lb	9 year old, 87 lb	2 year old, 30 lb	4 year old, 45 lb	6 year old, 50 lb
5 year old, 50 lb	7 year old, 64 lb	10 year old, 84 lb	7 year old, 70 lb	11 year old, 90 lb

Solution

Enter the age data into a list, L1, and the weight that corresponds to each age into another list, L2. Turn on Plot 1. Select *scatter plot* as the type of plot to display. Set the viewing window as follows:

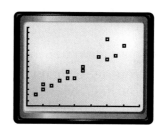

$$\text{Xmin} = 0 \quad \text{Xmax} = 14 \quad \text{Xscl} = 2$$
$$\text{Ymin} = 0 \quad \text{Ymax} = 140 \quad \text{Yscl} = 10$$

The relationship that a scatter plot displays is a *correlation*. A correlation can be positive or negative and weak or strong. Some scatter plots show no correlation.

A **positive correlation** means that as the horizontal axis values increase, so do the vertical axis values. A **negative correlation** means that as the horizontal axis values increase, the vertical axis values decrease.

When a scatter plot shows a correlation, a straight line can be drawn that best fits the set of data. This line, called a **line of best fit**, approximates a trend for the data in the scatter plot. For this reason, this line is often called a **trend line**. To find the best approximation for a line of best fit, use a graphing calculator.

Example 3

Graph a line of best fit for the scatter plot in Example 2.

Solution

The line of best fit is called a *linear regression*, which can be shown as an equation in $y = mx + b$ form. Since the data from Example 2 is already entered into your calculator in L1 and L2, you can find the equation from the home screen. Press [STAT] ▶ 4 [2nd] [L1] [,] [2nd] [L2] [ENTER] to find the equation of the linear regression to be $y = 7.88x + 14.02$.

To graph the line on the scatterplot, enter the equation into
Y1. You can also have the calculator enter the full equation
automatically. Press [Y=] and clear any stored equations. Then
press [VARS] 5 [▶] [▶] [ENTER]. The full equation is then entered.
Press [GRAPH] to graph both the scatter plot and line of best fit.

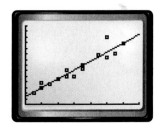

Example 4

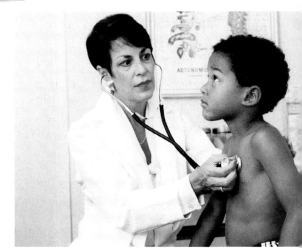

🌐nline **Personal Tutor at** mathmatters2.com

Use Examples 2 and 3 to answer the following.

a. Predict the weight of an eight-year-old child
visiting Dr. Warren for a checkup.

b. Name the type of correlation between age and
weight. Explain.

c. Which point lies farthest from the trend line?
What could account for this?

Solution

a. Locate age eight on the vertical axis. Move across
to the line, then down to the horizontal axis. The
weight is about 77 lb.

b. The trend line slopes upward to the right, so there is a positive correlation.
As a child gets older, the child's weight increases.

c. The point farthest from the trend line represents a ten-year-old child
who weighs 120 lb. This is a ten-year-old who is unusually tall or who
is overweight.

◥ TRY THESE EXERCISES

Use the scatter plot for Exercises 1–4.

1. During the month that had six days of rain, how
many days had above-average temperatures?

2. Find the range of days with above-average
temperatures.

3. Find the mode of the number of rainy days
each month.

4. Find the median number of days with above-
average temperatures.

5. Collect data from your classmates about the
amount of time each studied for the last test and
the grade each received. Display the data in a
scatter plot.

6. **GRAPHING** Graph a line of best fit for your scatter plot in Exercise 5.
Describe any correlation and draw a conclusion. Write a recommendation
for your classmates about time spent studying and expected grades.

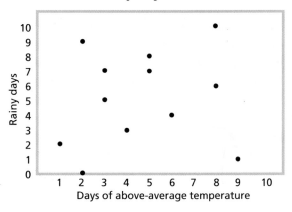

Use the scatter plot for Exercises 7–9.

TV Watching and Test Scores

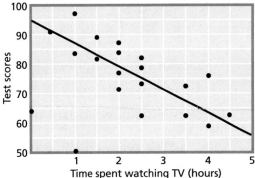

7. Predict the test score of a student who watched 3 h of television.

8. Does the graph show positive or negative correlation? Explain.

9. Which point lies farthest from the trend line? What could account for this?

MARKET RESEARCH Advertising is a major factor in a company's annual budget. The table shows several months' commercial time purchased and sales in millions of dollars.

5 min, 12.0 sales	40 min, 39.0 sales	21 min, 24.0 sales	10 min, 18.0 sales	30 min, 30.0 sales
42 min, 50.0 sales	33 min, 34.0 sales	7 min, 19.0 sales	11 min, 29.0 sales	21 min, 31.0 sales
13 min, 25.0 sales	38 min, 33.0 sales	15 min, 29.5 sales	17 min, 27.5 sales	22 min, 28.0 sales
25 min, 32.0 sales	28 min, 28.0 sales	33 min, 56.5 sales	34 min, 41.2 sales	11 min, 21.5 sales

10. Draw a scatter plot and line of best fit for the data.

11. Is there a positive or negative correlation between commercial time and sales?

12. Predict the company's sales if they have 50 min of commercials in 1 mo.

13. **WRITING MATH** Describe a relationship of data that shows a positive correlation. Describe another relationship that shows a negative correlation and another that shows no correlation. Provide sample data for each.

14. **DATA FILE** Refer to the data on cricket chirps/min on page 560. Make a scatter plot of the data, and show a trend line. Is there a correlation between the number of chirps a minute and temperature? Plot the chirps per minute on the horizontal axis.

■ EXTENDED PRACTICE EXERCISES

15. **ECONOMICS** Work with a partner. Choose Topic A or B. Research data to make a scatter plot. Explain any correlation and draw a conclusion.

   Topic A—prices for new cars and expected highway mileage (mi/gal)
   Topic B—population of a country and literacy rate (expressed as %)

16. **WRITING MATH** Write three paragraphs explaining when it is best to use a stem-and-leaf plot, a histogram and a scatter plot to display data.

■ MIXED REVIEW EXERCISES

Complete. (Basic math skills)

17. 4 m = __?__ mm

18. 80 cm = __?__ m

19. 45 ft = __?__ yd

20. 12 c = __?__ qt

21. 2 L = __?__ mL

22. 16 gal = __?__ qt

23. 2 km = __?__ m

24. 8 yd = __?__ ft

25. 300 dm = __?__ m

26. 2 yd = __?__ in.

27. 6,000 g = __?__ kg

28. 96 in. = __?__ ft

Review and Practice Your Skills

PRACTICE ◣ LESSON 1-3

Make a stem-and-leaf plot of each set of data.

1. Price of tickets: $23, $34, $27, $35, $31, $29, $42, $38

2. Miles traveled each day: 26 mi, 52 mi, 37 mi, 43 mi, 57 mi, 28 mi, 33 mi, 40 mi

3. The ages of the first 15 Presidents of the U.S. at the time of their first inauguration are listed below. Make a stem-and-leaf plot of the data.

   57   61   57   57   58   57   61   54   68   51   49   64   50   48   65

4. What was the most common age of the Presidents?

5. What is the median age of the Presidents?

6. What is the range in age of the Presidents?

Use the frequency table for Exercises 7–9.

7. Construct a histogram to display the data of TV use.

8. Which interval has the greatest frequency?

9. Which interval contains 10% of the data?

Television Use

Number of hours	Frequency
0-9	1
10-19	6
20-29	15
30-39	12
40-49	2
50-59	4

PRACTICE ◣ LESSON 1-4

Use the table of a basketball player's scoring average.

10. Make a scatter plot of the data. Plot the ages on the vertical axis.

11. What is the range of the players scoring average?

12. Does the scatter plot show a positive correlation, a negative correlation or no correlation?

Tell which of the following scatter plots would show a *positive correlation*, a *negative correlation* or *no correlation*.

13. price of a product, amount of tax on the product

14. height in inches, weight in pounds

15. population of a city, amount of rain

16. car speed, time needed to reach destination

Age	Scoring average
23	18
24	17.5
25	22.5
26	24
27	21.5
28	26
29	23.5
30	22.5
31	27.5
32	20.5

Use the scatter plot of home runs and runs batted in (RBI's).

17. Do players who hit a lot of home runs drive in a lot of runs?

18. Describe the correlation of the data?

19. If a player hit 25 home runs, about how many RBI's would they expect to have?

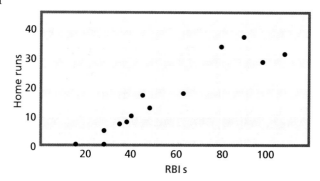

DATA FILE For Exercises 20–23, refer to the data on recent noteable earthquakes on page 564.

20. Organize the magnitude data into a stem-and-leaf plot.

21. Find the mode(s) of the data.

22. Find the median of the data.

23. Find the mean of the data.

Teenagers were polled about the number of evening meals they ate at home in a 4-wk period.

| 18 | 20 | 16 | 28 | 31 | 23 | 25 | 24 | 33 | 35 | 17 | 14 |

24. Organize the data into a stem-and-leaf plot. (Lesson 1-3)

25. What is the median number of meals eaten at home? (Lesson 1-2)

Mid-Chapter Quiz

POLITICAL SCIENCE A politician is running for reelection to Congress. She wants to know if people are likely to vote for her. What kind of sampling method is represented by each of these possibilities. (Lesson 1-1)

1. Ask all delegates at a political convention.

2. Ask all registered voters who live on randomly selected streets in her Congressional district.

3. Ask every tenth adult arriving at the county fair in her Congressional district.

Use the table for Exercises 4–8. (Lessons 1-2 through 1-4)

4. Find the mean, median, and mode of the data.

5. Make a frequency table for the data with intervals 1-3, 4-6, and so on.

6. How many families spent 10–14 days on vacation?

7. Organize the data in a stem-and-leaf plot.

8. Name any outliers, clusters or gaps in the data.

Number of Days Families Spend on Vacation per Year

7	10	4	5	9	14	18	9	6	3
2	8	3	1	4	17	7	14	1	9

Write *positive correlation, negative correlation* or *no correlation* for each. (Lesson 1-4)

9. As a person's age increases, the hours of sleep per night decreases.

10. As a person's age increases, the number of vacation days per year increases.

Problem Solving Skills: Coefficient of Correlation

The arrangement of points on a scatter plot provides a visual idea of the *degree of correlation* between two variables. This means you can tell if the relationship is positive or negative, strong or weak. Another way to tell the characteristics of a relationship is a **coefficient of correlation,** a statistical measure of how closely data fits a line.

A coefficient of correlation, *r*, is between 1 and −1. The closer a coefficient of correlation is to 1 or −1, the stronger the relationship between *x* and *y*. Visually, the points almost form a line. The closer a coefficient of correlation is to 0, the weaker the relationship is between *x* and *y*.

You can predict a coefficient of correlation by looking at data or a scatter plot. To be precise or confirm your prediction, use a graphing calculator to calculate linear regression and the coefficient. This strategy is **guess and check**.

Problem Solving Strategies

✔ Guess and check

Look for a pattern

Solve a simpler problem

Make a table, chart or list

Use a picture, diagram or model

Act it out

Work backwards

Eliminate possibilities

Use an equation or formula

Problem

RETAIL The manager of a concession stand orders bottled water and juice based on the daily temperature and number of bottles sold.

a. Is the correlation between temperature and combined sales positive or negative, strong or weak?

b. Predict the drink with a stronger correlation to temperature.

c. Use a graphing calculator to identify the coefficient of correlation for temperature and bottled water sales and the coefficient of correlation for temperature and juice sales.

d. Were your predictions in parts a and b correct?

Daily high temperature	Bottled water sales	Fruit juice sales
66°F	141	142
70°F	149	138
74°F	159	133
78°F	165	114
82°F	175	146
86°F	180	166
90°F	193	160
94°F	195	158
98°F	210	178

Solve the Problem

a. As temperature increases, sales of drinks increase. The correlation is strong and positive, so the coefficient of correlation will be close to 1.

b. As temperature increases, sales of bottled water increase the most.

c. Enter temperatures in a list, L1; water sales in a second list, L2; and juice sales in a third list, L3. To find the coefficient for temperature and water, use L1 and L2. Use L1 and L3 to find the coefficient for temperature and fruit juice. The screens at the right show that $r \approx 0.995$ for water sales and $r \approx 0.732$ for juice sales.

d. The correlation is strong and positive. The coefficient *r* is positive and close to 1 for both. However, bottled water sales show a stronger relationship.

Match each scatter plot with a possible description. Describe its correlation.

1. *x*-value: hours of daylight
   y-value: hours after dark

2. *x*-value: amount of rainfall
   y-value: levels of area rivers

3. *x*-value: heart rate
   y-value: reading speed

Graph A

Graph B

Graph C

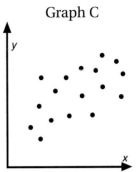

■ **PRACTICE EXERCISES**

Predict the characteristics of the correlation between the two variables. Find the coefficient of correlation to check your prediction.

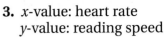

Five-step Plan

1 Read
2 Plan
3 Solve
4 Answer
5 Check

4.

x	1.0	2.0	2.5	3.1	4.2	5.0	7.0	8.3
y	1.1	3.9	5.4	7.1	10.6	13.2	19.1	22.1

5.

Year	1993	1994	1995	1996	1997	1998	1999	2000	2001
Number of evening papers	954	935	891	846	816	781	760	727	704

6.

Year	1984	1986	1988	1990	1992	1994	1996	1998	2000
Average MPG	17.4	17.4	18.8	20.3	21.0	20.8	21.2	21.6	22.0

WRITING MATH State whether you think each of the following sets of variables would show a *positive, negative,* or *zero* correlation. Explain your reasoning.

7. outside temperature and snow ski sales

8. an adult shoe size and annual salary

9. miles a car is driven and its gas mileage

10. a person's height and weight from birth to age 20

■ **MIXED REVIEW EXERCISES**

Write three equivalent ratios. (Basic math skills)

11. $\frac{3}{12}$ **12.** $\frac{5}{15}$ **13.** $\frac{20}{25}$ **14.** $\frac{1}{2}$ **15.** $\frac{7}{8}$ **16.** $\frac{4}{20}$

Simplify. (Basic math skills)

17. $-4 - (-6) + 12$

18. $13 + (-6) - 14$

19. $-3 + (-7) - 8 + 6$

20. $7 - (-4) - 16 + (-2)$

21. $-6 - (-5) - (-8)$

22. $15 - 8 - (-3) + (-9)$

23. $5 + (-8) - (-9) + 4$

24. $-9 + (-9) - (-9) + 9$

25. $8 - 7 - 6 - (-5) + 4$

Quartiles and Percentiles

Goals
- Identify quartiles and calculate percentiles.
- Create a box-and-whisker plot.

Applications Education, Market research, Statistics

EDUCATION The test scores for students in one class are given.

58	84	72	40	95	78	92	98	82	50
67	90	75	93	87	55	84	86	62	67

1. Find the median of the scores.

2. Is the median a good indicator of how well the class did on this test? Explain.

3. Find the median of all the scores below the median class score and the median of all the scores above the median class score.

4. How can using the three medians from Questions 1 and 3 give a better indication of how well the class did on the test?

5. Into how many equal parts do the three medians separate the scores?

◣ BUILD UNDERSTANDING

Another way to analyze data is by **quartiles**, three numbers that group the data into four equal parts. To find the quartiles, first determine the median, also called the *second quartile*. The *third quartile* is the median of the data above the second quartile. The *first quartile* is the median of the data below the second quartile.

Data that are greater than the third quartile are in the top quarter of the data set. Data that are less than the first quartile are in the bottom quarter. The **interquartile range** is the difference between the first and third quartiles.

> **Reading Math**
>
> The abbreviations Q_1, Q_2 and Q_3 are used to represent the first, second and third quartiles.

Example 1

Find the following for the data.

Age of employees: 18 29 56 42 58 31 40 28 37 46

a. median   **b.** first quartile   **c.** third quartile   **d.** interquartile range

Solution

Write the ages in order from least to greatest.

median of all data, or
second quartile
↓

18 28 29 31 37 40 42 46 56 58

↑ first quartile   ↑ third quartile

lower half   *upper half*

a. The median age is 38.5, halfway between the two middle ages. So the second quartile is 38.5.

b. There are five ages below 38.5. The middle of these is 29. So the first quartile is 29.

c. There are five ages above 38.5. The middle of these is 46. So the third quartile is 46.

d. The interquartile range is the difference between the first quartile and the third quartile, $46 - 29 = 17$. So the interquartile range is 17.

Check Understanding

In Example 1, suppose the lowest score, 18, is not listed in the data. Find the first quartile, median and third quartile.

Some graphs show the distribution of data related to measures in a data set. One such graph is a **box-and-whisker plot** that uses quartiles and a box to illustrate the interquartile range. Box-and-whisker plots are drawn using a number line; however, the number line does not have to be part of the completed plot.

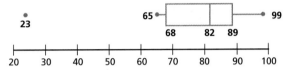

The lines that are drawn from the ends of boxes to the least and greatest values of the data are called **whiskers**. **Outliers**, which are marked with only a point, are data that are at least 1.5 times the interquartile range above the third quartile or at least 1.5 times the interquartile range below the first quartile.

Example 2

 Personal Tutor at <u>mathmatters2.com</u>

Make a box-and-whisker plot for the data.

Prices of Microwave Ovens (dollars)

225 257 175 300 265 185 229 235 299

Solution

175 185 225 229 235 257 265 299 300   Write in order from least to greatest.

Find the quartile values.

Q_2: The median of the data is 235.

Q_1: The median of the lower half of the data is midway between 185 and 225.

$$185 + 225 = 410$$
$$410 \div 2 = 205$$

Q_3: The median of the upper half of the data is midway between 265 and 299.

$$265 + 299 = 564$$
$$564 \div 2 = 282$$

Use points to mark these values below a number line. Complete the box and whiskers. No data is far from the rest of the data, so there are no outliers.

Prices of Microwave Ovens (dollars)

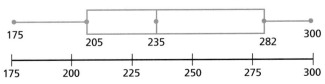

Technology Note

You can use a graphing calculator to create a box-and-whisker plot, which is also called a boxplot. Enter the data into a list, and select *boxplot* as the statistical plot to display.

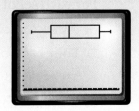

When you take a standardized test, your score is often reported in terms of percentiles. Your **percentile** indicates the percent of those who took the test who achieved at or below your score.

For example, if your score is in the 83rd percentile, approximately 83% of those who took the test had a score less than or equal to yours. Only 17% of those who took the test had a score greater than yours. So a percentile is a measure of a rank, or standing, within a group.

$$\text{Percentile} = \frac{\text{number of scores less than or equal to given score}}{\text{total number of scores}} \cdot 100$$

Example 3

EDUCATION Bernardo took a placement test in order to be accepted into a computer course. His score is 48th from the highest out of the 760 students who took the test. Find the percentile rank that Bernardo achieved.

Solution

The total number of students who took the test was 760. Bernardo was 48th. So there were 47 students who had higher scores than he did. The number of students who scored equal to or less than he did, was 760 − 47, or 713.

$$\frac{\text{number of scores less than or equal to given score}}{\text{total number of scores}} = \frac{713}{760} \approx 0.938$$

So, Bernardo's ranking is in the 94th percentile.

TRY THESE EXERCISES

1. The interquartile range is the difference of which quartiles?

2. Find the first quartile, the median and the third quartile of the science test scores shown. Then find the interquartile range.

3. Draw a box-and-whisker plot for the data in Exercise 2.

4. On a test, Cheung has the ninth highest test score. If there are 28 students who take the test, what is Cheung's percentile rank?

5. If you score in the 90th percentile on a test, how many people scored above you out of 20 students?

Science Test Scores

84	89	76	65	74
73	85	89	91	74
93	82	68	76	94
63	83	80	80	70

PRACTICE EXERCISES • For Extra Practice, see page 587.

For each set of data, find the first quartile, the median and the third quartile.

6. Test scores: 71, 86, 92, 53, 87, 76, 75, 84, 83

7. Points scored: 20, 16, 15, 13, 12, 19, 18, 20, 18, 10, 12, 14

8. Miles biked: 4.6, 8.3, 9.3, 7.2, 5.8, 8.7, 3.2, 5.9, 11.6

9. Bowling scores: 200, 114, 162, 260, 149, 140, 146, 125, 172

10–13. Create a box-and-whisker plot for each set of data in Exercises 6–9.

14. SPORTS Make a box-and-whisker plot for these bowling scores.
200 101 162 273 149 153 146
125 118 129 135 142 111 156

15. Exercise 9 gives Mario's bowling scores and Exercise 14 gives Shannon's bowling scores. Use their box-and-whisker plots to compare their scores. Write a statement that describes Mario's scores in terms of Shannon's scores.

16. DATA FILE Refer to the data on the ten hottest U.S. temperatures on record on page 577. Make a box-and-whisker plot to show the distribution of temperatures.

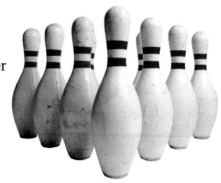

Find the percentile ranks for each of the following students' science midterm exam taken from the science test scores on the previous page.

	Test 1		
Apple, Felicia	93		
Bean, Jamal	91		
Carrot, Robert	76		
Ditto, Bethany	68		
Ever, Mark	80		
Floor, Martha	70		

17. Felicia   **18.** Martha   **19.** Mark

20. Bethany   **21.** Robert   **22.** Jamal

23. WRITING MATH Explain step-by-step how to solve the following problem. If there is any information not needed, explain how you know it is not needed.

Bly took a history exam and received a score of 27 out of 50. If out of 32 students in the class he ranked 19th, what is his percentile rank?

24. If you score at the 96th percentile on a test, how many people scored above you out of 25 students?

25. Explain the difference between the 90th percentile and the 10th percentile.

EXTENDED PRACTICE EXERCISES

26. If your score on a test is equal to the median of the scores, in what percentile rank is your score?

27. If exactly one-quarter of your class scored higher than you did on a test, what is your percentile ranking?

28. CRITICAL THINKING If your score on a test is equal to the mean of all the scores, can you determine your percentile ranking? Explain.

29. CHAPTER INVESTIGATION Organize your data to determine the information to highlight about the cafeteria. Calculate the mean, median, mode, range and percentiles that you can use to support your claims.

MIXED REVIEW EXERCISES

Multiply or divide. (Basic math skills)

30. $-72 \div (-4)$ **31.** $4(-3)$ **32.** $81 \div (-9)$

33. $-7(13)$ **34.** $-70 \div (-5)$ **35.** $-108 \div 12$

36. $5(-6)(4)$ **37.** $-96 \div (-2)$ **38.** $7(-6)(-5)$

Review and Practice Your Skills

PRACTICE ◤ LESSON 1-5

Predict the characteristics of the correlation between the two variables. Find the coefficient of correlation to check your prediction.

1.

Years since 1900	10	20	30	40	50	60	70	80	90	100
U.S. Population (millions)	92.2	106.0	123.2	132.2	151.3	179.3	203.3	226.5	248.7	281.4

2.

Average monthly temperature (°F)	8	14	29	37	45	52
Average monthly snowfall (inches)	22	17	12	6	2	0

State whether each of the following sets of variables would show a *positive*, *negative*, or *zero* correlation.

3. hours of TV watched per week, GPA

4. your age, the number of letters in your name

5. age of a car, the value of the car

6. height of a cylinder, the volume of the cylinder

PRACTICE ◤ LESSON 1-6

For each set of data, find the first quartile, the median, and the third quartile.

7. Number of sit-ups: 12   14   20   21   25   29   30   30   31   32   59   46

8. Number of vacation trips: 5   7   1   3   5   4   5   4   6   2   6   4

Make a box-and whisker plot for these exam scores. Label Q_1, Q_2, and Q_3.

9. 70   65   85   72   93   97   89   94   88

DATA FILE For Exercises 10–14, use the data on page 575 to make a box-and-whisker plot of the birth years of the states.

10. What is the median year?

11. What is Q_1, Q_2, and Q_3?

12. What is the interquartile range?

13. Does the birth of Texas fall between Q_1 and Q_3?

14. Does the birth of North Carolina fall between Q_1 and Q_3?

Find the percentile ranks for each of the following students' English final exam. Round to the nearest percent.

15. Arnold

16. Lynn

17. Bo

18. Kavita

Arnold	77
Bo	90
Ciana	100
Deb	68
Eduardo	80
Gayle	81
Kavita	78
Jose	61
Lynn	84
Mika	90
Naomi	89

19. Find the mean, median, and mode of the following test scores. (Lesson 1-2)

89   76   76   90   93   97   80   78   90   82   85

Use the box-and-whisker plot to answer the following questions. (Lesson 1-6)

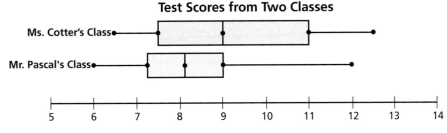

Test Scores from Two Classes

20. Which class had the higher median score?

21. Which class had its scores grouped more closely around its median?

22. Which class, as a whole, scored better on the test?

MathWorks
Workplace Knowhow

Career – Market Researcher

A market researcher plays a key role in a company's decisions. The results of the studies that a market researcher performs determine what new merchandise will be sold in different locations. It is critical that the data collected by a market researcher be accurate and easily interpreted. The table records the shirt sizes (collar and sleeve length) and pant sizes (waist and inseam) of ten men. The market researcher plans to use this data to recommend how many of each size of shirts and pants that a men's specialty shop should keep in stock.

1. Construct a frequency table of the collar size data listed in the table. What are the median and the mode for this data?

2. Draw a scatter plot to determine whether there is a positive or negative correlation between the waist size data (x-axis) and inseam data (y-axis). Is the coefficient of correlation closer to 1 or -1?

3. Design a stem-and-leaf plot of the data for collar size. Be sure to include a key that explains the plot.

Collar size (inches)	Sleeve length (inches)	Waist size (inches)	Inseam (inches)
15.5	35	34	32
16.0	34	33	30
17.0	36	36	35
17.0	36	37	38
14.5	32	32	33
16.5	36	35	34
15.5	34	35	33
16.5	34	36	35
16.5	33	35	35
17.5	37	38	37

4. Construct a box-and-whisker plot of the sleeve length data. Calculate the first, second and third quartiles as well as the range and the interquartile range.

1-7 Misleading Graphs and Statistics

Goals
- Recognize how a graph can be misleading.
- Identify the misleading use of the word "average."

Applications Insurance, Business, Advertising, Market research

One advantage of a statistical graph is that it can provide information at a glance. But sometimes this can be a disadvantage. There are ways to change the appearance of a graph to create different impressions.

1. Explain a time when you were convinced by a statistical graph.

2. Did you later find out that the graph was misleading?

3. How was the data presented that led you to misunderstand?

◼ BUILD UNDERSTANDING

Data is presented in many different formats. People see displays of data in the news, on television, in magazines, on financial reports, and in medical records. Much of this data is presented so that a judgement will be formed quickly without a lot of attention to detail.

Misleading data representation is data representation that leads to a false perception. One way to present correct data so that it is misinterpreted is to alter a scale or show only a certain segment of the results.

Example 1

BUSINESS Admissions data for a waterpark is shown.

a. Make a line graph of the data using a vertical scale from 500,000 to 530,000 with each unit equal to 5000.

b. Make a line graph of the same data using a vertical scale from 515,000 to 527,000 with each unit equal to 2000.

c. If you are a manager planning to include a graph in a report for investors, which graph would you use? Explain.

Month	Admissions
May	516,633
June	519,312
July	521,654
August	523,809
September	525,248

Solution

a.

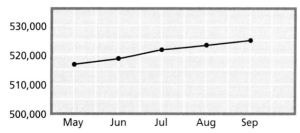

Waterpark Admissions

b.

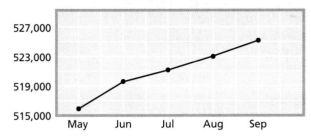

Waterpark Admissions

c. A manager would show the graph created in part b. This graph makes it appear that admissions have increased dramatically.

It is not uncommon for the word "average" to be used in place of any of the three measures of central tendency. However, when people read the word "average," they often think of *arithmetic average*. In these cases, the word "average" is used correctly, but the misleading aspect comes from the interpretation by the reader.

E x a m p l e   2 **Personal Tutor at** **mathmatters2.com**

ADVERTISING Furniture World has seven salespeople. The commissions they earned last week were $493, $283, $301, $299, $366, $304 and $299. The owner of the store places an ad in the newspaper for additional salespeople.

a. Which measure of central tendency does the owner use in the ad?

b. Does this ad give a fair picture of the salespeople's weekly commissions? Explain.

c. Which measure(s) of central tendency would be useful to a person considering a sales position at Furniture World? Explain.

> **EXPERIENCED SALESPEOPLE**
>
> wanted for furniture store.
> Earn an average weekly
> commission of $335.
> Apply in person at
>
> **FURNITURE WORLD**
> 219 Plainfield Tpke.

Solution

a. Determine the three measures of central tendency. Arranging the data in order simplifies finding the median and the mode.

$283 $299 $299 $301 $304 $366 $493

median: The middle commission of the seven is $301.

mode: The commission of $299 is the mode since it appears twice.

mean: Add the commissions and divide by 7.

$283 + $299 + $299 + $301 + $304 + $366 + $493 = $2345
2345 ÷ 7 = $335

The measure of central tendency used in the ad is the mean.

b. No. Only two of the seven commissions are greater than or equal to the mean. The other five commissions are not close to the mean.

c. The median and mode are useful because four commissions are equal to or near these measures.

> **Math: Who, Where, When**
>
> The Latin expression *caveat emptor* means "let the buyer beware." Remember this rule when making a decision based on survey data that has been interpreted for you.

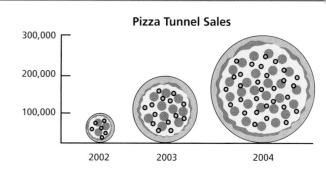

Pizza Tunnel Sales

300,000 —
200,000 —
100,000 —

2002　2003　2004

FOOD SERVICE The number of pizzas sold by the Pizza Tunnel chain for each of the three years since it opened is as follows: 2002–100,000; 2003–200,000; 2004–300,000. To show the sales, the chain's president made the graph shown.

1. Is the president justified in claiming the graph presents the data accurately? How might the graph be misleading?

2. How can the graph be drawn to avoid misleading anyone?

WEATHER The monthly rainfall, in centimeters, at Mountain Valley Resort last year is shown. A part of the resort's brochure is also shown.

J	F	M	A	M	J	J	A	S	O	N	D
0	2	7	12	18	24	25	25	22	6	3	0

3. Which measure of central tendency was used in the brochure?

4. Does the statement about rainfall in the brochure give a fair picture of weather conditions? Explain.

5. Which measure of central tendency would be a better indicator?

BEAUTIFUL SUMMER WEATHER

Average monthly rainfall less than 10 cm!

MUSIC A songwriter earned the following yearly totals in royalties over a 4-yr period.

1996—$35,000　1997—$35,800　1998—$37,250　1999—$38,000

6. Draw a graph that shows the actual trend in the writer's earnings.

7. Draw a graph so that the writer's earnings appear to be growing substantially.

MARKET RESEARCH The bar graph shows the average ratings from a taste test for ten brands of ketchup. This information is used by the makers of Brand G.

8. Approximately how many times as high is the tallest bar than the shortest bar?

9. Make a list of the ten brands of ketchup and the number of people who favored each brand. For example, 1600 people chose Brand A as their favorite.

10. Based on your numbers, how does the most favored brand compare with the least favored.

11. What caused the distortion in the graph?

12. Suppose Brands C and D want to speak out against Brand G's advertisement that included the bar graph above. Redraw the graph so that it is about the same overall size but gives a more accurate picture of the situation.

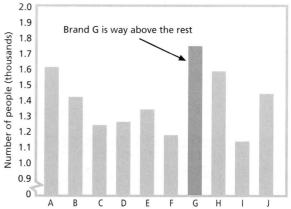

Taste Test of Ketchup Brands

Brand G is way above the rest

Number of people (thousands)

2.0　1.9　1.8　1.7　1.6　1.5　1.4　1.3　1.2　1.1　1.0　0.9　0

A　B　C　D　E　F　G　H　I　J

ENTERTAINMENT A movie theater charges $7 for a ticket. The theater manager is considering raising the ticket prices and asked 25 people to name the greatest amount they would be willing to pay to see a movie. The results are shown in the table.

Greatest Amount People Would Pay to See a Movie

Amount	Frequency
$20.00	3
15.00	4
12.00	3
10.00	5
8.00	7
7.00	1
5.00	1
2.00	1

13. In order to justify a large increase in the ticket price, which measure of central tendency might the theater manager use? Explain.

14. If the theater patrons saw this survey, which measure of central tendency might they point out to the manager to keep the price increase low? Explain.

15. Do you think the results of the manager's survey are biased?

16. How would you conduct your own survey differently to show the theater manager that the ticket price should not be increased?

 17. **WRITING MATH** Write a short paragraph explaining what to look for to be sure that you are drawing correct conclusions about data presented in a line graph.

EXTENDED PRACTICE EXERCISES

Gloria's Gloves Sales

Month	1	2	3	4	5	6
Sales (hundreds)	72	65	81	84	75	79

18. Use the data to create two graphs. In one, make it appear that sales fluctuated greatly from month to month. In the second, make it appear that sales varied little from month to month.

19. **CRITICAL THINKING** Name a reason why a marketer would use each of the graphs in Exercise 18 in different advertisements.

 20. **CHAPTER INVESTIGATION** Compare the different types of visual displays and your data. Choose the statistical graphs and the statistical measures that best represent your recommendations. Prepare the visuals and the summary of your market research.

MIXED REVIEW EXERCISES

Find the perimeter of each figure. (Basic geometry skills)

21.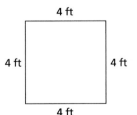
4 ft
4 ft   4 ft
4 ft

22.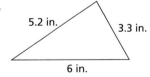
5.2 in.   3.3 in.
6 in.

23.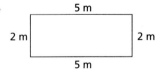
5 m
2 m   2 m
5 m

Determine whether each equation is *true* or *false*. (Basic math skills)

24. $3 \cdot 12 = 36$

25. $(-2)(6) = 12$

26. $\dfrac{-48}{12} = -4$

27. $(-5)(-9) = 45$

28. $72 \div 4 = 18$

29. $28 \div (-7) = -3$

30. $(-5)(-8) = 40$

31. $(-6)(-3) = -18$

32. $\dfrac{-81}{9} = -9$

1-8 Use Matrices to Organize Data

Goals
- ■ Organize and display data in matrices.
- ■ Peform basic operations using matrices.

Applications Business, Retail, Food service, Recreation

Divide the class into two groups. Answer Questions 1 and 2 in your group and Questions 3 and 4 as a class.

1. For all members of your group, record their hair color and the color of their shirt.

2. Decide how to organize the data so that it can be combined easily with the data collected by the other group. Do not discuss your ideas with the other group. Display your organized data.

3. Compare the sets of data, and discuss how compatible the displays are. Will it be simple to add the corresponding data together? Will it be simple to determine how many people are wearing red?

4. Discuss methods for organizing and displaying data that may make it simple to add data or to quickly identify a single fact.

◤ BUILD UNDERSTANDING

Data can be displayed in many different ways. A table and a spreadsheet are two common ways to organize and display data.

	Store 1	Store 2	Store 3
Cam Plus	77	15	42
Easy-Cam	19	89	21

Table

If you remove the labels from a table or spreadsheet, you have a rectangular array of numbers. By placing brackets around the numbers, you create a matrix. A **matrix** is a rectangular arrangement of data in rows and columns and enclosed by brackets.

	A	B	C
1	77	15	42
2	19	89	21

Spreadsheet

Each number in a matrix is an **element,** or *entry*. The sample matrix has elements 77, 15, 42, 19, 89 and 21.

The number of rows and columns in a matrix determine its dimensions. Since the sample matrix has 2 rows and 3 columns, its dimensions are 2 × 3, read 2 by 3.

$$\begin{bmatrix} 77 & 15 & 42 \\ 19 & 89 & 21 \end{bmatrix}$$

Matrix

Example 1

Give the dimensions of matrix A.

$$A = \begin{bmatrix} 3 & 1 & 16 & 4 \\ 0 & 1 & 3 & 2 \\ 21 & 3 & 6 & 0 \end{bmatrix}$$

Solution

The matrix has 3 rows and 4 columns, so its dimensions are 3 × 4.

Reading Math

The plural of matrix is *matrices*.

Capital letters are used to name matrices so that they can be referred to easily.

A matrix with the same number of rows and columns is a **square matrix**. When two matrices have equal dimensions, *corresponding elements* are the elements in the same position of each matrix. To find the sum or difference of two matrices with the same dimensions, add or subtract corresponding elements.

Example 2

Use matrices *A* and *B*.

$$A = \begin{bmatrix} 14 & 7 \\ 21 & 19 \\ 35 & 12 \end{bmatrix} \qquad B = \begin{bmatrix} 11 & 0 \\ 9 & 18 \\ 25 & 5 \end{bmatrix}$$

a. Find $A + B$.   **b.** Find $A - B$.

Math: Who, Where, When

In 1850, British mathematician James Sylvester (1814–1897) coined the term "matrix" from the Hebrew word "gematria," an ancient system that assigned numbers to the letters in Hebrew words. Sylvester and Arthur Cayley (1821–1895) developed the theory of matrices.

Solution

a. $A + B = \begin{bmatrix} 14 + 11 & 7 + 0 \\ 21 + 9 & 19 + 18 \\ 35 + 25 & 12 + 5 \end{bmatrix} = \begin{bmatrix} 25 & 7 \\ 30 & 37 \\ 60 & 17 \end{bmatrix}$

b. $A - B = \begin{bmatrix} 14 - 11 & 7 - 0 \\ 21 - 9 & 19 - 18 \\ 35 - 25 & 12 - 5 \end{bmatrix} = \begin{bmatrix} 3 & 7 \\ 12 & 1 \\ 10 & 7 \end{bmatrix}$

You can add and subtract matrices on a graphing calculator. First define the dimensions and the elements for each matrix. You can perform the basic operations at the home screen. The left and center figures below show the defining screens, and the right figure shows the home-screen operations.

   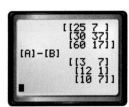

Example 3

Use a graphing calculator to find the sum and the difference of matrices *A* and *B*.

$$A = \begin{bmatrix} 5 & 5 \\ 55 & 19 \end{bmatrix} \qquad B = \begin{bmatrix} 3 & 19 \\ 13 & 2 \end{bmatrix}$$

Solution

Use the matrix feature to enter matrix A and matrix B. At the home screen calculate the sum and difference.

Example 4

online Personal Tutor at mathmatters2.com

	Brown	Black
Deck shoes	38	52
Sandals	49	70
Penny loafers	25	41

BUSINESS The spreadsheet shows the inventory levels at the Gold-n-Sole for two colors of three styles of shoes on May 1. During May, the manufacturer delivers the following quantities of brown and black shoes in each style.

deck shoes (25, 32)   sandals (30, 35)   penny loafers (18, 12)

During May, customers purchased brown and black shoes in the following quantities.

deck shoes (31, 19)   sandals (39, 43)   penny loafers (11, 17)

a. Write inventory matrix A.   **b.** Write delivery matrix B.

c. Write purchase matrix C.   **d.** Calculate $A + B - C$. What does this matrix represent?

Solution

a. $A = \begin{bmatrix} 38 & 52 \\ 49 & 70 \\ 25 & 41 \end{bmatrix}$ **b.** $B = \begin{bmatrix} 25 & 32 \\ 30 & 35 \\ 18 & 12 \end{bmatrix}$ **c.** $C = \begin{bmatrix} 31 & 19 \\ 39 & 43 \\ 11 & 17 \end{bmatrix}$

d. $A + B - C = \begin{bmatrix} 38 + 25 - 31 & 52 + 32 - 19 \\ 49 + 30 - 39 & 70 + 35 - 43 \\ 25 + 18 - 11 & 41 + 12 - 17 \end{bmatrix} = \begin{bmatrix} 32 & 65 \\ 40 & 62 \\ 32 & 36 \end{bmatrix}$

This matrix represents the inventory levels for each color shoe in each style at the end of May.

TRY THESE EXERCISES

Write each set of data as a matrix. Name its dimensions.

1.

	A	B	C
1		Adult	Child
2	Friday	106	255
3	Saturday	348	491
4	Sunday	196	304

2.

Age	A	B	C
Under 18	15	72	44
18-40	91	60	29
Over 40	17	19	88

3.

Distance	Number of parks	Distance	Number of parks
390	1	404	3
400	5	410	3
402	1	440	1

4. Can any of the matrices from Exercises 1–3 be added together? If so, find their sum. If not, explain why.

5. **WRITING MATH** Make a list of all the new terms presented in this lesson, and associate them to the matrix and its elements in Exercise 3.

PRACTICE EXERCISES   •   For Extra Practice, see page 588.

Use matrices E, F, G and H for Exercises 6–11.

6. Give the elements of E.

7. Name the dimensions of H.

8. Find $E + G$.   **9.** Find $G + F$.

10. Find $F - G$.   **11.** Find $F + G - E$.

$E = \begin{bmatrix} 12 & 9 \\ 8 & 6 \\ 20 & 13 \end{bmatrix}$ $F = \begin{bmatrix} 9 & 5 \\ 8 & 0 \\ 16 & 7 \end{bmatrix}$

$G = \begin{bmatrix} 26 & 14 \\ 9 & 12 \\ 5 & 25 \end{bmatrix}$ $H = \begin{bmatrix} -1 & 6 \\ 2 & -7 \end{bmatrix}$

RECREATION The 1999 membership in the CC Astronomy Club is listed by grade level and gender.

At beginning of year: seniors (117 males, 88 females)   juniors (91 males, 95 females)

Joined during the year: seniors (38 males, 35 females)   juniors (25 males, 40 females)

Stopped attending: seniors (13 males, 9 females)   juniors (6 males, 10 females)

12. Write initial membership matrix A.   **13.** Write new member matrix B.

14. Write stopped attending matrix C.   **15.** Calculate $A + B - C$.

16. What does the matrix in Exercise 15 represent? Explain.

BUSINESS Matrix W represents the inventories of three models of handheld vacuums in two warehouses at the beginning of April. Matrix V represents the number of vacuums received from the manufacturer during the month. Matrix Z represents vacuums shipped out during the same month.

$$W = \begin{bmatrix} 317 & 490 & 166 \\ 555 & 207 & 181 \end{bmatrix} \quad V = \begin{bmatrix} 52 & 70 & 48 \\ 88 & 86 & 66 \end{bmatrix} \quad Z = \begin{bmatrix} 114 & 98 & 50 \\ 61 & 90 & 77 \end{bmatrix}$$

17. At the beginning of April, how many vacuums, all models combined, were in stock in both warehouses?

18. Write a matrix that represents inventory of the three models at month's end.

RETAIL The matrix summarizes the sticker price of four car models at three automobile dealerships. Each dealer will add 8% in taxes to the sticker price.

19. Write a matrix giving the tax on each car.

20. Add the matrices to show the total prices.

$$\begin{bmatrix} 12{,}400 & 12{,}600 & 12{,}000 \\ 11{,}100 & 11{,}400 & 11{,}500 \\ 14{,}800 & 14{,}100 & 14{,}400 \\ 10{,}200 & 10{,}900 & 10{,}300 \end{bmatrix}$$

EXTENDED PRACTICE EXERCISES

CRITICAL THINKING Find the matrix M that makes each equation true.

21. $\begin{bmatrix} 19 & 44 & 30 \\ 62 & 91 & 55 \end{bmatrix} - M = \begin{bmatrix} 8 & 29 & 6 \\ 43 & 60 & 27 \end{bmatrix}$ **22.** $M + \begin{bmatrix} 29 & 63 \\ 77 & 49 \end{bmatrix} = \begin{bmatrix} 81 & 83 \\ 115 & 62 \end{bmatrix}$

23. WRITING MATH Is the addition of matrices commutative? (Hint: Does $A + B = B + A$ if A and B have the same dimensions?) Justify your answer.

MIXED REVIEW EXERCISES

Find the area of each figure. (Basic geometry skills)

24.

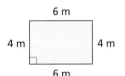

25.

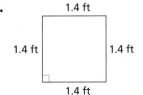

26.

DATA FILE **Refer to the data on top concert tours on page 563.** (Lesson 1-3)

27. Make a histogram of the data in intervals of $10 million.

28. Which concert is closest to the mean?

Chapter 1 Review

VOCABULARY ◤

Choose the word from the list that best completes each statement.

1. A number that is much greater or much less than the other values in a set of data is known as a(n) __?__ .

2. __?__ divide a set of data arranged in numerical order into four equal parts.

3. __?__ are isolated groups of values in a set of data.

4. The mean, median and mode are __?__ that represent a middle value of a data set.

5. A(n) __?__ records the number of times a response occurs.

6. A(n) __?__ uses a number line to show the distribution of data.

7. A rectangular arrangement of data in rows and columns enclosed in brackets is called a(n) __?__ .

8. A(n) __?__ is a representative part of a population.

9. The relationship of two sets of data can be shown using a(n) __?__ .

10. Survey results that do not truly represent the population are __?__ .

a.	biased
b.	box-and-whisker plot
c.	clusters
d.	frequency table
e.	interquartile range
f.	matrix
g.	measures of central tendency
h.	outlier
i.	percentile
j.	quartiles
k.	sample
l.	scatter plot

LESSON 1-1 ◤ Surveys and Sampling Methods, p. 6

▶ **Random, cluster, convenience,** and **systematic** sampling are methods of collecting data that can be used to make decisions about a population.

To identify the most popular current movie among teenagers, seven students seated at the same table in the school cafeteria were asked to name the most recent movie they had seen.

11. What kind of sampling method is this?

12. How might the results be biased?

LESSON 1-2 ◤ Measures of Central Tendency and Range, p. 10

▶ The **mean** is the sum of the values in a data set divided by the number of data. The **median** is the middle value of the data when the data are arranged in numerical order. The **mode** is the number that occurs most often in a set of data. There may be no mode or more than one mode.

▶ The **range** is the difference between the greatest and least values in a set of data.

Price of Shoes Sold at Shoe Mart

$49	$34	$37	$28	$39	$44	$34	$49
$52	$34	$37	$37	$49	$39	$34	$39

13. Find the mean, median, mode, and range of the data.

14. Would an ad stating "Most styles priced at $34" be misleading? Explain.

LESSON 1-3 ◣ Histograms and Stem-and-Leaf Plots, p. 16

▶ The frequency of data can be displayed in a **histogram**, a type of bar graph.

▶ Individual data items can be displayed in **stem-and-leaf plots**.

Daily High Temperature (°F)

80	91	68	92	86	69	71	75	90	86
70	91	83	81	79	78	99	76	80	86
90	71	79	86	90	76	84	78	88	81

15. Use a frequency table to make a histogram of the data using intervals of 5 starting with 65.

16. Make a stem-and-leaf plot of the data.

17. Locate the greatest gap in the data.

18. How many days had a high temperature below 80°?

LESSON 1-4 ◣ Scatter Plots and Lines of Best Fit, p. 20

▶ A **scatter plot** shows data grouped as ordered pairs and graphed as points on a grid. The points are not connected.

▶ A **line of best fit**, or **trend line**, on a scatter plot indicates either a **positive**, **negative**, or **no correlation** between items of data.

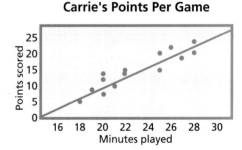
Carrie's Points Per Game

19. In the scatter plot, what is the correlation between points scored and minutes played?

20. Predict how many points Carrie would score if she played 24 min in one game.

21. Predict how many points Carrie would score if she played 30 minutes.

LESSON 1-5 ◣ Problem Solving Skills: Coefficient of Correlation, p. 26

▶ The coefficient of correlation is a statistical measure of how closely data fits a line.

Predict the characteristics of the correlation between the two variables. Find the coefficient of correlation to check your prediction.

22.

x	1.5	2.6	3	4	4.3	5	1	1	2	2	3.4	5.7	6	6
y	0.5	1.5	2.5	3	4	4	1	2	2	3.5	4	6	6	5.8

23.

x	4	7	8	5	3	1	6	9	2	10
y	5	3	1	7	5	6	2	3	7	1

State whether each of the following sets of variables would show a positive, negative, or zero correlation.

24. your height and month of your birth

25. temperature and the cost of heating bill

LESSON 1-6 ◣ Quartiles and Percentiles, p. 28

▶ **Box-and-whisker-plots** are used to display the distribution of data.

▶ A **percentile** represents the percent of scores at or below a particular score by dividing the number of scores less than or equal to a given score by the total number of scores.

26. Make a box-and-whisker plot of the data for Exercises 13–14.

27. What is the interquartile range of the data for Exercises 13–14?

28. Out of 30 students who took a test, Darius had the fifth highest score. What is his percentile rank?

LESSON 1-7 ◣ Misleading Graphs and Statistics, p. 34

▶ **Misleading data** is data that leads to a false perception.

Average Monthly Temp. (°F) at Sand City Resort

J	F	M	A	M	J	J	A	S	O	N	D
45°	50°	70°	82°	88°	97°	98°	95°	80°	50°	50°	40°

29. Which measure(s) of central tendency would make this beach resort seem a little too cold for swimming?

30. Which measure(s) of central tendency would the beach resort use to encourage people to come to the resort to swim?

LESSON 1-8 ◣ Use Matrices to Organize Data, p. 38

▶ A **matrix** is a rectangular arrangement of data in rows and columns enclosed by brackets.

$$A = \begin{bmatrix} 3 & 10 \\ 5 & 6 \\ 11 & 4 \\ 8 & 1 \\ 0 & 5 \end{bmatrix} \quad B = \begin{bmatrix} 7 & 13 \\ 6 & 6 \\ 9 & 2 \\ 0 & 3 \\ 7 & 2 \end{bmatrix}$$

31. Give the elements of A.

32. Name the dimensions of A.

33. Find $A + B$.

34. Find $A - B$.

CHAPTER INVESTIGATION

EXTENSION Present to the class your recommendations to the manager. Be sure to include advertisement suggestions, what to offer, and claims that the cafeteria can use to support your findings. After all groups have presented, as a class, discuss the differences and similarities in the findings and recommendations among the groups. What did each group do to get different results? How could the statistical graphs and measures have been used to mislead the students? How would the presentation differ if the class unites the data from each group and makes visual displays and recommendations?

Chapter 1 Assessment

To find out the most popular breed of dog, every tenth person who entered a pet shop one Saturday was asked to name one favorite breed.

1. What kind of sampling method is being used?

2. How might the results of this survey be biased?

The histogram shows the number of hours worked each week by part-time employees.

3. What is the most common number of hours worked?

4. How many employees work less than 20 h?

5. How many part-time employees are there?

A hotel charges the following daily rates for its different types of rooms.

$75 $120 $65 $84 $150 $79

6. Find the mean, median, mode, and range of these data.

7. The hotel used the median to advertise its average room rate. Is this misleading? Explain.

Hours Worked by Employees

(histogram: Frequency vs Hours — bars: 5-9 = 5, 10-14 = 6, 15-19 = 3, 20-24 = 8, 25-29 = 6, 30-34 = 10, 35-39 = 9)

The number of games won by a certain baseball team during the past eleven seasons is shown in the stem-and-leaf plot.

8. During how many seasons did the team win 50 or fewer games?

9. What is the range of the number of games won?

10. During how many seasons did the team win more than 60 games?

```
4 | 7  8  9  9
5 | 0  1  8  9
6 | 1  2
8 | 3
```

4|8 represents 48 games.

Use the scatter plot for Exercises 11–12.

11. Is there a positive or negative correlation between the year and distance?

12. Predict the winning distance for the 1996 long-jump.

13. On a test, Susan had the fourth highest score. If there were 30 students who took the test, what is Susan's percentile rank?

14. Predict the characteristics of the correlation between the two variables. Find the coefficient of correlation to check your prediction.

Yearly Long-Jump Winning Distances

(scatter plot: Feet vs Year, '85 to '96)

x	1	6	3	1	2	4	3	2	5	6
y	8	1	4	9	6	3	5	7	2	0

Use matrices A and B for Exercises 15–18.

15. Give the elements of A.

16. Give the dimensions of B.

17. Find A + B.

18. Find A − B.

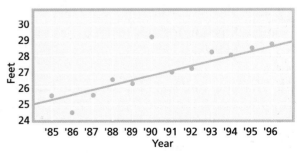

$$A = \begin{bmatrix} 10 & 3 \\ 8 & 9 \\ 3 & 7 \end{bmatrix} \quad B = \begin{bmatrix} 4 & -3 \\ -1 & 0 \\ 7 & 3 \end{bmatrix}$$

Standardized Test Practice

Part 1   Multiple Choice

Record your answers on the answer sheet provided by your teacher or on a sheet of paper.

1. You will be collecting information on how athletes at your school view good sportsmanship. Each of the coaches have sent you a roster. You plan to write the names on pieces of paper and select 20 names out of a hat to choose the students you will interview. What type of sampling is this method?   (Lesson 1-1)

   (A) cluster sampling

   (B) convenience sampling

   (C) random sampling

   (D) systematic sampling

2. What measure of central tendency is determined by finding the middle value when the data are arranged in numerical order? (Lesson 1-2)

   (A) mean   (B) median

   (C) mode   (D) range

3. You have test scores of 86, 87, and 90. You would like a mean score of 90 for the end of the semester. If all your tests are weighted the same and the highest score is 100, what is the lowest score you must get on the fourth and final test? (Lesson 1-2)

   (A) 95   (B) 97   (C) 99

   (D) It is not possible to get a mean score of 90.

4. The stem and leaf plot shows the number of minutes that music is played in one hour on nine radio stations. Find the range of the data. (Lesson 1-3)

3	2
4	1 1 2 2 4 9 9
5	0

   3|2 represents 32 min.

   (A) 9   (B) 18   (C) 42   (D) 50

Use the scatter plot for Questions 5–7.

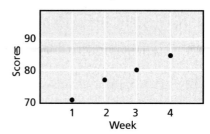

John's Quiz Scores

5. What kind of correlation does the data in the scatter plot show? (Lesson 1-4)

   (A) positive

   (B) negative

   (C) no correlation

   (D) cannot be determined

6. On a quiz, John's score was fifth from the highest in a class of 30 students. Find the percentile rank that John achieved. (Lesson 1-6)

   (A) 20th   (B) 83rd

   (C) 87th   (D) 95th

7. Which part of the graph is misleading? (Lesson 1-7)

   (A) x-axis

   (B) y-axis

   (C) title

   (D) The graph is not misleading.

8. Find the second quartile in the data. (Lesson 1-6)

   14.1, 16.4, 12.3, 10.9, 12.0, 9.2, 15.5

   (A) 10.9   (B) 12.0   (C) 12.2   (D) 12.3

Test-Taking Tip

Question 6
If you don't know how to solve a problem, eliminate the answer choices you know are incorrect and then guess from the remaining choices. Even eliminating one answer choice greatly increases your chance of guessing the correct answer.

Part 2 Short Response/Grid In

Record your answers on the answer sheet provided by your teacher or on a sheet of paper.

The graph shows the number of heads that resulted from 10 tosses of a coin for each student in a group of 30 students. Use the graph for Questions 9-14. (Lesson 1-2)

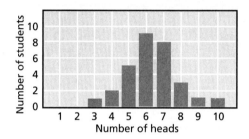

9. How many students got 4 heads in 10 coin tosses?

10. How many students got 2 tails in 10 coin tosses?

11. What is the mean of the data?

12. What is the mode of the data?

13. What is the median of the data?

14. What is the range of the data?

15. The table shows the average cost of a loaf of bread from 1960–2000. What type of relationship is shown from the data? (Lesson 1-4)

Year	1960	1970	1980	1990	2000
Cents	20	24	52	72	99

South Central High School's band and flag corps scored the following number of points during its performances this past season. Use the data for Questions 16–18. (Lesson 1-6)

27, 32, 6, 24, 29, 30, 8, 26, 30, 32

16. Find the first, second, and third quartiles of the data.

17. What is the interquartile range?

18. Name any outliers?

Use matrices A and B for Questions 19 and 20. (Lesson 1-8)

$$A = \begin{bmatrix} 5 & -3 \\ 8 & 2 \\ -1 & 5 \end{bmatrix} \quad B = \begin{bmatrix} 4 & 7 \\ 0 & -1 \\ 8 & -5 \end{bmatrix}$$

19. Find $A + B$.

20. Find $A - B$.

21. Name the dimensions of matrix K. (Lesson 1-8)

$$K = \begin{bmatrix} 2 & 7 & -10 \\ -9 & 3 & 6 \end{bmatrix}$$

Part 3 Extended Response

Record your answers on a sheet of paper. Show your work.

The table shows the personal income of individuals from 1975 to 2000. Use the data for Questions 22 and 23.

Year	Income
1975	$6,166
1980	$10,205
1985	$14,738
1990	$19,614
1995	$23,571
2000	$30,069

22. Make a scatter plot of the data. Sketch a trend line on your scatter plot. (Lesson 1-4)

23. Does the data show a positive or negative correlation? Find the coefficient of correlation. (Lesson 1-5)

24. Describe a relationship that would show a negative correlation. (Lesson 1-5)

25. A music group has designed a new marketing campaign. Attendances at their recent concerts are 125, 55, 65, 98, and 112. They also opened for a big name band at the state fair where the attendance was 55,000. In their new brochure, the band claims that "Average attendance at concerts is over 9,000 people. Evaluate this claim. (Lesson 1-7)

Foundations of Algebra

THEME: Population

In all aspects of life, algebra is the foundation for expressing known and unknown quantities. You have been using algebraic concepts since you first started learning mathematics. These concepts continue to be used throughout life in both personal and professional situations.

Expressions, variables, and exponents are a few of the many algebra topics that are used when studying population data. Although the term population refers to the organisms that live in a specific area, population is usually connected with the number of people that live in the world, a country, a state, or a city.

- The attendance at a sporting event is a population. **Concession stand operators** (page 61) use the buying patterns of the population that regularly attend events to make decisions about inventory and product availability.

- **Commercial aircraft designers** (page 81) use data about the average weight of passengers and cargo to establish size and weight restrictions of airplanes.

Math Online

mathmatters2.com/chapter_theme

Area and Populations of Ten Countries

Country	Capital	Area (square kilometers)	Population
Bangladesh	Dhaka	144,000	138,448,210
Brazil	Brasilia	8,511,965	182,032,604
China, People's Republic of	Beijing	9,596,960	1,286,975,468
India	New Delhi	3,287,590	1,049,700,118
Indonesia	Jakarta	1,919,440	234,893,453
Japan	Tokyo	377,835	127,214,499
Nigeria	Abuja	923,768	133,881,703
Pakistan	Islamabad	803,940	150,694,740
Russia	Moscow	17,075,200	144,526,278
United States	Washington, D.C.	9,629,091	290,342,554

Data Activity: Area and Population of Ten Countries

Use the table for Questions 1–5.

1. How many countries in the table have a population less than Russia?

2. In the table, what is the capital of the country with the largest population?

3. In the table, what is the capital of the country with the smallest population?

4. What is the approximate area per person in the United States?

5. Russia's population would have to increase by how many people to become ranked fifth most populous in the table?

CHAPTER INVESTIGATION

Population density is the number of people per square mile or square kilometer that live in a given location, such as a town, county, city, state, or country. For example, the population density of Indiana is 155. This means that for every square mile of Indiana, approximately 155 people live there.

Working Together

The population density of a state is not necessarily the same as the population density of a city within that state. Choose a state and a city within that state to compare population densities. Use the Chapter Investigation icons to guide your group.

Are You Ready?

Refresh Your Math Skills for Chapter 2

The skills on these two pages are ones you have already learned. Use the examples to refresh your memory and complete the exercises. For additional practice on these and more prerequisite skills, see pages 576–584.

ADDING AND SUBTRACTING FRACTIONS

You will perform many computations in this chapter. Remember when adding and subtracting fractions, you must find a common denominator.

Examples

$$3\frac{1}{4} = 3\frac{3}{12}$$
$$+ 1\frac{5}{6} = 1\frac{10}{12}$$
$$\overline{\qquad 4\frac{13}{12} = 5\frac{1}{12}}$$

$$16\frac{2}{5} = 16\frac{8}{20} = 15\frac{28}{20}$$
$$- 7\frac{3}{4} = 7\frac{15}{20} = 7\frac{15}{20}$$
$$\overline{\qquad\qquad\qquad\quad 8\frac{13}{20}}$$

Add or subtract.

1. $\frac{1}{3} + \frac{3}{8}$

2. $\frac{6}{7} + \frac{2}{9}$

3. $2\frac{4}{5} + 6\frac{8}{9}$

4. $3\frac{7}{8} + 1\frac{3}{4}$

5. $\frac{5}{8} - \frac{1}{3}$

6. $\frac{7}{9} - \frac{1}{2}$

7. $6\frac{2}{3} - 5\frac{8}{9}$

8. $3\frac{1}{2} - \frac{6}{7}$

9. $8\frac{4}{5} + 3\frac{1}{2}$

MULTIPLYING AND DIVIDING FRACTIONS

Common denominators are not needed when multiplying and dividing fractions. When multiplying fractions remember to find the product of all the denominators and the product of all the numerators.

Division problems are changed to multiplication problems by finding the reciprocal of the divisor.

Examples

$$2\frac{3}{8} \cdot \frac{5}{6} = \frac{19}{8} \cdot \frac{5}{6}$$
$$= \frac{95}{48}$$
$$= 1\frac{47}{48}$$

$$6\frac{2}{3} \div 1\frac{1}{2} = \frac{20}{3} \div \frac{3}{2}$$
$$= \frac{20}{3} \cdot \frac{2}{3}$$
$$= \frac{40}{9} = 4\frac{4}{9}$$

Multiply or divide.

10. $2\frac{1}{8} \cdot \frac{1}{4}$

11. $1\frac{6}{7} \cdot 1\frac{5}{6}$

12. $5\frac{2}{3} \cdot 4\frac{3}{5}$

13. $1\frac{1}{3} \div \frac{5}{6}$

14. $2\frac{3}{4} \div 1\frac{7}{8}$

15. $8\frac{3}{4} \div 4\frac{4}{9}$

16. $6\frac{2}{3} \cdot 1\frac{5}{8} \div 3\frac{1}{2}$

17. $4\frac{3}{4} \div 8\frac{2}{3} \cdot 7\frac{1}{5}$

18. $4\frac{2}{3} + \left(6\frac{1}{8} \cdot 1\frac{1}{2}\right)$

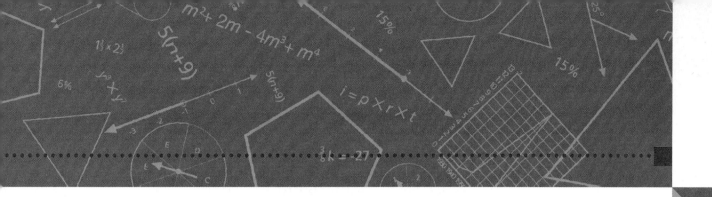

OPERATIONS WITH DECIMALS

You will also be working with decimals in this chapter. Refresh your memory of how to work with decimals.

Examples

$$\begin{array}{r} 5.6 \\ + \ 0.034 \\ \hline 5.634 \end{array} \qquad \begin{array}{r} 10.95 \\ - \ 4.063 \\ \hline 6.887 \end{array} \qquad \begin{array}{r} 6.4 \\ \times \ 0.07 \\ \hline 0.448 \end{array} \qquad \begin{array}{r} 2.6 \\ 4.6\overline{)11.96} \\ 92 \\ \hline 276 \\ 276 \\ \hline 0 \end{array}$$

Simplify.

19. $6.5 + 11.84$

20. $4 - 3.23$

21. $1.25 \cdot 0.9$

22. $0.088 \div 1.1$

23. $3.27 + 18.9$

24. $14.75 + 8.125$

25. $0.076 \cdot 1.5$

26. $0.1035 \div 0.23$

ABSOLUTE VALUE

The *absolute value* of a number is the distance the number is from zero on a number line.

Find each absolute value.

27. $|-3|$

28. $|7|$

29. $|0|$

30. $\left|-\dfrac{1}{3}\right|$

31. $|0.7|$

32. $|-4|$

33. $|1|$

34. $|-47|$

OPERATIONS WITH INTEGERS

The set $\{\ldots, -3, -2, -1, 0, 1, 2, 3, \ldots\}$ is the set of *integers*. Addition, subtraction, multiplication and division can be applied to integers.

To add numbers with the same signs, add their absolute values. The sign of the sum is the same sign as the numbers. To add numbers with different signs, subtract their absolute values. The sign of the sum is the sign of the number with the greater absolute value. To subtract a number, add its opposite.

The product or quotient of two numbers with the same sign is positive. The product or quotient of two numbers with different signs is negative.

Examples

$$4 + (-3) = 1 \qquad -3 - 8 = -3 + (-8) = -11 \qquad -4 \cdot 6 = -24 \qquad \dfrac{-18}{-6} = 3$$

Perform the indicated operation.

35. $5 + 7$

36. $17 - 20$

37. $8 - (-3)$

38. $6 + (-2)$

39. $-7 \cdot 5$

40. $\dfrac{-15}{3}$

41. $-11.7 - 3.9$

42. $\dfrac{-120}{-4}$

43. $12 - (-8) + 6$

44. $-3 \cdot (-3) \cdot 9$

45. $15 - 9 + (-27)$

46. $4 \cdot (-8) \cdot 12$

2-1 Real Numbers

Goals ■ Graph sets of numbers on a number line.
■ Evaluate expressions with absolute value.

Applications Weather, Sports, Population

Work in a group of two or three students.

1. Copy the number line. Place a point on the birth year of each family member. Label each point with the person's name.

2. If Joe's birth year were equal to 0, what numbers would correspond with the following events?

   a. mother's birth　**b.** father's birth　**c.** brother's birth　**d.** sister's birth

3. Redraw the time line using positive and negative numbers.

4. Construct two time lines using you and your family's birth years. One should use the actual years and another should use positive and negative numbers.

Joe's Family

Family member	Birth year
Father	1957
Mother	1959
Joe	1984
Sister	1988
Brother	1992

◤ BUILD UNDERSTANDING

The set of **integers** consists of the whole numbers and their opposites. **Opposites** of whole numbers are the same distance from zero but in the opposite direction.

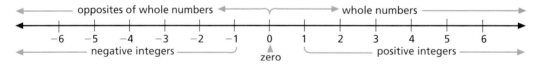

Positive integers are greater than zero. Negative integers are less than zero. Zero is neither positive nor negative. Zero is its own opposite.

A **rational number** is a number that can be expressed as a ratio of two integers a and b, where b is not equal to zero. This is usually written $\frac{a}{b}$, $b \neq 0$. The symbol $\neq$ means "is not equal to."

All rational numbers can be expressed as **terminating decimals** or **repeating decimals**. Some numbers, such as π (pi) and $\sqrt{2}$, are non-terminating and non-repeating decimals. Such numbers are called **irrational numbers**. Together the set of rational and irrational numbers make up the set of **real numbers**.

The number that corresponds to a point on a number line is called the **coordinate of the point**. Each real number corresponds to exactly one point on a number line. The point that corresponds to a number is called the **graph of the number** and is indicated on the number line by a solid dot. Each point on a number line corresponds to exactly one real number.

COncepts in MOtion
Animation
mathmatters2.com

Real Numbers

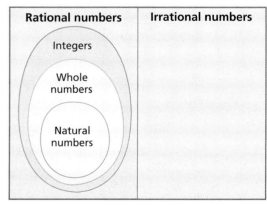

Example 1

Graph the set of numbers $\{0.25, -1, -2\frac{3}{4}, -4\}$ on a number line.

Solution

Draw a number line. Use a solid dot to graph each number.

Reading Math

A **set** is a collection of objects, such as a set of integers or the set of whole numbers. Objects in the set are called **members**, or **elements**, of the set.

A set is indicated by braces { } enclosing the names of set members.

Example 2

 Personal Tutor at mathmatters2.com

Use a number line to compare numbers. Replace each ■ with <, > or =.

a. -3 ■ 1 **b.** 2 ■ 0 **c.** $\frac{1}{4}$ ■ $-\frac{3}{4}$

Solution

Draw a number line and graph each number.

a. -3 is to the left of 1, so $-3 < 1$.

b. 2 is to the right of 0, so $2 > 0$.

c. $\frac{1}{4}$ is to the right of $-\frac{3}{4}$, so $\frac{1}{4} > -\frac{3}{4}$.

Example 3

Graph each set of numbers on a number line.

a. the integers from -3 to 2   **b.** the real numbers from -3 to 2

c. all real numbers less than or equal to 2   **d.** all real numbers greater than -1

Solution

a. The set consists of $-3, -2, -1, 0, 1,$ and 2. To graph the set, put a solid dot at each of these points on the number line.

b. The set consists of -3 and 2 and all real numbers between. Graph the set by drawing solid dots at -3 and 2 and connecting the two points.

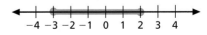

c. The set consists of 2 and all real numbers less than 2. Graph the set by drawing an arrow beginning at 2 and pointing to the left. To indicate that 2 is part of the set, draw a solid dot at 2.

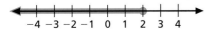

d. The set consists of all real numbers greater than -1. Graph the set by drawing an arrow beginning at -1 and pointing to the right. To indicate that -1 is not part of the set, draw an open circle at -1.

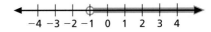

A **variable** is a symbol used to represent a number. Letters such as a and n can be used as variables.

The distance a number is from zero on the number line is the **absolute value** of the number. The absolute value of an integer a is written as $|a|$.

The **opposite of the opposite** property states if n is a real number then $-(-n) = n$.

Example 4

Evaluate each expression.

a. $-t$, when $t = 2.7$

b. $-(-d)$, when $d = -\frac{1}{3}$

c. $|k|$, when $k = -6.5$

d. $-|-n|$, when $n = 12$

Solution

a. Since $t = 2.7$, $-t = -2.7$.

b. Since $d = -\frac{1}{3}$, $-(-d) = -\left[-\left(-\frac{1}{3}\right)\right] = -\frac{1}{3}$.

c. Since $k = -6.5$, $|k| = |-6.5| = 6.5$.

d. Since $n = 12$, $-|-n| = -|-12| = -12$.

◤ TRY THESE EXERCISES

Graph the given sets of numbers on a number line.

1. $\{2.5, 0.5, -1, -2\}$

2. $\left\{2\frac{2}{3}, 1\frac{1}{3}, -1\frac{2}{3}, -2\frac{1}{3}\right\}$

3. $\{-3.75, -2.5, 0, 2.25, 5.75\}$

Use a number line to compare numbers. Replace each ▦ with $<$, $>$, or $=$.

4. 2 ▦ -2

5. 0 ▦ 3

6. $-2\frac{2}{3}$ ▦ $1\frac{1}{3}$

Graph each set of numbers on a number line.

7. whole numbers from 1 to 5

8. the integers from -5 to 1

9. all real numbers less than 4

Evaluate each expression.

10. $|n|$, when $n = 79$

11. $|x|$, when $x = 0$

12. $|w|$, when $w = -212$

13. $-r$, when $r = 3$

14. $-|-k|$, when $k = -2$

15. $-(-y)$, when $y = \frac{3}{4}$

◤ PRACTICE EXERCISES • For Extra Practice, see page 588.

Graph each set of numbers on a number line.

16. $\{-2, -0.75, 0, 1, 3\}$

17. $\left\{-5, -\frac{1}{2}, 0.25, 3\frac{1}{3}, 4.5\right\}$

18. the integers from -7 to -2

19. all real numbers greater than 6

20. all real numbers less than or equal to $-6\frac{1}{2}$

21. all real numbers between -3 and 3

22. WRITING MATH Explain what your graph looks like for Exercise 21. Did you use open or closed circles?

23. WEATHER In January the average temperature in Barrow, Alaska, is $-13°F$. In Fairbanks, Alaska, the average temperature in January is $-10°F$. Which place is warmer in January? Explain.

Replace each ■ with <, >, or =.

24. -3.5 ■ -2.8

25. $|-9|$ ■ 9

26. $5\frac{1}{3}$ ■ $5\frac{2}{5}$

27. -4 ■ $|-4|$

28. $|16|$ ■ $-|-20|$

29. $-|3|$ ■ 3

Evaluate each expression when $a = -4$, $b = 2\frac{1}{2}$, and $c = -8$.

30. $-a$

31. $-|-b|$

32. $-(-c)$

33. $-|a|$

34. $-\left(\dfrac{c}{a}\right)$

35. $-bc$

36. $|-ab|$

37. $-|abc|$

38. POPULATION Between 1990 to 2002, the population changed -5.9% in Philadelphia, 4.4% in Austin, and -4.7 in New Orleans. Which city had the greatest population decrease over the 12-yr period?

SPORTS Golf scores are based on a number called par. The object is to score as far under par as possible. So a score of -2 is two strokes under par. The table shows four players' scores for a round of golf.

Player	Final score
Marcus	-3
Darnesha	$+4$
Joelle	-1
Garrett	$+7$

39. Graph the set of scores on a number line.

40. Which golfer had the best score?

Draw a number line for each situation.

41. Jamar's test scores were 76, 79, 80, 90 and 82.

42. Raja's GPA (grade point average) ranges from 3.25 to 3.6.

43. The temperature in Wisconsin ranges from $-15°F$ to $105°F$.

■ EXTENDED PRACTICE EXERCISES

44. CRITICAL THINKING Explain what is meant by the statement "Every real number can be matched with a point on a number line."

Write *true* or *false* for each statement. For any false statement, give an example that proves the statement is false.

45. Between any two integers there is another integer.

46. Between any two rational numbers there is another rational number.

47. All negative numbers are rational numbers.

48. For any real number n where $n < 0$, $|n| = n$.

■ MIXED REVIEW EXERCISES

RECYCLING A condominium association is considering making recycling bins available to the home owners. They want to know if the residents will recycle. They conducted a survey by asking all the owners who attended a home owner's meeting one evening. (Lesson 1-1)

49. What method of sampling did they use?

50. Could the results of this survey be biased? How?

Order of Operations

Goals ■ Evaluate numerical expressions using order of operations.

Applications Part-time job, Fitness, Entertainment, Population

Determine where the symbols +, −, ×, and ( ) should be placed to make each sentence true.

1. 8 ■ 3 ■ 5 = 55

2. 5 ■ 7 ■ 6 ■ 5 = 65

3. 6 ■ 9 ■ 7 = 96

4. Did you use parentheses in any of the above questions? If so, were they necessary to make the sentence true? Explain.

◤ BUILD UNDERSTANDING

A **numerical expression** is two or more numbers joined by operations such as addition, subtraction, multiplication, and division. Parentheses also can be used in a numerical expression.

$$2 + 7 - 4 \qquad 2 \div 7 + 4 \qquad 5 + 2(7 - 4)$$

The number represented by the numerical expression is called its **value**. When you find the value of the numerical expression, you **simplify** the expression.

Example 1 nline **Personal Tutor at** mathmatters2.com

Simplify each numerical expression.

a. $9.64 - 3.2$ **b.** $\frac{1}{2}(12) + 7$ **c.** $(60 \div 3) \cdot 4$

Solution

a. $9.64 - 3.2 = 6.44$ **b.** $\frac{1}{2}(12) + 7 = 6 + 7$
$= 13$ **c.** $(60 \div 3) \cdot 4 = 20 \cdot 4$
$= 80$

Without parentheses or rules, it is possible to get two different answers for the same expression. For this reason, it is important to follow the **order of operations**.

Order of Operations	1. First, perform all calculations within parentheses and brackets.
	2. Then perform all calculations involving exponents.
	3. Next, multiply or divide in order from left to right.
	4. Finally, add or subtract in order from left to right.

An **exponent** tells how many times a number is used as a factor. The small number written to the upper right of the factor is the exponent.

Example 2

Simplify each numerical expression.

a. $(4 + 3) \cdot 5^2$

b. $55 - 7 - 12 \div 4$

c. $2^3 \cdot (6 - 3)$

Solution

a. $(4 + 3) \cdot 5^2 = 7 \cdot 5^2$
$= 7 \cdot 25$
$= 175$

b. $55 - 7 - 12 \div 4 = 55 - 7 - 3$
$= 48 - 3$
$= 45$

c. $2^3 \cdot (6 - 3) = 2^3 \cdot 3$
$= 8 \cdot 3$
$= 24$

An expression containing one or more variables is called a **variable expression**.

Variable expressions that involve multiplication can be written with or without the $\times$ sign, with the symbol $\cdot$ or with parentheses. Each of the variable expressions below represents the product of 6 and the variable m.

$6 \times m$ $6 \cdot m$ $6(m)$ $6m$

Similarly, the following expressions all represent division.

$n \div 6$ $6\overline{)n}$ $\dfrac{n}{6}$ $n/6$

To **evaluate** a variable expression, substitute a given number for each variable. Then simplify the numerical expression.

> **Reading Math**
>
> Read the expression $9y^2 \div 15$ as "nine y squared divided by 15."
>
> Read the expression $\frac{1}{2}(x - 5)$ as "one-half times the quantity x minus 5."

Example 3

Evaluate each variable expression when $n = 1.8$.

a. $7 + n^2$

b. $\dfrac{1}{3}n + 2.7$

c. $(4.6 - n) \div 4$

d. $n \cdot 3^2 + 45 \div 9$

Solution

In each expression, substitute 1.8 for n.

a. $7 + n^2 = 7 + (1.8)^2$
$= 7 + 3.24$
$= 10.24$

b. $\dfrac{1}{3}n + 2.7 = \dfrac{1}{3}(1.8) + 2.7$
$= 0.6 + 2.7$
$= 3.3$

c. $(4.6 - n) \div 4 = (4.6 - 1.8) \div 4$
$= 2.8 \div 4$
$= 0.7$

d. $n \cdot 3^2 + 45 \div 9 = 1.8 \cdot 3^2 + 45 \div 9$
$= 1.8 \cdot 9 + 45 \div 9$
$= 16.2 + 5$
$= 21.2$

Algeblocks are models that can be used to represent variable expressions.

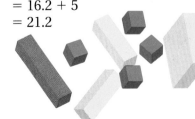

Example 4

MODELING Use Algeblocks to represent each variable expression.

a. $x + 4$ **b.** $3y$ **c.** $2x^2 - 3$ **d.** $-4x + y^2$

Solution

a. 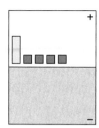   **b.** 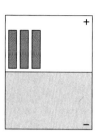   **c.** 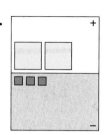   **d.**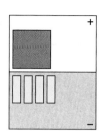

◤ TRY THESE EXERCISES

Simplify each numerical expression.

1. $8 - 2 \cdot 3$ **2.** $(102 + 42) \div 6^2$ **3.** $1.3 + (2.5 \div 1.25)$

4. $(14 - 4) \cdot 4 + 4^2$ **5.** $80 \div (3 + 7) - 1^5$ **6.** $10 + 8 \div 2 \cdot 5$

Evaluate each variable expression when $b = 25$.

7. $b + 4$ **8.** $\frac{1}{2}(b - 1)$ **9.** $10b + 20 \div 2^2$ **10.** $b + 5b$

11. CALCULATOR Some calculators follow order of operations, and some do not. Does your calculator follow order of operations? Use Exercise 6 to test your calculator. Explain your results.

12. PART-TIME JOB Ana works $3\frac{1}{2}$ h/day on Monday, Wednesday and Friday and 5 h on Saturday. Write and simplify a numerical expression for the number of hours that Ana works per week.

◤ PRACTICE EXERCISES • For Extra Practice, see page 589.

Simplify each numerical expression.

13. $4 \cdot 9 - 9 + 7$ **14.** $(20 + 8) \div 7 \cdot 6$ **15.** $1^3 \cdot 1 + 5 \cdot 2$

16. $\frac{1}{2}\left(\frac{1}{2} + \frac{5}{8}\right)$ **17.** $6.9 - 3.2 \cdot (10 \div 5)$ **18.** $[20 + (2 \cdot 8)] \div 3^2$

19. $7^2 + 81 \div 9 + 3$ **20.** $(120 + 42.5) \div (4^3 + 1)$ **21.** $14 - (6 + 3) \div 3 + 10$

Evaluate each expression when $w = 16$.

22. $30 - w$ **23.** $(w^2 + 0) + 7$ **24.** $\frac{w}{4} - w \div 8$ **25.** $\frac{1}{4}w \cdot (w \div 2)$

26. WRITING MATH Do the expressions $12(4 + 3)$ and $12 \cdot 4 + 3$ equal the same value? Explain why or why not.

27. FITNESS On Saturday Rosa rode her bike 15 mi. On Sunday she rode 6 mi less than she did on Saturday. Write and simplify a numerical expression to determine the number of miles she rode on both days.

28. ENTERTAINMENT Carla is selling tickets for the school play. Floor seats cost $5 and bleacher seats cost $4. She sold 46 floor tickets and 79 bleacher tickets. Write and simplify a numerical expression for the amount of money that Carla collected for the tickets.

29. FITNESS Javier ran for x miles on Monday, y miles on Tuesday and z miles on Wednesday. Write a variable expression for the average number of miles Javier ran Monday through Wednesday.

30. POPULATION The population of Wyoming, the least-populated U.S. state, can be represented by the variable expression $5x - 20{,}000 + (2800 \div 2)$ where $x = 100{,}000$. Evaluate the expression to find the population of Wyoming.

 MODELING Write the variable expression represented by each Basic Mat. Find the value of each expression if each x-block equals 2 and each y-block equals 4.

31. 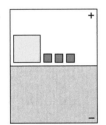   **32.** 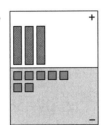   **33.**    **34.**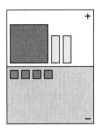

▚ EXTENDED PRACTICE EXERCISES

Evaluate each expression when $n = 10.2$, $p = 6$ and $r = 0.8$.

35. $n \div p^2$ **36.** $p(n + r)$ **37.** $n \cdot (p - r)$ **38.** $rp + n^3$

39. $\dfrac{n + p + r}{3p}$ **40.** $\dfrac{2}{3}n + \dfrac{1}{4}p$ **41.** $2n - 6r$ **42.** $8p \div (3r + n)$

 43. CHAPTER INVESTIGATION Use an almanac or the Internet to find the population and number of square miles or square kilometers of any state and a city in that state.

▚ MIXED REVIEW EXERCISES

Use the data in the table to answer the following questions. (Lesson 1-2)

Name	Test 1	Test 2	Test 3	Test 4	Final
Carol Jenn	89	76	90	85	92
Michael Sand	79	54	87	88	91
Morgan Tate	86	95	89	92	95

44. What is the range of the student's grades?

45. What is the overall average test score for these three students?

46. Which student would benefit the most if the teacher decides to weight the final test as twice the value of the other four tests? Explain.

Review and Practice Your Skills

Graph the given sets of numbers on a number line.

1. $\{4, 1.5, 0, -1, -5\}$

2. $\{-4, -2, 0, 2, 4\}$

3. $\left\{2\frac{1}{2}, \frac{1}{2}, 0, -1\frac{1}{2}, -2\frac{1}{2}\right\}$

4. $\{\sqrt{4}, \pi, \sqrt{25}\}$

5. $\{-5.25, -2.25, 1.75, 4.5\}$

6. even single digit numbers

Use a number line to compare numbers. Replace each ■ with <, > or = .

7. $0 \; ■ \; -3$

8. $|-6| \; ■ \; 6$

9. $-2 \; ■ \; -1$

10. $\frac{1}{3} \; ■ \; \frac{1}{2}$

11. $0.5 \; ■ \; 1.5$

12. $-7 \; ■ \; -10$

Graph each set of numbers on a number line.

13. all real numbers greater than 5

14. the integers between 0 and $|-3|$

15. the integers from -4 to 4

16. the odd numbers between 10 and 18

17. all real numbers less than -2

18. the number 5 and its opposite

Evaluate each expression.

19. $|x|$, when $x = 17$

20. $|y|$, when $y = -2$

21. $|r|$, when $r = -111$

22. $-w$, when $w = -4$

23. $-|d|$, when $d = 4$

24. $-(-n)$, when $n = \frac{2}{3}$

Simplify each numerical expression.

25. $6 + 3 \cdot 7$

26. $3^2 \div (9 - 6)$

27. $6 \cdot 2 - 4 \cdot 2$

28. $5^2 - 4 \cdot 4 + 5^2$

29. $3^4 - 3^3 \div 3^3 - 3^2$

30. $2(34 - 19)$

Evaluate each expression when $x = 8$.

31. $x - 2$

32. $3x + x \cdot 3^2$

33. $x^2 + x + 1$

Evaluate each variable expression when $a = 6$, $b = 4$ and $c = 3$.

34. $5(a - b)$

35. $5a - 5b$

36. $(5b - 2c) + c^2$

37. $\frac{1}{b}(6c - a)$

38. $b^3 - \frac{1}{3}c^3$

39. $10c \div a + b^2$

Evaluate each expression when $w = 2$, $x = 3$, $y = 12$ and $z = -10$.

40. $7(y - x)$

41. $y + x \cdot 5$

42. $16x - (4x + y)$

43. $w^2 + x \cdot y$

44. $[x + w\,(y)]x$

45. $y^2 - 2w + y$

46. $x(3 + w^2)$

47. $3\,|y|$

48. $3\,|z|$

Replace each ■ with <, > or =. (Lesson 2-1)

49. -3 ■ $|-3|$

50. $|3|$ ■ $|-3|$

51. $\dfrac{1}{2}$ ■ $\dfrac{1}{4}$

52. 0.2 ■ $.02$

53. -1 ■ $-\dfrac{4}{4}$

54. $\dfrac{1}{4}$ ■ 0.25

Graph each set of numbers on a number line. (Lesson 2-1)

55. the integers from -6 to -1

56. all real numbers less than or equal to $3\dfrac{1}{2}$

57. $\left\{-4, -2.5, -\dfrac{3}{4}, 2\dfrac{2}{3}, 3.5\right\}$

58. the integers between $|-2|$ and -4

Simplify each numerical expression. (Lesson 2-2)

59. $(15 \cdot 2) \div (5 \cdot 3)$

60. $16 \div 2 \cdot 8 + 4 - 4$

61. $3[2 + 4(4)]$

62. $7 \div 7 - 7 \cdot 7$

63. $2 \cdot 2 + 2 \div 2 - 2$

64. $|-2| - 2$

Math*Works*
Workplace Knowhow
Career – Concession Stand Operator

Concession stands provide refreshments for sports fans and participants. Concession stand operators strive to provide the highest quality food products as well as efficient service. As a concession stand operator plans for a homecoming football game, he or she must order appropriate quantities of food, drinks, utensils, and other supplies.

Use the table at the right for Exercises 1–3.

1. Complete the table below, indicating which years had increases and which years had decreases in the numbers of men, women, and youth attending the games.

Projected Attendance

Year	Men	Women	Youth (under 18)
2000 - 2001	240	113	147
2001 - 2002	244	58	226
2002 - 2003	291	76	193
2003 - 2004	302	94	193
2004 - 2005	257	125	122

Increases and Decreases in Attendance

Year	Men	Women	Youth	Total
2001 - 2002	+4	−55	+79	+28
2002 - 2003	+47	+18	−33	+32
2003 - 2004	■	■	■	■
2004 - 2005	■	■	■	■

2. Between which years was there an increase in the total population of those attending the games? Between how many years was there a decrease?

3. Is there a correlation between the increases and decreases of the individual groups and the total population? What factors might influence the increases or decreases?

2-3 Write Variable Expressions

Goals ■ Write variable expressions to represent word phrases.
■ Write word phrases to represent variable expressions.

Applications Part-time job, Space science, Photography, Population

HISTORY Diophantus of Alexandria is often known as the father of algebra. He lived from 200–284 A.D. The following inscription appears on his gravestone.

> This stone marks the grave of Diophantus.
> If you solve this riddle, you will know his age.
> He spent one-sixth of his life as a child,
> Then one-twelfth as a youth.
> He was married for one-seventh of his life.
> Five years after he married, his son was born.
> Fate overtook his beloved child; he died
> When he was half the age of his father.
> Four more years did the father live
> Before reaching the end of this life.

Translate the following lines of the inscription into a mathematical expression. Choose and identify a variable for each expression.

1. line 3
2. line 4
3. line 5
4. line 6
5. line 8
6. line 9

■ BUILD UNDERSTANDING

In mathematics, words and ideas often have to be translated into variable expressions. Many words and phrases suggest certain operations. Any variable can be used to represent a number.

	Word phrase	Variable expression
Addition	three *more than* a number	$3 + x$
	the *sum* of a number and -20	$n + (-20)$
	a number *increased* by eight	$y + 8$
Subtraction	the *difference* of a number and 20	$y - 20$
	seven *less than* a number	$p - 7$
	negative two *decreased* by a number	$-2 - m$
Multiplication	six *times* a number	$6a$
	the *product* of a number and 15	$15n$
	a number *multiplied* by -11	$-11t$
Division	the *quotient* of a number and -12	$\dfrac{t}{-12}$
	seven *divided* by a number	$\dfrac{7}{d}$

Example 1

Write each phrase as a variable expression.

a. a number increased by 33

b. negative three times a number

c. the difference of 15 and a number

d. the quotient of five and twice a number

Solution

Let n = a number.

a. $n + 33$

b. $-3n$

c. $15 - n$

d. $\dfrac{5}{2n}$

This process can be reversed to write word phrases for variable expressions.

Example 2

Translate each variable expression into a word phrase.

a. $4 + y$

b. $\dfrac{x}{20}$

c. $-\dfrac{1}{2}m$

d. $2n - 8$

Solution

a. the sum of four and a number

b. the quotient of a number and 20

c. negative one half times a number

d. twice a number decreased by eight

> **Check Understanding**
>
> Give a different possible solution for each variable expression in Example 2.

Example 3

PART-TIME JOB When Alicia babysits, she charges $2.50/h for each child age 2 and under and $2.00/h for each child above age 2. Write a variable expression to represent the amount of money Alicia earns each hour of babysitting.

Solution

Let x = the number of children age 2 and under. Let y = the number of children above age 2.

Think of a word phrase. *$2.50 times the number of children age 2 and under plus $2.00 times the number of children above age 2*

Write the variable expression. $2.50x + 2.00y$

Alicia earns $2.50x + 2.00y$ per hour for babysitting.

Write each phrase as a variable expression.

1. the product of seven and a number
2. one-half less than a number
3. negative two increased by a number
4. the quotient of 16 and a number
5. a number decreased by 22
6. four more than three times a number

Translate each variable expression into a word phrase.

7. $8x$

8. $-5 + y$

9. $\dfrac{-20}{x}$

10. $z - 3$

11. $\dfrac{1}{2}x$

12. $10 - 6m$

13. **SPACE SCIENCE** The time that a space shuttle is to blast off is referred to as t. For this situation, what does the expression $t - 9$ sec refer to?

■ **PRACTICE EXERCISES** • For Extra Practice, see page 589.

Write each phrase as a variable expression.

14. five more than a number
15. the difference of a number and 101
16. negative ten times a number
17. the quotient of negative six and a number
18. a number decreased by two
19. the sum of 32 and twice a number
20. one-third more than a number
21. the product of a number and 16
22. nine less than a number times six
23. five tenths increased by a number
24. the sum of 11 and five times a number
25. negative four decreased by a number
26. seven times a number divided by -13
27. the difference of eight and a number
28. one-half a number multiplied by 22
29. 30 times a number divided by -62

Translate each variable expression into a word phrase.

30. $m + 17$

31. $6 - f$

32. $-6y$

33. $\dfrac{1}{2} + n$

34. $-12 - v$

35. $\dfrac{y}{25}$

36. $2x + 6$

37. $\dfrac{-21}{3y}$

38. $0.8w$

39. $3x - 12$

40. $11s + \dfrac{1}{3}$

41. $\dfrac{-5x}{-11}$

 MODELING Write a word phrase for each variable expression that is represented by Algeblocks.

42.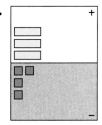

43.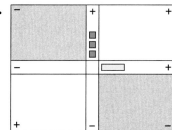

44.

DATA FILE Refer to the data on the top ten national parks on page 566. Let *n* equal the acreage of the Grand Canyon. Write a variable expression and a word phrase in terms of *n* for the acreage of the following parks.

45. Yellowstone Park

46. Gates of the Arctic

47. the Everglades

48. Katmai

Yellowstone Park

Write each phrase as a variable expression.

49. PHOTOGRAPHY Two more than half the number of pictures remain on the roll of film.

50. INDUSTRY Eight less than six times the number of boxes of books remain in stock.

51. POPULATION The approximate population of Montana is 101,000 less than twice the population of Wyoming.

52. HOBBIES To rent a paddle boat, it costs $10 the first hour and $3 for each additional half hour.

53. If a polygon has *n* sides, the sum of the measures of its interior angles is 180° times two less than the number of sides.

EXTENDED PRACTICE EXERCISES

WRITING MATH Write a real-life situation to represent each variable expression.

54. $y + 12$ **55.** $\dfrac{3n}{5}$ **56.** $\dfrac{1}{2}(b - 9)$ **57.** $5x$

58. Write a variable expression for the phrase *seven times the difference of four and twice a number*.

59. Write a word phrase for the expression $-12(3x + 9)$.

60. CRITICAL THINKING Write two phrases that are stated differently but equal the same value.

MIXED REVIEW EXERCISES

Use the data for Exercises 61–64.

48	64	26	39	22	45	56	67	43	26
41	35	39	42	68	65	52	58	48	33

61. Make a stem-and-leaf plot to display the data. (Lesson 1-3)

62. Name any outliers, clusters or gaps in the data. (Lesson 1-3)

63. Find the interquartile range of the data. (Lesson 1-4)

64. Suppose a data set includes the 20 elements above plus one more. The mean of this set is 46. What is the value of the 21st element? (Lesson 1-2)

Add and Subtract Variable Expressions

Goals ■ Simplify variable expressions.
■ Evaluate variable expressions

Applications Sports, Recycling, Population, Detective work

Marcus, Rob and Jiro went to a media store. Marcus bought 1 CD and 2 movies, Rob bought 3 CDs and 1 movie, and Jiro bought 2 CDs. Write a variable expression for the total amount spent on CDs and movies by each of the following. Let c = the price for each CD and m = the price for each movie.

1. Marcus   **2.** Rob   **3.** Jiro

4. Marcus and Rob   **5.** Rob and Jiro   **6.** Marcus, Rob, and Jiro

◨ BUILD UNDERSTANDING

The parts of a variable expression separated by addition or subtraction signs are called **terms** of the expression.

one term: $2x$ two terms: $3x + 6y$ three terms: $-4a + 5b - 7$

Terms that have identical variable parts are called **like terms**. Terms that have different variable parts are called **unlike terms**.

like terms: $2x$ and $4.5x$ unlike terms: $2a$ and $2\frac{1}{2}b$
$3st$ and $-\frac{1}{3}st$ $7xy$ and $7xz$
$2a^3b$ and $-9a^3b$ $8x^4y^2$ and $4x^3y^4$

To **simplify** a variable expression, perform as many of the indicated operations as possible. Use the distributive property in reverse order to simplify an expression that contains like terms. This process is called **combining like terms**. An expression is simplified when only unlike terms remain.

Think Back

The distributive property states that each factor outside parentheses is multiplied by each term within the parentheses.

$a(b + c) = ab + ac$

Example 1

Simplify.

a. $3x + 2x$ **b.** $-9n + n + 3m$ **c.** $12xy + 8xz + (-15xy)$

Solution

a. $3x + 2x = (3 + 2)x$ Use the distributive property.
$= 5x$

Another way to solve this example is in expanded form.

$3x + 2x = x + x + x + x + x$
$= 5x$

b. $-9n + n + 3m = (-9 + 1)n + 3m$ Use the distributive property.
$$= -8n + 3m$$

c. $12xy + 8xz + (-15xy) = 12xy + (-15xy) + 8xz$ Rewrite using the commutative property.
$$= [12 + (-15)]xy + 8xz$$ Use the distributive property.
$$= -3xy + 8xz$$

By applying the distributive property, the resulting sum inside the parentheses is the sum of the coefficients of the like terms. To simplify an expression involving subtraction, change subtraction to *addition of the opposite*.

CONcepts in MOtion
BrainPOP®
mathmatters2.com

E x a m p l e 2

Simplify.

a. $\dfrac{1}{3}a - \dfrac{2}{3}a$ **b.** $-3x + 8y - 6y$ **c.** $2.4st - (-7.1sy) - 10.8st$

Solution

a. $\dfrac{1}{3}a - \dfrac{2}{3}a = \dfrac{1}{3}a + \left(-\dfrac{2}{3}a\right)$ Change subtraction to addition of the opposite.

$$= \left[\dfrac{1}{3} + \left(-\dfrac{2}{3}\right)\right]a$$ Use the distributive property.

$$= -\dfrac{1}{3}a$$

b. $-3x + 8y - 6y = -3x + 8y + (-6y)$ Change subtraction to addition of the opposite.
$$= -3x + [8 + (-6)]y$$
$$= -3x + 2y$$

c. $2.4st - (-7.1sy) - 10.8st = 2.4st + 7.1sy + (-10.8st)$
$$= 2.4st + (-10.8st) + 7.1sy$$
$$= [2.4 + (-10.8)]st + 7.1sy$$
$$= -8.4st + 7.1sy$$

Check Understanding

Describe how to combine like terms without using the words "distributive property."

E x a m p l e 3

Evaluate each expression when $m = -7$ and $p = 12$.

a. $m + 2m$ **b.** $m - p + 3p$ **c.** $7m - 5m - p + 3p$

Solution

First, simplify each variable expression. Then substitute the appropriate value for each variable. Finally, simplify the expression using the order of operations.

a. $m + 2m = 3m$ **b.** $m - p + 3p = m + 2p$ **c.** $7m - 5m - p + 3p = 2m + 2p$
$$= 3(-7)$$ $$= -7 + 2(12)$$ $$= 2(-7) + 2(12)$$
$$= -21$$ $$= -7 + 24$$ $$= -14 + 24$$
$$= 17$$ $$= 10$$

Simplifying expressions can be modeled by Algeblocks. When equal numbers of the same block are on opposite sides of the mat, they are called **zero pairs**. To simplify an expression, remove zero pairs from the mat.

Example 4

MODELING Use Algeblocks to simplify each expression.

a. $-3x + 7x$

b. $2y - 4y + 2$

Solution

a.

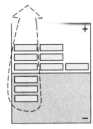 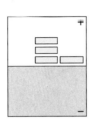

$-3x + 7x = 4x$

b.

 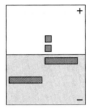

$2y - 4y + 2 = -2y + 2$

TRY THESE EXERCISES

Simplify.

1. $5t + 2t$

2. $-3r + r$

3. $-6t - 6t$

4. $-0.9m + 1.3n - 4.6m$

5. $\frac{3}{5}x - \frac{1}{10}x + \frac{1}{5}y$

6. $7xy - xy + xz$

7. $-3ab - 9ac + 11ab$

8. $pq - 6rs - 3pq + 6rs$

Evaluate each expression when $a = -4$ and $b = 11$.

9. $a - 6a$

10. $-b + 3b$

11. $a - b + 2b - 9a$

12. $\frac{1}{2}a + (-b) + \frac{1}{4}a$

 13. MODELING Use Algeblocks to simplify $-3x + 6x + 5 - 1$.

14. SPORTS A football team completed a pass for a gain of $5x$ yards. On the next play, they were penalized $2x$ yards. Write and simplify an expression for the yardage gained after these two plays.

PRACTICE EXERCISES • For Extra Practice, see page 590.

Simplify.

15. $8s + 6s$

16. $-2r + 7r$

17. $18k - 25k$

18. $4b - (-5b)$

19. $-\frac{1}{3}m + \frac{4}{9}m$

20. $3g - 11g + 16g$

21. $-7c - (-7c) + 7c$

22. $2.1x - 3.5y - 1.5x$

23. $11xy - 3xy + x$

24. $-36pq + 9pq - 3$

25. $15a - 2b + 3a - 6b$

26. $-7jk - (-9jk) + 8j$

27. $\frac{2}{3}d - \frac{5}{6}d + \frac{1}{3}d$

28. $ab - 22ab - 9ab$

29. $-0.7st + 9.4st - 0.4t$

30. $8m - 2n + 3m - 3n$

31. $a - 7b - a + b$

32. $3r - 6rs + 6r + r$

33. $\frac{7}{8}xy - \frac{1}{4}x + \frac{3}{4}xy$

34. $-c + 16cd - 12c + cd$

35. $7wz + 7w - wz - 3w + 11w$

Evaluate each expression when $x = -5$, $y = 4$ and $z = \frac{1}{2}$.

36. $2y - 6y$

37. $-5x + 11x$

38. $-3z - (-9z)$

39. $-3x + 4y - 2x$

40. $\frac{1}{2}y - \frac{3}{4}y - \frac{1}{2}z$

41. $4z + 4x - 6z - x$

42. $7y - 2x + 9y - x$

43. $3xy - 6xy + 2y$

44. RECYCLING During a recycling drive, the eighth grade collected $(2h + 9t)$ newspapers, the ninth grade collected $(h + 7t)$ newspapers and the tenth grade collected $4h$ newspapers. Write and simplify an expression for the total number of newspapers collected.

45. WRITING MATH Write a short paragraph explaining how to simplify a variable expression. Include examples.

46. POPULATION There are approximately $(1.7x + 0.2)$ million Americans between the ages of 15 and 19. There are approximately $(2x + 0.6)$ million Americans between the ages of 35 and 39. Find the difference in the population of these two age groups.

47. DETECTIVE WORK When part of a skeleton is found, detectives determine the height of the victim to help find the identity. If the victim is male and the length of the femur (large thigh bone) is known, the expression $69.089 + 2.238F$ can be used to find the height, where F is the length of the femur. Find the height of a male victim whose femur measures 41 cm.

MODELING Use Algeblocks to simplify each expression.

48. $5x - 2x$ **49.** $-2y + y - 3$ **50.** $3x - y - 6x + 2y$

DATA FILE For Exercises 51–54, use the data on sleep times on page 561. Let x equal the average hours of sleep per day for a human adult. Write a variable expression for the sleep time of the following creatures.

51. elephant   **52.** armadillo   **53.** sheep   **54.** pig

55. Write and simplify an expression in terms of x for the combined sleep time per day of the creatures in Exercises 51–54.

EXTENDED PRACTICE EXERCISES

56. Subtract $(4a + b) + (3a + 2b)$ from $6a + 8b + (-4b + 7a)$.

57. Find the sum of $4a - 3b$, $a - b$ and $a - 5b$.

58. CRITICAL THINKING On Monday Kaz did n sit-ups, on Tuesday he did $(n + 2)$ sit-ups, on Wednesday he did $(n + 5)$ sit-ups and on Thursday he did $(n + 9)$ sit-ups. If Kaz exercises every day and the pattern continues, how many sit-ups will he have done in all for the seven-day week?

MIXED REVIEW EXERCISES

Use the data for Exercises 59 and 60. (Lesson 1-4)

| 15 | 37 | 23 | 39 | 34 | 26 | 11 | 23 | 18 | 29 |
| 26 | 34 | 19 | 12 | 15 | 28 | 21 | 36 | 29 | 35 |

59. Create a box-and-whisker plot to display the data.

60. Find the interquartile range of the data.

Review and Practice Your Skills

PRACTICE ◣ LESSON 2-3

Write each phrase as a variable expression.

1. eight increased by a number
2. the sum of 11 and a number
3. five less than a number
4. some number less 19
5. the product of eight and y
6. the quotient of 15 and a number
7. four times a number decreased by seven
8. a number divided by three
9. one-half a number increased by six
10. a number increased by itself
11. two-thirds a number minus 14
12. the quotient of 45 and a number

Translate each variable expression into a word phrase.

13. $b - 6$
14. $3a + 4$
15. $-4 - k$
16. $\frac{1}{3}x$

17. $-7x$
18. $\frac{t}{8}$
19. $\frac{5x}{10}$
20. $\frac{x}{y}$

21. $6x - 11$
22. $\frac{3}{x}$
23. $n + (-15)$
24. $4x$
25. $2x - 7$
26. $x - 9$
27. $5xy$
28. $\frac{1}{2}x + 10$

PRACTICE ◣ LESSON 2-4

Simplify.

29. $x + x$
30. $4x - x$
31. $2y - 4y$
32. $3b - 5b + 10b$
33. $7c - (-4c)$
34. $x + 2y - x$
35. $\frac{1}{2}t - \left(-\frac{1}{2}t\right) + \frac{1}{4}t$
36. $6c - 2d + 7c - 5d$
37. $3xy - xy + 4x$
38. $7ab + 9bc + (-4ab)$
39. $5.2r - 3.1r + 7.5t$
40. $p + p - q - q$
41. $2z - 4y - 2z - 4y$
42. $\frac{1}{2}a - \frac{1}{4}a + \frac{2}{3}b$
43. $-3cd + 4d - 7d + 6cd$
44. $0.3x - 1.7y + 2.3y - 4.1x$
45. $n - 2m + 2n - m$
46. $\frac{3}{7}p - \frac{1}{14}p + \frac{2}{3}q - \frac{1}{2}q$
47. $7a + \frac{1}{3}b + \frac{2}{3}a - \frac{2}{3}b$
48. $\frac{11}{15}x + \frac{2}{9}z + \frac{2}{5}z - \frac{1}{3}x$
49. $19.3x - 17.7x - 3.7t + 5.9t$
50. $17x - (-23x) - 2y + 14y$
51. $2.6j - 4.8k - 4.3k - 0.2j$
52. $\frac{1}{3}x + \frac{3}{4}y + \frac{2}{5}x - \frac{1}{8}y$

Evaluate each expression when $w = 2$, $x = 3$, $y = 4$ and $z = 6$.

53. $-7x - y$
54. $-5z - (-3y)$
55. $-9w + 2y + 4w$
56. $\frac{1}{2}y + \frac{2}{3}z$
57. $y - x + 6z$
58. $4x - 3w + x - 2w$
59. $4w + 2y$
60. $z - 2w$
61. $-x - x$
62. $-\frac{1}{2}w + y$
63. $w + x - y + z$
64. $-w - x - y - z$
65. $4z + 4w - 4x$
66. $\frac{1}{3}(z + x)$
67. $\frac{1}{4}w + \frac{2}{3}z$
68. $2x - 4y + 7w - z$
69. $\frac{2}{7}x + \frac{z}{7} + \frac{w}{7}$
70. $1.2z + 7.9w - 0.7z - 2.9w$

Write each phrase as a variable expression. (Lesson 2-3)

71. the sum of five and twice a number

72. seven less than a number

73. 12 divided by a number

74. a number multiplied by -13

Translate each variable expression into a word phrase. (Lesson 2-3)

75. $8.7 - z$

76. $x - 270.5$

77. $\dfrac{x}{3}$

78. $2(x + 5)$

Simplify. (Lesson 2-4)

79. $-5x - (-5x) + 5x$

80. $-3xy - 4xy - 7xy$

81. $-2x + 2x$

82. $9x + 12 + x$

83. $15 + 14z - z$

84. $6n + 7m + 16n$

Evaluate each expression when $a = 3$, $b = 8$ and $c = 5$. (Lesson 2-1–Lesson 2-4)

85. $-3a + 2a + 4b$

86. $4b - 5b$

87. $\dfrac{1}{5}c - \dfrac{1}{3}a + \dfrac{1}{2}b$

88. $-b + a - (-c)$

89. $2ab - c$

90. $-2a + 5c + c$

Mid-Chapter Quiz

Graph the given sets of numbers on a number line. (Lesson 2-1)

1. $\left\{-\dfrac{1}{2}, 0, \dfrac{1}{2}, \dfrac{3}{2}\right\}$

2. $\{-3.25, -1.5, -0.75, 1.75\}$

3. all real numbers greater than -2

4. integers between -2 and 7

Simplify each numerical expression. (Lesson 2-2)

5. $(7 + 21) \div 2^2$

6. $4 \cdot 6 - 3 + 11$

7. $3.4 - 2(0.8 - 1.3)$

8. $3 \cdot 5 - 2 \div 4$

9. $(5^2 \div 5) + 5 \cdot 2^3$

10. $\dfrac{2}{3} \cdot 3^2 - 6$

Write each word phrase as a variable expression. (Lesson 2-3)

11. seven more than a number

12. the quotient of a number and 11

13. one-third of a number

14. the difference of -44 and a number

Evaluate each variable expression when $x = -4$, $y = -10$ and $z = 3.5$. (Lesson 2-4)

15. $2y - 10$

16. $-|x|$

17. $\dfrac{yz}{x}$

18. $(x + y) - z$

19. $(x - y)z$

20. $-4z - \left(\dfrac{2y}{x}\right)$

21. $x - 2y$

22. $|y + 3z|$

Simplify. (Lesson 2-4)

23. $4x + 8x$

24. $7y - 15y$

25. $-\dfrac{2}{3}b + \dfrac{1}{6}b$

26. $1.6a - 3.5b + 2.7a$

27. $-4x - (-4x) + 4x$

28. $5ab - 7a - 11ab + b$

Multiply and Divide Variable Expressions

Goals   ■ Simplify variable expressions.

■ Evaluate variable expressions.

Applications   Part-time job, Weather, Engineering, Spreadsheets

Use Algeblocks to multiply the variable expression $2(x - 3)$.

1. On a Quadrant Mat, place 2 unit blocks on the positive part of the horizontal axis.

2. Place 1 x-block in the positive part of the vertical axis and 3 unit blocks in the negative part of the vertical axis.

3. Form rectangular areas in all the quadrants that are bounded by Algeblocks.

4. Read the answer from the Quadrant Mat.

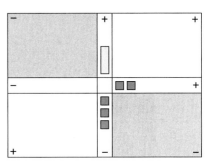

◣ BUILD UNDERSTANDING

To multiply variable expressions, use the distributive property.

Example 1

Simplify.

a. $4(m + 7)$ **b.** $-8(4.5 - rs)$ **c.** $-(2x + y)$

Solution

a. $4(m + 7) = 4m + 4(7)$ Use the distributive property.
$\qquad\qquad\quad = 4m + 28$

b. $-8(4.5 - rs) = -8(4.5) - (-8)(rs)$ Use the distributive property.
$\qquad\qquad\quad\ = -36 - (-8rs)$ To subtract, add the opposite.
$\qquad\qquad\quad\ = -36 + 8rs$

c. $-(2x + y) = (-1)(2x + y)$ Apply the multiplication property of -1.
$\qquad\qquad\ = (-1)(2x) + (-1)(y)$ Use the distributive property.
$\qquad\qquad\ = -2x + (-y) = -2x - y$

Example 1, part c, is an illustration of the **property of the opposite of a sum**. The negative sign on the outside of parentheses indicates you are to multiply each term by -1.

Property of the Opposite of a Sum	For all real numbers a and b, $-(a + b) = -1(a + b)$ $\qquad\qquad\quad = -a + (-b)$

To divide a variable expression, divide each term in the numerator by the denominator.

Example 2

Simplify.

a. $\dfrac{2x + 6}{2}$

b. $\dfrac{-7y - 16}{-4}$

c. $\dfrac{3.6 + 1.2rs}{0.6}$

Solution

a. $\dfrac{2x + 6}{2} = \dfrac{2x}{2} + \dfrac{6}{2}$

$= x + 3$

b. $\dfrac{-7y - 16}{-4} = \dfrac{-7y}{-4} - \dfrac{16}{-4}$

$= \dfrac{7}{4}y + 4$

c. $\dfrac{3.6 + 1.2rs}{0.6} = \dfrac{3.6}{0.6} + \dfrac{1.2rs}{0.6}$

$= 6 + 2rs$

> **Think Back**
>
> Recall division of signed numbers.
>
> $\dfrac{12}{4} = 3 \qquad \dfrac{-12}{4} = -3$
>
> $\dfrac{12}{-4} = -3 \qquad \dfrac{-12}{-4} = 3$

Example 3

Evaluate each expression when $a = 6$, $b = -5$, and $c = \dfrac{1}{3}$.

a. $-2(a + 7)$

b. $\dfrac{4b - 12}{-4}$

c. $9c(a - 10)$

Solution

First substitute the appropriate value for each variable. Then simplify the expression using the order of operations.

a. $-2(a + 7) = -2(6 + 7)$

$= -2(13)$

$= -26$

b. $\dfrac{4b - 12}{-4} = \dfrac{4(-5) - 12}{-4}$

$= \dfrac{-20 - 12}{-4}$

$= \dfrac{-32}{-4}$

$= 8$

c. $9c(a - 10) = 9\left(\dfrac{1}{3}\right)(6 - 10)$

$= 3(-4)$

$= -12$

Example 4

PART-TIME JOB Katie works $(9x + 6)$ hours every week. She works 3 days for the same amount of time each day. How many hours does Katie work in one day?

Solution

$\dfrac{9x + 6}{3} = \dfrac{9x}{3} + \dfrac{6}{3}$ number of hours per week
number of days per week

$= 3x + 2$

Katie works $(3x + 2)$ hours in one day.

Simplify.

1. $2(5 + s)$

2. $-3\left(x + \dfrac{1}{6}\right)$

3. $6(3.1 - 4b)$

4. $-8(x - y)$

5. $\dfrac{2d - 6}{2}$

6. $\dfrac{-15j + 30}{5}$

7. $\dfrac{-4.2w - 3.5}{0.7}$

8. $\dfrac{12x + 15y}{-3}$

Evaluate each expression when $x = -3$, $y = 7$, and $z = 0.5$.

9. $3(z - 0.7)$

10. $-\dfrac{1}{2}(11 - y)$

11. $\dfrac{2x - 9}{3}$

12. $\dfrac{3y + x}{-9}$

13. MODELING Use Algeblocks to find the product $-4(y + 2)$.

14. PART-TIME JOB Tim earns \$3.50/h plus tips as a waiter after school. Write and simplify an expression for the amount of money he earns in 5 h.

15. WRITING MATH Describe how you would simplify $-7(x + y)$.

PRACTICE EXERCISES • For Extra Practice, see page 590.

Simplify.

16. $3(a + 11)$

17. $-5(m + 3)$

18. $16(rs - 5)$

19. $\dfrac{1}{4}(-36 - x)$

20. $-17(2a + b)$

21. $0.4(0.8h - 0.6k)$

22. $-2(-5n + 8p)$

23. $-3(-3q - 4r)$

24. $-8(-7h + 5j)$

25. $7(-8q - 7r)$

26. $11(-10s - 11t)$

27. $-\dfrac{7}{8}(-16c + 40d)$

28. $\dfrac{-6a + 42}{6}$

29. $\dfrac{24b + 8}{-4}$

30. $\dfrac{14x - 84}{-7}$

31. $\dfrac{3.5 - 4.0p}{0.5}$

32. $\dfrac{-36r - 48s}{12}$

33. $\dfrac{-20m + 50n}{-5}$

34. $\dfrac{56j - 48k}{8}$

35. $\dfrac{-9q + 9r}{-3}$

36. $\dfrac{2.7c - 3.3d}{-0.3}$

37. $\dfrac{-6d - 3g}{-3}$

38. $-17s + \dfrac{16t}{4}$

39. $\dfrac{169w - 39v}{-11}$

Evaluate each expression when $a = \dfrac{1}{3}$, $b = -5$ and $c = 6$.

40. $-9(a - 27)$

41. $8(-b - 6)$

42. $\dfrac{c - 21}{3}$

43. $\dfrac{6a + 14}{2}$

44. $a(-9 + c)$

45. $-0.5(b - 2c)$

46. $\dfrac{-9a - b}{5}$

47. $\dfrac{3c - 2b}{7}$

MODELING Use Algeblocks to find each product.

48. $3(x - 4)$

49. $-2(y + 1)$

50. $2(x - y)$

51. $-1(2x + 3)$

52. Write and simplify an expression for the area of the figure shown.

2x + 3

15

53. Most movie screens in movie theaters have a length of $(x + 14)$ feet and a height of x feet. Find the area of a movie screen if $x = 19$.

54. WEATHER The formula $C = \dfrac{5}{9}(F - 32)$ relates Celsius and Fahrenheit temperatures. The temperature in Celsius is C, and the degrees in Fahrenheit is F. Find the number of degrees Celsius that is equal to a temperature of $50°F$.

SPREADSHEET Cells of a spreadsheet can include variable expressions. Cell A2 contains the number of people that are registered to play basketball.

	A	B	C	D
1	number of players	uniform costs	number of teams	
2				
3				
4				
5				

55. Write an expression for cell B2 that will compute the total cost of uniforms at $21/uniform.

56. Write an expression for cell C2 that will compute the number of teams that can be formed if each team must include at least 10 people.

57. Suppose the value in cell A2 is 127. Find the total cost of uniforms.

58. Find the number of teams that can be formed with 127 total players. Will there be exactly 10 members on each team?

59. ERROR ALERT Ermano simplified the expression $-5(2x - 3y + 4xy)$. His work is shown. Is he correct? Explain.

$$-5(2x - 3y + 4xy) = -10x + 15y + 4xy$$
$$= 5xy + 4xy$$
$$= 9xy$$

EXTENDED PRACTICE EXERCISES

Simplify.

60. $-\dfrac{2}{5}(10xy + 25x - 15y)$

61. $41(-9a + 25b - c + 12d)$

62. $\dfrac{-21f - 49g + 7h - 21k}{-7}$

63. ENGINEERING Before building a house, an engineer calculates the amount of material to be removed from the ground in order to pour the foundation. The formula to find the amount of earth to remove for the shape shown is $\dfrac{d(a + b)}{2} \cdot L$. Find the amount of earth to be removed when $a = 43$ m, $b = 39$ m, $d = 7$ m and $L = 32$ m.

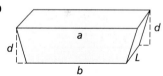

64. CRITICAL THINKING If $3(x + y) = 3x + 3y$, is $3 + (x \cdot y) = (3 + x)(3 + y)$ true? Experiment by substituting values for x and y. Explain your conclusion.

65. CHAPTER INVESTIGATION Calculate the population density of the state and city you chose by dividing the population by its area in square miles or kilometers.

MIXED REVIEW EXERCISES

Refer to the histogram to answer the questions. (Lesson 1-3)

66. Which interval contains the heights of the greatest number of people?

67. Name an interval that contains about 25% of the people.

68. How many sixteen-year-olds are between 5 ft 3 in. and 5 ft 7 in. tall?

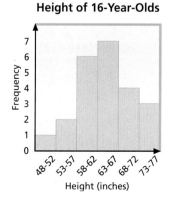

Height of 16-Year-Olds

Simplify Variable Expressions

Goals ■ Add, subtract, multiply and divide to simplify variable expressions.

Applications Sports, Finance, Photography, Fashion, Population

---

For each step below, list the rule from the order of operations or the property that is used to simplify the expression.

Rule or Property

1. $4(x + 6) + 2^3 = 4(x + 6) + 8$ _____

2. $\quad\quad\quad\quad = 4(x) + 4(6) + 8$ _____

3. $\quad\quad\quad\quad = 4x + 24 + 8$ _____

4. $\quad\quad\quad\quad = 4x + 32$ _____

5. Use order of operations to simplify $6(3 - y) - 4^2$.

◣ BUILD UNDERSTANDING

In Lesson 2-4, variable expressions were simplified by addition and subtraction. In Lesson 2-5, variable expressions were simplified by multiplication and division. Together these four operations can be combined to simplify a variable expression. Recall the order of operations when simplifying a variable expression with more than one operation.

Think Back

Recall the **order of operations.**

1. Perform all operations within parentheses and brackets.

2. Then perform all calculations involving exponents.

3. Multiply or divide in order from left to right.

4. Add or subtract In order from left to right.

Example 1

Simplify.

a. $3x + 2(5x - 1)$ **b.** $9 - 4(x - 4)$

Solution

a. $3x + 2(5x - 1) = 3x + 2(5x) + 2(-1)$ Use the distributive property.

$\quad\quad\quad\quad\quad\quad = 3x + 10x + (-2)$ Simplify.

$\quad\quad\quad\quad\quad\quad = 13x - 2$ Combine like terms.

b. $9 - 4(x - 4) = 9 + (-4)[x + (-4)]$ Rewrite subtraction as addition of the opposite.

$\quad\quad\quad\quad\quad\quad = 9 + (-4)(x) + (-4)(-4)$ Use the distributive property.

$\quad\quad\quad\quad\quad\quad = 9 - 4x + 16$ Simplify.

$\quad\quad\quad\quad\quad\quad = (9 + 16) - 4x$ Use the associative property to combine like terms.

$\quad\quad\quad\quad\quad\quad = 25 - 4x$ Simplify.

Example 2

Simplify.

a. $2(x + 3) + 3(x - 4)$

b. $3(ab - a) - 3(b + a)$

Solution

a. $2(x + 3) + 3(x - 4) = 2(x) + 2(3) + 3(x) + 3(-4)$ Use the distributive property.

$$= 2x + 6 + 3x + (-12)$$

$$= (2x + 3x) + [6 + (-12)]$$ Use the associative property to combine like terms.

$$= 5x + (-6)$$

$$= 5x - 6$$

b. $3(ab - a) - 3(b + a) = 3(ab) + 3(-a) - 3(b) - 3(a)$ Use the distributive property.

$$= 3ab + (-3a) - 3b - 3a$$

$$= 3ab + [(-3a) - 3a] - 3b$$ Combine like terms.

$$= 3ab + (-6a) - 3b$$

$$= 3ab - 6a - 3b$$

In Lesson 2-3, you translated word phrases into variable expressions. This skill is useful when solving real-life problems.

Example 3 **Personal Tutor at** mathmatters2.com

SPORTS A baseball team buys 12 bats. Some are wooden and some are aluminum. Wooden bats cost $36 each, and aluminum bats cost $43 each. Write and simplify a variable expression for the total cost of the baseball bats.

Solution

Let n = the number of wooden bats. If n of the 12 bats are wooden, then $12 - n$ are aluminum. Write an expression for the amount spent on each type of bat.

    wooden: $36(n)$

    aluminum: $43(12 - n)$

Find the total cost.

$$36n + 43(12 - n) = 36n + 43(12) + 43(-n)$$

$$= 36n + 516 + (-43n)$$

$$= -7n + 516$$

The total cost of the baseball bats is $(-7n + 516)$ dollars.

Simplify.

1. $5x + 7(-2x + 3)$
2. $7y - 2(8 + 3x)$
3. $-3n + 3(-n + 2m)$
4. $15u + \frac{1}{3}(12u - 9n)$
5. $6(c - 9) - (2c + 30)$
6. $2(6x - 2y) + 3(x + y)$
7. $(a - 2) - 1.5(a + b)$
8. $7(gh + 8g) - 4(g + 5.2)$
9. $8(3x + y) - (-x + 21)$

10. **WRITING MATH** Explain how you used the order of operations in Exercise 9.

11. **FINANCE** A cellular phone company sells stock for $83 a share. A cable company sells stock for $56 a share. Between the two companies, Felipe buys 15 shares of stock. Write and simplify a variable expression for the total amount that Felipe spent on the stocks.

◣ **PRACTICE EXERCISES** • For Extra Practice, see page 591.

Simplify.

12. $6x + 12(10 - x)$
13. $m + 5(m - 1)$
14. $8y - (2y - 3)$
15. $-3(x - 1) + 9$
16. $\frac{3}{4}(4n - 8) + 6n$
17. $6(x + 4) - 2x$
18. $\frac{1}{3}(6b + 9) + b$
19. $2(1.5d + 4) - 3.7$
20. $(n - 6) - 2(3n + 4)$
21. $3(x - 4) + 6(2x + 7)$
22. $-4(3w - 9) + 9(w - 2)$
23. $-\frac{1}{2}(y + 6) + \frac{2}{3}(6y - 9)$
24. $5(3 - 2.1x) - 5(2 + 0.9x)$
25. $11(2f - 8) - 10(-f + 6)$
26. $3.5(6a - 6) + 9(3a - 2)$
27. $-9(3t - 2) + 5(-4t + 8)$
28. $6(2n + 7) - 3(-7n - 4)$
29. $\frac{2}{3}(9 - 6w) + 7(2 - w)$
30. $2(12a - b) + (-a + b)$
31. $6(p + 3q) - (7p + 4q)$
32. $4(ab - 5a) - (9ab - 6a)$
33. $-3(r + 2s) + 2(-3r - s)$
34. $7(wz + w) - 2(wz + z)$
35. $16(-c - \frac{1}{4}cd) - (cd + 12d)$

36. **PHOTOGRAPHY** Queisha used a roll of 25 pictures at her family reunion. Some of the pictures were panoramic, and the rest were standard size. It costs $0.19 to develop standard size and $0.32 to develop panoramic. Write and simplify a variable expression for the total cost of developing the film.

37. **FASHION** Phoebe is the buyer for women's clothing at a department store. This month she buys 200 women's jeans. Some are slim fit, and the rest are relaxed fit. The manufacturer cost of the slim fit jeans is $18, and the manufacturer cost of the relaxed fit jeans is $21. Write and simplify an expression for the total amount spent on the jeans.

38. Find the perimeter of a rectangle with a length of $(3x - 2)$ in. and a width of $(x + 7)$ in.

Match each expression with the equivalent expression in simplified form.

39. $5(a + b) - 6(a - b)$
    a. $10a + 13b$

40. $2(6a + b) - (a + 3b)$
    b. $-11a + b$

41. $3(a + 2b) + 7(a + b)$
    c. $11a - b$

42. $-7(3a + 2b) - 5(-2a - 3b)$
    d. $13a + 3b$

43. $4(4a - 6b) - 3(a - 9b)$
    e. $-a + 11b$

44. DATA FILE Refer to the data on calories spent per minute on page 568. Twice a week Kuan goes to the gym to walk around the track and then swim. He spends 90 minutes working out. Kuan weighs about 150 lb. Write and simplify a variable expression for the number of calories that Kuan burns each day that he works out at the gym.

45. POPULATION In 1970 the population of New Jersey was approximately $10a + 1.7b$. From 1970 to 1980, the population of New Jersey increased by 28.1%. From 1980 to 1990, the population increased by 16.2%. Write and simplify a variable expression for the population of New Jersey in 1990.

46. YOU MAKE THE CALL Alicia and Frank each simplified the variable expression $3(-4g + 6f) - 3(g + 2f)$. Explain where Alicia and Frank each made a mistake. Then simplify and write the variable expression correctly.

Alicia's work	Frank's work
$3(-4g + 6f) - 3(g + 2f)$	$3(-4g + 6f) - 3(g + 2f)$
$= -12g + 6f - 3g - 6f$	$= 12g + 18f - 3g + 6f$
$= -15g$	$= -9g + 24f$

◤ EXTENDED PRACTICE EXERCISES

Simplify.

47. $3(-4a + 7b) - 9(8a - 3b) + 4(-a - 4b)$

48. $\frac{1}{3}(6p - 18q) - \frac{3}{4}(-24q - 16p) + \frac{3}{5}(15p - 40q)$

49. $-11(ab - 3a + 7b) + 3(-13ab + 6b - 8a) - 4(-12ab - 14a + 6b)$

50. SPORTS Kashauna made 11 baskets at the basketball game. Some were 3-point baskets, some were 2-point baskets and the rest were 1-point free throws. Write and simplify a variable expression for the total number of points that Kashauna had in the basketball game.

51. CRITICAL THINKING Simplify the expression $-4[3[2(x + 1)]]$.

◤ MIXED REVIEW EXERCISES

Use the scatter plot to answer the following questions. (Lesson 1-4)

52. How many weekly quizzes were given?

53. What is the least possible range of the average scores?

54. What is the greatest possible range of the average scores?

55. Is there a positive or negative correlation between the scores and the number of quizzes given?

56. Using one sentence, describe the trend in the data.

57. In what scoring group is the median weekly average score?

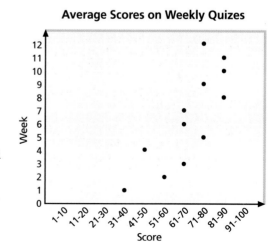

Average Scores on Weekly Quizzes

Review and Practice Your Skills

PRACTICE ◣ LESSON 2-5

Simplify each variable expression.

1. $-(6c - d)$

2. $\dfrac{3x + 12}{3}$

3. $\dfrac{-3v - 16}{-2}$

4. $-7(y - 4z)$

5. $\dfrac{-5}{6}(-12e + 36f)$

6. $\dfrac{1.6 - 3.2w}{0.8}$

7. $\dfrac{4x - 8}{4}$

8. $\dfrac{21w + 14y}{-7}$

9. $-(a + 2b)$

10. $\dfrac{-(7w - 4)}{3}$

11. $-3(6bc - 7)$

12. $\dfrac{1}{5}(-x - 25)$

13. $\dfrac{-44s - 66t}{11}$

14. $\dfrac{-8(x + 1.5)}{3}$

15. $\dfrac{132c - 60}{12}$

Evaluate each expression when $x = 7$, $y = 4$, and $z = -1$.

16. $x - (y - z)$

17. $x^2 + 2x + y$

18. $(x + y)(y - z)$

19. $3(x - y)$

20. $\dfrac{5x - 21}{2}$

21. $\dfrac{12z + 10y}{7}$

22. $\dfrac{-16z + 20}{-y}$

23. $\dfrac{-6x - 5y}{2}$

24. $\dfrac{1}{2} - (7z + x)$

25. $\dfrac{3(x - y)}{z^2}$

26. $\dfrac{4z - 4y}{y}$

27. $\dfrac{x^2 - y}{9z}$

PRACTICE ◣ LESSON 2-6

Simplify.

28. $3(a + b) + b$

29. $8(x + 2y) - x$

30. $5(x + y) + 4(x + y)$

31. $5(a + 2) + 10(a + 3)$

32. $12(m - 3) - (4m - 2)$

33. $9(b + 3) - 4(a - 2)$

34. $3 + 2(8 + 4c) - 2c$

35. $-\dfrac{1}{5}(25x - 5y) + (-x + y)$

36. $-8(3 + a) + 7(6 + 2a)$

37. $0.4(4 - 2x) - 5(2 + 0.1x)$

38. $\dfrac{3}{4}(16x - 8y) + (-x - 2y)$

39. $4(2x + 3) + 5(x - 1)$

40. $9(xy + x) - 3(xy + x)$

41. $-(a - b) - a$

42. $\dfrac{1}{3}(6x - 9y) + \dfrac{1}{4}(16x - 8y)$

43. Find the perimeter of a rectangle with a length of $\left(\dfrac{1}{4}x + 6z\right)$ cm and a width of $\left(x - \dfrac{1}{3}z\right)$ cm.

Match each expression with the equivalent expression in simplified form.

44. $3(3p + q) - 3(2p - 3q)$ a. $12p - 3q$

45. $-5(3p + 2q) + 2(6p - q)$ b. $3p + 12q$

46. $4(2p + q) - (-6p + q)$ c. $-12p - 3q$

47. $2(4p - 3q) + (4p + 3q)$ d. $-3p - 12q$

48. $-6(p + 2q) + 3(-2p + 3q)$ e. $14p + 3q$

Simplify. (Lesson 2-1–Lesson 2-6)

49. $-(3x - 5) + x$

50. $(c + 7) + 8(2c - 1)$

51. $(r + 3) - 2(7 - r)$

52. $\frac{1}{4}t - \left(-\frac{1}{2}t\right) + t^2$

53. $(5b - 7) - (5b - 7)$

54. $9 + 4(y - 3) - y$

55. $-3(x - 7)$

56. $\frac{1}{3}(-15x - 5)$

57. $4(4t + 6)$

58. $\dfrac{14r + 12}{-2}$

59. $\dfrac{100ab - 10}{10}$

60. $\dfrac{-81 - 9x}{9}$

Evaluate each expression when $a = \dfrac{1}{2}$, $b = -3$, **and** $c = 6$. (Lesson 2-1–Lesson 2-6)

61. $a(8b + 10)$

62. $b^2 + (3c - b)$

63. $a(c - b)$

64. $\dfrac{7c - 12}{b}$

65. $\dfrac{-7b - c}{a}$

66. $\dfrac{-c\,|b|}{a}$

MathWorks Career – Commercial Aircraft Designer
Workplace Knowhow

Commercial aircraft designers must consider many factors when engineering a safe airplane. One of the most critical factors is the amount of weight that will be carried. This influences everything from the types of seats that will be installed, to the material used to construct the floor of the passenger seating area, to the amount of thrust required from the engines.

Airline Travelers

Age Group (years)	Average Weight (pounds)	Percent of Travelers
0 - 9	35	7%
10 - 14	85	8%
15 - 19	125	12%
20 - 39	135	28%
40 and above	155	45%

Use the table for Exercises 1–5.

1. If x number of people board an airplane, write a variable expression for the number of people 20–39 years old who board the flight.

2. Using the answer from Exercise 1, write an expression for the total weight of the 20–39 year-old passengers on this flight.

3. Write variable expressions for the total weight of each age category in the table.

4. Combine the variable expressions from Exercises 2 and 3 to express the total weight of all persons on board the aircraft in simplified form.

5. If the maximum capacity of a Boeing B-777 is 279 people, what is the total passenger weight?

2-7 Properties of Exponents

Goals
- Choose appropriate units of measure.
- Evaluate variable expressions.

Applications Biology, Finance, Computers, Population

BIOLOGY During a lab experiment, Kayla observes that a single-cell organism divides into two organisms every hour.

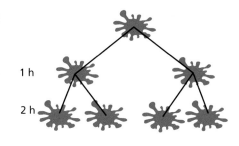

1 h

2 h

1. Copy and complete the table to determine the number of organisms after 5 h.

Hours	1	2	3	4	5
Numbers of organisms					

2. What pattern do you notice every hour?

3. Rewrite the number of organisms each hour by repeatedly multiplying by the same number. What pattern do you notice?

4. Predict the number of organisms in the sixth, seventh and eighth hours.

5. How many hours will it take until the number of organisms is more than 2000?

◣ BUILD UNDERSTANDING

A shortcut to writing a number that is repeatedly multiplied by itself is called **exponential form**. A number written in exponential form has a base and an exponent.

$$a \cdot a \cdot a \cdot a \cdot a - a^5$$

The **base** tells what factor is being multiplied. The **exponent** tells how many equal factors there are. The expression a^5 is read as "a to the fifth power." Any number raised to the first power is that number.

$$n^1 = n, \text{ so } 8^1 = 8$$

Reading Math

Read xy^2 as 'x times y squared.'

Read $(xy)^2$ as 'the quantity x times y, squared.'

E x a m p l e 1   **Personal Tutor at mathmatters2.com**

Evaluate each expression. Let $x = 4$ and $y = -3$.

a. x^2 **b.** y^3 **c.** xy^2

Solution

a. $x^2 = 4^2$
$= 4 \cdot 4$
$= 16$

b. $y^3 = (-3)^3$
$= (-3)(-3)(-3)$
$= -27$

c. $xy^2 = (4)(-3)^2$
$= (4)(-3)(-3)$
$= (4)(9) = 36$

To find the product of $3^2 \cdot 3^4$, you can write out the factors for each term and then write the product using exponents.

$$3^2 \cdot 3^4 = 3 \cdot 3 \cdot 3 \cdot 3 \cdot 3 \cdot 3 = 3^6$$

A shorter method is to use the **product rule**. To multiply numbers with the same base, write the base raised to the sum of the exponents.

$$a^m \cdot a^n = a^{m+n} \qquad\qquad 3^2 \cdot 3^4 = 3^{2+4} = 3^6$$

Apply the product rule to an exponential number raised to an exponent.

$$(5^2)^4 = 5^2 \cdot 5^2 \cdot 5^2 \cdot 5^2 = 5^{2+2+2+2} = 5^8$$

This relationship is described by the **power rule**. To raise an exponential number to an exponent, multiply exponents.

$$(a^m)^n = a^{mn} \qquad\qquad (5^2)^4 = 5^{2\cdot4} = 5^8$$

When a product is raised to an exponent, each factor is raised to that exponent.

$$(2y^2)^3 = (2y^2)(2y^2)(2y^2) = (2)(2)(2)(y^2)(y^2)(y^2) = 8y^6$$

The result is described by the **power of a product rule**. To find the power of a product, find the power of each factor and multiply.

$$(ab)^m = a^m b^m \qquad\qquad (2y^2)^3 = (2^3)(y^2)^3 = 8y^6$$

Properties of Exponents for Multiplication	For all real numbers a and b, if m and n are integers, then $a^m \cdot a^n = a^{m+n}$ $\qquad$ $(a^m)^n = a^{mn}$ $\qquad$ $(ab)^m = a^m b^m$

Example 2

Simplify.

a. $x^3 \cdot x^5$ $\qquad\qquad$ **b.** $(5r^4)^2$ $\qquad\qquad$ **c.** $(u^3v)^4$

Solution

a. $x^3 \cdot x^5 = x^{3+5}$
$\qquad = x^8$

b. $(5^2)(r^4)^2 = 25r^{4\cdot2}$
$\qquad\qquad = 25r^8$

c. $(u^3v)^4 = u^{3\cdot4}v^{1\cdot4}$
$\qquad\qquad = u^{12}v^4$

Study the following examples.

$$34 \div 32 = \frac{3^4}{3^2} = \frac{3 \cdot 3 \cdot 3 \cdot 3}{3 \cdot 3} = 3^2 \qquad\qquad \left(\frac{2}{5}\right)^3 = \frac{2}{5} \cdot \frac{2}{5} \cdot \frac{2}{5} = \frac{2^3}{5^3}$$

These examples illustrate two properties of exponents for division. One property is the **quotient rule**. To divide numbers with the same base, write the base with the difference of the exponents.

$$a^m \div a^n = \frac{a^m}{a^n} = a^{m-n}, a \neq 0 \qquad\qquad \frac{3^4}{3^2} = 3^{4-2} = 3^2$$

Another property is the **power of a quotient rule**. To find the power of a quotient, find the power of each number and divide.

$$\left(\frac{a}{b}\right)^m = \frac{a^m}{b^m}, b \neq 0 \qquad\qquad \left(\frac{2}{5}\right)^3 = \frac{2^3}{5^3}$$

Properties of Exponents for Division	For all real numbers a and b, if m and n are integers, then
	$\dfrac{a^m}{a^n} = a^{m-n}$, if $a \neq 0$ $\left(\dfrac{a}{b}\right)^m = \dfrac{a^m}{b^m}$, if $b \neq 0$

Example 3

Simplify.

a. $\dfrac{t^5}{t^2}$, $t \neq 0$

b. $\left(\dfrac{d}{2}\right)^3$

c. $\left(\dfrac{x^3}{x^2}\right)^4$, $x \neq 0$

Solution

a. $\dfrac{t^5}{t^2} = t^{5-2}$

$\qquad = t^3$

b. $\left(\dfrac{d}{2}\right)^3 = \dfrac{d^3}{2^3}$

$\qquad = \dfrac{d^3}{8}$

c. $\left(\dfrac{x^3}{x^2}\right)^4 = (x^{3-2})^4$

$\qquad = (x^1)^4$

$\qquad = x^4$

◼ TRY THESE EXERCISES

Evaluate each expression when $x = 5$ and $y = -2$.

1. y^2

2. x^4

3. $x^2 y$

4. xy^3

Simplify.

5. $c^4 \cdot c^7$

6. $x \cdot x^9$

7. $(d^5)^2$

8. $(gh^2)^3$

9. $\dfrac{v^7}{v^3}$, $v \neq 0$

10. $\left(\dfrac{c}{3}\right)^3$

11. $z^5 \div z$, $z \neq 0$

12. $\left(\dfrac{t^6}{t^4}\right)^3$, $t \neq 0$

 13. WRITING MATH Explain why it is necessary to include $v \neq 0$ in Exercise 9.

◼ PRACTICE EXERCISES • For Extra Practice, see page 591.

Evaluate each expression when $a = 3$ and $b = -4$.

14. a^3

15. $a^2 - b^2$

16. b^4

17. $\left(\dfrac{b}{7}\right)^2$

18. $4a^2 b$

19. $\dfrac{a^8}{a^7}$

20. $(2 + b)^3$

21. $(a^2 - 5)^2$

Simplify.

22. $c^4 \cdot c^3$

23. $(x^3)^5$

24. $\dfrac{y^9}{y^4}$, $y \neq 0$

25. $g(g^5)$

26. $\dfrac{r^{10}}{r^9}$, $r \neq 0$

27. $j^{10} \cdot j^2$

28. $(z^4)^4$

29. $\left(\dfrac{0}{d}\right)^6$, $d \neq 0$

30. $\left(\dfrac{m}{4}\right)^3$

31. $\dfrac{15w^7}{3w^5}$, $w \neq 0$

32. $3(y^3)^4$

33. $(d^2)(d^3)(d^4)$

34. FINANCE If \$500 is invested in a savings account at 5% interest, it will take a little more than 14 years to double (with no deposits or withdrawals). What is the value of the account in 70 years?

COMPUTERS A byte is the amount of memory in a computer needed to store a single character. A kilobyte (K) is equivalent to 2^{10} or 1024 bytes. A megabyte (MB) is equivalent to 2^{20} or 1,048,576 bytes. A gigabyte (GB) is equivalent to 2^{30} bytes.

35. If a computer disk has 800K of memory, how many bytes does it contain?

36. A word processing program requires 2500K of memory to run. How many megabytes does it require?

37. How many times larger is a megabyte than a kilobyte?

Simplify. You may need to use more than one of the properties.

38. $(x^2 y)^3$

39. $\left(\dfrac{x^8}{x^2} \right)^3$, $x \neq 0$

40. $m^2(m^3 n^4)$

41. $\dfrac{f^2 g^2}{f}$, $f \neq 0$

42. $\dfrac{25 z^8}{50 z^6}$, $z \neq 0$

43. $\left(\dfrac{r}{s^2} \right)^5$, $s \neq 0$

44. $(b^3 \cdot b^4)^3$

45. $\dfrac{v^2 (v^5)}{v^3}$, $v \neq 0$

POPULATION In 1994 the population of the U. S. was about 250 million people. The expression $250(1.01)^x$ can be used to represent the U.S. population where x represents the number of years since 1994. Determine the approximate U.S. population in the following years. Round to the nearest tenth.

46. 1998

47. 2002

48. 2005

EXTENDED PRACTICE EXERCISES

Find each missing exponent.

49. $a^? \cdot a^2 = a^6$

50. $\dfrac{c^?}{c^4} = c^3$, $c \neq 0$

51. $\dfrac{x^7 y^?}{x^2 y^3} = x^5 y^9$, $x \neq 0$ and $y \neq 0$

Write *true* or *false*. Give examples to support your answer.

52. If a and b are integers and $a < b$, then $a^2 < b^2$.

53. If a and b are integers and $a < b$, then $a^3 < b^3$.

54. CRITICAL THINKING A certain type of bacteria doubles in number every 35 min. At 10:00 A.M. there were 16,384 bacteria. At what time had there been exactly half that amount?

MIXED REVIEW EXERCISES

Simplify each numerical expression. (Lesson 2-2)

55. $18 - 5 \cdot 3$

56. $14 \cdot (28 - 4^2)$

57. $6.4 + 9.6 \div 0.3$

58. $26 - 12 \div 2^2 + 16$

59. $10 + 8 - 16 \div 2$

60. $81 \div (3 + 6) - 6$

Draw a number line for each situation. (Lesson 2-1)

61. The number of miles Antjuan walked each day this week is 3.5, 1.4, 2.6, 3.7, 4, 2.8 and 2.

62. Mitch was thinking of a number. Five people made guesses. They were off by -2, 3, 7, -5 and -1.

2-8 Zero and Negative Exponents

Goals
- Write numbers using zero and negative integers as expon▪
- Write numbers in scientific notation.

Applications Physics, Astronomy, Population

Use grid paper. Copy and label the figures shown.

1. Describe the relationship between 2^4 and 2^3.

2. How are 2^3 and 2^2 related? How are 2^3 and 2^1 related?

3. Write a statement summarizing the pattern.

4. What repeating pattern of shapes do you notice in the figures?

5. Following the pattern, draw the next figure after the 2 by 1 rectangle.

6. What is the area of the figure you drew? What power of 2 do you think this figure represents?

7. Draw the next three figures in the pattern that represent 2^{-1}, 2^{-2} and 2^{-3}.

8. What is the area of each figure?

9. Complete: $2^{-n} = \dfrac{\blacksquare}{\blacksquare}$

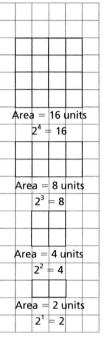

Area = 16 units
$2^4 = 16$

Area = 8 units
$2^3 = 8$

Area = 4 units
$2^2 = 4$

Area = 2 units
$2^1 = 2$

◥ BUILD UNDERSTANDING

Any nonzero real number raised to the zero power has a value that is equal to 1.

Zero Property of Exponents	For any nonzero real number a, $$a^0 = 1$$

Exponents can be negative numbers. The quotient property of exponents can help you understand the meaning of negative exponents.

Check Understanding

Write each expression as a fraction.

a. 5^{-2} b. 7^{-5}

c. $(-2)^{-4}$ d. x^{-3}

$$\frac{4^3}{4^5} = \frac{4 \cdot 4 \cdot 4}{4 \cdot 4 \cdot 4 \cdot 4 \cdot 4} = \frac{1}{4 \cdot 4} = \frac{1}{4^2}$$

Using the quotient property, $\dfrac{4^3}{4^5} = 4^{3-5} = 4^{-2} = \dfrac{1}{4^2}$.

Property of Negative Exponents	For any nonzero real number a, if n is a positive integer, $$a^{-n} = \frac{1}{a^n}$$

Example 1

Simplify each expression.

a. $x^7 \div x^{-3}$

b. $c^5 \cdot c^{-2}$

c. $(y^3)^{-4}$

Solution

a. $x^7 \div x^{-3} = x^{7-(-3)}$
$= x^{10}$

b. $c^5 \cdot c^{-2} = c^{5+(-2)}$
$= c^3$

c. $(y^3)^{-4} = y^{3(-4)}$
$= y^{-12}$ or $\dfrac{1}{y^{12}}$

Example 2

Evaluate each expression when $m = 2$ and $s = -3$.

a. m^{-5}

b. $(s^2)^{-3}$

c. $m^3 s^{-2}$

Solution

a. $m^{-5} = (2)^{-5}$
$= \dfrac{1}{2^5} = \dfrac{1}{32}$

b. $(s^2)^{-3} = s^{-6} = (-3)^{-6}$
$= \dfrac{1}{(-3)^6} = \dfrac{1}{729}$

c. $m^3 s^{-2} = m^3 \cdot \dfrac{1}{s^2} = \dfrac{m^3}{s^2}$
$= \dfrac{(2)^3}{(-3)^2} = \dfrac{8}{9}$

Any number in decimal form such as 26.376, 0.0067 or 200,000,000 is in **standard form**. In order to be able to write and compute with very large and small numbers, the number can be written in **scientific notation**. A number written in scientific notation has two factors. The first factor is greater than or equal to 1 and less than 10. The second factor is a power of 10.

standard form	*scientific notation*
0.0000268	$2.68 \cdot 10^{-5}$
370,000,000	$3.7 \cdot 10^8$

Calculators use scientific notation to display very large and very small numbers. On some calculators you can convert between scientific notation and standard form by changing the mode.

Mental Math Tip

To multiply by 10^n when n is a positive integer, move the decimal point n places to the right.

$5.18 \cdot 10^7 = 51,800,000$

To multiply by 10^n when n is a negative integer, move the decimal point n places to the left.

$9.2 \cdot 10^{-4} = 0.00092$

Example 3

a. Write 2,674,000 in scientific notation.

b. Write $6.3 \cdot 10^{-5}$ in standard form.

Solution

a. $2,674,000 = 2,674,000.$

2.674000   Move the decimal point so the first factor is between 1 and 10.

$= 2.674 \cdot 10^6$ The exponent of 10 is the number of places that the decimal point moved.

b. $6.3 \cdot 10^{-5} = 0.000063$ Move the decimal point to the left 5 places.

$= 0.000063$

Example 4

Technology Note

The first three lines of the calculator screen show Example 4 in normal mode.

The last three lines show it in scientific mode.

PHYSICS The speed of light is $3.00 \cdot 10^5$ km/sec. How far does light travel in 1 h? Write the answer in scientific notation.

Solution

Find the number of seconds in 1 h.

$$1 \text{ h} = 60 \text{ min} - 60(60) \text{ sec} = 3600 \text{ sec} - (3.6 \cdot 10^3) \text{ sec}$$

To find the distance light travels in 1 h, multiply.

$$(3.00 \cdot 10^5)(3.6 \cdot 10^3) = (3.00 \cdot 3.6)(10^5 \cdot 10^3)$$
$$= (10.8)(10^{5+3})$$
$$= 10.8 \cdot 10^8$$
$$= 1.08 \cdot 10^9 \quad \text{Remember that the first factor must be less than 10.}$$

Light travels $1.08 \cdot 10^9$ km in 1 h.

◤ TRY THESE EXERCISES

Simplify.

1. $x^4 \div x^7$ **2.** $b^{-5} \cdot b^9$ **3.** $(w^5)^{-4}$ **4.** $d^3 \cdot d^{-8}$

Evaluate each expression when $m = -4$ and $n = 5$.

5. m^{-2} **6.** n^{-4} **7.** $m^0 n^{-2}$ **8.** $n^3 \cdot n^{-6}$

Write each number in scientific notation.

9. 29,000,000 **10.** 0.0039 **11.** 0.0000808

Write each number in standard form.

12. $7.6 \cdot 10^8$ **13.** $9.5 \cdot 10^{-3}$ **14.** $4.03 \cdot 10^{-6}$

15. POPULATION The most highly populated city in the world is Tokyo, Japan. Its population is about 27,000,000. Express the population of Tokyo in scientific notation.

◤ PRACTICE EXERCISES • For Extra Practice, see page 592.

Simplify.

16. $z^{-10} \div z^2$ **17.** $y^{-4} \cdot y^{-6}$ **18.** $(w^3)^{-3}$ **19.** $a^8 \cdot a^{-8}$

20. $x^5 \div x^8$ **21.** $b^{-3} \cdot b^3$ **22.** $\dfrac{p^2}{p^5}$ **23.** $\left(\dfrac{a}{a^3}\right)^6$

Evaluate each expression when $s = -1$ and $t = 4$.

24. s^{-3} **25.** t^{-5} **26.** $(st)^{-2}$ **27.** $t^6 \cdot t^{-4}$

28. $\dfrac{s^2}{s^7}$ **29.** $(t^2)^{-1}$ **30.** $\dfrac{t^{-3}}{t^{-2}}$ **31.** $\left(\dfrac{s^2}{s^7}\right)^3$

Write each number in scientific notation.

32. 8450

33. 0.04658

34. 0.0000317

35. 15,000,000

36. 0.0096

37. 3,658,000

Write each number in standard form.

38. $2.01 \cdot 10^{-4}$

39. $5.3 \cdot 10^{8}$

40. $8.6 \cdot 10^{-6}$

41. $3.907 \cdot 10^{3}$

42. $9.98 \cdot 10^{-5}$

43. $6.0046 \cdot 10^{7}$

DATA FILE For Exercises 44 46, refer to the data on American pet ownership on page 560. Write the number of each pet in scientific notation.

44. reptiles

45. dogs

46. cats

47. ASTRONOMY Mercury travels around the sun at a speed of approximately $1.72 \cdot 10^{5}$ km/h. At this rate, how far does Mercury travel in 30 days?

48. PHYSICS The wavelength of red light is about $6.5 \cdot 10^{-5}$ cm. Express this distance in meters.

Write in order from least to greatest.

49. $2^3 \cdot 2^{-2}; (6 \cdot 3)^0; \left(\dfrac{1}{3}\right)^{-1}$

50. $(-4)^{-2}(-4); (-5)^2(-5); (-5)(-5)^{-2}$

EXTENDED PRACTICE EXERCISES

51. CRITICAL THINKING You know that $2^2 = 4$ and $2^{-2} = \dfrac{1}{4}$. What do you think is the value of $\dfrac{1}{2^{-2}}$? If a is a nonzero real number and n is an integer, what does $\dfrac{1}{a^{-n}}$ represent?

Use the results of Exercise 51. Rewrite each with positive exponents.

52. $\dfrac{1}{m^{-3}}$

53. $\dfrac{x^{-2}}{y^{-2}}$

54. $\dfrac{a^4}{b^{-3}}$

55. $\dfrac{m^{-7}}{n^3}$

56. CHAPTER INVESTIGATION Compare the population density of the state and city you chose. Explain similarities and differences.

MIXED REVIEW EXERCISES

Simplify. (Lessons 2-4 and 2-5)

57. $-5v + 6w - 8w$

58. $\dfrac{1s}{2} - 9s + 6s$

59. $-11tw - (6w)$

60. $1.4x + 0.69x$

61. $\dfrac{3}{4}(-8r + 16)$

62. $\dfrac{28f + 16g}{-4}$

63. $-0.5(17v - 8)$

64. $\dfrac{32q + 14r}{8}$

65. $\dfrac{1}{4}(8x - 6y + 12)$

66. DATA FILE Refer to the data on record transportation speeds on page 573. Find the record speed of the combat jet fighter in ft/s. (Hint: 1 mi = 5280 ft) (Prerequisite Skill)

Review and Practice Your Skills

Evaluate each expression when $x = 2$ and $y = -3$.

1. $(x^2y)^2$

2. $5xy^2$

3. $\left(\dfrac{y}{4}\right)^2$

4. $4x^2y$

5. $y^4 - x^3$

6. $-x^3$

7. $\dfrac{x^5}{x^4}$

8. $(5 + y)^5$

9. $\dfrac{x^3}{y^2}$

10. $x^3 - y^2$

11. $2\left(\dfrac{y^5}{y^3}\right)$

12. $(x + 4)^3$

Simplify.

13. $a^2 \cdot a^5$

14. $(c^2)^4$

15. $(c^4d)^2$

16. $\dfrac{g^7}{g^3}$, $g \neq 0$

17. $4(x^3)^2$

18. $f^5 \cdot f^3$

19. $\left(\dfrac{2}{y}\right)^4$, $y \neq 0$

20. $(3r^3)^3$

21. $x \cdot x$

22. $y^4 \div y$, $y \neq 0$

23. $\left(\dfrac{m^3}{m}\right)^4$, $m \neq 0$

24. $(-2x^2)^3$

25. $\left(\dfrac{a^4}{a}\right)^3$, $a \neq 0$

26. $b^3 \cdot b^8$

27. $(4x^2)^3$

28. $(x^2y^4)^5$

29. $-3(d^2e)^3$

30. $\left(\dfrac{-2}{n}\right)^3$, $n \neq 0$

Simplify.

31. $y^{-3} \cdot y^{-5}$

32. $b^{-4} \cdot b^4$

33. $r^0 \cdot b^{-2}$

34. $x^3 \div x^9$

35. $\dfrac{m^2}{m^7}$, $m \neq 0$

36. $(p^{-3})^{-2}$

Evaluate each expression when $f = 2$ and $g = -4$.

37. f^{-4}

38. $(f^3)^{-2}$

39. $g^0 \cdot f^3$

40. $\dfrac{f^{-2}}{f^{-4}}$

41. $(g^2)^{-1}$

42. $g^4 \cdot f^{-4}$

Write each number in scientific notation.

43. 1,000,000

44. 45,000

45. 3334

46. 0.00505

47. 0.00013

48. 0.0147

Write each number in standard form.

49. $7.2 \cdot 10^{-3}$

50. $9.45 \cdot 10^6$

51. $1.2 \cdot 10^{-5}$

52. $3.03 \cdot 10^{-2}$

53. $5.5 \cdot 10^4$

54. $7.1 \cdot 10^7$

Graph the given sets of numbers on a number line. (Lesson 2-1)

55. $\left\{-\dfrac{2}{3}, 0, \dfrac{1}{2}, \dfrac{3}{2}\right\}$

56. $\{-1.5, -0.5, 0, 2.5\}$

57. the integers from -3 to 5

58. all real numbers less than 2

59. all real numbers between -4 and 4

60. all real numbers greater than 6.

Simplify each numerical expression. (Lesson 2-2)

61. $3 \cdot 4 - 7$

62. $4(7 - 9) + 6$

63. $-\dfrac{1}{2} + \dfrac{4}{3} \cdot 3^2$

64. $7 - 6 \div 2 + 11$

65. $6^2 \div 2 + 14$

66. $11 - (7 - 3) \div 2^2$

Translate each variable expression into a word phrase. (Lesson 2-3)

67. $x + 7$

68. $\dfrac{4}{a}$

69. $-3y$

70. $\dfrac{1}{2}x + 15$

71. $-0.5 - y$

72. $\dfrac{4x}{3}$

Simplify. (Lesson 2-4–Lesson 2-8)

73. $b + b$

74. $b \cdot b$

75. $(2ab)(a^2b)$

76. $4(x - 10) + 2(x + 4)$

77. $\dfrac{10(x + 3)}{5}$

78. $-7a - 5 - 2a$

79. $-(a - 3) - 2a - 11$

80. $8 + 2^3 \div 4$

81. $5 + 3 \cdot 3^2$

82. $(2ab)^2 + 3a^2b^2$

83. $(2a^2b)^3$

84. $x^4 \cdot x^{-3}$

85. $7(x - 4) - 6(2x + 8) + 3(-4x + 1)$

86. $-4(2a - b) + 6(4a + 2b) - 3(-a + 7b)$

True or False. (Lesson 2-1–Lesson 2-8)

87. $|-3| = -3$

88. $2x^2 = (2x)^2$

89. $\dfrac{f^6}{f^4} = f^{-2}$

90. $4.28 \cdot 10^4 = 42,800$

91. $2x^2 + 2x^2 = 4x^2$

92. $2f^{-2} = \dfrac{1}{2f^2}$

93. $9^0 = 1$

94. $0.0000714 = 7.14 \cdot 10^5$

95. $(4x^2)^3 = 64x^5$

96. "six less than a number" is the expression $6 - x$.

97. "a number squared times two" is the expression $2x^2$.

Evaluate each expression when $x = 2$, $y = -5$ and $z = 3$. (Lesson 2-1–Lesson 2-8)

98. $xy - z$

99. $6(z - x) + y^2$

100. $2x - 3y + z$

101. $z^2 + 2x - y$

102. $2|y| - z$

103. $\dfrac{1}{2}(2y - 4z)$

104. $\dfrac{1}{5}y - \dfrac{2}{3}z + \dfrac{3}{2}x$

105. $-9(x + y) - 3z$

106. $-6|x| - 4z$

107. $\dfrac{-(xy - z)}{26}$

108. $\dfrac{12z + 20x}{4}$

109. $0.4x - 1.5y + 0.2z$

110. $-4x^3$

111. $\dfrac{(-2z)^4}{18z^0}$

112. $4x^2 \cdot (x^3)^2$

Problem Solving Skills: Find a Pattern

Some problems are solved by recognizing a pattern, while other problems require that you extend the pattern to find the solution. This strategy is called **look for a pattern**, and it is used with many different types of problems.

A set of numbers that is arranged according to a pattern is called a **sequence**. Each number of the sequence is called a **term**. By figuring out the pattern, you can predict the next term. Making a table often helps you see a pattern, and it is frequently used with this strategy.

Problem Solving Strategies

Guess and check

✔ Look for a pattern

Solve a simpler problem

Make a table, chart or list

Use a picture, diagram or model

Act it out

Work backwards

Eliminate possibilities

Use an equation or formula

Problem

 Personal Tutor at mathmatters2.com

FINANCE Sun Li invests $2,000 in a mutual fund. The value of the investment will double every 6 yr. How long will it take for the investment to be worth $16,000?

Solve The Problem

Create a table to find a pattern in the value of the investment. Let $2000 be the first term of the pattern.

Years Invested	Value of Investment
0	$2000
6	$4000
12	$8000
18	$16,000

After 18 yr, the investment will be worth $16,000.

TRY THESE EXERCISES

The first term in a sequence and the pattern are given. Write the next five terms of each sequence.

1. 5; add 7

2. 0; subtract 5

3. −60; divide by 3

4. $4x$; multiply by −2

5. $2x − 3$; add $x + 5$

6. $32x^4$; divide by $2x$

Use a calculator to find the first four products. Look for a pattern. Predict the fifth product. Use a calculator to check your answers.

7. $23 \cdot 101 =$
$24 \cdot 101 =$
$25 \cdot 101 =$
$26 \cdot 101 =$

8. $202,202 \cdot 9 =$
$202,202 \cdot 8 =$
$202,202 \cdot 7 =$
$202,202 \cdot 6 =$

9. $3703 \cdot \ \ 3 =$
$3703 \cdot \ \ 6 =$
$3703 \cdot \ \ 9 =$
$3703 \cdot 12 =$

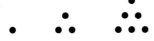

Five-step Plan
1. Read
2. Plan
3. Solve
4. Answer
5. Check

Draw the next three figures. Describe the mathematical pattern.

10.

11.

Describe the pattern in each sequence. Then write the next three terms.

12. 4, 8, 12, 16, . . .

13. 1, 3, 6, 10, 15, . . .

14. 1, 2, 4, 8, 16, . . .

15. 81, 27, 9, 3, . . .

PART-TIME JOB Mato works 40 h/wk and earns \$12/h. When he works overtime, he earns one and a half times his regular rate for each hour over 40 h. Calculate the amount that Mato earns when he works the following hours in one week.

16. 42 h

17. 46 h

18. 43.5 h

19. ADVERTISING Jonica works at a shoe store. During a back-to-school sale, the discount on any purchase is $5x$% where x is the number of shoes purchased up to 8 pairs. Jonica makes a sign to hang in the window to display how the sale works. Create a sample advertisement displaying the pattern of the discount when 1 to 8 pairs of shoes are purchased.

20. WRITING MATH Explain why the store in Exercise 19 places a limit on the number of shoes that can be purchased with this specific discount.

21. FINANCE Marcy puts \$500 in a savings bond. The amount in the bond will increase by half of the amount every 5 yr. How much will the savings bond be worth after 15 yr?

NUMBER SENSE A *perfect number* is a number equal to the sum of its factors excluding the number itself. For example, 6 is a perfect number since $6 = 1 + 2 + 3$. Determine if each number is a perfect number. Justify each answer.

22. 10

23. 28

24. 48

25. The *Fibonacci sequence,* named after Leonardo Fibonacci, is a sequence of numbers where each term is the sum of the previous two terms. Write the next six numbers in the sequence.

$$1, 1, 2, 3, 5, \ldots$$

◼ MIXED REVIEW EXERCISES

Replace each ◼ with <, > or =. (Lesson 2-1)

26. 6 ◼ −6

27. 0 ◼ −1

28. $-1\frac{1}{5}$ ◼ −1.2

29. WEATHER The temperature at the top of Mt. Washington is −25°F. The temperature at the base of the mountain is −17°F. Which temperature is colder? (Lesson 2-1)

Evaluate each expression. (Lesson 2-1)

30. $|y|$, when $y = -42$

31. $-|s|$, when $s = -18$

32. $-(-j)$, when $j = -\dfrac{4}{5}$

Chapter 2 Review

VOCABULARY ◣

Choose the word from the list that best completes each statement.

1. Two different integers that are the same distance from zero on the number line but in the opposite direction are __?__.

2. A(n) __?__ is a symbol used to represent a number.

3. The distance a number is from zero on the number line is the __?__ of the number.

4. The number that corresponds to a point on a number line is called the __?__ of the point.

5. A(n) __?__ is two or more numbers joined by operations such as addition, subtraction, multiplication and division.

6. To __?__ a variable expression, perform as many of the indicated operations as possible.

7. The __?__ of a number written in exponential form tells how many equal factors are being multiplied.

8. The __?__ are a set of rules that specify which operations must precede other operations in order to properly evaluate expression.

9. The __?__ of $1.2 \cdot 10^5$ is 120,000.

10. A set of numbers arranged according to a pattern is called a(n) __?__.

a.	absolute value
b.	base
c.	coordinate
d.	exponent
e.	numerical expression
f.	opposites
g.	order of operations
h.	scientific notation
i.	sequence
j.	simplify
k.	standard form
l.	variable

LESSON 2-1 ◣ Real Numbers, p. 52

▶ The set of **integers** consists of the whole numbers and their **opposites**.

▶ The set of **rational** and **irrational numbers** make up the set of **real numbers**.

Graph each set of numbers on a number line.

11. real numbers less than 1

12. real numbers greater than or equal to -4

Replace each ▓ with $<$, $>$, or $=$.

13. $2\frac{3}{8}$ ▓ $2\frac{2}{5}$

14. -5.3 ▓ 3.1

15. $|7|$ ▓ -7

16. Evaluate $-|x|$, when $x = -8$.

LESSON 2-2 ◣ Order of Operations, p. 56

▶ When you find the **value** of a **numerical expression**, you **simplify** the expression.

▶ An expression containing one or more **variables** is called a **variable expression**. To **evaluate** a variable expression, substitute a given number for each variable, then simplify the numerical expression.

Simplify each numerical expression.

17. $42 - 2 \cdot 5 + (8 - 2)$

18. $2[18 - (5 + 3^2) \div 7]$

Evaluate each expression when $x = 12$.

19. $x + 5$

20. $\frac{1}{3}x - 1$

21. $6x - 7$

22. $\frac{x}{2} + 10$

23. $(x)(x)$

LESSON 2-3 ◣ Write Variable Expressions, p. 62

▶ Many words and phrases suggest certain operations. Any variable can be used to represent a number.

Write each phrase as a variable expression.

24. six more than a number

25. nine less than a number

26. the product of a number and eight

27. the difference of four and twice a number

Translate each variable expression into a word phrase.

28. $5 - x$

29. $\frac{w}{3}$

30. $3n + 4$

31. $6x - 1$

LESSON 2-4 ◣ Add and Subtract Variable Expressions, p. 66

▶ To **simplify** a variable expression, use the properties and **combine like terms**.

▶ An expression is simplified when only like terms remain.

Simplify.

32. $6rs + rs - rt$ **33.** $-2 - mn + 3.2mn + 3$ **34.** $-2cd - (-3cd)$ **35.** $\frac{1}{2}xy + \frac{1}{2}xy - 3xy - x$

Evaluate each expression when $x = -2$, $y = 5$, and $z = \frac{1}{3}$.

36. $-3x + 7x$

37. $-11y - (-9y)$

38. $6y - 7y + 15z$

39. $9z - 2y - 3z + 5y$

LESSON 2-5 ◣ Multiply and Divide Variable Expressions, p. 72

▶ To multiply variable expressions, use the distributive property.

▶ To divide variable expressions, divide each term in the numerator by the denominator.

Simplify.

40. $-2(mn + 3.2)$

41. $-(c + 3d)$

42. $2(xy + y)$

43. $-3(2y - 4x)$

44. $\frac{12x - 16}{2}$

45. $\frac{-9y + 9x}{-3}$

46. $\frac{25n - 15m}{-5}$

47. $\frac{4x + y}{2}$

48. Evaluate $\frac{-2x - y}{5}$ when $x = 5$ and $y = -6$.

LESSON 2-6 ◣ Simplify Variable Expressions, p. 76

▶ Use the **order of operations** when simplifying variable expressions.

Simplify.

49. $2x + 3(4x - 1)$

50. $8 - 2(x - 5)$

51. $8(x - 1) - 5x$

52. $3(2x - 5) - 4(3x - 7)$

Simplify.

53. $4(x + 5) + 2(x - 1)$

54. $-(x - 9) - 2(-3x + 8)$

55. $\frac{1}{3}(6x + 21) - \frac{3}{4}(4x - 12)$

56. $4(xy - 6x) - 4(x - xy)$

57. Find the perimeter of a rectangle with a length of $(2x - 7)$ ft and a width of $(x + 3)$ ft.

LESSON 2-7 ◤ Properties of Exponents, p. 82

▶ A number written in exponential form has a **base** and an **exponent**.

▶ Use the properties of exponents for multiplication and division to simplify expressions.

Evaluate each expression when $m = 5$ and $n = -2$.

58. m^3

59. $m^2 - n^3$

60. $\left(\dfrac{m}{8}\right)^2$

61. mn^2

Simplify.

62. $x^2 \cdot x^5$

63. $(a^2)^4$

64. $\dfrac{27a^3}{9a}$

65. $\left(\dfrac{27w^3}{9w}\right)^2, w \neq 0$

LESSON 2-8 ◤ Zero and Negative Exponents, p. 86

▶ For every nonzero real number a, $a^0 = 1$ and $a^{-n} = \dfrac{1}{a^n}$.

▶ A number written in **scientific notation** has two factors. The first factor is greater than or equal to 1 and less than 10. The second factor is a power of 10.

Evaluate each expression when $r = -2$ and $s = 2$.

66. r^0

67. r^{-2}

68. $r^0 s^{-3}$

69. $s^2 \cdot s^{-5}$

70. Write 3,500,000 in scientific notation.

71. Write $7.5 \cdot 10^{-4}$ in standard form.

72. Write $2.4 \cdot 10^8$ in standard form.

73. Write 0.0000546 in scientific notation.

LESSON 2-9 ◤ Problem Solving Skills: Find a Pattern, p. 92

▶ A set of numbers that is arranged according to a pattern is called a **sequence**. Each number of the sequence is called a **term**.

▶ Some problems are solved by recognizing a pattern, and then extending the pattern to find the solution.

Write the next three terms in each sequence.

74. 95, 91, 87, 83, ■, ■, ■

75. $1, \dfrac{1}{3}, \dfrac{1}{9}, \dfrac{1}{27}$, ■, ■, ■

76. $-4, 8, -16, 32$, ■, ■, ■

77. 7, 18, 16, 27, 25, ■, ■, ■

CHAPTER INVESTIGATION

EXTENSION Find the population density of your town or city. Compare the population, number of square miles, and population density of your city with those of a city similar in population. Write a paragraph explaining the similarities and differences of these two cities. Include possible reasons for any differences in population density.

Chapter 2 Assessment

Graph each set of numbers on a number line.

1. all real numbers greater than -5

2. all real numbers less than or equal to -2

Evaluate each expression when $r = -4$.

3. $|r|$

4. $-|-r|$

5. $-r$

Simplify each numerical expression.

6. $20 + 8 \times 2$

7. $(300 + 42) \div 3^2$

8. $(18 - 2) \cdot 2 + 4^2$

Evaluate each variable expression when $c = 25$.

9. $2c - 15$

10. $\frac{1}{2}(c + 13)$

11. $100c \div 5^2$

Write each phrase as a variable expression.

12. the quotient of six and three times a number

13. a number decreased by 18

Translate each variable phrase into a word phrase.

14. $5x$

15. $12 + 2x$

16. $\frac{2}{-7}$

Simplify.

17. $5x + 7x$

18. $5mn + 3mn + (-2mn)$

19. $-3x - 5x + 8y$

20. $2(x + 4)$

21. $-4(2 - xy)$

22. $\frac{14x + 7}{7}$

23. $\frac{-6y + 24}{-3}$

24. $5(rs + s) - 3(rs + r)$

25. $6(y - 3) - 4y$

Simplify.

26. $x^3 x^5$

27. $(n^2)^3$

28. $\frac{16b^6}{4b^2}, b \neq 0$

29. $(3y^2)^3$

30. $x^{-8} \div x^5$

31. $y^{-2} \cdot y^{-5}$

32. $\frac{a^{-3}}{a^{-6}}$

33. $d^5 \div d^9$

Evaluate each expression when $a = 2$ and $b = 24$.

34. b^0

35. a^{-3}

36. $(a^3)^{-2}$

Write each number in scientific notation.

37. 0.00385

38. 9,380,000,000

39. 0.0543

Write the next three terms in each sequence.

40. 101, 100, 98, 95, 91, ▪, ▪, ▪

41. $x, 2x + 1, 3x + 2, 4x + 3$, ▪, ▪, ▪

42. David won $4000 in a contest. If the first day he spends half of this amount and then each day there after spends half of what is left, how long will it take before he only has about $4.00 left?

Standardized Test Practice

Part 1 Multiple Choice

**Record your answers on the answer sheet
provided by your teacher or on a sheet of paper.**

1. The data shows the heights, in inches, of plants
grown for a science experiment. What is the
median height of the plants? (Lesson 1-2)

| 29, 22, 27, 31, 19, 25, 22, 28 |

  Ⓐ 22 in.   Ⓑ 25 in.
  Ⓒ 26 in.   Ⓓ 29 in.

2. Kayla took last year's shot statistics for her
school's basketball team and created the
scatter plot below. What type of relationship
do the data show? (Lesson 1-4)

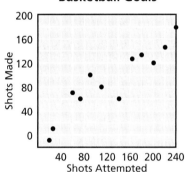

Basketball Goals

  Ⓐ positive
  Ⓑ negative
  Ⓒ even
  Ⓓ odd

3. Use the box-and-whisker plot to determine
what percent of the highest recorded wind
speeds in the U.S. range from 40 to 70 mph.
(Lesson 1-6)

Highest Recorded Wind Speeds (mph) in the U.S.

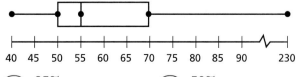

  Ⓐ 25%   Ⓑ 50%
  Ⓒ 75%   Ⓓ 100%

4. If $a = -3$ and $b = 3$, then which of the
following statements is *false*? (Lesson 2-1)

  Ⓐ $|a| > 2$
  Ⓑ $|a| = |b|$
  Ⓒ $|b| < 2$
  Ⓓ $|a| = b$

5. The Rockies scored four more runs than the
Cubs scored. Which expression represents the
number of runs the Cubs scored if the Rockies
scored n runs? (Lesson 2-3)

  Ⓐ $n + 4$
  Ⓑ $4n$
  Ⓒ $4 - n$
  Ⓓ $n - 4$

6. Simplify $\dfrac{40x - 16}{-8}$. (Lesson 2-5)

  Ⓐ $32x + 8$
  Ⓑ $-5x + 2$
  Ⓒ $-5x - 16$
  Ⓓ $5x - 2$

7. Which expression is equivalent to
$2(y + 7) - 3(y + 7)$? (Lesson 2-6)

  Ⓐ $-y - 7$
  Ⓑ $2y + 7$
  Ⓒ $-y + 7$
  Ⓓ $-y + 14$

8. Simplify $\left(\dfrac{n^7}{n^5}\right)^4$, $n \neq 0$. (Lesson 2-7)

  Ⓐ n^6 Ⓑ n^8
  Ⓒ n^{13} Ⓓ n^{23}

9. A human blinks about $6.25 \cdot 10^6$ times a year.
What is this number in standard notation?
(Lesson 2-8)

  Ⓐ 625,000
  Ⓑ 6,250,000
  Ⓒ 62,500,000
  Ⓓ 625,000,000

Part 2   Short Response/Grid In

Record your answers on the answer sheet provided by your teacher or on a sheet of paper.

10. The stem-and-leaf plot shows the number of points scored by the Grizzlies in each of their basketball games this season. In how many games did they score at least 30 points? (Lesson 1-3)

Stem	Leaf
1	8 9
2	0 2 3 3 6 8 8 9
3	0 1 4 4 5 6 8 9
4	0 1 2

1|8 represents 18 points.

11. Find $A - B$ if $A = \begin{bmatrix} -2 & 1 \\ 5 & -3 \end{bmatrix}$ and $B = \begin{bmatrix} 4 & 6 \\ -1 & 2 \end{bmatrix}$. (Lesson 1-8)

12. Write an algebraic expression for the phrase *three times the difference of m and n increased by twice p*. (Lesson 2-3)

13. Write an expression in simplest form for the perimeter of the figure. (Lesson 2-4)

14. What is the area of a triangle with a base of 8 and a height of $3x - 6$? (Lesson 2-5)

15. Simplify the expression $(7y^3)^2$. (Lesson 2-7)

Test-Taking Tip   Ⓐ Ⓑ Ⓒ Ⓓ

Question 18
If a problem seems difficult, don't panic. Reread the question slowly and carefully. Always ask yourself, "What have I been asked to find?" and "What information will help me find the answer?"

16. Evaluate $\dfrac{a^2}{b^{-3}}$ if $a = 3$ and $b = -2$. (Lesson 2-8)

17. What is the eighth term in the sequence 8, 7, 14, 13, 26, …? (Lesson 2-9)

Part 3   Extended Response

Record your answers on a sheet of paper. Show your work.

18. Anthony wants to change his cellular phone carrier. Before he signs an agreement, he compares the different service plans available.

Plan	Monthly Fee	Cost per minute
A	$5.95	$0.30
B	$12.95	$0.10

a. Suppose x represents the monthly fee, and y represents the number of minutes. Write an algebraic expression cost per month for any one of these phone plans. (Lesson 2-3)

b. Which plan should Anthony pick if he makes an average of 150 minutes worth of calls each month? Explain. (Lesson 2-2)

19. The figures are formed using toothpicks. If each toothpick is a unit, then the perimeter of the first figure is 4 units.

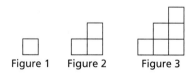

Figure 1   Figure 2   Figure 3

a. Copy and complete the table below. (Lesson 2-9)

Figure Number	1	2	3	4	5	6
Perimeter	4	8				

b. Write an expression for the perimeter of figure n? (Lesson 2-3)

c. Use the expression you wrote in part b to find the perimeter of Figure 20. (Lesson 2-1)

3

Equations and Inequalities

THEME: Physics

Any time you are dealing with relationships, you can use equations and inequalities to describe and study the interactions of cause-and-effect connections.

When you hear the word "physics" you may think of simple machines, acoustics, gravity, movement, or other natural and material phenomena, but it should also make you think of mathematics, especially algebra. Physics is the science of matter and energy and how they interact. These interactions can be expressed using equations and inequalities.

- **Mechanical engineers** (page 113) design and develop equipment. Much of their work involves using formulas associated with physics, or discovering formulas that can be applied to the physical world.

- Passenger safety is one of the primary goals of **automobile designers** (page 131). They study the physical principles that relate to acceleration, force, and mass to determine how cars react when a collision occurs.

Math nline

mathmatters2.com/chapter_theme

Planet Table

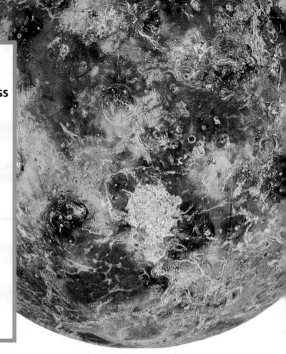

Planet	Length of year (Earth days)	Length of day (Earth hours)	Approximate mass (kilograms)
Mercury	87.969	1,407.6	$3.30 \cdot 10^{23}$ kg
Venus	224.7	8,532.5	$4.87 \cdot 10^{24}$ kg
Earth	365.26	23.934	$5.97 \cdot 10^{24}$ kg
Moon			$7.35 \cdot 10^{22}$ kg
Mars	686.98	24.62	$6.42 \cdot 10^{23}$ kg
Jupiter	4,330.6	9.92	$1.90 \cdot 10^{27}$ kg
Saturn	10,747	10.5	$5.69 \cdot 10^{26}$ kg
Uranus	30,588	17.24	$8.69 \cdot 10^{25}$ kg
Neptune	59,800	16.11	$1.02 \cdot 10^{26}$ kg
Pluto	90,591	153.3	$1.25 \cdot 10^{22}$ kg

Data Activity: Planet Table

Newton's Universal Law of Gravitation states that all objects in the universe attract all other objects. It also states that the strength of the force of attraction depends on the masses of the two objects. The mass of an object is a measure of how much material it contains.

Use the table for Questions 1–4.

1. Which planet exerts the greatest force of attraction?

2. About how many times greater is the force of attraction on Saturn than on Earth?

3. Which planet has a year that is about 84 times as long as a year on Earth?

4. About how many more hours are there in a Mars year than in an Earth year?

CHAPTER INVESTIGATION

Gravity is the force of attraction between any two pieces of matter. The force of attraction between you and the Earth is what allows you to stand still and not float away. The gravitational field strength of the Earth is the force of gravity exerted on an object with a mass of 1 kg. So a more massive object will have a greater gravity force acting on it.

Working Together

In the 1800's, scientists measured the strength of the Earth's gravity field with a simple pendulum. The pendulum was a heavy ball suspended on a long wire string. This procedure produced some of the most accurate measurements of the Earth's gravity field. Use the Chapter Investigation icons to measure the strength of the Earth's gravity field using a pendulum.

3 Are You Ready?

Refresh Your Math Skills for Chapter 3

The skills on these two pages are ones you have already learned. Use the examples to refresh your memory and complete the exercises. For additional practice on these and more prerequisite skills, see pages 576–584.

SIGNED NUMBERS

In this chapter you will solve equations that include signed numbers. It is important to understand how these numbers affect the final solution of the equation.

Examples Add a positive and negative number:

$5 + (-8) = -3$

Add two negative numbers:

$-3 + (-6) = -9$

Subtract a negative number from a positive number:

$8 - (-4) = 8 + 4 = 12$

Subtract a negative number from a negative number:

$-5 - (-4) = -5 + 4 = -1$

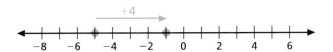

Add or subtract.

1. $-4 + 6$ **2.** $-7 - 9$ **3.** $5 - (-12)$

4. $-13 + (-3)$ **5.** $7 - (-4)$ **6.** $-8 + 15$

7. $-\dfrac{4}{5} - \dfrac{3}{15}$ **8.** $\dfrac{1}{2} - \dfrac{5}{8}$ **9.** $-\dfrac{3}{4} - \left(-\dfrac{7}{16}\right)$

10. $3.2 + (-0.5)$ **11.** $-2.8 + (-1.4)$ **12.** $4 - (-7.8)$

13. $-3 + (-5) + 6$ **14.** $5 - (-8) + (-3)$ **15.** $-4 - 7 + (-2)$

RATIOS AND RATES

Ratios and rates are found in many formulas.

Examples A **ratio** compares two quantities by division:

$$\dfrac{\text{number of flowers picked}}{\text{total number of flowers in the garden}}$$

A **rate** is a special ratio that compares different units:

$$\dfrac{\text{number of cans of soup}}{\text{cost of soup can}}$$

A **unit rate** is a comparison to 1 unit:

$$\frac{\text{number of miles traveled}}{1 \text{ h}} \text{ (miles per hour)}$$

A **proportion** is an equation stating that two ratios are equal. You can use cross products to check if the ratios are equal.

$$\frac{5}{10} = \frac{3}{6}$$

$$5 \cdot 6 = 10 \cdot 3$$

$$30 = 30$$

Identify each relation as a ratio, rate, or unit rate. Be as specific as possible.

16. $\dfrac{\text{number of questions answered correctly}}{\text{total number of questions}}$

17. $\dfrac{\text{number of miles traveled}}{1 \text{ gal of gas}}$

18. $\dfrac{\text{number of pages printed}}{90 \text{ sec}}$

19. $\dfrac{\text{number of sticks of gum}}{\$0.50}$

20. $\dfrac{\text{number of words typed}}{1 \text{ min}}$

21. $\dfrac{\text{number of pages for the index}}{\text{total number of pages in book}}$

Determine if each pair of ratios form a proportion. Write = or ≠.

22. $\dfrac{5}{6} \ \blacksquare \ \dfrac{25}{30}$

23. $\dfrac{3}{16} \ \blacksquare \ \dfrac{9}{48}$

24. $\dfrac{46}{60} \ \blacksquare \ \dfrac{4}{5}$

25. $\dfrac{17}{38} \ \blacksquare \ \dfrac{68}{152}$

26. $\dfrac{19}{26} \ \blacksquare \ \dfrac{38}{54}$

27. $\dfrac{7}{12} \ \blacksquare \ \dfrac{56}{96}$

28. $\dfrac{58}{104} \ \blacksquare \ \dfrac{29}{52}$

29. $\dfrac{4}{14} \ \blacksquare \ \dfrac{7}{17}$

SQUARES AND SQUARE ROOTS

Squares and square roots are found in many equations and formulas.

Examples   Find the square.   Find the square roots.

   a. $7^2 = 7 \cdot 7 = 49$ **a.** $\sqrt{64} = 8$ $8 \cdot 8 = 64$

   b. $(-6)^2 = (-6) \cdot (-6) = 36$ **b.** $-\sqrt{25} = -5$ $5 \cdot 5 = 25$

Find each square.

30. 9^2

31. $(-4)^2$

32. 12^2

33. $(-18)^2$

34. 22^2

35. $(-15)^2$

36. 13^2

37. $(-25)^2$

Find each square root. Use a calculator or table of squares. If necessary, round to the nearest ten-thousandth.

38. $\sqrt{196}$

39. $\sqrt{58}$

40. $\sqrt{96}$

41. $\sqrt{74}$

42. $\sqrt{1024}$

43. $\sqrt{67}$

44. $\sqrt{295}$

45. $\sqrt{430}$

Equations and Formulas

Goals
- Determine if a number is a solution of an equation.
- Solve an equation or formula.

Applications Travel, Safety, Physics

Work with a partner.

1. Make a set of ten number cards by writing one of these numbers on each card.

   -30 -15 -10 4 21 36 48 60 75 100

2. Make a set of ten expression cards by writing each expression on a card.

$16 - 12$	$40 \div (-4)$	$-20 + 5$	$16 + 5$	$-3 \cdot 10$
$40 - 4$	$12 \cdot 4$	$30 + 30$	$100 - 25$	$-60 + 160$

3. Shuffle the two sets of cards together; then arrange the cards face down in four rows of five cards.

4. The first player turns two cards face up and determines if the expressions are equal. If they are equal, the player keeps the cards. If the expressions on the cards are not equal, they are placed face down in their original position. The next player takes a turn. Play continues until all cards have been taken. The player with the greatest number of cards wins.

5. Work with your partner to make different sets of number and expression cards. Then play the new game.

◣ BUILD UNDERSTANDING

An **equation** is a statement in which two numbers or expressions are equal. Equations can be true, false or open. If both sides of an equation have the same numerical expression, the equation is *true*. If the numerical expressions on both sides of an equation are not equal, the equation is *false*.

true: $-8 + 2 = -6$ false: $-8 + 2 = -10$

$-6 = -6$ $-6 \neq -10$

An equation that contains one or more variables is a type of *open sentence*. An open sentence can be true or false depending upon the values that are substituted for the variables. The value that makes the equation true is called the **solution of the equation**.

equation: $x - 3 = 11$

solution: $x = 14$

Check Understanding

Tell whether each equation is true, false or open.

1. $x + 9 = 5$
2. $16 + 23 = 39$
3. $7y = 40$
4. $8(-6) = -54$

Example 1

Determine if 2 is a solution of each equation.

a. $7w + 4 = 18$

b. $8x - 1 = 4x + 3$

c. $y^2 + 5 = 9$

d. $-6a^2 = 12a$

Solution

Substitute 2 for the variable in each equation.

a. $7w + 4 = 18$

$7(2) + 4 \overset{?}{=} 18$

$14 + 4 \overset{?}{=} 18$

$18 = 18$ So 2 is a solution.

b. $8x - 1 = 4x + 3$

$8(2) - 1 \overset{?}{=} 4(2) + 3$

$16 - 1 \overset{?}{=} 8 + 3$

$15 \neq 11$ So 2 is not a solution.

c. $y^2 + 5 = 9$

$(2)^2 + 5 \overset{?}{=} 9$

$4 + 5 \overset{?}{=} 9$

$9 = 9$ So 2 is a solution.

d. $-6a^2 = 12a$

$-6(2^2) \overset{?}{=} 12(2)$

$-6(4) \overset{?}{=} 24$

$-24 \neq 24$ So 2 is not a solution.

To **solve an equation**, find all the values of the variable that make the equation true.

Example 2

Use mental math to solve each equation.

a. $p + 3 = 7$

b. $3w = 24$

Solution

a. $p + 3 = 7$ What number added to 3 equals 7?

You know that $4 + 3 = 7$, so $p = 4$.

b. $3w = 24$ Three times what number equals 24?

You know that $3 \cdot 8 = 24$, so $w = 8$.

A **formula** is an equation stating a relationship between two or more quantities. For example, the distance that you travel is equal to your average speed, or rate, multiplied by the time you spend traveling at that rate.

$$distance = rate \cdot time$$

$$d = rt$$

Often you can evaluate a variable in a formula using given information. For example, suppose a car travels at 55 mi/h for 3 h.

$$d = rt$$

$$d = 55 \text{ mi/h} \cdot 3\text{h}$$

$$= 165 \text{ mi}$$

The distance traveled is 165 mi.

Example 3   Online Personal Tutor at mathmatters2.com

    a. TRAVEL Mitch traveled for 120 mi in 2 h. What was his average speed?

    b. The formula for the perimeter of a triangle is $P = a + b + c$, where a, b and c are the lengths of the sides. The perimeter of a triangle is 25 cm. The length of one side is 5 cm, and another side is 10 cm. What is the length of the third side?

Solution

    a. $d = rt$

        $120 = r \cdot 2$ Two times what number is 120?

        Since $60 \cdot 2 = 120$, then $r = 60$.

        Mitch's average speed was 60 mi/h.

    b. $P = a + b + c$

        $25 = 5 + 10 + c$ What number added to 10 and 5 is 25?

        Since $5 + 10 + 10 = 25$, $c = 10$.

        The length of the third side is 10 cm.

◤ TRY THESE EXERCISES

Determine if -1, 1, or 4 is a solution of each equation.

1. $y + 7 = 6$ **2.** $2x - 5 = 3$ **3.** $8m - 4 = 4$

Use mental math to solve each equation.

4. $t - 1 = \dfrac{1}{2}$ **5.** $6w = 36$ **6.** $\dfrac{1}{3}b = 7$

7. The perimeter of a triangle is 1.8 m. The length of one side is 0.4 m, and another side is 0.8 m. Find the length of the third side.

8. The area of a rectangle is 18 ft², and the width is 3 ft. Use the formula $A = lw$ to find the length.

9. Caroline traveled for 67.5 mi in $2\dfrac{1}{2}$ h. Find her average speed.

◤ PRACTICE EXERCISES • For Extra Practice, see page 592.

Which of the given values is a solution of the equation?

10. $c + 4 = 2$; $-4, -2, 2$ **11.** $8x - 3 = 21$; $-3, 2, 3$

12. $m^2 - 8 = 41$; $-7, -6, 7$ **13.** $3x = 2\dfrac{1}{4}$; $\dfrac{1}{2}, \dfrac{2}{3}, \dfrac{3}{4}$

14. $2w + 2 = 3$; $0.1, 0.5, 1.5$ **15.** $\dfrac{2}{5}k = 14$; $28, 35, 70$

16. $3x^2 - 1 = 26$; $-4, -3, 3$ **17.** $10a + 10 = 10$; $-2, 0, 1$

18. $-\dfrac{2}{3} + m = -\dfrac{1}{2}$; $\dfrac{1}{3}, \dfrac{1}{6}, -\dfrac{1}{6}$ **19.** $0.5c^2 + 1 = 9$; $-4, 8, 4$

20. WRITING MATH Are all equations formulas? Are all formulas equations? Use an example to explain.

Use mental math to solve each equation.

21. $a - 2.5 = 0$ **22.** $\frac{1}{4}x = 12$ **23.** $8r = -48$ **24.** $-10 + w = -6$

DATA FILE Refer to the data on cricket chirps in relation to temperature on page 560. Use the formula $C = \frac{5}{9}(F - 32)$, where C is degrees Celsius and F is degrees Fahrenheit, to find the temperature in degrees Celsius for the following number of cricket chirps.

25. 64 chirps/min **26.** 160 chirps/min **27.** 98 chirps/sec

SAFETY The formula $d = \frac{s + s^2}{20}$ is used to determine the approximate stopping distance (d) in feet for a car traveling s mi/h on a dry road. Find the approximate stopping distance for each car traveling at the following speeds.

28. 20 mi/h **29.** 35 mi/h **30.** 40 mi/h **31.** 55 mi/h

▰ EXTENDED PRACTICE EXERCISES

Solve each equation. If no solution exists, explain why.

32. $-x = 13$ **33.** $|b| = 8$ **34.** $-p = -1$ **35.** $|x| = -3$

PHYSICS The formula for the amount of energy used by an appliance is $E = P \cdot T$, where E is energy used in kilowatt-hours (kWh), P is the power usage in watts/hour and T is the number of hours the appliance is used. (Hint: 1 kilowatt-hour = 1000 watts)

Appliance	Power usage (watts per hour)
Hair dryer	1000
Microwave	700
Television	200
Refrigerator	620
Stereo	110
100-watt bulb	100

36. The Changs watch their color TV an average of 4 h/day. Approximately how many kilowatt-hours of energy do they use each day watching TV?

37. Mario listens to his stereo an average of 2 h/day. Approximately how many kilowatt-hours of energy does Mario use each day listening to his stereo?

38. If the rate charged for electricity is $0.11/kWh, what is the cost of operating a refrigerator for the month of April? Assume the refrigerator is always running.

39. CHAPTER INVESTIGATION Set up a pendulum as shown. Measure the length L of the string.

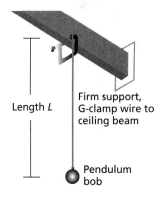

Length L

Firm support, G-clamp wire to ceiling beam

Pendulum bob

▰ MIXED REVIEW EXERCISES

Identify each of these numbers as an integer, a rational number, or an irrational number. (Lesson 2-1)

40. -34 **41.** $3.101001\ldots$ **42.** $\sqrt{7}$ **43.** $\frac{5}{16}$

Simplify each numerical expression. (Lesson 2-2)

44. $16 - (4 + 3) \cdot 2$ **45.** $8^2 - (3 \cdot 4) \div 2$ **46.** $84 \div 6 \cdot \frac{2}{3}$

47. $(9 - 2)^2 - (5 \cdot 3)$ **48.** $6 + 3 \cdot 2 - 8$ **49.** $\frac{1}{2}(7 - 3) \cdot 8 \div 4$

50. $[36 - (2 + 4)] \div 3$ **51.** $7.2 - (3 + 8) + 12.3$ **52.** $(180 \div 9) \div 4 + 2^3$

3-2 One-Step Equations

Goals
- Solve one-step equations.
- Solve formulas for a given variable.

Applications Physics, Packaging, Finance

Use Algeblocks and a Sentence Mat to solve the equation $x + 3 = 7$.

1. Sketch or show the equation on a Sentence Mat.

2. To isolate the x-block, create a zero pair by adding the opposite of 3, or -3, to each side of the Mat.

3. Simplify each side of the Mat by removing zero pairs.

4. Read the answer from what remains on the Mat.

5. Use Algeblocks and a Sentence Mat to solve the equation $x - 2 = 3$. Sketch or show each step.

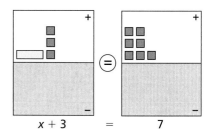

$x + 3 \quad = \quad 7$

◤ BUILD UNDERSTANDING

When two expressions are equal, if you add the same number to each expression, the resulting sums will be equal. This is called the **addition property of equality**.

Addition Property of Equality	If $a = b$, then $a + c = b + c$ and $c + a = c + b$.

COncepts in MOtion

Animation
mathmatters2.com

For example, if $x + 1 = 4$, then $x + 1 + 5 = 4 + 5$.

You can also multiply two equal expressions by the same number. The resulting products will be equal. This is called the **multiplication property of equality**.

Multiplication Property of Equality	If $a = b$, then $a \cdot c = b \cdot c$ and $c \cdot a = c \cdot b$.

For example, if $2x = 10$, then $3(2x) = 3(10)$.

If you cannot solve an equation mentally, use these properties to perform the same operations on both sides of the equation. When the variable is alone on one side, you have solved the equation.

Example 1

Solve each equation. Check the solution.

a. $x - 3 = 5$ **b.** $y + 2.7 = 6.1$ **c.** $6 - z = 14$

Think Back

The sum of a number and its opposite is 0.

$-5 + 5 = 0$

$\frac{2}{3} + \left(-\frac{2}{3}\right) = 0$

Solution

Since the sum of a number and its opposite is zero, use the Addition Property of Equality. To check each solution, substitute the solution into the original equation.

a. $x - 3 = 5$

$x - 3 + 3 = 5 + 3$ To isolate x, add 3 to both sides.

$x + 0 = 8$

$x = 8$

Check $x - 3 = 5$

$8 - 3 \stackrel{?}{=} 5$

$5 = 5$ ✔

b. $y + 2.7 = 6.1$

$y + 2.7 + (-2.7) = 6.1 + (-2.7)$ Add the opposite of 2.7 to both sides.

$y + 0 = 3.4$

$y = 3.4$

Check $y + 2.7 = 6.1$

$3.4 + 2.7 \stackrel{?}{=} 6.1$

$6.1 = 6.1$ ✔

c. $6 - z = 14$

$6 + (-6) - z = 14 + (-6)$

$0 + (-z) = 8$

$-z = 8$ The opposite of what number is 8?

$z = -8$ The opposite of 8 is -8.

Check $6 - z = 14$

$6 - (-8) \stackrel{?}{=} 14$

$6 + 8 \stackrel{?}{=} 14$

$14 = 14$ ✔

To solve an equation involving multiplication or division of a variable, perform the inverse operation.

COncepts in MOtion
Interactive Lab
mathmatters2.com

Example 2

Solve each equation. Check the solution.

a. $\dfrac{x}{4} = 2.5$

b. $-7s = |-35|$

Solution

a. $\dfrac{x}{4} = 2.5$ $\dfrac{x}{4}$ means $\left(\dfrac{1}{4}\right)x$.

$4\left(\dfrac{x}{4}\right) = (2.5)4$ Multiply both sides by 4, the reciprocal of $\dfrac{1}{4}$.

$1x = 10$

$x = 10$

Check $\dfrac{x}{4} = 2.5$

$\dfrac{10}{4} \stackrel{?}{=} 2.5$

$2.5 = 2.5$ ✔

b. $-7s = |-35|$ The absolute value of -35 is 35.

$\dfrac{-7s}{-7} = \dfrac{35}{-7}$ Divide both sides by -7.

$1s = -5$

$s = -5$

Check $-7s = |-35|$

$-7(-5) \stackrel{?}{=} 35$

$35 = 35$ ✔

Sometimes you must solve for a variable in a formula. To do so, solve for the indicated variable by performing opposite operations. Addition and subtraction are opposite operations, as are multiplication and division.

Example 3

Solve each formula for the indicated variable.

a. $d = rt$, solve for t

b. $P = a + b + c$, solve for c

Solution

a. $d = rt$

$\dfrac{d}{r} = \dfrac{rt}{r}$ *Divide both sides by r.*

$\dfrac{d}{r} = t$

b. $P = a + b + c$

$P - a = a - a + b + c$ *Subtract a from both sides.*

$P - a = b + c$ *Simplify.*

$P - a - b = b - b + c$ *Subtract b from both sides.*

$P - a - b = c$ *Simplify.*

◥ TRY THESE EXERCISES

State the operation and number you would use to solve each equation.

1. $3x = 12$

2. $m - 17 = 20$

3. $\dfrac{m}{8} = 1.6$

4. $y + 2 = 1$

5. $n + 3 = 0$

6. $7 - w = 12$

Solve each equation. Check the solution.

7. $4x = 48$

8. $x + 5 = 14$

9. $\dfrac{1}{2}c = 2.3$

10. $8 - z = 2$

11. $\dfrac{2}{3} + g = \dfrac{1}{10}$

12. $1.5d = 30$

13. $P = 4s$, solve for s

14. $A = bh$, solve for h

15. $I = prt$, solve for r

◥ PRACTICE EXERCISES • For Extra Practice, see page 593.

Solve each equation. Check the solution.

16. $6w = 12$

17. $b + 8 = 29$

18. $\dfrac{3}{4}b = -8$

19. $2.5 - t = 3.4$

20. $z - 16 = 5$

21. $12k = 1.44$

22. $-4d = 1$

23. $5n = -45$

24. $1.5 - c = 3$

25. $p - 3.5 = 10$

26. $\dfrac{5}{8}n = 15$

27. $-52 = 13h$

28. $-\dfrac{x}{7} = -4$

29. $-8c = 100$

30. $\dfrac{2}{3}n = -6$

31. $3.5 - x = 6$

32. $-8 = 4y$

33. $11 + c = 4\dfrac{1}{2}$

34. $-1 = \dfrac{3}{4}x$

35. $-\dfrac{x}{2} = 1\dfrac{1}{4}$

36. $12 + 22 + n = 10$

37. $-\dfrac{r}{5} = -13$

38. $-3.9q = -0.3$

39. $-\dfrac{4}{5}w = |-16|$

40. $e - 7 = |-12|$

41. $-\dfrac{2}{5} = \dfrac{3}{7}d$

42. $t + 3 = |-4|$

43. WRITING MATH Write a paragraph explaining the different ways you know to solve an equation. Use examples.

44. PHYSICS The formula $F = ma$ is used to find the force applied to an object, where F is the force, m is the mass and a is the acceleration of the object. Solve the equation for m. Then solve the equation for a.

45. YOU MAKE THE CALL Kenny solved the equation $-\frac{2}{5}x = 14$. His work is shown. Is he correct? Explain why or why not.

$$-\frac{2}{5}x = 14$$
$$\frac{5}{2}\left(-\frac{2}{5}\right)x = (14)\frac{5}{2}$$
$$x = 35$$

46. PACKAGING A case of canned pineapples weighs 18 lb and contains 24 cans. A store manager wrote the equation $24x = 18$. What does x represent? Solve the equation.

47. FINANCE The formula $I = prt$ is used to calculate the simple interest on money in a savings account. The amount of interest is represented by I, p is the principal or amount of money in the account, r is the interest rate, and t is the time in years that the money is in the account. Find the interest rate if $500 earns $162.50 in interest after 5 yr.

MODELING Use Algeblocks to model and solve each equation.

48. $x + 3 = -2$ **49.** $y - 4 = 3$ **50.** $x - 6 = -3$

Translate each word phrase into an equation. Then solve the equation.

51. A number increased by six is ten.   **52.** Three times a number is -21.

53. One-third of a number is 24.   **54.** Eight decreased by a number is 13.

■ EXTENDED PRACTICE EXERCISES

Complete.

55. If $3a + 4 = 10$, then $6a + 8 = $ ___?___.   **56.** If $8w - 6 = 9$, then $4w - 3 = $___?___.

57. CRITICAL THINKING Suppose $x + a = b$. What happens to x if a increases and b remains the same?

58. If $x + 8 = 17$, find the value of $5x$.   **59.** If $w - 3 = 21$, find the value of $3w - 100$.

■ MIXED REVIEW EXERCISES

Refer to the histogram for Exercises 60–62. (Lesson 1-3)

60. People of which age interval use their membership for emergency service most often?

61. Which intervals show equal frequencies?

62. Name the interval that contains 5% of the drivers.

Evaluate each expression. (Lesson 2-1)

63. $|g|$ when $g = 4$ **64.** $|f|$ when $f = -6$

65. $|r|$ when $r = -3.8$ **66.** $|q|$ when $q = 7.146$

67. $|h|$ when $h = 85$ **68.** $|m|$ when $m = -\frac{3}{4}$

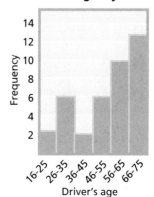

Drivers Using Car Club Membership for Emergency Service

Review and Practice Your Skills

Which of the given values is a solution of the equation?

1. $x + 5 = 3$; $-4, 2, -2$

2. $w + 0.9 = 8.3$; $17.3, 7.4, 9.2$

3. $b^2 - 7 = 29$; $-6, 7, 6$

4. $3x + 5 = 26$; $7, 8, 9$

5. $y - 2\frac{3}{4} = 9$; $6\frac{1}{4}, 12\frac{1}{4}, 11\frac{3}{4}$

6. $1.5c = 45$; $3, 0.3, 30$

7. $\frac{m}{8} = 0$; $0, 1, 8$

8. $d^2 - 7 = 9$; $-4, 0, 4$

Find the unknown side of each triangle. Use the formula $A = \frac{1}{2}bh$.

9. The area of the triangle is 24 in.^{2} and the base is 12 in. Find the height.

10. The area of the triangle is 40 m^2 and the height is 4 m. Find the base.

11. The area of the triangle is 27 ft^2 and the base is 9 ft. Find the height.

Use mental math to solve each equation.

12. $3m = -24$

13. $\frac{24}{x} = 12$

14. $-2 + y = -11$

15. $7b = 1$

16. $x + 4 = 9$

17. $4a = 48$

18. $\frac{x}{3} = 3$

19. $11 - y = 2$

Solve each equation. Check the solution.

20. $m + 3 = -5$

21. $3x = -21$

22. $-\frac{3}{4}y = 9$

23. $\frac{b}{4} = -8$

24. $a - 9 = -2$

25. $-4 + n = -3$

26. $x + 5 = -\frac{2}{3}$

27. $13b = -143$

28. $9x = -207$

29. $-\frac{4}{5} = -\frac{5}{4}P$

30. $a - 2.7 = -4.32$

31. $-60 = \frac{2}{3}x$

32. $-5x = 75$

33. $\frac{2}{3}v = 24$

34. $\frac{f}{-9} = 207$

35. $r - 17 = 15$

36. $x + 11 = 3$

37. $-\frac{1}{4} = 8c$

38. $-18 = 3y$

39. $5.2 - a = 3$

40. $\frac{x}{-9} = -3$

41. $2 = 4y$

42. $b - 6 = |-13|$

43. $\frac{3}{5}a = -\frac{2}{5}$

44. $-\frac{x}{6} = -15$

45. $0.7y = -4.9$

46. $|15| = -3a$

47. $-\frac{x}{2} = 2\frac{1}{2}$

48. $R = 3v$, solve for v

49. $I = prt$, solve for t

Translate each word phrase into an equation. Then solve the equation.

50. A number decreased by four is ten.

51. Seven times a number is -35.

52. One-fourth of a number is -16.

53. Ten increased by a number is 27.

Use mental math to solve each equation. (Lesson 3-1)

54. $|x| = -5$ **55.** $\dfrac{100}{x} = 20$ **56.** $x + 5 = 11$

57. $\dfrac{25}{x} = -5$ **58.** $7 - y = 3$ **59.** $\dfrac{1}{3}x = -9$

Solve each equation. Check the solution. (Lesson 3-2)

60. $A = \dfrac{1}{2}bh$, solve for h **61.** $2\pi r = C$, solve for r

62. $\dfrac{w}{4} = 15$ **63.** $A = B + C + D$, solve for C

64. $\dfrac{1}{4}x = 9$ **65.** $w - (-3) = 13$

MathWorks — Career – Mechanical Engineer
Workplace Knowhow

Mechanical engineers influence the design and development of modern technological equipment. The development of the automobile has been influenced by many physics applications. One of those applications is the source of power for the automobile, the internal combustion engine. The amount of power (measured in units called horsepower) produced by an engine whose pistons rotate a single crankshaft is given by the following equation.

$H = \dfrac{PAnl\omega}{396{,}000}$, where H = horsepower (hp), P = average piston pressure (lb/in.²),

A = piston area (in.²), n = number of pistons, l = piston stroke length, (in.) and ω = crankshaft revolutions per minute (rpm)

1. Solve the formula for n.

2. Calculate the number of pistons if the average piston pressure is 120 lb/in.², the piston area is 10 in.², the stroke length is 3.3 in., and the engine is producing 150 hp at 2500 rpm.

3. Determine what piston area is required for the engine to continue producing 150 hp at 2500 rpm if the piston pressure and stroke length remain the same as given in Exercise 2 and there are 8 pistons turning the crankshaft.

4. Assuming that all other variables remain constant, what do you predict will happen to the horsepower if the piston area doubles?

5. Calculate the horsepower from the engine when the piston area doubles its value from Exercise 3. Assume that the piston pressure, the stroke length, the number of pistons and the crankshaft revolutions per minute remain the same as in Exercise 2. Does this confirm your prediction in Exercise 4?

3-3 Problem Solving Skills: Model Algebra

A **model** is a physical or numerical representation of a real-life situation. There are various types of models. Algeblocks are a type of physical model in which blocks are used to represent variables and numbers. A **mathematical model** uses numbers, usually in a table or list, to describe a situation. An **algebraic model** is a mathematical model that includes a variable and is often written as an expression. A **rule** is an equation or a formula that represents a model.

Problem Solving Strategies

Guess and check

Look for a pattern

Solve a simpler problem

Make a table, chart or list

✔ Use a picture, diagram or model

Act it out

Work backwards

Eliminate possibilities

Use an equation or formula

Problem

 Personal Tutor at mathmatters2.com

COMMUNICATIONS A phone company charges $7.00 each month plus $0.10 per call.

a. Make a mathematical model of the billing system.

b. Write an algebraic model of the billing system.

c. Write a rule for the billing system.

Solve the Problem

a. Make a table to represent a mathematical model of this situation. Numbers are used to represent the billing system.

Number of calls	0	10	20	30	40	50
Monthly bill	$7.00	$8.00	$9.00	$10.00	$11.00	$12.00

b. To write an algebraic model, choose a variable to represent an element of the situation. Let c represent the number of calls per month. The algebraic model of the billing system is $7 + 0.10c$.

c. To write a rule, choose another variable to represent the entire situation. Let m represent a monthly bill. The rule for the billing system is $m = 7 + 0.10c$.

TRY THESE EXERCISES

MODELING Use Algeblocks to model each equation. Do not solve.

1. $2x - 6 = 4$ **2.** $4x + 2 = 10$ **3.** $9 = 3 - 3y$ **4.** $-8 - 2x = -6$

PHYSICS On Pluto the weight of an object is $\frac{1}{25}$ the weight of an object on Earth.

5. Make a mathematical model by using five different weights on Earth, E, to find the corresponding weight on Pluto, P.

6. Write an algebraic model.

7. Write a rule.

For each situation, set up a mathematical model (table), write an algebraic model and a rule.

Five-step
Plan

1 Read
2 Plan
3 Solve
4 Answer
5 Check

8. **ENTERTAINMENT** It costs $10 to become a member at a movie rental store and $2 to rent a movie.

9. **FITNESS** Elonzo starts a jogging program. He begins by running 1 mi and adds $\frac{1}{2}$ mi each week for the next 6 wk.

10. A photographer charges $35 for a sitting and $10 for each 8×10 sheet produced.

11. Barry opens a savings account with a deposit of $300. He puts $50 of each bi-weekly paycheck in this savings account.

12. Mika designs and sells quilts to schools with the school name, logo and mascot. She charges an initial fee of $650, which includes creating the design and the first 20 quilts. Each additional quilt costs $15.

PART-TIME JOB During the summer, Kristen runs a day camp at her house for neighborhood kids. She charges each family $15/day plus $5 for each child in the family that attends the camp.

13. Make a mathematical model for the amount each family pays per day.

14. Write an algebraic model for the amount each family pays per day.

15. Write a rule for the amount each family pays per day.

16. If 3 families attend her camp, set up a mathematical model for the amount that Kristen earns each day.

17. Write an algebraic model for Exercise 16.

18. Write a rule for Exercise 16.

19. Do you think the method of payment that Kristen uses is fair? Explain why or why not.

20. **WRITING MATH** Write a paragraph explaining the differences among a mathematical model, an algebraic model and a rule. Use examples to support your explanation.

MIXED REVIEW EXERCISES

Simplify. (Lesson 2-4)

21. $15a - 7a$

22. $4n - (-2n)$

23. $-17k + 31k$

24. $4.8x + 3.5x - 2x$

25. $-\frac{2}{3}b + \frac{5}{8}b + \frac{1}{2}b$

26. $5d(6) - 22d$

27. $3.1g + 1.2h - 1.8g + 2.7h$

28. $\frac{1}{2}(6x + 14y + 3x - 2y)$

29. $8s - 4t + 7t - 9s$

Simplify. (Lesson 2-7)

30. $g^3 \cdot g^6$

31. $(h^4)^3$

32. $\frac{z^9}{z^5}, z \neq 0$

33. $2(f^3)^7$

34. $\frac{q^5}{q^2}, q \neq 0$

35. $m^2\left(\frac{m^6}{m^3}\right), m \neq 0$

Equations with Two or More Operations

Goals ■ Solve two-step equations and formulas.

Applications Physics, Mechanics, Sports, Modeling

In Figure A, each bag contains the same unknown weight of nails. Each single nail weighs 1 oz. The two bags with the three additional nails weigh 21 oz.

1. Estimate the weight of each bag of nails. Check your estimate. Continue until you find the value of x.

2. If the bags are combined into one larger bag weighing the same as the two bags, will the scale remain balanced? Explain.

3. Write an equation to describe the scale in Figure B.

4. Test the solution you found in Question 1 in this equation. Does it satisfy the equation?

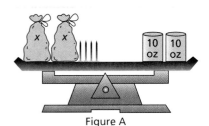

Figure A

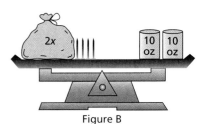

Figure B

◥ BUILD UNDERSTANDING

Some equations involve multiplication or division and addition or subtraction. To solve these **two-step equations**, use the addition property of equality first. Then use the multiplication property.

COncepts in MOtion
Animation
mathmatters2.com

Example 1

Solve each equation. Check the solution.

a. $\dfrac{x}{3} + 4 = 9$

b. $5a - 15 = 5$

Solution

a.
$$\frac{x}{3} + 4 = 9$$
$$\frac{x}{3} + 4 + (-4) = 9 + (-4) \qquad \text{Add } -4 \text{ to both sides.}$$
$$\frac{x}{3} = 5$$
$$3\left(\frac{x}{3}\right) = 3(5) \qquad \text{Multiply both sides by 3.}$$
$$x = 15$$

Check $\dfrac{x}{3} + 4 = 9$
$$\frac{15}{3} + 4 \overset{?}{=} 9$$
$$5 + 4 \overset{?}{=} 9$$
$$9 = 9 \quad ✔$$

b.
$$5a - 15 = 5$$
$$5a - 15 + 15 = 5 + 15 \qquad \text{Add 15 to both sides.}$$
$$5a = 20$$
$$\frac{5a}{5} = \frac{20}{5} \qquad \text{Divide both sides by 5.}$$
$$a = 4$$

Check $5a - 15 = 5$
$$5(4) - 15 \overset{?}{=} 5$$
$$20 - 15 \overset{?}{=} 5$$
$$5 = 5 \quad ✔$$

Sometimes it is necessary to simplify the equation before you can solve it. To simplify, combine like terms or distribute over parentheses.

Example 2

Solve each equation. Check the solution.

a. $7x + 6 - 2x + 3 = 34$

b. $2(x + 7) = 4(x - 1)$

Solution

a.

$$7x + 6 - 2x + 3 = 34$$

$$(7x - 2x) + (6 + 3) = 34 \qquad \text{Combine like terms.}$$

$$5x + 9 = 34$$

$$5x + 9 - 9 = 34 - 9 \qquad \text{Subtract 9 from both sides.}$$

$$5x = 25$$

$$\frac{5x}{5} = \frac{25}{5} \qquad \text{Divide both sides by 5.}$$

$$x = 5$$

Check $7x + 6 - 2x + 3 = 34$

$$7(5) + 6 - 2(5) + 3 \stackrel{?}{=} 34$$

$$35 + 6 - 10 + 3 \stackrel{?}{=} 34$$

$$34 = 34 \quad \checkmark$$

b.

$$2(x + 7) = 4(x - 1)$$

$$2x + 14 = 4x - 4 \qquad \text{Distribute 2 and 4.}$$

$$2x - 2x + 14 = 4x - 2x - 4 \qquad \text{Subtract } 2x \text{ from both sides.}$$

$$14 = 2x - 4$$

$$14 + 4 = 2x - 4 + 4 \qquad \text{Add 4 to both sides.}$$

$$18 = 2x$$

$$\frac{18}{2} = \frac{2x}{2} \qquad \text{Divide both sides by 2.}$$

$$9 = x$$

Check $2(x + 7) = 4(x - 1)$

$$2(9 + 7) \stackrel{?}{=} 4(9 - 1)$$

$$2(16) \stackrel{?}{=} 4(8)$$

$$32 = 32 \quad \checkmark$$

Formulas that involve two operations can be solved in a similar manner.

Example 3

Solve the formula $T = \dfrac{n}{4} + 40$ for n.

Solution

$$T = \frac{n}{4} + 40$$

$$T - 40 = \frac{n}{4} + 40 - 40 \qquad \text{Subtract 40 from both sides.}$$

$$T - 40 = \frac{n}{4}$$

$$4(T - 40) = 4\left(\frac{n}{4}\right) \qquad \text{Multiply both sides by 4.}$$

$$4(T - 40) = n$$

Example 4

MODELING Solve the equation $2y - 1 = 5$ using Algeblocks.

Solution

Step 1 Model the equation on a Sentence Mat.

Step 2 Add 1 to both sides of the Mat. Then remove zero pairs.

Step 3 There are two y-blocks. Make the 6 unit blocks into two equal groups.

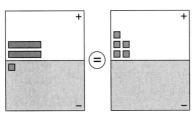 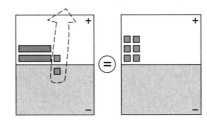 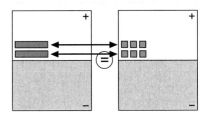

Read the answer for $1y$ from the Mat, $y = 3$.

◼ TRY THESE EXERCISES

Solve each equation. Check the solution.

1. $3x + 2 = 14$
2. $\frac{m}{2} - 4 = 8$
3. $2w - 7 = 25$
4. $-\frac{1}{4}n + 5 = 2$
5. $3(t - 2) = 15$
6. $-6x + 8 = 4x - 42$
7. $2(y + 7) = 3(y - 1)$
8. $P = 2l + 2w$, solve for w
9. $y = mx + b$, solve for x

10. **WRITING MATH** When solving two-step equations, why do you undo addition and subtraction before multiplication and division? Explain.

11. In a circle graph, all sections or percentages must add up to 100%. Write an equation for the circle graph shown. Solve the equation to determine the percent of each section.

12. **MODELING** Use Algeblocks to model and solve $3x + 4 = -2$.

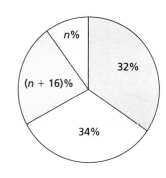

◼ PRACTICE EXERCISES • For Extra Practice, see page 593.

Solve each equation. Check the solution.

13. $8m - 1 = 31$
14. $\frac{w}{3} + 4 = 10$
15. $-6g + 8 = -46$
16. $\frac{x}{2} - 4 = 5$
17. $3p - p + 1 = -9$
18. $5a - 10 = 0$
19. $5q + 3q - 1 = -33$
20. $\frac{3}{4}n + 2 = 8$
21. $-4m - 10 = -6$
22. $7 = \frac{3}{8}t - 2$
23. $4r - 2.5 = 6.3$
24. $-2y - 2y + 3 = -13$
25. $4(b + 6) = -20$
26. $1.5z - 7 - 3.5z = 3$
27. $-12g + 8 + 3g + 3 = 29$
28. $6c - 2 = c + 13$
29. $8(x - 1) = 0$
30. $3(2w + 5) = 4$
31. $4(c + 3) = -3(c - 2)$
32. $2(9 - d) = 4$
33. $\frac{1}{2}(14k - 8) = 10$
34. $y = \frac{x}{3} - 20$, solve for x
35. $p = \frac{7}{8}r + 20$, solve for r
36. $4(w - 2) = 16v$, solve for w

PHYSICS The formula $F = \frac{9}{5}C + 32$ is used to convert a temperature in Celsius to Fahrenheit. Round to the nearest tenth.

37. Solve the equation for C.

38. Convert 27°C to Fahrenheit.

39. Convert 82°F to Celsius.

40. Convert −5°C to Fahrenheit.

MODELING Use Algeblocks to model and solve each equation.

41. $3x - 5 = 7$

42. $4y - 2 = 2$

43. $2x + 3 = -3$

44. In the figure shown, $\angle ABD$ and $\angle CBD$ are supplementary angles, so $m\angle ABD + m\angle CBD = 180°$. Write and solve an equation to find the measure of each angle.

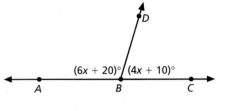

45. Complementary angles are angles whose sum is 90°. The $m\angle EFG$ is $(10x + 1)°$ and $m\angle GFH$ is $(8x - 1)°$. Find the measure of each angle if $\angle EFG$ and $\angle GFH$ are complementary angles.

46. **MECHANICS** Jontay had his car repaired, and the bill was $235. The cost for parts was $127. The cost of labor was $27/h. Write and solve an equation to find the total number of hours spent on repairing Jontay's car.

47. **SPORTS** The length of a football field is 67 yd greater than the width. The perimeter of the field is 346 yd. Find the length and width of a football field.

Translate each sentence into an equation. Then solve the equation.

48. Twice a number increased by four is 20.

49. Half a number decreased by five is −12.

50. Three more than half a number is two less than the number.

EXTENDED PRACTICE EXERCISES

51. Find the measure of each angle of the triangle shown.

Solve each equation. Check the solution.

52. $\frac{1}{2}(6t - 20) + 15 = -5 + 2(t + 3)$

53. $2(3 - k) = 3(2 + k)$

54. $4[1 - 3(x + 2)] + 2x = 2(x + 2)$

55. $2[3(g+3) - (g + 2)] = 5(g + 1) + 4$

56. **CHAPTER INVESTIGATION** Set the ball of the pendulum swinging through an arc of about 5°. Find the time t it takes the pendulum to make 50 swings. A swing is counted each time the ball moves back and forth completely.

MIXED REVIEW EXERCISES

Use the stem-and-leaf plot for Exercises 57–62. (Lesson 1-3)

57. Find any outliers in the data.

58. Find the median.

59. Find any clusters or gaps.

60. Find the mode.

61. Find the range.

62. Find the mean.

4	3 5
5	0
8	2 3 3 5 7 9 9
9	0 1 1 1 2 5 6 8 9
10	1 2 5 5 8
12	9

4 | 3 represents 43.

Review and Practice Your Skills

PRACTICE ▟ LESSON 3-3

FINANCE Luis opens a savings account with a deposit of $90. He puts $30 in each week from tips he has earned as a waiter.

1. Make a mathematical model for the amount he has in the bank for 1, 2, 3, 4, 5, and 6 weeks after opening his account.

2. Write an algebraic model for the amount he has in the bank.

3. Write a rule for the amount he has in the bank after a certain number of weeks.

TRANSPORTATION Car Rental Company A charges $45/day plus $0.10/mi. A competitor, Company B, charges $35/day plus $0.12/mi. Answer Exercises 4–9 based on a one day rental.

4. Make a mathematical model for Company A.

5. Make a mathematical model for Company B.

6. Write an algebraic model for Company A.

7. Write an algebraic model for Company B.

8. Write a rule for Company A.

9. Write a rule for Company B.

10. If you were going to travel about 300 mi each day which company would you prefer to use? Why?

PRACTICE ▟ LESSON 3-4

Solve each equation. Check the solution.

11. $7r + 5 = 26$
12. $4t - 6 = 14$
13. $-8 - 2w = 11$
14. $5(x - 5) = 10$
15. $12 - 3m = 24$
16. $-2(r + 9) = -28$
17. $-8 = 12 - 4x$
18. $\dfrac{x}{4} - 6 = 13$
19. $112 = 12 + 8y$
20. $(w - 3)5 + 7 = 2$
21. $\dfrac{b}{13} + 6 = 15$
22. $-1 = \dfrac{h}{11} - 14$
23. $-m + 3(m + 1) = 11$
24. $\left(\dfrac{1}{4}\right)x - 9 = 5$
25. $-9 = \dfrac{w}{4} - 2$

Translate each sentence into an equation. Then solve the equation.

26. Twice a number decreased by three is 21.

27. Two more than three-fourths a number is eight.

28. Four more than a number divided by three is ten.

29. Five times a number decreased by ten is zero.

30. Twice a number decreased by seven is 25.

Solve each equation. Check the solution. (Lessons 3-1 and 3-2)

31. $\frac{1}{3}x = 10$

32. $7r = -49$

33. $3.1 + b = 0$

34. $-9 + x = -4$

35. $\frac{a}{4} = -2$

36. $-25 = 5c$

37. Igashu has a lawn mowing business. He deposited $200 in a savings account and adds $50 a week. Set up a mathematical model, write an algebraic model and a rule. (Lesson 3-3)

Solve the equation. Check the solution. (Lesson 3-4)

38. $3x - 7 = 11$

39. $\frac{a}{2} - 3 = 12$

40. $4y - 2 = 14$

41. $3(y + 2) = 18$

42. $-3y - 4y + 6 = -8$

43. $11b - 5b + 7 = 19$

44. $-2x - 9 = 3$

45. $5(c + 4) = -(c + 4)$

46. $4(x + 3) = -3(x - 2)$

Mid-Chapter Quiz

Which of the given values is a solution of the equation? (Lesson 3-1)

1. $m - 3 = 6$; $-3, 3, 9$

2. $2x^2 + 1 = 33$; $-4, 4, 8$

3. $\frac{1}{3}b = -3$; $-9, -6, 3, 9$

4. $2.4y = 14.4$; $1.2, 2.4, 6.0$

5. $\frac{5}{6}r - \frac{1}{6} = 4$; $1, 2, 3, 4, 5$

6. $|x + 3| = 7$; $-10, -4, 4, 10$

7. A rectangle has an area of 70 ft^2 and width of 7 ft. Find its length.

Solve each equation. Check the solution. (Lessons 3-2 and 3-4)

8. $3x = 36$

9. $|5p| = 30$

10. $12p = -72$

11. $\frac{3}{5} - n = \frac{1}{3}$

12. $-4b = -38$

13. $-1.7n = -51$

14. $2\frac{1}{2}x = 7\frac{1}{2}$

15. $9y = 171$

16. $\frac{m}{5} - 7 = -2$

17. $26 - 17z = -8$

18. $13(y + 6) = 39$

19. $5x - 2 = -20 + 7x$

20. $-4(z - 1) = 6z$

21. $0.9x - 8.1 = 2.7$

22. $\frac{1}{2}x + 3 = x - 2$

23. $4c + 9c + 5 = 18$

Translate each word phrase into an equation. Then solve the equation. (Lesson 3-2)

24. The difference between a number and seven is negative four.

25. The product of 2.5 and a number is 20.

MUSIC Two weeks before a concert, a piano player begins to practice his repertoire 2 h/day plus $\frac{1}{2}$ h more each day until the day of the concert. (Lesson 3-3)

26. Make a mathematical model for the amount the pianist practices per day.

27. Write an algebraic model for the amount the pianist practices per day.

28. Write a rule for the amount the pianist practices per day.

3-5 Proportions

Goals ■ Write and solve proportions.

Applications Art, Machinery, Manufacturing, Business, Physics, Retail

ART Artists often create scale drawings in order to portray a real-life object. A *scale drawing* is a drawing that represents an object. The *scale* of the drawing is the ratio of the size of the drawing to the actual size of the object.

1. Use a centimeter ruler to measure the length of the car.

2. Use a centimeter ruler to measure the height of the car.

3. The scale of this drawing is 1 cm : 2 ft. This means that each centimeter of the drawing represents 2 ft of the actual object. Find the actual length of the car.

4. Find the actual height of the car.

5. Explain how you can use an equation for Questions 3 and 4.

■ BUILD UNDERSTANDING

Recall that a **proportion** is an equation stating that two ratios are equivalent. There are three different ways to write a proportion.

$$a \text{ is to } b \text{ as } c \text{ is to } d \qquad a : b = c : d \qquad \frac{a}{b} = \frac{c}{d}$$

In any of these forms, b and c are called the **means** of the proportion and a and d are called the **extremes**. In a proportion, the **cross-products** are equal. You can use cross-products to solve a proportion.

$$\frac{a}{b} = \frac{c}{d}, \text{ so } ad = bc$$

Check Understanding

Explain how you would set up and solve the following proportions:

1. 25 is to x as 40 is to 16

2. $-36 : 24 = 27 : x$

Example 1

Solve the proportion $\frac{9}{12} = \frac{21}{n}$; $n \neq 0$. Check the solution.

Solution

$$\frac{9}{12} = \frac{21}{n}$$

$$9 \cdot n = 12 \cdot 21 \qquad \text{Find the cross-products.}$$

$$9n = 252$$

$$\frac{9n}{9} = \frac{252}{9} \qquad \text{Divide both sides by 9.}$$

$$n = 28$$

Check

$$\frac{9}{12} = \frac{21}{n}$$

$$\frac{9}{12} \overset{?}{=} \frac{21}{28} \qquad \text{Reduce each ratio.}$$

$$\frac{3}{4} = \frac{3}{4} \checkmark$$

CŎncepts in MƆtion
Interactive Lab
mathmatters2.com

Example 2

MACHINERY A machine produces 45 hinges in 6 min. How many minutes does it take the machine to produce 75 hinges?

Solution

Set up a proportion. Let $x =$ the amount of time to produce 75 hinges.

$$\frac{\text{number of hinges}}{\text{time}} \qquad \frac{45}{6} = \frac{75}{x} \qquad \frac{\text{number of hinges}}{\text{time}}$$

$$45 \cdot x = 6 \cdot 75 \qquad \text{Find the cross-products.}$$

$$45x = 450$$

$$\frac{45x}{45} = \frac{450}{45}$$

$$x = 10$$

It takes the machine 10 min to produce 75 hinges.

Some proportions involve variable expressions as the terms of the ratios. To solve these equations, find the cross-products. Then solve for the variable.

> **Problem Solving Tip**
>
> When solving a word problem that requires you to write a proportion, checking the solution by substitution will only tell you if you solved the proportion correctly.
>
> Check your answer for reasonableness. Since 45 hinges are produced in 6 min, 90 hinges would be produced in 12 min. So 75 hinges in 10 min is reasonable.

Example 3

Solve each proportion. Check the solution.

a. $\dfrac{36}{n+2} = 4; \; n + 2 \neq 0$

b. $\dfrac{3y+3}{6} = \dfrac{2y-1}{2}$

Solution

a. $\dfrac{36}{n+2} = 4$

$\dfrac{36}{n+2} = \dfrac{4}{1}$ Write 4 as $\frac{4}{1}$.

$36 \cdot 1 = 4(n+2)$ Find the cross-products.

$36 = 4n + 8$

$36 - 8 = 4n + 8 - 8$ Subtract 8 from both sides.

$28 = 4n$

$\dfrac{28}{4} = \dfrac{4n}{4}$

$7 = n$

b. $\dfrac{3y+3}{6} = \dfrac{2y-1}{2}$

$2(3y+3) = 6(2y-1)$ Find the cross-products.

$6y + 6 = 12y - 6$

$6y - 6y + 6 = 12y - 6y - 6$

$6 = 6y - 6$

$6 + 6 = 6y - 6 + 6$

$12 = 6y$

$\dfrac{12}{6} = \dfrac{6y}{6}$

$2 = y$

Check $\dfrac{36}{n+2} = 4$

$\dfrac{36}{7+2} \overset{?}{=} 4$

$\dfrac{36}{9} \overset{?}{=} 4$

$4 = 4 \; ✔$

Check $\dfrac{3y+3}{6} = \dfrac{2y-1}{2}$

$\dfrac{3 \cdot 2 + 3}{6} \overset{?}{=} \dfrac{2 \cdot 2 - 1}{2}$

$\dfrac{6+3}{6} \overset{?}{=} \dfrac{4-1}{2}$

$\dfrac{9}{6} \overset{?}{=} \dfrac{3}{2}$ Reduce each ratio.

$\dfrac{3}{2} = \dfrac{3}{2} \; ✔$

Solve each proportion. Assume no denominator is 0. Check the solution.

1. $\dfrac{16}{n} = \dfrac{36}{27}$

2. $\dfrac{x}{9} = \dfrac{50}{15}$

3. $11 : 6 = m : 18$

4. $\dfrac{15}{r-3} = 3$

5. $\dfrac{x+4}{-5} = 10$

6. $\dfrac{-4z}{z+4} = -3$

7. $\dfrac{x}{x+2} = \dfrac{3}{5}$

8. $\dfrac{x+6}{3} = \dfrac{5x}{9}$

9. $\dfrac{b-4}{4} = \dfrac{b+4}{8}$

10. **WRITING MATH** Explain why it is necessary in Exercises 1–9 to assume that the denominator can't be zero.

11. A company produced 260 turbine engines. Of these, 15% are for motor boats. How many engines are for the motor boats? (Hint: The ratio represented by 15% is 15 to 100.)

12. **MANUFACTURING** Of all computer chips manufactured, 3% are defective. If a company produces 367 chips, approximately how many are defective?

Solve each proportion. Assume no denominator is 0. Check the solution.

13. $\dfrac{18}{48} = \dfrac{12}{n}$

14. $\dfrac{21}{z} = \dfrac{77}{33}$

15. $85 : 100 = c : 20$

16. $96 : 69 = 32 : x$

17. $\dfrac{f}{44} = \dfrac{24}{66}$

18. 21 is to k as 15 is to 25

19. $\dfrac{99}{18} = \dfrac{p}{8}$

20. $38 : g = 57 : 18$

21. $\dfrac{7.5}{21.2} = \dfrac{y}{106}$

22. $-6 = \dfrac{p+9}{11}$

23. $\dfrac{9}{w+3} = \dfrac{1}{6}$

24. $\dfrac{k-13}{9} = 2$

25. $\dfrac{u-3}{u} = \dfrac{2}{5}$

26. $\dfrac{9}{5} = \dfrac{3n}{8-n}$

27. $\dfrac{36}{t+2} = \dfrac{4}{3}$

28. $\dfrac{22}{a+3} = \dfrac{11}{12}$

29. $\dfrac{x}{-9} = \dfrac{x-6}{-27}$

30. $\dfrac{4}{5} = \dfrac{3}{m-2}$

31. $\dfrac{3y+4}{y} = 11$

32. $\dfrac{c-2}{3} = \dfrac{5c-2}{3}$

33. $\dfrac{20}{4} = \dfrac{n}{n-4}$

34. $\dfrac{t+3}{2} = \dfrac{t-3}{3}$

35. $\dfrac{2m+1}{9} = \dfrac{12-7m}{6}$

36. $\dfrac{6a-8}{7} = \dfrac{-40+4a}{-4}$

37. **YOU MAKE THE CALL** Erica and Jenisa solved the proportion $\dfrac{t-4}{-6} = \dfrac{3}{2}$. Erica's solution is $t = 5$, and Jenisa solution is $t = -5$. Who is correct? Explain Erica's or Jenisa's possible mistake.

38. An automobile manufacturer employs 21,550 people. Of these, 68% are production workers. How many production workers does the company employ?

39. **BUSINESS** For one year, a company's income was $820,650. The company's expenses for that year were $754,998. What percent of the company's income was profit? (Hint: income − expenses = profit)

40. PHYSICS A cylindrical water tank is 10-ft high. When the water is 4-ft deep, the volume of water is 75 ft³. How much water is there in a full tank?

41. WRITING MATH When solving a proportion involving a percent, explain how to set up each ratio.

DATA FILE For Exercises 42–44, refer to the data on the leading causes of death for men and women on page 569.

42. Approximately what percent of people who die from heart disease are men?

43. Approximately what percent of people who die from cancer are women?

44. Approximately what percent of people who die from accidents are women?

Use the scale drawing of a bus for Exercises 45–47. The scale is 1 cm : 2 m.

45. Find the actual height of the bus.

46. Find the actual length of the bus.

47. How many buses of this size could be parked along one side of a street that is 78 m long?

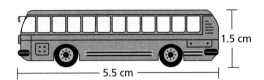

1.5 cm

5.5 cm

EXTENDED PRACTICE EXERCISES

Solve each proportion. Check the solution.

48. $\dfrac{-9h - 17}{2} = -3h + 5$

49. $\dfrac{5(x + 3)}{4} = \dfrac{3(x + 8)}{2}$

50. RETAIL The manager of a clothing store is having no success selling a particular shirt at $35.99. She decides to lower the price to $23.39. Write and solve a proportion to determine the percent of the discount. (Hint: Find the amount of the discount first.)

51. Keandre deposits $540 in a bank account paying simple interest. Exactly one year later, he closes the account and withdraws $564.30. At what percent is interest paid on his account during the year?

MIXED REVIEW EXERCISES

52. Find the median, mode, mean and range of the time that 15 students spent studying for a math test. Round to the nearest hundredth. (Lesson 1-2)

Hours Spent Studying for Math Test

4.3	2.8	3.1	1.9	2.0
0.5	3.6	4.2	2.3	2.1
3.1	2.5	1.8	3.4	3.1

Graph each set of numbers on a number line.
(Lesson 2-1)

53. integers from −4 to 3

54. real numbers from −2 to 5

55. integers from −3 to 8

56. real numbers from −4 to 0

57. real numbers greater than or equal to −3

58. real numbers less than or equal to 2

59. real numbers less than 3

60. real numbers greater than −4

Graph Inequalities on a Number Line

Goals
- Determine if a number is a solution of an inequality.
- Graph the solution of an inequality on a number line.

Applications Business, Government, Sports, Physics, Geography

---

BUSINESS In a company meeting, Kiesha presented the number line shown.

1. Kiesha said that sales this year are less than last year's sales. Name a possible amount of this year's sales.

2. She said that she expects next year's sales to be better than last year's. Name a possible amount of what Kiesha expects for next year's sales.

Last Year's Sales (dollars)

```
◄──┼──────┼──────◆──────┼──────┼──►
   0          1,000,000    2,000,000
```

3. In Question 1, could you have named other amounts? How many others? Where are the points corresponding to these amounts located on the number line shown?

4. In Question 2, could you have named other amounts? How many others? Where are the points corresponding to these amounts located on the number line shown?

◤ BUILD UNDERSTANDING

An **inequality** is a mathematical sentence containing one of these symbols: $<$, $>$, $\neq$, $\leq$, or $\geq$. Like an equation, an inequality may be true, or false, or an open sentence.

true: $-4 \leq -4$ *false:* $-4 < -4$ *open sentence:* $m < -4$

An inequality is an open sentence that is neither true nor false until the variables are replaced by specific values. A value of the variable that makes such an inequality true is called a **solution of the inequality**. Inequalities like these often have an infinite number of solutions.

Inequality: $x \leq -3$

Solution: any real number less than or equal to -3

Example 1   **Personal Tutor at mathmatters2.com**

Determine whether -7 is a solution of $x = 10$, $x < 10$ or $x > 10$.

Solution

Substitute -7 for x in each equation and inequality.

$x = 10$	$-7 = 10$	false
$x < 10$	$-7 < 10$	true
$x > 10$	$-7 > 10$	false

So -7 is a solution of $x < 10$.

Reading Math

Read the following inequality symbols as:

$<$	"is less than"
$>$	"is greater than"
$\leq$	"is less than or equal to"
$\geq$	"is greater than or equal to."
$\neq$	"is not equal to"

Besides −7, there are many other numbers that are solutions of $x < 10$. In fact, the solution of $x < 10$ includes all real numbers less than 10.

The solution of an inequality can be represented on a number line. Graph points using a solid circle. To show that a point is not included, use an open circle. Then shade the number line to include all other solutions of the inequality.

Reading Math

Words such as "at least," "at most," "greater than," "less than," "maximum" and "minimum" often indicate that an inequality will be used in solving a problem.

Example 2

Graph the solution of each inequality on a number line.

a. $x < 5$ **b.** $n \geq -3$

Solution

a. Place an open circle on 5 to indicate that 5 is not a solution. Then shade to the left of 5 to show that all real numbers less than 5 are solutions of the inequality.

b. Place a solid circle on −3 to indicate that −3 is a solution of the inequality. Then shade to the right of −3 to show that all real numbers greater than −3 are also solutions of the inequality.

Inequalities can be used to model real-world situations. Choose a variable to represent the solution. Then express the situation as an inequality.

Example 3

GOVERNMENT To vote in the U.S., a citizen must be 18 years of age or older.

a. Write an inequality that describes the age in years of voters in the U.S.

b. Graph the solution of the inequality on a number line.

Solution

a. Let $a =$ the age of U.S. voters. The inequality $a \geq 18$ represents the age of U.S. voters.

b. Place a solid circle on 18 to show that 18 is a solution. Shade to the right to show that all ages greater than or equal to 18 are solutions of the inequality.

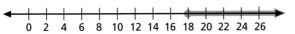

Example 4

State the inequality that is represented on each number line.

a.

b.

Solution

a. Since the circle on −4 is open, −4 is not a solution. The line is shaded to the right; therefore, all real numbers greater than −4 are solutions of the inequality. So $x > -4$.

b. Since the circle on 12 is closed, 12 is a solution. The line is shaded to the left; therefore, all real numbers less than or equal to 12 are solutions of the inequality. So $y \leq 12$.

◤ TRY THESE EXERCISES

Determine whether each number is a solution of $y = 2.8$, $y < 2.8$ or $y > 2.8$.

1. 2.5   **2.** 3.1   **3.** 2.8   **4.** −2.8

Graph the solution of each inequality on a number line.

5. $x > 4$ **6.** $b \leq -1$ **7.** $n > \dfrac{1}{2}$ **8.** $-8 \geq z$

State the inequality that is represented on each number line.

9.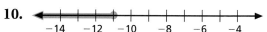

10.

SPORTS In every basketball game this season, Colin has scored at least 13 points.

11. Write an inequality to describe the number of points Colin scored each game.

12. Graph the solution on a number line.

◤ PRACTICE EXERCISES • For Extra Practice, see page 594.

Determine whether each number is a solution of $d = -\dfrac{1}{3}$, $d < -\dfrac{1}{3}$ or $d > -\dfrac{1}{3}$.

13. $-\dfrac{1}{2}$ **14.** $\dfrac{1}{3}$ **15.** $-\dfrac{1}{3}$ **16.** −1

17. WRITING MATH Describe two different ways you can show that your answers for Exercises 13–16 are correct.

Graph the solution of each inequality on a number line.

18. $a \geq 6$ **19.** $g < -2$ **20.** $m \leq 2.5$ **21.** $n > 0$

22. $b \leq -\dfrac{3}{4}$ **23.** $d > -4.5$ **24.** $6 \geq x$ **25.** $h < 3$

26. $\dfrac{3}{2} < y$ **27.** $h \geq 0.8$ **28.** $-1 \leq k$ **29.** $9 > t$

30. $h \geq 18$ **31.** $-20 < x$ **32.** $r \leq -\dfrac{1}{4}$ **33.** $2.2 > n$

State the inequality that is represented on each number line.

34.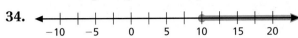

35.

36.

37.

38. To get an A in math class for the semester, Natasha must score at least 94% on the exam. Write an inequality that describes the score she must receive.

39. **PHYSICS** Water boils at 212°F. Write an inequality that describes the temperature of water that is not boiling.

GEOGRAPHY The Dead Sea, located in both Israel and Jordan, has the lowest continental altitude. It is 1312 ft below sea level.

40. Write an inequality that describes the elevations in feet of all other places in the world.

41. Graph the solution on a number line.

DATA FILE Refer to the data on the ten largest lakes of the world on page 565.

42. Write an inequality that describes the area in square miles of all the lakes in the world in relation to the Caspian Sea.

43. Write an inequality that describes the length in kilometers of all the lakes in the world in relation to the Caspian Sea.

Satellite image of the Dead Sea

EXTENDED PRACTICE EXERCISES

State whether each statement is *always*, *sometimes*, or *never* true for real numbers *a*, *b* and *c*. Justify each *sometimes* or *never* answer.

44. $a < a + 1$

45. $b > b - 1$

46. If $a < b$, then $b < a$

47. If $a \leq b$, then $b \leq a$

48. **CRITICAL THINKING** If $a > 0$ and $b > -2$, is it true that $a > b$? Explain.

MIXED REVIEW EXERCISES

Use a calculator. Identify each number as a repeating decimal, terminating decimal, or an irrational number. (Lesson 2-1)

49. 3.846

50. 5.989898...

51. 0.2222

52. 1.632632...

53. $\dfrac{3}{7}$

54. $\sqrt{54}$

55. $\sqrt{33.64}$

56. $\dfrac{9}{15}$

Simplify. (Lessons 2-7 and 2-8)

57. $z^{-5} \cdot z^2$

58. $(t^{-4})^5$

59. $r^4 \cdot r^{-2}$

60. $\dfrac{x^4}{x}$

61. $p^5 \div p^4$

62. $\left(\dfrac{c^3}{c^5}\right)^4$

63. $(v^{-5})^5$

64. $y^4 \cdot y^5$

65. **DATA FILE** Refer to the data on weekly time spent traveling on page 574. How much time is spent shopping and visiting friends and family during the week? Express your answer in minutes. (Prerequisite Skills)

Review and Practice Your Skills

PRACTICE ◣ LESSON 3-5

Solve each proportion. Assume no denominator is 0. Check the solution.

1. $\dfrac{21}{27} = \dfrac{x}{18}$

2. $\dfrac{3}{x} = \dfrac{5}{9}$

3. $\dfrac{10}{b} = \dfrac{20}{28}$

4. $\dfrac{5}{4} = \dfrac{10}{c}$

5. 4 is to x as 8 is to 24

6. $\dfrac{9}{16} = \dfrac{b}{32}$

7. $\dfrac{4}{9} = \dfrac{36}{x}$

8. $\dfrac{215}{5} = \dfrac{y}{8}$

9. $\dfrac{7}{1} = \dfrac{y}{7}$

10. $\dfrac{6}{14} = \dfrac{7}{x-3}$

11. $y:5 = 22:10$

12. $\dfrac{14}{10} = \dfrac{5+x}{x-3}$

13. $\dfrac{5}{x+3} = \dfrac{3}{2x-8}$

14. $\dfrac{3}{3a+2} = \dfrac{3}{8}$

15. $\dfrac{8}{z-5} = \dfrac{3}{4}$

16. $\dfrac{x-4}{4} = \dfrac{x+4}{8}$

17. $\dfrac{y-13}{9} = 2$

18. $\dfrac{a}{9} = \dfrac{a-6}{27}$

19. $\dfrac{b-2}{3} = \dfrac{5b-2}{3}$

20. $\dfrac{a+4}{-1} = \dfrac{a+4}{5}$

21. $\dfrac{c}{c+2} = \dfrac{3}{5}$

22. **BUSINESS** For one month, a company's income was $12,450. The company's expenses for that month were $7,470. What percent of the company's income was profit? (Hint: income − expenses = profit).

23. **RETAIL** A shoe store advertises a running shoe on sale for $38.40. The regular price on the running shoes is $48.00. What is the discount percent on the running shoes?

PRACTICE ◣ LESSON 3-6

Graph the solution of each inequality on a number line.

24. $x < 5$

25. $a \geq -6$

26. $g < -1$

27. $-10 < y$

28. $h > -6$

29. $x \leq 10$

30. $-2 > b$

31. $\dfrac{5}{2} \leq y$

32. $t < 6$

33. $t \geq -4$

34. $a \geq 0$

35. $3.2 > n$

State the inequality that is represented on each number line.

36.

37.

38.

Solve each equation. Check the solution. (Lessons 3-2 and 3-4)

39. $\frac{6}{5}y = 36$

40. $n - 3 = -11$

41. $3 + d = 7.5$

42. $3y - 8 = 43$

43. $\frac{2}{5} + x = \frac{3}{4}$

44. $-4 + 6w = 38$

Solve each proportion. Assume no denominator is 0. Check the solution. (Lesson 3-5)

45. $\frac{2}{8} = \frac{x}{40}$

46. $\frac{17}{4} = \frac{17}{y}$

47. $\frac{3}{5} = \frac{6}{y - 4}$

Graph the solution to the inequality on a number line. (Lesson 3-6)

48. $x > |-5|$

49. $x \le -2$

50. $x \ge 6.5$

MathWorks Career – Automobile Designer
Workplace Knowhow

Automobile design requires the application of many physical principles. For passenger safety, one of the most important principles to understand is how automobiles respond when they collide. To prevent serious injury, the air bag acts as a cushion of air between the passenger and the dashboard, steering wheel, and door.

The following exercises analyze a situation in which an automobile strikes a stationary object head on and a passenger in the front seat is thrown toward the dashboard. The following equations will be useful in this analysis.

Force = mass · acceleration

Acceleration = $\frac{\text{[final velocity − initial velocity]}}{\text{time}}$

Force is measured in newtons (N), mass is measured in kilograms (kg), acceleration is measured in meters per second per second (m/sec²), velocity is measured in meters per second (m/scc), and time is measured in seconds (sec).

1. A car traveling 13.4 m/sec (30 mi/h) strikes a utility pole and comes to rest in 2 sec. With how much force does a 54.4-kg (120-lb) person in the car strike the dashboard in the absence of an airbag?

2. If there are 4.448 N per pound of force, how many pounds of force were exerted on the person in Exercise 1?

3. An airbag is to be deployed in 1 sec to prevent the passenger described in Exercise 1 from striking the dashboard. The force exerted by the airbag against the passenger must be equivalent to or greater than the force with which the passenger would strike the dashboard. Assuming that the airbag weighs 2.3 kg (5 lb), what is the velocity of the airbag when it reaches the force calculated in Exercise 1?

4. Convert the velocity value in Exercise 3 from m/sec to mi/h. (1 mi/h = 0.4470 m/sec)

3-7 Solve Inequalities

Goals
- Solve and graph inequalities on a number line.
- Solve problems involving inequalities.

Applications Communications, Health, Fitness, Hobbies, Safety, Business

---

COMMUNICATIONS Two competing telephone companies offer two different plans. National Telephone Company charges $4.95/mo plus $0.10/call. American Long Distance charges $6.50/mo and $0.05/call.

1. Copy and complete the tables for each company.

2. For what number of calls is National a better deal than American?

3. Write an algebraic expression to represent the monthly charge for each phone company. Let c = the number of calls.

4. Compare the two expressions from Question 3 using < or > if a customer makes 25 calls/mo.

National Telephone Company

Number of calls	Monthly charge
5	
10	
15	
20	
30	
40	

American Long Distance

Number of calls	Monthly charge
5	
10	
15	
20	
30	
40	

◣ BUILD UNDERSTANDING

To **solve an inequality**, find all values of the variable that make the inequality true. You can use techniques similar to those used to solve an equation. For example, if you add the same number to each side of an inequality, the order of the inequality remains the same. This is called the **addition property of inequality**.

Addition Property of Inequality	If $a < b$, then $a + c < b + c$ and $c + a < c + b$. If $a > b$, then $a + c > b + c$ and $c + a > c + b$.

E x a m p l e 1

Solve and graph $n + 5 \geq 8$.

Solution

$$n + 5 \geq 8$$

$$n + 5 + (-5) \geq 8 + (-5) \qquad \text{Add } -5 \text{ to both sides.}$$

$$n \geq 3$$

Graph the solution. Use a closed circle to show that 3 is a solution.

You can also multiply or divide each side of an inequality by the same number, but it is very important to be aware of the sign of the number. The **multiplication and division properties of inequality** state the following:

If you multiply or divide each side of an inequality by the same positive number, the order of the inequality remains the same.

If you multiply or divide each side of an inequality by the same negative number, the order of the inequality is reversed.

Multiplication and Division Properties of Inequality	For all real numbers a, b and c: If $a > b$ and $c > 0$, then $ac > bc$ and $\dfrac{a}{c} > \dfrac{b}{c}$. If $a < b$ and $c > 0$, then $ac < bc$ and $\dfrac{a}{c} < \dfrac{b}{c}$. If $a > b$ and $c < 0$, then $ac < bc$ and $\dfrac{a}{c} < \dfrac{b}{c}$. If $a < b$ and $c < 0$, then $ac > bc$ and $\dfrac{a}{c} > \dfrac{b}{c}$.

COncepts in MOtion
Animation
mathmatters2.com

Example 2

Solve and graph each inequality.

a. $3x \leq 9$

b. $-\dfrac{1}{4}y < 2$

Solution

a. $3x \leq 9$

$\dfrac{3x}{3} \leq \dfrac{9}{3}$ Divide both sides by 3.

$x \leq 3$

Graph $x \leq 3$.

b. $-\dfrac{1}{4}y < 2$

$-4\left(-\dfrac{1}{4}y\right) > 2(-4)$ Multiply both sides by -4 and reverse the inequality.

$y > -8$

Graph $y > -8$.

Example 3

Solve and graph $-2x - 3 < -1$.

Solution

$-2x - 3 < -1$

$-2x - 3 + 3 < -1 + 3$ Add 3 to both sides.

$-2x < 2$

$\dfrac{-2x}{-2} > \dfrac{2}{-2}$ Divide both sides by -2 and reverse the order of the inequality.

$x > -1$

Graph $x > -1$.

Check Understanding

In the inequality $-\dfrac{r}{3} > -1$, do you reverse the inequality for the solution? Why or why not?

Example 4

Online Personal Tutor at mathmatters2.com

FITNESS Nita is training for a bike race. She plans to ride at least 100 mi/wk. One wheel broke on Monday, so she only rode 7 mi that day. What is the least number of miles she must average daily for the next 6 days to reach her goal?

Solution

Write and solve an inequality that represents the situation. Let m = number of miles Nita rides each day. Let $6m$ = number of miles for the next 6 days.

$$6m + 7 \geq 100$$
$$6m + 7 - 7 \geq 100 - 7$$
$$6m \geq 93$$
$$\frac{6m}{6} \geq \frac{93}{6}$$
$$m \geq 15.5$$

Nita must average at least 15.5 mi each day for the next 6 days.

TRY THESE EXERCISES

Solve and graph each inequality.

1. $z - 4 \leq -2$

2. $3m \geq 12$

3. $y + 9 < 3$

4. $\frac{x}{4} > -3$

5. $-\frac{2}{3}t \leq -8$

6. $8 - 2p \leq 10$

7. $-9 > 4n + 7$

8. $3x - 14 < 7$

9. $-\frac{w}{3} + 5 \geq -1$

10. WRITING MATH Explain each step you used to solve and graph the inequality in Exercise 9.

11. HEALTH Julio is on an 1800-calorie-a-day diet. He consumed 450 calories at breakfast. What must Julio's average caloric intake be for lunch and dinner in order for him to maintain his diet?

PRACTICE EXERCISES • For Extra Practice, see page 595.

Solve and graph each inequality.

12. $2 - c > -2$

13. $2 \geq -\frac{1}{2}x$

14. $d - 9 < -7$

15. $-4x \leq -12$

16. $5 > b + 11$

17. $\frac{3}{4}h \leq -6$

18. $-14 + p > -21$

19. $14 \leq -4r$

20. $6n + 5 < -25$

21. $7k + 4 \geq -10$

22. $4z + 8 > -2$

23. $5 + 4y < 37$

24. $14 < 5e + 4$

25. $4 - 3q > 13$

26. $16 \geq 3c + 7$

27. $\frac{2}{3}b - 3 \leq 1$

28. $10 \leq -2x + 5$

29. $0.8x - 7 \leq 0.2$

30. $3 - 9t \geq 30$

31. $2 - \frac{1}{4}a \geq 4$

32. $17 \geq 5 - 6x$

33. $3x - 6x \geq 12$

34. $4(c + 1) > -3$

35. $2p + 3.7 < p - 1.5$

36. $1 + m \leq 3 - 2m$

37. $5x + 4 \geq 3x + 2$

38. $2(x + 1) < 4(x + 3)$

39. HOBBIES Parker wants to read a 150-page book over 5 days. He plans to read 30 pages each day. However, on the first day he read 38 pages. What is the least number of pages Parker must average for the next 4 days to reach his goal?

40. SAFETY The maximum load for an elevator is 2400 lb. A crate weighing 300 lb is put on the elevator. Suppose the average weight of a passenger is 150 lb. How many passengers can get on the elevator with the crate?

Write and solve an inequality for each statement.

41. Negative six is less than 18 times a number.

42. A number decreased by seven is at least negative nine.

43. Eight increased by four times a number is at most −12.

44. Three-fourths of a number, decreased by 11, is greater than 4.

45. Sara can afford no more than 36 yd of fencing for her yard. What is the largest length of x?

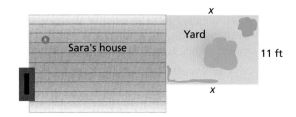

EXTENDED PRACTICE EXERCISES

46. BUSINESS Hoshi begins her own company selling music CDs. The start-up cost of her business is $4500. It costs Hoshi $8.50 for each CD, which she will sell for $15.00. Write and solve an inequality to show the number of CDs she must sell to earn a profit of at least $500.

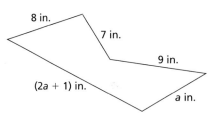

47. In order for the perimeter of the polygon shown to be no more than 50 in., what values can a have?

48. WRITING MATH Write a problem that could be solved using $3c + 25 \le 100$.

49. CRITICAL THINKING If $a > b$ and $c > d$, then is $a - c > b - d$ *sometimes*, *always* or *never* true? Use examples to justify your answer.

50. If $m \ge n$ and $-m \ge -n$, what do you think is true about m and n?

MIXED REVIEW EXERCISES

Write each number in scientific notation. (Lesson 2-8)

51. 7,382,047   **52.** 0.00826   **53.** 38,471,900

Write each number in standard form. (Lesson 2-8)

54. $3.2 \cdot 10^5$ **55.** $5.697 \cdot 10^{-4}$ **56.** $3.2168 \cdot 10^8$

Simplify each variable expression. (Lesson 2-5)

57. $5(2z + 8)$ **58.** $-3(3b + 1)$ **59.** $\frac{1}{2}(2f - 6g)$

60. $\dfrac{(5 - 12x)}{3}$ **61.** $\dfrac{(-4a - 12)}{2}$ **62.** $\dfrac{(6j - 30k)}{3}$

3-8 Equations with Squares and Square Roots

Goals
- Solve equations involving squares.
- Solve equations involving square roots.

Applications Physics, Safety, Engineering, Mechanics

PHYSICS The period of a pendulum is the time it takes the pendulum to complete a full swing (back and forth). The period is found using the formula $T = 2\pi \cdot \sqrt{\dfrac{L}{981}}$, where T is the period and L is the length of the pendulum.

1. Use a ruler, a piece of string and a weight (such as a washer, ring or binder clip) to construct a pendulum similar to the one shown.

2. Copy and complete the table. Use the formula to find the period.

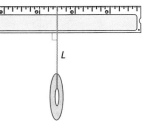

Length of pendulum (centimeters)	10	20	30	40	50
Period (seconds)					

3. Determine the length of the pendulum if the period is 1 sec.

BUILD UNDERSTANDING

To solve an equation, perform the inverse of the operation that is applied to the variable to both sides of the equation. The inverse of squaring a number is taking its square root. The inverse of taking the square root of a number is squaring that number.

Recall that when you multiply two radical expressions with the same radicand, the product is the radicand.

$$\sqrt{x} \cdot \sqrt{x} = x \qquad \sqrt{2x} \cdot \sqrt{2x} = 2x \qquad \sqrt{x+5} \cdot \sqrt{x+5} = x+5 \qquad \sqrt{\tfrac{2}{5}} \cdot \sqrt{\tfrac{2}{5}} = \tfrac{2}{5}$$

To solve an equation with x^2 on one side of the equation and a non-negative real number on the other side take the square root of both sides. Recall that every positive real number has a positive square root and a negative square root.

To solve an equation with $\sqrt{x}$ on one side of the equation and a real number on the other side, square both sides of the equation.

> **Think Back**
>
> An expression such as $\sqrt{36}$ is called a **radical**.
>
> $\sqrt{}$ is the **radical symbol**.
>
> The number under the radical symbol is the **radicand**.

Example 1

Solve each equation. Check the solutions.

a. $x^2 = 36$

b. $\sqrt{x} = 4$

Solution

a.
$$x^2 = 36$$
$$\sqrt{x^2} = \sqrt{36} \qquad \text{Find each square root.}$$
$$x = 6 \quad \text{or} \quad x = -6 \qquad \text{Include the positive and negative solutions.}$$

Check

$x^2 = 36$	$x^2 = 36$
$6^2 = 36$	$(-6)^2 = 36$
$36 = 36$ ✔	$36 = 36$ ✔

b. $\sqrt{x} = 4$

$(\sqrt{x})^2 = 4^2$ Square both sides.

$x = 16$

Check

$\sqrt{x} = 4$

$\sqrt{16} = 4$ ✔

Example 2

Solve each equation. Check the solutions.

a. $x^2 = 1.21$

b. $\sqrt{x} = \dfrac{3}{4}$

> **Check Understanding**
>
> Explain why part a of Example 1 has two solutions and part b has only one solution.

Solution

a. $x^2 = 1.21$

$\sqrt{x^2} = \sqrt{1.21}$ Find each square root.

$x = 1.1 \text{ or } x = -1.1$

Check

$x^2 = 1.21 \qquad x^2 = 1.21$

$(1.1)^2 \stackrel{?}{=} 1.21 \qquad (-1.1)^2 \stackrel{?}{=} 1.21$

$1.21 = 1.21$ ✔ $1.21 = 1.21$ ✔

b. $\sqrt{x} = \dfrac{3}{4}$

$(\sqrt{x})^2 = \left(\dfrac{3}{4}\right)^2$ Square both sides.

$x = \dfrac{9}{16}$

Check

$\sqrt{x} = \dfrac{3}{4}$

$\sqrt{\dfrac{9}{16}} \stackrel{?}{=} \dfrac{3}{4} \qquad \sqrt{\dfrac{9}{16}} = \dfrac{\sqrt{9}}{\sqrt{16}}$

$\dfrac{3}{4} = \dfrac{3}{4}$ ✔

Often equations with squares or square roots require more than one step. First isolate the term or expression with the variable. Then perform the inverse operation.

Example 3   Online **Personal Tutor at mathmatters2.com**

Solve each equation. Check the solutions.

a. $x^2 - 6 = 19$

b. $\sqrt{g + 4} = 5$

Solution

a. $x^2 - 6 = 19$

$x^2 - 6 + 6 = 19 + 6$ Add 6 to both sides.

$x^2 = 25$

$\sqrt{x^2} = \sqrt{25}$ Find each square root.

$x = \pm 5$

Check

$x^2 - 6 = 19 \qquad x^2 - 6 = 19$

$5^2 - 6 \stackrel{?}{=} 19 \qquad (-5)^2 - 6 \stackrel{?}{=} 19$

$25 - 6 \stackrel{?}{=} 19 \qquad 25 - 6 \stackrel{?}{=} 19$

$19 = 19$ ✔ $19 = 19$ ✔

b. $\sqrt{g + 4} = 5$

$(\sqrt{g + 4})^2 = 5^2$ Square both sides.

$g + 4 = 25$

$g + 4 - 4 = 25 - 4$ Subtract 4 from both sides.

$g = 21$

Check

$\sqrt{g + 4} = 5$

$\sqrt{21 + 4} \stackrel{?}{=} 5$

$\sqrt{25} \stackrel{?}{=} 5$

$5 = 5$ ✔

Example 4

PHYSICS In the formula $d = 16t^2$, d is the distance in feet traveled by a freely falling object dropped from a resting position. The time of the fall in seconds is t. Find the time in seconds for an object to fall 484 ft.

Solution

$$d = 16t^2$$
$$484 = 16t^2 \qquad \text{Substitute 484 for } d.$$
$$\frac{484}{16} = \frac{16t^2}{16}$$
$$30.25 = t^2$$
$$\sqrt{30.25} = \sqrt{t^2}$$
$$\pm\, 5.5 = t$$

Since t represents time, the solution cannot be negative. So it takes 5.5 sec for an object to fall 484 ft.

◼ TRY THESE EXERCISES

Solve each equation. Check the solutions.

1. $x^2 = 25$ **2.** $\sqrt{x} = 9$ **3.** $y^2 = 7$ **4.** $n^2 = 2.25$

5. $z^2 = \dfrac{49}{100}$ **6.** $\sqrt{m} = \dfrac{1}{4}$ **7.** $\sqrt{d+3} = 2$ **8.** $89 = x^2 + 8$

9. $2w^2 = 288$ **10.** $\sqrt{2x+1} = 5$ **11.** $0 = 2.4b^2$ **12.** $\sqrt{j} - 3 = 1$

13. WRITING MATH Explain the difference between Exercises 7 and 12. What did you do differently in solving each?

14. Using the formula $d = 16t^2$, find the time in seconds for an object to fall 96 ft.

◼ PRACTICE EXERCISES • For Extra Practice, see page 595.

Solve each equation. Check the solutions.

15. $\sqrt{x} = 9$ **16.** $\sqrt{m} = -11$ **17.** $y^2 = 64$ **18.** $h^2 = 169$

19. $\sqrt{a} = 2$ **20.** $\sqrt{b} = 0.3$ **21.** $59 = c^2$ **22.** $d^2 = \dfrac{1}{4}$

23. $m^2 = -1.8$ **24.** $6 = \sqrt{n-3}$ **25.** $a^2 - 1 = 15$ **26.** $\sqrt{m-4} = 0$

27. $279 = x^2 + 23$ **28.** $3n^2 = 300$ **29.** $5\sqrt{p} = 30$ **30.** $\sqrt{s+9} = 5$

31. $\sqrt{2k} - 5 = 7$ **32.** $2w^2 = 1.62$ **33.** $4y^2 = 200$ **34.** $\sqrt{3h} = \dfrac{1}{3}$

35. $\sqrt{3t-5} = 5$ **36.** $5 = \sqrt{2(x+1)}$ **37.** $x^2 - 3 = 102$ **38.** $9 + 3n^2 = 21$

39. SAFETY In the formula $b = 0.06v^2$, b is the distance in feet needed to stop a car after the brakes are applied. The speed at which a car is traveling when the brakes are applied in miles per hour is v. If a car traveled 96 ft after the brakes were applied, how fast was the car going when the driver applied the brakes?

40. The perimeter of the rectangle shown is 42 yd. Find the value of x.

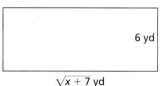

6 yd

$\sqrt{x + 7}$ yd

41. Heta has three identical square pieces of fabric with a total area of 588 in.² Write and solve an equation to find the dimensions of each piece of fabric.

42. ENGINEERING A circular pipe has an inner radius of r and a cross-sectional area of A. The outer radius R of the pipe is given by $R = \sqrt{\dfrac{A}{\pi} + r^2}$. Find the inner radius of a pipe with an outer radius of 2 cm and a cross-sectional area of 1.75π cm².

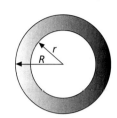

43. Solve the equation $T = 2\pi \cdot \sqrt{\dfrac{L}{981}}$ for L.

CALCULATOR Solve each equation. Then use a calculator to approximate the solutions to the nearest thousandth.

44. $a^2 = \dfrac{49}{10,000}$ **45.** $k^2 = 0.0144$ **46.** $67 = c^2$ **47.** $7f^2 = 1.75$

48. $x^2 - 8 = 30$ **49.** $328 = 4p^2$ **50.** $3m^2 + 2 = 50$ **51.** $117 = 6y^2 - 9$

52. MODELING What equation is modeled on the Sentence Mat? Solve for x.

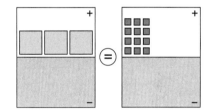

EXTENDED PRACTICE EXERCISES

Solve each radical equation.

53. $\sqrt{x + 7} = \sqrt{3x - 19}$

54. $\sqrt{5n + 2} = \sqrt{-2n + 23}$

55. MECHANICS An engine piston is acted on by a force F producing pressure P. The diameter d of the piston is given by $d = \sqrt{\dfrac{5F}{4P}}$. Find the force in pounds that produces a pressure 80 lb/in.² on a piston with a diameter of 3 in.

56. Use the formula $d = 16t^2$ to find the time in seconds (t) for an object to fall 200 ft, 400 ft and 600 ft. Round each answer to the nearest hundredth.

57. CRITICAL THINKING Compare your results for Exercise 56. Does it take twice as long for an object to fall twice as far? Is the difference in time between 600 ft and 400 ft the same as the difference between 400 ft and 200 ft? What conclusions can you draw?

58. CHAPTER INVESTIGATION Calculate the acceleration due to Earth's gravity using the formula $g = \dfrac{4\pi^2 L}{t}$, where L is the length of the pendulum string in feet and t is the time in seconds for 50 complete swings.

MIXED REVIEW EXERCISES

Write each phrase as a variable expression. (Lesson 2-3)

59. seven more than a number

60. the difference between a number and 143

61. five times a number divided by -12

62. the product of 14 and a number

Evaluate each expression when $x = 3$, $y = -4$, and $z = \dfrac{1}{4}$. (Lesson 2-4)

63. $5x + 3y$ **64.** $2(x + y)^2$ **65.** $z(3y - 4x)$ **66.** $2xy + 8z$

Chapter 3 Review

VOCABULARY

Choose the word from the list that best completes each statement.

1. A value of the variable that makes an equation true is called a(n) __?__ of the equation.

2. When you add the same number to each side of an inequality, the result is a(n) __?__ inequality.

3. An equation stating the relationship between two or more variable quantities is called a(n) __?__.

4. Squaring a number and finding that number's square root are __?__ operations.

5. In the equation $\frac{a}{b} = \frac{c}{d}$, a and d are called the __?__.

6. A mathematical sentence that states that "a is greater than or equal to -6" is an example of a(n) __?__.

7. One way to solve a proportion is to write an equation using the __?__.

8. A(n) __?__ uses a variable or variables to describe a real-life situation.

9. The statement $-8 + 2 = 6$ is an example of a false __?__.

10. The statement "1 is to 2 as 5 is to 10" is an example of a(n) __?__.

a. addition property of equality

b. algebraic model

c. cross products

d. equation

e. equivalent

f. extremes

g. formula

h. inequality

i. inverse

j. means

k. proportion

l. solution

LESSON 3-1 ◣ Equations and Formulas, p. 104

▶ An **equation** is a statement that two numbers or expressions are equal.

▶ To **solve an equation**, find all the values of the variable that make the equation true.

▶ A **formula** is an equation stating a relationship between two or more variable quantities.

Determine if 7, 12, or 21.5 are solutions of each equation.

11. $2p - 3 = 11$

12. $-4x = 6$

13. $9 = \frac{z}{3} + 5$

14. What is the weight (F) of a man of 100-kg mass on a mountain near the equator where $g = 9.76$ m/s^2? Use the formula $F = mg$.

For Exercises 15–16, use the cylinder at the right.

15. The formula for the volume of a cylinder is $V = \pi r^2 h$. In this formula, V is the volume of the cylinder, r is the radius of the base, and h is the height of the cylinder. Find the volume of the cylinder.

16. The formula for the surface area of a cylinder is $SA = 2\pi rh + 2\pi r^2$. In this formula, SA is the surface area, r is the radius of the base, and h is the height of the cylinder. Find the surface area of the cylinder.

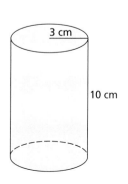

3 cm

10 cm

LESSON 3-2 ◾ One-Step Equations, p. 108

▶ When two expressions are equal, if you add or multiply the same number to each expression the resulting expressions will be equal.

Solve each equation. Check the solution.

17. $x + 7 = 3$

18. $-4m = 3.2$

19. $\frac{2}{3}y = 6$

Translate each word sentence into an equation. Then solve the equation.

20. A number decreased by -12 is two.

21. One-fourth a number is -15.

LESSON 3-3 ◾ Problem Solving Skills: Model Algebra, p. 114

▶ An **algebraic model** is a variable representation of a situation.

▶ A **mathematical model** uses numbers to describe a situation.

▶ A **rule** is an equation or formula that represents a model.

For each situation, set up a mathematical model (table), write an algebraic model and a rule.

22. Nicole starts a savings program. She begins by saving $10 and adds $5 each week for 8 weeks.

23. When an American alligator is hatched, it is about 8 in. long. It grows about 12 in. each year.

24. The cost of renting a car is $25 per day plus $0.10/mi.

25. Devin owes the library $1.80 for a previous overdue book. If he does not return the book today, he will be fined an additional $0.30 per day.

LESSON 3-4 ◾ Equations With Two or More Operations, p. 116

▶ To solve **multi-step equations**, first use the addition property of equality. Then use the multiplication property.

Solve each equation. Check the solution.

26. $\frac{c}{2} + 8 = 21$

27. $25 = 40n + 5$

28. $2z - 11 + 9z = 11$

29. $\frac{a}{8} + 21 = 14$

30. $10y - 2 = 8y - 1$

31. $4(2a - 1) = -10(a - 5)$

LESSON 3-5 ◾ Proportions, p. 122

▶ A **proportion** is an equation stating that two ratios are equivalent.

▶ In a proportion, the **cross-products** are equal.

Solve each proportion. Assume no denominator is 0. Check the solution.

32. $\frac{12}{22} = \frac{30}{x}$

33. $\frac{y}{8} = \frac{65}{52}$

34. $\frac{(n - 2)}{3} = \frac{(n + 2)}{6}$

35. The Harrisons' car requires 4 gal to travel 120 mi. How many gallons will they need to travel 300 mi?

LESSON 3-6 ◣ Graph Inequalities on a Number Line, p. 126

▶ An **inequality** is a mathematical sentence that contains one of these symbols: $\neq, <, >, \leq,$ or $\geq$.

▶ The **solution of an inequality** can be represented on a number line.

▶ An open circle on the graph indicates that the number is not part of the solution. A solid circle on the graph indicated that the number is part of the solution.

Graph the solution of each inequality on a number line.

36. $m < -3$ **37.** $4 \geq z$ **38.** $-5.2 \leq x$

39. $t > 3$ **40.** $-3.8 > h$ **41.** $a \leq 2.3$

LESSON 3-7 ◣ Solve Inequalities, p. 132

▶ To solve an inequality, find all the values of the variable that make the inequality true.

▶ Use the addition, multiplication, and division properties to solve inequalities.

Graph the solution of each inequality on a number line.

42. $3m \leq -3$ **43.** $4z - 4 > 4$ **44.** $-5 \leq \frac{x}{2} + 1$

45. $-7t + 19 < -16$ **46.** $-2 - \frac{d}{5} \leq -3$ **47.** $3(s - 7) + 9 < 21$

LESSON 3-8 ◣ Equations with Squares and Square Roots, p. 136

▶ The inverse of squaring a number is finding the square root.

▶ The inverse operation of taking the square root of a number is squaring that number.

Solve each equation. Check the solutions.

48. $y^2 = 49$ **49.** $\sqrt{x} = 25$ **50.** $m^2 + 3 + 19$

51. $\sqrt{x} + 2 = 10$ **52.** $4 = \sqrt{4(x - 5)}$ **53.** $4 + y^2 = 40$

54. Using the formula $d = 16t^2$, find the time in seconds for an object to fall 144 ft.

55. Meterologists can use the formula $t = \frac{d^3}{216}$ to estimate the amount of time t a storm of diameter d will last. Suppose the eye of a hurricane, which causes the greatest amount of destruction, is 9 mi in diameter. How long will the worst part of the hurricane last? Round to the nearest tenth of an hour.

CHAPTER INVESTIGATION

EXTENSION The weight of an object is the measure of the gravitational force on an object. The formula $F = mg$ represents the weight of object (F) as a product of its mass (m) and the strength of the Earth's gravity (g). Find three different objects of varying weights. Measure the weight of each object. Use the formula $F = mg$ to find the mass of each object.

Chapter 3 Assessment

Which of the given values is a solution of the equation?

1. $-3x = 12$; 2, -4, -2

2. $11 = 3 + \frac{c}{2}$; 16, 18, 20

3. $-\frac{b}{6} = 6$; 236, 26, 36

Solve each equation. Check the solution.

4. $d - 9 = 9$

5. $48 = -8n$

6. $w - 7.3 = -6.2$

7. $\frac{x}{6} = -2$

8. $5g - 13 = 32$

9. $-\frac{2}{3}n + 10 = 24$

10. $0 = 0.5t - 1$

11. $8g - 10 + g = 71$

12. $35 = 3x + 1 - 5x$

Solve each proportion. Assume no denominator is 0. Check the solution.

13. $\frac{6}{21} = \frac{28}{x}$

14. $\frac{t}{45} = \frac{8}{40}$

15. $\frac{y - 5}{7} = \frac{y + 11}{3}$

Solve and graph each inequality.

16. $7d \geq 21$

17. $3f - 2 < 4$

18. $\frac{-x}{2} + 1 < 3$

19. $-10 \leq 3k + 8$

20. $-x < 2$

21. $-5 \geq -z$

Solve.

22. If it is 95° Fahrenheit, what is the temperature in Celsius? Use the formula $F = \frac{9}{5}C + 32$.

23. Use the formula $p = \frac{a}{2} + 110$ to find the value of p when $a = 32$.

Solve each equation. Check the solutions.

24. $x^2 = 100$

25. $m^2 = 0.36$

26. $w^2 = \frac{9}{196}$

27. $\sqrt{y} = 225$

28. $\sqrt{4n} = 8$

29. $\sqrt{y + 9} = 3$

30. $\sqrt{3(w - 2)} = 1$

31. $\sqrt{y} - 7 = 2$

32. $11 + 9j^2 = 15$

Set up a mathematical model, then write an algebraic model and a rule for each situation.

33. Karl has been adding 12 cards to his baseball card collection each week for the past 5 weeks. He started with 60 cards.

34. A long distance phone call from the motel costs $3.50 plus $0.25 for each minute.

35. Kendra began with $20 worth of animal feed. Every minute for 5 minutes she gave $1.50 worth of feed to the animals.

Standardized Test Practice

Part 1 Multiple Choice

Record your answers on the answer sheet
provided by your teacher or on a sheet of paper.

1. The stem-and-leaf plot below lists the cost of
   various bicycle helmets. Which statement
   about the data is true? (Lesson 1-3)

2	1 3 4 4 5 7
3	0 0 0 1 5 9
4	3 9 9

   2|1 represents $21.

   A The median and mean are equal.
   B The median and mode are equal.
   C The median is greater than the mode.
   D The mode is greater than the mean.

2. The box-and-whisker plot below represents
   the high temperatures in degrees Fahrenheit
   for various cities. Which statement about the
   data is true? (Lesson 1-6)

   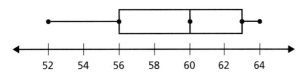

   A The median is 60°.
   B The first quartile is 52°.
   C The interquartile range is 12°.
   D 64° is an outlier.

3. Which statement is *not* true? (Lesson 2-1)
   A $|-3| > -3$

   B $-\dfrac{3}{2} < -1$

   C $\dfrac{1}{4} > \dfrac{1}{3}$

   D $-2\dfrac{2}{3} < 1\dfrac{3}{4}$

4. What is the value of $(14 - 4) \div 2 + 3^2$?
   (Lesson 2-2)
   A 14   B 19
   C 35   D 64

5. Which expression is equivalent to
   $3(2x - y) + 4(x + 2y) - 3(x + y)$?
   (Lesson 2-6)
   A $7x + 8y$ B $7x + 2y$
   C $9x$ D $9y$

6. An online music store charges $12 for
   each CD and $5 per order for shipping and
   handling. If c represents the number of CDs
   ordered, which rule can be used to determine
   the cost of the order? (Lesson 3-3)
   A $12c + 5$
   B $5c + 12$
   C $12c - 5$
   D $5c - 12$

7. What is the solution of $\dfrac{2x + 3}{6} = \dfrac{x - 4}{4}$?
   (Lesson 3-5)
   A 18   B $\dfrac{1}{2}$ C $-\dfrac{1}{2}$ D -18

8. Which graph represents the solution of
   $-2x + 3 \le 5$? (Lesson 3-7)

   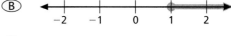

9. What is the solution of $\sqrt{x + 4} = 6$?
   (Lesson 3-8)
   A 2   B 32   C 36   D 40

Test-Taking Tip

Questions 7 and 9
Some multiple-choice questions ask you to solve an equation
or inequality. You can check your solution by replacing the
variable in the equation or inequality with your answer. The
answer choice that results in a true statement is the correct
answer.

Preparing for the Standardized Tests
For test-taking strategies and more
practice, see pages 627-644.

Part 2 Short Response/Grid In

**Record your answers on the answer sheet
provided by your teacher or on a sheet of paper.**

**The scatter plot shows the relationship between
the foot length and the height of a sample of
male students in the 11th grade. Use the scatter
plot for Questions 10–12.**

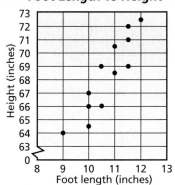

Foot Length vs Height

10. What is the median of the heights of the
males? (Lesson 1-2)

11. What is the mode of the foot lengths?
(Lesson 1-2)

12. Does the scatter plot show a positive
correlation, a negative correlation, or
no correlation? (Lesson 1-4)

13. Write four less than half of a number as a
variable expression. (Lesson 2-3)

14. Evaluate $2.5x + 7y - 3z$ when $x = 3$, $y = 4$,
and $z = -9$. (Lesson 2-4)

15. Simplify $(p^3)^2 \cdot p^{11}$. (Lesson 2-7)

16. The diameter of a red blood cell is about
$7.4 \cdot 10^{-4}$ cm. Write this number in standard
form. (Lesson 2-8)

17. What is the next term in the sequence?
(Lesson 2-9)

$$10, 6, 2, -2, -6, \ldots$$

18. The formula that can be used to change
degrees Celsius to degrees Fahrenheit
is $F = \frac{9}{5}C + 32$. What is the Fahrenheit
temperature that is equivalent to 30°C?
(Lesson 3-1)

19. If $x + 2 = 6$, what is the value of $4x$?
(Lesson 3-2)

20. Autumn withdrew an amount of money from
her bank account. She spent one fourth for
gasoline and had $90 left. How much money
did she withdraw? (Lesson 3-4)

21. Mr. Rollins drove 174 mi in 3 h. At that rate,
how long will it take him to travel 290 mi?
(Lesson 3-5)

22. Write an inequality represented by the
number line. (Lesson 3-6)

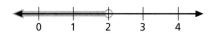

Part 3 Extended Response

**Record your answers on a sheet of paper.
Show your work.**

23. Mrs. Anderson and her family are going to
the zoo. She buys the same number of adult
tickets as child tickets. Adult tickets cost $12
and child tickets cost $8. She also buys a
T-shirt for $15. She spends $75 altogether.
Write an equation for this situation. Solve the
equation showing your work. Explain the
meaning of the solution. (Lesson 3-4)

24. Leon is driving his car at a rate of 7 mi every
10 min. Hana is driving her car at a rate of
20 mi every 25 min. Who is driving his or her
car faster? If the speed limit is 50 mi/h, is
either vehicle exceeding the speed limit?
Explain. (Lesson 3-5)

25. Explain why $x^2 = -4$ has no solution.
(Lesson 3-8)

mathmatters2.com/standardized_test

Probability

THEME: Games

At some point during the last week, have you said, "Chances are...?" Chances are the answer to that question is yes. Chance, also referred to as "likelihood," means the probability something will occur. Many events in your life are affected by probability.

People often find taking a chance as a form of entertainment. From children to adults, people enjoy playing card games, board games, electronic games, and sports games. Knowing the probability of winning a game makes game play exciting.

- **Baseball players** (page 157) are successful based on their game statistics. Coaches use the calculated probability of getting a hit to help establish a batting lineup.

- **Board Game designers** (page 177) use probability to build games of strategy, knowledge and nonsense. The board layout, game pieces and winning strategy must come together using sample spaces, permutations and probability.

Math Online

mathmatters2.com/chapter_theme

Nested Shapes

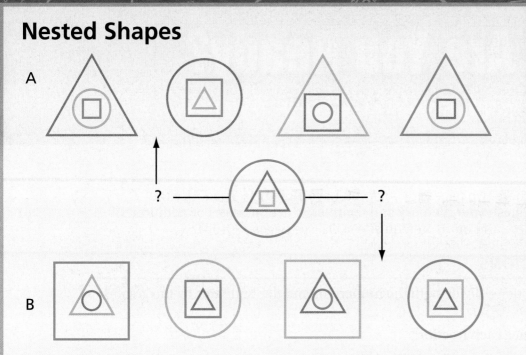

A

B

Data Activity: Nested Shapes

Use the figures above for Questions 1–5.

1. Identify the three shapes, colors and sizes that make up each nested figure.

2. Explain any similarities among the figures in group A.

3. Explain any similarities among the figures in group B.

4. Does the figure in the middle belong in group A or in group B?

5. Name some board games you have played. How was the game board marked? Did geometric shapes appear on the board or on the playing pieces? Did geometric shapes affect the way in which the game was played?

CHAPTER INVESTIGATION

Many popular games rely on the principles of probability and chance to determine outcomes. For example, in card games, being dealt a particular hand can be thought of as an event. As such, that event has a certain likelihood of occurring, which is its probability.

Working Together

Design your own game of chance and calculate the probabilities of various outcomes within the game. Use the Chapter Investigation icons throughout the chapter to guide you in the design of your game.

The skills on these two pages are ones you have already learned. Use the examples to refresh your memory and complete the exercises. For additional practice on these and more prerequisite skills, see pages 576–584.

REDUCING FRACTIONS

To reduce fractions, divide both the numerator and denominator by the greatest common factor (GCF).

Examples **Reduce each fraction.**

$\frac{18}{32} = \frac{9}{16}$ The GCF is 2.

$\frac{20}{32} = \frac{5}{8}$ The GCF is 4.

Reduce each fraction.

1. $\frac{9}{12}$

2. $\frac{24}{32}$

3. $\frac{28}{32}$

4. $\frac{28}{56}$

5. $\frac{25}{80}$

6. $\frac{15}{48}$

7. $\frac{28}{64}$

8. $\frac{18}{30}$

9. $\frac{27}{72}$

10. $\frac{35}{42}$

11. $\frac{15}{60}$

12. $\frac{91}{156}$

AREA

Examples Use the following formulas to find the area of a figure.

Figure	Area formula
square	$A = s^2$
rectangle	$A = b \cdot h$
triangle	$A = \frac{1}{2}b \cdot h$
parallelogram	$A = b \cdot h$
circle	$A = \pi r^2$

Find the area of each figure. Round to the nearest hundredth.

13. 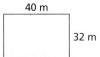 40 m, 32 m

14. 2 ft

15. 16 cm, 21 cm

16. 7 cm, 8 cm

Find the area of each figure with the given dimensions. Round to the nearest hundredth.

17. square: $s = 5$ ft

18. triangle: $b = 6.2$ cm, $h = 3.2$ cm

19. circle: $r = 3$ in.

20. parallelogram: $b = 2$ m, $h = 6$ m

21. rectangle: $b = \frac{2}{3}$ yd, $h = \frac{3}{4}$ yd

22. circle: $d = 6$ cm

SETS AND VENN DIAGRAMS

It will be helpful to review sets and Venn diagrams so that you can use them to compute probability.

Examples   The following lists several symbols used as set notation.

U	Universal set, or all the elements.	$U = \{1, 2, 3, 4, 5, 6\}$
{ }	Roster notation	$\{1, 2, 3, 4, 5, 6\}$
$\in$	"is an element of"	$3 \in \{1, 2, 3\}$
$\notin$	"is not an element of"	$8 \notin \{1, 2, 3, 4\}$
$\subseteq$	"is a subset of"	$\{1, 2, 3\} \subseteq \{1, 2, 3, 4, 5, 6\}$
$\nsubseteq$	"is not a subset of"	$\{1, 2, 3\} \nsubseteq \{1, 2, 4, 6\}$
$\varnothing$	the null set, which has no elements	{ }

Venn diagrams can help you see the sets and their relationships more clearly.

Complement of a set: The subset of all elements of a universal set that are not elements of subset A.

$$A = \{1, 2, 3, 4\} \qquad A' = \{5, 6\}$$

Union of two sets: All the elements of two sets.

$$A = \{1, 2, 3, 4\}, B = \{4, 5, 6\}$$

$$A \cup B = \{1, 2, 3, 4, 5, 6\}$$

Intersection of two sets: The elements that two sets have in common.

$$A = \{1, 2, 3, 4\}, B = \{4, 5, 6\}$$

$$A \cap B = \{4\}$$

Use roster notation to represent the sets named when $U = (1, 2, 3, 4, 5, 6, 7, 8, 9)$, $A = \{1, 4, 5, 7\}$, $B = \{2, 3, 5, 8\}$, and $C = \{5, 7, 8, 9\}$.

23. $A \cup B$ **24.** $A \cup C$ **25.** $B \cup C$

26. $A \cap B$ **27.** $A \cap C$ **28.** $B \cap C$

29. A' **30.** B' **31.** C'

Use the diagram to find the set named. List the elements in roster notation.

32. $A \cup B$ **33.** $A \cup C$ **34.** A'

35. $B \cup C$ **36.** $A \cap B$ **37.** B'

38. $A \cap C$ **39.** $B \cap C$ **40.** $(A \cup B)'$

4-1 Experiments and Probabilities

Goals ■ Use experiments to collect data.
■ Use data to find experimental probabilities.

Applications Music, Market research, Games, Statistics, Probability

---

When a coin is tossed, it will either land heads up or tails up. Prepare a table to tally the number of times a coin lands heads up and tails up.

Event	Tally	Results
Heads up	■	■
Tails up	■	■

1. Toss a coin 20 times. Record the results in the table.

2. Find the ratio of the number of times the coin lands tails up to the total number of times tossed. How does your ratio compare with the ratios of other groups?

3. Find the total number of times a coin lands tails up for your entire class, and divide by the total number of times a coin was tossed. What is the ratio? Is it more likely that a tossed coin will land heads up or tails up?

◤ BUILD UNDERSTANDING

Most real-world situations involve more than one possible outcome. Often these outcomes are not equally likely. While you cannot be sure of the outcome in advance, an experiment can help you find the likelihood of one particular outcome occurring.

An **experiment** is an activity that is used to produce data that can be observed and recorded. The **relative frequency** of an outcome compares the number of times the outcome occurs to the total number of observations.

The **experimental probability** represents an estimate of the likelihood of an event, E, or desired outcome.

Experimental Probability $P(E) = \dfrac{\text{number of observations favorable to } E}{\text{total number of observations}}$

Suppose you want to find whether a music department will be a successful addition to a store. If you ask 10 of the store's customers about the proposed music department, you would not have enough data to make a reliable prediction. But if you ask 2000 customers each week for a month, your prediction will be more accurate.

	Number of positive responses	Total number questioned	Relative frequency of positive responses
Week 1	384	2000	$\frac{384}{2000} = 0.192$
Week 2	420	2000	$\frac{420}{2000} = 0.21$
Week 3	396	2000	$\frac{396}{2000} = 0.198$
Week 4	412	2000	$\frac{412}{2000} = 0.206$

From the results in the table, the experimental probability that a customer will shop at the music department is about 0.20, or 1 out of 5.

Example 1 **Personal Tutor at** mathmatters2.com

MARKET RESEARCH The table shows the number of customers and skate rentals at a roller-skating rink during a week of summer vacation.

	Customers	Pairs of skates rented
Monday	192	130
Tuesday	328	212
Wednesday	296	222
Thursday	325	195
Friday	456	292

a. What is the experimental probability that a customer will rent skates on a Wednesday?

b. What is the experimental probability that a customer will rent skates on a Thursday?

Solution

For both parts a and b, use the experimental probability formula.

$$P(\text{customers renting skates}) = \frac{\text{pairs of skates rented}}{\text{number of customers}}$$

a. $P(\text{customers renting skates on Wednesday}) = \frac{222}{296} = 0.75$
The probability of a customer renting skates on Wednesday is 0.75, or $\frac{3}{4}$. This means that for every four customers on a given Wednesday, three of them will probably rent skates.

b. $P(\text{customers renting skates on Thursday}) = \frac{195}{325} = 0.6$
The probability of a customer renting skates on Thursday is 0.6, or $\frac{3}{5}$.

> ### Check Understanding
>
> Relative frequencies allow you to make statements that estimate probabilities.
>
> What can you conclude if the relative frequency of an outcome is 1? 0?

Example 2

According to the table, what is the probability that a July customer is over 40 years old?

Customers by Age

	12-18	19-25	26-40	41-55	Over 55
June	961	930	749	711	220
July	812	748	819	507	164
August	645	702	736	499	217

Solution

To find the experimental probability, find the total number of July customers and the number of July customers over 40 years old.

$$P(\text{July customers over 40}) = \frac{\text{July customers over 40 years old}}{\text{total number of July customers}}$$

$$= \frac{507 + 164}{812 + 748 + 819 + 507 + 164}$$

$$= \frac{671}{3050}$$

$$= 0.22$$

The probability that a customer is over 40 years old in July is 0.22.

Some probability problems can be solved by using an *area model*. Suppose that a region *A* contains a smaller region *B*. The probability (*P*) that a randomly chosen point in *A* is in *B* is given by:

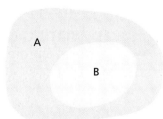

$$P = \frac{\text{area of } B}{\text{area of } A}$$

mathmatters2.com/extra_examples

Example 3

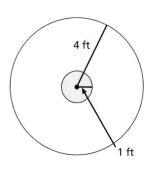

What is the probability that a thrown dart will hit the yellow region?

Solution

Find the areas of the yellow region and the entire region and divide.

$$P(\text{yellow}) = \frac{\text{area of yellow region}}{\text{area of entire region}}$$

$$= \frac{\pi \cdot 1^2}{\pi \cdot 4^2} = \frac{\pi \cdot 1}{\pi \cdot 16} = \frac{1}{16}$$ Use the formula for area of a circle, $A = \pi r^2$.

The probability that the dart will hit the yellow region is $\frac{1}{16}$.

◤ TRY THESE EXERCISES

1. At a fork in a hiking trail, 82 people take the lake trail while 43 people take the mountain trail. What is the probability of a hiker taking the lake trail?

2. **RETAIL** In June, 243 customers at an optician's office buy contact lenses and 207 buy glasses. Find the probability that a customer will buy glasses.

3. The ratio of boys to girls in a high school is 2 : 3. What is the probability that the first student to arrive at school one day is a boy? a girl?

4. What is the probability of a correct answer if 54 are correct out of 180?

Find the probability that a point selected at random lies in the shaded region. Round to the nearest hundredth.

5. 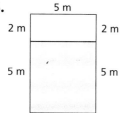   6. 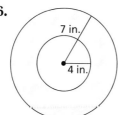   7. 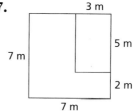   8.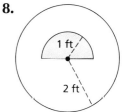

◤ PRACTICE EXERCISES • For Extra Practice, see page 596.

9. In a poll conducted to find students' favorite sports, 119 vote for football, 135 vote for baseball, 126 vote for volleyball and 160 vote for other sports. What is the probability that a student chooses baseball?

10. **STATISTICS** A survey of 500 students shows that the probability that a student's birthday is in April is 0.15. Of the boys surveyed, 29 have April birthdays. How many girls surveyed have April birthdays?

 11. **WRITING MATH** In Example 2, the manager can use probability to plan a promotional campaign targeted at the age of the customer who visits the store the most in July. Choose either Example 1 or 3, and describe a benefit of knowing the probability.

12. A circle with a 7-cm radius lies within a rectangle that is 22 cm by 21 cm. What is the probability that a point in the rectangle lies within the circle?

Use the table for Exercises 13–15.

13. On Friday what is the probability that a person in the library is a graduate student?

14. On Saturday what is the probability that a person using the library is a professor?

15. On Sunday what is the probability that a person using the library is a professor?

People Using the College Library

	Undergraduate students	Graduate students	Professors
Friday	737	105	33
Saturday	588	132	30
Sunday	448	91	36

Find the probability that a point selected at random lies in the shaded region.

16.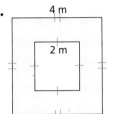

17.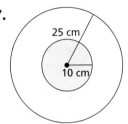

18.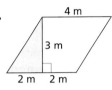

19.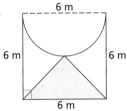

20. **MUSIC** Of the first 1200 people to buy tickets to a concert, 840 do not want the most expensive seats. What is the probability that the next ticket buyer wants one of the most expensive seats?

EXTENDED PRACTICE EXERCISES

21. **DATA FILE** Refer to the data on pet ownership on page 561. What is the probability of a household not owning a cat or a dog?

22. **WRITING MATH** Write a probability problem that can be solved by using an area model. Include a solution.

23. At a picnic, a man loses his wallet in a square field that is 50 m on each side. Each person at the picnic searches an area 10 m². Assuming someone finds the wallet, what is the probability that the man who lost the wallet will be the one who finds it?

MIXED REVIEW EXERCISES

24. **DATA FILE** Refer to the U.S. population data on page 574. Make a line graph of only the census years and make it appear that the population varied greatly from decade to decade. (Lesson 1-7)

25. **DATA FILE** Refer to the data on wind speed on page 565. Make a frequency table for the following wind speeds (in miles per hour) using the Beaufort Scale intervals. (Lesson 1-3)

11	4	10	8
6	34	14	6
15	23	33	51
36	9	8	9
30	5	49	12
2	22	17	10

Problem Solving Skills: Simulations

When you physically go through the motions described in a problem or you use objects to represent elements of the problem, you are using the **act it out** problem solving strategy.

In many cases, you can act out a complex probability problem through a simulation. A **simulation** is a model of a situation in which you carry out the trials, collect data and calculate probabilities. Simulations can be done using coins, number cubes, spinners or any method that involves random outcomes.

The manipulative you use should have the same number of outcomes as the number of possible outcomes. Computers and graphing utilities are useful when you need to produce many random numbers in a short amount of time.

Problem Solving Strategies

Guess and check

Look for a pattern

Solve a simpler problem

Make a table, chart or list

Use a picture, diagram or model

✔ Act it out

Work backwards

Eliminate possibilities

Use an equation or formula

Problem

PART-TIME JOB Two-thirds of the seniors at Central High School have part-time jobs. If three seniors are polled, what is the probability that at least one of them has a part-time job?

a. Design an experiment to simulate the situation.

b. Preform 20 trials of the experiment, and calculate the probability.

Solve the Problem

a. To model a ratio of $\frac{2}{3}$, use a number cube. Since there are six outcomes on a number cube, write a ratio equivalent to $\frac{2}{3}$ with a denominator of 6.

$$\frac{2}{3} = \frac{4}{6}$$

Let 1, 2, 3 and 4 represent having a part-time job.
Let 5 and 6 represent not having a part-time job.
Since there are three seniors polled, roll three number cubes at a time.

b. The following results indicate the three numbers rolled for each trial. The asterisk (*) indicates trials containing at least one of the numbers 1, 2, 3, or 4.

*251	*213	*264	556	*343	*422	*152	*125	555	*562
*361	*661	*242	*354	*246	*265	*445	*546	*221	*345

Since 18 of the 20 trials resulted in at least one student that works part-time,

$$P(\text{at least 1 student works part-time}) = \frac{18}{20} = \frac{9}{10} = 90\%.$$

The simulation shows that if three seniors are polled at Central High School, the probability that at least one of them has a part-time job is 90%.

Five-step Plan

1 Read
2 Plan
3 Solve
4 Answer
5 Check

◼ TRY THESE EXERCISES

Describe a model that could be used to simulate each situation.

1. The Bowers family is planning to have 4 children. Find the probability that there will be 2 boys and 2 girls.

2. At one university, a freshman has a 4 in 5 chance of returning his sophomore year. If 5 freshman are polled, what is the probability that 4 will return?

3. **CALCULATOR** Use a random number generator to find the probability in Exercise 2.

◼ PRACTICE EXERCISES

Describe a model for each situation. Perform 25 trials of the simulation and find the indicated probability.

4. **SPORTS** Gordon has a batting average of .250. In his next 10 times at bat, what is the probability that he gets exactly 3 hits?

5. **MEDICINE** A medicine has a 3 in 5 chance of curing the condition for which it is prescribed. If 2 patients are chosen at random to use the medicine, what is the probability that it will cure their condition?

6. A litter has 6 solid-colored kittens who are either black or white. What is the probability that there are 3 black and 3 white kittens?

7. **SPORTS** The likelihood that Tia will make a free throw in a basketball game is $\frac{2}{3}$. In her next 5 free throws, what is the probability that she will make 3?

8. At a stable, $\frac{1}{2}$ of the horses are brown and $\frac{1}{6}$ of the horses have braided manes. Design a model that uses two simulations to find the probability that a given horse is brown and has a braided mane.

9. **YOU MAKE THE CALL** To simulate a situation with two equally likely outcomes, Alex said you could only use a device with two outcomes, such as a coin. Jolon said you could also use a device that has more than two outcomes, such as a number cube. With whom do you agree? Explain.

10. **WRITING MATH** Explain an advantage of using the act it out strategy to solve a problem.

◼ MIXED REVIEW EXERCISES

Evaluate each expression when $a = \frac{1}{3}$, $b = -2$, and $c = 3$. (Lesson 2-5)

11. $6a + 5b$

12. $-2(4a - 3b)$

13. $\dfrac{(0.4a - 0.2b)}{0.2c}$

14. $3(4b - 4c)$

15. $\dfrac{(-5b - 2c)}{-4a}$

16. $\dfrac{(3c + 5b)}{2}$

17. $-2(a + b + c)$

18. $\dfrac{(4.8c - 2.4b)}{1.2a}$

Write each number in standard form. (Lesson 2-8)

19. $4.08 \cdot 10^4$

20. $6.22 \cdot 10^8$

21. $9.17 \cdot 10^{-5}$

22. $2.01 \cdot 10^4$

23. $3.2987 \cdot 10^6$

24. $1.7803 \cdot 10^{-5}$

25. $6.59 \cdot 10^{-4}$

26. $7.668 \cdot 10^7$

Review and Practice Your Skills

PRACTICE ◢ LESSON 4-1

Find the probability that a point selected at random lies in the shaded region.

1.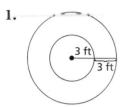

2.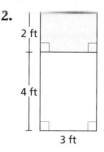

3.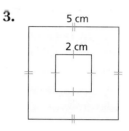

4.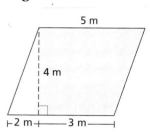

Use the table for Exercises 5–7.

5. On Thursday what is the probability that a person using the gym is not a freshman?

6. On Saturday what is the probability that a person using the gym is a freshman?

7. On Sunday what is the probability that a person using the gym is a sophomore?

People Using Gym

	Freshmen	Sophomores	Others
Tuesday	53	30	15
Thursday	62	37	21
Saturday	21	45	25
Sunday	19	41	30

8. There are 22 marbles in a bag. Fifteen are red and 7 are blue. What is the probability that a red marble will be selected?

9. In a poll conducted to find a student's favorite fruit juice, 107 vote for orange juice, 91 vote for apple juice, 120 vote for grape juice and 61 vote for pineapple juice. What is the probability that a randomly-chosen student chose apple juice?

10. A circle with an 8-cm radius lies within a rectangle that is 18 cm by 22 cm. What is the probability that a point in the rectangle lies within the circle?

PRACTICE ◢ LESSON 4-2

Describe a model for each situation. Perform 25 trials of the simulation and find the indicated probability.

11. Each box of pancake mix contains 1 of 6 different baseball cards. Assuming that the company has evenly distributed the cards among the boxes, what is the probability that you will find all 6 cards if you buy 6 boxes of pancake mix?

12. A litter has 5 puppies. What is the probability that there are 3 female puppies and 2 male puppies?

13. A husband and wife want to have 3 children. Find the probability that they will have 3 boys or 3 girls.

14. A baseball player has a batting average of 0.333. This means that he has a hit approximately 33.3% of the time that he is at bat. In his next 10 at bats, what is the probability that he gets exactly 3 hits?

Find the probability that a point selected at random lies in the shaded area.
(Lesson 4-1)

15.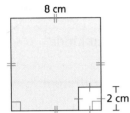
8 cm
2 cm

16.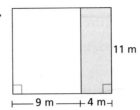
11 m
9 m   4 m

17.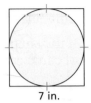
7 in.

Use the graph for Exercises 18–22. (Lessons 4-1 and 4-2).

18. What is the area of the triangle?

19. What is the area of the rectangle?

20. What is the probability that a point selected at random inside the rectangle is in the triangle?

21. What is the probability that a point selected at random inside the rectangle is not in the triangle?

22. What do you notice about your answer to Exercises 20 and 21?

MathWorks Career – Baseball Player
Workplace Knowhow

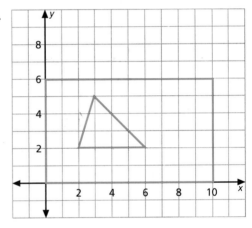

On average, a baseball player receives 20 pitches in four at bats during each game. During spring training a certain player swings and hits the ball every 40 out of 100 pitches. Of the pitches he hits, 13 are fly outs or ground outs, 20 are foul balls and 7 are base hits. Of these base hits, two are home runs. Use the spring training statistics to answer the following questions concerning the player's regular season.

1. What is the probability the player will hit a fly out or a ground out on a pitch?

2. Predict how many foul balls the player will hit per game.

3. Predict how many base hits the player will hit during the regular 162-game season.

4. At the All-Star break (after 81 games), the player has hit 34 home runs. By how many home runs is he exceeding or falling short of his expected number of home runs at this point in the season based on his spring training statistics? Are his spring training home run statistics an accurate prediction of the number of home runs he will hit in the regular season? Explain.

4-3 Sample Spaces and Theoretical Probability

Goals
■ Determine sample spaces using various methods.
■ Find theoretical probabilities.

Applications Sports, Food service, Games, Number theory

A game has cards marked with either a circle or a square.

1. If the shapes can be colored red, blue or green, how many different kinds of cards can there be?

2. If the shapes can be colored red, blue, green or yellow, how many different kinds of cards can there be?

3. How can you be sure that you have counted each kind of card?

■ BUILD UNDERSTANDING

When evaluating probability, you must know all the different things that can happen in, or outcomes of, a situation. Any one of the possible outcomes or combination of possible outcomes of an experiment is considered an **event**.

The **sample space** for a probability experiment is the set of all possible outcomes of the experiment. Often ordered pairs are used to organize and show a sample space.

Example 1 Personal Tutor at mathmatters2.com

In an experiment, a coin is tossed and a number cube is rolled. How many possible outcomes are there?

Solution

The sample space can be shown as a set of ordered pairs. For the coin, let H represent heads and T represent tails. For the number cube, use the number for each face: 1, 2, 3, 4, 5, 6. Then list the possible outcomes.

$(H, 1)$	$(H, 2)$	$(H, 3)$	$(H, 4)$	$(H, 5)$	$(H, 6)$
$(T, 1)$	$(T, 2)$	$(T, 3)$	$(T, 4)$	$(T, 5)$	$(T, 6)$

There are 12 possible outcomes.

Another way to organize and show a sample space is with a diagram. One common diagram used is a **tree diagram**. It is called a tree diagram since you list one part of an event and then add branches to show all the outcomes involving that part of the event. When the entire sample space is complete, the diagram looks similar to a tree.

Example 2

FOOD SERVICE A pizza parlor offers three sizes of pizza: large (L), medium (M), and small (S). It also offers three toppings: cheese (C), peppers (P), and onions (O). How many different pizzas with one topping are available? Use a tree diagram to solve the problem.

Solution

Use the tree diagram to count the total number of combinations.

The pizza parlor sells nine different one-topping pizzas.

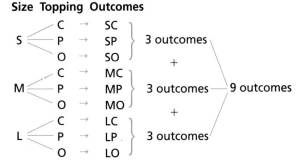

In many problems, the sample space is too large to list as a set of ordered pairs or to show as a tree diagram. According to the **fundamental counting principle**, if there are two or more stages of an activity, the total number of possible outcomes is the product of the number of possible outcomes for each stage of the activity.

Example 3

RETAIL A retail store sells shirts in 8 different sizes. For each size, there is a choice of 5 colors. For each color, there is a choice of 6 patterns. How many different kinds of shirts does the store have?

Solution

You must consider size, color, and pattern. Multiply the number of possible outcomes for each stage.

| size | · | color | · | pattern | = | possible outcomes |

$$8 \cdot 5 \cdot 6 = 240$$

The store has 240 different shirts.

Probability is not always determined from the observed outcomes of an experiment. The **theoretical probability** of an event, E, can be assigned using the formula below.

You may think of theoretical probabilities of events in a sample space as representing the results of an *ideal* experiment.

> **Check Understanding**
>
> What is the theoretical probability of getting heads on one toss of a coin?
>
> What is the experimental probability of getting heads if heads comes up 60 times in 100 tosses?

| **Theoretical Probability** | $P(E) = \dfrac{\text{number of favorable outcomes}}{\text{number of possible outcomes}}$ |

In Example 3, the number of different kinds of shirts represents the number of possible outcomes. If the store has exactly one of each kind and you are calculating the theoretical probability of selling a particular kind of shirt, the denominator would equal 240.

Example 4

A card is picked at random from a set of twelve marked with the numbers 1 through 12. Find *P*(odd number greater than 5).

Solution

There are twelve possible outcomes: 1, 2, 3, 4, 5, 6, 7, 8, 9, 10, 11, 12.
There are six odd numbers: 1, 3, 5, 7, 9, 11.
Only three odd numbers are greater than 5: 7, 9 and 11.
So there are three favorable outcomes.

$$P(\text{odd number greater than 5}) = \frac{3}{12} \qquad \frac{\text{favorable outcomes}}{\text{total possible outcomes}}$$

$$= \frac{1}{4} = 0.25$$

So the probability of drawing an odd number greater than 5 is 0.25.

◥ TRY THESE EXERCISES

1. **SPORTS** A store sells baseball bats in 3 different weights and 4 different lengths. The bats can be either wooden or aluminum. How many different types of bats does the store sell?

2. Show the sample space when a nickel and a dime are tossed. Use ordered pairs.

A number cube is rolled, and a spinner labeled A through D is spun.

3. Use a tree diagram to show the sample space.

4. How many possible outcomes are there?

5. Find *P*(1, A).   6. Find *P*(odd number, D).

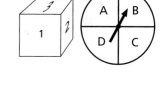

Use the spinner for Exercises 7 and 8.

7. In the spinner, what is *P*(multiple of 6)?

8. What is *P*(multiple of 10, less than 50)?

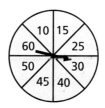

◥ PRACTICE EXERCISES • For Extra Practice, see page 596.

9. **MUSIC** At an audition for a three-piece band, there are 8 guitarists, 4 bass players and 5 drummers. How many different ways can a guitarist, bass player and a drummer be selected?

10. A spinner with four equal sections labeled 1 through 4 is spun, and a coin is tossed. Show the sample space using ordered pairs and a tree diagram.

Use both spinners for Exercises 11–15.

11. How many possible outcomes are there?

12. Find *P*(number less than 4, vowel).

13. Find *P*(number greater than or equal to six, G).

14. Find *P*(3, D).   15. Find *P*(odd number, B).

16. WRITING MATH Explain the similarities and differences between using ordered pairs and tree diagrams to list the sample space of an experiment.

17. A menu has a choice of 2 soups, 5 main dishes and 3 desserts. How many different three-course meals are possible?

18. A combination lock has 3 dials, each numbered from 1 to 8. How many different ways can the lock be set?

19. GAMES A card is picked at random from a full deck of 52 playing cards. What is the probability that it will have a value greater than nine? (Assume that aces are worth more than nines.)

20. WRITING MATH Describe when it is helpful to list the outcomes in a sample space and when it is more convenient to use the fundamental counting principle without identifying the outcomes.

21. A store sells both oil-based and latex house paints. Each type of paint is available in 12 colors and in 3 different can sizes. How many different choices of paint does the store carry?

22. DATA FILE Refer to the data on popular symphony orchestras on page 562. How many ways can a trio of trombonists be selected so that each orchestra is represented?

◼ EXTENDED PRACTICE EXERCISES

CRITICAL THINKING A number cube is rolled three times.

23. Find *P*(all sixes).   **24.** Find *P*(all even numbers).   **25.** Find *P*(no sixes).

26. How many five-digit numbers can be made using the digits 1, 2, 3, 4 and 5? (Assume that a digit can be used more than once.)

27. NUMBER THEORY A number cube is tossed twice, and each outcome is written down in order as a two-digit number. What is the probability that the two-digit number is a perfect square?

28. CHAPTER INVESTIGATION Choose a format for your chance game. This may include drawing cards from a deck, rolling a number cube, spinning a spinner, or some other event. Your game should include at least two events.

◼ MIXED REVIEW EXERCISES

Solve each equation. Check the solution. (Lesson 3-2)

29. $4z = 28$

30. $\frac{1}{3}d = -6$

31. $-3a = 18$

32. $\frac{f}{8} = -6$

33. $-2.5g = -5$

34. $\frac{y}{0.4} = -3.2$

Solve each proportion. Check the solution. (Lesson 3-5)

35. $\frac{15}{45} = \frac{n}{30}$

36. $\frac{g}{8} = \frac{7}{2}$

37. $\frac{3.6}{5.4} = \frac{h}{0.9}$

38. $\frac{4u}{8} = \frac{24}{16}$

39. $\frac{6x}{9} = \frac{24}{18}$

40. $\frac{25}{6+b} = \frac{10}{2}$

4-4 Probability of Compound Events

Goals
- ■ Find probabilities of compound events.
- ■ Explore mutually exclusive compound events.

Applications Games, Probability, Number sense

The Venn diagram shows even numbers and multiples of 5.

1. Draw a Venn diagram showing {numbers greater than 12} and {prime numbers} for whole numbers 1 through 20.

2. Draw a Venn diagram showing {multiples of 4} and {prime numbers} for the numbers 1 through 20.

3. What is the difference between your two Venn diagrams?

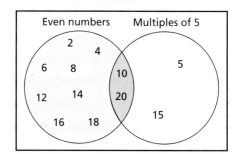

◣ BUILD UNDERSTANDING

A **compound event** is made up of two or more simpler events. Problems dealing with compound events ask for the probability of one event *and* another event occurring. The probability of event *A* and event *B* is written *P*(*A* and *B*). Other problems ask for the probability of one event *or* another event occurring. The probability of event *A* or event *B* is written *P*(*A* or *B*).

Sometimes two events are mutually exclusive. Events that are **mutually exclusive** are events that cannot occur at the same time. When two events *A* and *B* are mutually exclusive, the probability of *A* or *B* can be found using the formula *P*(*A* or *B*) = *P*(*A*) + *P*(*B*).

Example 1

Two number cubes are rolled. Find *P*(the sum is 7 or 11).

Solution

List the sample space for the experiment. There are 36 possible outcomes.

(1, 1) (1, 2) (1, 3) (1, 4) (1, 5) (1, 6)	(2, 1) (2, 2) (2, 3) (2, 4) (2, 5) (2, 6)
(3, 1) (3, 2) (3, 3) (3, 4) (3, 5) (3, 6)	(4, 1) (4, 2) (4, 3) (4, 4) (4, 5) (4, 6)
(5, 1) (5, 2) (5, 3) (5, 4) (5, 5) (5, 6)	(6, 1) (6, 2) (6, 3) (6, 4) (6, 5) (6, 6)

Six outcomes have a sum of 7: (1, 6); (2, 5); (3, 4); (4, 3); (5, 2); (6, 1).

$$P(\text{sum of } 7) = \frac{6}{36}$$

Two outcomes have a sum of 11: (5, 6); (6, 5).

$$P(\text{sum of } 11) = \frac{2}{36}$$

Since the sum cannot be 7 and 11 at the same time, both sums are mutually exclusive.

$$P(\text{sum of 7 or 11}) = \frac{6}{36} + \frac{2}{36} = \frac{8}{36} = \frac{2}{9}$$

Events that are not mutually exclusive can happen at the same time. The next example shows how to find the probability of two such events.

Example 2

Two number cubes are rolled. Find the probability that the sum of the numbers rolled is even or a multiple of 3.

Solution

Refer to the sample space of Example 1.
The events are not mutually exclusive because a sum can be both even and a multiple of 3. Of the 36 possible outcomes, 18 are even sums.

$$P(\text{even}) = \frac{18}{36} = \frac{1}{2}$$

Sums of 3, 6, 9 and 12 are multiples of 3. There are 12 sums that are multiples of 3.

$$P(\text{multiple of 3}) = \frac{12}{36} = \frac{1}{3}$$

However, sums that are even and a multiple of 3 have been counted twice. These are 6 and 12. There are six pairs with these sums.

$$P(\text{even and multiple of 3}) = \frac{6}{36} = \frac{1}{6}$$

Subtract the probability of the sums that have been counted twice.

$$P(\text{even or multiple of 3}) = \frac{1}{2} + \frac{1}{3} - \frac{1}{6} = \frac{3}{6} + \frac{2}{6} - \frac{1}{6} = \frac{4}{6} = \frac{2}{3}$$

The probability of an even sum or a sum that is a multiple of 3 is $\frac{2}{3}$.

> **Problem Solving Tip**
>
> When listing the outcomes in a sample space, double check them to make sure you have accounted for all possibilities and that none of them have been repeated.

When two events A and B are not mutually exclusive, the probability of A or B can be found using the formula: $P(A \text{ or } B) = P(A) + P(B) - P(A \text{ and } B)$.

> **COncepts in MOtion**
>
> BrainPOP®
>
> mathmatters2.com

Example 3

GAMES A card is drawn at random from a standard deck of 52 playing cards. Find the probability that the card is a heart or an ace.

Solution

The events are not mutually exclusive. A card can be both a heart and an ace.

$$P(\text{heart or ace}) = P(\text{heart}) + P(\text{ace}) - P(\text{heart and ace})$$

Of the 52 cards, there are 13 hearts; so $P(\text{heart}) = \frac{13}{52}$.

Of the 52 cards, there are 4 aces; so $P(\text{ace}) = \frac{4}{52}$.

There is 1 heart that is an ace, so $P(\text{heart and ace}) = \frac{1}{52}$.

$$P(\text{heart or ace}) = \frac{13}{52} + \frac{4}{52} - \frac{1}{52}$$
$$= \frac{16}{52}$$
$$= \frac{4}{13}$$

> **Check Understanding**
>
> A card is drawn at random from a standard deck of 52 cards. Find $P(\text{red and 5})$.

The probability that the card drawn is a heart or an ace is $\frac{4}{13}$.

1. Two coins are tossed. Find the probability that the coins show 2 tails or 2 heads.

2. Two number cubes are rolled. Find the probability that the numbers rolled are doubles and have a sum that is a multiple of 4.

3. Two number cubes are rolled. Find the probability that the numbers rolled are doubles or have a sum of 4.

4. A card is drawn from a standard deck of 52 cards. Find the probability that it is a red card or a jack.

A card is drawn from a standard deck of 52 cards. Find each probability.

5. P(red and face card)   6. P(black or face card)   7. P(red or club)

8. **WRITING MATH** Explain why it is necessary to involve subtraction when finding the probability of two events that are not mutually exclusive.

■ **PRACTICE EXERCISES** • For Extra Practice, see page 597.

9. Two number cubes are rolled. Find the probability that the sum of the numbers is 5 or 6.

10. A card is drawn at random from a standard deck of 52 cards. Find the probability that the card is a 10 or a jack.

11. Two number cubes are rolled. Find the probability that the sum of the numbers is odd and less than 6.

12. A card is drawn from a standard deck of 52 cards. Find the probability that the card is a diamond or a face card.

The spinner shown is spun one time. Find each probability.

13. P(blue or 7)

14. P(blue and an even number)

15. P(green or less than 6)

16. P(yellow or greater than 5)

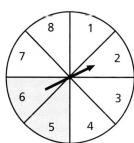

17. **GAMES** While playing a board game, Yasuhiro's piece is seven blocks away from landing on the home space. What is the probability that he will miss the home space on his next turn if two number cubes are rolled?

18. **DATA FILE** Refer to the data on trips to foreign destinations on page 574. What is the probability that an American visited Mexico or Asia?

19. Three coins are tossed. Draw a tree diagram of the outcomes, and find the probability that the coins show three heads or three tails.

20. **NUMBER SENSE** Six coins are tossed. Use the fundamental counting principle to find the probability of all heads or all tails.

A box contains 9 ball-point pens and 6 felt-tip pens. Of the ball-point pens, 4 are blue, 2 are red and 3 are black. Of the felt-tip pens, 3 are blue, 1 is red and 2 are black. A pen is picked at random from the box. Find each probability.

21. P(ball-point pen or blue)

22. P(felt-tip pen or red)

23. P(black ball-point pen)

24. P(red or ball-point pen)

25. YOU MAKE THE CALL In a recent county election, 70% of the voters surveyed said they voted for a county auditor, but only 61% actually did. On the same survey, 50% of the voters surveyed said they voted for a county clerk, but only 43% actually did. Mary says that the probability of someone lying on this survey is 9% + 7% = 16%. Do you agree or disagree? Explain.

26. WRITING MATH Describe the difference between events that are mutually exclusive and events that are not mutually exclusive. Give an example of each type, and describe how the probability of each can be found.

EXTENDED PRACTICE EXERCISES

CRITICAL THINKING A photo album contains pictures of Joe or Catina, or both Joe and Catina. Joe is in 25 of the pictures, and Catina is in 30 of them. Joe and Catina are together in 15 of the pictures.

27. How many pictures are in the photo album?

28. If a picture is selected at random, what is the probability that it shows only Joe or only Catina?

The **complement** of the event A is the event *not A*. For example, if you roll two number cubes, the event *the sum is 3* has the complement *the sum is not 3*.

29. Does $P(A$ or complement of $A) = P(A) + P$(complement of A)? Explain.

30. What does $P(A) + P$(complement of A) equal?

31. Two number cubes are rolled. Find the probability that the sum of the numbers is neither a multiple of 3 nor a multiple of 4.

MIXED REVIEW EXERCISES

Find the mean, median, and mode of each set of numbers. Round to the nearest tenth. (Lesson 1-2)

32. 4, 8, 12, 16, 15, 13, 7, 9, 12, 15, 14

33. 1, 2, 1, 2, 1, 2, 3, 2, 1, 2, 3

34. 26, 28, 31, 42, 37, 26, 28, 31, 26, 19, 24

35. 6, 8, 12, 6, 12, 7, 9, 10, 12, 8, 6

36. 52, 54, 56, 57, 54, 53, 51, 50, 58, 59, 55

Simplify each numerical expression. (Lesson 2-2)

37. $5 + 3 \cdot 9$

38. $3(8 - 6) \div 2$

39. $2^3 - (8 - 3) \div 5$

40. $\frac{2}{3}(6^2 + 3)$

41. $12 \cdot 8 + 2$

42. $12(8 + 2)$

43. $16 \div 4 + 1 - 3^2$

44. $(2 + 3)^2 - 8 \cdot 7$

45. $35 \div 3^2 + 4 - 7$

46. $5^2 + 64 \div 8 - 2$

47. $2.1 \cdot 3.4 \div 1.5$

48. $6^2 - 32 \div 8 + 6$

Review and Practice Your Skills

PRACTICE ◣ LESSON 4-3

A number cube is rolled and a spinner with sections A, B, C, D and E is spun. Each section of the spinner is the same size.

1. How many possible outcomes are there?

2. Find $P(C, 2)$.

3. Find $P(B, \text{even number})$.

4. Find $P(\text{number less than 4, consonant})$.

5. Find $P(\text{vowel, factor of 4})$.

6. At an audition for a brass trio there are 7 trumpets, 5 french horns and 4 trombones. How many different ways can a trumpeter, french hornist and a trombonist be selected?

7. Find the probability of drawing a red 9 from a standard deck of playing cards.

8. A combination lock has 4 dials, each numbered from 1 to 9. How many different ways can the lock be set?

9. A spinner with 5 equal sections labeled 1 through 5 is spun and a coin is tossed. Show the sample space using ordered pairs and a tree diagram.

10. A menu has a choice of 2 soups, 4 main dishes and 3 desserts. How many different three-course meals are possible?

A number cube is rolled two times.

11. Find $P(\text{all fours})$.

12. Find $P(\text{all odd numbers})$.

13. Find $P(\text{no fours})$.

PRACTICE ◣ LESSON 4-4

A card is drawn at random from a standard deck of 52 cards.

14. Find the probability that the card is a 2, 3 or 4.

15. Find the probability that the card is a heart or a face card.

16. Find the probability that the card is a spade and an even number.

A box contains 8 regular marbles and 7 shooter marbles. Of the regular marbles, 3 are blue, 2 are green and 3 are yellow. Of the shooter marbles, 2 are blue, 4 are green and 1 is yellow. Find each probability.

17. $P(\text{regular marble or green})$

18. $P(\text{shooter marble or blue})$

19. $P(\text{green shooter marble})$

20. $P(\text{yellow or regular marble})$

Two number cubes are rolled.

21. Find the probability that the sum of the numbers is 10 or 11.

22. Find the probability that the sum is odd and greater than 5.

23. Find the probability that the number cubes rolled are doubles and the sum is greater than 4.

Find the probability that a point selected at random is in the shaded region. Round to the nearest hundredth. (Lesson 4-1)

24.
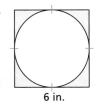
6 in.

25.
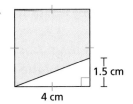
1.5 cm
4 cm

26.
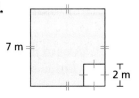
7 m
2 m

List all the elements of the sample space as ordered pairs for each of the following experiments. (Lesson 4-3)

27. You flip a dime and a nickel.

28. You spin each of these spinners once.

Find each probability. (Lesson 4-4)

29. Guessing incorrectly on a multiple-choice question with 5 choices.

30. Rolling 2 number cubes and getting a multiple of 3.

Mid-Chapter Quiz

Find the probability that a point selected at random lies in the shaded region. (Lesson 4-1)

1.

4 ft
2 ft

2.
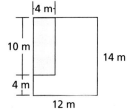
4 m
10 m
14 m
4 m
12 m

3.
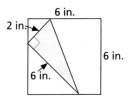
6 in.
2 in.
6 in.
6 in.

4. A box contains 12 tennis balls, 15 golf balls, 9 ping-pong balls and 6 baseballs. What is the probability that a golf ball is chosen? (Lesson 4-1)

5. A glass has a 5 in 6 chance of breaking if dropped onto a wooden floor. If 3 glasses are dropped, what is the probability that just one of them breaks? Describe a model and perform 25 trials of the simulation. (Lesson 4-1)

6. A family wants to adopt 1 puppy and 1 kitten from an animal shelter. They have narrowed their choices down to 6 puppies and 4 kittens. How many possible different dog and cat duos are possible? (Lesson 4-2)

A number cube is rolled and a letter from the alphabet is randomly selected. (Lesson 4-3)

7. How many possible outcomes are there?

8. Find $P(6, Z)$.

9. Find P(even number, vowel).

10. Find P(number less than 4, B).

4-5

Independent and Dependent Events

Goals
- Find probabilities of dependent events.
- Find the probability of independent events.

Applications Government, Health, Sports, Games

Refer to the set of cards. Assume they are placed face down.

1. A card is selected at random, then replaced. Another card is selected. Make a tree diagram of the sample space.

2. Use the tree diagram to find P(B, then L).

3. A card is selected at random, but it is not replaced. Then another card is selected. Make a tree diagram for the sample space.

4. Use the tree diagram to find P(B, then L).

5. Compare your results from Questions 2 and 4. What do you notice?

◤ BUILD UNDERSTANDING

Two events are **independent** if the result of the second event is not affected by the result of the first event.

If A and B are independent events, the probability of both events occurring is the product of the probabilities of the individual events.

$$P(A \text{ and } B) = P(A) \cdot P(B)$$

Sometimes $P(A \text{ and } B)$ is written $P(A, \text{ then } B)$ to emphasize that A and B do not characterize a single event.

Example 1

A bag contains 3 red marbles, 4 green marbles and 5 blue marbles. One marble is taken at random and then replaced. Then another marble is taken at random.

Find the probability that the first marble is red and the second marble is blue.

Solution

Because the first marble is replaced, the sample space of 12 marbles does not change for each event. The two events are independent.

$$P(\text{red, then blue}) = P(\text{red}) \cdot P(\text{blue}) \qquad \frac{\text{red marbles}}{\text{total marbles}} \cdot \frac{\text{blue marbles}}{\text{total marbles}}$$

$$= \frac{3}{12} \cdot \frac{5}{12}$$

$$= \frac{15}{144} = \frac{5}{48}$$

The probability of picking red, then blue is $\frac{5}{48}$.

Two events are **dependent** if the result of one event *is* affected by the result of another event. If *A* and *B* are dependent,

$$P(A \text{ and } B) = P(A) \cdot P(B, \text{ given that } A \text{ occurred}).$$

Example 2

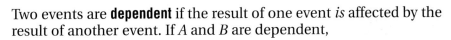 **Personal Tutor at** mathmatters2.com

A bag contains 3 red marbles, 4 green marbles, and 5 blue marbles. One marble is taken at random from the bag. It is not replaced. Then another marble is taken at random.

Find the probability that the first marble is red and the second marble is blue.

Solution

Because the first marble is not replaced, the second event is dependent on the first event. Find the probability of the first event.

$$P(\text{red}) = \frac{3}{12} = \frac{1}{4} \qquad \frac{\text{red marbles}}{\text{total marbles}}$$

On the next selection, there are only 11 marbles. Assuming that a red marble was removed, there are still 5 marbles in the bag.

$$P(\text{blue after red}) = \frac{5}{11} \qquad \frac{\text{blue marbles}}{\text{remaining marbles}}$$

Multiply the two probabilities together.

$$P(\text{red, then blue}) = \frac{1}{4} \cdot \frac{5}{11}$$
$$= \frac{5}{44}$$

The probability of picking red, then blue is $\frac{5}{44}$.

Example 3

A bag contains 3 red marbles, 4 green marbles, and 5 blue marbles. Two marbles are taken at random from the bag.

Find the probability that both marbles are blue.

Solution

Consider that one of the marbles must be taken before the other. Since the first marble is not replaced, the second event is dependent on the first event.

$$P(\text{first blue marble}) = \frac{5}{12} \qquad \frac{\text{blue marbles}}{\text{total marbles}}$$

$$P(\text{second blue marble}) = \frac{5-1}{12-1} = \frac{4}{11} \qquad \frac{\text{remaining blue marbles}}{\text{remaining marbles}}$$

Multiply the two probabilities together.

$$P(\text{blue, then blue}) = \frac{5}{12} \cdot \frac{4}{11} = \frac{20}{132} = \frac{5}{33}$$

The probability of picking blue, then blue is $\frac{5}{33}$.

4-6 Permutations of a Set

Goals ■ Find the number of permutations of a set.

Applications Cooking, Travel, Music, Sports, Games

Work with a partner. Use coins to help you answer Questions 1–3.

1. Suppose you have a penny, nickel, dime and quarter in your pocket. If you take out two coins, how many different amounts of money might you have in your hand?

2. How can you be sure you have listed all the different amounts of money?

3. Suppose you have the same coins as in Question 1. In how many different ways could you take out two coins and give one each to two friends?

■ BUILD UNDERSTANDING

A word is a set of letters arranged in a definite order. For example, *charm* and *march* are different words even though they contain the same letters. An arrangement of items in a particular order is called a **permutation**. Order is important in permutations.

You can use the fundamental counting principle to find the number of possible permutations of a set of items. Each time you pick an item for a position in a permutation, there will be one less item to pick from for the next position of the permutation.

n! **(*n* factorial)**	The number of permutations of *n* different items is $n(n-1)(n-2) \ldots (2)(1)$ and is written as $n!$

Example 1

How many different "words" can be made with the letters a, b, c, d, and e if all the letters are used?

(The "words" do not have to make sense in English or any other language.)

Solution

Find the number of permutations of five letters.

$$\text{number of permutations} = 5! = 5 \cdot 4 \cdot 3 \cdot 2 \cdot 1 = 120$$

There are 120 different "words" that can be made using five letters.

> **Technology Note**
>
> Most advanced calculators have a factorial function key. You can often find it in the probability menu.

Sometimes you may want to use only part of a set. The number of permutations of n different items taken r items at a time and with no repetitions, is written $_nP_r$. This is read as "the number of permutations of n items taken r at a time." You can use the formula below to find the number of permutations when only part of the set is used.

$_nP_r$

$$_nP_r = \frac{n!}{(n-r)!}$$

where n is the number of different items and r is the number of items taken at a time.

By definition, $n!$ is the number of permutations of n items taken n at a time, so

$$_nP_n = \frac{n!}{(n-n)!} = \frac{n!}{0!}.$$

This means that $0! = 1$.

Example 2

How many different "words" can be made with the letters a, b, c, d and e if only three different letters are used in each word?

Technology Note

Most advanced calculators have a permutation function key. It is usually symbolized as $_nP_r$.

For example, entering the sequence 5 $_nP_r$ 3 calculates $\frac{5!}{2!} = 60$.

Solution

There are five letters to be taken three at a time. So $n = 5$ and $r = 3$ in the permutation formula.

$$_nP_r = \frac{n!}{(n-r)!} \qquad _5P_3 = \frac{5!}{(5-3)!}$$

$$= \frac{5 \cdot 4 \cdot 3 \cdot \cancel{2} \cdot \cancel{1}}{\cancel{2} \cdot \cancel{1}} \qquad \text{Cancel common factors to simplify.}$$

$$= 60$$

There are 60 "words" that can be made.

Example 3

COOKING A sandwich shop offers 6 toppings for their submarine sandwiches: ham, turkey, cheese, lettuce, tomatoes and onions. How many different ways can the ingredients of a sandwich be arranged if exactly 2 ingredients are chosen from the menu?

Solution

There are six different menu items to be taken two at a time. So $n = 6$ and $r = 2$ in the permutation formula.

$$_nP_r = \frac{n!}{(n-r)!} \qquad _6P_2 = \frac{6!}{(6-2)!}$$

$$= \frac{6 \cdot 5 \cdot \cancel{4} \cdot \cancel{3} \cdot \cancel{2} \cdot \cancel{1}}{\cancel{4} \cdot \cancel{3} \cdot \cancel{2} \cdot \cancel{1}} \qquad \text{Cancel common factors to simplify.}$$

$$= 30$$

There are 30 different ways that the ingredients can be arranged.

Calculate each of the following permutations.

1. $_5P_2$ **2.** $_8P_2$ **3.** $_4P_1$ **4.** $_7P_5$

5. In how many different ways can you arrange the letters m, n, o, p, q and r?

6. SPORTS There are 7 finalists in a swimming race. Medals are awarded for first, second and third place. In how many different ways can the medals be awarded?

7. Find the number of 5-letter "words" that can be formed from the letters of the word "horse."

8. A professor is grading 8 term papers in order of merit. If none of the papers are of equal merit, in how many different orders can the papers be graded?

Calculate each of the following permutations.

9. $_6P_2$ **10.** $_{11}P_9$ **11.** $_5P_0$ **12.** $_8P_8$

13. MUSIC A five-piece band consists of a guitarist, a bass player, a trumpeter, a saxophone player and a pianist. If three of the band members each play a solo in an introduction to a song, how many different permutations of three band members can be used?

14. ENTERTAINMENT A radio disc jockey has a set of 9 songs to play. In how many different ways can the disc jockey play the next 3 songs?

15. How many different four-digit numbers can be formed from the digits 4, 5, 6 and 7 if each digit can be used only once?

16. In how many different orders could a poet recite 5 poems from a collection of 9 poems?

17. Find the number of 7-letter "words" that can be formed from the letters of the word "Florida."

18. Dermarr and Luke are two of the 12 actors auditioning for a show. The order in which the actors will perform is picked at random. What is the probability that Dermarr will perform first and Luke will perform second?

19. TRAVEL The towns of Anderson and Mt. Washington are connected by 7 roads. Mt. Washington and Hyde Park are connected by 5 roads. In how many different ways can a tourist travel from Anderson to Hyde Park via Mt. Washington?

20. DATA FILE Refer to the data on threatened and endangered species on page 567. Find the number of permutations of types of species listed three at a time. What is the probability that the group consists of snails, insects, and birds listed in that order?

21. How many 5-digit zip codes can be formed using odd digits if digits can be reused? How many can be formed if digits cannot be reused?

22. How many different ways can 6 classes be scheduled in a 6-period day?

23. **GAMES** Nakeisha selects seven tiles from the group of letters to begin a Scrabble® game. How many different ways can she arrange her 7 tiles?

24. **WRITING MATH** Describe a situation in which the number of outcomes is given by $_6P_3$.

25. The lunch menu at a school cafeteria lists 3 soups, 6 sandwiches and 4 fruits. In how many different ways can a student choose a lunch that consists of a soup, a sandwich, and a fruit?

26. **ERROR ALERT** Manuel is asked to determine how many different 2-digit numbers can be formed from the digits 3, 4, 5 and 6 if a digit can appear only once. He reasons that since there are 4 numbers, there are $4 \cdot 4 = 16$ possible 2-digit numbers. What mistake has Manuel made, and what is the correct answer?

◤ EXTENDED PRACTICE EXERCISES

27. In how many ways can the five CD-ROMs of an encyclopedia be placed next to each other on a shelf if the first CD-ROM is always on the extreme left?

28. Damon needs to name the vertices of a triangle and wants to use the letters of his name. In how many ways can he do this?

29. Krystal wants to collect all 5 prizes given in a box of cereal. If her parents buy a box of cereal a week, what is the probability that she will get all five prizes in five weeks?

30. **CRITICAL THINKING** If the digits 1–9 are equally likely to be selected, what is the probability that a 3-digit number is a palindrome? A *palindrome* is a word or number that reads the same backwards and forwards.

31. How many even numbers can be formed from the digits 1, 2, 3, 4 and 5 if each digit is used once in each number?

◤ MIXED REVIEW EXERCISES

Graph the solution of each inequality on a number line. (Lesson 3-6)

32. $b \leq 7$	33. $a > -3.5$	34. $c < 1.5$
35. $3 \leq h$	36. $2 \geq g$	37. $l \leq 3$
38. $p \geq -3$	39. $r < -2.5$	40. $m < -4$
41. $d \geq -2$	42. $-3 < e$	43. $4 > j$

Solve each inequality. (Lesson 3-7)

44. $4 + c > -2$	45. $-3x \leq -12$	46. $10 \geq g + 4$
47. $\frac{r}{2} - 6 < -2$	48. $7k + 8 > 3$	49. $1.4f - 3 < -0.2$
50. $3 + \frac{1}{4}g \geq 7$	51. $3r - 5 \leq 4$	52. $3(2.5h - 1) \geq 12$
53. $8 \geq 0.3s - 7$	54. $4(3h + 1) < 6h - 4$	55. $2(8 + a) > 3(4 + a)$

Review and Practice Your Skills

A bag contains 5 quarters, 4 dimes and 3 nickels. Coins are taken at random and then replaced. Find each probability.

1. P(quarter, then dime)
2. P(dime, then dime)
3. P(nickel, then quarter)
4. P(nickel, then dime)
5. P(quarter, then quarter)
6. P(dime, then quarter)

A box contains 6 yellow, 4 purple and 3 pink marbles. Marbles are taken at random and not replaced. Find each probability.

7. P(yellow, then pink)
8. P(purple, then yellow)
9. P(pink, then purple)
10. P(yellow, then yellow)
11. P(pink, then pink)
12. P(purple, then purple)

A drawer contains 3 pairs of black socks, 2 pairs of brown socks and 4 pairs of white socks. One sock is taken at a time at random and not replaced. Find each probability.

13. P(black, then black)
14. P(white, then black)
15. P(brown, then brown)
16. P(brown, then white)
17. P(white, then white)
18. P(black, then brown)

Calculate each of the following permutations.

19. $_7P_3$
20. $_8P_4$
21. $_5P_2$
22. $_3P_2$
23. $_9P_3$
24. $_5P_3$
25. $_5P_1$
26. $_8P_6$
27. $_{11}P_6$

28. How many different 4-digit numbers can be formed from the digits, 2, 3, 4 and 5 if each digit can be used only once?

29. The art club will select 4 of 10 students to be on a committee for the positions of president, vice president, secretary and treasurer. How many different committees are possible?

30. A school fair will have a snack bar with items from Brazil, Ethiopia, Greece, Haiti, Italy, Korea, Mexico and Thailand. How many ways can the flags of 4 of these nations be displayed in a circular arrangement around the snack bar?

31. A store owner received a shipment of picture frames in 7 new styles. How many different ways can the owner display 3 of the new styles on a shelf?

32. How many different ways can you arrange the letters l, m, n, o and p?

PRACTICE ◣ LESSON 4-1–LESSON 4-6

A number cube is rolled and a spinner labeled A through H is spun. (Lesson 4-3)

33. How many possible outcomes are there?

34. Find P(6, H).

35. Find P(even, vowel).

36. Find P(number less than 3, C).

37. Find P(number greater than or equal to 2, G).

A card is drawn at random from a standard deck of 52 cards. (Lesson 4-4)

38. Find the probability that the card drawn is a face card.

39. Find the probability that the card drawn is a diamond or a face card.

40. Find the probability that the card drawn is a club and not a face card.

41. A bag of marbles contains 4 green marbles, 5 red marbles and 7 yellow marbles. Find the probability that when three marbles are taken and not replaced you get one of each color. (Lesson 4-5)

Calculate each of the following permutations. (Lesson 4-6)

42. $_5P_3$ **43.** $_6P_3$ **44.** $_7P_5$ **45.** $_4P_2$

MathWorks
Workplace Knowhow

Career – Board Game Designer

Board game designers must decide the logic behind the game, the layout, the pieces to be included and the possible ways to win the game. When a board game is created, the designer must know the probability of certain events occurring. Finding these probabilities will determine the rules of the game. A new board game has two number cubes. Each number cube shows numbers 1 through 6. Players get an extra turn if they roll the same number on each number cube.

1. Determine the number of possible outcomes one could roll with the number cubes.

2. Use a tree diagram to show the sample space.

3. What is the probability of getting an extra turn by rolling the same number on both number cubes?

4. What is the probability of rolling a 3 or 5 on one of the number cubes?

5. What is the probability of rolling a 3 and a 5?

4-7 Combinations of a Set

Goals ▪ Find the number of combinations of a set

Applications Safety, Sports, Games, Probability, Landscaping

Work in groups of four. Record your answers.

1. From your group of four, select three officers: a president, a vice president and a treasurer. List all the different possibilities.

2. From your group of four, select a committee of three. List all possibilities.

3. Are your lists the same for Questions 1 and 2? Can you find a relationship between the two lists? Explain.

4. From a group of five, how many ways can you select three officers?

◼ BUILD UNDERSTANDING

In the last lesson, you learned how to use the fundamental counting principle to find the number of possible permutations of a set of items. In this case, the order of the items is important. Sometimes, however, the order of items is *not* important. A set of items in which order is not important is called a **combination**. For example, *acb* and *bac* are both combinations of the letters *a*, *b* and *c*.

The number of combinations of *n* items taken *r* items at a time is written $_nC_r$. You can use the formula below to find the number of possible combinations when only part of a set is used.

$_nC_r$

$$_nC_r = \frac{n!}{(n - r)!\, r!}$$

where *n* is the number of different items and *r* is the number of items taken at time.

Example 1

Calculate the combinations.

a. $_4C_2$

b. $_6C_4$

Solution

a. $_4C_2 = \dfrac{4!}{(4 - 2)!\, 2!}$

$= \dfrac{4 \cdot 3 \cdot 2 \cdot 1}{(2 \cdot 1)(2 \cdot 1)}$

$= \dfrac{12}{2}$

$= 6$

b. $_6C_4 = \dfrac{6!}{(6 - 4)!\, 4!}$

$= \dfrac{6 \cdot 5 \cdot 4 \cdot 3 \cdot 2 \cdot 1}{(2 \cdot 1)(4 \cdot 3 \cdot 2 \cdot 1)}$

$= \dfrac{30}{2}$

$= 15$

Technology Note

Many calculators have a combination function key symbolized $_nC_r$.

For example, entering the sequence 5 [MATH] ▶ ▶ ▶ 3 1 will calculate $\dfrac{5!}{4!1!} = 5.$

Example 2

How many different ways can a 2-person committee be chosen from 8 people if there are no restrictions?

Solution

So $n = 8$ and $r = 2$ in the combination formula.

$$_nC_r = \frac{n!}{(n-r)!\,r!}$$

$$_8C_2 = \frac{8!}{6!\,2!}$$

$$= \frac{8 \cdot 7 \cdot \cancel{6} \cdot \cancel{5} \cdot \cancel{4} \cdot \cancel{3} \cdot \cancel{2} \cdot \cancel{1}}{(\cancel{6} \cdot \cancel{5} \cdot \cancel{4} \cdot \cancel{3} \cdot \cancel{2} \cdot \cancel{1})(2 \cdot 1)}$$ Cancel common factors to simplify.

$$= \frac{56}{2}$$

$$= 28$$

There are 28 ways to select the committee.

> ### Reading Math
>
> Read the combination $_8C_2$ in Example 2 as "combinations of eight, taken two at a time."

Example 3   **Personal Tutor at** mathmatters2.com

A random drawing is held to determine which 2 of the 6 members of the math club will be sent to a regional math contest.

a. How many different pairs of two could be sent to the contest?

b. If the members of the club are James, Dontae, Colin, Daphne, Nashota and Nicole, what is the probability that James and Nashota will be picked?

Solution

a. There are six people to be picked two at a time, and the order is not important. So $n = 6$ and $r = 2$ in the combination formula.

$$_nC_r = \frac{n!}{(n-r)!\,r!}$$

$$_6C_2 = \frac{6!}{(6-2)!\,2!}$$

$$= \frac{6 \cdot 5 \cdot \cancel{4} \cdot \cancel{3} \cdot \cancel{2} \cdot \cancel{1}}{(\cancel{4} \cdot \cancel{3} \cdot \cancel{2} \cdot \cancel{1})(2 \cdot 1)}$$ Cancel common factors to simplify.

$$= \frac{30}{2}$$

$$= 15$$

There are 15 different groups of two members.

b. The group of James and Nashota is one of the 15 possible combinations.

$$P(\text{James and Nashota}) = \frac{\text{favorable outcomes}}{\text{possible outcomes}}$$

$$= \frac{1}{15}$$

The probability that James and Nashota will be picked is $\frac{1}{15}$.

TRY THESE EXERCISES

Calculate each combination.

1. $_5C_3$

2. $_8C_4$

3. $_5C_5$

4. $_7C_5$

5. SPORTS A doubles team is to be picked at random from the 9 members of a tennis team. If Marciann and Leah are two of the members, what is the probability that they will both be picked?

6. There are 7 different points on a circle. How many straight lines can be drawn through pairs of these points?

7. There are six 1-qt cans, each containing a different fruit juice. How many different types of fruit punch could be obtained by mixing 3 full containers together?

8. LANDSCAPING How many different flower arrangements can be made from 6 different types of flowers if each arrangement contains 3 types?

PRACTICE EXERCISES • For Extra Practice, see page 598.

Calculate each combination.

9. $_{11}C_8$

10. $_4C_2$

11. $_3C_0$

12. $_9C_1$

13. A deck of 26 alphabet cards marked A through Z is shuffled, and 2 cards are dealt. What is the probability that the cards are X and Y?

14. A committee of three people is to be selected at random from a group of 24 people that includes Santiago, Tina, and Rashad. Find the probability that the committee will consist of Santiago, Tina, and Rashad.

15. A basketball coach wants to know how many different 5-member teams she can play from a roster of 10 team members.

16. Each of five friends gives Mary his or her favorite book. Mary only has time to read 3 of the books. In how many ways can she select 3 of the 5 books? If you are one of the friends, then what is the probability that your book will be one of the three books read? (Hint: To help you find the probability, list the sample space.)

17. SAFETY In how many ways can a 6-person neighborhood nightwatch be selected from 12 men and 8 women who are on the safety patrol?

18. There are 5 points in a plane, no 3 of which are collinear. If 2 noncollinear points determine a line, how many lines are determined?

19. WRITING MATH Explain the differences and similarities between permutations and combinations of a set of items. Give an example of each.

20. GAMES In the game of *bridge*, a hand has 13 cards. Determine the number of bridge hands possible from a standard deck of 52 cards. Write the answer in scientific notation.

NEIGHBORHOOD CRIME WATCH

We immediately report all SUSPICIOUS PERSONS and activities to our Police Dept.

21. **ENTERTAINMENT** A popular touring band has 20 songs. How many combinations of songs can the band play in their opening 3-song set?

22. There are 12 people at a party. If each person shakes hands with every other person, how many handshakes are exchanged?

23. A standard deck of 52 cards is shuffled, and 4 cards are turned over. What is the probability that the 4 cards all have the same value? (Hint: Use the combination value as the denominator in the probability fraction.)

24. How many triangles are determined by 9 points if none of the 9 points lie on a straight line?

25. A radio station chooses 22 postcards from the many listeners who enter a monthly contest. In addition to their prize, 4 of those chosen will also receive backstage passes. In how many different ways can the past winners be selected?

◼ EXTENDED PRACTICE EXERCISES

26. **CRITICAL THINKING** Simplify the expression $_nC_{n-1}$.

27. Eldora is told to study 10 questions for an exam. Three of the questions will be randomly selected for the exam. Unfortunately, she only has time to study 3 of the questions. What is the probability that the 3 questions on the test are the 3 that she has studied?

28. **YOU MAKE THE CALL** Suppose that 4 officers will be selected. The first person will be the president, the second the vice-president, the third the secretary and the fourth the treasurer. Tia says that to solve this problem she would use a permutation and not a combination. Do you agree or disagree? Explain.

29. **CHAPTER INVESTIGATION** Create a set of rules for your game. Include a list of the significant outcomes along with their respective probabilities.

◼ MIXED REVIEW EXERCISES

Solve each equation. Check the solution. (Lesson 3-4)

30. $8 + 4c = 10$
31. $\frac{x}{2} - 7 = 3$
32. $6a - 4 = 20$
33. $12 + 2d = 17$
34. $3(x + 4) = 6$
35. $2(k - 3) = 8$

Solve each equation. Check the solution. (Lesson 3-8)

36. $\sqrt{f} = 8$
37. $x^2 = 75$
38. $b^2 - 2 = 7$
39. $3 + \sqrt{h} = 9$
40. $g^2 = \frac{1}{9}$
41. $\sqrt{4x + 3} = 6$

DATA FILE **Refer to the data on target heart rates on page 571.** (Lesson 3-3)

42. How many times does a 30 year-old's heart beat per minute if she achieves a target heart rate of 50%.

43. Write an algebraic model of the target heart rate for a 30-year old where t is the target heart rate and p is the percent of the average maximum heart rate.

44. How many times does a 30 year-old's heart beat per minute if she achieves a target heart rate of 65%?

Chapter 4 Review

VOCABULARY ◣

Choose the word from the list that best completes each statement.

1. A(n) __?__ is an activity that is used to produce data that can be observed and recorded.

2. When you use objects to represent elements of the problem, you are using the __?__ strategy.

3. A(n) __?__ contains all the possible outcomes of an experiment.

4. Any outcome or combination of possible outcomes of an experiment is considered a(n) __?__.

5. Problems dealing with __?__ may ask for the probability of one event *and/or* another occurring.

6. Two events that cannot occur at the same time are called __?__.

7. When one event affects the outcome of another event, the events are called __?__.

8. The number of ways a group of children can stand in line is an example of a(n) __?__.

9. The fundamental counting principle uses __?__ to determine the number of possibilities.

10. A(n) __?__ can be used to systematically list all possible outcomes.

a. act it out
b. addition
c. combination
d. compound events
e. dependent
f. event
g. experiment
h. multiplication
i. mutually exclusive
j. permutation
k. sample space
l. tree diagram

LESSON 4-1 ◣ Experiments and Probabilities, p. 150

▶ An **experimental probability** is an estimate of the likelihood of an event, *E*, or desired outcome. It can be expressed using the formula

$$P(E) = \frac{\text{number of observations favorable to } E}{\text{total number of observations}}.$$

11. Last month a music store sold 1178 cassette tapes, 2574 CDs and 1968 DVDs. What is the relative frequency of cassette tape sales?

12. In a random survey of 240 students, 168 said they would vote for Maria for class president. What is the probability that a student will vote for Maria?

The enrollment at Washington High School is given below.

Washington High School Enrollment

	Freshmen	Sophomores	Juniors	Seniors
Male	55	50	48	62
Female	43	62	47	53

13. What is the probability that a student picked at random is a junior?

14. What is the probability that a student picked at random is a male?

LESSON 4-2 ◣ Problem Solving Skills: Simulations, p. 154

▶ A **simulation** is a model of a problem that is easier to implement than the actual problem.

15. At a grocery store, $\frac{1}{3}$ of all the bread is whole wheat and $\frac{1}{2}$ of all the bread is dated to be sold by the middle of the week. Design a simulation using a number cube and a coin to find the probability that a given loaf of bread is whole wheat and dated to be sold by the middle of the week.

16. Victor has 6 ties. He works at the mall every Friday, Saturday and Sunday. Each work day he chooses a tie at random to wear for his job. Design a simulation to find the probability that Victor will wear a different tie each of the three work days.

17. A restaurant chain gives one of three action figures at random with every child's meal. Design a simulation to find the probability that a child will get all three action figures if he or she buys five of these meals.

LESSON 4-3 ◣ Sample Spaces and Theoretical Probability, p 158

▶ The **sample space** for a probability is the set of all possible outcomes of the experiment.

▶ The **fundamental counting principle** states that to find the number of possible outcomes for an activity, multiply the number of possible outcomes for each stage of the activity.

▶ The **theoretical probability** of an event, E, can be assigned using the formula
$$P(E) = \frac{\text{number of favorable outcomes}}{\text{number of possible outcomes}}.$$

18. A company has printed eleven new books about American artists. Each book is available in hardcover or paperback and in regular type or large type. How many different books were printed?

19. A car comes in two or four doors, a four or six-cylinder engine, and eight exterior colors. How many of these cars are available?

20. A spinner with six equal sections marked A through F is spun and a number cube is rolled. Find P(vowel, number greater than 3).

LESSON 4-4 ◣ Probability of Compound Events, p. 162

▶ If A and B are **mutually exclusive events**, they cannot occur at the same time. $P(A \text{ or } B) = P(A) + P(B)$.

▶ If A and B are **not mutually exclusive events**, they can occur at the same time. $P(A \text{ or } B) = P(A) + P(B) - P(A \text{ and } B)$.

21. Two number cubes are rolled. Find the probability that the sum of the numbers rolled is 7 or greater than 10.

22. A card is drawn at random from a standard deck of 52 cards. Find the probability that it is a red card or a king.

23. The numbers 1 through 30 are each written on a separate piece of paper. If one piece of paper is picked at random, what is the probability that the number is divisible by 2 or 3?

LESSON 4-5 ◣ Independent and Dependent Events, p. 168

▶ Two events are **independent** if the outcome of one does not affect the outcome of the other. If A and B are independent events, $P(A \text{ and } B) = P(A) \cdot P(B)$.

▶ Two events are **dependent** if the outcome of one affects the outcome of the other. If A and B are dependent events, $P(A \text{ and } B) = P(A) \cdot P(B, \text{ given } A)$.

A box contains 2 green cards, 3 red cards and 5 blue cards. Cards are picked one at a time, then replaced. Find each probability.

24. P(blue, then red)

25. P(green, then blue)

Two cards are drawn from a deck of ten cards numbered 1 through 10. Once a card is selected, it is not replaced. Find each probability.

26. P(two even numbers)

27. P(two numbers greater than 4)

LESSON 4-6 ◣ Permutations of a Set, p. 172

▶ A **permutation** is an arrangement of items in a particular order.

$n!$ (**n factorial**) The number of permutations of n different items is $n(n - 1)(n - 2) \ldots (2)(1)$ and is written as $n!$

$_nP_r$ $_nP_r = \dfrac{n!}{(n-r)!}$ where n is the number of different items and r is the number of items taken at a time.

28. In how many different ways can 8 paintings be awarded first, second, and third prizes?

29. How many ways can you arrange the letters in the word *math*?

30. There are 9 players on a baseball team. How many ways can the coach pick the first 4 batters?

LESSON 4-7 ◣ Combinations of a Set, p. 178

▶ A **combination** is a set of items in which order is not important.

$_nC_r$ $_nC_r = \dfrac{n!}{(n - r)!r!}$, where n is the number of different items and r is the number of items taken at a time.

31. How many different pairs of students can be chosen from a class of 28?

32. How many ways can you choose 3 CDs out of 10 CDs to take on a trip?

33. Pizza Palace offers 8 different toppings for its pizza. How many four-topping pizzas does the Pizza Palace offer?

CHAPTER INVESTIGATION

EXTENSION Present your game to the rest of the class and play a few rounds as a demonstration. Explain your reasoning for the assignment of point values to the various outcomes as well as why you have chosen the particular format. Discuss how probability plays a key role in the outcome of your game.

Chapter 4 Assessment

Answer each question.

1. In a random survey of students, 152 students could swim and 48 could not. What is the probability that a student can swim?

2. Suppose you spin two spinners, one marked 1 through 4 and the other marked A through E. How many possible outcomes are there?

3. In a random sample of 1,000 voters in Kerr City, 525 said they would vote for Higgins for mayor. If there are 122,680 voters in the city, how many might be expected to vote for Higgins?

4. An ice cream store sells 24 different flavors and offers a choice of 3 sizes of cones and 5 types of sprinkles. How many choices of a cone with sprinkles are there?

5. Each card in a set is marked with a letter, a number, and a shape. The letters that can be used are A, B, C, D; the numbers that can be used are 1 to 5; and the shapes are a square, circle, and triangle. A card is selected at random from a box that contains one of each possible card. Find the probability that the card shows an even number and a circle.

6. Find the probability that a point selected at random lies in the shaded region.

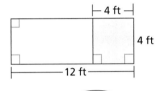

The spinners are both spun for Exercises 7–10.

7. Find $P(D, 1)$.

8. Find $P(\text{not } E, 4 \text{ or } 5)$.

9. Find $P(D, \text{odd number})$.

10. Find $P(\text{vowel, number greater than 2})$.

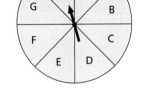

11. Two number cubes are rolled. Find the probability that the sum of the numbers rolled is 6 or 10.

12. Find the probability that the sum of the numbers rolled on two number cubes is less than 6 or a multiple of 3.

A bag contains 7 green marbles, 3 red marbles and 5 yellow marbles.

13. Find $P(\text{yellow, then green})$ if you take one marble, replace it, and then take another marble.

14. Find $P(\text{red, then green})$ if you take one marble and then take another marble without replacing the first.

15. How many ways can a group of 5 books be lined up on a shelf it there are 9 books to choose from?

16. How many ways can 5 runners be chosen from a track team that has 10 members?

Standardized Test Practice

Part 1 Multiple Choice

Record your answers on the answer sheet provided by your teacher or on a sheet of paper.

1. Joannie works for a marketing research company. The company acquires names and phone numbers by purchasing a list of all subscribers to a certain magazine. Then the telemarketers call everyone on the list. This is an example of what type of sampling? (Lesson 1-1)

   Ⓐ convenience
   Ⓑ random
   Ⓒ systematic
   Ⓓ none of these

Use the scatter plot for Exercises 2 and 3.

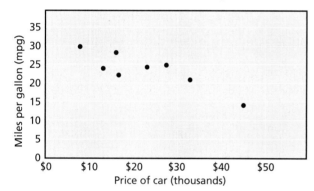

Price of Cars and Miles per Gallon

2. What kind of correlation does the data represent? (Lesson 1-4)

   Ⓐ positive correlation
   Ⓑ negative correlation
   Ⓒ no correlation
   Ⓓ cannot be determined

3. Estimate the coefficient of correlation. (Lesson 1-5)

   Ⓐ 0.90
   Ⓑ 0.50
   Ⓒ 0
   Ⓓ −0.90

4. A dust particle has a mass of 0.00000081 g. What is this number in scientific notation? (Lesson 2-8)

   Ⓐ $81 \cdot 10^8$
   Ⓑ $8.1 \cdot 10^7$
   Ⓒ $81 \cdot 10^{-8}$
   Ⓓ $8.1 \cdot 10^{-7}$

5. Find the next term in the sequence. (Lesson 2-9)

   $$2, -6, 18, -54, \ldots$$

   Ⓐ −162
   Ⓑ −90
   Ⓒ 90
   Ⓓ 162

6. The perimeter of a rectangle can be found using the formula $P = 2l + 2w$. Solve the formula for l. (Lesson 3-4)

   Ⓐ $l = P - w$
   Ⓑ $l = w - P$
   Ⓒ $z = \frac{1}{2}P - w$
   Ⓓ $l = \frac{P - w}{2}$

7. Solve $\frac{2x - 3}{3} = \frac{3x + 1}{4}$. (Lesson 3-5)

   Ⓐ $x = 17$
   Ⓑ $x = 15$
   Ⓒ $x = -15$
   Ⓓ $x = -17$

8. The square root of a number is 4. What is the number? (Lesson 3-8)

   Ⓐ 16
   Ⓑ 2
   Ⓒ −2
   Ⓓ −16

9. How many outcomes are there for rolling a number cube and spinning the spinner at the right? (Lesson 4-3)

   Ⓐ 36
   Ⓑ 24
   Ⓒ 12
   Ⓓ 10

10. What is the probability of drawing a club or an ace out of a standard deck of 52 cards? (Lesson 4-4)

    Ⓐ $\frac{1}{13}$
    Ⓑ $\frac{4}{13}$
    Ⓒ $\frac{17}{52}$
    Ⓓ 0

11. The forecast predicts a 40% chance of rain on Wednesday and a 60% chance of rain on Thursday. If these probabilities are independent, what is the chance that it will rain on both days? (Lesson 4-5)

    Ⓐ 2.4%
    Ⓑ 20%
    Ⓒ 24%
    Ⓓ 100%

Part 2 Short Response/Grid In

Record your answers on the answer sheet provided by your teacher or on a sheet of paper.

Use the following data for Questions 12–14.

Number of Daily Service Calls by Town and Country Heating in November

20	1	20	45	32	31
41	24	3	20	32	37
48	51	37	36	4	6
29	54	44	28	3	38
20	8	38	27	29	30

12. Find the range of the data. (Lesson 1-2)

13. Find the mean of the data. (Lesson 1-2)

14. Find the first, second, and third quartiles of the data. (Lesson 1-6)

15. Trevor scored 12 points less than half Emma's total in Laser tag. If Emma scored x points, what expression can be used to describe Trevor's points? (Lesson 2-3)

16. Simplify: $4(x - 2) + \frac{1}{2}(2x - 6)$. (Lesson 2-6)

17. The sum of twice a number and 6 is 10. What is the number? (Lesson 3-4)

18. The time it takes for a falling object to travel a certain distance d is given by the equation $t = \sqrt{\frac{d}{16}}$, where t is in seconds and d is in feet. If you dropped a ball from a window 28 ft above the ground, how long will it take for the ball to reach the ground? (Lesson 3-8)

Two cubes are rolled. The numbers on the number cube are multiplied. Use this information to answer Questions 19–22.

19. How many outcomes are there in a sample space? (Lesson 4-3)

20. What is the probability of rolling a product of 2? (Lesson 4-3)

21. Is the probability of an odd product less than, equal to, or greater than the probability of an even product? (Lesson 4-3)

22. What is the probability of getting an odd or an even product? (Lesson 4-4)

23. The pentatonic scale has five notes: C#, D#, F#, G#, and A#. These are the black keys on a piano. How many different five-note sequences can be written if each note is used only once? (Lesson 4-6)

24. How many different two-topping pizzas can be made with the toppings sausage, hamburger, ham, peppers, onions, mushrooms, olives, and pineapple? (Lesson 4-7)

Part 3 Extended Response

Record your answers on a sheet of paper. Show your work.

25. Design a spinner with four sections so that no two sections have an equal probability. Explain your answer. (Lesson 4-1)

26. Jasmine usually makes $\frac{1}{3}$ of her free-throw shots. Devise a simulation to determine the probability that Jasmine will make her next two free-throw shots. (Lesson 4-2)

Test-Taking Tip

Questions 23 and 24
If you have time at the end of a test, go back to check your calculations and answers. If the test allows you to use a calculator, use it to check your calculations.

LOGIC AND GEOMETRY

THEME: Navigation

The study of geometry involves the properties, measurements, and relationships of points, lines, angles, planes, and solids. Navigation combines these geometric elements together with geography and logical reasoning. Think back to the last time you read a map or gave someone directions to your home. You used geometry, measurement, and logic. Not only is navigation used by people to get from here to there, but it is also used by many in their professions.

- **Cattle ranchers** (page 201) use navigation to herd their cattle to the best grazing and watering locations. They must do this without crossing land that is treacherous to the animals.

- **Ship captains** (page 221) determine course, speed, and effects of the weather to transport their cargo. Ship captains use compasses and anemometers that measure relationships among the geometric elements found in nature and the location of the stars.

Math Online

mathmatters2.com/chapter_theme

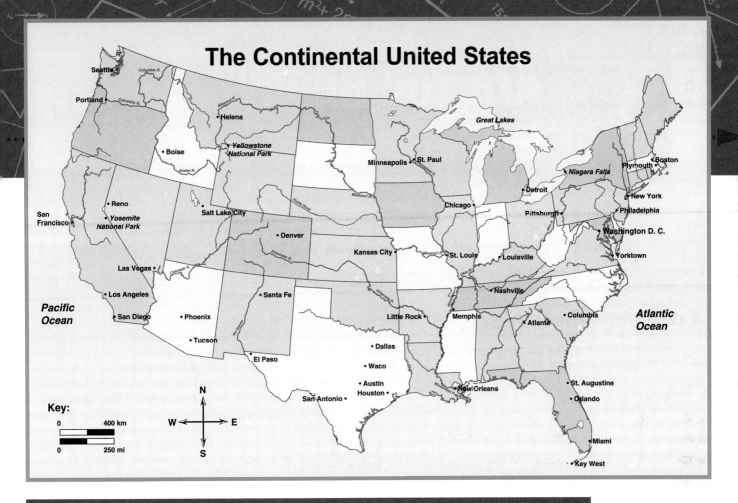

The Continental United States

Key:

| 0 | 400 km |
| 0 | 250 mi |

N
W ← → E
S

Data Activity: The United States

Use the map for Questions 1–4.

1. On this map, 250 mi is equal to how many inches?

2. What is the approximate distance in miles between New York and Kansas City?

3. Pedro took a trip from Chicago to El Paso. He drove at a rate of 65 mi/h. To the nearest hour, how long was the trip?

4. What direction would you travel from Salt Lake City to Yosemite National Park?

CHAPTER INVESTIGATION

A compass is a navigational tool that works in conjunction with the Earth's magnetic fields, particularly the *magnetic north pole*. A compass is a mechanical device that has a magnetic needle, which pivots from its center and points toward magnetic north.

Working Together

If a compass is not available, the sun and a stick can be used to identify which direction is north, south, east and west. Use the Chapter Investigation icons to build a compass to use in nature.

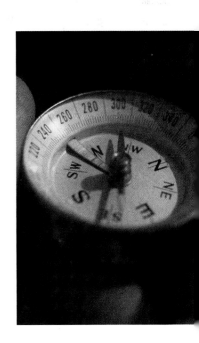

The skills on these two pages are ones you have already learned. Use the examples to refresh your memory and complete the exercises. For additional practice on these and more prerequisite skills, see pages 576–584.

CLASSIFYING TRIANGLES

In this chapter you will work with triangles. It is helpful to be able to recognize different types of triangles.

Examples

Triangles can be classified by the lengths of their sides.

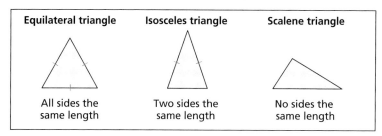

| Equilateral triangle | Isosceles triangle | Scalene triangle |
| All sides the same length | Two sides the same length | No sides the same length |

Triangles can also be classified by the measure of their angles.

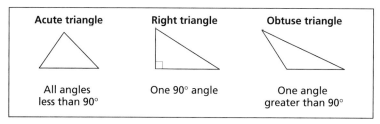

| Acute triangle | Right triangle | Obtuse triangle |
| All angles less than 90° | One 90° angle | One angle greater than 90° |

Classify each triangle both by side and by angle measurements.

1.

2.

3.

4.

5.

6.

7.

8.

SUM OF ANGLES

In working with triangles and other polygons, it is helpful to know that the sum of the angles of any triangle is always 180°.

Examples Find the measure of the missing angle measure.

$$45 + 78 + x = 180$$
$$123 + x = 180$$
$$x = 57$$

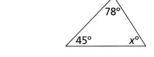

Find the measure of each missing angle.

9.

10.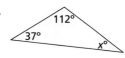

11.

12. What are the measures of the angles in an equilateral triangle?

13. What are the measures of the angles in a right isosceles triangle?

14. If one angle of a right triangle measures 30°, what are the measures of the other two angles?

PARALLELOGRAMS

Examples Parallelograms that have unique characteristics are known by specific names.

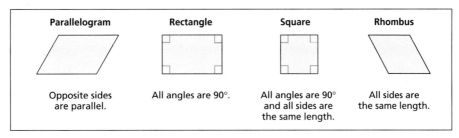

Parallelogram	Rectangle	Square	Rhombus
Opposite sides are parallel.	All angles are 90°.	All angles are 90° and all sides are the same length.	All sides are the same length.

Write the most common name of each shape—parallelogram, rectangle, square or rhombus.

15.

16.

17.

18.

19.

20.

Elements of Geometry

Goals
- Identify fundamental geometric concepts.
- Identify and use basic geometric postulates.

Applications Construction, Art, Photography, Navigation

Draw three dots spread apart that do not lie in a straight line. Label them points *A*, *B*, and *C*.

1. How many straight lines can be drawn through point *A*?

2. How many straight lines can be drawn that will pass through both points *B* and *C*?

3. How many straight lines can be drawn that will pass through all three points?

◤ BUILD UNDERSTANDING

Geometry (from the Greek words *geo*, meaning "earth," and *metria*, meaning "measurement") is the study of points in space. In geometry, *point*, *line*, and *plane* are basic terms.

A **point** (*P*) is a location in space having no dimensions. Every other geometric figure is composed of sets of points. A **line** ($\overleftrightarrow{TB}$) is a set of points that extends infinitely in two opposite directions. A **plane** (*M*) is a flat surface that extends endlessly in all directions. **Space** is the set of all points.

Collinear points (*T*, *B*, and *A*) lie on the same line, while **noncollinear points** (*A*, *K*, and *C*) do not lie on the same line. **Coplanar points** (*J*, *K*, and *C*) lie in the same plane, while **noncoplanar points** (*J*, *K*, *C*, and *B*) do not lie in the same plane.

Example 1   **Personal Tutor at** mathmatters2.com

In the figures shown, name the following.

a. two collinear points
b. three noncollinear points
c. three coplanar points
d. four noncoplanar points

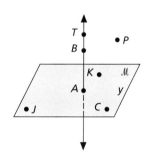

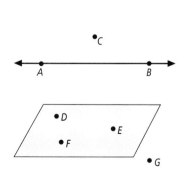

Solution

a. *A* and *B* lie on the same line, so they are collinear.

b. *A*, *B*, and *C* are noncollinear since there is no line that contains all three points.

c. *D*, *E*, and *F* lie in the same plane, so they are coplanar.

d. *D*, *E*, *F*, and *G* are noncoplanar since they do not lie in the same plane.

The **intersection** of two figures is the set of points that both figures share. Two lines intersect in a point. A plane and a line can intersect in a point or a line. A plane and a line may not intersect at all.

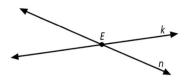

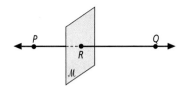

Check Understanding

Describe the intersection of plane $\mathcal{M}$ and $\overrightarrow{PQ}$. Draw a figure to represent the other two possible intersections of a plane and a line.

A **line segment** is a part of a line consisting of two endpoints and all points that lie between these two endpoints. **Congruent line segments** have the same measure. The **midpoint of a segment** is the point that divides the segment into two congruent segments. The symbol $\cong$ means "is congruent to." A **bisector of a segment** is any line, segment, ray, or plane that intersects the segment at its midpoint.

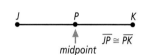
$\overline{JP} \cong \overline{PK}$
midpoint

Example 2

In the figure, $\overleftrightarrow{XY}$ bisects $\overline{MN}$. Name two congruent line segments.

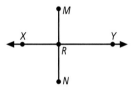

Solution

Since $\overleftrightarrow{XY}$ bisects $\overline{MN}$, point R is the midpoint of $\overline{MN}$. Therefore, segments $\overline{MR}$ and $\overline{RN}$ have the same length, or $\overline{MR} \cong \overline{RN}$.

There are certain statements about the relationships between *points, lines* and *planes* that are assumed to be true. These assumptions are called **postulates**.

Point, Line and Plane Postulates	*Postulate 1* Through any two points, there is exactly one line.
	Postulate 2 Through any three noncollinear points, there is exactly one plane.
	Postulate 3 If two points lie in a plane, then the line joining them lies in that plane.
	Postulate 4 If two planes intersect, then their intersection is a line.

Example 3

State which postulate is illustrated in each figure.

a.

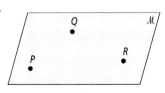

b.

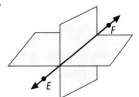

Solution

a. Points P, Q and R are noncollinear and are all contained in plane $\mathcal{M}$, so the first figure illustrates *Postulate 2*.

b. The second figure shows two planes intersecting at $\overleftrightarrow{EF}$, which illustrates *Postulate 4*.

Refer to the figure.

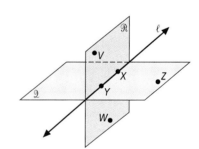

1. Name three points that determine plane $\mathcal{J}$.
2. Name the intersection of planes $\mathcal{J}$ and $\mathcal{K}$.
3. How many lines do points A and B determine?
4. In how many planes shown is $\overleftrightarrow{XY}$ contained?
5. Does $\overrightarrow{AC}$ lie on plane $\mathcal{J}$ or plane $\mathcal{K}$?

Draw a figure to illustrate each situation.

6. Points A, B, and C are noncollinear.
7. Points W, X, Y, and Z are noncoplanar.
8. Line m intersects plane $\mathcal{L}$ at point Q.
9. Planes J and K intersect at line t.

10. **PHOTOGRAPHY** Photographers often use tripods, three-legged stands, to hold their cameras steady. Which postulate does this illustrate? Why do you think they do not use stands with four legs?

Refer to the figure.

11. Name two points that determine line l.
12. Name three points that determine plane $\mathcal{R}$.
13. Name the intersection of plane $\mathcal{R}$ and plane $\mathcal{Q}$.
14. Name three lines that lie in plane $\mathcal{Q}$.

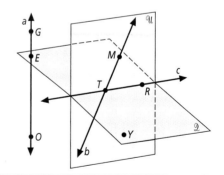

State whether each statement is *true* or *false*. If the statement is false, explain why.

15. Points G and O determine line a.
16. Points M and T determine plane $\mathcal{U}$.
17. The intersection of lines b and c is point T.
18. The intersection of planes $\mathcal{U}$ and $\mathcal{Q}$ is points R and T.
19. Points G, E, and O are collinear.
20. Points M, T, R, and E are coplanar.

21. **CONSTRUCTION** To make sure a wall being built is straight, a mason will place two sticks in the ground flush with the wall and pull a rope tight between the sticks. Which postulate does the mason apply to this situation?

22. **NAVIGATION** A forest ranger is leading a group of hikers through the woods to visit three noncollinear locations. Describe the plane that contains these three points. Is the ground a good model for a plane? Explain.

Refer to the figure. Point *E* is not in plane *M*.

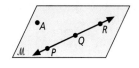

23. Which postulate states that exactly one plane contains *A*, *P*, and *Q*?

24. Which postulate states that exactly one line contains *A* and *R*?

25. If points *A* and *P* are in plane *M*, which postulate states that $\overleftrightarrow{AP}$ is in plane *M*?

ART Many works of artist Piet Mondrian consist only of rectangles and parts of rectangles. Refer to the geometric representation of one of his works in the figure shown.

26. Name all the points that are collinear with points *A* and *D*.

27. Identify the intersection of $\overline{LN}$ and $\overline{RM}$.

28. Identify the intersection of $\overline{KQ}$ and $\overline{EP}$.

29. Name all segments shown for which point *L* is one endpoint.

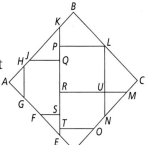

■ EXTENDED PRACTICE EXERCISES

30. **MODELING** Use a sheet of paper and a cube to demonstrate how the intersection of a plane and a cube could be a point, a line segment, or a square region.

31. **CRITICAL THINKING** What do you think the intersection of three planes might look like? Think of a corner of a room to help you visualize the answer.

32. **WRITING MATH** Explain why it is important to include the word *noncollinear* in the statement of *Postulate 2*.

33. **DATA FILE** Refer to the data on road congestion at ten major cities on page 575. Use an atlas to locate three of these cities that are approximately collinear. List the three cities.

34. **CHAPTER INVESTIGATION** On a sunny morning, select a level area and place a 4-ft stick in the ground before noon. Mark the tip of the shadow and label it point *W*. Wait until after noon when the shadow creates a straight line with point *W* in the opposite direction. Mark the tip of the new shadow point *E*.

■ MIXED REVIEW EXERCISES

Use the frequency table for Exercises 35–38. (Lesson 1-3).

35. How many more flights head to Chicago than to Boston each day?

36. The number of flights to Los Angeles each day is the same as the combined number of flights to which two cities?

37. How many fewer flights are there to Denver than to St. Louis each day?

38. Suppose you took a random survey of people in this airport asking each person to what city they were flying. The results showed that more people were flying to Los Angeles than Chicago. Would you consider the survey to be invalid? Explain. (Lesson 1-1)

Departing flights by destination	Tally	Frequency
Boston	|||| |||| |||| |||| |||	23
Los Angeles	|||| |||| |||| |||| |||| |||| |||| ||||	39
Chicago	|||| |||| |||| |||| |||| |||| |||| |||| |||| ||	47
Baltimore	|||| |||| ||||		16
Cincinnati	|||| |||	8
Denver	|||| |||| ||	12
St. Louis	|||| |||| ||||		16

5-2 Angles and Perpendicular Lines

Goals ■ Identify and use perpendicular lines.
■ Identify and use angle relationships.

Applications Health, Physics, Paper folding, Navigation

Use a file folder or a piece of paper, a straightedge and a pencil.

1. Draw ∠*ABC* on the file folder or paper.

2. Position the folder so *m*∠*ABC* is 0°, 90°, and then 180°.

3. Position the folder so *m*∠*ABC* is greater than 0° but less than 90°.

4. Work with a partner. Position your two papers so the sum of the measures of the angles together is 90°, then 180°.

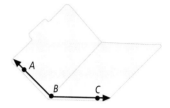

■ BUILD UNDERSTANDING

A **ray** is part of a line that begins at one endpoint and extends without end in one direction. If point *B* is between points *A* and *C*, then $\overrightarrow{BA}$ and $\overrightarrow{BC}$ are **opposite rays**.

An **angle** is the figure formed by two rays that have a common endpoint. The endpoint is called the **vertex** (plural: *vertices*) of the angle. Each ray forms a *side* of the angle. The size of an angle is measured in units called **degrees**.

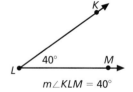

Two angles whose measures have a sum of 90° are called **complementary angles**. Two angles whose measures have a sum of 180° are called **supplementary angles**.

  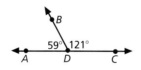

Think Back

Recall that *acute angles* measure less than 90°, *right angles* measure exactly 90°, *obtuse angles* measure between 90° and 180° and *straight angles* measure exactly 180°.

Example 1

Use the three angles shown to name:

a. two complementary angles

b. two supplementary angles

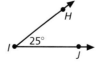

Solution

a. *m*∠*HIJ* + *m*∠*SRT* = 25° + 65° = 90°, so the angles are complementary.

b. *m*∠*SRT* + *m*∠*DEF* = 65° + 115° = 180°, so the angles are supplementary.

Two angles that have a common vertex and share a common side, but do not overlap are called **adjacent angles**, such as ∠1 and ∠2.

Congruent angles have the same angle measure.

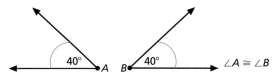

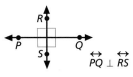

∠A ≅ ∠B

Perpendicular lines are two lines that intersect to form right angles.

$\overleftrightarrow{PQ} \perp \overleftrightarrow{RS}$

When two lines intersect, the angles that are not adjacent to each other are called **vertical angles**. Vertical angles are congruent. In the figure, ∠1 and ∠3 are vertical and ∠2 and ∠4 are vertical.

Example 2

In the figure, $\overline{AD} \perp \overline{CF}$ and ∠FGE and ∠BGC are vertical angles.

a. Name all right angles.   **b.** Find $m\angle AGB$.   **c.** Find $m\angle EGF$.

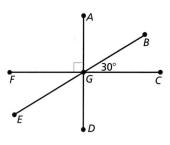

Solution

a. $\overline{AD} \perp \overline{CF}$. So ∠AGF, ∠AGC, ∠CGD, and ∠DGF are all right angles.

b. ∠AGB and ∠BGC are complementary angles. So $m\angle AGB = 90° − m\angle BGC = 90° − 30° = 60°$.

c. ∠BGC and ∠EGF are vertical angles, so $m\angle BGC = m\angle EGF = 30°$.

**COncepts
in MOtion**
Animation
mathmatters2.com

The **bisector of an angle** is a ray that divides the angle into two congruent adjacent angles. $\overrightarrow{LN}$ is the bisector of ∠KLM.

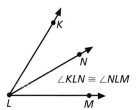

∠KLN ≅ ∠NLM

Example 3

GEOMETRY SOFTWARE Use geometry software to bisect an angle and verify that the bisector divides the angle into two congruent adjacent angles.

Solution

Step 1 Construct an angle and display the measure of the angle.

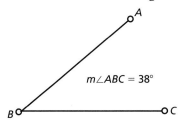

$m\angle ABC = 38°$

Step 2 Using the construct option, bisect the angle.

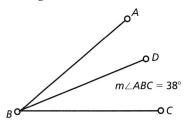

$m\angle ABC = 38°$

Step 3 Display the measures of the newly created adjacent angles. Notice they have the same measure.

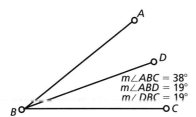

$m\angle ABC = 38°$
$m\angle ABD = 19°$
$m\angle DBC = 19°$

Step 4 Keep all of the measures displayed and move the original angle. Observe how the angle measurements change.

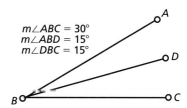

$m\angle ABC = 30°$
$m\angle ABD = 15°$
$m\angle DBC = 15°$

◤ TRY THESE EXERCISES

Find the measure of the complement and supplement of each angle.

1. $m\angle A = 14°$ **2.** $m\angle B = 47°$ **3.** $m\angle 1 = 30°$ **4.** $m\angle DEG = 89°$

In the figure shown, $\overline{FG} \perp \overline{BE}$.

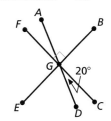

5. Name all right angles.

6. Name a pair of vertical angles.

7. Find $m\angle AGB$.

8. Find $m\angle BGC$.

Find the value of x in each figure. Justify your answer.

9.
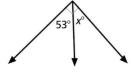

10.
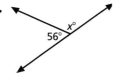

11.
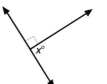

12.
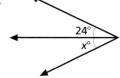

◤ PRACTICE EXERCISES • For Extra Practice, see page 599.

For Exercises 13–15, use the figure shown.

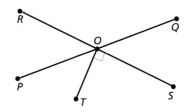

13. Name a pair of perpendicular segments.

14. Name two adjacent complementary angles.

15. Name two pairs of vertical angles.

In the figures shown, $\overrightarrow{OX}$ and $\overrightarrow{OY}$ are opposite rays. Find m$\angle XOZ$.

16.
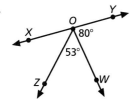

17.
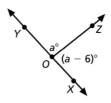

18.
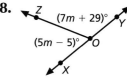

19. PHYSICS According to the *law of reflection*, when a ray of light is reflected from a plane surface, the *angle of incidence* is congruent to the *angle of reflection*. Copy the figure shown, and give as many angle measures as you can.

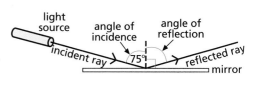

In the figures shown, ∠RST is a right angle. Find m∠RSP.

20.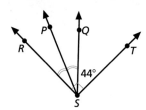

21.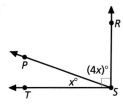

22.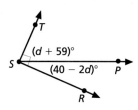

23. **PAPER FOLDING** Draw an angle of any size on a sheet of paper. Fold the paper so one side of the angle lies directly on top of the other. Unfold the paper. What does the crease in the paper represent?

24. **WRITING MATH** Look up the words *complementary* and *supplementary* in a dictionary. Describe why you think they are used in reference to angles.

HEALTH An orthopedist adjusts the position of a patient's crutch handles so that the hand pieces allow a 30° elbow angle.

25. Measure the indicated elbow angle with a protractor.

26. Find the supplement of this angle.

27. Should the hand piece be lowered or raised to obtain the proper elbow angle?

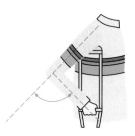

EXTENDED PRACTICE EXERCISES

NAVIGATION Submarines use a periscope to see above the water.

28. Describe any perpendicular lines if ∠2 and ∠5 are right angles.

29. If ∠2 and ∠5 are right angles, estimate $m\angle4$, $m\angle1$ and $m\angle3$.

CRITICAL THINKING Classify each statement as *true* or *false*. If the statement is false, explain why.

30. Two vertical angles may also be adjacent.

31. The complement of an acute angle is an obtuse angle.

32. Two vertical angles may be complementary.

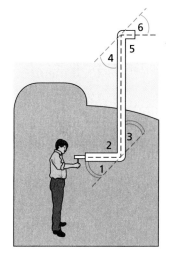

MIXED REVIEW EXERCISES

Solve each inequality. Graph each solution on a number line.
(Lessons 3-6 and 3-7)

33. $4x < -12$

34. $v + -18 \geq -5$

35. $4 - w \leq 36$

36. $\frac{3}{5}y > 21$

37. $3d - 9 \geq -15$

38. $16 - \frac{1}{5}f > -1$

39. Miranda spent at least 120 min on homework this week. She spent the same amount of time on her homework Monday, Tuesday, Wednesday and Thursday. She did not study on Friday or Saturday, but she spent 60 min on homework on Sunday. Find the least amount of time Miranda spent on her homework on Monday. (Lesson 3-7)

Add. (Basic Math Skills)

40. $\frac{2}{3} + \frac{1}{3}$

41. $\frac{3}{5} + \frac{7}{10}$

42. $8 + \frac{1}{4} + 6\frac{1}{3}$

Review and Practice Your Skills

PRACTICE ◤ LESSON 5-1

1. How many points determine a plane?

2. Name three undefined terms.

Refer to the figure for Exercises 3–10.

3. Name two points coplanar with *H*, *D*, and *E*.

4. Name three points on plane $\mathcal{I}$.

5. Name the intersection of planes $\mathcal{I}$ and $\mathcal{J}$.

Tell whether each statement is *true* or *false*.

6. $\overline{BA}$ is a bisector of $\overline{DF}$.

7. $\overline{BA}$ intersects $\overline{DF}$.

8. *A*, *C*, and *B* are coplanar.

9. Three points determine a line.

10. If $\overline{BE} \cong \overline{EG}$, then *E* is the midpoint of $\overline{BG}$.

Draw a figure to illustrate each situation.

11. Line *m* intersects plane $\mathcal{X}$ at point *B*.

12. The midpoint of $\overline{AB}$ is *C*. Line *CD* bisects $\overleftrightarrow{AB}$.

PRACTICE ◤ LESSON 5-2

Use the figure shown for Exercises 13–17.

13. Find $m\angle ABF$ if $m\angle ABE = 140°$ and $\overrightarrow{BF}$ bisects $\angle ABE$.

14. Name the angle vertical to $\angle ABD$.

15. Name all right angles.

16. Name two angles supplementary to $\angle EBC$.

17. Find the measure of $\angle ABD$.

Find the measures of the complement and supplement of each angle.

18. $m\angle 1 = 23°$	19. $m\angle 2 = 77°$
20. $m\angle 3 = 89°$	21. $m\angle 4 = 59°$
22. $m\angle 5 = 13°$	23. $m\angle 6 = 60°$
24. $m\angle 7 = 12°$	25. $m\angle 8 = 45°$
26. $m\angle 9 = 70°$	27. $m\angle 10 = 31°$
28. $m\angle 11 = 55°$	29. $m\angle 12 = 5°$

Use the figure to match each term with the correct symbol. (Lessons 5-1 and 5-2)

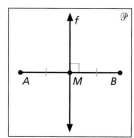

30. midpoint   **a.** $\mathcal{P}$

31. bisector   **b.** f

32. plane   **c.** A, M, B

33. collinear points   **d.** $\perp$

34. $f \blacksquare \overline{AB}$ **e.** M

Find the value of x in each figure. List a vocabulary word to justify each answer. (Lessons 5-1 and 5-2)

35.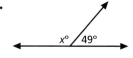
36.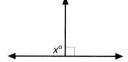
37.
38.

MathWorks Career – Cattle Rancher
Workplace Knowhow

A cattle rancher is responsible for the well-being of the cattle he or she is raising. Part of providing for the animals requires knowing the best grazing and watering locations. In the mountains it can be difficult to move animal herds, as certain areas are too steep or treacherous for cattle to negotiate. In the diagram, the ranch is at point A. The best grazing fields are at point C. However, because a straight path from A to C is too steep, they travel first to point B at an angle of 70° to the vertical line. At point B, they turn 50° to the left and head up the mountain to point C. After grazing, they walk down the mountain, reaching the river at point D.

Refer to the diagram for Exercises 1–4.

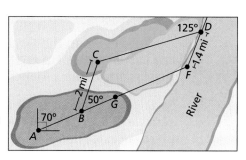

1. What acute angle is formed by the herd's path approaching the river and the river's shore?

2. If the herd retraces its path on the way home, it moves from D to C, turns left and continues to B, and turns right at B and returns toward A. At what angle must the herd turn right at B in order to head directly home?

3. Point G is the midpoint of $\overline{AF}$. If $GF = 2DF$, find AF.

4. The length of CD is 3.7 mi and $AB = 1.8$ mi. Find the length of the entire trip.

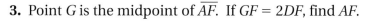

5-3 Parallel Lines and Transversals

Goals
- Identify angles formed by parallel lines and transversals.
- Identify and use properties of parallel lines.

Applications Construction, Safety, Navigation, Music

Use line *m* and point *P* to construct congruent angles.

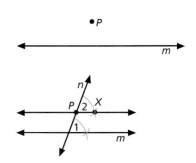

1. Draw line *m* and point *P* as shown. Through *P* draw any line *n* that intersects line *m*. Label ∠1.

2. At *P*, using line *n* as one side, construct ∠2 so that ∠2 ≅ ∠1. Label the intersection of the two arcs point *X*.

3. Draw and label $\overleftrightarrow{PX}$. What seems to be true about the relationship between $\overleftrightarrow{PX}$ and line *m*?

■ BUILD UNDERSTANDING

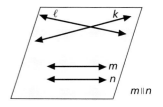

Coplanar lines may intersect in a point. Coplanar lines that do not intersect are called **parallel lines**. In the figure, lines *k* and *l* intersect. Lines *m* and *n* have no points in common, so they are parallel lines.

Planes that do not intersect are called **parallel planes**. The top and bottom of a cube are contained in two parallel planes. Noncoplanar lines that do not intersect and are not parallel are called **skew lines**.

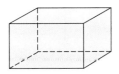

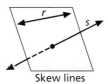

Skew lines

A **transversal** is a line that intersects each of two other coplanar lines in different points to produce **interior** and **exterior angles**. In the figure, transversal *t* intersects lines *j* and *k*.

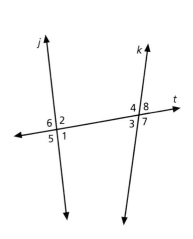

Two nonadjacent interior angles on opposite sides of a transversal are called **alternate interior angles**. The two pairs of alternate interior angles are ∠1 and ∠4, and ∠2 and ∠3.

Interior angles on the same side of a transversal are called **same-side interior angles**. In the figure, the two pairs of same-side interior angles are ∠1 and ∠3, and ∠2 and ∠4.

Two nonadjacent exterior angles on opposite sides of the transversal are called **alternate exterior angles**. The two pairs of alternate exterior angles are ∠5 and ∠8, and ∠6 and ∠7.

Two angles in corresponding positions relative to two lines cut by a transversal are called **corresponding angles**. In the figure, the four pairs of corresponding angles are ∠1 and ∠7, ∠2 and ∠8, ∠3 and ∠5, and ∠4 and ∠6.

Example 1

 Personal Tutor at mathmatters2.com

Refer to the figure to name the following.

a. all pairs of alternate interior angles

b. all pairs of alternate exterior angles

c. all pairs of same-side interior angles

d. all pairs of corresponding angles

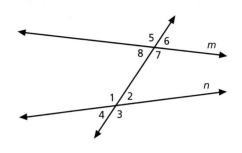

Solution

a. ∠1 and ∠7, ∠2 and ∠8   b. ∠5 and ∠3, ∠6 and ∠4

c. ∠2 and ∠7, ∠1 and ∠8   d. ∠1 and ∠5, ∠2 and ∠6, ∠3 and ∠7, ∠4 and ∠8

When a transversal intersects two parallel lines, the angles formed have special relationships, which are postulates in geometry.

Parallel Line Postulates	*Postulate 5* If two parallel lines are cut by a transversal, then corresponding angles are congruent.
	Postulate 6 If two lines are cut by a transversal so that corresponding angles are congruent, then the lines are parallel.

From these postulates, it can be shown that the following statements are true.

Statement 5A If two parallel lines are cut by a transversal, then alternate interior angles are congruent.

Statement 5B If two parallel lines are cut by a transversal, then alternate exterior angles are congruent.

Statement 6A If two lines are cut by a transversal so that alternate interior angles are congruent, then the lines are parallel.

Statement 6B If two lines are cut by a transversal so that alternate exterior angles are congruent, then the lines are parallel.

Example 2

In the figure, $\overleftrightarrow{AB} \parallel \overleftrightarrow{CD}$. **Name the postulate or statement that gives the reason why each statement is true.**

a. ∠2 ≅ ∠6   b. ∠4 ≅ ∠5   c. ∠1 ≅ ∠8

Solution

a. ∠2 and ∠6 are corresponding angles, so *Postulate* 5 says that they are congruent angles.

b. ∠4 and ∠5 are alternate interior angles, so *Statement 5A* says that they are congruent angles.

c. ∠1 and ∠8 are alternate exterior angles, so *Statement 5B* says that they are congruent angles.

Refer to the figure to name a pair of each type of angle.

1. alternate interior angles   2. alternate exterior angles

3. same-side interior angles   4. corresponding angles

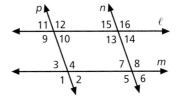

In the figure, line *PQ* ∥ line *RS*. Justify each statement.

5. ∠5 ≅ ∠1   6. ∠8 ≅ ∠1

7. ∠4 ≅ ∠5   8. ∠2 ≅ ∠6

In the figure, *m*∠5 = 120°. Find each measure.

9. *m*∠4   10. *m*∠2

11. *m*∠6   12. *m*∠7

◤ PRACTICE EXERCISES • For Extra Practice, see page 600.

Refer to the figure to classify each pair of angles.

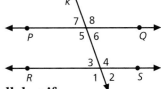

13. ∠1 and ∠9   14. ∠9 and ∠3

15. ∠1 and ∠5   16. ∠3 and ∠10

17. ∠15 and ∠6   18. ∠8 and ∠13

19. ∠5 and ∠16   20. ∠7 and ∠13

In the figure, $\overleftrightarrow{PQ}$ ∥ $\overleftrightarrow{RS}$ and *m*∠5 = 105°. Find each measure.

21. *m*∠1   22. *m*∠2   23. *m*∠4

24. *m*∠6   25. *m*∠7   26. *m*∠8

State whether the lines cut by the transversal are parallel, not parallel or if there is not enough information. Justify your answer if they are not parallel.

27. 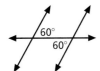   28. 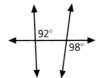   29.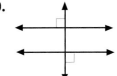

30. **CONSTRUCTION** The top of a house is shown. Use the figure to name all skew line segments, parallel line segments, and parallel planes.

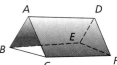
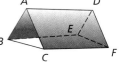

31. **WRITING MATH** Describe an object or objects that model each of the following: parallel lines, parallel planes, and skew lines.

Find each unknown angle measure.

32.   33. 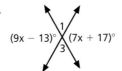   34.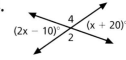

35. SAFETY To avoid possible collisions, airplanes headed eastbound are assigned an altitude level that is an odd number of thousands of feet while westbound planes fly at an even number of thousands of feet. What type of lines are modeled?

36. GEOMETRY SOFTWARE Use geometry software to draw three lines that are all parallel to each other. Then draw a transversal that cuts each of the three lines. Use the measure tool to find the measurements of the angles created. What do you notice about corresponding, alternate interior, and alternate exterior angles?

NAVIGATION Sailboats cannot sail directly into the wind. Instead, sailors use a technique called *tacking*, where the boat sails a series of short paths. In the figure, the boat is tacking first 45° left of the wind, then 45° right of the wind and so on.

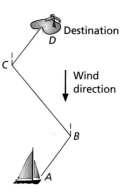

37. What appears to be true of the relationship between $\overline{AB}$ and $\overline{CD}$?

38. If the relationship in Exercise 37 is true, what does $\overline{BC}$ represent?

39. What type of angles are represented by $\angle DCB$ and $\angle ABC$?

■ EXTENDED PRACTICE EXERCISES

Classify each statement as *true* or *false*. If the statement is false, explain why.

40. Two skew lines never intersect.

41. Two lines in the same plane are sometimes skew lines.

42. If a transversal is perpendicular to one of two parallel lines, then it is perpendicular to the other line.

43. MUSIC The word *parallel* is used in music to describe songs that have consistency, such as harmony with parallel voices. Describe two other uses of the word *parallel* in real world applications.

■ MIXED REVIEW EXERCISES

A bag contains 3 white marbles, 6 blue marbles, 2 green marbles and 1 red marble. Find the probability of the following events. (Lessons 4-1 and 4-5)

44. *P*(drawing one red marble)

45. *P*(drawing 1 green marble, then a second green marble without replacement)

46. *P*(drawing 1 green marble, then a second green marble with replacement)

47. Describe an event in relation to this bag of marbles that has the probability of 1.

48. Describe an event in relation to this bag of marbles that has the probability of 0.

5-4 Properties of Triangles

Goals ■ Classify triangles according to their sides and angles.
■ Identify and use properties of triangles.

Applications Travel, Interior design, Navigation

On a sheet of paper, trace the triangle shown.

1. Use a protractor to find the measure of each angle of the triangle. What is the sum of the three angles?

2. Which side is longest? Which angle is largest?

3. Which side is shortest? Which angle is smallest?

4. Based on these results, describe what you think are some properties of triangles. Do you think they apply to all triangles?

5. Draw a triangle of your own, and test these properties.

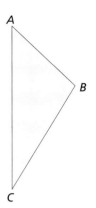

◥ BUILD UNDERSTANDING

A **triangle** is a closed plane figure formed by three line segments joining three noncollinear points. Each point is called a **vertex**. Each vertex corresponds to an angle of the triangle. Each of the line segments that joins the vertices is called a side.

Think Back

Equal number of tick marks on the sides of a figure indicate congruence.

Vertices: Points P, Q and R

Sides: $\overline{PQ}$, $\overline{QR}$ and $\overline{RP}$

Angles: $\angle P$, $\angle Q$ and $\angle R$

A triangle is named by its vertices. The figure is named triangle PQR, usually written $\triangle PQR$.

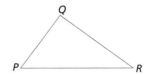

Two sides of a triangle that have the same length are **congruent sides**. Two angles of a triangle that have the same measure are **congruent angles**.

A triangle may be classified by its number of congruent sides.

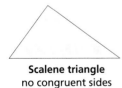

Scalene triangle
no congruent sides

Isosceles triangle
at least two congruent sides

Equilateral triangle
all three sides congruent

A triangle may also be classified by its angles.

Acute triangle
three acute angles

Obtuse triangle
one obtuse angle

Right triangle
one right angle

Isosceles triangle
at least two
congruent angles

Equiangular triangle
all three angles congruent

The following properties are true for all triangles.

Properties of Triangles	*Property 1* The sum of the angles of a triangle is 180°.
	Property 2 The sum of the lengths of any two sides is greater than the length of the third side.
	Property 3 The longest side is opposite the largest angle, and the shortest side is opposite the smallest angle.

Example 1

State whether it is possible to have a triangle with sides of the given lengths.

a. 14, 9, 6   **b.** 8, 5, 3

Solution

a. It is possible because $14 + 9 = 23$, $9 + 6 = 15$ and $14 + 6 = 20$. In each case, the sum of any two sides is greater than the third side.

b. It is not possible because $3 + 5 = 8$, which is not greater than the length of the remaining side, which is 8.

Check Understanding

For example 1, part b, use a centimeter ruler to illustrate the truth of *Property 2*.

In the figure, $\angle LMR$ is called an **exterior angle** of $\triangle LMN$. Notice that $m\angle LMR = m\angle L + m\angle N$.

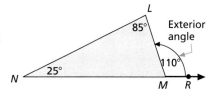

For isosceles triangles, the two angles opposite the congruent sides, called **base angles**, are congruent.

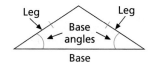

These two properties are summarized below.

More Properties of Triangles	*Property 4* If one side of a triangle is extended, then the exterior angle formed is equal to the sum of the two remote interior angles of the triangle.
	Property 5 If two sides of a triangle are congruent, then the angles opposite those sides are congruent.

Example 2   Personal Tutor at **mathmatters2.com**

Find the values of *a* and *b*.

a.

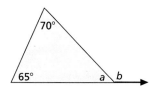

b.

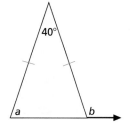

Solution

a. According to *Property 1*, $a + 70° + 65° = 180°$.
$$a = 180° - 70° - 65° = 45°$$

Using *Property 4*, $b = 70° + 65° = 135°$.

b. Since the triangle is isosceles, the base angles are congruent.

$$a + a + 40° = 180° \qquad b = 40° + 70°$$
$$2a = 140° \qquad b = 110°$$
$$a = 70°$$

◤ TRY THESE EXERCISES

State whether it is possible to have a triangle with sides of the given lengths.

1. 12, 7, 6   **2.** 13, 7, 5   **3.** 11, 14, 16   **4.** 1, 2, 3

Find the unknown angle measures in each figure.

5. 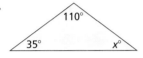   **6.**   **7.**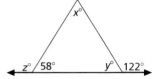

8. WRITING MATH Explain how to find the measure of one angle of a triangle when you are given the other two angle measures.

◤ PRACTICE EXERCISES • For Extra Practice, see page 600.

State whether it is possible to have a triangle with sides of the given lengths.

9. 11, 32, 21   **10.** 0.2, 0.3, 0.5   **11.** 3, 6, 8   **12.** 56, 32, 85

Use the figure to classify each triangle by its sides.

13. △ABC   **14.** △BDC

15. △AEB   **16.** △BCE

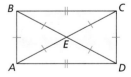

Find the unknown angle measures in each figure.

17.   **18.**   **19.**

20. A triangle has angle measures of $4x°$, $(2x - 30)°$ and $(2x - 30)°$. Classify the triangle, and find its angle measures.

21. TRAVEL At what angle (opposite McDonald Blvd.) do Jasmine Ave. and Rogers Rd. intersect? What property of triangles does this model?

In the figure, $\overrightarrow{AB} \parallel \overrightarrow{CD}$. Find each measure.

22. $m\angle 1$ **23.** $m\angle 2$ **24.** $m\angle 3$

25. $m\angle 4$ **26.** $m\angle 5$ **27.** $m\angle 6$

28. $m\angle 7$ **29.** $m\angle 8$ **30.** $m\angle 9$

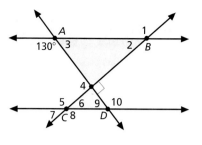

31. WRITING MATH Briefly describe the different types of triangles in this lesson. Draw a right isosceles triangle, and explain how you know it is both isosceles and right.

32. INTERIOR DESIGN Felicity is redesigning her kitchen and wishes to have a large window installed that is in the shape of an isosceles triangle. If two of the sides of the window are 8 ft long, can the base be 18 ft long? Explain.

33. NAVIGATION The Bermuda Triangle is the area of the Atlantic Ocean famous for causing navigational difficulties. It is an equilateral triangle formed by Bermuda, Puerto Rico and Ft. Lauderdale, Florida. The distance between vertices is about 1000 mi. Since the Bermuda Triangle is equilateral, is it necessarily equiangular? Explain.

Bermuda Triangle

◣ EXTENDED PRACTICE EXERCISES

34. CRITICAL THINKING Explain why a triangle cannot have more than one obtuse angle.

35. An exterior angle of a triangle measures $(14x + 43)°$, and the remote interior angles both measure $(13x − 11.5)°$. Classify the triangle, and find the measure of the angles.

Complete each statement with *always*, *sometimes*, or *never*.

36. An isosceles triangle is ___?___ scalene.

37. An exterior angle of a triangle is ___?___ acute.

38. In a right triangle, the acute angles are ___?___ complementary.

39. CHAPTER INVESTIGATION Connect points W and E. This is your east-west line. Draw the perpendicular bisector of $\overline{EW}$, making it the same length as $\overline{EW}$. Label the top endpoint of the bisector N and the bottom endpoint S. This is your north-south line.

◣ MIXED REVIEW EXERCISES

Simplify. (Lessons 2-7 and 2-8)

40. $3x^2 + 7x − x^2$ **41.** $5(7t^5)$ **42.** $2(r^2)^3$

43. $5s(s − 5)$ **44.** $\dfrac{14v^3}{7v}$ **45.** $\dfrac{(36s − 21s^2t)}{3s}$

46. $3t(4t^4)^2$ **47.** $\dfrac{x^2(3x − x^3)}{x}$ **48.** $(3x^8)(2x^{-4})(x^{-4})$

Review and Practice Your Skills

Assume line ℓ || line m and line n || line p. Classify each pair of angles.

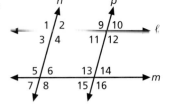

1. ∠1 and ∠4

2. ∠1 and ∠7

3. ∠3 and ∠6

4. ∠11 and ∠15

5. ∠2 and ∠10

6. ∠12 and ∠13

7. ∠11 and ∠13

8. ∠14 and ∠16

9. ∠2 and ∠7

10. ∠8 and ∠13

In the figure, line ℓ || line m and $m\angle 3 = 120°$. Find each measure.

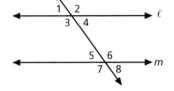

11. $m\angle 1$

12. $m\angle 2$

13. $m\angle 4$

14. $m\angle 6$

15. $m\angle 7$

16. $m\angle 8$

State whether the lines cut by the transversal are parallel, not parallel or if there is not enough information. Justify your answer if they are parallel.

17.
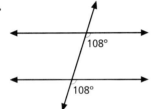

18.
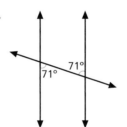

19.
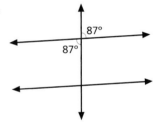

State whether it is possible to have a triangle with sides of the given lengths.

20. 7, 11, 14

21. 10, 10, 10

22. 0.4, 0.5, 0.9

23. 1, 3, 6

24. 30, 40, 50

25. 30, 30, 60

In the figure, $\overrightarrow{EF} \parallel \overrightarrow{GH}$.

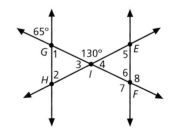

26. $m\angle 1$

27. $m\angle 2$

28. $m\angle 3$

29. $m\angle 4$

30. $m\angle 5$

31. $m\angle 6$

32. $m\angle 7$

33. $m\angle 8$

Draw a figure to illustrate each triangle. (Lesson 5-4)

34. Draw an acute triangle.

35. Draw an obtuse triangle.

36. Draw an isosceles right triangle.

37. Draw an equilateral triangle.

Use the figure to determine whether each statement is *true* or *false*.
(Lessons 5-1 and 5-3)

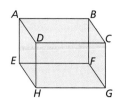

38. $\overleftrightarrow{AD} \parallel \overleftrightarrow{FG}$

39. $\overleftrightarrow{AD}$ and $\overleftrightarrow{CG}$ are skew lines.

40. Plane *ABCD* and plane *BCGF* are parallel.

Find the measure of the complement and supplement of each angle. (Lesson 5-2)

41. $m\angle 1 = 41°$ **42.** $m\angle 2 = 60°$ **43.** $m\angle 3 = 14°$

44. $m\angle 4 = 10°$ **45.** $m\angle 5 = 82°$ **46.** $m\angle 6 = 33°$

Find the unknown angle measures in each figure. (Lesson 5-4)

47. 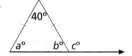   **48.**   **49.**   **50.**

Mid-Chapter Quiz

For exercises 1–6, use the figure shown. (Lesson 5-1)

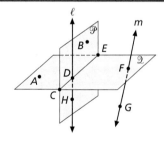

1. Name two points that determine line ℓ.

2. Name three points that determine plane $\mathscr{P}$.

3. Name two collinear points on line *m*.

4. Name three coplanar points in plane $\mathscr{Q}$.

5. Name the intersection of planes $\mathscr{P}$ and $\mathscr{Q}$.

6. How many lines do points *A* and *B* determine?

Use the figure shown to find the measure of the angles and make an observation about their relation to each other. (Lesson 5-2)

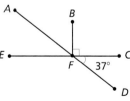

7. $\angle EFA$, $\angle AFB$ 8. $\angle EFB$, $\angle BFC$

In the figure, $\overleftrightarrow{JK} \parallel \overleftrightarrow{LM}$. Classify each pair of angles and find their measures. (Lessons 5-2 and 5-3)

9. $\angle 10$ and $\angle 14$ 10. $\angle 6$ and $\angle 7$

11. $\angle 12$ and $\angle 13$ 12. $\angle 9$ and $\angle 16$

State whether it is possible to have a triangle with sides of the given lengths. (Lesson 5-3)

13. 10, 9, 8   14. 21, 4, 6   15. 7, 7, 7   16. 100, 51, 49

5-5

Congruent Triangles

Goals ■ Use postulates to identify congruent triangles.

Applications Engineering, Art, Recreation

Use the three line segments to construct a triangle.

1. With a straightedge, draw line *m*.

2. Choose point *X* on line *m*. Set a compass for the length of $\overline{AB}$. With *X* as the center, draw an arc that intersects line *m*. Label that point *Z*.

3. Set the compass for the length of $\overline{CD}$. With *X* as the center, draw an arc above line *m*.

4. Set the compass for the length of $\overline{EF}$. With *Z* as the center, draw an arc that intersects the arc drawn in Step 3. Label that intersection point *Y*.

5. Connect points *X* and *Y* and points *Z* and *Y*.

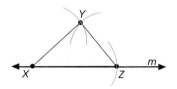

■ BUILD UNDERSTANDING

If you trace $\triangle KLM$ and slide it over $\triangle PQR$, the vertices of one triangle match, or correspond to, the vertices of the other triangle. The angles and sides of one triangle match, or correspond to, the angles and sides of the other.

Corresponding sides	Corresponding angles
$\overline{LK}$ and $\overline{QP}$	$\angle K$ and $\angle P$
$\overline{LM}$ and $\overline{QR}$	$\angle L$ and $\angle Q$
$\overline{KM}$ and $\overline{PR}$	$\angle M$ and $\angle R$

Corresponding parts have equal measures and, therefore, are congruent.

Two triangles are **congruent** if their vertices can be matched so that corresponding parts of the triangles are congruent.

Example 1

Name the congruent sides and angles of $\triangle KLM$ and $\triangle PQR$ shown above.

Solution

$$\overline{LK} \cong \overline{QP} \qquad \angle K \cong \angle P$$

$$\overline{LM} \cong \overline{QR} \qquad \angle L \cong \angle Q$$

$$\overline{KM} \cong \overline{PR} \qquad \angle M \cong \angle R$$

Triangles *KLM* and *PQR* are congruent, written $\triangle KLM \cong \triangle PQR$.

> **Problem Solving Tip**
>
> To name two congruent triangles, match the vertices of one triangle to the corresponding vertices of the other triangle.

To show that two triangles are congruent, you need only show that certain combinations of at least three of the corresponding parts are congruent.

Side-Side-Side Postulate (SSS)	If three sides of one triangle are congruent to three corresponding sides of another triangle, then the triangles are congruent.

△GHI ≅ △JKL by the Side-Side-Side Postulate.

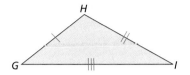 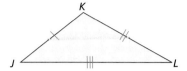

Side-Angle-Side Postulate (SAS)	If two sides and the included angle of one triangle are congruent to two corresponding sides and the included angle of another triangle, then the triangles are congruent.

The **included angle** for the two sides of a triangle is the angle formed by the two sides.

△PQR ≅ △BCD by the Side-Angle-Side Postulate.

Angle-Side-Angle Postulate (ASA)	If two angles and the included side of one triangle are congruent to two corresponding angles and the included side of another triangle, then the triangles are congruent.

The **included side** for two angles of a triangle is the side whose endpoints are the vertices of the angles.

△ABC ≅ △DEF by the Angle-Side-Angle Postulate.

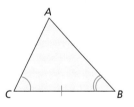 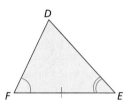

E x a m p l e 2   🌐nline **Personal Tutor at** mathmatters2.com

State whether each pair of triangles is congruent. If a pair is congruent, name the congruence and the appropriate postulate.

a. 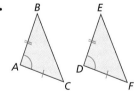 **b.** 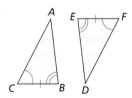 **c.**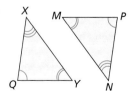

Solution

a. △ABC ≅ △DEF by the Side-Angle-Side Postulate (SAS)

b. △CAB ≅ △FDE by the Angle-Side-Angle Postulate (ASA)

c. not necessarily congruent

The order of the vertices in naming a triangle congruence is not important as long as the corresponding vertices match. For example, stating that △ABC ≅ △DEF is the same as stating △BCA ≅ △EFD.

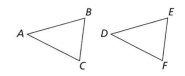

1. In the figure, △ABC ≅ △DEF. Name the corresponding, congruent parts.

State whether each pair of triangles is congruent. If a pair is congruent, name the congruence and the appropriate postulate.

2. 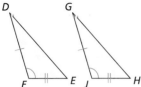 3. 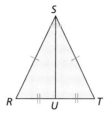 4.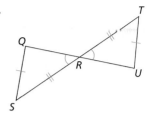

For Exercises 5–8, use △ABC.

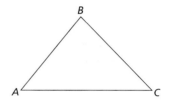

5. Which angle is included between $\overline{AB}$ and $\overline{BC}$?

6. Which angle is included between $\overline{AC}$ and $\overline{AB}$?

7. Which side is included between ∠A and ∠B?

8. Which side is included between ∠B and ∠C?

■ **PRACTICE EXERCISES** • **For Extra Practice, see page 601.**

9. If △RST ≅ △XYZ, name all the corresponding congruent parts.

State whether the pair of triangles is congruent by SAS, ASA or SSS.

10. 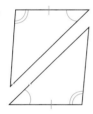 11. 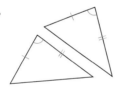 12.  13.

14. **ENGINEERING** The bridge shown uses a triangular truss design. In the figure, △RSU ≅ △TSU. Name the corresponding congruent parts.

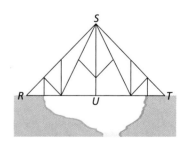

Use the figure of the bridge to answer the following.

15. Which angle is included between $\overline{SR}$ and $\overline{RU}$?

16. Which angle is included between $\overline{SU}$ and $\overline{UT}$?

17. Which side is included between ∠SRU and ∠RUS?

18. Which side is included between ∠UST and ∠STU?

19. **RECREATION** Are the two halves of the kite congruent triangles? If so, which postulate guarantees their congruence?

20. WRITING MATH To apply the Angle-Side-Angle Postulate, what information do you need to know?

21. ART In the design pattern, quadrilateral *ABCD* is a square and points *W, X, Y* and *Z* are the midpoints of the sides. Can you use this information to conclude that $\triangle WAX \cong \triangle YCZ$? If so, which postulate allows you to draw this conclusion?

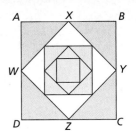

Find the value of *x* for each pair of congruent triangles.

22.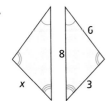

23.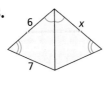

24.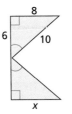

25. YOU MAKE THE CALL Gabe claims that the two triangles in the figure are not congruent because the side marked is not included between the two congruent angles. Latravis says that since the sum of the angles of a triangle is 180°, the triangles are congruent by ASA. Who is correct and why?

EXTENDED PRACTICE EXERCISES

26. WRITING MATH Compare and contrast the Side-Side-Side, Side-Angle-Side and Angle-Side-Angle Triangle Congruence Postulates.

27. CRITICAL THINKING If three angles of a triangle are congruent to three corresponding angles of another triangle, are the two triangles necessarily congruent? Explain.

28. GEOMETRY SOFTWARE Use geometry software to determine if SSA is a valid method of proving triangle congruence. Fix the lengths of two sides, and then fix the size of one of the two angles that is not included between your two fixed sides. Does this result in only one triangle, or could you have more than one? Is SSA a valid Triangle Congruence Postulate?

Boston, Massachusetts

29. DATA FILE Refer to the data on the ten windiest U.S. cities on page 577. Use an atlas to find which of the cities creates an isosceles triangle with Dodge City, KS, and Goodland, KS. (Hint: The base is 143 mi, and the sides are about 580 mi.)

MIXED REVIEW EXERCISES

Find the next three terms for each sequence. (Lesson 2-9)

30. $-32, -16, -8, -4, -2, \ldots$ **31.** $5, -25, 125, -625, 3125, \ldots$ **32.** $\dfrac{3}{4}, \dfrac{1}{2}, \dfrac{1}{3}, \dfrac{2}{9}, \ldots$

33. $1, 1, 2, 3, 5, \ldots$ **34.** $-2, -2, -4, -6, -10, \ldots$ **35.** $4, 8, 13, 19, 26, \ldots$

Find the 100th term for each sequence. (Lesson 2-9)

36. $0, 3, 6, 9, 12, \ldots$ **37.** $-\dfrac{1}{2}, \dfrac{1}{2}, -\dfrac{1}{4}, \dfrac{1}{4}, -\dfrac{1}{6}, \dfrac{1}{6}, \ldots$ **38.** $x^2, 2x^3, 3x^4, 4x^5, 5x^6, \ldots$

Quadrilaterals and Parallelograms

Goals
 ■ Classify different types of quadrilaterals.
 ■ Identify and use properties of parallelograms.

Applications Construction, Civil engineering, Navigation

Refer to parallelogram *ABCD*.

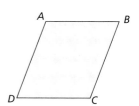

1. Use a protractor to measure each of the angles of *ABCD*.

2. What do you notice about the relationship between ∠*A* and ∠*C* and the relationship between ∠*B* and ∠*D*?

3. Use a ruler to measure the length of each side of *ABCD*.

4. What do you notice about the relationship between the different sides?

◥ BUILD UNDERSTANDING

A **quadrilateral** is a closed plane figure that has four sides. A **parallelogram** is a quadrilateral with two pairs of parallel sides. The relationships between different kinds of quadrilaterals are shown in the diagram.

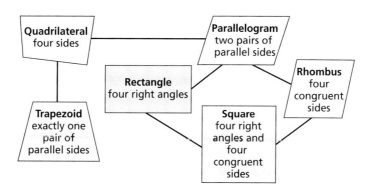

The parts of a quadrilateral have special names. In parallelogram *ABCD*, angles such as ∠*A* and ∠*C* are **opposite angles**. Angles such as ∠*A* and ∠*B* are **consecutive angles**. Sides such as $\overline{AB}$ and $\overline{CD}$ are **opposite sides**. Sides such as $\overline{AB}$ and $\overline{BC}$ are **consecutive sides**.

There are several important properties of parallelograms.

Properties of Parallelograms	*Property 1* The opposite sides of a parallelogram are congruent.
	Property 2 The opposite angles of a parallelogram are congruent.
	Property 3 The consecutive angles of a parallelogram are supplementary.
	Property 4 The sum of the angle measures of a parallelogram is 360°.

Example 1

In parallelogram *DEFG*, $m\angle D = 60°$. Find $m\angle E$, $m\angle F$, and $m\angle G$.

Solution

$\angle E$ and $\angle D$ are consecutive angles, so they are supplementary.

$$m\angle E + m\angle D = 180°$$
$$m\angle E + 60° = 180°$$
$$m\angle E = 180° - 60° = 120°$$

$\angle F$ and $\angle D$ are opposite angles, so they are congruent.
$$\angle F \cong \angle D, \text{ so } m\angle F = 60°$$

$\angle E$ and $\angle G$ are opposite angles, so they are also congruent.
$$\angle E \cong \angle G, \text{ so } m\angle G = 120°$$

> **Reading Math**
>
> The symbol $\square$ is used to represent a parallelogram.
>
> The expression "$\square DEFG$" is read "parallelogram *DEFG*."

Another group of properties of parallelograms involves diagonals.

Diagonals of Parallelograms	*Property 5* The diagonals of a parallelogram bisect each other.
	Property 6 The diagonals of a rectangle are congruent.
	Property 7 The diagonals of a rhombus are perpendicular.

Example 2

Figure *ABCD* is a square. Name all the pairs of congruent segments.

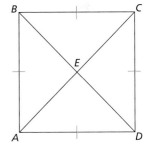

Solution

Since *ABCD* is a square, by definition $\overline{AB} \cong \overline{BC} \cong \overline{CD} \cong \overline{DA}$.

Since a square is also a rectangle, by *Property 6*, $\overline{AC}$ and $\overline{BD}$ are congruent.

Since the figure is parallelogram, by *Property 5*, $\overline{AE} \cong \overline{CE}$ and $\overline{BE} \cong \overline{DE}$.

For $\overline{AB}$, the notation *AB* (no segment bar) means the measure of $\overline{AB}$. When you state that two segments are congruent, write $\overline{AB} \cong \overline{CD}$. Use an equal symbol to state that the measure of two segments are equal, $AB = CD$.

Example 3   **Personal Tutor at** mathmatters2.com

CONSTRUCTION A builder is adding a handrail to a staircase. The handrail is supported by vertical posts called *balusters* and must be constructed so that $\overline{PQ} \parallel \overline{SR}$. If the baluster at the bottom step is 3 ft high, what should be the height of the other balusters?

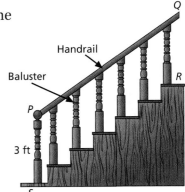

Solution

Since all of the balusters must be vertical, $\overline{PS} \parallel \overline{QR}$. It is given that $\overline{PQ} \parallel \overline{SR}$, so *PQRS* is a parallelogram. *Property* 1 says that opposite sides of a parallelogram are congruent. Since $PS = 3$ ft, $QR = 3$ ft. By similar reasoning, the height of each other baluster also must be 3 ft.

TRY THESE EXERCISES

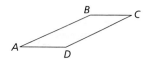

In □ABCD, m∠A = 42°. Find each measure.

1. m∠B   2. m∠C   3. m∠D

4. Name all the pairs of congruent segments in □EFGH. Justify each answer.

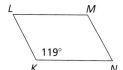

Find the length of each segment.

5. $\overline{EH}$ 6. $\overline{HG}$ 7. $\overline{DG}$

8. $\overline{EG}$ 9. $\overline{FD}$ 10. $\overline{FH}$

PRACTICE EXERCISES • For Extra Practice, see page 601.

Find the unknown angle measures in each parallelogram.

11. m∠D   12. m∠M

13. m∠B   14. m∠N

15. m∠C   16. m∠L

Find each unknown measure in □ABCD.

17. BC

18. CD

19. m∠B

20. m∠C

21. m∠D

Name all the types of quadrilaterals with the following properties.

22. Opposite sides are congruent.   23. Consecutive sides are perpendicular.

24. Diagonals bisect each other.   25. Opposite angles are congruent.

26. Diagonals are congruent.   27. Diagonals are perpendicular.

28. Consecutive angles are congruent.   29. All four angles are congruent.

30. **YOU MAKE THE CALL** A rectangle is defined as a quadrilateral with four right angles. Sying says this means that a rectangle is not a parallelogram, but Hal says a rectangle is a parallelogram. Who is correct and why?

31. **CIVIL ENGINEERING** In Center City, all lettered streets are parallel to each other, and all numbered streets are parallel to each other. The corner of F Street and 6th Street has an angle measure of 121°. What are the angle measures at the other three corners bounded?

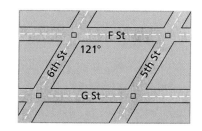

32. **WRITING MATH** Given that quadrilateral RSTU is a parallelogram, list as many facts about the angles, sides and diagonals of quadrilateral RSTU as you can.

Refer to □ABCD to answer the following.

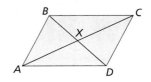

33. If $AB = 10$, then $DC = $ ___?___ .

34. If $AC = 15$, then $AX = $ ___?___ .

35. If $m\angle CBA = 111°$, then $m\angle ADC = $ ___?___ .

36. If $m\angle BAD = 69°$, then $m\angle BCD = $ ___?___ .

Quadrilateral *PARK* is a parallelogram. Tell if each statement is justified.

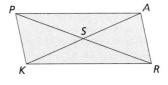

37. $\overline{PS} \cong \overline{KS}$

38. $\overline{PK} \cong \overline{RA}$

39. $\overline{PA} \perp \overline{AR}$

40. $\overline{PK} \parallel \overline{AR}$

41. $SR = \frac{1}{2}PR$

42. $m\angle ARP = \frac{1}{2}m\angle ARK$

43. $\angle KPA \cong \angle RAP$

44. $\angle PAR \cong \angle RKP$

45. NAVIGATION Four buoys have been placed at the vertices of a parallelogram to outline the course for a yacht race. If the boats start northwest of buoy *R*, what is the sum of angle measures of the four turns they will have to make? Do you think turn *T* or turn *U* will be more difficult? Explain.

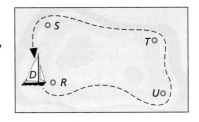

EXTENDED PRACTICE EXERCISES

Use the properties you have learned to complete each statement.

46. If a quadrilateral is a parallelogram, then ___?___ .

47. If ___?___ , then the quadrilateral is a parallelogram.

48. Show why a parallelogram having four angles of equal measure must be a rectangle. (Hint: Let $x = $ the measure of any angle.)

49. CRITICAL THINKING The adjustable ironing board is built so that when open, $\overline{AO} \cong \overline{OC} \cong \overline{BO} \cong \overline{OD}$. Explain why the ironing surface is parallel to the floor.

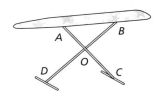

MIXED REVIEW EXERCISES

Solve each equation. Check the solution. (Lesson 3-4)

50. $3x + 12 = -9$

51. $4d - 19 = 28$

52. $\frac{2}{5} - \frac{1}{10}s = 3\frac{1}{2}$

53. $\frac{f}{5} - 9 = -4$

54. $-\frac{1}{2}g + 15 = 13$

55. $-4t + -12 = -20$

56. $2y - 17 = 212$

57. $\frac{3}{4}x = 7.5$

58. $-5 - 4a = 27$

59. $\frac{2}{3}x - 9 = 7 + \frac{1}{3}x$

60. $7t + 110 = 319$

61. $-y + 0.55 = 3.75$

62. RETAIL Maska has a scratch card for a clothing store. The sales clerk will scratch the card to reveal the discount Maska can take on his purchases. Maska wants to buy a coat that regularly costs $175. He can only afford to pay $125. What is the least discount Maska needs in order to buy the coat? Round to the nearest tenth of a percent. (Lesson 3-7)

Review and Practice Your Skills

PRACTICE ◤ LESSON 5-5

State whether each pair of triangles is congruent by SAS, SSS, or ASA.

1.

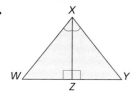

2.

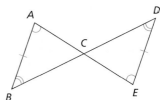

3.

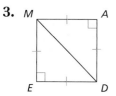

4.

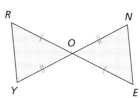

5.

6.

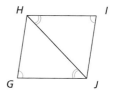

△*MNO* ≅ △*STU*. Fill in the blanks.

7. ∠*M* ≅ ___?___

8. ∠*O* ≅ ___?___

9. ∠*N* ≅ ___?___

10. $\overline{MN}$ ≅ ___?___

11. $\overline{NO}$ ≅ ___?___

12. $\overline{MO}$ ≅ ___?___

PRACTICE ◤ LESSON 5-6

Refer to parallelogram DARK.

13. If *DA* = 8, then *RK* = ___?___.

14. If *DR* = 18, then *DX* = ___?___.

15. If *m*∠*RAD* = 81°, then *m*∠*DKR* = ___?___.

16. If *m*∠*ADK* = 60°, then *m*∠*DKR* = ___?___.

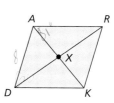

Fill in the blanks for each parallelogram.

17. If *m*∠*KRA* = 128°, then *m*∠*ABK* = ___?___.

18. If *BARK* is a rhombus, then *m*∠*KXB* = ___?___.

19. If *m*∠*KBA* = 128°, then *m*∠*BAR* = ___?___.

20. If *BR* = 12, then *BX* = ___?___.

21. $\overline{FI}$ ≅ ___?___

22. $\overline{FY}$ ≅ ___?___

23. ∠*FDN* ≅ ___?___

24. ___?___ and ___?___ are supplementary to ∠*DFI*.

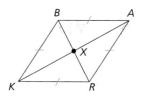

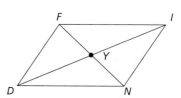

State whether each pair of triangles is congruent by SSS, SAS, or ASA. (Lesson 5-5)

25.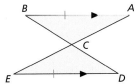
26.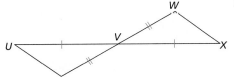
27.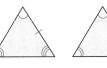

In the figure, $\overline{YN} \parallel \overline{ED}$ and $m\angle 4 = 82°$. Find each measure. (Lesson 5-3)

28. $m\angle 1$

29. $m\angle 2$

30. $m\angle 3$

31. $m\angle 5$

32. $m\angle 6$

33. $m\angle 7$

34. $m\angle 8$

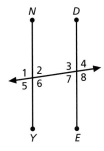

MathWorks Career – Ship Captain

Workplace Knowhow

Ship captains are in command of a water vessel, such as a deep-sea merchant ship, tugboat, ferry, yacht, cruise ship or other waterborne craft. The captain determines the course and speed, monitors the vessel's position, directs the crew and maintains logs of the ship's movements and cargo. When a ship sails from Cape Verde, Africa, to Cape Hatteras, North Carolina, to the southern coast of Newfoundland, to Lisbon, Spain, and back to Cape Verde, it sails approximately in the shape of a parallelogram.

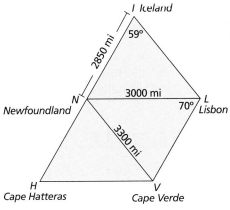

Refer to the diagram for Exercises 1–5.

1. Through what angle must the ship's captain turn the ship when arriving at the south coast of Newfoundland from Cape Hatteras to head toward Lisbon?

2. If the boat is headed toward Lisbon from Newfoundland and at Lisbon turns north to travel toward Iceland, through what angle must he turn the ship at Lisbon?

3. What is the approximate distance between Cape Verde and Cape Hatteras?

4. What is the maximum distance in miles between Iceland and Lisbon?

5. There is a very tiny island located at the intersection of the diagonals of the parallelogram formed by Cape Verde, Cape Hatteras, Newfoundland, and Lisbon. What is the distance between Newfoundland and this island?

5-7 Diagonals and Angles of Polygons

Goals
- Classify polygons according to their sides.
- Find the sum of the angle measures of polygons.

Applications Safety, Hobbies, Nature

With a compass, protractor and straightedge, you can construct a regular hexagon.

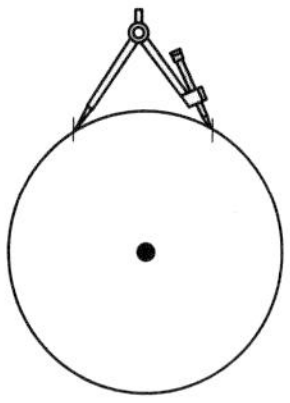

1. Use the compass to construct a circle on a sheet of paper.
2. With the same setting from Step 1, set the point of the compass on the circle and strike an arc on the circle.
3. Place the point of the compass on your arc and strike a new arc. Repeat this around the circle until you reach your first arc.
4. Use the straightedge to connect the points where the arcs intersect the sides of the circle. The resulting figure is a regular hexagon.

Build Understanding

A **polygon** is a simple, closed plane figure formed by joining three or more line segments at their endpoints. Each segment, or **side**, of the polygon intersects exactly two other segments, one at each endpoint. The point at which the endpoints meet is called a **vertex** of the polygon. Polygons are named by their number of sides.

Name of polygon	Number of sides
Triangle	3
Quadrilateral	4
Pentagon	5
Hexagon	6
Heptagon	7
Octagon	8
Nonagon	9
Decagon	10
n-gon	*n*

A polygon is **convex** if each line containing a side has no points in the interior of the polygon. A polygon in which a line that contains a side of the polygon also contains a point in its interior, is called **concave**.

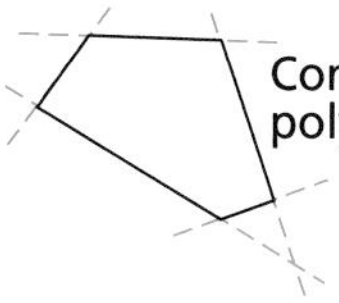
Convex polygon

Concave polygon

A polygon that has all sides congruent and all angles congruent is called a **regular polygon**.

Check Understanding

How are the number of sides and the number of vertices of a polygon related?

Example 1

Classify each polygon by its number of sides. State whether it is convex or concave, regular or not regular.

a.
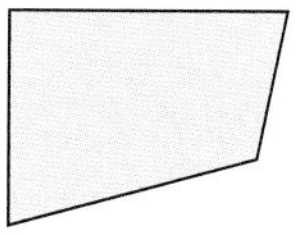

b.
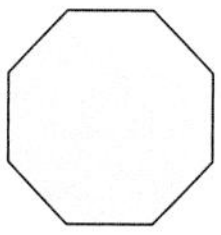

c.
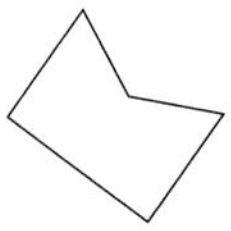

d.
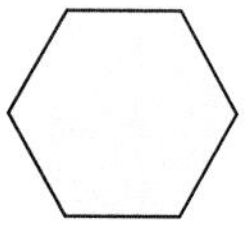

Solution

a. The figure has four sides, so it is a quadrilateral. It is convex, but since not all sides and angles are congruent, it is not regular.

b. The figure has eight congruent sides and eight congruent angles. It is a convex, regular octagon.

c. The figure has five sides of unequal length. There are two lines containing sides with points in the interior, so it is a concave pentagon that is not regular.

d. The figure has six congruent sides and six congruent angles. It is a convex, regular hexagon.

A **diagonal** of a polygon is a segment that joins two vertices but is not a side. You can find the sum of the angles of any convex polygon by drawing the diagonals from any vertex. The diagonals separate the interior of the polygon into nonoverlapping triangular regions.

The sum of the measures of the interior angles of the polygon is the product of the number of triangles formed and 180°.

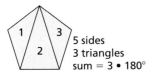 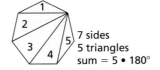 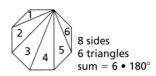

The number of triangles is two less than the number of sides of the polygon.

Angle Sum of a Polygon	The sum of the interior angle measures of a convex polygon with n sides is $(n - 2)180°$.

Example 2

Online **Personal Tutor** at **mathmatters2.com**

Find the sum of the interior angles of each convex polygon.

a. pentagon

b. heptagon

Solution

a. A pentagon has 5 sides.
$(n - 2)180°$ Substitute 5 for n.
$(5 - 2)180° = 3 \cdot 180°$
$\qquad\qquad = 540°$

b. A heptagon has 7 sides.
$(n - 2)180°$ Substitute 7 for n.
$(7 - 2)180° = 5 \cdot 180°$
$\qquad\qquad = 900°$

You can find the measure of each interior angle of a regular polygon when you know the number of sides. A regular polygon of n sides has n angles of equal measure. Divide the sum of the angles by the number of angles.

Angle Measure of a Regular Polygon	The measure of each interior angle of a regular polygon with n sides is given by the formula $\dfrac{(n - 2)180°}{n}$.

Example 3

NATURE A beehive is a large array of regular hexagons. Find the measure of an interior angle of each regular hexagon.

Solution

Use the formula for the interior angle measure of a regular hexagon.

$$\frac{(n-2)180°}{n} = \frac{(6-2)180°}{6} = \frac{4 \cdot 180°}{6} = \frac{720°}{6} = 120°$$

TRY THESE EXERCISES

Classify each polygon by its number of sides. Tell whether it is convex or concave, regular or not regular.

1.   **2.**   **3.**

Find the sum of the interior angles of each convex polygon.

4. octagon **5.** decagon **6.** nonagon **7.** 13-gon

Find the measure of an interior angle of each regular polygon.

8. pentagon **9.** decagon **10.** triangle **11.** 15-gon

PRACTICE EXERCISES • For Extra Practice, see page 602.

Sketch each polygon.

12. a regular triangle **13.** a convex quadrilateral **14.** a concave quadrilateral

For each convex polygon, find the sum of the interior angles.

15. quadrilateral **16.** triangle **17.** 20-gon **18.** heptagon

19. Find the measure of an interior angle of a regular 24-sided polygon.

20. What is the measure of an interior angle of a regular 45-gon?

21. HOBBIES Masaru is designing a quilt that will contain the pattern shown. Classify the polygon containing the cloth pattern.

22. ERROR ALERT Angela calculates the sum of the interior angles of a regular 13-gon by multiplying 13 by 180° to arrive at 2340°. What mistake has Angela made, and what is the correct answer?

SAFETY Name the shape of each sign. Is it a polygon?

23.   **24.**   **25.**   **26.**

Classify each statement as *true* or *false*. If the statement is false, explain why.

27. The sum of the measures of the interior angles of an 11-gon is 1820°.

28. The measure of an interior angle of a regular 18-gon is 160°.

29. An interior angle of a regular polygon cannot have a measure of 148°.

30. The sum of the interior angle measures of a polygon cannot be 1500°.

Sketch the polygon described by each set of characteristics.

31. A quadrilateral with both pairs of opposite sides lying along parallel lines.

32. A concave hexagon with at least three sides the same length.

33. A convex pentagon with exactly two right angles.

◣ EXTENDED PRACTICE EXERCISES

CRITICAL THINKING At each vertex of any convex polygon, there is a pair of supplementary angles. So a convex polygon of *n* vertices has *n* pairs of supplementary angles, one interior angle and one exterior angle at each vertex.

34. Write an expression for the sum of all the interior and exterior angles of a polygon with *n* sides.

35. Show that the sum of the exterior angles of a convex polygon is 360° by solving the equation:

   (sum of interior angles) + (sum of exterior angles) = $n(180°)$.

36. **WRITING MATH** Sketch a concave heptagon and a convex heptagon. Describe in your own words the difference between a concave and a convex polygon.

37. If the measure of an interior angle of a regular polygon is 168°, how many sides does the polygon have?

◣ MIXED REVIEW EXERCISES

Gina's Pizzeria offers the following toppings for pizzas: pepperoni, bacon, hamburger, sausage, peppers, onions, and tomatoes. (Lessons 4-3 and 4-7)

38. How many unique pizzas can be created using three toppings? None of the toppings are duplicated.

39. How many unique two-topping pizzas can be created?

40. How many unique four-topping pizzas can be created?

41. Of all the three-topping pizzas possible, what is the probability of a customer ordering a pizza with pepperoni and tomatoes?

42. If Gina's only allows up to five toppings on each pizza, how many different five-topping pizzas can be created?

Properties of Circles

Goals
- ■ Understand relationships among parts of a circle.
- ■ Identify and use properties of circles.

Applications Market research, Food service, Art, Recreation, Navigation

Use a compass, a ruler and a calculator.

1. On a sheet of paper, draw a circle with a radius of 2 in. Then keeping the same center, draw a circle with a radius of 4 in.

2. Study the two circles. How do you think the circumferences of the two circles compare?

3. How do you think the areas of the two circles compare?

4. Recall that the formulas for the area and circumference of a circle are $A \approx 3.14 \cdot r^2$ and $C \approx 3.14 \cdot d.$ Use a calculator to find the areas and circumferences of the two circles.

5. How do the circles compare in terms of area and circumference? Was your guess correct?

■ BUILD UNDERSTANDING

A **circle** is the set of all points in a plane that are a given distance from a fixed point in the plane. The fixed point is called the **center** of the circle. The given distance is the **radius** (plural: *radii*). A radius is a segment that has one endpoint at the center and one on the circle.

A **chord** is a segment with both endpoints on the circle.

A **diameter** is a chord that passes through the center of the circle. The length of a diameter is twice the length of a radius.

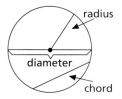

Example 1

Use letters to name the parts of circle P.

  a. three radii   **b.** a diameter   **c.** two chords

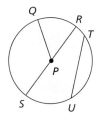

Solution

  a. $\overline{PQ}$ or $\overline{QP}$

     $\overline{RP}$ or $\overline{PR}$

     $\overline{SP}$ or $\overline{PS}$

  b. $\overline{RS}$ or $\overline{SR}$

  c. $\overline{TU}$ or $\overline{UT}$

     $\overline{RS}$ or $\overline{SR}$

Reading Math

The symbol ⊙ is used to name a circle by its center.

The expression "⊙P" is read "circle P."

A **central angle** of a circle is an angle with its vertex at the center of a circle. In ⊙D, ∠ADB is a central angle.

An **arc** is a section of the circumference of a circle.

A **semicircle** is an arc of a circle with endpoints that are the endpoints of a diameter. Three letters are used to name a semicircle. Read $\widehat{ABC}$ as "arc ABC" or "semicircle ABC."

A **minor arc** is an arc that is smaller than a semicircle. A minor arc is named by its two endpoints. Read $\widehat{AB}$ as "arc AB."

A **major arc** is an arc that is larger than a semicircle. A major arc is named by three points, the first and last being its endpoints. Read $\widehat{ACB}$ as "arc ACB."

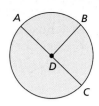

COncepts in MOtion
Animation
mathmatters2.com

Example 2

Identify the following parts of ⊙G.

a. a minor arc

b. a major arc

c. a semicircle

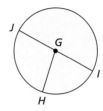

Solution

a. $\widehat{JH}$ or $\widehat{HJ}$, $\widehat{HI}$ or $\widehat{IH}$

b. $\widehat{HJI}$ or $\widehat{IJH}$, $\widehat{HIJ}$ or $\widehat{JIH}$

c. $\widehat{IHJ}$ or $\widehat{JHI}$

Each minor arc is associated with a central angle that is said to intercept the minor arc. The measure of a minor arc is defined to be the same as the measure of its central angle. In ⊙N, $m\angle LNM = m\widehat{LM} = 75°$.

The measure of a major arc is found by subtracting the measure of its related minor arc from 360°. In ⊙N, $m\widehat{LPM} = 360° - m\widehat{LM} = 360° - 75° = 285°$.

The measure of a semicircle, or half of a circle, is 180°.

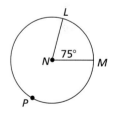

Example 3

MARKET RESEARCH The results of a survey on favorite types of music are shown. Find the measure of the arc for the part of the circle representing each type of music.

Solution

The arc for classical music is a semicircle, so it measures 180°.

The arc for blues measures 55°.

The arc for country measures 180° − 55°, or 125°.

An **inscribed angle** is an angle whose vertex lies on the circle and whose sides contain chords of the circle. In the figure, ∠ACB is an inscribed angle. The measure of an inscribed angle is one-half the measure of the arc it intercepts.

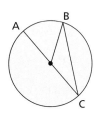

Example 4

Identify the following for ⊙N.

a. an inscribed angle   **b.** a central angle

c. m∠ONQ   **d.** m∠OPQ

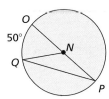

Solution

a. ∠OPQ   **b.** ∠ONQ or ∠QNP

c. 50°   **d.** 25°

TRY THESE EXERCISES

Identify the following for ⊙L.

1. a radius   **2.** a chord

3. a major arc   **4.** a minor arc

5. $m\widehat{MPN}$ **6.** $m\widehat{PN}$

7. $m\widehat{POM}$ **8.** $m\widehat{MON}$

9. an inscribed angle   **10.** m∠MNO

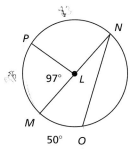

PRACTICE EXERCISES • For Extra Practice, see page 602.

In the figure, m∠KHJ = 60° and m∠JHI = 45°. Find each measure.

11. $m\widehat{JK}$ **12.** $m\widehat{IJ}$

13. $m\widehat{IK}$ **14.** $m\widehat{ILK}$

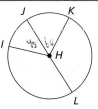

In ⊙M, $\overline{NO}$ and $\overline{PQ}$ are diameters. Identify the following.

15. four semicircles   **16.** four central angles

17. $m\widehat{NQ}$ **18.** $m\widehat{PNQ}$

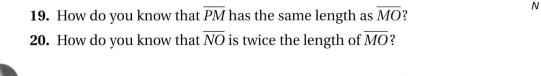

19. How do you know that $\overline{PM}$ has the same length as $\overline{MO}$?

20. How do you know that $\overline{NO}$ is twice the length of $\overline{MO}$?

21. WRITING MATH Discuss how an arc of a circle is similar to a line segment and how it is different.

22. FOOD SERVICE A pizza company cuts each pizza into six approximately-equal sized slices. Find the measure of the arc formed by the crust of a piece of pizza. Does this measure depend on the size of the pizza? Explain.

23. ART An artist's design for a floor display is shown. What is the measure of a central angle formed by two consecutive outer points on the large star?

Classify each statement as *true* or *false*. If the statement is false, explain why.

24. A chord of a circle contains exactly two points on the circle.

25. The measure of a radius is always one-half the measure of a chord.

26. If a central angle and an inscribed angle are congruent, their arcs are congruent.

GEOMETRY SOFTWARE Use geometry software to draw a circle and construct two chords to form an inscribed angle.

27. Use the measure tools to verify the relationship between the intercepted arc and the measure of the inscribed angle.

28. Do you think the measure of the inscribed angle will increase if the size of the circle is increased? Explain why or why not, and use the software to check.

▰ EXTENDED PRACTICE EXERCISES

29. NAVIGATION A *compass rose* is used to determine the heading, or direction, of a boat at sea. Find the measure of the minor arc formed by the headings 212° SSW and 12° NNE.

30. CRITICAL THINKING If you double the measure of a minor arc, will the measure of the related major arc be doubled? Will the measure of the central angle be doubled? Explain.

31. RECREATION The Jones family is having a circular pool installed that has a radius of 14 ft. The pool will have a wooden deck hanging over the edge as shown in the figure. What is the length of the arc covered by the deck? Round your answer to the nearest tenth. (Hint: Find the pool's circumference and set up a proportion.)

32. CHAPTER INVESTIGATION Draw the circle formed by points *N*, *S*, *E* and *W*. Draw four radii to bisect each of the four regions of your compass. Label the endpoints of the radii *NE*, *SE*, *SW* and *NW* as appropriate to complete your compass.

▰ MIXED REVIEW EXERCISES

Use the spinner shown and a number cube. (Lessons 4-4 and 4-5)

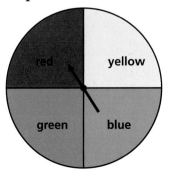

33. *P*(spinning blue, then red)

34. *P*(spinning green and rolling an even number)

35. *P*(spinning blue and rolling an odd number)

36. *P*(spinning red or green and rolling a prime number)

37. *P*(spinning blue or green and rolling a 6)

38. *P*(spinning yellow and rolling a number less than 7)

Review and Practice Your Skills

PRACTICE ■ LESSON 5-7

Sketch each polygon.

1. a regular quadrilateral

2. a concave pentagon

3. a regular pentagon

4. a concave hexagon

For each regular polygon, find the sum of the interior angles and the measure of an interior angle.

5. 12-gon

6. heptagon

7. 40-gon

8. octagon

Classify each polygon by its number of sides. Tell whether it is convex or concave, regular or not regular.

9.

10.

11.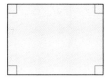

12.

PRACTICE ■ LESSON 5-8

Match the following for circle *P*.

13. a minor arc   **a.** $\angle QPR$

14. a chord   **b.** $\overset{\frown}{QS}$

15. a radius   **c.** $\overset{\frown}{RSQ}$

16. an inscribed angle   **d.** $\overline{QS}$

17. a major arc   **e.** $\angle QSR$

18. a central angle   **f.** $\overline{SR}$

19. a diameter   **g.** $\overline{QP}$

In circle *K*, find each measure.

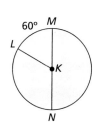

20. $m\angle LKM$

21. $m\,\overset{\frown}{LNM}$

22. $m\,\overset{\frown}{MLN}$

23. $m\,\overset{\frown}{LN}$

24. $m\angle LKN$

25. $m\angle MKN$

Fill in the blanks. In the figure, $\overleftrightarrow{AB}$ bisects $\overline{CD}$. (Lesson 5-1)

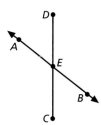

26. ___?___ ≅ ___?___

27. The intersection of two lines is a ___?___, and the intersection of two planes is a ___?___.

In the figure, $\ell \parallel m$, classify each pair of angles as congruent, supplementary, or neither. (Lessons 5-3)

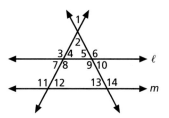

28. ∠1 and ∠2

29. ∠7 and ∠12

30. ∠3 and ∠7

31. ∠4 and ∠12

32. ∠11 and ∠14

33. ∠11 and ∠7

34. ∠6 and ∠9

35. ∠2 and ∠5

State whether it is possible to have a triangle with sides of the given lengths. (Lesson 5-4)

36. 1, 2, 3

37. 7, 8, 11

38. $\dfrac{1}{2}, \dfrac{2}{3}, \dfrac{3}{4}$

39. 14, 17, 25

40. 2, 4, 6

41. 37, 49, 90

Find the value of x in each figure. (Lesson 5-4)

42.

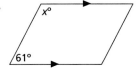

43.

44.

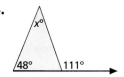

State whether each pair of triangles is congruent by SSS, SAS, or ASA. (Lesson 5-5)

45.

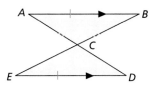

46.

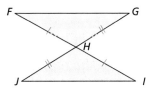

47.

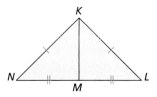

Classify each statement as *true* or *false*. (Lessons 5-2, 5-7 and 5-8)

48. A diameter in a circle is a chord.

49. A radius is a chord.

50. An inscribed angle is equal to the measure of the arc it intercepts.

51. The diagonals of a rectangle are congruent.

52. A rhombus is a square.

53. A square is a rhombus.

54. Two angles whose measures have a sum of 180° are called complementary angles.

Problem Solving Skills: Circle Graphs

An effective problem solving strategy is to **use a picture, diagram or model**. A **circle graph** is one way to display data to make comparisons. The whole circle represents 100% of the data. Each part, or percent of the data, is represented by a **sector**. Since there are 360° in a circle, the sum of the measures of the central angles for the sectors is 360°.

Problem Solving Strategies

Guess and check

Look for a pattern

Solve a simpler problem

Make a table, chart or list

✔ Use a picture, diagram or model

Act it out

Work backwards

Eliminate possibilities

Use an equation or formula

Problem

ENTERTAINMENT A recent poll surveyed students about their favorite form of entertainment. Of the 950 students surveyed, 229 chose movies, 209 said music, 176 liked TV, 155 said sports, 73 said video games, 61 chose the Internet, and 47 said dancing. Make a circle graph of the data.

Solve the Problem

Step 1 Write each response as a percent of the entire survey, 950.

Step 2 Find the number of degrees in the central angle by multiplying by 360°.

Movies	$229 \div 950 \approx 24\%$	$0.24 \cdot 360° \approx 86°$
Music	$209 \div 950 = 22\%$	$0.22 \cdot 360° \approx 79°$
Watching TV	$176 \div 950 \approx 19\%$	$0.19 \cdot 360° \approx 68°$
Sports	$155 \div 950 \approx 16\%$	$0.16 \cdot 360° \approx 58°$
Video games	$73 \div 950 \approx 8\%$	$0.08 \cdot 360° \approx 29°$
Internet	$61 \div 950 \approx 6\%$	$0.06 \cdot 360° \approx 22°$
Dancing	$47 \div 950 \approx 5\%$	$0.05 \cdot 360° = 18°$

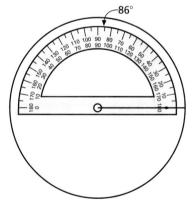

Step 3 Draw a circle with a compass. To construct the central angles for the sectors, draw any radius. Place a protractor along the radius with 0 at the circle's center. Measure and mark 86°. Draw another radius from the center to this point.

Step 4 Place the protractor along the new radius. Measure and mark 79°. Draw another radius to this point. Draw the rest of the central angles this way. Label each sector and title the graph.

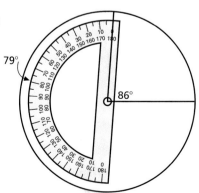

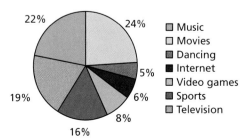

Favorite Forms of Entertainment

- Music
- Movies
- Dancing
- Internet
- Video games
- Sports
- Television

Make a circle graph for each set of data. Write a title for your graph.

Five-step Plan
1. Read
2. Plan
3. Solve
4. Answer
5. Check

1. Travel to Work in U.S.

Means of travel	%
Drive alone	76
Carpool	12
Bus or train	5
Walk only	3
Work at home	3
Other	1

2. The Earth's Surface

Element	%
Oxygen	47
Silicon	28
Aluminum	8
Iron	5
Calcium	4
Potassium	3
Sodium	3
Magnesium	2

3. Monthly Budget

Expenses	%
Car	15
Travel	19
Food	26
Clothing	23
Savings	12
Miscellaneous	5

Make a circle for each set of data. Begin by finding each sector's percent.

4. Cruise Line Destinations

Destination	Number of cruises
Acapulco, Mexico	47
Anchorage, Alaska	28
Honolulu, Hawaii	53
Vancouver, Canada	14
Melbourne, Australia	18

5. University Degrees

Major	Number of students
Education	822
Computers	650
Engineering	510
Psychology	423
Math	119
Law	100

6. INDUSTRY A car manufacturer produces 735 cars in one month. Of these, 215 are economy cars, 190 sports cars, 145 minivans, 130 luxury cars and 55 full-size vans. Make a circle graph representing the information.

7. NAVIGATION In Pierre Mondrin's career in exploration, he sailed to Antarctica 12 times, the South China Sea 9 times, Alaska 8 times and the Mediterranean Sea 2 times. Make a circle graph representing Pierre's favorite places to explore.

8. WRITING MATH Describe the advantages of using a circle graph to represent data. How does it make comparing information easier?

9. DATA FILE Refer to the data on water usage for a family on page 567. Make a circle graph of the information presented. Round each percent to the nearest tenth.

Simplify. (Lesson 2-6)

10. $14g - 5(g - 7)$

11. $\frac{3}{4}(d + 21)$

12. $\frac{3}{5}(15j - 3)$

13. $-3(n - 1) + 6(4 - n)$

14. $4(2.5k + 1.7) - 3.6k$

15. $3.5(6 + 6a) + 0.5(4 + 9a)$

16. Jamesha has measured phone service. She pays $0.06/call for the first 60 calls and $0.05/call for any additional calls. Her base fee is $5.40. Write and simplify a variable expression for the total cost of 150 calls.

Chapter 5 Review

VOCABULARY

Choose a word from the list to complete each statement.

1. Two lines that intersect to form right angles are called ___?___.

2. A(n) ___?___ is a quadrilateral with two pairs of parallel sides.

3. A(n) ___?___ is a line that intersects two other coplanar lines in different points.

4. A(n) ___?___ is formed by two rays that have a common endpoint.

5. A segment with both endpoints on a circle is called a(n) ___?___.

6. A triangle with all three sides congruent is called ___?___.

7. A ray that divides an angle into two congruent adjacent angles is the ___?___ of the angle.

8. If the line containing a side of a polygon contains points in the interior of the polygon, the polygon is called ___?___.

9. A(n) ___?___ is a statement that is assumed to be true.

10. If two adjacent angles form a right angle, the angles are called ___?___.

a.	angle
b.	bisector
c.	chord
d.	complementary
e.	concave
f.	equilateral
g.	parallelogram
h.	perpendicular
i.	postulate
j.	skew
k.	supplementary
l.	transversal

LESSON 5-1 ▶ Elements of Geometry, p. 192

▶ The **midpoint** of a segment is the point that divides the segment into two **congruent segments**. A **bisector** of a segment is any line, segment, ray, or plane that intersects the segment at its midpoint.

Draw a figure to illustrate each situation.

11. Points C, D, and E are noncollinear.

12. Points P, Q, R and S are noncoplanar.

13. Line t intersects plane $\mathcal{H}$ at point R.

14. Planes $\mathcal{P}$ and $\mathcal{Q}$ intersect at line m.

15. You can think of your classroom as a model of six planes: the ceiling, the floor, and the four walls. Find two planes that do *not* intersect.

LESSON 5-2 ▶ Angles and Perpendicular Lines, p. 196

▶ Two angles whose measures have a sum of 90° are called **complementary angles**. Two angles whose measures have a sum of 180° are called **supplementary angles**.

Find the measure of the complement and supplement of each angle.

16. $m\angle A = 34°$

17. $m\angle B = 49°$

18. $m\angle C = 87°$

19. $m\angle D = 12°$

20. If $\overline{BD}$ is the bisector of $\angle ABC$, what is the value of x?

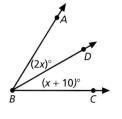

LESSON 5-3 ◣ Parallel Lines and Transversals, p. 202

▶ When two parallel lines are cut by a **transversal**, pairs of **corresponding angles**, **alternate interior angles**, and **alternate exterior angles** formed are congruent.

Refer to the figure to name two pairs of each type of angle.

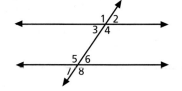

21. corresponding angles   **22.** alternate interior angles

23. alternate exterior angles   **24.** same-side interior angles

25. Refer to the figure. If $m\angle 1 = 120°$ and $m\angle 8 = 120°$, are the lines cut by the transversal parallel? Justify your answer.

LESSON 5-4 ◣ Properties of Triangles, p. 206

▶ The sum of the **angles** of a **triangle** is 180°. The sum of the lengths of any two **sides** of a triangle is greater than the length of the third side.

State whether it is possible to have a triangle with sides of the given lengths.

26. 13, 19, 8   **27.** 9, 5, 4   **28.** 11, 22, 33

Find the unknown angle measures in each figure.

29.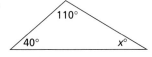

30.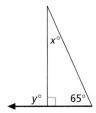

31.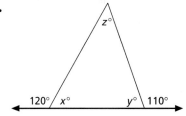

LESSON 5-5 ◣ Congruent Triangles, p. 212

▶ To show that two triangles are **congruent** you can use the SSS, SAS, or ASA postulates.

State whether $\triangle ABC \cong \triangle DEF$. If so, name the appropriate postulate.

32.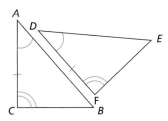

33.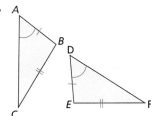

34.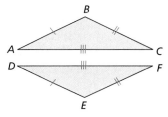

LESSON 5-6 ◣ Quadrilaterals and Parallelograms, p. 216

▶ A **quadrilateral** is a closed plane figure that has four sides.

▶ **Parallelograms** have pairs of congruent opposite sides and opposite angles.

Use $\square ABCD$ to find each unknown measure.

35. $\overline{AD}$ **36.** $m\angle A$ **37.** $m\angle D$

Tell whether each statement is *true* or *false*.

38. All squares are parallelograms.

39. All parallelograms are rectangles.

40. A trapezoid can never be a parallelogram.

LESSON 5-7 ◣ Diagonals and Angles of Polygons, p. 222

▶ The sum of the angle measures of a polygon with n sides is $(n - 2)180°$.

▶ The measure of each interior angle of a regular n-gon is $\frac{(n - 2)180°}{n}$.

41. Find the sum of the interior angle measures of an 11-gon.

42. Find the sum of the interior angles of a polygon with 15 sides.

43. Find the measure of each interior angle of a regular octagon.

LESSON 5-8 ◣ Properties of Circles, p. 226

▶ An **arc** is an unbroken part of a **circle**. A **central angle** has its vertex at the center of a circle. An **inscribed angle** has its vertex on the circle.

Identify the following for ⊙N.

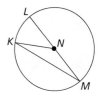

44. a central angle

45. an inscribed angle

46. major arc

47. minor arc

48. a radius

49. a chord

LESSON 5-9 ◣ Problem Solving Skills: Circle Graphs, p. 232

▶ In a circle graph the whole circle represents 100% of the data.

▶ Each percent of the data is represented by a sector of the circle.

Make a circle graph of each set of data.

50. Possible Points on a Test

Section	Number of points
True/False	48
Fill-in	25
Essay	60
Matching	50
Multiple choice	17

51. Ocean Surface Areas

Ocean	Area (square miles)
Pacific	64,186,300
Atlantic	33,420,000
Indian	28,350,500
Arctic	5,105,700

CHAPTER INVESTIGATION

EXTENSION Notice the objects and landmarks that surround your compass. Create a map of the area. Include at least four objects or landmarks, a scale and a directional key (N-S-E-W). Present your map to the class. Note geometric properties, such as collinear and noncollinear points, polygons and parallel lines.

Chapter 5 Assessment

Refer to the figure.

1. Name three points that determine plane $\mathcal{M}$.
2. Name the intersection of planes $\mathcal{M}$ and $\mathcal{N}$.
3. Name three lines that lie in plane $\mathcal{N}$.

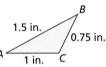

For exercises 4–8, use the figure shown to name each.

4. a pair of perpendicular segments
5. two adjacent complementary angles
6. two pairs of vertical angles
7. two pairs of adjacent supplementary angles
8. the measure of the complement of $\angle CDE$

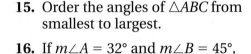

In the figure, $k \parallel m$ and $m\angle 4 = 125°$. Find each measure.

9. $m\angle 1$ 10. $m\angle 2$ 11. $m\angle 3$

12. $m\angle 5$ 13. $m\angle 6$ 14. $m\angle 7$

15. Order the angles of $\triangle ABC$ from smallest to largest.

16. If $m\angle A = 32°$ and $m\angle B = 45°$, what is the measure of $\angle C$?

17. If $\triangle RST \cong \triangle ABC$, name all the corresponding, congruent parts.

18. State whether the pair of triangles is congruent by SSS, SAS, or ASA.

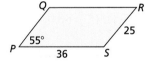

Find each unknown measure of the parallelogram.

19. $\overline{QR}$ 20. $\overline{PQ}$

21. $\angle S$ 22. $\angle R$

23. Sketch a concave pentagon.

24. Find the sum of the interior angles of a 14-gon.

25. Find the measure of an interior angle of a regular 36-gon.

Refer to the figure to name the following.

26. a diameter   27. a chord

28. a minor arc   29. an inscribed angle

30. Find the approximate number of degrees of each central angle of a circle graph that could represent each category in the table shown.

Craven Junior High Students

Grade	Percent
Sixth	18%
Seventh	25%
Eighth	27%
Ninth	30%

Standardized Test Practice

Part 1 | Multiple Choice

Record your answers on the answer sheet provided by your teacher or on a sheet of paper.

1. According to the box-and-whisker plot, between which two numbers will you find half of the data? (Lesson 1-6)

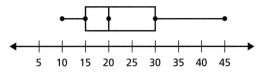

   10 and 15   (B) 10 and 30
   (C) 15 and 30   (D) 20 and 30

2. The student council of South High School is planning to sell school T-shirts. They hire a company to make the T-shirts. The company charges $40 to prepare the design and $8 for each T-shirt. If t represents the number of T-shirts, which expression represents the total cost to make the T-shirts? (Lesson 2-3)

   (A) $8t + 40$ (B) $40t + 8$
   (C) $40 + 8 + t$ (D) $t(40 + 8)$

3. Which expression is equivalent to $\dfrac{-45t + 60s}{-15}$? (Lesson 2-5)

   (A) $-3t + 4s$ (B) $3t - 4s$
   (C) $-3t + 60s$ (D) $3t - 60s$

4. Which equation has a solution that is equivalent to the solution of $4x + 6 = -18$? (Lesson 3-4)

   (A) $4x = -24$ (B) $4x = -12$
   (C) $x + 6 = -4.5$ (D) $x + 6 = 4.5$

5. Which inequality has the solution represented on the graph? (Lesson 3-7)

   (A) $-2x - 4 < -2$
   (B) $-2x - 4 > -2$
   (C) $-2x + 4 < 2$
   (D) $-2x + 4 > 2$

6. Brian has 10 rock CDs, 5 jazz CDs, and 8 country CDs. If Brian picks one CD at random to play, what is the probability that he will choose a CD that is *not* jazz? (Lesson 4-3)

   (A) $\dfrac{5}{23}$ (B) $\dfrac{5}{18}$
   (C) $\dfrac{5}{15}$ (D) $\dfrac{18}{23}$

7. Tomaso drops a penny in a pond, and then he drops a nickel in the pond. What is the probability that both coins land with tails showing? (Lesson 4-4)

   (A) $\dfrac{1}{8}$ (B) $\dfrac{1}{4}$ (C) $\dfrac{1}{2}$ (D) $\dfrac{3}{4}$

8. An ice cream store has 20 flavors. Abigail wants to buy three different flavors of ice cream to take home for a birthday party. Which expression can be used to determine how many ways she can choose the flavors of ice cream? (Lesson 4-7)

   (A) $\dfrac{17!}{20!\,3!}$ (B) $\dfrac{20!}{17!\,3!}$
   (C) $\dfrac{17!}{3!}$ (D) $\dfrac{20!}{17!}$

9. Point C is the midpoint of $\overline{AB}$. Point D is the midpoint of $\overline{AC}$. Which of the following segments is the longest? (Lesson 5-1)

   (A) $\overline{AC}$ (B) $\overline{AD}$ (C) $\overline{BD}$ (D) $\overline{CD}$

10. Which is *not* a postulate that can be used to prove two triangles are congruent? (Lesson 5-5)

   (A) Angle-Angle-Side
   (B) Angle-Side-Angle
   (C) Side-Angle-Side
   (D) Side-Side-Side

Test-Taking Tip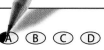

Question 6
Read each question carefully. Be sure you understand what the question asks. Look for words like *not*, *estimate*, and *approximately*.

Preparing for the Standardized Tests
For test-taking strategies and more
practice, see pages 627–644.

Part 2 Short Response/Grid In

**Record your answers on the answer sheet
provided by your teacher or on a sheet of paper.**

11. Find $R + S$. (Lesson 1-8)

$$R = \begin{bmatrix} -2 & 5 & 0 \\ -8 & -1 & 4 \end{bmatrix} \quad S = \begin{bmatrix} -4 & 7 & -5 \\ 1 & -3 & 2 \end{bmatrix}$$

12. Simplify $4(q + 2p) - (8q + 3p)$.
(Lesson 2-6)

13. What is the value of $(-1)^{21}$? (Lesson 2-8)

14. The day after a hurricane, the barometric
pressure in a coastal town had risen to 29.7 in.
of mercury, which is 2.9 in. of mercury higher
than the pressure when the eye of the
hurricane passed over. Write an equation
to represent the situation.
(Lesson 3-2)

15. When 2000 lb of paper are recycled, 17 trees
are saved. How many trees would be saved if
5000 lb of paper are recycled? (Lesson 3-5)

16. Solve $\sqrt{4x + 1} = 5$. (Lesson 3-8)

17. A poll is taken to determine whether
registered voters plan to vote for the school
levy. The results of the poll are given below.
What is the experimental probability that a
registered voter is undecided?
(Lesson 4-1)

Response	Yes	No	Undecided
Men	83	70	92
Women	103	52	100

18. There are 17 floats in the Founder's Day
Parade. If the queen's float must be last, how
many ways can the organizer pick the first
two floats in the parade? (Lesson 4-6)

19. Find the value of y in the figure.
(Lesson 5-4)

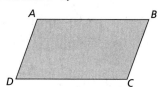

20. Suppose $\triangle DEF \cong \triangle JKL$. Name the congruent
sides of the two triangles. (Lesson 5-5)

21. In quadrilateral $MNOP$, $m\angle M = 73°$,
$m\angle N = 101°$, and $m\angle O = 84°$. Find the
measure of $\angle P$. (Lesson 5-6)

22. What is the sum of the interior angles of a
convex hexagon? (Lesson 5-7)

23. $\overline{AD}$ is a diameter of circle D. If the measure of
$\angle ADC$ equals 45°, what is the measure of $\overset{\frown}{BC}$?
(Lesson 5-8)

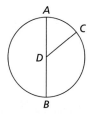

Part 3 Extended Response

**Record your answers on a sheet of paper.
Show your work.**

24. Draw a set of parallel lines and a transversal
where the same side interior angles are
congruent. Explain why the lines are parallel
and why the same side interior angles are
congruent. (Lesson 5-2)

25. Write six true statements about parallelogram
$ABCD$. (Lesson 5-6)

26. The results of a student survey about favorite
foods are as follows: 23 chose pizza, 14 chose
hamburgers, 18 chose spaghetti, and 9 chose
fish. Make a circle graph that reflects the data.
Label each sector with the favorite food, the
percentage of students the sector represents,
and the measure of the central angle.
(Lesson 5-9)

Math Online
mathmatters2.com/standardized_test

Graphing Functions

THEME: Business

Have you ever thought of someday owning your own business and being your own boss? Business owners must understand how to solve equations that represent relationships between factors such as cost and profit, cost and reliability, and time and quantity. Businesses often use graphs to visually represent these relationships. As a business owner, an analysis of financial information each year will help you plan for future income and spending.

- One responsibility of **music store owners** (page 253) is to display product on the store floor so that customers are encouraged to make purchases.

- **Restaurateurs** (page 273) use previous sales and inventory costs to place orders for supplies, determine staffing needs, and estimate weekly profits.

Math Online

mathmatters2.com/chapter_theme

Sales of Cars in the U.S., 1995-2001

Year	Sales (thousands)
1995	8687
1996	8527
1997	8273
1998	8142
1999	8697
2000	8852
2001	8422

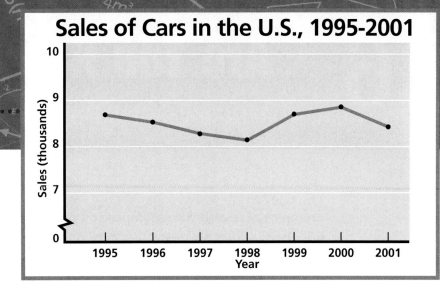

Sales of Cars in the U.S., 1995-2001

Data Activity: Sales of Cars in the U.S., 1995 -2001

Use the table and graph for Questions 1–5.

1. Describe the change in car sales from 1995-2001.

2. Did you look to the table or graph first to answer Question 1? Why?

3. Which year had the greatest sales? Which year had the least sales?

4. Use the graph to predict the sales for the year 2006, 2007, and 2008.

5. If possible, check the accuracy of your predictions by using current data available.

CHAPTER INVESTIGATION

Most companies, especially those that have stockholders or investors, compile an annual report of profit and loss information. This report includes financial statements, graphs, and other data for the past year, as well as any significant plans for the future. It is important to understand the language used in a business plan. Two terms frequently used are *sales revenue* and *operating costs*. *Sales revenue* is the amount of money received for products or services sold to customers. Sales revenue can also be referred to as income. *Operating cost* is the amount of money that is needed to run the business. This amount includes all monies used to produce, market, and sell a product or service.

Working Together

Use the library or the Internet to get an annual report of a company. Locate information in the report about the company's operating costs, sales revenue and operating profit/loss. Draw graphs that illustrate the finances of your selected company. Use the Chapter Investigation icons to guide your group through the analysis of the annual report.

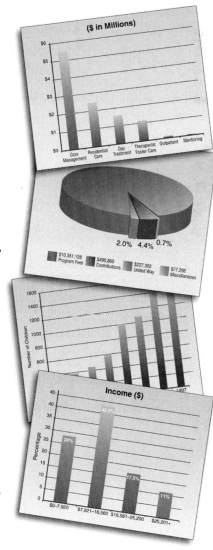

Are You Ready?

Refresh Your Math Skills for Chapter 6

The skills on these two pages are ones you have already learned. Use the examples to refresh your memory and complete the exercises. For additional practice on these and more prerequisite skills, see pages 576–584.

THE COORDINATE PLANE

In this chapter you will learn to graph equations of several types. It is helpful to review the basics of graphing on a coordinate plane.

Example

Give the coordinates (x, y) of point L and two points in Quadrant I.

The coordinates of point L are $(-6, -4)$. Points E and F are both in Quadrant I.

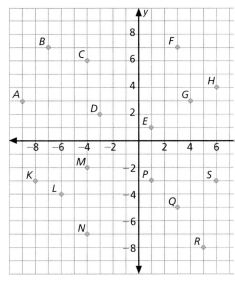

Give the coordinates of the indicated points.

1. point A

2. two points in Quadrant II

3. point S

4. two points in Quadrant III

Name each point.

5. the point at $(-4, 6)$

6. the point at $(3, -5)$

7. the point at $(3, 7)$

8. the point at $(-4, -2)$

EVALUATE EXPRESSIONS

To evaluate an expression, substitute the given value for each variable and then follow the order of operations.

Example

Evaluate the expression $3x + 5$ when $x = 4$. $\quad 3x + 5$

$$3 \cdot 4 + 5 = 12 + 5 = 17$$

Evaluate each expression for the given value of the variable.

9. $1 + 7a$, when $a = 2$

10. $-4t$, when $t = -10$

11. $\dfrac{x - 8}{2}$, when $x = -4$

12. $\dfrac{-10 + y}{-4}$, when $y = -6$

13. $2(p + 9)$, when $p = 12$

14. $3(r - 20)$, when $r = 15$

15. $x^2 + 14$, when $x = 3$

16. $16 - d^2$, when $d = 5$

17. $8y - y^2$, when $y = 2$

18. $\dfrac{24 + b}{3b}$, when $b = 3$

19. $\dfrac{4 - x}{x - 4}$, when $x = -1$

20. $k(2k - 7)$, when $k = -5$

21. $\dfrac{7v - 6}{-v}$, when $v = 2$

22. $c^2 - 4c$, when $c = 9$

23. $4x - x^2 + 21$, when $x =$

Use Multiplication and Division to Solve Equations

Recall that to solve an equation, you must isolate the variable on one side of the equal symbol. Use inverse operations to make 1 the variable. If the operation on the variable is multiplication, then divide to solve the equation. If the operation on the variable is division, then multiply to solve the equation.

Example

Solve each equation.

$$7y = -35$$

$$\frac{7y}{7} = \frac{-35}{7} \quad \text{Divide both sides by 7.}$$

$$y = -5$$

$$\frac{x}{4} = 5$$

$$4 \cdot \frac{x}{4} = 5 \cdot 4 \quad \text{Multiply both sides by 4.}$$

$$x = 20$$

Solve each equation. Check the solution.

24. $-5x = 5$

25. $32b = 192$

26. $\frac{a}{7} = 2$

27. $\frac{3}{5}t = 6$

28. $-24h = 120$

29. $17x = 51$

30. $\frac{z}{12} = -4$

31. $8 = -\frac{2}{3}x$

32. $18k = 90$

33. $96y = 288$

34. $\frac{d}{-13} = 5$

35. $\frac{2}{9}x = -6$

36. $-43t = 86$

37. $\frac{r}{9} = -11$

38. $63c = -7$

Patterns and Sequences

In this chapter, you will learn how to write a function rule from a table of x- and y-values. To do this, it will be important to recognize the pattern in the x-values and y-values.

Example

State the next three numbers in the sequence. $\qquad$ 1, 2, 4, 8, 16, . . .

To get the next number in the sequence, you must multiply each number by 2.

The next three numbers in the sequence are 32, 64, and 128.

Explain each pattern and state the next three numbers in the sequence.

39. 6, 12, 18, 24, 30, . . .

40. 1, 4, 9, 16, 25, . . .

41. $-6, -3, 0, 3, 6, 9, . . .$

42. 5, 6, 9, 10, 13, 14, . . .

43. $1, \frac{1}{2}, \frac{1}{4}, \frac{1}{8}, \frac{1}{16}, . . .$

44. 4, 12, 36, 108, 324, . . .

45. 1, 3, 6, 10, 15, 21, . . .

46. 0.2, 0.4, 0.8, 1.4, 2.2, 3.2, . . .

47. 1, 1, 2, 3, 5, 8, 13, . . .

48. 4, 7, 13, 25, 49, . . .

6-1 Distance in the Coordinate Plane

Goals
- Use the distance formula to find the distance between two points.
- Use the midpoint formula.

Applications Geography, Market research, Community service, Architecture

GEOGRAPHY Directions on the map of Four-World Fun Park are provided by first giving the units to move east or west and then the units to move north or south.

1. Give directions from the main gate to Tiger Trek.

2. Give directions from the main gate to Glacier Climb.

3. If you walk east 2 units from the main gate and then south 4 units, where are you?

4. Between which two regions do you travel if you walk directly from the restaurant to Lion's Lair?

Four-World Fun Park

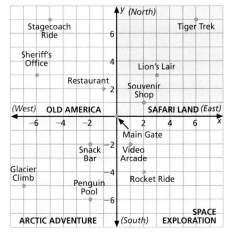

BUILD UNDERSTANDING

On the **coordinate plane**, two number lines are drawn perpendicular to each other and form four **quadrants**. The **x-axis** is the horizontal number line, and the **y-axis** is the vertical number line. Points in the plane are **ordered pairs**. It is important that the order of the coordinates is stated as (x, y). The point $(0, 0)$ is the **origin**, which is where the x-axis and the y-axis intersect.

When two points lie on a line that is parallel to the x-axis or y-axis, find the distance between the points by calculating the absolute value of the difference between the x-coordinates or the y-coordinates, whichever is appropriate.

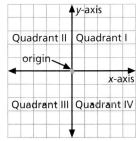

Example 1

Use the graph to calculate the length of each line segment.

a. BC **b.** PQ

Solution

a. Points $B(-7, 2)$ and $C(4, 2)$ are on a line parallel to the x-axis. So use the x-coordinates.

$$|-7 - 4| = |-11| = 11$$
The distance between B and C is 11 units.

b. Points $P(7, 6)$ and $Q(7, -2)$ are on a line parallel to the y-axis. So use the y-coordinates.

$$|6 - (-2)| = |8| = 8$$
The distance between P and Q is 8 units.

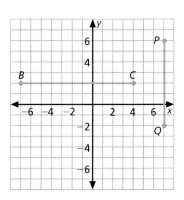

To find the distance between two points that do not lie on a line parallel to the x-axis or the y-axis, use the coordinates of the ordered pairs in the distance formula.

Distance Formula	$d = \sqrt{(x_2 - x_1)^2 + (y_2 - y_1)^2}$ for any points (x_1, y_1) and (x_2, y_2).

COncepts in MOtion
Interactive Lab
mathmatters2.com

Example 2

Find the distance between $M(4, -6)$ and $N(-2, 3)$. Round to the nearest tenth.

Solution

$$MN = \sqrt{(x_2 - x_1)^2 + (y_2 - y_1)^2}$$
$$MN = \sqrt{(-2 - 4)^2 + (3 - (-6))^2} \quad x_2 = -2, x_1 = 4, y_2 = 3, y_1 = -6$$
$$= \sqrt{(-6)^2 + (9)^2} \quad \text{Simplify.}$$
$$= \sqrt{36 + 81}$$
$$= \sqrt{117} \approx 10.817 \quad \text{Approximate.}$$

The distance between M and N is approximately 10.8 units.

Check Understanding

Does it make a difference in the distance formula which ordered pair is used for (x_1, y_1)? What order is important when you substitute the coordinates?

The midpoint of a segment is the point halfway between the endpoints of that segment. To find the coordinates of a midpoint, use the midpoint formula.

Midpoint Formula	$M = \left(\dfrac{x_1 + x_2}{2}, \dfrac{y_1 + y_2}{2} \right)$ for any points (x_1, y_1) and (x_2, y_2).

Example 3

MARKET RESEARCH In marketing class, Chris tracks the constant rise in gasoline prices. He uses a coordinate plane to represent his data. Find the midpoint of his results.

Solution

Use the midpoint formula to find midpoint M of $\overline{AB}$. The endpoints of $\overline{AB}$ are (1, 1) and (9, 5).

$$M = \left(\frac{1 + 9}{2}, \frac{1 + 5}{2} \right)$$
$$= \left(\frac{10}{2}, \frac{6}{2} \right)$$
$$= (5, 3)$$

So $M(5, 3)$ is the midpoint of $\overline{AB}$.

Gasoline Prices

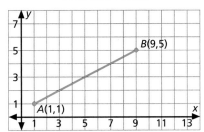

The distance and midpoint formulas can be used with geometric figures that are drawn in the coordinate plane.

Example 4

 Personal Tutor at mathmatters2.com

The endpoints of a diameter of ⊙P are $(-1, 10)$ and $(4, 6)$.

a. Find the length of the diameter.   **b.** Find the center of the circle.

Solution

a. Use the distance formula.

$$d = \sqrt{(x_2 - x_1)^2 + (y_2 - y_1)^2}$$

$$d = \sqrt{(-1 - 4)^2 + (10 - 6)^2}$$

$$d = \sqrt{(-5)^2 + (4)^2}$$

$$d = \sqrt{25 + 16}$$

$$= \sqrt{41} \approx 6.403$$

The diameter of ⊙P is approximately 6.4 units.

b. Use the midpoint formula.

$$M = \left(\frac{x_1 + x_2}{2}, \frac{y_1 + y_2}{2}\right)$$

$$M = \left(\frac{-1 + 4}{2}, \frac{10 + 6}{2}\right)$$

$$M = \left(\frac{3}{2}, \frac{16}{2}\right)$$

$$M = \left(\frac{3}{2}, 8\right)$$

The center of ⊙P is $(1.5, 8)$.

TRY THESE EXERCISES

Use the graph to calculate the length of each segment.

1. $\overline{AB}$ **2.** $\overline{PQ}$

Find the distance between the points. Round to the nearest tenth.

3. $D(-8, -3)$, $E(4, 2)$ **4.** $X(-5, 2)$, $Y(-2, 5)$

5–6. Find the midpoint of the segments in Exercises 3 and 4.

7. A circle has a diameter with endpoints $(4, 7)$ and $(10, 11)$. Find the length of the diameter, length of the radius and coordinates of the center.

8. The vertices of a triangle are $A(2, 3)$, $B(5, 7)$ and $C(8, 3)$. Find the length of each side. What type of triangle is $\triangle ABC$?

9. Name the midpoints of each side of the triangle in Exercise 8.

PRACTICE EXERCISES • For Extra Practice, see page 603.

Use the graph to calculate the length of each segment.

10. $\overline{AB}$ **11.** $\overline{BD}$ **12.** $\overline{DE}$ **13.** $\overline{CF}$

14–17. Find the midpoints of the segments in Exercises 10–13.

18. YOU MAKE THE CALL Su says to double the length of a line segment on a coordinate plane, double the coordinates of each endpoint. Do you agree or disagree? Explain.

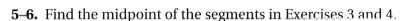

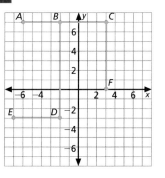

Find the distance between the points. Round to the nearest tenth.

19. $A(6, 2)$, $B(5, -1)$

20. $C(-2, -1)$, $D(-2, 9)$

21. $E(-8, 3)$, $F(4, -2)$

22. $G\left(\dfrac{2}{3}, -5\right)$, $H(0, 6)$

23. $A(2, 6)$, $B(-1, 5)$

24. $C(-1, -2)$, $D(9, -2)$

25. $E(3, 3)$, $F(4, 5)$

26. $G(2, 8)$, $H(10, -6)$

27. $X(0, 0)$, $Z\left(\dfrac{-1}{2}, \dfrac{6}{7}\right)$

Use the given endpoints of each circle's diameter. Find each circle's center and the lengths of its diameter and radius.

28. $(0, 0)$, $(0, -8)$

29. $(10, -2)$, $(8, -8)$

30. $(1, 1)$, $(6, 6)$

31. $(-2, -4)$, $(10, 5)$

32. $(12, -3)$, $(0, -8)$

33. $(11, -2)$, $(1, -10)$

34. The vertices of a quadrilateral are $S(2, 3)$, $T(5, 7)$, $U(11, 8)$ and $V(8, 4)$. Find the length of each side. What type of quadrilateral is this?

35. COMMUNITY SERVICE The layout of the Fairlands Park can be represented by the coordinates $M(4, 1)$, $N(4, 22)$ and $O(17, 22)$. Assume that each square unit of the coordinate plane represents 1 mi². How much land does the clean-up committee need to cover if they clean the entire park?

36. WRITING MATH The midpoint formula uses the basic math skill of calculating an average. Compare and contrast the average of a set of numbers to the midpoint of a segment.

EXTENDED PRACTICE EXERCISES

37. Ramón is a meteorologist tracking a storm system over the western U.S. On a coordinate grid over a U.S. map, the straight path of the storm is from $(-190, 47)$ in Washington to $(-17, 31)$ in Nevada. Ramón divides the path into four equal sections. Name the coordinates of the three points that divide the storm into four equal sections.

38. WRITING MATH Three points are marked on the coordinate plane: $P(-3, -5)$, $Q(-2, 2)$, and $R(6, 4)$. Name the point that is closer to Q. Explain step-by-step how to determine if point P or R is closer to Q.

39. CRITICAL THINKING The center of $\odot S$ is at $(-2, 6)$. A diameter of that circle has an endpoint at $(4, -2)$. Find the other endpoint of that diameter.

40. CHAPTER INVESTIGATION Draw a diagram that shows only the first and fourth quadrants of a coordinate plane. Label four intervals on the horizontal axis as 1st Q, 2nd Q, 3rd Q and 4th Q. Q stands for Quarter. The first quarter of a year is January–March, the second quarter is April–June and so on. Label the vertical axis as dollar amounts.

MIXED REVIEW EXERCISES

41. DATA FILE Refer to the data on shopping day preferences on page 571. Make a circle graph representing the information. (Lesson 5-9)

Solve each equation. Check the solution. (Lesson 3-4)

42. $6b + 8 = -4$

43. $\dfrac{q}{3} + 6 = 5$

44. $12r - 7 = -20$

45. $12 - 5z = 17$

6-2 Slope of a Line

Goals
- Find the slope of a line.
- Identify horizontal and vertical lines.

Applications Business, Science, Transportation

Create a spreadsheet like the one shown.

1. Write ten sets of two ordered pairs. Choose a variety of values including zero, fractions and decimals for both the x- and y-elements. Include one set where the x-elements of both ordered pairs are equal. Include another set where the y-elements are equal.

	A	B	C	D	E	F	G
	x_1	y_1	x_2	y_2	change in x	change in y	ratio of change
1							
2							
3							

2. Enter your ordered pairs from Question 1 into the spreadsheet so that each set of ordered pairs fills Columns A through D.

3. Format Column E to find the difference of the data in Columns A and C.

4. Format Column F to find the difference of the data in Columns B and D.

5. Format Column G to divide Column F by Column E.

6. Does any data cause a formula error? Describe the cause of the errors.

◼ BUILD UNDERSTANDING

Often the meaning of a math term is related to the meaning of the word in everyday life. This is the case with the concept of slope as it refers to the steepness or slant of a line. The **slope** of a segment is the ratio of its change in vertical distance compared to its change in horizontal distance. The variable m is usually used to represent slope.

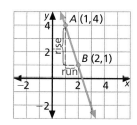

$$m = \frac{\text{rise}}{\text{run}} = \frac{\text{change in } y\text{-coordinates}}{\text{change in } x\text{-coordinates}}$$

Using a plane, count the units of rise and run between any two points on a line to find its slope. Conversely, given a point on a line and the slope of that line, you can locate other points on the line.

Example 1

Find the slope of $\overleftrightarrow{AB}$.

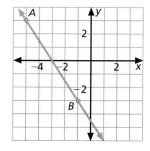

Solution

$$m = \frac{\text{rise}}{\text{run}} = \frac{\text{down } 6}{\text{right } 4} = \frac{-6}{4} = -\frac{3}{2}$$

The slope of $\overleftrightarrow{AB}$ is $-\frac{3}{2}$.

Check Understanding

Name another point on $\overleftrightarrow{AB}$ that is not shown in the graph for Example 1.

Example 2

Graph the line that passes through the point (2, 1) and has a slope of $\frac{5}{4}$.

Problem Solving Tip

A line with a positive slope line slants upward to the right. Its slope is a positive fraction, so think of it as:

$$\frac{\text{move up}}{\text{move right}} \text{ or } \frac{\text{move down}}{\text{move left}}$$

A line with a negative slope line slants downward to the right. Its slope is a negative fraction, so think of it as:

$$\frac{\text{move up}}{\text{move left}} \text{ or } \frac{\text{move down}}{\text{move right}}$$

Solution

First plot the point (2, 1). Since the slope is $\frac{5}{4}$, the change of rise over run equals a positive ratio.

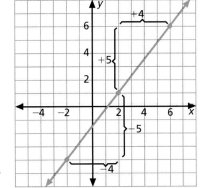

To locate another point, place your pencil at (2, 1). Rise 5 units up and run 4 units right.

To locate a third point on the line, you can rise 5 units and run 4 units again. Or go back to the point (2,1), and rise 5 units down and run 4 units left.

You can also find the slope of a line algebraically by using the following formula.

Slope of a Line	$m = \dfrac{y_2 - y_1}{x_2 - x_1}$, given two points (x_1, y_1) and (x_2, y_2) on a line.

All points on a horizontal line have equal y-coordinates. The numerator of the slope formula simplifies to zero. So a *horizontal line has a slope of 0.* You can describe a horizontal line by the equation $y = b$. The value of b is the value of all y-coordinates of points on the horizontal line.

Horizontal Line	A horizontal line containing the point (a, b) is described by the equation $y = b$.

All points on a vertical line have equal x-coordinates. So the denominator of the slope simplifies to zero. Since division by 0 is undefined, the *slope of a vertical line is undefined.* You can describe a vertical line by the equation $x = c$. The value of c is the value of every x-coordinate of points on the vertical line.

Vertical Line	A vertical line containing the point (c, d) is described by the equation $x = c$.

Example 3

Is the line containing the given points horizontal or vertical? Name the slope.

a. $(6, -2)$ and $(-1, -2)$

b. $(3, -5)$ and $(3, 0)$

c. $\left(0, \frac{1}{2}\right)$ and $\left(\frac{1}{2}, \frac{1}{2}\right)$

Solution

a. The y-coordinates are equal, so it is a horizontal line that has a slope of 0.

b. The x-coordinates are equal, so it is a vertical line that has an undefined slope.

c. The y-coordinates are equal, so it is a horizontal line that has a slope of 0.

Find the slope of a line that passes through the given points. Graph each line.

a. $(0, 0)$ and $(3, 4)$ **b.** $(-2, 3)$ and $(4, 0)$ **c.** $(1, -1)$ and $(-3, -1)$ **d.** $(2, 0)$ and $(2, 5)$

Solution

Substitute the coordinates in the slope formula.

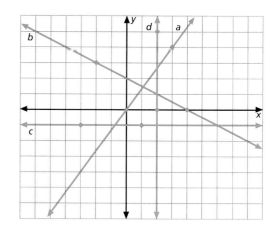

a. $m = \dfrac{4 - 0}{3 - 0} = \dfrac{4}{3}$ $m = \dfrac{y_2 - y_1}{x_2 - x_1}$

b. $m = \dfrac{0 - 3}{4 - (-2)} = \dfrac{-3}{6} = -\dfrac{1}{2}$

c. $m = \dfrac{-1 - (-1)}{-3 - 1} = \dfrac{0}{-4} = 0$

d. $m = \dfrac{5 - 0}{2 - 2} = \dfrac{5}{0}$ Slope is undefined.

To graph each line, plot both points. Use the slope to verify other points on each line.

▊ TRY THESE EXERCISES

Find the slope of each line shown. Note that each graph contains two lines.

1.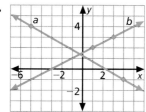

2.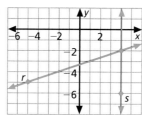

3.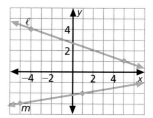

Graph the line that passes through the given point and has the given slope.

4. $(3, 5), m = -\dfrac{4}{5}$ **5.** $(-2, 1), m = \dfrac{3}{4}$ **6.** $(4, -1), m = 2$

7. $(1, -3), m = -3$ **8.** $(3, 9), m$ is undefined   **9.** $(1, -1), m = 0$

Find the slope of the line containing the given points. Name any vertical and horizontal lines.

10. $(2, 3), (8, 6)$ **11.** $(-4, 3), (-8, 6)$ **12.** $(2, 2), (2, -1)$

▊ PRACTICE EXERCISES • For Extra Practice, see page 603.

Find the slope of each line segment.

13. $\overline{AB}$ **14.** $\overline{CD}$

15. $\overline{EF}$ **16.** $\overline{GH}$

17. $\overline{IJ}$ **18.** $\overline{KL}$

19. $\overline{MN}$ **20.** $\overline{OP}$

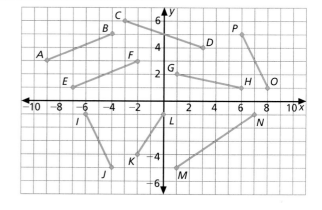

Graph a line using the given information.

21. $(1, -3)$, $m = -\dfrac{1}{3}$ 22. $(-2, -3)$, $m = \dfrac{2}{3}$ 23. $(5, 0)$, m is undefined

24. $(0, 0)$, $m = 0.5$ 25. $(0, 0)$, $m = -4$ 26. $\left(\dfrac{1}{2}, \dfrac{1}{2}\right)$, m is undefined

27. $(3, -1)$, $m = \dfrac{-5}{3}$ 28. $(-5, -5)$, $m = 0$ 29. $(-4, 1)$, $m = \dfrac{1}{4}$

30. Find the slope of the line passing through points $(-1, 2)$ and $(3, -2)$.

31. What equation describes a vertical line that passes through $P(4, 0)$?

32. What equation describes a horizontal line that passes through $P(-4, 2)$?

33. **WRITING MATH** Explain how you can tell just from the coordinates of points on a line whether the line is horizontal or vertical.

Find the slope of each line.

34.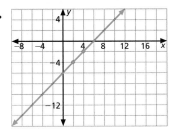

35.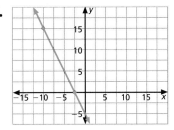

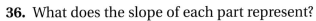

EXTENDED PRACTICE EXERCISES

BUSINESS TRAVEL Kerrie drove from her home to a business meeting and returned in one day. The graph plots her time against her distance from home. The trip is divided into five parts as shown in the graph.

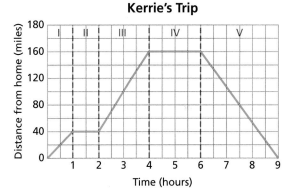

36. What does the slope of each part represent?

37. Find the average speed of section I and section III?

38. Explain which section of the graph represents Kerrie's time in the meeting.

39. **CRITICAL THINKING** What does the negative slope in section V represent?

40. Give one possible situation that explains the graph in section II.

MIXED REVIEW EXERCISES

Use the histogram. (Lesson 1-3)

41. How many students read 7–9 books?

42. How many students read 13–15 books?

43. How many more students read 0–3 books than read 16–18 books?

44. How many fewer students read 4–6 books than read 10–12 books?

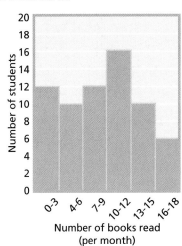

Review and Practice Your Skills

Find the distance between the points. Round to the nearest tenth.

1. $C(-5, -6)$, $D(7, -1)$ **2.** $T(2, 9)$, $W(15, 1)$ **3.** $E(-4, 0)$, $F(12, -3)$

4. $A(-2, 8)$, $B(5, -6)$ **5.** $X(10, 3)$, $Y(1, -9)$ **6.** $G\left(-\dfrac{1}{2}, \dfrac{3}{4}\right)$, $H(6, -8)$

7–12. Find the midpoints of the segments in Exercises 1–6.

Use the graph to calculate the length of each segment.

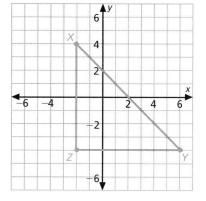

13. $\overline{XY}$

14. $\overline{YZ}$

15. $\overline{XZ}$

16. What kind of a triangle is $\triangle XYZ$?

17–19. Find the coordinate of the midpoint of the segments in Exercises 13–15.

Use the given endpoints of each circle's diameter. Find each circle's center and the lengths of its diameter and radius.

20. $(-5, 2)$, $(7, 2)$ **21.** $(3, 12)$, $(3, -2)$ **22.** $(-10, 4)$, $(3, -9)$

23. $(4, 0)$, $(-4, 4)$ **24.** $(4, 9)$, $(-6, 9)$ **25.** $(5, -3)$, $(5, 6)$

Find the slope of each line shown. Note that each graph contains two lines.

26.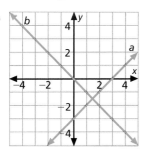

27.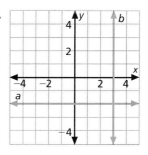

28.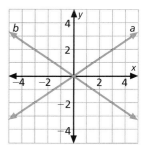

Graph a line using the given information.

29. $(1, 3)$, $m = \dfrac{3}{4}$ **30.** $(0, 2)$, $m = -2$ **31.** $(3, 0)$, $m = -\dfrac{1}{3}$

32. $(-1, 2)$, $m = -3$ **33.** $(1, 2)$, m is undefined   **34.** $(-2, -2)$, $m = 0$

35. Find the slope of the line passing through points $(-2, 4)$ and $(5, 10)$.

36. What equation describes a vertical line that passes through $P(2, 5)$?

37. What equation describes a horizontal line that passes through $P(-3, 4)$?

Use the given endpoints of each circle's diameter. Find each circle's center and the length of its diameter and radius. (Lesson 6-1)

38. $(0, 0)$, $(-8, 8)$

39. $\left(\frac{1}{2}, -\frac{2}{3}\right)$, $(0, 0)$

40. $(4.5, -6)$, $(-1, 0.5)$

41. $(-2, -7)$, $(-1, 4)$

42. $(-6, 3)$, $(14, 3)$

43. $(-8, 3)$, $(-7, 5)$

Find the distance between the points. Round to the nearest tenth. (Lesson 6-1)

44. $A(0, -4)$, $B(2, 8)$

45. $C(5, 10)$, $D(-3, -7)$

46. $E(0, -7)$, $F(-1, -1)$

47. $G(0, -4)$, $H(2, -2)$

48. $I(2, 1)$, $J(7, 9)$

49. $K\left(\frac{4}{3}, \frac{6}{5}\right)$, $L(-9, 15)$

Graph a line using the given information. (Lesson 6-2)

50. $(-1, 2)$, $m = -1$

51. $(-2, -3)$, $m = \frac{2}{5}$

52. $(0, -4)$, $m = -\frac{1}{3}$

53. $(1, 0)$, $m = -2$

54. $(2, 2)$, $m = \frac{1}{3}$

55. $(4, -3)$, $m = 0$

MathWorks — Career – Music Store Owner
Workplace Knowhow

Owners of specialty stores purchase merchandise, manage employees, prepare and evaluate financial documents and determine the layout of the store. The layout of the merchandise can affect the store's success. The owner of a music store needs to determine the best layout for the various sections of music on the showroom floor. The figure shows the general layout for the showroom, but more specific information is needed to determine exactly where the center of each section should be located. The center of the store is at (0 ft, 0 ft). The center of the rock music section is at (0 ft, −45 ft). The center of the classical music section should be 30 ft from the back wall of the store, along the y-axis. The central point in the New Age section is at (−37.5 ft, 0 ft).

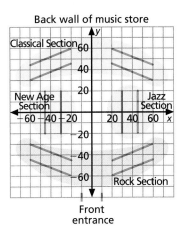

1. What is the distance from the Rock section center to the Classical section center?

2. The centers of the New Age and Jazz sections will be on the x-axis. The centers of both sections are equidistant from the center of the Rock section, with the Jazz and New Age on opposite sides of the store. Give the coordinates of the center of the Jazz section.

3. The slanted lines located in the Classical and Rock sections represent shelves. What is the slope of the shelves on the left-hand side of the Classical section? How does this slope compare to the slope of the shelves on the right-hand side of the Rock section?

4. What is the slope of the lines representing the shelves in the New Age and Jazz sections?

6-3

Write and Graph Linear Equations

Goals
- Write equations of lines using slope, intercepts and points.
- Graph a line given the equation.

Applications Finance, Sports, Transportation, Recreation

Work with a partner.

1. Study the graphs at the right and their corresponding equations.

2. Find the slope of each line.

3. Find the point where each line crosses the *y*-axis.

4. Compare these with the equations for the lines. What do you notice?

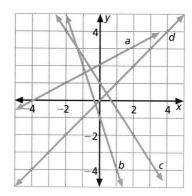

a. $y = \dfrac{1}{2}x + 2$

b. $y = -3x - 1$

c. $y = -\dfrac{3}{2}x + 1$

d. $y = x$

◼ BUILD UNDERSTANDING

A **linear equation** is an equation whose graph is a straight line. If the line intersects the *y*-axis, the point of intersection is the **y-intercept**.

When the equation of a line is written in the form $y = mx + b$, you can quickly determine the slope and *y*-intercept of a line. Then you can use this information to graph the line.

Slope-Intercept Form	$y = mx + b$ where *m* is the slope and *b* is the *y*-intercept.

COncepts in MOtion

BrainPOP®

mathmatters2.com

Example 1

Name the slope and *y*-intercept for the line with the given equation. Graph each line on a coordinate plane.

    a. $y = 3x - 2$ **b.** $3x + 2y = 6$

Solution

Rewrite the equation so that the operation sign is addition.

$$y = 3x + (-2)$$

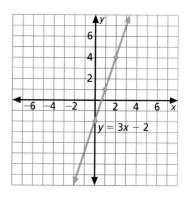

a. The slope is the coefficient of the *x*-term. The *y*-intercept is the constant. So the slope is 3 and the *y*-intercept is −2.

To graph the line, plot the point (0, −2). Then use the slope as $\dfrac{3}{1}$ to plot two or three additional points.

b. Rewrite the equation in slope-intercept form. Solve for y.

$$3x + 2y = 6$$

$$3x - 3x + 2y = -3x + 6 \qquad \text{Subtract } 3x \text{ from both sides.}$$

$$2y = -3x + 6$$

$$\frac{2y}{2} = \frac{-3x + 6}{2} \qquad \text{Divide each term by 2.}$$

$$y = -\frac{3}{2}x + 3$$

The slope is $-\dfrac{3}{2}$ and the y-intercept is 3.

To graph the line, plot the point $(0, 3)$. Then use the slope, $-\dfrac{3}{2}$, to plot two or three additional points.

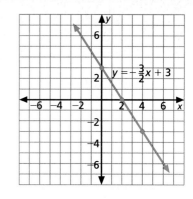

To write the equation of a line when given two points, first calculate the slope. To find the value of b, substitute the slope and the coordinates of one point into $y = mx + b$. Then use slope-intercept form to write the equation of the line.

Problem Solving Tip

Since the x-coordinate for every point on the y-axis is 0, the coordinates of the y-intercept are $(0, b)$.

E x a m p l e 2

Write an equation of a line using the information given.

a. $(7, 1)$, y-intercept is -1 **b.** $m = 3$, point on line is $(4, -1)$

Solution

a. The y-intercept coordinates are $(0, -1)$. Use ordered pairs to calculate the slope.

$$m = \frac{1 - (-1)}{7 - 0} = \frac{2}{7} \qquad \text{Use slope formula.}$$

Substitute $m = \dfrac{2}{7}$ and $b = -1$. The equation of the line is $y = \dfrac{2}{7}x - 1$.

b. To find the value of b, use the slope and the point given in $y = mx + b$

$$y = mx + b \qquad \text{Use slope-intercept form.}$$

$$-1 = 3(4) + b \qquad m = 3, x = 4, y = -1$$

$$-1 - 12 = b$$

$$-13 = b$$

Substitute $m = 3$ and $b = -13$. The equation of the line is $y = 3x - 13$.

To write the equation for a line given the graph of the line, use the graph to get information that will help you find the slope and y-intercept.

E x a m p l e 3   nline **Personal Tutor** at **mathmatters2.com**

Use the graph to write the equation of lines a and b.

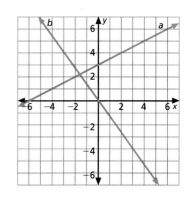

Solution

a. Choose two points on the line, and count rise units and run units. $m = \dfrac{1}{2}$. The line crosses the y-axis at $(0, 3)$. The equation is $y = \dfrac{1}{2}x + 3$.

b. The rise is -4 units and the run is 3 units, so $m = -\dfrac{4}{3}$. The line crosses the y-axis at $(0, 0)$. The equation is $y = -\dfrac{4}{3}x$.

You can write the equation of a line using point-slope form if you know the slope and the coordinates of any point on the line. You can also use this form if you know any two points on the line.

Point-Slope Form	$y - y_1 = m(x - x_1)$ where m is the slope and (x_1, y_1) is a point on the line.

Example 4

Write the equation of the line using the given information.

a. slope is $\frac{4}{5}$, point on line is $(3, 0)$

b. points on line are $(-1, 5)$ and $(1, -3)$

Solution

a. Use $m = \frac{4}{5}$, $x_1 = 3$ and $y_1 = 0$.

$$y - y_1 = m(x - x_1)$$
$$y - 0 = \frac{4}{5}(x - 3)$$
$$y = \frac{4}{5}x - \frac{12}{5}$$

b. Find the slope.

$$m = \frac{-3 - 5}{1 - (-1)} = -\frac{8}{2} = -4$$

Select either point to use. Substitute.

Use $m = -4$, $x_1 = 1$ and $y_1 = -3$.

$$y - y_1 = m(x - x_1)$$
$$y - (-3) = -4(x - 1)$$
$$y + 3 = -4x + 4$$
$$y + 3 - 3 = -4x + 4 - 3$$
$$y = -4x + 1$$

◥ TRY THESE EXERCISES

Use the graph to write an equation for each line a–f.

1. line d **2.** line e **3.** line f

4. line a **5.** line b **6.** line c

7. Graph $y = -2x + 5$ using the slope and y-intercept.

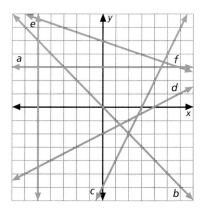

Write the equation of each line using the given information.

8. $m = 3$, $(1, 5)$ **9.** $(3, -6)$, $(1, -4)$ **10.** $m = 0$, $(0, 0)$

◥ PRACTICE EXERCISES • For Extra Practice, see page 604.

Identify the slope and y-intercept for each line. Graph each line.

11. $-5x + 4y = -8$ **12.** $y = -x + 1$ **13.** $y = 5x + 3$ **14.** $y = -8$

15. WRITING MATH Explain how to use the slope and the y-intercept of a line to graph the line. Explain what to do if the line is in the form $ax + by = c$.

Write the equation of each line using the given information.

16. $m = -4$, $(6, 2)$

17. $(9, 3)$, $(-2, -2)$

18. $(1, 1)$, $b = 5$

19. $(9, 5)$, $(5, 6)$

20. $m = -\dfrac{5}{2}$, $(3, 0)$

21. m is undefined, $(0, 0)$

22. $m = 4$, $(-1, 2)$

23. $m = -1$, $(3, -4)$

24. $(-1, 1)$, $(1, 5)$

25. $(4, 2)$, $(-1, 7)$

26. $m = -\dfrac{1}{3}$, $(1, 5)$

27. $m = 0$, $(3, -4)$

28. $(2, 3)$, $(1, 1)$

29. $(3, -6)$, $(1, -4)$

30. $(-1, 0)$, $b = 3$

31. $m = \dfrac{1}{2}$, $(-2, -7)$

32. $(0, -1)$, $(1, 2)$

33. $m = \dfrac{5}{2}$, $(0, -3)$

34. **RETAIL** Bulk dog food sells for \$1.50/lb. Write an equation to represent the cost in terms of the number of pounds purchased. Find the cost of 3.5 lb of food. Explain the meaning of slope for this problem.

SCIENCE An approximate formula for a barometric reading is $p = 760 - 0.09h$, where p is measured in millimeters and h is the altitude and is measured in meters. The formula is for altitudes less than 500 m.

35. What does the slope represent? What is implied by the slope being negative?

36. Explain the meaning of slope for this problem.

37. If there is a change of altitude from 100 m to 250 m, what is the change in the barometric reading?

◢ EXTENDED PRACTICE EXERCISES

38. Write the equation of a line that passes through all points where the x-coordinate equals the y-coordinate.

39. **CRITICAL THINKING** A linear equation is in *standard form* when it is written as $Ax + By = C$ where A, B and C are real numbers and A and B are not both zero. Use the standard form and solve for y. Make a general statement that summarizes how to find the slope and y-intercept for an equation in the form $Ax + By = C$.

40. **CHAPTER INVESTIGATION** Find the data on your company's operating costs for each quarter of the year. Plot four ordered pairs, (Quarter, Operating cost). Draw a line connecting each quarter to represent the increase, decrease or steady pace of the operating costs. Describe the operating costs by quarters and by the entire year.

◢ MIXED REVIEW EXERCISES

Graph the solution of each inequality on a number line. (Lesson 3-6)

41. $a \leq -4$

42. $b > -2$

43. $c < 3$

44. $d \leq 2$

45. $e > 4$

46. $f \geq 1$

47. $g \leq -1$

48. $h > -5$

Find the measure of the complement and the supplement of each angle. (Lesson 5-2)

49. $m\angle 18°$

50. $m\angle 71°$

51. $m\angle 83°$

52. $m\angle 50°$

6-4 Write and Graph Linear Inequalities

Goals
- Write linear inequalities in two variables.
- Graph linear inequalities in two variables on the coordinate plane.

Applications Business, Market research, Inventory

Use the graph to answer Questions 1–3.

1. What is the relationship between the x- and y-coordinate for each point on the line? Write the equation of the line.

2. What is the relationship between the x-coordinate and the y-coordinate for each point in the blue region?

3. What is the relationship between the x-coordinate and the y-coordinate for each point in the unshaded region?

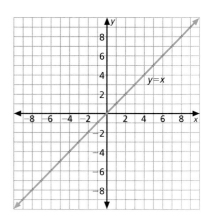

◤ BUILD UNDERSTANDING

The graphed line above separates the coordinate plane into two regions. The region on either side of a line is called an **open half-plane**. The line separating the half-planes forms the **boundary**, or edge, of each half-plane.

When the equal symbol in a linear equation is replaced with an inequality symbol ($>, <, \geq, \leq, \neq$) the result is a **linear inequality**. Any ordered pair of real numbers that makes an inequality true is a **solution of the inequality**. The graph of all the solutions is the **graph of an inequality** and shows the boundary, with either a solid line or dashed line, along with a shaded region.

To determine if an ordered pair is a solution, substitute its values into the linear inequality. When the simplified inequality is true, the ordered pair is a solution and thus part of the shaded region. When the simplified inequality is false, the ordered pair is not part of the solution and thus not part of the shaded region.

Example 1

Tell whether the ordered pair is a solution of the inequality.

a. $(2, 6)$; $y \geq 2x - 1$ **b.** $(-2, 4)$; $y < 2x - 1$

Solution

a. Substitute 2 for x and 6 for y in the inequality.

$$y \geq 2x - 1$$
$$6 \overset{?}{\geq} 2(2) - 1$$
$$6 \geq 3$$

Since the inequality is true, the ordered pair $(2, 6)$ is a solution of $y \geq 2x - 1$.

b. Substitute -2 for x and 4 for y in the inequality.

$$y < 2x - 1$$
$$4 \overset{?}{<} 2(-2) - 1$$
$$4 < -5$$

Since the inequality is false, the ordered pair $(-2, 4)$ is not a solution of $y < 2x - 1$.

The graph at the right shows the solution of $y \geq 2x - 1$. The solution includes the blue shaded half-plane and the line $y = 2x - 1$. Whether the line is included in the solution depends on the inequality symbol. A solution that includes the line is called a **closed half-plane**.

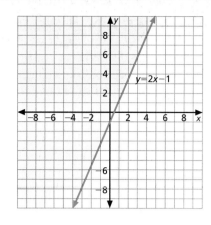

The symbols $>$ and $<$ mean the boundary is not part of the solution and the line is graphed as a dashed line. The symbols $\geq$ and $\leq$, mean the boundary is part of the solution and the line is graphed as a solid line.

To graph an inequality with two variables, first graph the equation related to the inequality. Determine if the line is graphed as a solid or dashed rule. Use a *test point* to determine which region is shaded.

A **test point** is a point that does not lie on the boundary, but rather above or below it. Substitute the coordinates of the ordered pair into the inequality. If the simplified inequality is true, shade the half-plane where the test point lies. If the simplified inequality is false, shade the other half-plane. If $(0, 0)$ is not on the boundary, use it as a test point.

Example 2   Personal Tutor at mathmatters2.com

Graph $2x + y \leq -4$.

Solution

Graph the equation $2x + y = -4$ as the boundary.

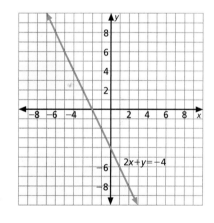

Rewrite the equation in slope-intercept form.

$$2x + y = -4$$
$$2x - 2x + y = -2x + (-4)$$
$$y = -2x + (-4)$$

The slope is -2 and the y-intercept is -4.

Since the inequality symbol includes "equal to," use a solid line. Then choose and test a point: $(0, 0)$.

$$2x + y \overset{\leq}{} -4$$
$$2(0) + 0 \overset{?}{\leq} -4 \quad \text{Use } x = 0 \text{ and } y = 0.$$
$$0 \leq -4$$

Since $(0, 0)$ does not satisfy the inequality, shade the region that does not include $(0, 0)$. The solution is the lower half-plane together with the boundary.

The statements in the box below summarize the shading of open and closed half-planes.

Forms of Inequalities	Shade the region above a dashed line if $y > mx + b$.
	Shade the region above a solid line if $y \geq mx + b$.
	Shade the region below a dashed line if $y < mx + b$.
	Shade the region below a solid line if $y \leq mx + b$.

Example 3

BUSINESS A company pays 32% of sales in taxes. The amount of taxes paid is modeled by the equation $y = 0.32x$, where x is the amount of sales and y is the amount of taxes. Determine when taxes are less than or equal to 32% of sales.

Solution

A graphing calculator can help you graph the equation. Enter the equation $y = 0.32x$ and graph the line.

Since the problem is stated *less than or equal to*, the inequality is $y \leq 0.32x$ and the line is displayed as a solid line.

Choose a test point: (1, 0).

Determine if (1, 0) is a solution.

$$y < 0.32x$$
$$0 \overset{?}{<} 0.32(1) \quad \text{Use } x = 1 \text{ and } y = 0.$$
$$0 < 0.32$$

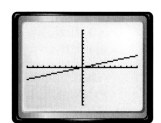

Since (1, 0) is below the line, shade the lower half-plane.

The line and the shaded region of the graph show when taxes are less than or equal to 32% of sales.

◥ TRY THESE EXERCISES

Tell if each ordered pair is a solution of the inequality.

1. $(4, 1); y > x - 5$

2. $(-1, 5); y \leq 4x - 3$

3. $(-3, -1); y \geq x + 4$

4. $(1, 1); y < 3x$

Determine whether solutions of the inequality are *above* or *below* the boundary. State if the boundary line is included.

5. $y - x < 0$

6. $y > 3x - 2$

7. $2x - y \geq 4$

8. $-2y > x + 1$

9. $2x + y \geq 4$

10. $3y - x > 4$

Graph each inequality.

11. $y < x + 4$

12. $y \geq 3x - 1$

13. $2x - y \leq 2$

14. $x + y < 6$

◥ PRACTICE EXERCISES • For Extra Practice, see page 604.

Determine whether solutions of the inequality are *above* or *below* the boundary. State if the boundary line is included.

15. $y + x < 0$

16. $2x + y \leq 4$

17. $3y - x > 6$

18. $-2y < x + 3$

Graph each inequality.

19. $y < 4$

20. $x \geq -1$

21. $y < -x$

22. $y > 0.5x + 1$

23. $y > -x + 2$

24. $x + y \geq 3$

25. $2x + y \leq 1$

26. $y \leq 4x + 2$

27. $y \leq \dfrac{1}{2}x + 1$

28. $y < -\dfrac{1}{3}x - 1$

29. $y < -\dfrac{3}{2}x$

30. $y - \dfrac{2}{3}x > 3$

Write each statement as an inequality. Then graph the inequality.

31. The y-coordinate of a point is at most 3 more than the x-coordinate.

32. The sum of the x-coordinate and the y-coordinate is less than 8.

33. The y-coordinate of a point less 3 times the x-coordinate is at least 2.

Write an inequality for each graph.

34. 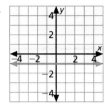   **35.** 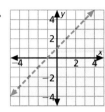   **36.**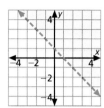

37. MARKET RESEARCH Suppose a company plans to increase sales by $125,000 each year for the next ten years. Sales are now at $1 million. Write and graph an inequality that represents the future sales for the company if sales increase by at least $125,000/yr. Use an appropriate scale for the y-axis.

38. WRITING MATH Use an example to explain the steps you would use to check a solution for an inequality.

■ EXTENDED PRACTICE EXERCISES

39. Use a graphing calculator to graph $y = |x|$.

Use your graph from Exercise 39 to determine which ordered pairs are solutions to the inequality $y \geq |x|$.

40. $(0, 1)$ **41.** $(-2, 5)$ **42.** $(3, -3)$ **43.** $(2, 10)$

44. BUSINESS A company wants to have sales such that the exports are half of the imports. Write an equation that represents the number of exports in terms of the number of imports. Make a graph to show when the number of exports is not more than half the imports.

45. CHAPTER INVESTIGATION Create a coordinate plane like the one you created in Lesson 6-1. Find the data on your company's sales revenue for each quarter of the year. Plot four ordered pairs as (Quarter, Sales revenue). Draw a line connecting each quarter to represent the increase, decrease or steady pace of sales. Describe the sales by quarters and by the entire year.

■ MIXED REVIEW EXERCISES

A group of lettered cards contains one A, two Q's, three K's and four Z's. Cards are picked one at a time, then replaced. Find each probability. (Lesson 4-5)

46. P(Q, then K)   **47.** P(A, then Q)   **48.** P(K, then A)

Simplify. (Lesson 2-8)

49. $p^{12} \div p^3$ **50.** $\left(\dfrac{z}{z^4}\right)^3$ **51.** $\dfrac{m^{-2}}{m^{-6}}$ **52.** $\dfrac{k^3}{k^8}$

53. $\dfrac{f^{-3}}{f^{-4}}$ **54.** $a^{-6} \cdot a^{-3}$ **55.** $(r^2)^{-4}$ **56.** $\left(\dfrac{x}{x^4}\right)^5$

Review and Practice Your Skills

Identify the slope and *y*-intercept for each line.

1. $y - x + 2$

2. $y - 2$

3. $y - 2x$

4. $-3x + 6y = -18$

5. $2y = 10x - 12$

6. $y = \dfrac{x}{3} + 1$

Write the equation of each line using the given information.

7. $(0, 4), m = \dfrac{2}{3}$

8. $(2, 3), m = 0$

9. $(2, -2), (-4, 6)$

10. $(1, 2), (-2, 5)$

11. $(5, -1), (-3, 3)$

12. $(3, 3), (3, 7)$

13. $m = 1, (3, -5)$

14. $(-2, 5), (10, 6)$

15. $(0, -1), (-1, -7)$

16. $m = \dfrac{2}{3}, (2, -4)$

17. $m = 0, (-5, 5)$

18. $(-3, 8), (-1, -3)$

19. $(-1, 3), b = 4$

20. $(0, -4), (11, 3)$

21. $m = \dfrac{1}{3}, (6, -2)$

22. $(-1, -1), (-5, -5)$

23. $m = -1, (-6, 1)$

24. $(0, 3), (-4, 0)$

25. $(12, -2), (9, 3)$

26. $m = \dfrac{3}{2}, (5, 0)$

27. $(4, -3), b = -1$

Tell if each ordered pair is a solution of the inequality.

28. $(5, -2); y \le 4x + 1$

29. $(4, 1); y \ge 3x - 4$

30. $(0, -3); x + y \ge -2$

31. $(-1, -4); y < -\dfrac{1}{2}x - 3$

32. $(-7, 3); 2x - y < 2$

33. $\left(-\dfrac{1}{2}, \dfrac{1}{3}\right); 4x - 3y \ge -3$

Determine whether solutions of the inequality are *above* or *below* the boundary. State if the boundary line is included.

34. $y < 3x + 4$

35. $2y \ge -3x + 12$

36. $y > 4$

37. $-y - x \le -7$

38. $9y > 12x - 3$

39. $-2y \ge 3x + 3$

Graph each inequality.

40. $y < 2x - 1$

41. $y \le -3x - 3$

42. $y > 2x$

43. $y > \dfrac{1}{3}x$

44. $x < -1$

45. $2x + 2y \ge 4$

Write an inequality for each graph.

46.
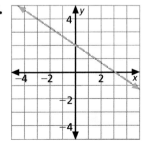

47.
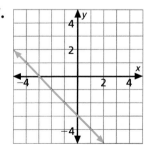

48.
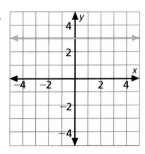

Find the distance between the points. Round to the nearest tenth. (Lesson 6-1)

49. $A(1, 5)$, $B(-1, -5)$ **50.** $X(8, 2)$, $Y(-3, 2)$ **51.** $M(0, 0)$, $N(5, 12)$

52. Find the slope of a line passing through $(-1, 4)$ and $(5, 12)$. (Lesson 6-2)

Write the equation of a line using the given information. (Lesson 6-3)

53. $(2, -2)$, $m = -2$ **54.** $(-1, 1)$, $m = \dfrac{1}{2}$ **55.** $(-2, 4)$, $(0, -6)$

Identify the slope and y-intercept for each line. Graph each line. (Lesson 6-3)

56. $y = \dfrac{x}{2} - 3$ **57.** $y = x + 1$ **58.** $y = -2x$

Graph each inequality. (Lesson 6-4)

59. $y < 2x - 4$ **60.** $y \geq x + 2$ **61.** $2x - 3y \leq 12$

The vertices of a triangle are $A(1, 7)$, $B(7, 9)$ and $C(9, 3)$. Find the following. (Lesson 6-4)

62. $AB = \underline{\ ?\ }$.   **63.** $BC = \underline{\ ?\ }$.   **64.** $AC = \underline{\ ?\ }$.

65. What type of triangle is $\triangle ABC$?

Mid-Chapter Quiz

Find the distance between the points. Round to the nearest tenth. (Lesson 6-1)

1. $P(-4, 2)$, $Q(-1, 6)$ **2.** $A(3, 7)$, $B(10, 2)$ **3.** $F(-1, -2)$, $G(0, 3)$

4. $X(-8, -4)$, $Y(2, 9)$ **5.** $H(0, 11)$, $I(12, -4)$ **6.** $C(5, -1)$, $D(-5, 3)$

Find the slope and y-intercept of a line using the information given. Identify any vertical or horizontal lines. (Lessons 6-2 and 6-3)

7. $(-1, 3)$, $(7, 5)$ **8.** $(4, 4)$, $(4, -2)$ **9.** $(2.9, 7.2)$, $(5.4, 12.2)$

10. $2x + 5y = 25$ **11.** $-x = -2y + 6$ **12.** $\dfrac{3}{4} = \dfrac{1}{8}y - \dfrac{3}{2}x$

Graph each line using the given information. (Lesson 6-3)

13. $(1, 2)$, $m = 4$ **14.** $(7, -2)$, $m = -1$ **15.** $(3, 12)$, $(-2, 2)$

16. $\left(\dfrac{3}{2}, 2\right)$, $m = -2$ **17.** $(1, 10)$, $(-2, 1)$ **18.** $(-6, 4)$, $(3, 1)$

Determine whether solutions of the inequality are *above* or *below* the boundary. State if the boundary line is included. (Lesson 6-3)

19. $y + 2x \leq 4$ **20.** $y > \dfrac{1}{3}x - 4$ **21.** $y \geq -\dfrac{3}{4}x + 2$

Graph each inequality. (Lesson 6-4)

22. $y < 5$ **23.** $y \geq -x - 1$ **24.** $x - y > 3$

6-5 Linear and Nonlinear Functions

Goals
- Graph linear functions.
- Identify the domain and range of a function.

Applications Machinery, Travel, Temperature

Use the graph for Questions 1–4.

1. Determine whether the coordinates of P and Q are solutions of the equation $y = 2x + 3$.

2. Give the coordinate pairs for three other points that are solutions of $y = 2x + 3$.

3. Do you think there are points on the line that are not solutions of $y = 2x + 3$? Explain.

4. Are there any breaks in the line? Explain.

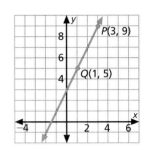

◣ BUILD UNDERSTANDING

When an equation in two variables has a relationship where each x-coordinate is paired with exactly one y-coordinate, the equation represents a **function**. You can represent functions using a mapping, a function table or a function rule written in **function notation**, $f(x)$.

An example of a function rule is $x - 3$. Its function notation is $f(x) = x - 3$. However, you can use the equation $y = x - 3$ to represent the function. This equation expresses y as a function of x. The set of all possible values of x is called the **domain** of the function. The set of all possible values of y is called the **range**, or *values of the function*.

> ### Think Back
>
> A mapping is a visual representation of how the x elements and $f(x)$ elements are paired.
>
> A function table is a table of x and $f(x)$ values.

Example 1

RECREATION A block-party association needs 12 packages of buns for a party. Write an equation where h is the number of hotdog bun packages and r is the number of hamburger bun packages. Make a function table. Graph the data.

Solution

The relationship of the number of each package is a function, written $f(r) = 12 - r$. As an equation, it is written $h = 12 - r$. Select domain values for a function table. Find the range.

Use the horizontal axis to represent hotdog buns and the vertical axis to represent hamburger buns.

The graph shows 13 possible combinations.

r	h
0	12
1	11
2	10
.	.
.	.
.	.
10	2
11	1
12	0

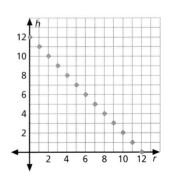

In the situation in Example 1, the domain is whole numbers from 0 to 12. Negative and fractional values do not make sense when purchasing packages of buns.

In other mathematical situations the domain is not restricted. In such cases, the domain of the function is all real numbers. Ordered pairs are graphed and the points are connected to show that the graph is **continuous**. This shows that for all real numbers, *every point* on the line is a solution of the equation.

All functions can be graphed using a function table. Set up a table. Choose a value for x, and find the corresponding $f(x)$ value. Once you have a function table, you are able to graph the function.

Example 2

Graph each function when the domain is the set of real numbers.

a. $f(x) = 4x - 3$ **b.** $f(x) = -4$ **c.** $2x + y = 4$

Solution

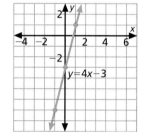

a. Make a table to show the ordered pairs. Choose at least three values of x from the domain. Calculate $f(x)$. Then graph the points that correspond to the ordered pairs, $(x, f(x))$. Draw a line to connect the points.

x	f(x)
1	1
0	-3
-1	-7

You can also think of the function as $y = 4x - 3$. Draw the graph using the slope and y-intercept.

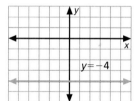

b. This function is written as $y = -4$. Recall that the graph of an equation in this form is a horizontal line. In this case, graph the horizontal line $y = -4$.

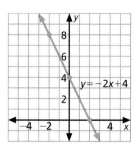

c. Make a function table or solve the equation for y. Then graph using the slope and y-intercept.

$$2x + y = 4$$
$$2x - 2x + y = -2x + 4$$
$$y = -2x + 4$$

The slope is -2 and the y-intercept is 4.

A function that can be represented by a linear equation, where the domain is all real numbers, is called a **linear function.** Not all functions are linear functions. You can tell from looking at a graph whether the function is linear or nonlinear. Recall that the word linear means straight line.

To test whether a graph is a function, use the **vertical line test.** The vertical line test states that if a vertical line drawn anywhere on the graph crosses the graph no more than once, then the graph represents a function.

Example 3

Online **Personal Tutor** at mathmatters2.com

Determine if each graph represents a function. Explain.

a. 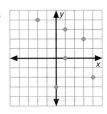   b. 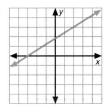   c.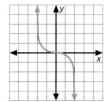

Solution

a. 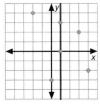   b. 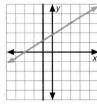   c.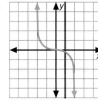

  not a function   function   function

Example 4

Make a table or use your graphing utility to graph the nonlinear function $f(x) = x^3 - 1$.

Solution

To use your graphing utility, enter $y = x^3 - 1$ at the equation screen and then graph. Notice the graph is a curve.

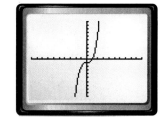

Set up a table that includes positive and negative values for x. Substitute each x-value into the equation and solve for the $f(x)$ value.

x	0	−1	1	2	−2
f(x)	−1	−2	0	7	−9

Plot each point on a coordinate plane. Since there are no values for which the function is not defined, the domain is the set of real numbers. The function is continuous.

◤ TRY THESE EXERCISES

Determine if each is a function.

1.
x	1	1	0	2	10
y	5	−4	−2	0	10

2. $f(x) = 3x$

3.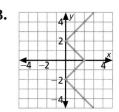

4.
x	0	−1	2	12	−1
y	5	0	−2	1	5

5. $y = x^2 + 1$

6. The area of a circle is a function of the length of the radius. Let y represent the area and r the length of the radius. Graph the area function $y = \pi r^2$ when the domain is {1, 2, 3, 4, 5}. Round y-values to the nearest whole number.

Graph each function for the given domain.

7. $y = -4x - 1$; $\{-2, -1, 0, 1, 2\}$

8. $f(x) = \frac{1}{2}x + 3$; $\{-6, -4, 0, 2, 3\}$

9. $f(x) = -3$; {all real numbers}

10. $y = x^2 - 2$; {all real numbers}

Determine if each graph represents a function. Explain.

11.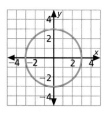

12.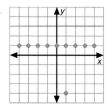

13.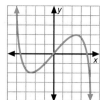

Graph each function.

14. $y = |x| + 3$

15. $y = x^2 - 1$

16. $y = \frac{3}{2}x - 2$

17. $y = \sqrt{x}$

18. $y = -x - 5$

19. $2\sqrt{x} + y = 0$

20. $y = -0.5x^2$

21. $x - y = 2$

22. WRITING MATH Explain the difference between "y as a function of x" and "x as a function of y." Include examples to show the difference.

23. BUSINESS In Jawaun's computer repair business, he charges $20 for making a house call and $12 per hour. Write a function for Jawaun's total charges, c, in terms of time, t. Is this a linear or nonlinear function?

■ **EXTENDED PRACTICE EXERCISES**

24. CHEMISTRY According to a scientific supply catalog, the cost of a chemical in cents C is represented by the equation $C = 3V$, where V is the amount of liquid in milliliters. Graph the function to determine the cost of 15 mL and 60 mL of the chemical.

25. Use the graph from Exercise 24 to find how much of the chemical can be purchased for $0.39.

26. The greatest integer function is another type of nonlinear function that models real-world situations, such as postage rates. The function notation for the *greatest integer function* is $f(x) = [x]$, and it means $f(x)$ equals the greatest integer that does not exceed the value of x. For example, when $x = 11.23$, $f(x) = 11$; when $x = 0.346$, $f(x) = 0$. Make a table of at least 12 ordered pairs. Graph the function $f(x) = [x]$ when the domain is greater than 0.

■ **MIXED REVIEW EXERCISES**

Which of the given values is a solution of the equation? (Lesson 3-1)

27. $5d + 6 = -9$; $-5, -3, 3$

28. $12r - 3 = 21$; $2, 3, 7$

29. $e^2 - 4 = 12$; $-4, -3, 4$

Solve each inequality. (Lesson 3-7)

30. $5 + d > 3$

31. $3 + 4z < 27$

32. $4p \geq -18$

33. $2a + 4.7 \leq a - 1.3$

34. $12c - 6 > 6c + 9$

35. $5r - 3 < 3r + 8$

6-6 Graph Quadratic Functions

Goals
- Identify points given the graph of a quadratic function.
- Graph simple quadratic functions.

Applications Sports, Physics, Agriculture

Use a graphing calculator for Questions 1–7.

1. Graph $y = x$.

2. Turn off the graph in Question 1. Graph $y = x^2$.

3. Describe the difference between the two graphs?

4. Turn on the graph in Question 1 so that you can view both graphs on the same coordinate plane, and check your answer to Question 3.

5. Look at the form of the following equations. Which equations do you think have a graph similar to the graph in Question 1? Explain your choices.

$$y = x + 3 \qquad y = x^2 + 1$$
$$y = 3x^2 \qquad y = 2x$$

6. Which of the equations above do you think has a graph similar to the graph in Question 2? Explain your choices.

7. Graph the equations in Question 5, and check your answers.

◤ BUILD UNDERSTANDING

Not all equations with two variables represent linear functions. Linear functions are represented by first-degree equations written in the standard form $Ax + By = C$. A first-degree equation is an equation whose variable terms have only exponents of one.

A **quadratic function** is a nonlinear function that when written in standard form is a second-degree equation, which means that it has one squared term.

Quadratic Function	$y = Ax^2 + Bx + C$, where A, B, and C are real numbers and $A \neq 0$.

If the domain of a quadratic function is not specified, it is understood that the domain is the set of real numbers. The graph of a quadratic function is a curve known as a *parabola*. Not all parabolas represent functions since parabolas can open upward, downward, to the right or to the left.

To graph quadratic functions, make a table of ordered pairs. Use enough points to get an accurate picture of the graph.

Example 1

Graph $y = x^2 - 2$.

Solution

Make a table with at least five ordered pairs. Include both positive and negative values for x.

x	−2	−1	0	1	2
y	2	−1	−2	−1	2

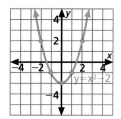

Graph the points that correspond to the ordered pairs in the table.

Draw a smooth curve through the points.

Graphs of quadratic functions are used to solve many types of problems. Use the graph of the function to find maximum or minimum points.

Technology Note

If you use a graphing calculator for Example 2 you may need to adjust your viewing window to see the maximum point of the graph.

Example 2

PHYSICS A golf ball is hit into the air with an initial velocity of 144 ft/sec. The height h, in feet, of the ball above the ground is modeled by the equation $h = -16t^2 + 144t$ where t is in seconds. Use a graph to find the time when the following occurs.

a. the ball reaches 224 ft

b. the ball reaches its maximum height

c. the ball hits the ground

Solution

Make a table of ordered pairs. Then draw a graph. Let the horizontal axis represent time and the vertical axis represent height.

t	0	1	2	3	4	5	6	7	8	9
h	0	128	224	288	320	320	288	224	128	0

a. From the table and graph, you see when $t = 2$, $h = 224$ on the way up. On the way down, the ball is 224 ft above the ground when $t = 7$. So, the ball has a height of 224 ft twice, at 2 sec and again at about 7 sec.

b. The maximum height of the ball is the h-value at the top of the curve. Maximum height occurs at approximately 4.5 sec.

c. The ball hits the ground when $h = 0$. So the ball hits the ground after 9 sec.

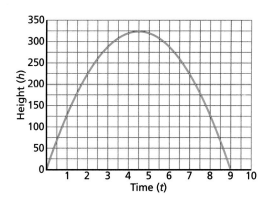

TRY THESE EXERCISES

Graph each function for the domain of real numbers.

1. $y = x^2 + 2$

2. $y = -2x^2 - 5$

3. $y = 3x^2 - x + 3$

4. WRITING MATH Write a few sentences describing the similarities among the graphs of the quadratic equations in Exercises 1–3.

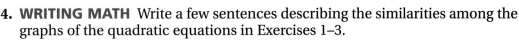

SPORTS An arrow is shot into the air. The height of the arrow is a function of the time that the arrow is in the air. The equation that represents this function is $h = -16t^2 + 112t$, where h is the height in feet and t is the seconds the arrow is in the air.

5. Make a table of ordered pairs. Graph the points, and draw a smooth curve through the points.

6. When does the arrow reach a height of 96 ft?

7. When does the arrow reach its maximum height?

8. When does the arrow hit the ground?

Complete each ordered pair so that it corresponds to a point on the graph.

9. $(-1, ?)$

10. $(3, ?)$

11. $(?, 1)$

12. $(?, 5)$

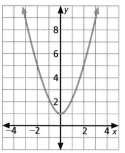

◥ PRACTICE EXERCISES • For Extra Practice, see page 605.

Graph each function for the domain of real numbers.

13. $y = x^2 + 4$

14. $y = 3x^2 - 6$

15. $y = 2x^2$

16. $y = x^2 - 2x + 3$

17. $y = -x^2 - 4$

18. $y = 2x^2 - 1$

AGRICULTURE Suppose you have 100 ft of fence for a rectangular garden.

19. Name three possible lengths and widths of the garden to be fenced in.

20. Write an equation for the area.

21. Graph the area as a function of the width.

22. What values make sense for the domain?

23. What is the maximum area possible?

Complete each ordered pair so that it corresponds to a point on the graph.

24. $(1, ?)$

25. $(0, ?)$

26. $(?, -2)$

27. $(?, -1)$

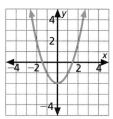

PHYSICS The height of a softball hit into the air is given by the equation $h = -4.9t^2 + 24t + 3$, where h is the height in feet and t is the time in seconds.

28. Graph the function.

29. At what time does the disc reach its maximum height? (Remember that time is measured in quarter seconds.)

30. How many seconds is the ball in the air?

31. **GRAPHING** Use a graphing utility to graph the equations below on a coordinate plane.

$$y = x^2 \qquad y = x^2 + 1 \qquad y = x^2 - 1 \qquad y = x^2 + 2 \qquad y = x^2 - 2$$

32. Compare the graphs in Exercise 31. What type of changes occur in the graph based on the value of the constant term being positive or negative?

33. Name the function of the graph of $y = x^2$ translated 10 units up. Name the function of the graph of $y = x^2$ translated 10 units down.

34. **GRAPHING** Use a graphing calculator to graph the equations below on the same coordinate plane.

$$y = x^2 \qquad y = (x + 1)^2 \qquad y = (x - 1)^2 \qquad y = (x + 2)^2 \qquad y = (x - 2)^2$$

35. Compare the graphs in Exercise 34. Does it matter whether the value of the constant term is positive or negative?

36. What function would a graph the same as the graph of $y = x^2$ but 10 units to the left of it represent? 10 units to the right of it?

37. **GRAPHING** Use a graphing calculator to graph the equations below on the same coordinate plane.

$$y = x^2 \qquad\qquad y = -x^2 \qquad\qquad y = 2x^2$$
$$y = -2x^2 \qquad\qquad y = 0.5x^2 \qquad\qquad y = -0.5x^2$$

38. Compare the graphs in Exercise 37 to others in the form $y = ax^2$. Does it matter if the value of the coefficient of x is positive or negative?

39. Compare the graphs in Exercise 37 to others in the form $y = ax^2$. Does it matter if the value of the coefficient of x is between 0 and 1, or is greater than 1?

Use what you discovered in Exercises 31–39 to graph each quadratic function.

40. $y = x^2 + 5$ 41. $y = 3x^2 - 2$ 42. $y = -3x^2$ 43. $y = (x - 2)^2$

44. $y = x^2 - 4$ 45. $y = 2x^2 - 7$ 46. $y = -2x^2$ 47. $y = 2(x - 2)^2$

◣ EXTENDED PRACTICE EXERCISES

48. **CRITICAL THINKING** For the graph of a quadratic function $y = ax^2 + bx + c$, find the x-coordinate of the vertex in terms of a and b.

49. Use your answer to Exercise 48. Write the coordinates of the vertex of any quadratic function $y = ax^2 + bx + c$ in terms of a and b.

50. **CHAPTER INVESTIGATION** Use your graphs from Lessons 6-3 and 6-4 to determine which quarters were profitable and which had a loss. Brainstorm ideas about business decisions and planning based on similar graphs.

◣ MIXED REVIEW EXERCISES

Find the unknown angle measures in each parallelogram. (Lesson 5-6)

51. $m\angle A$

52. $m\angle B$

53. $m\angle C$

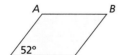

54. $m\angle Q$

55. $m\angle R$

56. $m\angle S$

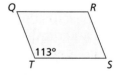

Write each phrase as a variable expression. (Lesson 2-3)

57. a number multiplied by 27

58. the quotient of seven and a number

59. the product of a number and 13

60. the difference of a number and three.

Review and Practice Your Skills

Graph each function for the given domain.

1. $y = 2x - 3$; $\{-4, -3, -1, 0, 2\}$

2. $f(x) = \frac{1}{2}x$; $\{-4, -2, 0, 2, 4\}$

3. $f(x) = -3$; $\{-3, -1, 1, 3\}$

4. $y = -3x^2$; {all real numbers}

Determine if each graph represents a function. If not, explain why.

5.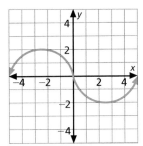

6.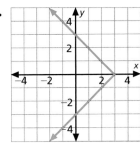

7.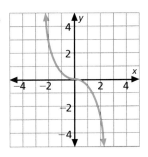

Graph each function.

8. $y = |x + 3|$

9. $y = \frac{x}{2} + 2$

10. $y = -|x|$

11. $y = 4\sqrt{x}$

12. $y = x^2 + 2$

13. $x - y = 3$

Graph each function for the domain of real numbers.

14. $y = x^2$

15. $y = x^2 - 3$

16. $y = -3x^2$

17. $y = \frac{1}{3}x^2$

18. $y = -x^2 - 3$

19. $y = x^2 + 3$

Complete each ordered pair so that it corresponds to a point on the graph.

20. $(?, 0)$

21. $(1, ?)$

22. $(-2, ?)$

23. $(?, 3)$

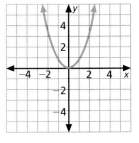

PHYSICS A ball is dropped from the top of a building 30-m tall. The height of the ball is a function of the time that the ball is in the air. The equation that represents this function is $h = -4.9t^2 + 30$ where h is the height in meters and t is the time in seconds the ball is in the air.

24. Copy and complete the table of ordered pairs. Round to the nearest tenth.

t	0	0.5	1	1.5	2	2.2	2.4
h	■	■	■	■	■	■	■

25. When does the ball reach a height of 10.4 m?

26. When does the ball hit the ground?

27. Is the ball above or below ground at $t = 3$ sec? Justify your answer.

Find the midpoints of the segments with the following endpoints. (Lesson 6-1)

28. $A(-7, 3)$, $B(0, -1)$

29. $C\left(4, \dfrac{1}{2}\right)$, $D\left(\dfrac{1}{8}, -6\right)$

30. $F(-1, -1)$, $G(6, -8)$

Identify the slope and y-intercept of each line. (Lesson 6-3)

31. $y = -x$

32. $y = \dfrac{x}{3} - 3$

33. $2x - y = 4$

34. $5x - 3y = 15$

Write an inequality for each graph. (Lesson 6-4)

35.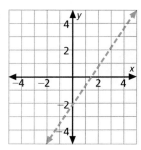

36.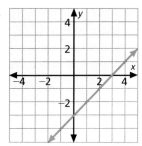

37.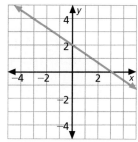

38. Graph the nonlinear function $y = x^2 + 3$. (Lesson 6-6)

MathWorks Career – Restaurateur
Workplace Knowhow

A restaurateur sells and serves food to the public. Restaurateurs are responsible for hiring employees, purchasing food, paying bills and satisfying the customers. A restaurateur has determined that the following equation roughly approximates the amount of profit he can make, based on x number of customers.

$$\text{Profit} = 25x - 1000$$

1. Graph the profit equation and determine the "break-even" point. The "break-even" point is the minimum number of customers required to obtain a positive profit.

2. Suppose that on a given day there were 50 customers and the restaurateur made a profit of $300. This profit is more than what is expected from 50 customers. Shade the region of the graph where the profit is greater than what is normally expected for a given number of customers. What circumstances might occur to allow profits to be higher than expected?

3. Write the profit equation above using function notation, $f(x)$. What are the domain and range of this profit function?

4. Suppose that the profit function is $f(x) = 25x^2 - 1000$. Graph this function and determine the break-even point. Plot the graph using x-values 0 through 10.

Problem Solving Skills: Patterns and Functions

Functions form patterns that can be summarized using function rules. An equation with two variables is another way to write a function rule. For example, for the function rule $x + 3$, you can write it as $y = x + 3$. An effective strategy to organize and classify the domain and range values is to **make a table**. In this case, a function table can lead you to a function rule.

Problem Solving Strategies

Guess and check

Look for a pattern

Solve a simpler problem

✔ Make a table, chart or list

Use a picture, diagram or model

Act it out

Work backwards

Eliminate possibilities

Use an equation or formula

Problem

 Personal Tutor at mathmatters2.com

Darnell is considering a new long-distance telephone provider. The new provider charges $1.10 for placing a call and $0.08 for each minute. His current provider charges $0.12/min, and the cost of his calls range from $2.50 to $7.50. Compare providers' costs.

Solve the Problem

Use a table to find a pattern or function rule for the new service. The function is $t = 1.10 + 0.08m$. Use the lower price (2.50), and solve for m. Then use the upper price (7.50), and solve for m.

	Base charge	Cost of minutes	Total cost
One-minute call	1.10	$0.08 \cdot 1 = 0.08$	1.18
Two-minute call	1.10	$0.08 \cdot 2 = 0.16$	1.26
Three-minute call	1.10	$0.08 \cdot 3 = 0.24$	1.34
Any number of minutes, m	1.10	$0.08 \cdot m = 0.08m$	$1.10 + 0.08m$

$$2.50 = 1.10 + 0.08m \qquad 7.50 = 1.10 + 0.08m$$
$$1.40 = 0.08m \qquad\qquad 6.40 = 0.08m$$
$$17.5 = m \qquad\qquad\quad 80 = m$$

With the new provider, you can talk from 17.5 min to 80 min for the price range.

To find the minutes that Darnell currently talks, divide the cost by the minutes.
$$2.50 \div 0.12 = 20.8 \qquad 7.50 \div 0.12 = 62.5$$

Currently, Darnell can talk from 20.8 min to 62.5 min. for the price range.

◢ TRY THESE EXERCISES

Match each function table with the appropriate function rule.

1.

x	5	−10	0	2	4
y	3	8	−2	0	2

2.

x	0	6	−12	30	3
y	0	4	−8	20	2

a. $f(x) = \dfrac{2}{3}x$

b. $f(x) = 2x^3$

c. $f(x) = |x| - 2$

Five-step Plan

1 Read
2 Plan
3 Solve
4 Answer
5 Check

Use each domain to find ordered pair solutions of each equation.

3. $y = 2x - 1$; $\{4, -2, 0, 1, 2\}$

4. $y = -3x + 5$; $\{-5, -3, 0, 2, 4\}$

Write a function rule for each function table.

5.

x	0	1	16	100	$\frac{9}{4}$
y	0	1	4	10	$\frac{3}{2}$

6.

x	10	-6	-1	25	0
y	-2	-18	-13	13	-12

7. If a car is moving at a constant speed of 50 mi/h, the function $y = 50x$ gives the distance y, in miles, that it travels in x hours. Graph the function when the domain is $\left\{\frac{1}{4}, \frac{1}{2}, 1, 1\frac{1}{2}, 2\right\}$.

8. **NUMBER SENSE** Is the relationship that matches a number with its positive square root a function? Explain.

Let the domain be the set of real numbers. Give the range of the function.

9. $y = |x|$

10. $y = -|x|$

11. $y = x$

12. $y = -x$

13. **ECONOMICS** Use the current cost of a loaf of bread. Due to an average rate of inflation of 4% each year, the cost of food increases accordingly. What is the estimated cost of a loaf of bread 10 yr from today?

14. **WRITING MATH** Compare *range* as used in this lesson to how it is used in statistics in Chapter 1. How are the meanings alike? How are they different?

BUSINESS The cost of parking is $1.50 for the first hour and $0.75 for each additional hour.

15. Write a function to represent the cost of parking x hours.

16. What values make sense for the domain?

17. What meaning does the y-intercept have for this problem?

18. How much will a five-hour stay cost?

19. For how long can you park if you have $18.50?

20. **DATA FILE** Refer to the data on the age of Internet users on page 571. Write a function that represents the number of Internet users of age 9–17 for a given group of people.

■ **MIXED REVIEW EXERCISES**

Calculate each of the following permutations. (Lesson 4-6)

21. $_{12}P_6$

22. $_{14}P_3$

23. $_8P_3$

24. $_{14}P_6$

Solve each equation. Check the solutions. (Lesson 3-8)

25. $\sqrt{p} = 6$

26. $z^2 = 81$

27. $b^2 + 4 = 53$

28. $\sqrt{s + 6} = 8$

6-8

Direct Variation

Goals
- Solve problems involving direct variation function.
- Solve problems involving direct square variation functions.

Applications   Finance, Biology, Physics, Business

Use the table to answer Questions 1–5.

1. Let m = the number of muffins, and let C = the cost. Find the ratio $\dfrac{C}{m}$ for each pair of entries. What do you discover?

2. As the number of muffins increases, what happens to the cost?

3. When the number of muffins doubles, what happens to the cost?

4. Use a proportion to find the cost of 7 muffins.

5. The cost of the muffins is a function of the number of muffins. Write the function rule using the variables used in Question 1.

Number of muffins	Cost
5	$3.45
10	$6.90
12	$8.28
20	$13.80
24	$16.56

■ BUILD UNDERSTANDING

Some linear functions have a relationship that when you divide the range (ouput) value by its corresponding domain (input) value, the quotient is the same. This constant quotient means that "y varies directly as x." Such a function is a **direct variation**. The constant k is called the **constant of variation**.

Direct Variation	$y = kx$, where $k \neq 0$.

Direct variation problems can be solved by writing and solving an equation or by writing and solving a proportion.

Example 1

PHYSICS   The distance a spring stretches varies directly as the weight applied to it. A 12-kg weight stretches the spring 15 cm. How many centimeters will a 20-kg weight stretch the spring?

Solution

The weight applied is x, and the length the spring stretches is y.

Equation Method

$$y = kx \longrightarrow 15 = k(12)$$
$$\frac{15}{12} = k$$
$$\frac{5}{4} = k$$

Substitute 20 for x and $\frac{5}{4}$ for k in $y = kx$.

$$y = \frac{5}{4}(20) = 25$$

Proportion Method $\dfrac{\text{known length}}{\text{known weight}} = \dfrac{\text{new length}}{\text{new weight}}$

$$\dfrac{15}{12} = \dfrac{y}{20}$$

$$15(20) = 12y \qquad \text{Cross-products are equal to each other.}$$

$$300 = 12y$$

$$25 = y \qquad \text{Divide both sides of equation by 12.}$$

The length the spring stretches is 25 cm when 20 kg of weight is applied.

In some cases one quantity varies directly as the square of another. For example, the area of a circle varies directly as the square of the radius, assuming $\pi = 3.14$.

Area of a Circle

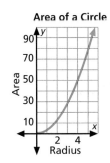
Area of a Circle

Radius	1	2	3	4	5
Area	3.14	12.56	28.26	50.24	78.50

The graph shows that the relationship between the area and the radius is a quadratic function. This is an example of a **direct square variation** stated as "y varies directly as x^2," or "y is directly proportional to x^2."

Direct Square Variation	$y = kx^2$, where $k \neq 0$.

Check Understanding

Direct variation is a special case of a linear function $y = mx + b$, where $b = 0$. When a direct variation is graphed, how does the value of the constant of variation relate to the slope of the graph? Through what point on the coordinate plane will the graph of a direct variation always pass?

Example 2

 Personal Tutor at mathmatters2.com

PHYSICS The distance a car rolls down a hill from a stand still position varies directly as the square of the amount of time it is rolling. If the car rolls 15 m downhill in 4 sec, how far will it roll in 10 sec?

Solution

Substitute known values of x and y into the equation to find k. The time the car is rolling is x, and the distance the car rolls is y.

$$y = kx^2$$

$$15 = k(4)^2$$

$$15 = 16k$$

$$\dfrac{15}{16} = k \approx 0.9375$$

Substitute 10 for x and 4.0625 for k into the equation $y = kx^2$.

$$y = 0.9375(10)^2$$

$$= 0.9375(100)$$

$$= 93.75$$

The car will roll approximately 93.75 m in 10 sec.

1. Assume that y varies directly as x. When $x = 24$, $y = 16$. Find y when $x = 42$.

2. Assume that y varies directly as x. When $x = 3$, $y = 63$. Find y when $x = 5$.

3. **FINANCE** The annual simple interest on a loan varies directly as the amount of the loan. The interest in one year on a $400 loan is $56. Find the interest in one year on a loan of $650.

4. **WRITING MATH** Use examples to explain how you know whether the relationship between the area and the radius of a circle is a direct variation or a direct square variation.

5. **AGRICULTURE** An acre is equivalent to 4840 yd². A chain is a special measure of length equal to 22 yd. How many square chains are equivalent to 1 acre?

6. A map is scaled so that 1 in. = 2.5 mi. How far apart are two cities if they are 6 in. apart on the map?

◤ PRACTICE EXERCISES • For Extra Practice, see page 606.

7. Assume that y varies directly as x. When $x = 9$, $y = 30$. Find y when $x = 15$.

8. Assume that y varies directly as x. When $x = 21$, $y = 12$. Find y when $x = 35$.

9. Assume that y varies directly as x. When $x = 6$, $y = 198$. Find y when $x = 2$.

10. Assume that y varies directly as x. When $x = 10$, $y = 10$. Find y when $x = 7$.

Find the constant of variation for each.

11. y = days, x = weeks

12. y = days, x = hours

13. y = yards, x = miles

14. y = in.², x = ft²

15. **SPACE** The weight of an object on the moon varies directly with its weight on Earth. A person who weighs 141 lb on Earth weighs only 23.5 lb on the moon. About how much do you weigh on the moon if you weigh 109 lb on Earth?

16. A length of 50 ft is equivalent to a length of 1524 cm. How many centimeters are equivalent to 125 ft?

17. **PHYSICS** The distance an object falls from a given height varies directly as the square of the time the object falls. A ball falls about 312 m in 8 sec. How far did it fall during the first 4 sec?

18. **PART-TIME JOB** For every pound of cherries she picks, Hanna gets paid $1.80. Write a function rule relating her earnings in dollars to the number of pounds she picks. Use E to represent her earnings and p to represent the number of pounds she picks. Draw a graph that shows the relationship. Use it to find how many pounds of cherries Hanna must pick to earn $6.

19. WRITING MATH Write a problem that involves the relationship between two sets of data that increase or decrease together at a constant rate. Provide a solution to your problem.

20. The table shows the cost of operating a refrigerator. Write the rule for the function that describes the relationship of the cost of operating the refrigerator and the amount of time the refrigerator is in operation. What is the constant of variation?

Refrigerator Operating Costs

Cost (cents)	3	6	9	12	15
Time (hours)	1	2	3	4	5

Identify each relationship as direct variation, direct square variation or neither.

21. the perimeter of a square and the measure of its side

22. the area of a square and the measure of its side

23. the volume of a cube and the measure of its side

24. the surface area of a cube and the measure of an edge

■ EXTENDED PRACTICE EXERCISES

Assume that y varies directly as x^2.

25. How is y affected when x is doubled? When x is tripled?

26. How is y affected when x is multiplied by k?

27. CRITICAL THINKING If $\dfrac{y}{x} = \dfrac{a}{b}$, show that $\dfrac{y}{x} = \dfrac{y+a}{x+b}$.

28. If $y = kx^2$ and $k < 0$, does y increase or decrease as x increases for $x \geq 0$?

29. If $y = kx^2$ and $k > 0$, does y increase or decrease as x increases for $x \geq 0$?

30. If $y = kx^2$ and $x < 0$, does y increase or decrease as k increases for $k \geq 0$?

31. If $y = kx^2$ and $x > 0$, does y increase or decrease as k decreases for $k \geq 0$?

32. CHAPTER INVESTIGATION Look over your graphs, the list of decision ideas and other information you have about your company. Name as many relationships as you can find that you think are examples of direct variation. Discuss your ideas with the class.

■ MIXED REVIEW EXERCISES

Make a stem-and-leaf plot for each set of data. (Lesson 1-3)

33. Number of long distance calls per month
    4   18   29   27   20   31   19   17
   12   18   24   27   19   15   8

34. Number of beagles entered in AKC shows
   52   48   65   104   73   81   85   67
   58   65   76   103   88   46   92

Simplify each variable expression. (Lesson 2-5)

35. $-3(b + 4)$ **36.** $6(a - 13)$ **37.** $\dfrac{72c - 18d}{9}$ **38.** $\dfrac{15r + 6t}{-3}$

Evaluate each expression when $a = 0.5$, $b = -2$, and $c = 3$. (Lesson 2-5)

39. $-3(a - 2)$ **40.** $\dfrac{6b}{-3}$ **41.** $\dfrac{-2a + 4b}{6}$ **42.** $c(6a - 4b)$

Review and Practice Your Skills

Match each function table with the appropriate function rule.

1. $f(x) = x^2$

a.

x	0	1	2	3	4
y	0	$\frac{1}{2}$	1	$\frac{3}{2}$	2

2. $f(x) = \frac{1}{2}x$

b.

x	0	1	2	3	4
y	−3	−2	−1	0	1

3. $f(x) = x - 3$

c.

x	0	1	2	3	4
y	0	1	4	9	16

Use each domain to find ordered pair solutions of each equation.

4. $y = 4x - 1$; {−4, −2, 0, 2, 4}

5. $y = x - 5$; {−4, −2, 0, 5, 10}

6. $y = \frac{2}{3}x$; {−6, −3, 0, 3, 6}

7. $y = -|x| + 2$; {−6, −3, 0, 3, 6}

Write a function rule for each function table.

8.

x	2	3	4	5	6
y	8	11	14	17	20

9.

x	0	1	2	3	4
y	0	3	6	9	12

10.

x	2	3	4	5	6
y	2	7	14	23	34

11.

x	−4	−2	0	2	4
y	4	2	0	2	4

12. Assume y varies directly as x. When $x = 4$, $y = -12$. Find y when $x = -7$.

13. Assume y varies directly as x. When $x = 5$, $y = 15$. Find y when $x = 8$.

14. Assume y varies directly as x. When $x = 3$, $y = 18$. Find y when $x = 6$.

15. Assume y varies directly as x. When $x = -2$, $y = -13$. Find y when $x = 6$.

16. Assume y varies directly as x. When $x = -3$, $y = -17$. Find y when $x = 4$.

17. Assume y varies directly as the square of x. When $x = 4$, $y = -16$. Find y when $x = -2$.

18. Assume y varies directly as the square of x. When $x = -5$, $y = 75$. Find y when $x = 8$.

19. Suppose the price of a pizza varies directly with the square of its diameter. At the Pizza Inn an 8 in. pizza cost $6.00. How much would a 13 in. pizza cost?

20. The refund r you get varies directly with the number n of cans you recycle. If you receive a $6.50 refund for 150 cans, how much should you receive for 500 cans?

Match the slope with the graph shown. (Lesson 6-2)

21.

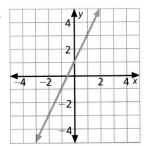

22.

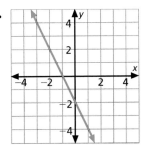

a. $\dfrac{1}{2}$

b. 2

c. -2

d. $-\dfrac{1}{2}$

23.

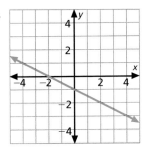

24.

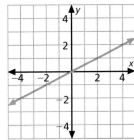

Determine if the following graphs are functions. If not, explain why. (Lesson 6-5)

25.

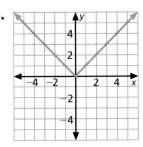

26.

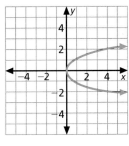

27.

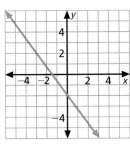

28.

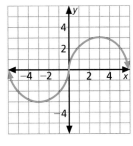

Make a function table for the following function rules. (Lesson 6-7)

29. $y = \dfrac{x}{2} + 3$, {2, 4, 6, 8}

30. $y = x^2 - 3$, {2, 4, 6, 8}

Use the given endpoints of each circle's diameter. Find each circle's center and the lengths of its diameter and radius. (Lesson 6-1)

31. $(-1, 7), (4, -3)$ **32.** $(0, 1), (-3, 7)$ **33.** $(10, 7), (-7, 10)$ **34.** $(4, 6), (-1, 1)$

6-9 Inverse Variation

Goals
- Solve problems involving inverse variation functions.
- Solve problems involving inverse square variation functions.

Applications Music, Physics, Industry, Travel

On grid paper, draw as many rectangles as you can that have an area of 80 square units.

1. Copy and complete the table of values for the length x and the width y of your rectangles. Write pairs in order of increasing values of x.

Rectangles with Area of 80

length x					
width y					

2. As x increases, does y increase or decrease? Does x vary directly as y? Explain.

3. Draw a graph to show the lengths and widths of all rectangles whose area is 80 square units. Is the graph linear?

◤ BUILD UNDERSTANDING

Some functions have this relationship: as x increases in value, y decreases in value. This relationship is an **inverse variation** if the product of the two values remain constant. For example, the time it takes to travel a given distance is inversely related to velocity.

> **Inverse Variation** $y = \dfrac{k}{x}$ or $xy = k$, where $k \neq 0$ and $x \neq 0$.

When you describe an inverse variation, you say "y varies inversely as x," or "y is inversely proportional to x." The constant of variation is k.

Example 1

MUSIC If the tension in a guitar string is constant, the frequency of a note varies inversely as the length of the string. When the string is 75 cm, the frequency is 512 hertz (Hz). Find the length of the string that produces a note at a frequency of 640 Hz.

Solution

Substitute the length of string for x and the frequency of note for y.

$$xy = k$$
$$(75)(512) = k$$
$$38,400 = k$$

Substitute 640 for y and 38,400 for k into the equation $xy = k$.

$$x(640) = 38,400$$
$$x = \frac{38,400}{640} = 60$$

The length of the string that produces a note at a frequency of 640 Hz is 60 cm.

When you stand a certain distance from a light source and then move twice as far away, the intensity of the light is one–fourth as great. If you move three times as far away, the intensity is one–ninth as great. This is an example of **inverse square variation**, stated as "y varies inversely as x^2," or "y is inversely proportional to x^2."

Inverse Square Variation	$y = \dfrac{k}{x^2}$ or $x^2y = k$, where $k \neq 0$ and $x \neq 0$.

Example 2

PHYSICS The intensity of light varies inversely as the square of the distance from the source. At a distance of 8 ft from the source, a light meter measures 24 units of intensity. How many units of intensity will the light meter measure at a distance of 32 ft from the source?

Solution

Substitute known values for x and y in the equation to find k. The distance from the light is x, and the intensity measured by the light meter is y.

$$x^2y = k$$
$$(8)^2(24) = k \qquad x = 8, y = 24$$
$$1536 = k$$

Substitute 32 for x and 1536 for k into the equation $x^2y = k$.

$$(32)^2(y) = 1536$$
$$1024y = 1536$$
$$y = \frac{1536}{1024} = 1.5$$

The light meter will measure 1.5 units of intensity at 32 ft.

Check Understanding

Suppose that y varies inversely as x. When $x = n$, $y = m$.

Find y when $x = m$.

Example 3

 Personal Tutor at mathmatters2.com

PHYSICS The force needed to pry open a crate varies inversely as the length of the crowbar used. When the length is 2 m, the force needed is 12 newtons. What force is needed if the crowbar is 1.6 m long?

Solution

Substitute 2 for x and 12 for y.

$$xy = k$$
$$2(12) = k = 24$$

Use $k = 24$ and $x = 1.6$ to find the force needed.

$$xy = k$$
$$1.6y = 24$$
$$y = \frac{24}{1.6} = 15$$

The force needed if the crowbar is 1.6 m long is 15 newtons.

1. Assume that y varies inversely as x. When $x = 3$, $y = 8$. Find y when $x = 4$.

2. Assume that y varies inversely as x. When $x = 9$, $y = 12$. Find y when $x = 6$.

TRAVEL The table shows the time taken to travel 100 yd at various speeds. The time taken varies inversely as the speed. If t represents the time in seconds and r represents the rate in yards per second, the function can be represented by the rule $t = \dfrac{100}{r}$.

Time Taken to Travel 100 yd

Time (sec)	2	4	5	8	10	12.5	20	25	50
Rate (yd/sec)	50	25	20	12.5	10	8	5	4	2

3. As the amount of time decreases, does the rate of speed increase or decrease?

4. What amount of time does it take to travel 100 yd at a rate of 8 yd/sec?

5. What rate of speed is needed if you want to travel 100 yd in 20 sec?

Identify each relationship as a direct variation or an inverse variation.

6. the area of a circle and its radius

7. the speed of a vehicle and the time it takes to travel 1 mi

8. **WRITING MATH** Write a paragraph describing the difference between direct variation and inverse variation. Give an example of each.

9. Assume that y varies inversely as x. When $x = 10$, $y = 64$. Find y when $x = 2$.

10. Assume that y varies inversely as x. When $x = 2.5$, $y = 6$. Find y when $x = 20$.

11. Assume that y varies inversely as x. When $x = 10$, $y = 1$. Find y when $x = 4$.

12. Assume that y varies inversely as x. When $x = 4$, $y = 18$. Find y when $x = 3$.

13. **ENGINEERING** The time needed to fill a water tank varies inversely as the square of the diameter of the pipe used to fill it. A pipe with a diameter of 2 cm takes 10 min to fill the tank. How long does it take to fill the tank with a pipe having a diameter of 5 cm?

14. **INDUSTRY** The volume of gas varies inversely as the pressure applied to it. Air pressure at sea level is 1 atmosphere. If a balloon rises to a point where the air pressure is 0.8 atmosphere, by what percent will its volume increase?

15. **TRAVEL** The amount paid by each member of a group chartering a bus varies inversely as the number of people in the group. When there are 30 people in the group, the cost per person is $20. What is the cost per person when there are 25 people in the group?

16. **MUSIC** The frequency of a note varies inversely as the length of a guitar string. When a string is 45 cm long, it produces a frequency of 826 Hz. What frequency will a string produce when it is 70 cm long?

Identify each relationship as a *direct variation* or an *inverse variation*.

17. the capacity of a paint can and the number of cans needed to paint a wall

18. the weight of an object and the force needed to lift it

19. the loudness of a sound and the distance from the sound source

20. the length of a rope and its weight

21. **PHYSICS** The force of attraction between two magnets varies inversely as the square of the distance between them. When two magnets are 3 cm apart, the force is 49 newtons. How great is the force when the magnets are 21 cm apart?

22. **BUSINESS** The manager of a lumber store schedules 6 employees to take inventory in an 8-hr work period. The manager assumes all employees work at the same rate. If 2 employees call in sick, how many hours will 4 employees need to take inventory?

◼ EXTENDED PRACTICE EXERCISES

When $x = 4$, $y = 20$. Find y when $x = 100$ for each variation.

23. y varies directly as x

24. y varies directly as x^2

25. y varies inversely as x

26. y varies inversely as x^2

27. **CRITICAL THINKING** Suppose that y varies inversely as x and x varies inversely as z^2. What is the relationship between y and z?

28. A triangle has an area of 48 cm². Write an equation that describes how the length of the base varies in relation to the height of the triangle.

29. A cone has a volume of 44 cm³. Write an equation that describes how the height varies in relation to the radius of the cone.

30. **CHAPTER INVESTIGATION** Look over your graphs, the list of decision ideas and other information you have about your company. Name as many relationships as you can find that you think are examples of indirect variation. Discuss your ideas with the class.

◼ MIXED REVIEW EXERCISES

In the figure, $m\angle ABC = 42°$ and $m\angle CBD = 75°$. Find each measure. (Lesson 5-8)

31. $m\widehat{AC}$

32. $m\widehat{CD}$

33. $m\widehat{ACD}$

34. $m\widehat{AED}$

Find the value of x in each figure. (Lesson 5-2)

35.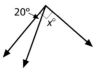

36.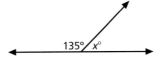

37.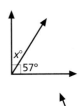

38.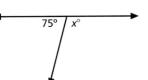

39.

40.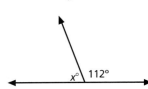

Chapter 6 Review

VOCABULARY

Choose the word from the list that best completes each statement.

1. On a coordinate plane, the horizontal axis is called the __?__.

2. A region on one side of a line is called an __?__.

3. The __?__ of a segment is the ratio of its change in vertical distance compared to its change in horizontal distance.

4. When each x-coordinate is paired with exactly one y-coordinate, the graph represents a __?__.

5. The __?__ is the point where the graph of a line crosses the y-axis.

6. If the ratio of each member of the range to the corresponding member of the domain is constant, the function is called a __?__.

7. The vertical axis on a coordinate plane is called the __?__.

8. Use a(n) __?__ to determine which half-plane should be shaded in the graph of a linear inequality.

9. The __?__ is used to determine if a graph is a function.

10. For the inequality $y < 3x + 4$, the line $y = 3x + 4$ is called the __?__ of each half-plane.

a. boundary

b. direct variation

c. function

d. inverse variation

e. open half-plane

f. slope

g. test point

h. vertical line test

i. x-axis

j. x-intercept

k. y-axis

l. y-intercept

LESSON 6-1 ◤ Distance in the Coordinate Plane, p. 244

▶ To find the distance or midpoint between any two points, (x_1, y_1) and (x_2, y_2), use these formulas: **Distance Formula**, $d = \sqrt{(x_2 - x_1) + (y_2 - y_1)^2}$, and

Midpoint Formula, $M = \left(\dfrac{x_1 + x_2}{2}, \dfrac{y_1 + y_2}{2} \right)$.

Use the graph to calculate the length of each segment. Round to the nearest tenth if necessary.

11. $\overline{AB}$

12. $\overline{BC}$

13. $\overline{AC}$

14. $\overline{EF}$

15. $\overline{EC}$

16. $\overline{CF}$

17. $\overline{GB}$

18. $\overline{GD}$

19. $\overline{BD}$

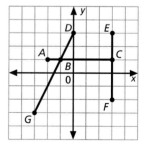

Find the midpoint of each segment. Use the graph.

20. $\overline{EF}$

21. $\overline{GD}$

22. $\overline{GB}$

LESSON 6-2 ◤ Slope of a Line, p. 248

▶ To find the **slope** of a line given two points (x_1, y_1), and (x_2, y_2) on the line, use the formula $m = \dfrac{y_2 - y_1}{x_2 - x_1}$.

23. Find the slope of the line that passes through $A(-3, -1)$ and $B(-5, 7)$.

24. Graph the line that passes through $(-2, -3)$ and has a slope of $\dfrac{2}{5}$.

Find the slope of a line that passes through the given points.

25. (1, −6), (3, 6)

26. (−2, −4), (3, 6)

27. (0, 5), (−3, 0)

28. (0, 5), (12, −4)

29. (−5, 8), (−5, −10)

30. (−10, −8), (3, 7)

LESSON 6-3 Write and Graph Linear Equations, p. 254

- **Slope intercept form** of a line is $y = mx + b$, where m = slope and b = y-intercept.
- **Point-slope form** of a line is $y - y_1 = m(x - x_1)$, where m = slope and (x_1, y_1) is a point on the line.

Write an equation of each line using the given information.

31. $m = -\frac{2}{5}$, $b = 3$

32. $m = -3$, (−1, 2)

33. (−6, 1), (−3, 4)

34. $m = 3$, (2, −5)

35. (4, 7), (−8, −5)

36. $m = \frac{4}{5}$, $b = -3$

37. (7, −3), (2, 7)

38. (2, −4), (−5, 10)

39. $m = -\frac{1}{2}$, $b = 2$

40. $m = -2$, (2, 0)

41. $m = \frac{2}{3}$, $b = 4$

42. $m = \frac{3}{2}$, (−4, 6)

LESSON 6-4 Write and Graph Linear Inequalities, p. 258

- The **graph of an inequality** is the set of all ordered pairs that make the inequality true. Any one of these pairs is a solution of the inequality.

Tell if each ordered pair is a solution of the inequality.

43. (2, 5); $y < 2x + 2$

44. (1, 4); $y \leq x - 4$

45. (0, −3); $2x - y \geq 4$

Write an inequality for each graph.

46.

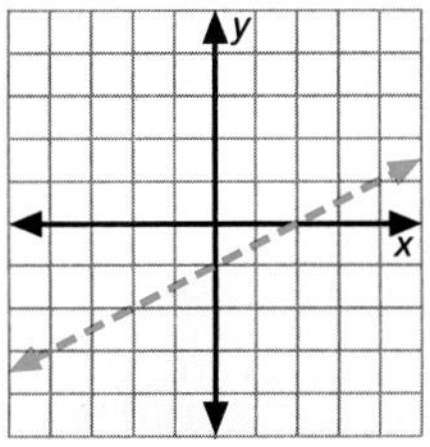

47.

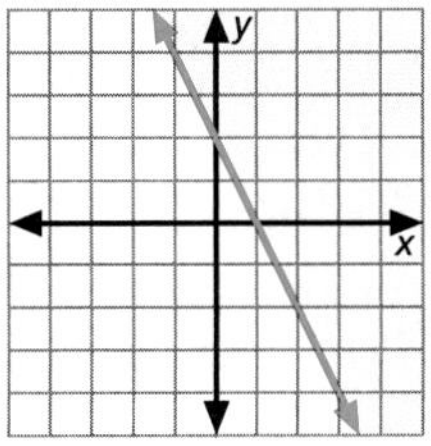

48.

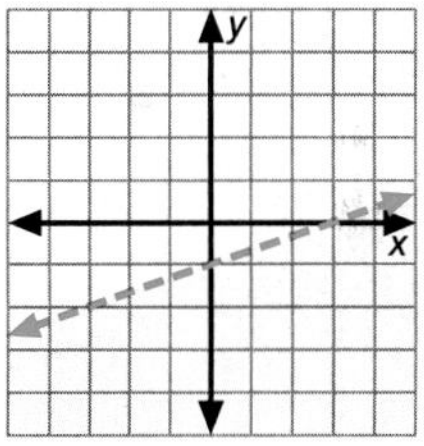

LESSON 6-5 Linear and Nonlinear Functions, p. 264

- A **function** that can be represented by a linear equation where the domain is all real numbers is called a linear function.

Graph each function for the given domain.

49. $y = x - 1$; all real numbers

50. $y = x^2 - 1$; {−2, −1, 0, 1, 2}

Determine if each graph represents a function. Explain.

51.

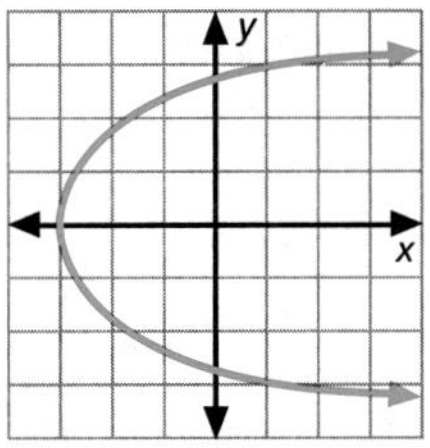

52.

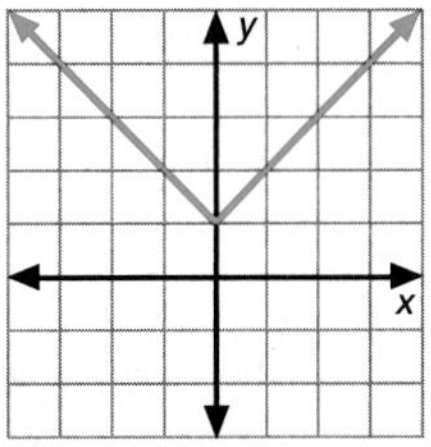

53.

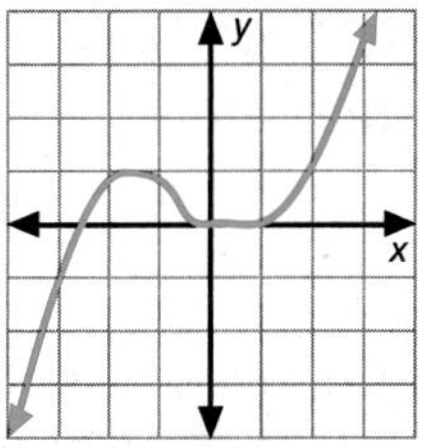

LESSON 6-6 ◢ Graph Quadratic Functions, p. 268

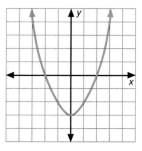

▶ A **quadratic function** is a nonlinear function whose graph is a curve known as a *parabola*. A quadratic function is given by the equation $y = Ax^2 + Bx + C$, where A, B, and C are real numbers and $A \neq 0$.

Complete each ordered pair so that it corresponds to a point on the graph.

54. $(1, ?)$ **55.** $(2, ?)$ **56.** $(0, ?)$

LESSON 6-7 ◢ Problem Solving: Patterns and Functions, p. 274

▶ An effective strategy to organize and classify the domain and range values is to **make a table**.

Use each domain to find ordered pair solutions of each equation.

57. $y = x - 4$; $\{-3, -2, 0, 1, 2\}$ **58.** $y = 2x + 2$; $\{-2, -1, 0, 1, 2\}$

Write a function rule for each function table.

59.

x	−3	0	2	3
y	18	0	8	18

60.

x	−12	0	4	16
y	−3	0	1	4

LESSON 6-8 ◢ Direct Variation, p. 276

▶ A **direct variation** is a function in the form $y = kx$, where $k \neq 0$.

▶ A **direct square variation** is in the form $y = kx^2$, where $k \neq 0$.

Assume y varies directly as x.

61. When $x = 4.2$, $y = 21$. Find y when $x = 3$.   **62.** When $x = 9$, $y = 54$. Find y when $x = 3.6$.

63. When $x = 14$, $y = 49$. Find y when $x = 4$.   **64.** When $x = 5.2$, $y = 20.8$. Find y when $x = 6.4$.

LESSON 6-9 ◢ Inverse Variation, p. 282

▶ An **inverse variation** is a function in the form $y = \dfrac{k}{x}$, where $k \neq 0$ and $x \neq 0$.

▶ An **inverse square** variation is in the form $y = \dfrac{k}{x^2}$ or $x^2 y = k$, where $k \neq 0$.

Assume y varies inversely as x.

65. When $x = 15$, $y = 8$. Find y when $x = 16$.   **66.** When $x = 5$, $y = 10$. Find y when $x = 25$.

67. When $x = \dfrac{1}{2}$, $y = 6$. Find y when $x = 12$.   **68.** When $x = 16$, $y = \dfrac{7}{8}$. Find y when $x = 2$.

CHAPTER INVESTIGATION

EXTENSION Use your graphs to write equations that model each quarter of the operating costs graph. Also write equations that model each quarter of the sales revenue graph. Select another group that researched a different company than your group. Trade sets of equations. Use the other group's equations for operating costs, and draw a graph that includes all four quarters. Do the same with the sales revenue equations. Compare your group's graphs with the other group's graphs. Discuss reasons for any differences among the graphs.

Chapter 6 Assessment

Use the graph to calculate the length of each segment.

1. $\overline{KL}$
2. $\overline{MN}$
3. $\overline{OP}$

4. Find the slope of each segment in Exercises 1–3.

5. Find the midpoint of OP.

6. Graph the line that passes through $(1, -2)$ and has a slope of 3.

7. Find the slope of a line containing the points $(0, 5)$ and $(3, -7)$.

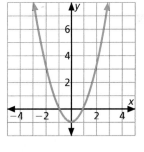

Identify the slope and y-intercept for each line. Graph each line.

8. $y = 2x + 1$
9. $y = 3x + 2$
10. $x + 2y = 4$

Write an equation of each line from the given information.

11. $m = -6, b = 1$
12. $m = 3, (-2, 4)$
13. $(3, -4), (8, -3)$

Tell if each ordered pair is a solution of the inequality.

14. $(1, 4); y > 2x - 2$
15. $(3, 3); y \le x - 3$
16. $(-6, 0); 3x - y - 12$

17. Graph the inequality $y > 2x - 4$.

18. Graph $y = -3$ for the domain of real numbers.

19. Graph $y = x + 2$ for the domain of real numbers.

Complete each ordered pair so that it corresponds to a point on the graph.

20. $(-2, ?)$

21. $(0, ?)$

22. $(?, 8)$

23. Graph $y = x^2 - 2$ for the domain of real numbers.

Use each domain to find ordered pair solutions of each equation.

24. $y = x - 3; \{-4, -2, 0, 2, 4\}$
25. $y = 2x + 3; \{-2, -1, 0, 1, 2\}$

26. Use the table to write a function rule.

x	2	4	6	8	10
f(x)	5	9	13	17	21

27. The distance a truck needs to reach a full stop varies directly as the square of its speed. From the speed of 40 mi/h, the truck needs a distance of 86 ft to stop. What distance will it need to stop from a speed of 30 mi/h?

28. The time it takes to build a wall varies inversely as the number of people doing the job. If it takes 28 h for 12 people, how long would it take 16 people?

Standardized Test Practice

Part 1 | Multiple Choice

Record your answers on the answer sheet provided by your teacher or on a sheet of paper.

1. Which sentence is *not* true? (Lesson 2-1)
   - Ⓐ All natural numbers are whole numbers.
   - Ⓑ Every whole number is a natural number.
   - Ⓒ Natural numbers are positive numbers.
   - Ⓓ Zero is neither positive nor negative.

2. Ms. Lee is planning a business trip for which she needs to rent a car. The car rental company charges $36 per day plus $0.50 per mile over 100 mi. Suppose Ms. Lee rents the car for 5 d and drives 180 mi. Which expression can be used to determine how much Ms. Lee must pay the car rental company? (Lesson 2-2)
   - Ⓐ $0.5(36) + 5(180)$
   - Ⓑ $5(36) + 0.5(180)$
   - Ⓒ $0.5(36) + 5(180 - 100)$
   - Ⓓ $5(36) + 0.5(180 - 100)$

3. There are ten socks in a drawer: 2 yellow, 2 green, 2 blue, 2 white, and 2 red. If you pull out one sock and then another sock without replacing the first, what is the probability of choosing two blue socks? (Lesson 4-5)
   - Ⓐ $\frac{1}{90}$
   - Ⓑ $\frac{1}{9}$
   - Ⓒ $\frac{1}{5}$
   - Ⓓ $\frac{2}{9}$

4. What is the value of y? (Lesson 5-4)
   - Ⓐ 35°
   - Ⓑ 72.5°
   - Ⓒ 107.5°
   - Ⓓ 125°

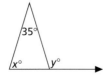

5. Which theorem or postulate can be used to prove that the two triangles are congruent? (Lesson 5-5)
   - Ⓐ AAS
   - Ⓑ ASA
   - Ⓒ SAS
   - Ⓓ SSS

6. What is the value of a in the parallelogram? (Lesson 5-6)

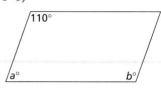

   - Ⓐ 20
   - Ⓑ 45
   - Ⓒ 70
   - Ⓓ 110

7. Find the sum of the interior angles of a 25-gon. (Lesson 5-7)
   - Ⓐ 4140°
   - Ⓑ 4320°
   - Ⓒ 4500°
   - Ⓓ 4680°

8. Sam plotted his house, school, and library on a coordinate plane. What is the shortest distance from his house to the library? (Lesson 6-1)

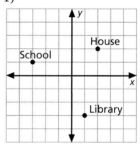

   - Ⓐ 5.8 units
   - Ⓑ 5.1 units
   - Ⓒ 4.9 units
   - Ⓓ 3.2 units

9. Raini is buying chocolate-covered pretzels from a bulk-foods store. The pretzels (P) are $1.49 per pound. Which equation represents the cost (C)? (Lesson 6-3)
   - Ⓐ $C = 1.49 - P$
   - Ⓑ $C = 1.49 + 1.49P$
   - Ⓒ $C = \frac{1.49}{P}$
   - Ⓓ $C = 1.49P$

Test-Taking Tip

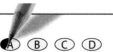

If you're having difficulty answering a question, mark it in your test booklet and go on to the next question. Make sure that you also skip the question on your answer sheet. At the end of the test, go back and answer the questions that you skipped.

Preparing for the Standardized Tests
For test-taking strategies and more practice, see pages 627–644.

Part 2 | Short Response/Grid In

Record your answers on the answer sheet provided by your teacher or on a sheet of paper.

10. The formula for Ohm's Law is $E = IR$, where E represents voltage measured in volts, I represents current measured in amperes, and R represents resistance measured in ohms. Suppose a current of 0.25 ampere flows through a resistor connected to a 12-volt battery. What is the resistance in the circuit? (Lesson 3-1)

11. The expected increase of a population of organisms is directly proportional to the current population. If a sample of 360 organisms increases by 18, by how many will a population of 9,000 increase? (Lesson 3-5)

12. If you spin the arrow on the spinner below, what is the probability that the arrow will land on an even number? (Lesson 4-1)

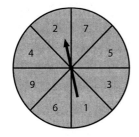

13. The two triangles are congruent. What is the value of x? (Lesson 5-5)

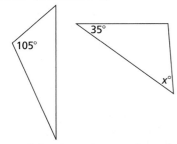

14. The sum of the interior angles of a regular polygon is 900°. Which polygon is it? (Lesson 5-7)

15. What is the slope of the line that passes through (3, 5) and (−2, −6)? (Lesson 6-2)

16. The cost of bananas varies directly with their weight. The cost of $3\frac{1}{2}$ lb of bananas is $1.12. What is the cost in dollars of $4\frac{1}{4}$ lb of bananas? (Lesson 6-8)

Part 3 | Extended Response

Record your answers on a sheet of paper. Show your work.

17. Pedro wants to construct a triangle. (Lesson 5-4)
    a. He has three pieces of wood. The lengths are 6 in., 8 in., and 16 in. Is it possible for Pedro to form a triangle using these three lengths? Explain.
    b. Two of the three sides of the triangle are 10 in. and 12 in. long. Write an inequality that expresses the possible lengths of the third side.
    c. Pedro constructs a right triangle with a base of 8 in. and a height of 12 in. What is the area of the triangle?

18. In the figure, B is the midpoint of $\overline{AE}$. $\overline{AC} \parallel \overline{DE}$. (Lesson 5-5)

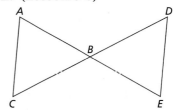

    a. Which pairs of angles are congruent? Explain.
    b. Determine whether the two triangles in the figure are congruent. Explain.

19. The table shows the loss in billions of dollars each year due to computer viruses. Determine whether the data is linear or nonlinear. Explain. (Lesson 6-2)

Year	2000	2001	2002	2003	2004
Loss (billions)	0.1	0.3	0.7	1.4	2.7

Coordinate Graphing and Transformations

Just as a map provides directions for travelers, a building blueprint provides instructions for a construction crew. Building a structure requires that architects describe the concept and specifications on paper in an established format. Construction crews interpret the blueprints and build the structure accordingly.

Architects and draftpersons use the coordinate plane to document the measurements and layout of structures and landscaped areas. They create plans for structures such as houses, office complexes, shopping malls, gardens, and monuments.

- **Campus facilities managers** (page 305) use transformations to design the layout of buildings and courtyards. Often they create attractive areas on a campus where the structures and landscaping are of a similar style.

- **Architects** (page 323) incorporate rotations, translations, and dilations of basic geometric shapes to design the interior and exterior of a building.

Math Online

mathmatters2.com/chapter_theme

World's Ten Highest Dams

Name	River, State and Country	Structural Height		Gross Reservoir Capacity		Year Completed
		feet	meters	thousands of acre feet	millions of cubic meters	
Rogun	Vakhsh, Tajikistan	1099	335	9,404	11,600	1985
Nurek	Vakhsh, Tajikistan	984	300	8,512	10,500	1980
Grande Dixence	Dixence, Switzerland	935	285	324	400	1962
Inguri	Inguri, Georgia	892	272	801	1,100	1984
Vaiont	Vaiont, Italy	859	262	137	169	1961
Manuel M. Torres	Grijalva, Mexico	856	261	1,346	1,660	1981
Tehri	Bhagirathi, India	856	261	2,869	3,540	UC
Alvaro Obregon	Mextiquic, Mexico	853	260	n.a.	n.a.	1926
Mauvoisn	Drance de Bagnes, Switzerland	820	250	146	180	1957
Alberto Lleras	Orinoca, Columbia	797	243	811	1,000	1989

Data Activity: World's Ten Highest Dams

Use the table for Questions 1–4.

1. How old is the highest dam in Switzerland?

2. Replace each ▧ with >, ≥, < or ≤ for the structural height of each set of dams.
   a. Nurek ▧ Tehri
   b. Vaiont ▧ Alberto Lleras
   c. Alvaro Obregon ▧ Rogun
   d. Mauvoism ▧ Inguri

3. When comparing any two dams, would it be correct to assume that the higher dam has the greater reservoir capacity? Explain.

4. Is the ratio of feet to meters consistent in the structural height column? Explain.

CHAPTER INVESTIGATION

One of the primary uses of geometry in the real world is architecture. Early in the planning stages of a project, an architect will draw a blueprint showing the layout of the floor plan.

Working Together

Use grid paper, a straightedge, a compass and any other necessary tools to create a blueprint for the floor plan of your dream house. This will be a diagram showing the size and location of the different rooms in the house. Use the Chapter Investigation icons to guide you in the creation of your blueprint.

The skills on these two pages are ones you have already learned. Use the examples to refresh your memory and complete the exercises. For additional practice on these and more prerequisite skills, see pages 576–584.

GRAPHING ON THE COORDINATE PLANE

In this chapter you will graph lines and figures on the coordinate plane. It is helpful to be able to identify the points on a graph.

Example   Graph $\triangle ABC$ with vertices $A(-3, -3)$, $B(-1, 2)$ and $C(8, 3)$.

- Locate point A. Start at $(0, 0)$. Go 3 units to the left and 3 units down.

- Locate point B. Start at $(0, 0)$. Go 1 unit to the left and 2 units up.

- Locate point C. Start at $(0, 0)$. Go 8 units to the right and 3 units up.

- Connect the points.

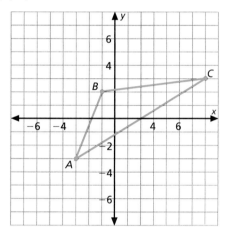

Use grid paper to draw each figure.

1. rectangle $PRST$ with vertices $P(-1, 2)$, $R(7, 2)$, $S(7, -2)$ and $T(-1, -2)$

2. triangle FGH with vertices $F(-8, 8)$, $G(-2, 7)$ and $H(-7, 4)$

3. parallelogram $ABCD$ with vertices $A(3, 2)$, $B(8, 2)$, $C(6, -1)$ and $D(1, -1)$

4. square $WXYZ$ with vertices $W(1, 3)$, $X(3, 6)$, $Y(6, 4)$ and $Z(4, 1)$

5. triangle LMN with vertices $L(-2, 2)$, $M(5, 8)$ and $N(4, 0)$

6. parallelogram $ABCD$ with vertices $A(-5, 2)$, $B(-4, 7)$, $C(2, 7)$ and $D(1, 2)$

7. trapezoid $MNOP$ with vertices $M(5, -2)$, $N(5, 1)$, $O(8, 2)$ and $P(8, -5)$

8. quadrilateral $STUV$ with vertices $S(-4, 3)$, $T(1, 2)$, $U(3, -5)$ and $V(-6, -1)$

9. rectangle $HIJK$ with vertices $H(-5, 6)$, $I(5, 6)$, $J(5, 4)$ and $K(-5, 4)$

10. triangle QRS with vertices $Q(-3, 0)$, $R(3, 3)$ and $S(1, -5)$

Write the coordinates for each point on the graph.

11. A 12. B 13. C 14. D

15. E 16. F 17. G 18. H

MIDPOINT FORMULA

Being able to apply the midpoint formula will help you as you learn about reflections across an axis or a given line.

Example   Find the midpoint of $\overline{ST}$.

The midpoint formula is $M = \left(\dfrac{x_1 + x_2}{2}, \dfrac{y_1 + y_2}{2} \right)$.

The endpoints of $\overline{ST}$ are $(-4, 3)$ and $(3, 8)$.

$M = \left(\dfrac{-4 + 3}{2}, \dfrac{3 + 8}{2} \right) = \left(\dfrac{-1}{2}, \dfrac{11}{2} \right)$

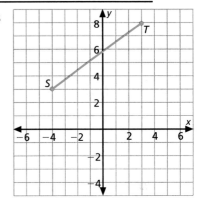

Find the midpoint of each line segment.

19.
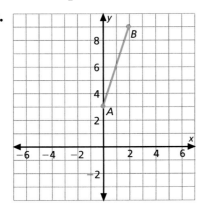

20.
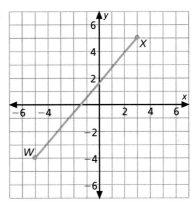

21. $C(8, 4), D(-2, 6)$ **22.** $R(-3, 5), S(6, -2)$ **23.** $K(8, 0), L(4, -1)$

24. $Y(-7, -4), Z(-8, 5)$ **25.** $D(0, 5), E(-4, 9)$ **26.** $J(3, -6), K(0, -3)$

DRAW AND MEASURE ANGLES

In this chapter you will use a protractor to rotate figures. It will be helpful to practice using a protractor to draw and measure angles.

Use a protractor to measure $\angle XYZ$ in each figure.

27.
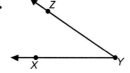

28.
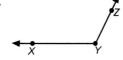

29.

Use a protractor to draw angles with the given measure.

30. 90°   **31.** 35°   **32.** 120°   **33.** 15°   **34.** 160°

Translations in the Coordinate Plane

Goals ■ Describe and graph translation images on a coordinate plane.

Applications Music, Architecture, Sports, Photography

Use grid paper, a ruler, and scissors.

1. Use the ruler to draw an isosceles triangle on grid paper. Make the base 4 units and the height 5 units. Cut out the triangle and label the vertices *A*, *B* and *C*.

2. Draw a coordinate plane. Label each axis from −12 to 12.

3. Place the triangle in the third quadrant so that the coordinates of each vertex are integers. Trace it and label it Triangle 1. Record the coordinates of the vertices.

4. Carefully slide the triangle up 7 units and to the right 8 units. Trace the triangle and label it Triangle 2. Record the coordinates of the vertices.

5. Compare the *x*-coordinates and *y*-coordinates of both triangles. What do you notice?

◥ BUILD UNDERSTANDING

A **translation**, or *slide*, of a figure produces a new figure exactly like the original. The new figure is the **image** of the original figure, and the original figure is the **preimage**. A move like a translation is called a **transformation** of a figure.

As a figure is translated, you can imagine all its points sliding along a plane at once in the same direction and for the same distance. Therefore, the sides and angles of an image are equal in measure to the sides and angles of its preimage. Also, each side of an image is parallel to the corresponding side of its preimage. An image and its preimage are congruent figures.

Example 1

Graph the image of △*ABC* with vertices *A*(1, −4), *B*(2, −2), and *C*(5, −3) under a translation of 7 units up and 3 units left.

Solution

First graph △*ABC*. To slide the image up 7 units, add 7 to each *y*-coordinate. To slide the image left 3 units, subtract 3 from each *x*-coordinate. Graph △*A'B'C'*.

$$A(1, -4) \rightarrow A'(1 - 3, -4 + 7) = A'(-2, 3)$$

$$B(2, -2) \rightarrow B'(2 - 3, -2 + 7) = B'(-1, 5)$$

$$C(5, -3) \rightarrow C'(5 - 3, -3 + 7) = C'(2, 4)$$

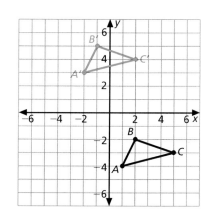

A translation can be described by a rule stating the number of units to the left or right and the number of units up or down.

Reading Math

In describing a translation, an image is said to be translated "under" a given translation. This does not necessarily mean that the image is located beneath the preimage on the plane.

Example 2

Write the rule that describes the translation of $\triangle RST$ to $\triangle R'S'T'$.

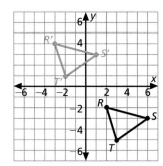

Solution

Look for a pattern between the x-coordinates and y-coordinates of each vertex of the preimage ($\triangle RST$) and the x-coordinates and y-coordinates of each vertex of the image ($\triangle R'S'T'$).

$R(2, -2)$ $S(6, -3)$ $T(3, -5)$

$R'(-3, 4)$ $S'(1, 3)$ $T'(-2, 1)$

To find the x-coordinate of each image vertex, you must subtract 5 from the x-coordinate of each preimage vertex. To find the y-coordinate of each image vertex, you must add 6 to the y-coordinate of each preimage vertex.

The rule $(x, y) \rightarrow (x - 5, y + 6)$ describes the translation of $\triangle RST$ 5 units to the left and 6 units up.

Example 3   **Personal Tutor at mathmatters2.com**

SPORTS Coach Higgins diagrams a play for his basketball team. Each of the five players is represented by the letters A, B, C, D and E. Draw the image under the given translations. Which player ends up going to the basket?

$A(-4, -3)$ under a translation of 5 right and 11 up.

$B(-4, 8)$ under a translation of 0 left or right and 7 down.

$C(3, 5)$ under a translation of 6 left and 7 down.

$D(3, -3)$ under a translation of 3 left and 1 down.

$E(5, 7)$ under a translation of 0 left or right and 0 up or down.

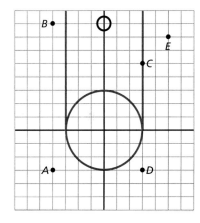

Solution

Apply each translation to graph the five images.

$A(-4, -3) \rightarrow A'(-4 + 5, -3 + 11) = A'(1, 8)$

$B(-4, 8) \rightarrow B'(-4 + 0, 8 - 7) = B'(-4, 1)$

$C(3, 5) \rightarrow C'(3 - 6, 5 - 7) = C'(-3, -2)$

$D(3, -3) \rightarrow D'(3 - 3, -3 - 1) = D'(0, -4)$

$E(5, 7) \rightarrow E'(5 + 0, 7 + 0) = E'(5, 7)$

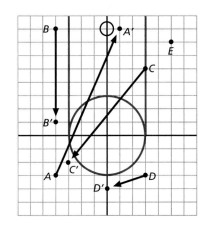

Each player's translation is drawn in the figure. Player A ends up going to the basket.

Math Online mathmatters2.com/extra_examples

Graph the image of rectangle DEFG under the given translations.

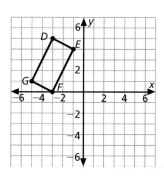

1. 4 units right   **2.** 7 units down

3. 1 unit up and 6 units right   **4.** 5 units right and 5 units down

5. What would the coordinates of rectangle *DEFG* be under a translation of 557 units to the left and 159 units down?

Write the rule that describes each translation.

6.

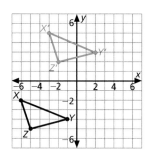

7.

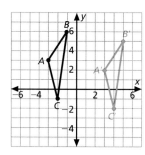

8.

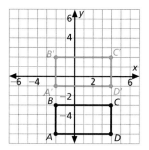

Pentagon *HIJKL* has vertices *H*(−1, 5), *I*(2, 6), *J*(5, 5), *K*(3, 2), and *L*(1, 2). Graph the pentagon and its image under the given translations.

9. 4 units left   **10.** 7 units down

11. 3 units up   **12.** 5 units right

13. 7 units left and 6 units down   **14.** 3 units up and 3 units right

15. Is pentagon *HIJKL* a regular pentagon? Explain.

Write the rule that describes each translation.

16.

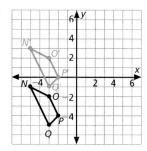

17.

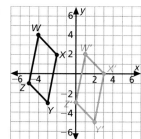

18.

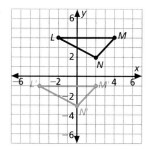

19. WRITING MATH Suppose a preimage is in Quadrant III of the coordinate plane. Describe a translation under which the image is in Quadrant I.

20. MUSIC The diagram shows the musical scale that Jontay uses when he practices the flute each day. Copy the diagram, and sketch the next 16 notes that Jontay will play if he repeats what is shown. Describe how this is an example of a translation.

21. ARCHITECTURE An architect designs a deck for a client. To access the deck, there is a glass sliding door. Describe how the sliding door is an example of a translation.

22. PHOTOGRAPHY When you take a picture with a camera, a shutter opens to expose the film to light. The amount of time that the shutter remains open is known as the *shutter speed*. To illustrate motion in a photograph, a photographer can use a long shutter speed. This suggests a translated image. Sketch a picture that demonstrates a translation in a photograph.

23. ERROR ALERT In describing $\triangle ABC$ with vertices $A(1, 4)$, $B(4, -2)$, and $C(7, 9)$ under a translation of 2 units up and 4 units right, Mandy arrives at $\triangle A'B'C'$ with vertices $A'(3, 8)$, $B'(6, 2)$, and $C'(9, 13)$. What mistake has Mandy made, and what are the correct vertices?

■ EXTENDED PRACTICE EXERCISES

24. CRITICAL THINKING Triangle QRS with vertices $Q(-4, 7)$, $R(-1, 3)$ and $S(-5, 2)$ is translated using the rule $(x, y) \rightarrow (x + 7, y - 1)$ to create $\triangle Q'R'S'$. Triangle $Q'R'S'$ is then translated using the rule $(x, y) \rightarrow (x - 4, y - 7)$ to create $\triangle Q''R''S''$. In words, describe the position of $\triangle Q''R''S''$ in relation to $\triangle QRS$ in the coordinate plane.

25. In the figure shown, is the transformation $\triangle ABC \rightarrow \triangle A'B'C'$ a translation? Explain why or why not.

26. GEOMETRY SOFTWARE Use geometry software to graph several polygons. Then use the software's transformation functions to translate the images in different directions.

27. CHAPTER INVESTIGATION On grid paper, draw the general layout and shape of the main floor of your house. It can be a rectangle, U shape or any other design. Make the blueprint large enough to fill most of the page.

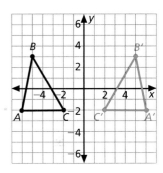

■ MIXED REVIEW EXERCISES

Find each probability. (Lesson 4-4)

28. Two number cubes are tossed. What is the probability that the sum of the numbers is 3 or 4?

29. One card is drawn at random from a standard deck of 52 cards. What is the probability that the card is a diamond or a king?

30. Two number cubes are tossed. What is the probability that both cubes show an odd number?

Solve. (Lesson 3-2)

31. $8p = 32$

32. $6r = 3$

33. $z - 8 = 4$

34. $\frac{1}{2}f = 5$

35. $k - 6 = |-2|$

36. $0.3m = 1.8$

37. $8.2 + n = 4.7$

38. $\frac{t}{4} = 1.6$

39. $g + 3 = |-17|$

40. $2d = -8$

41. $-3c = -18$

42. $k + 3 = 11$

Reflections in the Coordinate Plane

Goals
- Graph reflection images on a coordinate plane.
- Identify lines of reflection.

Applications Art, Landscaping, Navigation, Architecture

Use grid paper, a ruler and food coloring or paint.

1. On grid paper, draw a coordinate plane.

2. Place a small drop of food coloring or paint in the Quadrant I. Carefully fold the paper in half along the *y*-axis.

3. Unfold your paper. There should now be a shape on both sides of the *y*-axis.

4. Describe the shapes on each side of the fold line.

■ BUILD UNDERSTANDING

Translations are one type of transformation that can be applied to figures in the coordinate plane to produce an image that is exactly like its preimage. Another type of transformation that yields a congruent figure is a **reflection**, or *flip*. Under a reflection, a figure is *reflected*, or *flipped*, across a *line of reflection*.

When you reflect a point across the *y*-axis, the *y*-coordinate remains the same, but the *x*-coordinate is made its opposite. The reflection of the point (x, y) across the *y*-axis is the point $(-x, y)$.

When you reflect a point across the *x*-axis, the *x*-coordinate remains the same, but the *y*-coordinate is made its opposite. The reflection of the point (x, y) across the *x*-axis is $(x, -y)$.

Example 1

Graph the image of $\triangle DEF$ with vertices $D(4, 5)$, $E(5, 1)$ and $F(1, 3)$ reflected across the *x*-axis.

Solution

First graph $\triangle DEF$. Multiply the *y*-coordinate of each vertex by -1.

$$D(4, 5) \rightarrow D'(4, -5)$$

$$E(5, 1) \rightarrow E'(5, -1)$$

$$F(1, 3) \rightarrow F'(1, -3)$$

Graph $\triangle D'E'F'$.

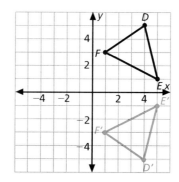

Check Understanding

Name the coordinates of $\triangle DEF$ in Example 1 when it is reflected across the *y*-axis.

Figures can be reflected over lines other than the *x*-axis or *y*-axis. When a point (x, y) is reflected across the line $y = x$, the image is the point (y, x). When a point is reflected across the line $y = -x$, the image is the point $(-y, -x)$.

Example 2

 Personal Tutor at mathmatters2.com

Graph the image of $\triangle JKL$ with vertices $J(-4, 2)$, $K(-2, -1)$, and $L(-5, -3)$ under a reflection across the line $y = x$.

Solution

Graph $\triangle JKL$ and the line $y = x$. Transpose the *x*-coordinate and *y*-coordinate of each vertex using the rule $(x, y) \rightarrow (y, x)$.

$$J(-4, 2) \rightarrow J'(2, -4)$$

$$K(-2, -1) \rightarrow K'(-1, -2)$$

$$L(-5, -3) \rightarrow L'(-3, -5)$$

Graph $\triangle J'K'L'$.

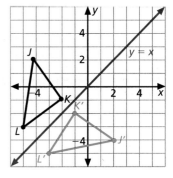

Each point of a reflection image is the same distance from the line of reflection as the corresponding point of its preimage. In other words, the line of reflection lies directly in the middle between the image and its preimage. You can use this fact to help find reflection lines.

Example 3

ART Mani is designing a pattern for a picture she is going to paint. In the pattern, there is a triangle and its reflection image. Triangle QRS has vertices $Q(-7, 1)$, $R(-3, -2)$ and $S(-4, -4)$; and its image has vertices $Q'(5, 1)$, $R'(1, -2)$ and $S'(2, -4)$. Graph the triangles and find the line of reflection.

Check Understanding

To find a line of reflection, is it necessary to find the midpoint of each segment that connects the vertices of a figure and its reflected image? If not, how many midpoints is it necessary to find? Explain.

Solution

First graph $\triangle QRS$ and $\triangle Q'R'S'$. Imagine line segments connecting each pair of corresponding vertices. Use the midpoint formula to find the midpoint of each of these three line segments. Recall the midpoint formula.

$$M = \left(\frac{x_1 + x_2}{2}, \frac{y_1 + y_2}{2} \right)$$

For $\overline{QQ'}$, $M = \left(\frac{-7 + 5}{2}, \frac{1 + 1}{2} \right) = \left(\frac{-2}{2}, \frac{2}{2} \right) = (-1, 1)$

For $\overline{RR'}$, $M = \left(\frac{-3 + 1}{2}, \frac{-2 + (-2)}{2} \right) = \left(\frac{-2}{2}, \frac{-4}{2} \right) = (-1, -2)$

For $\overline{SS'}$, $M = \left(\frac{-4 + 2}{2}, \frac{-4 + (-4)}{2} \right) = \left(\frac{-2}{2}, \frac{-8}{2} \right) = (-1, -4)$

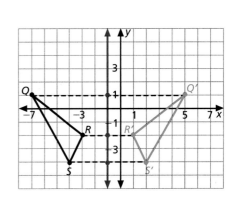

Draw a line through the midpoints. The equation of the line of reflection is $x = -1$.

Graph △XYZ and its image under the given reflection.

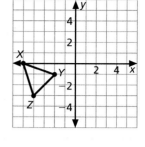

1. y-axis   2. x-axis   3. $y = x$

4. $y = -x$ 5. $x = 1$ 6. $y = 3$

7. Graph △MNO with vertices $M(-1, 3)$, $N(3, 3)$ and $O(4, -1)$ and its reflected image △M'N'O' with vertices $M'(-2, 2)$, $N'(-2, -2)$ and $O'(2, -3)$ on a coordinate plane. Then graph and identify the line of reflection.

8. The reflected image of $\overline{JK}$ whose endpoints are $J(3, 1)$ and $K(3, -10)$ is $\overline{J'K'}$ with endpoints $J'(-7, 1)$ and $K'(-7, -10)$. What is the line of reflection?

9. **WRITING MATH** Suppose a preimage is contained entirely in Quadrant IV. Describe how to graph its reflected image across the x-axis. In which quadrant is the image located?

Graph quadrilateral HIJK and its image under the given reflection.

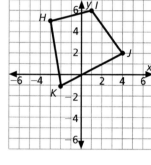

10. y-axis   11. x-axis   12. $y = x$

13. $y = -x$ 14. $x = 6$ 15. $y = -4$

Give the coordinates of the image of each point under a reflection across the given line.

16. $(5, -3)$; x-axis   17. $(4, 0)$; $y = -x$ 18. $(-2, -7)$; y-axis

19. $(-6, 8)$; $y = x$ 20. $(0, -6)$; x-axis   21. $(3, -3)$; $y = -x$

Copy each figure and its reflected image. Identify the line of reflection.

22.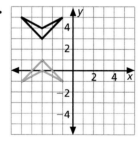

23.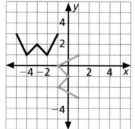

24. **NAVIGATION** A periscope uses two mirrors positioned parallel to each other at 45° angles to allow you to see above the line of sight. Explain how a periscope is an example of a reflection.

25. **WRITING MATH** Suppose you are given △ABC on the left side of a vertical line m. Explain how to find the reflection of △ABC across line m. Include a diagram with your explanation.

26. LANDSCAPING A landscaping company is sketching a preliminary design for a garden that will be on both sides of a trail in a park. The corners of one side are $(-3, -4)$, $(-3, 7)$, $(-9, -4)$ and $(-9, 7)$. It is to be reflected across the y-axis. What are the coordinates of the other side of the garden?

Copy each diagram onto grid paper. Then sketch the image of each set of squares under a reflection across line l.

27.

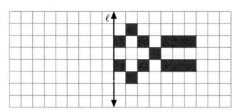

28.

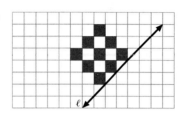

29. ARCHITECTURE In designing the twin towers of an office complex, an architect makes a sketch on grid paper. The vertices of one of the towers are $(-10, 0)$, $(-5, 0)$, $(-7.5, 14)$, $(-5, 11)$, and $(-10, 11)$. If he wishes to reflect the tower across the line $x = -1$, what will the vertices of the image be? Draw both the image and preimage on grid paper.

EXTENDED PRACTICE EXERCISES

30. GEOMETRY SOFTWARE Use geometry software to graph the lines $y = -3$ and $x = 4$, on separate coordinate planes. Draw a figure on each plane, and use the software's tools to reflect it across the line. Write a rule for each reflection.

31. CRITICAL THINKING Suppose $\triangle ABC$ is reflected across the y-axis. What does the reflection of this image across the y-axis look like?

32. DATA FILE Refer to the data on the most visited sites in the National Park System on page 566. Use an atlas to locate three of the parks, and form a triangle by connecting the vertices on a copy of the atlas or map. Draw a line on the copy of the map and reflect the triangle across the line.

Blue Ridge Parkway, North Carolina

MIXED REVIEW EXERCISES

Find the mean, median, mode, and range of each set of football data. (Lesson 1-2)

33. Rushing yards (per game)
208 94 132 169 152 145
98 145 152 165 212

34. Passing yards (per game)
122 138 46 154 92 131 197
119 185 131 182 75 105

35. Field goals (lengths in yards)
26 26 32 50 43 32 20
35 41 32 18 27 28 39

36. Kickoff returns (lengths in yards)
8 16 9 91 10 12 18 33
6 87 7 13 17 22 26 20

Find the measure of the complement and supplement of each angle. (Lesson 5-2)

37. $m\angle 31°$ **38.** $m\angle 15°$ **39.** $m\angle 79°$ **40.** $m\angle 58°$

41. $m\angle 22°$ **42.** $m\angle 4°$ **43.** $m\angle 84°$ **44.** $m\angle 61°$

Review and Practice Your Skills

Graph the image of △*LMN* under the given translations.

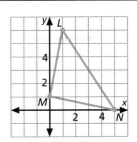

1. 3 units right

2. 6 units down

3. 2 units up and 5 units left

4. 3 units right and 3 units down

Write the rule that describes each translation.

5.

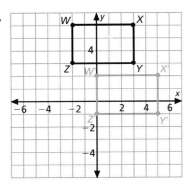

6.

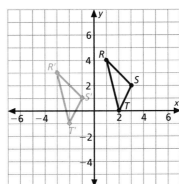

7.

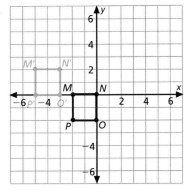

Triangle *ABC* has vertices *A* (−2, 1), *B* (−1, 0) and *C* (−3, −3). What are the coordinates of the vertices of the image under each translation?

8. 5 units right

9. 2 units up

10. 6 units left and 5 units up

11. 4 units right and 9 units down

Graph quadrilateral *WXYZ* and its image under the given reflection.

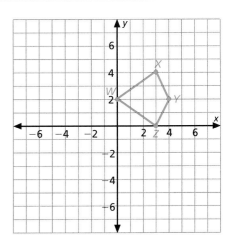

12. *y*-axis

13. *x*-axis

14. $y = x$

15. $y = -x$

16. $y = -2$

17. $x = 5$

Give the coordinates of the image of each point under a reflection across the given line.

18. (4, 1); *x*-axis

19. (−4, 4); *y*-axis

20. (5, 2); $y = x$

21. (1, 6); $y = -x$

22. (−2, 2); *x*-axis

23. (−3, 0); *y*-axis

Write the rule that describes each translation. (Lesson 7-1)

24.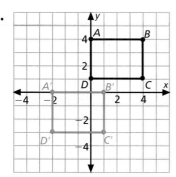

25.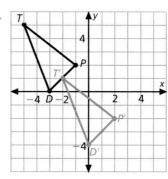

26.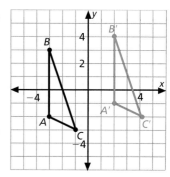

Give the coordinates of the image of each point under a reflection across the given line. (Lesson 7-2)

27. $(5, 4)$; x-axis

28. $(-1, -3)$; y-axis

29. $(0, -6)$; $y = -x$

30. $(2, -4)$; $y = x$

31. $(0, 3)$; $y = 1$

32. $(-12, 6)$; x-axis

Math*Works* Career – Campus Facilities Manager
Workplace Knowhow

A campus facilities manager is involved in many aspects of a college or university. They supervise landscaping, maintenance, renovations and new building design. A campus facilities manager is coordinating the design of a new residence complex consisting of four buildings that enclose a courtyard. The campus facilities manager decided to design the four new buildings using the mathematical principles of translation and reflection. The figure shows the location of the first residence building and the courtyard. Complete the following exercises to determine the shapes and locations of the other three buildings.

1. Reflect Building *ABCD* across the line $x = 2$ to determine the location of Building 2, $A'B'C'D'$.

2. Across which line would you reflect Building 2 so the majority of Building 3 ($A''B''C''D''$) would be in the first quadrant on the coordinate grid? Reflect Building 2 across that line.

3. Describe how you would create Building 4 so that the new residence complexes enclose the courtyard. Create Building 4 through a transformation.

4. Would the reflection of Building 1 across the line $y = -x$ create Building 3? Would the reflection of Building 2 across the line $y = x$ create Building 4? Explain.

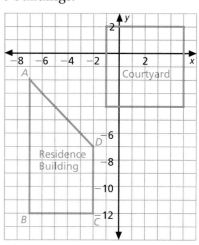

7-3 Rotations in the Coordinate Plane

Goals
- Graph rotated images in the coordinate plane.
- Identify centers, angles and directions of rotations.

Applications Machinery, Art, Architecture, Recreation

Refer to the photograph for Questions 1–3.

1. From the side at which the photograph was taken, does this Ferris wheel turn clockwise or counterclockwise around its center? Explain.

2. What fractional part of a turn does it take for a rider to get from the bottom to the top of a Ferris wheel?

3. How many degrees does a Ferris wheel turn while a rider goes completely around one time?

BUILD UNDERSTANDING

You have learned about two types of transformations – translations and reflections. A third transformation that produces a new figure exactly like the original is a **rotation**, or *turn*. Under a rotation, a figure is rotated, or turned, about a given point. A complete turn is 360°, a half turn is 180° and a quarter turn is 90°.

The description of a rotation includes three pieces of information.

1. The **center of rotation**, or point about which the figure is rotated

2. The amount of turn expressed as a fractional part of a whole turn, or as the **angle of rotation** in degrees

3. The direction of rotation – clockwise or counterclockwise

When you rotate a point 180° clockwise about (0, 0), both the x-coordinate and the y-coordinate become their opposites.

Check Understanding

How many degrees are in a three-quarter turn? How many degrees are in a full turn?

Example 1

Draw the image of $\triangle QRS$ with vertices $Q(2, 4)$, $R(1, 1)$ and $S(4, 1)$ under a rotation of 180° clockwise about $(0, 0)$.

Solution

The rotation is 180°, so the x-coordinate and y-coordinate become their opposites. Multiply each coordinate by -1.

$$Q(2, 4) \rightarrow Q'(2(-1), 4(-1)) = Q'(-2, -4)$$

$$R(1, 1) \rightarrow R'(1(-1), 1(-1)) = R'(-1, -1)$$

$$S(4, 1) \rightarrow S'(4(-1), 1(-1)) = S'(-4, -1)$$

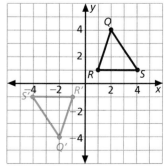

Use these rules when the angle of rotation is 90°, 180°, and 360° and the center is (0, 0).

Rules of Rotations about (0, 0)	90° clockwise $(x, y) \rightarrow (y, -x)$
	90° counterclockwise $(x, y) \rightarrow (-y, x)$
	180° clockwise or counterclockwise $(x, y) \rightarrow (-x, -y)$
	360° clockwise or counterclockwise $(x, y) \rightarrow (x, y)$

Example 2

MACHINERY A large mechanical shovel is used to move gravel. If the coordinates of the shovel are $A(0, 0)$, $B(-2, 2)$, $C(-4, 4)$, and $D(-4, 2)$ after being rotated clockwise 90° about (0, 0), what were the coordinates in its original position?

Solution

To find the coordinates of the shovel in its original position, rotate its current image 90° counterclockwise about the origin. Do so by multiplying the y-coordinate of each point by -1. Then transpose the x- and y-coordinates.

$$A(0, 0) \rightarrow A'(0, 0(-1)) = A'(0, 0)$$

$$B(-2, 2) \rightarrow B'(-2, 2(-1)) = B'(-2, -2)$$

$$C(-4, 4) \rightarrow C'(-4, 4(-1)) = C'(-4, -4)$$

$$D(-4, 2) \rightarrow D'(-4, 2(-1)) = D'(-4, -2)$$

Check Understanding

What rotation would result in a figure fitting back on itself?

To rotate figures when the rules stated above do not apply, use a protractor, compass and ruler. Use a protractor to measure angles where the center of rotation is the vertex. The compass and ruler are used for measuring distances.

Example 3

Draw the image of $\triangle ABC$ after a 120° turn clockwise about point P.

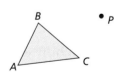

Solution

Step 1: Draw a segment from vertex C to point P.

Step 2: Use a protractor to draw a ray from point P that creates a 120° angle with $\overline{CP}$.

Step 3: Use a compass to measure the length of $\overline{CP}$.

Step 4: Use this measure to locate point C' on the ray drawn in Step 2. Label point C'.

Repeat steps 1–4 to locate A' and B'. Draw $\overline{A'B'}$, $\overline{B'C'}$, and $\overline{C'A'}$.

The rotated image is $\triangle A'B'C'$.

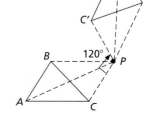

To find the center of rotation, draw segments that join corresponding vertices and find each midpoint. Construct a perpendicular bisector to each segment. Locate the point of intersection of the three perpendicular bisectors.

Example 4

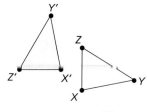

For the image of △XYZ, identify the center of rotation, the angle of rotation and the direction of rotation.

Solution

Draw a segment connecting each pair of corresponding vertices (shown in red). Construct the perpendicular bisectors of $\overline{XX'}$, $\overline{YY'}$ and $\overline{ZZ'}$ (shown in blue). Label the point where the bisectors intersect as *T*. Draw a segment that connects two corresponding vertices to *T*. Then measure the angle formed by these segments.

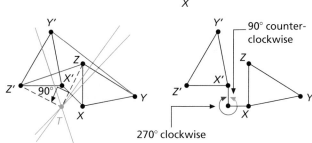

Point *T* is the center of rotation, and the angle rotation is either 90° counterclockwise or 270° clockwise.

TRY THESE EXERCISES

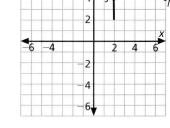

Graph the flag containing points *J, K, L* and *M* on a coordinate plane. Graph its image under the given rotation about zero.

1. 90° clockwise   2. 180° counterclockwise

3. 135° counterclockwise 4. 220° counterclockwise

5. **WRITING MATH** Write step-by-step instructions for locating the center of rotation of any image and its preimage.

Copy △XYZ and its rotation image on grid paper.

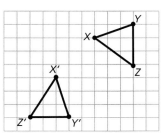

6. Identify the center of rotation of the image. Label it point *T*.

7. Identify the angle of rotation and the direction of rotation of the image.

8. Identify another angle and direction of rotation of the image.

PRACTICE EXERCISES • For Extra Practice, see page 608.

On grid paper, graph each figure and its image under the given rotation.

9. *A*(0, 0), *B*(1, 2), *C*(3, 3) and *D*(2, 1); 60° clockwise about (0, 0)

10. *X*(−3, 1), *Y*(−3, 2) and *Z*(−1, 1); 180° counterclockwise about (0, 0)

11. *D*(−2, 2), *E*(2, 2), *F*(2, −2) and *G*(−2, −2); 100° counterclockwise about (0, 0)

12. *R*(−3, 5), *S*(2, 0) and *T*(−1, −3); 360° clockwise about (0, 0)

Trace each figure and its rotation image. Identify the center of rotation, the angle of rotation and the direction of rotation.

13.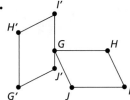

14.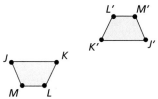

15. **ART** Pentagon *OLCEK* represents one-fourth of a finished display space for an art festival. Find the vertices of the rotation image of pentagon *OLCEK* after a 90° clockwise turn about (0, 0). (Each grid block is 100-ft long.)

16. Complete the display in Exercise 15. Draw the preimage and three rotation images of 90° clockwise about (0, 0).

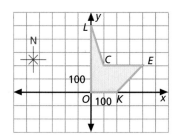

Refer to the figure shown.

17. Which triangle is the rotation image of △*A* about (−2, 0)?

18. Which triangle is the translation image of △*A*?

19. Which triangle is the rotation image of △*A* after a turn of 180° clockwise about (0, 0)?

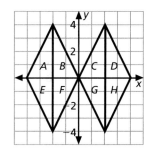

■ EXTENDED PRACTICE EXERCISES

20. **CRITICAL THINKING** Does the order that transformations are performed affect the image produced? If it does affect the image, sketch an example.

    a. translation, then translation
    b. translation, then reflection
    c. translation, then rotation
    d. reflection, then rotation

21. The gondola in an amusement park ride begins its swing at *P* and takes 6 sec to swing counterclockwise from *P* to *Q* and back to *P*. Is the gondola swinging clockwise or counterclockwise after swinging for 43 sec?

22. **ARCHITECTURE** Use a protractor to estimate the angle of rotation that the Tower of Pisa structure has undergone over the years.

Tower of Pisa, Italy

■ MIXED REVIEW EXERCISES

State whether the pair of triangles is congruent by SAS, ASA, or SSS. Otherwise, write *not congruent*. (Lesson 5-5)

23.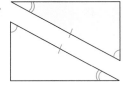

24.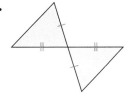

25.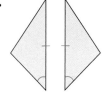

Find the distance between each pair of points. Round to the nearest tenth. (Lesson 6-1)

26. (4, 2) and (3, 6)

27. (2, 1) and (0, 4)

28. (−2, 3) and (−4, 2)

Line Symmetry and Rotational Symmetry

Goals   ■ Identify lines of symmetry.
             ■ Identify order of rotational symmetry.

Applications   Industry, Weather, Architecture, Number theory, Art

---

Work with a partner. You will need scissors and a mirror.

1. Draw and cut out a scalene triangle, an isosceles triangle and an equilateral triangle.

2. Place a mirror on each triangle so that the part in front of the mirror along with the reflection in the mirror forms the original figure. For each you may find one way, more than one way or no way.

3. For each way you found in Step 2 to form the original triangle, have your partner hold the mirror in place while you a draw a line on the triangle along the bottom edge of the mirror.

4. How many lines did you draw on each triangle?

■ BUILD UNDERSTANDING

Many figures are drawn so that a line can divide them exactly into two equal, overlapping parts. A plane figure has **line symmetry** (sometimes called *reflection symmetry*) if you can divide it along a line into two parts that are mirror images of each other. The line is called a **line of symmetry**. Some figures have one line of symmetry. Others have two or more, and still others have none.

Check Understanding

Draw all the types of triangles that have line symmetry. Show the lines of symmetry.

Example 1

Trace each figure, and draw all its lines of symmetry. If a figure has no lines of symmetry, write *none*.

a.   b.    c.

Solution

a.   b.    c.

two lines of symmetry   one line of symmetry   none

Example 2

Half of a figure and its line of symmetry is shown. Complete the figure by drawing the other half.

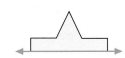

Solution

Draw a reflection of the figure across the line of symmetry. The completed figure is the same on each side of the line of symmetry.

Sometimes a figure does not have line symmetry. However, a figure that has a point about which it is rotated so that it fits exactly over its original position more than once during a complete turn has **rotational symmetry**.

If you rotate the figure shown about point *T* one complete turn, it will fit over the original position of the figure 3 times. Therefore, the figure has rotational symmetry, and its *order of rotational symmetry* is 3.

Example 3

Give the order of rotational symmetry for each figure.

a.

b.

c.

Solution

a. The figure fits over its original position 2 times during a complete turn, so its order of rotational symmetry is 2.

b. The figure fits over its original position 5 times during a complete turn, so its order of rotational symmetry is 5.

c. The figure fits over its original position 4 times during a complete turn, so its order of rotational symmetry is 4.

Example 4

INDUSTRY Give the order of rotational symmetry for each industrial tool.

a. the blade of a circular saw with 100 teeth

b. the large wheel of a steamroller

Solution

a. During a complete turn, the blade will fit over its original position 100 times, so its order of rotational symmetry is 100.

b. During a complete turn, the wheel will always look like its original position, so its order of rotational symmetry is infinite.

Trace each figure, and draw all lines of symmetry. If applicable, write *none*.

1.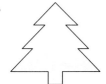

2.

3.

Give the order of rotational symmetry for each figure.

4.

5.

6.

 7. **WRITING MATH** Is it possible for a figure to have both line symmetry and rotational symmetry? If so, give an example.

Tell whether each dashed line is a line of symmetry. If not, trace the line and one side of the figure. Complete the drawing so that it has line symmetry.

8.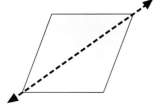

9.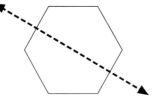

10.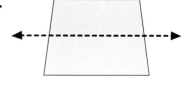

Give the order of rotational symmetry for each figure.

11.

12.

13.

14. **WEATHER** Describe the symmetries of the snowflake shown.

 15. **WRITING MATH** Explain why every circle has infinitely many lines of symmetry.

16. **ARCHITECTURE** Draw the front of a building that has line symmetry. Explain the position of the line of symmetry.

17. NUMBER THEORY The numbers 121 and 1221 are called *palindromes* because they read the same forward or backward. Using block numbers, find a palindrome that has line symmetry. Find one with rotational symmetry.

ART Describe the symmetries of each woven basket design.

18.  **19.**  **20.**

21. List four capital letters that have a horizontal line of symmetry. List four capital letters that have a vertical line of symmetry.

EXTENDED PRACTICE EXERCISES

Draw a polygon with exactly the number of lines of symmetry given.

22. 1 line   **23.** 2 lines   **24.** 3 lines   **25.** 4 lines

Draw a polygon with the order of rotational symmetry given.

26. order 2   **27.** order 3   **28.** order 4   **29.** order 5

30. CRITICAL THINKING How many lines of symmetry does a regular polygon with n sides have?

31. What is the order of rotational symmetry of a regular polygon with n sides?

 32. CHAPTER INVESTIGATION Draw the designs of the rooms and hallways of your house on grid paper. Label the different types of rooms including any special features. Include transformational principles such as two rooms being mirror reflections of each other.

MIXED REVIEW EXERCISES

Calculate each combination. (Lesson 4-7)

33. $_7C_3$ **34.** $_{14}C_5$ **35.** $_{12}C_4$ **36.** $_{13}C_6$

37. Eight students are running for Student Council advisors. Three students will be elected. How many different combinations of students can be elected?

38. Twelve cards numbered 1 to 12 are placed in a bag. If 5 cards are drawn at random, how many different combinations of cards can be drawn?

Find the measure of an interior angle of each regular polygon. (Lesson 5-7)

39.   **40.**   **41.**   **42.**

Review and Practice Your Skills

PRACTICE ◣ LESSON 7-3

On grid paper, graph each figure and its image under the given rotation.

1. $A(2, 3)$, $B(4, 6)$, $C(8, 6)$, $D(7, 3)$; 90° clockwise about $(0, 0)$

2. $Q(-6, 2)$, $R(-4, 6)$, $S(-2, 2)$; 180° clockwise about $(0, 0)$

3. $M(1, -1)$, $N(4, -2)$, $O(4, -5)$, $P(1, -4)$; 120° clockwise about $(0, 0)$

4. $S(-4, -1)$, $T(-2, -4)$, $U(-5, -7)$; 60° clockwise about $(0, 0)$

5. Draw the image of $\triangle XYZ$ after a turn of 120° clockwise about point P.

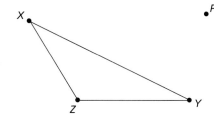

Trace each figure and its rotated image. Identify the center of rotation, the angle of rotation and the direction of rotation.

6.

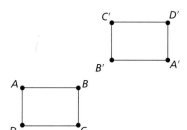

7.

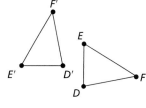

PRACTICE ◣ LESSON 7-4

Trace each figure, and draw all lines of symmetry. If applicable, write _none_.

8.

9.

10.

11.

12.

13.

Give the order of rotational symmetry for each figure.

14.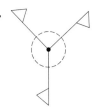

15.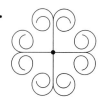

16.

Use the figure shown for Exercises 17–21. (Lesson 7-1–Lesson 7-3)

17. Which triangle is the rotation image of triangle 1 about the point $(-3, 0)$?

18. Which triangle is the reflection image of triangle 1 across the x-axis?

19. Which triangle is the reflection image of triangle 1 across the y-axis?

20. Which triangle is the rotation image of triangle 1 180° clockwise about $(0, 0)$?

21. Which triangle is the reflection image of triangle 1 across the line $x = -3$?

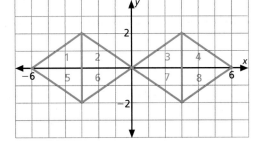

Trace each figure, and draw all lines of symmetry. If applicable, write *none*. (Lesson 7-4)

22.  23. 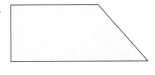 24.

Mid-Chapter Quiz

Rectangle $ABCD$ has vertices $A(-5, 2)$, $B(1, 2)$, $C(1, -1)$ and $D(-5, -1)$. **Graph the rectangle and its image under the given translations.** (Lesson 7-1)

1. 3 units up

2. 7 units right

3. 1 unit left and 2 units down

4. 4 units down and 3 units right

Triangle EFG has vertices $E(3, 2)$, $F(6, -3)$ and $G(1, -4)$. **Graph the triangle and its images under the given reflection from the original position.** (Lesson 7-2)

5. y-axis

6. x-axis

7. $y = x$

8. $y = -x$

9. $y = 2$

10. $x = 4$

Graph each figure and its image under the given rotation about $(0, 0)$. (Lesson 7-3)

11. $H(2, 3)$, $I(4, 3)$, $J(4, 1)$, and $K(2, 1)$; 90° clockwise

12. $L(-4, -2)$, $M(-2, 3)$, and $N(-1, -1)$; 180° counterclockwise

Draw all lines of symmetry and give the rotational symmetry for each figure. (Lesson 7-4)

13.   14.   15.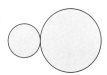

7-5 Dilations in the Coordinate Plane

Goals
■ Draw dilation images on a coordinate plane.
■ Determine the scale factor of dilations.

Applications Photography, Entertainment, Architecture, Cooking

Draw △*RST* and point *C* nearby the triangle.

1. Draw $\overrightarrow{CR}, \overrightarrow{CS}, \overrightarrow{CT}$. Use a ruler to measure $\overline{CR}, \overline{CS}$ and $\overline{CT}$ to the nearest millimeter.

2. Calculate the values $(2 \cdot CR), (2 \cdot CS)$ and $(2 \cdot CT)$.

3. Use a ruler to find point R' on $\overrightarrow{CR}$ so that $\overline{CR'} = 2 \cdot CR$. Locate points S' and T' in the same manner. Draw △$R'S'T'$.

4. Compare corresponding angle measures in △*RST* and △$R'S'T'$. What do you notice?

5. Compare the measures of corresponding sides in △*RST* and △$R'S'T'$. What do you notice?

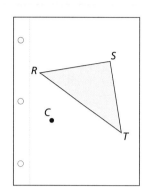

■ BUILD UNDERSTANDING

A transformation that produces an image that is the same shape as the original figure but a different size is a **dilation**. A dilation used to create an image that is larger than the preimage is an **enlargement**, while one that is used to create a smaller figure is a **reduction**. A preimage and its dilation image are similar.

The description of a dilation includes the **scale factor** and the **center of dilation**. Each length in the image is equal to the corresponding length of the preimage multiplied by the scale factor.

The distance from the center of the dilation to each point on the image is equal to the distance from the center of dilation to each corresponding point on the preimage times the scale factor.

COncepts in MOtion
Animation
mathmatters1.com

Example 1

Draw the dilation of △*ABC* with vertices $A(-1, -1)$, $B(0, 2)$ and $C(1, -2)$ with the center of dilation at the origin and a scale factor of 3.

Solution

Graph △*ABC*. Multiply the *x*-coordinates and *y*-coordinates of each vertex by the scale factor of 3.

$$A(-1, -1) \rightarrow A'(-1 \cdot 3, -1 \cdot 3) = A'(-3, -3)$$

$$B(0, 2) \rightarrow B'(0 \cdot 3, 2 \cdot 3) = B'(0, 6)$$

$$C(1, -2) \rightarrow C'(1 \cdot 3, -2 \cdot 3) = C'(3, -6)$$

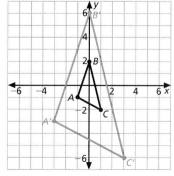

When the center of dilation is a vertex of the preimage, the corresponding vertex of the dilation image is the same point.

Example 2

Draw the dilation image of △*EFG* with the center of dilation at *E* and a scale factor of $\frac{1}{2}$.

Solution

The distance from the center of dilation, *E*, to *F* is 6 units. So the distance from *E* to *F′* is $\frac{1}{2} \cdot$ 6, or 3 units.

The distance from *E* to *G* is 8 units. So the distance from *E* to *G′* is $\frac{1}{2} \cdot$ 8, or 4 units. Points *E* and *E′* coincide.

Example 3 **Personal Tutor at** mathmatters2.com

PHOTOGRAPHY Dilations are used when pictures are "blown up" and made larger. Suppose that a film negative is originally $\frac{2}{3}$ in. by 1 in.

a. If the negative makes a 4 in. by 6 in. print, what is the scale factor of the dilation?

b. If a print of the negative is blown up by a scale factor of 24, what are the dimensions of the print?

Check Understanding

What positive scale factors indicate an enlargement? What positive scale factors indicate a reduction?

Solution

a. To determine the scale factor, compare the print length and width to the film negative length and width.

$$\frac{4}{\left(\frac{2}{3}\right)} = 6 \qquad\qquad \frac{6}{1} = 6$$

The length and width of the film negative are multiplied by 6 to achieve the print size, so the scale factor is 6.

b. To find the size of the print, multiply the length and width of the negative by the scale factor.

$$\left(\frac{2}{3}\right) \cdot 24 = 16 \qquad\qquad 1 \cdot 24 = 24$$

The print is 16 in. by 24 in.

◤ TRY THESE EXERCISES

Copy △*TBA* on grid paper. Draw each dilation.

1. scale factor 2, center (0, 0)

2. scale factor $\frac{2}{3}$, center (0, 0)

3. scale factor 3, center *T*

4. scale factor 2.5, center *A*

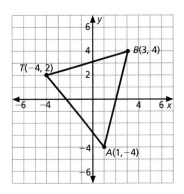

For Exercises 5–8, give the scale factor and center for each dilation.

5. Triangle *ABC* has vertices *A*(2, 1), *B*(3, −5), and *C*(6, 4). Its dilation image is △*A'B'C'* with vertices *A'*(6, 3), *B'*(9, −15), and *C'*(18, 12).

6. Rectangle *QRST* has vertices *Q*(2, 1), *R*(6, 1), *S*(2, 7) and *T*(6, 7). Its dilation image is rectangle *Q'R'S'T'* with vertices *Q'*(2, 1), *R'*(4, 1), *S'*(2, 4), and *T'*(4, 4).

7.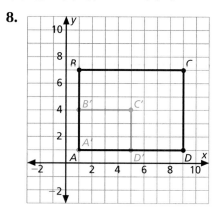

8.

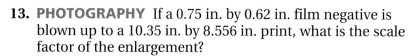

PRACTICE EXERCISES • For Extra Practice, see page 609.

For Extra Practice, see page 609.

Copy parallelogram *ABCD* on grid paper. Draw each dilation.

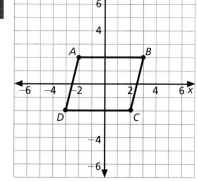

9. scale factor 2, center (0, 0)

10. scale factor 0.75, center *A*

11. scale factor 1, center (0, 0)

12. scale factor 4, center *C*

13. **PHOTOGRAPHY** If a 0.75 in. by 0.62 in. film negative is blown up to a 10.35 in. by 8.556 in. print, what is the scale factor of the enlargement?

14. An equilateral triangle with 4-in. sides undergoes a dilation with a scale factor of 2.8. In the image, what are the lengths of the sides? What are the measures of the angles of the image?

15. **ENTERTAINMENT** The projection of a film on a movie screen is a dilation. Most movies are recorded on 35-mm wide film. If the width of a movie screen is 12 m, what is the enlargement scale factor?

16. **ARCHITECTURE** An architect is designing a shopping center. A shop's dimensions on the blueprint are 19 cm by 25 cm. Using a scale factor of 1 cm : 2 m, what are the shop's dimensions in meters?

17. **WRITING MATH** Write a paragraph describing how dilations are similar to translations, reflections and rotations and how they differ.

Give the scale factor and center for each dilation represented by the vertices.

18. *C*(0, 0), *D*(0, 4), *E*(6, 0), and *C'*(0, 0), *D'*(0, 10), *E'*(15, 0)

19. *S*(0, 0), *T*(0, −3), *U*(−3, −6), and *S'*(0, 0), *T'*(0, −1), *U'*(−1, −2)

20. CRITICAL THINKING Is it possible for a scale factor to be a negative number? Explain why or why not.

21. GARDENING Troy made a scale drawing of the plan for his garden. It will be a rectangle measuring 18 ft by 12 ft. On the scaled version, it measures 8 in. on the longer sides. What is the measure of each of the shorter sides?

▰ EXTENDED PRACTICE EXERCISES

For each pair of figures, describe the two transformations used to create each image.

22.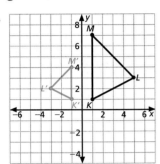

23.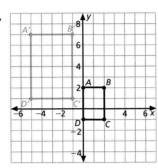

24.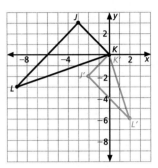

25. GEOMETRY SOFTWARE Use Cabri Jr. to draw △*GHI* with vertices *G*(0, 0), *H*(0, 6), and *I*(3, 0) on a coordinate plane. Use the software's tools to dilate the triangle about the origin with scale factors of 2 and 0.5. Find the area of △*GHI*, its enlargement, and its reduction.

26. Draw rectangle *TUVW* with vertices *T*(1, 1), *U*(1, 4), *V*(7, 4), and *W*(7, 1) on a coordinate plane. Draw the dilation images of the rectangle with the center of dilation at *T* and scale factors of 3 and $\frac{1}{3}$. Find the area of the rectangle, its enlargement and its reduction.

27. DATA FILE Refer to the data on balls used in various sports on page 572. A movie creates large-scale models of certain balls to create the illusion of tiny people. If the scale factor used is 45, what is the height (diameter) of a tennis ball? a baseball? a basketball?

28. CHAPTER INVESTIGATION Decide how large your house will be by choosing a scale factor for the blueprint to the actual house. Label the sizes of the rooms and halls on the blueprint, and calculate the square footage of the whole floor of your house.

▰ MIXED REVIEW EXERCISES

Simplify each numerical expression. (Lesson 2-2)

29. $5 - 3 \cdot 2 - 4$

30. $3(8 + 2) \div 5$

31. $3^2 - 4 + 8 \div 2$

32. $13 - 4 \cdot 9 + 3 \cdot 2$

33. $20 - (4 + 3) \div 2 + 6$

34. $5 + (3 - 1)^2 \div 3 + 8$

35. $16 \div 4 - (2 + 5)^2$

36. $1 + 6^2 \cdot \frac{1}{4}(9 - 5)^2$

37. $6 + (4 - 6)^3 + 2 - 8 \cdot 2$

Identify the slope and *y*-intercept for each line. (Lesson 6-3)

38. $y = 4x - 2$

39. $y = 3x + 5$

40. $y = \frac{1}{2}x + 2$

41. $2x - 2y = 8$

42. $-3x + y = 4$

43. $5x + 2y = 6$

44. $y = \frac{2}{3}x - 4$

45. $-x + 2y = 10$

46. $4x + 5y = -15$

Some problems are solved by recognizing a pattern, while other problems require that you extend the pattern to find the solution. This strategy is called **look for a pattern**, and it is used with many different types of problems. Look for a pattern that is numerical, visual or behavioral. By figuring out the pattern, you can predict the next element or figure out all the elements.

A **tessellation** is a repeating pattern of one or more figures that completely covers a plane without gaps or overlaps. Tessellations are also called *tilings*. A set of figures that can be used to create a tessellation is said to *tessellate*. You can find tessellations in art, nature and everyday life.

Problem Solving Strategies

Guess and check

✔ Look for a pattern

Solve a simpler problem

Make a table, chart or list

Use a picture, diagram or model

Act it out

Work backwards

Eliminate possibilities

Use an equation or formula

Problem

Construct a tessellation using regular octagons and squares as tessellating figures.

Solve the Problem

Step 1 Draw a regular octagon.

Step 2 Draw 4 squares.

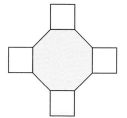

Step 3 Complete the tessellation.

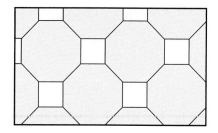

◥ TRY THESE EXERCISES

Concepts in MOtion
Animation
mathmatters2.com

Determine if each polygon can be a tessellating figure. Write *yes* or *no*.

1. equilateral triangle

2. regular octagon

3. regular hexagon

4. isosceles triangle

5. Draw the tessellation for one of the polygons to which you answered "yes" in Exercises 1–4.

6. **NATURE** Describe the tessellating figure in a honeycomb beehive.

7. Draw a tessellation whose tessellating figure is composed of regular hexagons, squares, and equilateral triangles.

Five-step Plan

1 Read
2 Plan
3 Solve
4 Answer
5 Check

Determine if each polygon can be a tessellating figure. Write *yes* or *no*.

8. regular pentagon

9. square

10. circle

11. parallelogram

12. HOBBIES Draw a tessellation in which the tessellating figure is a puzzle piece.

13. WRITING MATH Describe how tessellations are used in everyday life. Sketch some of the examples.

Use each figure to create a tessellation on dot or grid paper.

14.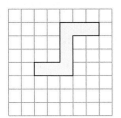

15.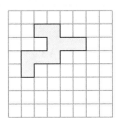

16.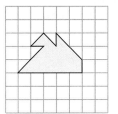

17. ARCHITECTURE The floor of the Taj Mahal in India contains many beautiful tile patterns. Describe the pattern and the tessellation of one of the tile patterns in the Taj Mahal shown at the right.

18. Use isometric grid paper to create a tessellation using large and small equilateral triangles.

19. MODELING Fold patty paper to create a tessellation design, or fold the paper into a grid and draw a tessellation on it. Name the shapes that appear in the tessellation.

20. CRITICAL THINKING Draw a tessellation using a tessellating figure of a quadrilateral that contains no right angles.

■ **MIXED REVIEW EXERCISES**

Solve each proportion. Check the solution. (Lesson 3-5)

21. $\dfrac{7}{8} = \dfrac{m}{32}$

22. $\dfrac{6}{5} = \dfrac{42}{g}$

23. $\dfrac{13}{20} = \dfrac{r}{4}$

24. $\dfrac{1}{3} = \dfrac{f}{21}$

25. $\dfrac{2s}{3} = \dfrac{24}{12}$

26. $\dfrac{7}{8} = \dfrac{4p}{56}$

27. $\dfrac{15}{25} = \dfrac{5}{2z}$

28. $\dfrac{10}{8} = \dfrac{x}{0.8}$

29. During one flight, 32 passengers chose *In Flight* magazine to read, 16 chose a newspaper, 18 chose a book and 8 chose nothing to read. What is the probability that a passenger chose to read *In Flight*? (Lesson 4-1)

30. On the same flight, 24 passengers chose to drink apple juice, 17 chose soda, 19 chose coffee and 14 chose iced tea. What is the probability that a passenger chose iced tea to drink? (Lesson 4-1)

Review and Practice Your Skills

PRACTICE ◥ LESSON 7-5

Copy △ABC on grid paper. Draw each dilation.

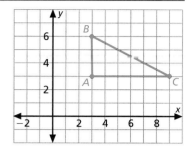

1. scale factor 2, center (0, 0)

2. scale factor $\frac{1}{2}$, center (0, 0)

3. scale factor 2, center B

4. scale factor $\frac{2}{3}$, center A

Give the coordinates of each dilated image.

5. scale factor 4, center (0, 0)

6. scale factor $\frac{1}{2}$, center (0, 0)

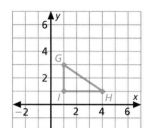

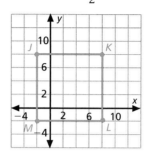

7. scale factor 1.5, center (0, 0)

8. scale factor $\frac{1}{2}$, center S

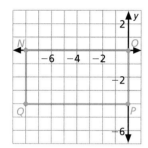

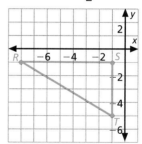

PRACTICE ◥ LESSON 7-6

Determine if each polygon can be a tessellating figure. Write *yes* or *no*.

9.

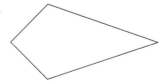

10.

11.

Determine if each figure can be a tessellating figure. If so, create a tessellation on dot or grid paper.

12.

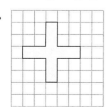

13.

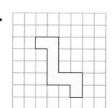

14.

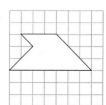

15. Graph △XYZ with vertices X(0, 0), Y(2, −3) and Z(6, −3) and its reflection over the y-axis. (Lesson 7-2)

16. Graph △ABC with vertices A(2, −2), B(4, −2) and C(2, −6) and its image with a 180° rotation about (0, 0). (Lesson 7-3)

Trace each figure, and draw all lines of symmetry. If applicable, write *none*. (Lesson 7-4)

17.

18.

19.

20. Gayle has a film negative that is 1.25 in. by 1.75 in. To get a print blown up to a 5 in. by 7 in., what is the scale factor of the enlargement? (Lesson 7-5)

MathWorks Career – Architect
Workplace Knowhow

Architects design buildings and other structures. They are often involved in all phases of development, from the initial discussion with the client through the entire life of the facility. Their duties require skills such as design, engineering, communication and supervision. Architects often begin the design of a group of buildings by choosing a basic geometric shape and repeating it in the design. Rotation, translation and dilation of the chosen shape results in buildings of various shapes and sizes. The steps below outline the process of designing a library. Use one grid to construct all three buildings, allowing both the x- and y-axes to extend from −15 to +15 units.

1. Rotate Building 1 90° counterclockwise around the origin to determine the location of Building 2, A′B′C′D′. What are the coordinates of A′, B′, C′ and D′?

2. Draw a line of symmetry on Building 1. A flagpole, which is 5 units from Building 1 along the line of symmetry, is located between Buildings 1 and 2. Draw an "x" at the location of the flagpole. What are its coordinates?

3. Building 3, figure A″B″C″D″ is three times the size of Building 2. To determine its location, draw the dilation of A′B′C′D′ with the center of dilation at (11, 10) and translate the dilated figure to the left 3 units. Give the coordinates of A″, B″, C″ and D″.

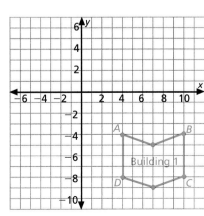

Chapter 7 Review

VOCABULARY

Choose the word from the list that best completes each statement.

1. An image produced by a flip across a line is a(n) ___?___.

2. Under a(n) ___?___, a figure is turned about a given point.

3. A transformation that produces an image that is the same shape as the original figure but a different size is a(n) ___?___.

4. A plane figure has ___?___ if you can divide it along a line into two parts that are mirror images of each other.

5. A slide transformation is a(n) ___?___.

6. A(n) ___?___ occurs when the image is smaller than the original.

7. A repeating pattern that covers a plane without gaps or overlaps is called a(n) ___?___.

8. A plane figure has ___?___ if you can rotate it so that it fits exactly over its original position more than once in a complete turn.

9. In an enlargement, the ___?___ is greater than one.

10. In a transformation, the original figure is called a(n) ___?___.

a.	dilation
b.	enlargement
c.	line symmetry
d.	preimage
e.	reduction
f.	reflection
g.	rotation
h.	rotational symmetry
i.	scale factor
j.	tessellation
k.	transformation
l.	translation

LESSON 7-1 ◣ Translations in the Coordinate Plane, p. 296

▶ A move like a **translation**, or slide, is called a **transformation** of a figure. The original figure is the **preimage**. The new figure is the **image** of the original figure.

▶ A translation can be described by a rule stating the number of units to the left or right and the number of units up or down.

11. Triangle STU has vertices $S(-2, 0)$, $T(2, -2)$, and $U(21, 24)$. Graph the image of $\triangle STU$ under a translation 2 units left and 4 units up.

12. Triangle ABC has vertices $A(-5, -2)$, $B(-2, 3)$, and $C(-2, -3)$. Graph the image of $\triangle ABC$ under a translation 1 unit right and 3 units up.

Write the rule that describes each translation.

13.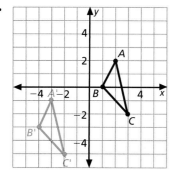

14.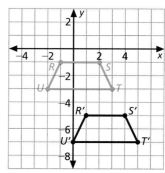

LESSON 7-2 ◣ Reflections in the Coordinate Plane, p. 300

▶ Under a **reflection**, a figure is reflected, or flipped, across a *line of reflection*. A reflection is a transformation that yields a congruent figure.

15. Graph the image of parallelogram *ABCD* under a reflection across the line $y = -x$.

16. Graph the image of parallelogram *ABCD* under a reflection across the *x*-axis.

17. Graph the image of parallelogram *ABCD* under a reflection across the *y*-axis.

18. Graph the image of parallelogram *ABCD* under a reflection across the line $y = x$.

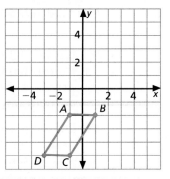

LESSON 7-3 ◣ Rotations in the Coordinate Plane, p. 306

▶ A rotation is a transformation that includes the following.

   a. The **center of rotation**, or point about which the figure is rotated.

   b. The amount of turn expressed as a fractional part of a whole turn or as the **angle of rotation** in degrees.

   c. The direction of rotation—clockwise or counterclockwise.

19. Triangle *FGH* has vertices $F(0, 0)$, $G(5, -2)$, and $H(4, -5)$. Graph the rotation image of $\triangle FGH$ under a turn of 90° clockwise about $(0, 0)$.

20. Triangle *VWX* has vertices $V(-4, 2)$, $W(-2, 4)$, and $X(2, 1)$. Graph the rotation image of $\triangle VWX$ under a turn of 180° clockwise about $(0, 0)$.

21. Triangle *ABC* has vertices $A(-5, 3)$, $B(-2, 5)$, and $C(-3, 2)$. Graph the rotation image of $\triangle ABC$ under a turn of 90° counterclockwise about $(0, 0)$.

LESSON 7-4 ◣ Line Symmetry and Rotational Symmetry, p. 310

▶ A figure has **line symmetry** if, when you fold it along a line, one side fits exactly over the other side. The line is called a **line of symmetry**.

▶ A figure has **rotational symmetry** if, when you turn it about a point, the figure fits exactly over its original position more than once before the turn is completed.

Trace each figure, and draw all lines of symmetry. If applicable, write *none*.

22.

23.

24.

Give the order of rotational symmetry for each figure.

25.

26.

27.

LESSON 7-5 ◼ Dilations in the Coordinate Plane, p. 316

▶ A **dilation** is a transformation that produces the same shape but a different size image of a figure.

▶ An **enlargement** is an image that is larger than its preimage. A **reduction** is an image that is smaller than its preimage.

▶ Each length in the image is equal to the corresponding side of the preimage multiplied by the **scale factor**. The distance from the **center of dilation** to each point on the image is equal to the distance from the **center of dilation** to each corresponding point on the preimage times the scale factor.

28. Draw the dilation image of parallelogram *MNOP* with the center of dilation at (0, 0) and a scale factor of 2.

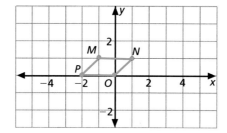

29. Triangle *ABC* has vertices *A*(−4, 1), *B*(−3, 4), and *C*(−1, 1). Its dilation image is △*A′B′C′* with vertices *A′*(−8, 2), *B′*(−6, 8), and *C′*(−2, 2). What is the scale factor and center of the dilation?

30. A photograph is 8 in. by 10 in. It is being reduced by a scale factor of $\frac{3}{4}$. What are the dimensions of the new photograph?

31. A projector focuses an image from film that is 16 millimeters wide onto a screen that is 1.5 meters wide. What is the scale factor?

LESSON 7-6 ◼ Problem Solving Skills: Tessellations, p. 320

▶ A **tessellation**, or **tiling**, is a repeating pattern of figures that completely covers a plane without gaps or overlaps.

Use each figure to create a tessellation on dot or grid paper.

32.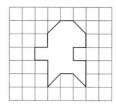

33.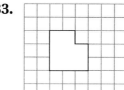

34.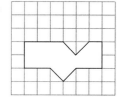

CHAPTER INVESTIGATION

EXTENSION Present your blueprint to the class. Describe your reasons for the types of rooms you have included and the way in which they are oriented. Also describe any geometric principles that you have employed in the creation of the blueprint. These may include any transformational principles used, different types of polygons and shapes, a tessellated floor pattern and more.

Chapter 7 Assessment

Use the figure for Exercises 1–3.

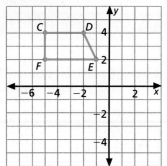

1. Graph the image of trapezoid *CDEF* under a translation 3 units to the right and 4 units down.

2. Graph the image of trapezoid *CDEF* under a reflection across the line $y = x$.

3. Graph the rotation image of trapezoid *CDEF* after a 90° turn counterclockwise about (0, 0).

4. Triangle *T′U′V′* is the reflected image of △*TUV*. Graph the line of reflection.

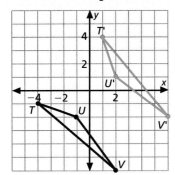

5. Triangle *A′B′C′* is the rotation image of △*ABC*. Identify the center of rotation, *T*, the angle of rotation and the direction of rotation.

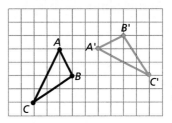

6. How many lines of symmetry does the figure below have?

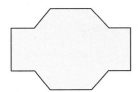

7. What is the order of rotational symmetry of the figure below?

8. Write the rule that describes the translation of the black preimage to the blue image.

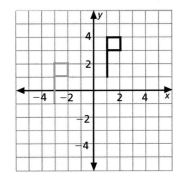

9. Draw the dilation image of △*WXY* with the center of dilation at the origin and a scale factor of $\frac{1}{3}$.

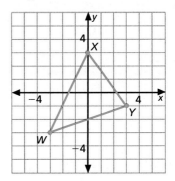

10. Use the figures to create a tessellation on dot or grid paper.

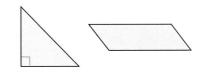

Standardized Test Practice

Record your answers on the answer sheet
provided by your teacher or on a sheet of paper.

1. The staff at a research company is paid to
   survey shoppers walking past them at the
   mall. What kind of sampling is this?
   (Lesson 1-1)
   - (A) cluster sampling
   - (B) convenience sampling
   - (C) random sampling
   - (D) systematic sampling

2. Simplify $\frac{-6x + 3}{-3}$. (Lesson 2-5)
   - (A) $2x + 1$
   - (B) $2x - 1$
   - (C) $3x + 1$
   - (D) $3x - 1$

3. What is the next term in the sequence?
   (Lesson 2-9)

   $$2, 3, 5, 8, 13, 21, \ldots$$

   - (A) 24
   - (B) 28
   - (C) 30
   - (D) 34

4. If $3a = -6$, what does $2a$ equal?
   (Lesson 3-2)
   - (A) -6
   - (B) -4
   - (C) 4
   - (D) 6

5. Solve $\sqrt{d} = 4$. (Lesson 3-8)
   - (A) $d = 16$ or -16
   - (B) $d = 2$ or -2
   - (C) $d = 16$
   - (D) $d = 2$

6. What is the probability of an event that is
   certain to happen? (Lesson 4-1)
   - (A) 0
   - (B) 0.5
   - (C) 1
   - (D) depends on the circumstances

7. If you rolled a number cube five times and
   got a 6 each time, what is the theoretical
   probability of rolling a 6 on the sixth roll?
   (Lesson 4-3)
   - (A) 1
   - (B) $\frac{1}{6}$
   - (C) $\frac{1}{36}$
   - (D) 0

8. What is the total measure of all the interior
   angles in a polygon with 14 sides?
   (Lesson 5-7)
   - (A) 2160°
   - (B) 2520°
   - (C) 4320°
   - (D) 5140°

9. $\triangle ABC \cong \triangle DEF$. The vertices of $\triangle ABC$ are
   $A(-1, 3)$, $B(2, 7)$ and $C(2, -1)$. Find DF.
   (Lesson 6-1)
   - (A) $\sqrt{5}$
   - (B) $\sqrt{8}$
   - (C) 5
   - (D) 8

10. The brightness of a light bulb varies inversely
    as the square of the distance from the source.
    If a light bulb has a brightness of 300 lumens
    at 2 ft, what is its brightness at 10 ft?
    (Lesson 6-7)
    - (A) 120 lumens
    - (B) 60 lumens
    - (C) 12 lumens
    - (D) 6 lumens

11. Which of the following
    figures is *not* a rotation of
    the figure at the right?
    (Lesson 7-3)

    - (A)
    - (B)
    - (C)
    - (D)

12. Which figure has exactly two lines of
    symmetry? (Lesson 7-4)
    - (A)
    - (B)
    - (C)
    - (D)

Preparing for the Standardized Tests
For test-taking strategies and more
practice, see pages 627–644.

Part 2 Short Response/Grid In

Record your answers on the answer sheet
provided by your teacher or on a sheet of paper.

13. The mean of r and s is 25. The mean of r, s,
and t is 30. What is the value of t?
(Lesson 1-2)

14. On a car trip, Kwan drove 50 mi more than
half the number of miles Molly drove.
Together they drove 425 mi. How many
miles did Kwan drive? (Lesson 3-4)

15. A bag contains 2 red, 6 blue, 7 yellow, and
3 orange marbles. Once a marble is selected,
it is not replaced. If two marbles are picked
at random, what is the probability that
both marbles are yellow? (Lesson 4-5)

16. Find the value of x in the figure.
(Lesson 5-4)

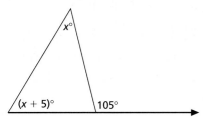

Use parallelogram *ABCD* for Exercises 17 and 18.

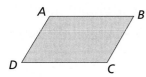

17. If $AB = 7$ and $BC = 3$, find DC. (Lesson 5-6)

18. If $m\angle A = 110°$, find $m\angle B$. (Lesson 5-6)

19. The weight of an object on the moon varies
directly with its weight on Earth. With all of
his equipment, an astronaut weighed 420 lb
on Earth, but weighed only 70 lb on the
moon. Suppose an object weighs 144 lb
on Earth. Find its weight on the moon.
(Lesson 6-8)

20. The vertices of $\triangle ABC$ are $A(-2, 3)$, $B(1, 4)$,
and $C(2, -3)$. Find the coordinates of
$\triangle A'B'C'$ if $\triangle ABC$ is translated 3 units
down and 4 units to the right. (Lesson 7-1)

21. Which capital letters of the alphabet produce
the same letter after being rotated 180°?
(Lesson 7-3)

Part 3 Extended Response

Record your answers on a sheet of paper.
Show your work.

22. The drawing shows the pattern for the right
half of a shirt. Copy the pattern onto grid
paper. Then draw the outline of the pattern
after it has been flipped over a vertical line
to show the left half of the pattern. Label it
"Left Front." Describe the relationship
between the right and left fronts of the
pattern in mathematical terms. (Lesson 7-2)

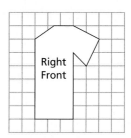

23. Jamal is designing a floor for a customer.
The customer wants to use three geometrical
shapes to tessellate the floor. Create a design
that will work and explain how you can use
your geometry skills to know that the figures
tessellate. (Lesson 7-6)

Test-Taking Tip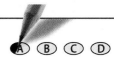

Question 20
Most standardized tests allow you to write in the test booklet.
You may sketch the figure and its translation on a coordinate
plane. Doing so will help keep you from making careless errors.

Systems of Equations and Inequalities

The point of intersection in two or more lines signifies a relationship that can indicate where supply equals demand, the location of a ball and a receiver, or number of calories burned equals calories consumed.

Systems of equations and inequalities play an important role in sports. The point or the region of intersection may represent a winning play, a maximum performance, or a dynamic combination of players. Equations and inequalities can represent many aspects of the sports game and the solution to the system is a means to develop the ultimate goal: teamwork and victory.

- Athletes, such as **runners** (page 343), can use equations and systems of equations to monitor their running times so that their performance consistently improves.

- **Coaches** (page 361) can use equations and systems of equations to represent characteristics of individual performers and then plan a winning strategy.

Math Online

mathmatters2.com/chapter_theme

All Time Winter Olympics Medal Standings, 1924-2002

Rank	Nation	Gold	Silver	Bronze	Total
1	Norway	94	94	75	263
2	Soviet Union (1956-1988)	78	57	59	194
3	United States	69	72	52	193
4	Austria	41	57	64	162
5	Finland	42	51	49	142
6	Germany (1928-36, 1992-)	47	46	32	125
7	East Germany (1956-1988)	43	39	36	118
8	Sweden	39	30	39	108
9	Switzerland	32	33	38	103
10	Canada	31	31	27	89

Data Activity: Winter Olympics Medal Standings

Use the table for Questions 1–3.

1. Let gold = 3 points, silver = 2 points and bronze = 1 point. Who has more points, Switzerland or Sweden?

2. Let gold = 5 points, silver = 3 points and bronze = 1 point. Who has more points, the United States or Austria?

3. Draw a bar graph using three of the countries in the table. Use different colors to represent gold, silver and bronze.

CHAPTER INVESTIGATION

In the Central High School League, there are ten football teams. The top four teams advance to the league tournament. To determine the top four teams, points are awarded to each team for a win. Each team also receives points whenever a team it beats wins a game or has already won a game.

Working Together

The first table indicates who played each week, and the winning team is circled. Teams are numbered 1 through 10. The second table shows the top four teams and their total points for the year. Use the tables to determine the number of points awarded for winning a game and the number of points awarded whenever a team it beats wins a game or has already won a game. Use the Chapter Investigation icons to guide your group.

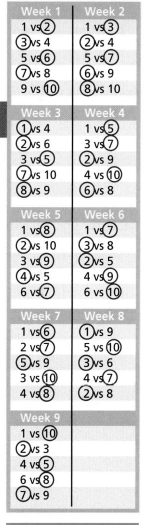

Week 1	Week 2
1 vs ②	1 vs ③
③ vs 4	② vs 4
5 vs ⑥	5 vs ⑦
⑦ vs 8	⑥ vs 9
9 vs ⑩	⑧ vs 10

Week 3	Week 4
① vs 4	1 vs ⑤
② vs 6	3 vs ⑦
3 vs ⑤	② vs 9
⑦ vs 10	4 vs ⑩
⑧ vs 9	⑥ vs 8

Week 5	Week 6
1 vs ⑧	1 vs ⑦
② vs 10	③ vs 8
3 vs ⑨	② vs 5
④ vs 5	4 vs ⑨
6 vs ⑦	6 vs ⑩

Week 7	Week 8
1 vs ⑥	① vs 9
2 vs ⑦	5 vs ⑩
⑤ vs 9	③ vs 6
3 vs ⑩	4 vs ⑦
4 vs ⑧	② vs 8

Week 9	
1 vs ⑩	
② vs 3	
4 vs ⑤	
6 vs ⑧	
⑦ vs 9	

Team	Total Points
Team 7	198
Team 2	160
Team 10	104
Team 8	86

The skills on these two pages are ones you have already learned. Use the examples to refresh your memory and complete the exercises. For additional practice on these and more prerequisite skills, see pages 576–584.

RECIPROCALS

To find the *reciprocal* of a number, switch the numerator and denominator.

Example Write the reciprocal of $\frac{2}{3}$.

The reciprocal of $\frac{2}{3}$ is $\frac{3}{2}$.

Write the reciprocal of each number.

1. $\frac{1}{2}$ **2.** 5   **3.** $\frac{4}{5}$ **4.** $-\frac{2}{7}$

5. $\frac{4}{3}$ **6.** -6 **7.** $\frac{11}{12}$ **8.** $-\frac{9}{5}$

9. 12   **10.** $\frac{17}{15}$ **11.** $-\frac{1}{8}$ **12.** $\frac{99}{100}$

SLOPE OF A LINE

In this chapter you will solve systems of equations by graphing. It is helpful to be able to find the slope of a graphed equation.

Example What is the slope of the line that includes points $A(-3, -2)$ and $B(4, 6)$?

Use the slope formula, $m = \frac{y_2 - y_1}{x_2 - x_1}$, where (x_1, y_1) and (x_2, y_2) are points on the line.

$$m = \frac{y_2 - y_1}{x_2 - x_1}$$

$$m = \frac{6 - (-2)}{4 - (-3)}$$

$$m = \frac{8}{7}$$

Since the slope is positive, the line slants upward to the right.

A line with a negative slope runs downward to the right.

The slope of a vertical line is undefined. The slope of a horizontal line is 0.

Find the slope of a line that passes through the given points.

13. $(-4, 2), (3, -8)$ **14.** $(-5, -1), (0, 0)$ **15.** $(6, 2), (1, 4)$ **16.** $(8, -6), (4, -2)$

17. $(1, -5), (4, 0)$ **18.** $(-2, 1), (1, -2)$ **19.** $(3, -1), (3, 5)$ **20.** $(4, 4), (-1, -1)$

21. $(0, 0), (2, 5)$ **22.** $(-2, -4), (-3, 1)$ **23.** $(-4, -6), (-2, 1)$ **24.** $(0, 3), (6, 0)$

25. $(4, -1), (-2, 1)$ **26.** $(2, 9), (-4, 6)$ **27.** $(2, 5), (-4, 5)$ **28.** $(12, -6), (-3, 9)$

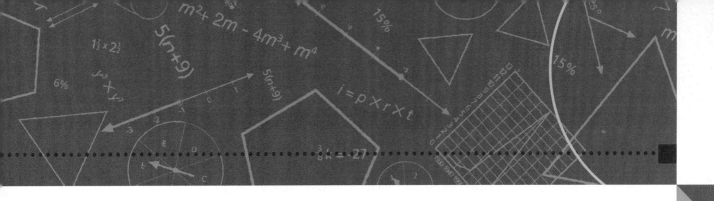

GRAPHING EQUATIONS

You can graph any linear equation by identifying the slope of the line and the y-intercept.

Example   Graph the equation $y = 2x - 3$.

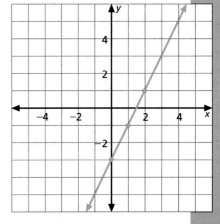

The slope is the coefficient of the x-term.

$$m = 2$$

The y-intercept is the constant term.

$$y\text{-intercept} = -3$$

Plot the point $(0, -3)$. Then use the slope of $\frac{2}{1}$ to plot two or three additional points.

Identify the slope and the y-intercept of each equation. Graph each equation.

29. $y = 4x - 3$

30. $y = \frac{1}{2}x + 4$

31. $y = -2x + 1$

32. $3x + y = 5$

33. $4y + 8x = 12$

34. $-3x + y = 6$

35. $4y - 3 = 2x$

36. $2x - 5 = -3y$

37. $3x + 2 = 3y$

GRAPHING INEQUALITIES

Example   Graph the inequality $y < x + 4$.

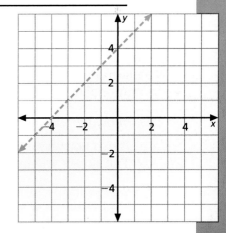

To graph the inequality, first graph the equation related to the inequality: $y = x + 4$.

Since the inequality symbol is $<$, the line is not included in the solution. Therefore, the line should be a dashed line.

Choose a test point to determine if you shade above or below the dashed line. Choose $(0, 0)$.

$$y < x + 4$$

$$0 < 0 + 4$$

$$0 < 4$$

Since the test point results in a true statement, shade the half-plane that includes $(0, 0)$ as a solution.

Graph each inequality.

38. $y > x - 2$

39. $y < 2x + 1$

40. $y \geq x - 4$

41. $y \leq 3x + 4$

42. $x + y > -3$

43. $2x \leq y - 4$

44. $3x - 4y \geq 8$

45. $y < 3 - 4x$

46. $3x - 2y > -6$

8-1 Parallel and Perpendicular Lines

Goals ■ Determine if two lines are parallel or perpendicular.
■ Write equations of parallel and perpendicular lines.

Applications Sports, Travel, Safety

GRAPHING Use a graphing calculator for Questions 1–6.

1. Enter and graph these linear functions one at a time on your graphing calculator. Graph the functions on the same coordinate plane.

   a. $y = 2x + 5$ **b.** $y = 2x + 1$ **c.** $y = 2x - 1$

2. Predict the location of the graph of $y = 2x - 6$.

3. Use your graphing calculator to check your prediction in Question 2.

4. Enter and graph $y = -\frac{1}{2}x + 3$ on the same graph as the linear functions in Questions 1 and 2. What appears to be the relationship of this line to the other four lines?

5. Predict the location of the graph of $y = -\frac{1}{2}x - 7$.

6. Use your graphing calculator to check your prediction in Question 5.

◤ BUILD UNDERSTANDING

The following relationships exist between parallel lines.

> If two nonvertical lines are parallel, they have the same slope.

> If two distinct lines have the same slope, they are parallel.

Recall that two rational numbers that have a product of 1 are called reciprocals. If two rational numbers have a product of -1, they are called **negative reciprocals**.

The following relationships exist between perpendicular lines.

> If two nonvertical lines are perpendicular, the product of their slopes is -1. The slopes are negative reciprocals of each other.

> If the slopes of two lines are negative reciprocals of each other, then the lines are perpendicular.

Example 1

For each line identified by two points, state the slope of a line parallel and the slope of a line perpendicular to it.

a. $A(6, 3)$ and $B(4, 6)$

b. $C(-1, -5)$ and $D(3, -2)$

c. $E(0, 5)$ and $F(-1, 6)$

d. $G(-4, -2)$ and $H(-5, -2)$

Solution

Recall that $m = \frac{y_2 - y_1}{x_2 - x_1}$. The slope of the parallel line is m, and the slope of the perpendicular line is $\frac{-1}{m}$.

	Line	∥ Line	⊥ Line
a.	$\frac{6-3}{4-6} = \frac{3}{-2} = -\frac{3}{2}$	$-\frac{3}{2}$	$\frac{2}{3}$
b.	$\frac{-2-(-5)}{3-(-1)} = \frac{3}{4} = \frac{3}{4}$	$\frac{3}{4}$	$-\frac{4}{3}$
c.	$\frac{6-5}{-1-0} = \frac{1}{-1} = -1$	-1	1
d.	$\frac{-2-(-2)}{-5-(-4)} = \frac{0}{-1} = 0$	0	undefined

Example 2

Determine if the graphs will show parallel or perpendicular lines, or neither.

a. $y = \frac{1}{3}x - 4$
  $y = -3x + 1$

b. $y = -2x + \frac{1}{2}$
  $6x + 3y = 9$

c. $y = 5x - 4$
  $2x - 4y = 8$

Solution

a. The lines are perpendicular since their slopes are negative reciprocals. $\frac{1}{3} \cdot (-3) = -1$

b. Write the second equation in slope-intercept form to find the slope.

  $6x + 3y = 9$
  $y = -2x + 3$ Subtract $6x$ and divide by 3.

  The lines are parallel since they have the same slope, $m = -2$.

c. Write the second equation in slope-intercept form.

  $2x - 4y = 8$
  $y = \frac{1}{2}x - 2$ Subtract $2x$ and divide by -4.

  The lines are neither parallel nor perpendicular since $m = 5$ in the first equation and $m = \frac{1}{2}$ in the second equation.

Technology Note

Use a graphing utility to check the solutions In Example 2. Graph each pair of lines on the same coordinate plane.

Example 3

Write an equation in slope-intercept form of a line that passes through (2, 4) and is parallel to the line $y = -\frac{1}{2}x + 1$.

Solution

The slope of the line is $-\frac{1}{2}$. Substitute $-\frac{1}{2}$ for m and the point (2, 4) in slope-intercept form to solve for b.

  $4 = -\frac{1}{2}(2) + b$ $y = mx + b$
  $4 = -1 + b$
  $5 = b$

Use the values of m and b to write the equation of the line, $y = -\frac{1}{2}x + 5$.

Example 4

Online Personal Tutor at mathmatters2.com

Write an equation in slope-intercept form of a line that passes through (6, 5) and is perpendicular to the line $-6x - 2y = 12$.

Solution

First write the equation in slope-intercept form.

$$-6x - 2y = 12$$
$$y = -3x - 6 \qquad \text{Add } 6x \text{ and divide by } -2.$$

The slope is -3. The slope of a line perpendicular to this line is $\frac{1}{3}$. Substitute $\frac{1}{3}$ for m and the point (6, 5) in slope-intercept form.

$$5 = \frac{1}{3}(6) + b \qquad y = mx + b.$$
$$5 = 2 + b$$
$$3 = b$$

The equation of the line is $y = \frac{1}{3}x + 3$.

TRY THESE EXERCISES

For each line identified by two points, state the slope of a line parallel and the slope of a line perpendicular to it.

1. $A(7, 2)$ and $B(5, 7)$
2. $C(1, 5)$ and $D(-2, 3)$
3. $E(0, -2)$ and $F(-3, -5)$

Determine if the graphs will show parallel or perpendicular lines, or neither.

4. $y = \frac{2}{3}x - 6$
$2x - 4 = 3y$

5. $y = \frac{1}{2}x + 4$
$4x - 3y = 12$

6. $2y = 10x - \frac{2}{5}$
$\frac{1}{5}x + y = 3$

Write an equation in slope-intercept form of a line passing through the given point and parallel to the given line.

7. $(2, -7); y = 3x - 1$
8. $(0, -5); 3x - 6y = 15$
9. $(2, -1); 7y = 5x - 3$

Write an equation in slope-intercept form of a line passing through the given point and perpendicular to the given line.

10. $(-3, 1); \frac{1}{3}x + y = 2$
11. $(3, 7); y = \frac{3}{4}x - 1$
12. $(4, -3); 2x - 7y = 12$

PRACTICE EXERCISES • For Extra Practice, see page 609.

For each line identified by two points, state the slope of a line parallel and the slope of a line perpendicular to it.

13. $M(-2, 3)$ and $N(2, 1)$
14. $P(-7, -3)$ and $Q(0, -1)$
15. $F(-3, 3)$ and $G(2, -4)$

16. $T(5, -1)$ and $U(-3, 1)$
17. $E(-4, 1)$ and $F(-4, 3)$
18. $G(-5, 2)$ and $H(6, 4)$

19. WRITING MATH Explain how you can tell from the coordinates of two points on a line whether the line is horizontal or vertical?

Determine if the graphs will show parallel or perpendicular lines, or neither.

20. $y = -7x + 1$
$-7x + y = 1$

21. $3x + 4y = 12$
$y = \dfrac{4}{3}x - 5$

22. $y = x + 11$
$x + y = 0$

23. $y = -4 - 8x$
$16x + 2y = 18$

24. $y - 3 = 0$
$y = 6$

25. $y = \dfrac{4}{3}x + \dfrac{1}{3}$
$3x - 4y = -20$

Write an equation in slope-intercept form of a line passing through the given point and parallel to the given line.

26. $(2, -1); y = -\dfrac{1}{2}x + 4$

27. $(-10, -8); 2x + 3y = -1$

28. $(5, 6); 3y + x = 3$

29. $(0, -6); -7x + y = 4$

30. $(1.5, -0.5); x + y = 0$

31. $(-3, 2); 3y = -5x + 15$

Write an equation in slope-intercept form of a line passing through the given point and perpendicular to the given line.

32. $(2, -1); 2x - 9y = 5$

33. $(0, -1); 5x - y = 3$

34. $(3, -3); 3x - 2y = -7$

35. $(-4, 0); 7y = -2x - 1$

36. $(5, -6); -y = -2x + 4$

37. $(-1, 0); -\dfrac{2}{3}x - y = \dfrac{1}{2}$

38. SPORTS The vertical drop of a ski slope is 1320 ft, and the horizontal distance traveled is 1 mi. What is the slope of the ski slope? (Hint: 1 mi = 5280 ft)

39. TRAVEL An aircraft takes off following the path shown on the map. Another plane has $y = \dfrac{1}{3}x + 5$ as the equation of its take-off path. Graph the path of the second plane. Will the planes crash? Explain.

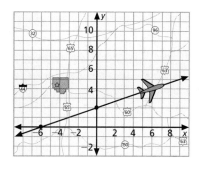

◣ EXTENDED PRACTICE EXERCISES

40. Plot the points $A(1, 2)$, $B(5, 7)$, $C(9, 2)$ and $D(5, -3)$. How many segments do the points determine? Draw the segments and find their slopes. What is true of the opposite sides and diagonals? What type of quadrilateral is this?

CRITICAL THINKING Three lines, l, m, and n, lie on the same coordinate plane. Lines l and m are perpendicular and intersect at the point $(0, 3)$. Lines m and n are also perpendicular. Line n is the graph of $y = \dfrac{3}{2}x + 4$.

41. Write the equation of line m.

42. Write the equation of line l.

43. Graph the three lines.

44. Describe the relationship of lines l and n.

◣ MIXED REVIEW EXERCISES

Find the value of y in each direct variation. (Lesson 6-8)

45. Assume that y varies directly as x. When $x = 3$, $y = 6$. Find y when $x = 7$.

46. Assume that y varies directly as x. When $x = 12$, $y = 7$. Find y when $x = 42$.

Find the value of x in each figure. (Lesson 5-2)

47.

48.

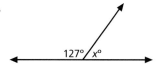

Solve Systems of Equations Graphically

Goals
- ■ Determine if an ordered pair is a solution of a system of equations.
- ■ Solve systems of linear equations graphically.

Applications   Sports, Safety, Economics

Use the graph.

1. An equation for line *a* is $x = 2$. Write three ordered pairs that are on line *a*. What do you notice?

2. An equation for line *c* is $y = 3$. Write three ordered pairs that are on line *c*. What do you notice?

3. Find an ordered pair that is a solution of both equations.

4. An equation for line *b* is $y = x$. Can you find an ordered pair that is a solution for all three equations? Why or why not?

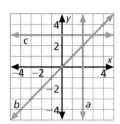

◥ BUILD UNDERSTANDING

Two (or more) linear equations with the same two variables form a **system of equations**. A **solution of a system** of equations is an ordered pair that makes both equations true.

Example 1

Determine whether the ordered pair $(-1, 6)$ is a solution of the system of equations.

$$2x + y = 4$$
$$x - y = -7$$

Solution

Substitute -1 for *x* and 6 for *y* in both equations.

$$2x + y = 4 \qquad\qquad\qquad x - y = -7$$
$$2(-1) + 6 \stackrel{?}{=} 4 \qquad\qquad -1 - 6 \stackrel{?}{=} -7$$
$$-2 + 6 \stackrel{?}{=} 4 \qquad\qquad\qquad -7 = -7 \;\checkmark$$
$$4 = 4 \;\checkmark$$

Since $(-1, 6)$ makes both equations true, it is a solution of the system of equations.

One way to solve a system of equations is to graph both equations on the same coordinate plane. All points of intersection are solutions of the system of equations. If the lines are parallel, there are no solutions.

Example 2

Solve each system of equations graphically. Check the solution.

a. $y = -2x + 4$
    $y = x - 2$

b. $y = 3x - 1$
    $y = 3x + 2$

Solution

a. First graph each equation using the slope m and the y-intercept b. Then read the solution from the graph.

$y = -2x + 4$ $b = 4; m = -2$ $y = mx + b$

$y = x - 2$ $b = -2; m = 1$

The solution is (2, 0), the point of intersection of the two lines.

To check the solution, substitute (2, 0) in each equation.

$y = -2x + 4$ $y = x - 2$

$0 \stackrel{?}{=} -2(2) + 4$ $0 \stackrel{?}{=} 2 - 2$

$0 = 0$ ✔ $0 = 0$ ✔

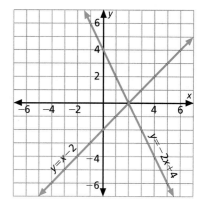

b. Since the graphs have the same slope, $m = 3$, but different y-intercepts, you do not need to graph the lines to know that they are parallel. Therefore, there is no solution.

COncepts in MOtion
Interactive Lab
mathmatters2.com

Example 3

Solve the system of equations.

$$x + 3y = 6$$
$$3x + 2y = 4$$

Solution

First write each equation in slope-intercept form, $y = mx + b$.

$x + 3y = 6$ $3x + 2y = 4$

$3y = -x + 6$ $2y = -3x + 4$

$y = -\dfrac{1}{3}x + 2$ $y = -\dfrac{3}{2}x + 2$

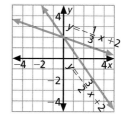

Then graph each equation using m and b.

$y = -\dfrac{1}{3}x + 2$ $b = 2$

                       $m = -\dfrac{1}{3}$

$y = -\dfrac{3}{2}x + 2$ $b = 2$

                       $m = -\dfrac{3}{2}$

The solution is (0, 2).

Check Understanding

Explain how you would check the solution of Example 3.

Graphing calculators can be used to find or check the solution of a system of equations. You can use the intersection or trace feature to estimate the coordinates of the point of intersection in a graph.

Example 4

GRAPHING Use a graphing calculator to solve the system of equations.

$$2x + y = 3$$
$$x - y = 3$$

Solution

First write the equations in slope-intercept form, $y = mx + b$.

$$2x + y = 3 \qquad x - y = 3$$

$$y = -2x + 3 \qquad y = x - 3$$

Then graph both equations on your graphing calculator.

To find the intersection of the two lines,

press $\boxed{\text{2nd}}$ $\boxed{\text{[CALC]}}$ 5 $\boxed{\text{ENTER}}$ $\boxed{\text{ENTER}}$ $\boxed{\text{ENTER}}$.

The point of intersection is $(2, -1)$.

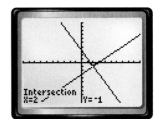

◥ TRY THESE EXERCISES

Determine if the given ordered pair is a solution of the system of equations.

1. $(2, 1)$; $3x - 2y = 6$
$\qquad\ x + y = 2$

2. $(2, 5)$; $-4x + y = -3$
$\qquad\ 2x + y = 9$

3. $(-1, 2)$; $y = 4x + 6$
$\qquad\ y = x + 3$

Solve each system of equations graphically. Check the solution.

4. $y = 2x - 6$
$\quad\ y = x - 4$

5. $2x + 3y = 3$
$\quad\ 3x + 4y = 5$

6. $y = -x - 1$
$\quad\ y = x + 3$

SPORTS Julian and Leticia are in a race together. Julian runs 8 ft/sec and Leticia runs 12 ft/sec. Suppose Leticia starts 10 ft behind Julian in the race.

7. Write an equation for both runners using $d = rt$, where d = distance, r = speed and t = time.

8. Graph both equations to determine when Leticia will catch Julian.

◥ PRACTICE EXERCISES • For Extra Practice, see page 610.

Determine if the given ordered pair is a solution of the system of equations.

9. $(2, -3)$; $2x - y = 10$
$\qquad\ x + 2y = -5$

10. $(1, 0)$; $3x + y = 3$
$\qquad\ x - 4y = 1$

11. $(0, 5)$; $y = 3x + 5$
$\qquad\ y = -3x + 5$

12. $(3, 5)$; $y = 4x - 7$
$\qquad\ y = -x + 8$

13. $(-2, 3)$; $y = 3x + 3$
$\qquad\ y = -x + 1$

14. $(0.5, 1)$; $2x + 3y = 4$
$\qquad\ -8x - 2y = -6$

Solve each system of equations graphically. Check the solution.

15. $y = -x + 7$
$y = 2x - 5$

16. $3x - 2y = 4$
$2x + y = 5$

17. $y = -3x + 4$
$y = -x + 2$

18. $y = x + 1$
$y = 2x - 3$

19. $2x - 2y = 4$
$x + y = 8$

20. $3y - x = 7$
$y - x = 1$

21. $y = -2x + 3$
$y = 2x - 5$

22. $3x - 2y = 4$
$2x + y = 5$

 GRAPHING Use a graphing calculator to solve each system of equations. Round to the nearest hundredth, if necessary.

23. $y = -x$
$y = -x + 7$

24. $y = -5x - 13$
$y = -4x - 10$

25. $2x + y = 16$
$x - 3y = 9$

26. $5x + 6y = 24$
$-7x + y = -29$

27. SAFETY The position of an accident on a highway is located at $P(4, -6)$ on a map grid. Three emergency vehicles travel on paths given by the following equations. Which of the vehicles will arrive exactly at the site of the accident?

    a. fire truck: $x - 12 = 2y$ **b.** ambulance: $y = -10 - x$ **c.** police car: $x - 5y = 34$

28. WRITING MATH Write a paragraph explaining why the intersection of two lines is the solution of a system of equations.

EXTENDED PRACTICE EXERCISES

Solve each system of equations. State the number of solutions.

29. $y = 4x - 2$
$x - 4y = 2$

30. $y = 7 - 3x$
$6x + 2y = 14$

31. $y = 2x - 8$
$2x - y = 10$

32. CRITICAL THINKING Use the results in Exercises 29–31 to describe how you can determine the number of solutions for a system of equations by comparing the slopes and the y-intercepts.

 33. YOU MAKE THE CALL Carla says she can determine the x-value of the point of intersection of any two lines, $y = m_1x + b_1$ and $y = m_2x + b_2$. She says that $x = \dfrac{b_2 - b_1}{m_1 - m_2}$. Test her conjecture with the two equations in Exercise 15.

34. ECONOMICS When Lorinda graduated from college, she was offered two jobs. One paid an annual salary of \$30,000 plus guaranteed increases of \$2000/yr. The other job paid an annual salary of \$25,000 plus guaranteed increases of \$2500/yr. Considering only this information, which job should Lorinda take? Explain why.

MIXED REVIEW EXERCISES

Find the distance between the points. Round to the nearest tenth. (Lesson 6-1)

35. $(-4, 2), (3, 8)$

36. $(-2, -5), (4, 6)$

37. $(8, 4), (3, -6)$

38. Judges at the county fair judge under a weighted system. Judge 1 has a weight of 1, Judge 2 has a weight of 2, Judge 3 has a weight of 3 and Judge 4 has a weight of 4. There are a total of 10 scores. Pies are scored from 1 to 6, with 6 being the most tasty. Determine the total scores on the score sheet. Which pie won the contest? (Lesson 1-2)

	Pie A	Pie B	Pie C	Pie D
Judge 1	2	5	6	4
Judge 2	4	3	5	3
Judge 3	3	5	2	1
Judge 4	2	4	2	5
Total score	■	■	■	■

Review and Practice Your Skills

For each line identified by two points, state the slope of a line parallel and a slope of a line perpendicular to it.

1. $A(1, 2)$ and $B(5, 7)$

2. $C(-3, -4)$ and $D(-1, 6)$

3. $E(1, 1)$ and $F(-2, 3)$

4. $G(0, 4)$ and $H(-2, 2)$

5. $I(1, 5)$ and $J(-6, 2)$

6. $K(2, 2)$ and $L(-2, 5)$

Determine if the graphs will show parallel or perpendicular lines, or neither.

7. $y = 2x - 2$
$2x + y = -5$

8. $3x + 2y = 4$
$y = \dfrac{-3}{2}x + 1$

9. $y - 2 = 0$
$y = 5$

10. $x + y = 12$
$-x + y = -1$

11. $3x - 4y = 1$
$y = -3x + 1$

12. $y = \dfrac{1}{2}x$
$y + 2x = 2$

Write an equation in slope-intercept form of a line passing through the given point and parallel to the given line.

13. $(3, -2); y - 3x = 2$

14. $(-3, 3); 2x + y = 4$

15. $(-1, 3); 2y = 4x - 8$

Write an equation in slope-intercept form of a line passing through the given point and perpendicular to the given line.

16. $(-3, -2); y = 3x + 2$

17. $(1, 3); x - 3y = 9$

18. $(-1, 4); y = 7 - 3x$

Determine if the given ordered pair is a solution of the system of equations.

19. $(1, 2); 2x + y = 5$
$x - 2y = -4$

20. $(1, 1); y = x$
$y = 2 - x$

21. $(-4, 3); y = \dfrac{1}{2}x + 5$
$2x + y = -5$

22. $(6, 5); y = x - 1$
$x + y = 11$

23. $(-3, 2); y = \dfrac{2}{3}x + 1$
$y = 2x - 2$

24. $(0, -6); y = x - 6$
$y = 4x + 6$

Solve each system of equations graphically. Check the solution.

25. $x - y = 2$
$2x + 3y = 9$

26. $y = x$
$x + y = 4$

27. $x + y = 6$
$x - y = 2$

28. $x + 2y = 7$
$y = 2x + 1$

29. $x + y = 3$
$y = x + 3$

30. $y = x - 4$
$2x + y = 5$

GRAPHING Use a graphing calculator to solve the system of equations.

31. $y = -x + 7$
$y = 2x - 5$

32. $3y = x + 7$
$y = x + 1$

33. $3x + y = 4$
$x + y = 2$

Determine if the given ordered pair is a solution of the system of equations.
(Lesson 8-2)

34. $(-2, -1); y = 2x + 3$
$y = -2x - 5$

35. $(2, -3); y = 2x - 5$
$x + y = 7$

36. $(4, 1); 4x + y = 17$
$2x + 4y = 12$

Solve each system of equations graphically. Check the solution. (Lesson 8-2)

37. $x + y = 5$
$x - 2y = -4$

38. $y = -3x$
$4x + y = 2$

39. $y = -x$
$x - y = 2$

Write an equation in slope-intercept form of a line passing through the given point and parallel to the given line. (Lesson 8-1)

40. $(-3, 1); y = -\dfrac{1}{2}x + 1$

41. $(-2, -1); x - y = 0$

42. $(1, 1); 2x + 3y = -2$

Write an equation in slope-intercept form of a line passing through the given point and perpendicular to the given line. (Lesson 8-1)

43. $(-3, 4); -3x + 2y = 3$

44. $(-6, -1); 2x + y = 3$

45. $(1, 2); y = \dfrac{x}{3} - 3$

Math*Works* Career – Runner
Workplace Knowhow

A runner has been practicing for an upcoming track and field meet in Europe, trying to improve her time in the mile run by 0.1 min every month. Unfortunately, 8 mo before the event, she sustained an ankle injury that did not allow her to run for 2 mo. When she began running again, she ran the mile in 4.5 min, 0.75 min slower than before her injury. Hoping to still participate in the track and field meet, she set a goal of improving her time by 0.2 min each month for the remaining 6 mo before the meet.

1. Write an equation in slope-intercept form for her time (t) to run 1 mi as a function of the number of months (m) of training, beginning with 6 mo before the track and field meet. Write one equation assuming that she did not have an injury and a second equation describing her performance with the injury.

2. Graph each of these equations on the same set of axes. Do these lines intersect? If so, where do they intersect? If not, why don't they intersect?

3. Draw the line $t = 3.75$ min on the graph that was drawn in Exercise 2. Where does this line intersect the line representing the runner's performance after her injury? What does the x-value of this point of intersection represent?

4. When the track and field meet begins, what speed will she have achieved in the mile run? If she had not injured her ankle, what speed would she have run at the start of the track and field meet?

8-3

Solve Systems by Substitution

Goals ■ Solve systems of equations using substitution.

Applications Transportation, Construction, Sports, Recreation

TRANSPORTATION The two ships shown on the map will stop in the middle of the strait in order to meet and exchange some cargo.

1. Make a guess at the coordinates of the meeting place.

2. Compare your solutions with those of other classmates.

3. Did you all agree on the coordinates? Why or why not?

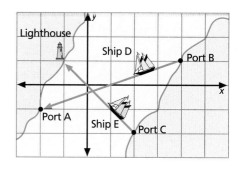

◤ BUILD UNDERSTANDING

When a system of equations is difficult to solve by reading the graph, you can use algebraic methods to solve it. One of these methods is **substitution**. In any algebraic method, you need to eliminate variables so you have one equation in one variable. Here are the steps to follow when using the substitution method.

1. Solve one of the equations for one variable. It will be in terms of the other.

2. Substitute that expression in the other equation and solve.

3. Substitute that value in one of the original equations and solve.

4. Check the solution in both of the original equations.

Example 1

Solve the system of equations. Check the solution.

$$x - 2y = 3$$
$$x + y = 6$$

Solution

$$x + y = 6 \qquad \text{Solve the second equation for } y \text{ in terms of } x.$$

$$y = -x + 6$$

$$x - 2y = 3 \qquad \text{Write the first equation.}$$

$$x - 2(-x + 6) = 3 \qquad \text{Substitute } (-x + 6) \text{ for } y.$$

$$x + 2x - 12 = 3 \qquad \text{Solve for } x.$$

$$3x - 12 = 3$$

$$3x - 12 + 12 = 3 + 12$$

$$\frac{3x}{3} = \frac{15}{3}$$

$$x = 5$$

> **Problem Solving Tip**
>
> When deciding which equation to solve for a variable, use the equation that has a variable with a coefficient of 1.

Choose one of the original equations.

$$x + y = 6$$
$$5 + y = 6 \qquad \text{Substitute 5 for } x.$$
$$-5 + 5 + y = 6 - 5 \qquad \text{Solve for } y.$$
$$y = 1$$

Check $x = 5$ and $y = 1$ in each original equation.

$$x - 2y = 3 \qquad\qquad x + y = 6$$
$$5 - 2(1) \stackrel{?}{=} 3 \qquad\qquad 5 + 1 \stackrel{?}{=} 6$$
$$3 = 3 \; ✔ \qquad\qquad\quad 6 = 6 \; ✔$$

The solution is (5, 1).

Check Understanding

In Example 1, if you solve for *x* instead of *y*, how does the solution change?

Example 2

 Personal Tutor at mathmatters2.com

CONSTRUCTION M & K Construction Company sends 29 workers out in two crews. One crew has 5 more than twice the number of workers in the other crew. How many workers are in each crew?

Solution

Define each of the variables. Write and solve a representative system of equations.

Let n = number of workers in larger crew

$\quad\;\; s$ = number of workers in smaller crew

There are 29 workers.

$$n + s = 29 \qquad \text{This equation is ready for substitution.}$$

The larger crew has 5 more than twice the smaller.

$$n = 2s + 5$$
$$n + s = 29$$
$$(2s + 5) + s = 29 \qquad \text{Substitute } (2s + 5) \text{ for } n.$$
$$3s + 5 = 29 \qquad \text{Solve for } s.$$
$$3s + 5 - 5 = 29 - 5$$
$$\frac{3s}{3} = \frac{24}{3}$$
$$s = 8$$

$$n + s = 29 \qquad \text{Substitute 8 for } s.$$
$$n + 8 = 29 \qquad \text{Solve for } n.$$
$$n + 8 - 8 = 29 - 8$$
$$n = 21$$

The larger crew has 21 workers.
The smaller crew has 8 workers.

Check

$$n + s = 29 \qquad\qquad n = 2s + 5$$
$$21 + 8 \stackrel{?}{=} 29 \qquad\qquad 21 \stackrel{?}{=} 2(8) + 5$$
$$29 = 29 \; ✔ \qquad\qquad 21 \stackrel{?}{=} 16 + 5$$
$$21 = 21 \; ✔$$

In general, if you solve a system of linear equations and the result is a true statement (an identity such as $-4 = -4$), the system has an infinite number of solutions. However, if the result is a fake statement (for example, $-4 = 5$), the system has no solution.

Example 3

Solve the system of equations.

$$x - 2y = -9$$
$$-2x + 4y = 13$$

Solution

$x - 2y = -9$	Solve for x.
$x = 2y - 9$	
$-2x + 4y = 13$	
$-2(2y - 9) + 4y \stackrel{?}{=} 13$	Substitute $(2y - 9)$ for x.
$-4y + 18 + 4y \stackrel{?}{=} 13$	
$18 \stackrel{?}{=} 13$	The variables cancel and leave a false statement.

Since $18 \neq 13$, the lines are parallel. There is no solution.

Check Understanding

If you solved the system of equations in Example 3 using graphs, what kind of lines would they be?

◣ TRY THESE EXERCISES

Solve each system of equations. Check the solution.

1. $2x - y = 14$
$x + y = 7$

2. $x + 2y = 9$
$x - y = 6$

3. $x + y = 7$
$3x - 2y = 6$

4. $x - 2y = 7$
$x - 2y = 7$

5. $x - 3y = -4$
$2x + 6y = 7$

6. $-5x + y = 3$
$20x - 4y = -2$

7. RECREATION During halftime, Joe bought 3 hot dogs and 4 drinks for $10. Cindy paid $5 for 1 hot dog and 3 drinks. Find the cost of each hot dog and each drink.

◣ PRACTICE EXERCISES • For Extra Practice, see page 610.

Solve each system of equations. Check the solution.

8. $x + y = 4$
$3x - 2y = 2$

9. $-2x + y = 1$
$x + y = 1$

10. $2x + y = 2$
$4x - 2y = 20$

11. $3x + 2y = 11$
$-2x + y = -5$

12. $2x + y = 7$
$x - y = 2$

13. $3x - y = 1$
$-12x + 4y = -3$

14. $x + 2y = 4$
$3x + y = 7$

15. $4x + y = 17$
$x + 2y = 6$

16. $x + y = 5$
$2x - y = 4$

17. $x + 5y = 9$
$3x - 2y = -7$

18. $5x + 2y = 10$
$x + y = 5$

19. $9x - 6y = 12$
$3x + 2y = 8$

20. $2x - 7y = -2$
$-4x + 14y = 3$

21. $-2x + y = -3$
$3x - 3y = -18$

22. $-x + 2y = -2$
$-2x + y = -3$

23. **WRITING MATH** Write an outline explaining the substitution method step-by-step. Use an example with your outline.

24. **SPORTS** On a cross-country team, there are 14 athletes. Three times the number of males is 2 more than twice the number of females. How many females are on the team?

25. **DATA FILE** Refer to the data on favorite pastimes on page 570. In a survey on favorite pastimes, x men and y women were surveyed. If the total number of men and women who chose watching TV was 785 and the total number who chose exercise was 160, how many men were surveyed?

26. The sum of Kelly's and Fernando's ages is 29. Kelly's age exceeds Fernando's by 5 yr. How old is each person?

Solve each system of equations by substitution.

27. $2x - y = 4$
$-4x + 2y = -8$

28. $2x + y = 3$
$-12x - 6y = -18$

29. $3x - y = -4$
$-12x + 4y = 8$

30. $-5x + 4y = -12$
$10x - 8y = -16$

31-34. **GRAPHING** Solve each system of equations in Exercises 27–30 using a graphing calculator.

■ EXTENDED PRACTICE EXERCISES

To solve a system of three equations in three variables, try to eliminate one variable and get two equations in two variables. Solve each system of equations.

35. $x + y - z = -4$
$x = -2y$
$y - z = -6$

36. $a + b + c = 1$
$b = 2c$
$-a + c = -5$

37. $r + s + t = 0$
$r = -3t$
$s + t = 3$

38. **CRITICAL THINKING** The sum of the ages of Latifa, Kyle and Luke is 38. Latifa is 6 yr older than Luke. Luke is 2 yr younger than Kyle. Find the age of each.

39. **CHAPTER INVESTIGATION** Choose two of the top four teams in the football league. Write and solve a system of equations using substitution to find the point value of the league standings.

■ MIXED REVIEW EXERCISES

Graph each function for the domain of real numbers. (Lesson 6-6)

40. $y = x^2 - 5$

41. $y = 2x^2 + 1$

42. $y = 3x^2$

43. $y = 2x^2 - 3$

44. $y = 3x^2 - 7$

45. $y = x^2 + 4$

46. A deck of 15 cards lettered A to O is shuffled and 4 cards are dealt. How many different combinations of cards could be dealt? (Lesson 4-7)

47. How many different 6-member volleyball teams can be chosen from a roster of 12 players if there are no restrictions? (Lesson 4-7)

48. Randy has to select a 6-digit number as his computer password. If no digits are repeated, how many possible passwords can he select? (Lesson 4-6)

8-4 Solve Systems by Adding, Subtracting, and Multiplying

Goals
- ■ Solve systems of equations by adding or subtracting.
- ■ Solve systems of equations by adding and multiplying.

Applications   Landscaping, Construction, Sports

NUMBER SENSE The addition property of opposites can help you explore other ways to solve a system of equations. Recall that the sum of opposites is always 0.

$$-x + x = 0 \qquad 5a + (-5a) = 0$$

For each of the following, explain what you would do to the first term to make the two terms opposites.

1. $m, -m$

2. r, r

3. $-2z, -4z$

4. $5g, -15g$

5. $3x, x$

6. $-q, q$

■ BUILD UNDERSTANDING

When solving a system of equations by an algebraic method, you need to eliminate one of the variables to get one equation in one variable. If the coefficients of one of the variables are opposites, add the equations to eliminate one of the variables. This is the addition method.

Example 1

Solve the system of equations. Check the solution.

$$2x + 3y = 6$$
$$-2x + y = 2$$

Solution

$$\begin{array}{l} 2x + 3y = 6 \\ \underline{-2x + y = 2} \\ 0 + 4y = 8 \end{array}$$

The x-coefficients are opposites.

Add the equations.

Solve for y.

$$\frac{4y}{4} = \frac{8}{4}$$

$$y = 2$$

$$2x + 3y = 6$$

Choose one of the original equations.

$$2x + 3(2) = 6$$

Substitute 2 for y.

$$2x + 6 = 6$$

Solve for x.

$$2x + 6 - 6 = 6 - 6$$

$$\frac{2x}{2} = \frac{0}{2}$$

$$x = 0$$

The solution is $(0, 2)$.

> **Check Understanding**
>
> In Example 1, if $-2x$ is changed to $2x$, what will you have had to multiply the second equation by to eliminate x?

Check

$$2x + 3y = 6$$
$$2(0) + 3(2) \stackrel{?}{=} 6$$
$$6 = 6 \ ✔$$

$$-2x + y = 2$$
$$-2(0) + 2 \stackrel{?}{=} 2$$
$$2 = 2 \ ✔$$

If the coefficients of one of the variables are the same, you can use the subtraction method to eliminate that variable.

Example 2

Solve the system of equations. Check the solution.

$$3x + 4y = 14$$
$$x + 4y = 10$$

Solution

$$
\begin{array}{rl}
3x + 4y = 14 \\
-(x + 4y = 10)
\end{array}
\longrightarrow
\begin{array}{rl}
3x + 4y = 14 \\
-x - 4y = -10
\end{array}
$$

The y-coefficients are the same.
Subtract. Distribute the negative over the equation.

$$2x + 0 = 4$$ Solve for x.

$$\frac{2x}{2} = \frac{4}{2}$$

$$x = 2$$

Choose one of the original equations.

$$x + 4y = 10$$ Substitute 2 for x.

$$2 + 4y = 10$$ Solve for y.

$$\frac{4y}{4} = \frac{8}{4}$$

$$y = 2$$

Check

$3x + 4y = 14$	$x + 4y = 10$
$3(2) + 4(2) \overset{?}{=} 14$	$2 + 4(2) \overset{?}{=} 10$
$6 + 8 \overset{?}{=} 14$	$2 + 8 \overset{?}{=} 10$
$14 = 14$ ✔	$10 = 10$ ✔

The solution is (2, 2).

Sometimes you will need to multiply one or both of the equations by a number to get coefficients of one of the variables to be opposites. This is the multiplication and addition method.

Example 3

Solve the system of equations.

$$2x + 3y = 8$$
$$x + y = -3$$

Solution

$$
\begin{array}{rl}
2x + 3y = 8 \\
-2(x + y = -3)
\end{array}
\longrightarrow
\begin{array}{rl}
2x + 3y = 8 \\
-2x - 2y = 6
\end{array}
$$

Multiply the second equation by -2.

$$0 + y = 14$$ Add.

$$y = 14$$

Choose one of the original equations.

$$x + y = -3$$ Substitute 14 for y.

$$x + 14 = -3$$ Solve for x.

$$x = -17$$

Be sure to check the solution.

The solution is $(-17, 14)$.

Check Understanding

In Example 3, what can you multiply the second equation by to eliminate y?

Example 4

Online Personal Tutor at mathmatters2.com

Kate has nickels and dimes in her pocket. There are 11 coins. The value of the coins is $0.95. How many of each kind of coin does Kate have?

Solution

Let n = the number of nickels and d = the number of dimes. The total number of coins is 11, so $n + d = 11$. The value of the coins is $0.95, so $0.05n + 0.10d = 0.95$.

$$-0.05(n + d = 11)$$
$$0.05n + 0.10d = 0.95$$

$$-0.05n - 0.05d = -0.55$$ Multiply the first equation by -0.05.
$$0.05n + 0.10d = 0.95$$ Add.

$$0 + 0.05d = 0.4$$ Solve for d.
$$\frac{0.05d}{0.05} = \frac{0.4}{0.05}$$
$$d = 8$$

Choose one of the original equations.

$$n + d = 11$$ Substitute 8 for d.
$$n + 8 = 11$$ Solve for n.
$$n = 3$$

Kate has 3 nickels and 8 dimes. Check the solution.

TRY THESE EXERCISES

Solve each system of equations. Check the solution.

1. $3x - y = 15$
$x + y = 1$

2. $-5a + 4b = -1$
$7a + 4b = 11$

3. $4x + 3y = 2$
$x - y = -10$

4. $5x - y = -23$
$3x - y = -15$

5. $j + 3k = 10$
$j + 2k = 7$

6. $7m - 5n = -2$
$-8m - n = 9$

7. LANDSCAPING A landscaping firm is designing a flower bed to border a rectangular pool. The perimeter of the pool is 32 m. Three times the width is the same as five times the length. What are the dimensions of the pool?

PRACTICE EXERCISES • For Extra Practice, see page 611.

Solve each system of equations. Check the solution.

8. $3x + y = 9$
$-3x + y = -3$

9. $-7x - 8y = 8$
$7x - 8y = 8$

10. $7a + 2b = 16$
$8a - 2b = 14$

11. $2r + 3s = 1$
$9r - 3s = 54$

12. $r + s = 4$
$2r + 3s = 8$

13. $2x + 3y = 8$
$2x + 3y = 8$

14. $3c - d = 1$
$c + 5d = 11$

15. $2x + 2y = 2$
$-3x + y = -3$

16. $-2x - y = 1$
$3x + 8y = 5$

17. $c - 2d = 10$
$2c + 5d = 11$

18. $2x + y = 6$
$-3x + 2y = -2$

19. $4x - 3y = 12$
$2x + 6y = 5$

20. $5x + 3y = 11$
$3x + 2y = 1$

21. $2r - 5s = 7$
$3r - 2s = -17$

22. $3a + 8b = 1$
$2a + 7b = 4$

23. WRITING MATH Write a brief paragraph explaining how you solved Exercise 22.

24. Yuki pays $5.05 for 3 muffins and 4 coffees. Dave pays $4.90 for 4 muffins and 2 coffees. How much does each item cost?

25. SPORTS Cara had a combined total of 64 hits during her junior and senior years of fastpitch softball. In her senior year, Cara had 5 fewer than twice as many hits as her junior year. How many hits did Cara have in her senior year?

26. CONSTRUCTION A work crew of 5 bricklayers and 3 carpenters earns $891 for a job. Another crew of 12 bricklayers and 4 carpenters earns $1748 on the same job. Find the wage of each type of worker.

27. DATA FILE Refer to the data on pizza toppings on page 568. Suppose that John ate x slices of pepperoni pizza and y slices of pizza with extra cheese. If the total number of calories is 576 and the total number of fat grams is 37, how many slices of each type of pizza did John eat?

■ EXTENDED PRACTICE EXERCISES

Solve each system of equations.

28. $-5x + 2y = 12$
$10x - 4y = -24$

29. $6x - 9y = 36$
$-2x + 3y = -12$

30. $8x - 2y = -10$
$-16x + 4y = 20$

Use the results of Exercises 28–30.

31. What do you notice about the relationship between the two equations in each of the systems? What do you notice about the number of solutions?

32. Solve the systems graphically. What do you notice about the graphs?

33. CRITICAL THINKING Write a statement that generalizes your observations.

34. CHAPTER INVESTIGATION Choose two of the top four football teams, other than the pair chosen in Lesson 8-3. Write and solve a system of equations using addition, subtraction or multiplication. Verify that this is the same answer as that in Lesson 8-3.

■ MIXED REVIEW EXERCISES

Find the slope of the line containing the given points. Identify any vertical or horizontal lines. (Lesson 6-2)

35. $(3, 4)$, $(1, -6)$

36. $(5, -2)$, $(3, 7)$

37. $(-3, -2)$, $(5, 3)$

38. $(6, -1)$, $(-3, -1)$

39. $(4, -4)$, $(7, 1)$

40. $(7, -2)$, $(7, 7)$

In the figure, $\overleftrightarrow{AB} \parallel \overleftrightarrow{CD}$. Find each measure. (Lesson 5-3)

41. $m\angle 1$

42. $m\angle 2$

43. $m\angle 3$

44. $m\angle 5$

45. $m\angle 6$

46. $m\angle 7$

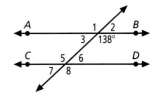

Review and Practice Your Skills

Use substitution to solve each system of equations. Check the solution.

1. $y - 2x$

    $2x + y = 8$

2. $y - 3x$

    $x + 2y = -21$

3. $x - 2y$

    $4x + 2y = 15$

4. $x = y + 4$

    $x + 3y = 16$

5. $y = x + 1$

    $2x - 3y = -5$

6. $x = 3 - 2y$

    $2x - 2y = 6$

7. $3x = y + 9$

    $2x - 4y = 16$

8. $-4x + 3y = -16$

    $-x + 2y = -4$

9. $\dfrac{x}{2} - y = \dfrac{5}{4}$

    $8x + 3y = 1$

10. $5x + 2y = 1$

    $3x + 4y = -5$

11. $6x - 3y = -9$

    $13x - 5y = -15$

12. $10x - 5y = 65$

    $10y - 5x = -55$

13. $y = -x + 5$

    $2x - y = 4$

14. $-2x = -y - 3$

    $3x = 3y - 18$

15. $y = 2x + 2$

    $4x - 2y = 3$

16. The sum of Sophia's and Austin's age is 32. Sophia's age exceeds Austin's age by 4 yr. How old is each person?

Use addition, subtraction or multiplication to solve each system of equations. Check the solution.

17. $-x + 2y = 3$

    $3x + 2y = -1$

18. $x - 5y = 7$

    $-9x - 10y = -8$

19. $-2x + 6y = 10$

    $2x - 9y = -19$

20. $-10x + 6y = 25$

    $2x + 9y = 12$

21. $4x - 9y = -1$

    $-8x + 9y = 1$

22. $7x + 4y = 6$

    $7x + 5y = -3$

23. $8x + 3y = -4$

    $6x + 5y = 8$

24. $8x - 3y = 18$

    $8x - 2y = 12$

25. $2x - \dfrac{1}{2}y = 3$

    $4x + \dfrac{1}{2}y = 9$

26. $-\dfrac{1}{2}x + y = 3$

    $\dfrac{1}{2}x - 3y = 1$

27. $9a + 7b = 4$

    $6a - 3b = 18$

28. $2x - 5y = -6$

    $6x - 6y = 18$

29. $x - y = 4$

    $-2x + 4y = 4$

30. $4x + 6y = 68$

    $2x - 5y = 10$

31. $4x + 5y = 34$

    $-3x + 5y = 6$

32. Samantha paid $6.00 for 3 hotdogs and 2 juices. Damian paid $9.75 for 5 hotdogs and 3 juices. How much does each item cost?

33. The measure of one angle is 15 more than twice the measure of its supplementary angle. Find the measure of each angle.

Write an equation in slope-intercept form of the line passing through the given point and perpendicular to the given line. (Lesson 8-1)

34. $(-2, -3)$; $y = 2x$

35. $(1, 4)$; $x + 3y = 6$

36. $(3, 1)$; $2y = 3x - 4$

GRAPHING **Use a graphing calculator to solve each system of equations.** (Lesson 8-2)

37. $y = \dfrac{2}{3}x$

$y = -\dfrac{2}{3}x + 4$

38. $y = \dfrac{1}{2}x - 2$

$-x + y = -5$

39. $-2x + y = 1$

$y = x + 3$

Use substitution to solve each system of equations. Check the solution. (Lesson 8-3)

40. $3x + 2y = 12$

$y = 2x - 1$

41. $2x + y = 6$

$x - 2y = 8$

42. $8x + y = 2$

$6x - 2y = 7$

Use addition, subtraction or multiplication to solve each system of equations. Check the solution. (Lesson 8-4)

43. $7x + 5y = 14$

$4x - 5y = 8$

44. $9x - 2y = 5$

$9x + y = 8$

45. $12x - y = 7$

$6x - y = 1$

Mid-Chapter Quiz

For each line identified by two points, state the slope of a line parallel and the slope of a line perpendicular to it. (Lesson 8-1)

1. $A(3, -5)$ and $B(-6, 0)$

2. $C(11, 4)$ and $D(9, -4)$

3. $E(0, 7)$ and $F(-4, -1)$

Determine if the given ordered pair is a solution of the system of equations. (Lesson 8-2)

4. $(1, 2)$; $y = -2x + 4$

$y = \dfrac{3}{2}x + \dfrac{1}{2}$

5. $(-6, 9)$; $y = -2x - 3$

$y + x = 3$

6. $(-4, -3)$; $y = x - 7$

$x - y = -1$

Solve each system of equations. Check the solution. (Lessons 8-3 and 8-4)

7. $y = 7x - 9$

$y = x + 3$

8. $4y + x = -2$

$x = -4y$

9. $2x - y = 2$

$2y - 4x = -4$

10. $2x + 4y = 8$

$x + y = 1$

11. $x + 3y = 25$

$y - x = -9$

12. $4x - 5y = -17$

$4x + 6y = 38$

13. $3x + 6y = -15$

$-x - 4y = 13$

14. $7x + 2y = 9$

$4x - 3y = -28$

15. $-2m - 5n = 2$

$5m + 2n = 16$

16. There are 37 trucks and cars waiting in line to pay a toll. The number of cars is 2 less than twice the number of trucks. How many cars and trucks are in line? (Lesson 8-4)

8-5 Matrices and Determinants

Goals ■ Find the determinant of a 2 × 2 matrix.
■ Solve systems of equations using determinants.

Applications Sports, Construction, Fitness

Work with a partner.

1. In how many ways can you solve a system of equations with two unknowns?

2. Solve the system of equations using each method learned in the previous three lessons.

$$2x - 3y = 4$$
$$-3x + 5y = 1$$

3. Did you arrive at the same solution using each method?

4. Which method do you prefer? Why?

5. Discuss with others the method they prefer.

◣ BUILD UNDERSTANDING

Recall that a matrix is a rectangular array of values. The matrix shown has elements 4, −1, 0, 3, 6, and 2, and its dimensions are 2 × 3. A matrix of n rows and m columns has dimensions $n \times m$.

$$\begin{bmatrix} 4 & -1 & 0 \\ 3 & 6 & 2 \end{bmatrix}$$

A **square matrix** has the same number of rows and columns. Associated with a 2 × 2 square matrix is a number called the determinant. The **determinant** of a matrix is symbolized by using vertical bars in place of the matrix brackets.

The value of a 2 × 2 determinant is the difference of the products of the diagonal entries. The determinant of a square matrix A is named det A.

Determinant $\quad \det A = \begin{vmatrix} a & b \\ c & d \end{vmatrix} = ad - bc$, where a, b, c and d are real numbers.

Example 1

Evaluate the determinant of matrix $A = \begin{bmatrix} 1 & 2 \\ -3 & 4 \end{bmatrix}$.

Solution

$$\det A = \begin{vmatrix} 1 & 2 \\ -3 & 4 \end{vmatrix}$$

$$= 1(4) - 2(-3) \qquad a = 1, b = 2, c = -3, d = 4$$

$$= 4 + 6$$

$$= 10$$

One of the important applications of determinants is in solving systems of equations. The method used is called **Cramer's rule**, after the Swiss mathematician Gabriel Cramer (1704–1752). Cramer's rule is demonstrated in Example 2. Notice that to use Cramer's rule, you must first write both equations in standard form, $Ax + By = C$.

Math: Who, Where, When

A method of detached coefficients similar to determinants has been traced to twelfth-century Chinese mathematicians. As the Chinese studied patterns involved in solving systems of equations, they discovered that they could arrange numerical coefficients in a square array similar to those on their counting board (the abacus).

Example 2

Solve the system of equations using the method of determinants.

$$2x - 3y = 4$$

$$-3x + 5y = 1$$

Solution

Write the coefficients of x and y in a determinant A.

y-coefficients

$$\det A = \begin{vmatrix} 2 & -3 \\ -3 & 5 \end{vmatrix}$$

x-coefficients

Write another determinant. Use A and replace the x-column with the constants from the equations. Label it A_x.

$$\det A_x = \begin{vmatrix} 4 & -3 \\ 1 & 5 \end{vmatrix}$$

Replace x-coefficients with constants.

Write a third determinant. Use A and replace the y-column with the constants from the equations. Label it A_y.

$$\det A_y = \begin{vmatrix} 2 & 4 \\ -3 & 1 \end{vmatrix}$$

Replace y-coefficients with constants.

If $A \neq 0$, the solution of the system is (x, y), where $x = \dfrac{A_x}{A}$ and $y = \dfrac{A_y}{A}$. Solve for x and y.

$$x = \frac{A_x}{A} = \frac{\begin{vmatrix} 4 & -3 \\ 1 & 5 \end{vmatrix}}{\begin{vmatrix} 2 & 3 \\ -3 & 5 \end{vmatrix}} = \frac{4(5) - (-3)(1)}{2(5) - (-3)(3)} = \frac{20 + 3}{10 - 9} = \frac{23}{1} = 23$$

$$y = \frac{A_y}{A} = \frac{\begin{vmatrix} 2 & 4 \\ -3 & 1 \end{vmatrix}}{\begin{vmatrix} 2 & -3 \\ -3 & 5 \end{vmatrix}} = \frac{2(1) - 4(-3)}{2(5) - (-3)(-3)} = \frac{2 + 12}{10 - 9} = \frac{14}{1} = 14$$

Check the solution.

$$2x - 3y = 4 \qquad\qquad\qquad -3x + 5y = 1$$

$$2(23) - 3(14) \stackrel{?}{=} 4 \qquad\qquad -3(23) + 5(14) \stackrel{?}{=} 1$$

$$46 - 42 \stackrel{?}{=} 4 \qquad\qquad\qquad -69 + 70 \stackrel{?}{=} 1$$

$$4 = 4 \;\checkmark \qquad\qquad\qquad\qquad 1 = 1 \;\checkmark$$

The solution is $(23, 14)$.

Example 3 **Personal Tutor at** mathmatters2.com

MOVIES Movie tickets to a matinee cost $7.25 for adults and $5.50 for students. A group of friends purchased 8 matinee tickets for $52.75. How many adult tickets and student tickets were purchased?

Solution

Let x represent the total number of adult tickets. Let y represent the total number of student tickets.

$x + y = 8$

$7.25x + 5.5y = 52.75$

Use the method of determinants to find x and y.

$$\det A = \begin{vmatrix} 1 & 1 \\ 7.25 & 5.5 \end{vmatrix}, \det A_x = \begin{vmatrix} 8 & 1 \\ 52.75 & 5.5 \end{vmatrix}, \det A_y = \begin{vmatrix} 1 & 8 \\ 7.25 & 52.75 \end{vmatrix}$$

$$x = \frac{A_x}{A} = \frac{\begin{vmatrix} 8 & 1 \\ 52.75 & 5.5 \end{vmatrix}}{\begin{vmatrix} 1 & 1 \\ 7.25 & 5.5 \end{vmatrix}} = \frac{8(5.5) - 1(52.75)}{1(5.5) - 1(7.25)} = \frac{-8.75}{-1.75} = 5$$

$$y = \frac{A_y}{A} = \frac{\begin{vmatrix} 1 & 8 \\ 7.25 & 52.75 \end{vmatrix}}{\begin{vmatrix} 1 & 1 \\ 7.25 & 5.5 \end{vmatrix}} = \frac{1(52.75) - 8(7.25)}{1(5.5) - 1(7.25)} = \frac{-5.25}{-1.75} = 3$$

So, there were 5 adult tickets and 3 student tickets purchased.

TRY THESE EXERCISES

Evaluate each determinant.

1. $\begin{vmatrix} 0 & 4 \\ -2 & 3 \end{vmatrix}$
2. $\begin{vmatrix} -7 & 7 \\ 3 & -3 \end{vmatrix}$
3. $\begin{vmatrix} 4.1 & -2.7 \\ 0.1 & -1.2 \end{vmatrix}$
4. $\begin{vmatrix} 6 & -1 \\ 4 & 0 \end{vmatrix}$

Solve each system of equations using the method of determinants.

5. $-4x + 5y = 2$
$-3x + 6y = -3$

6. $-5x + 5y = -5$
$3x - 4y = -1$

7. $3x + 5y = 9$
$-2x - 3y = 7$

8. $4x - y = 9$
$x - 3y = 16$

9. $3x - 5y = -23$
$5x + 4y = 11$

10. $3x + 2y = 24$
$15x - 2y = 48$

11. SPORTS A swim team has 52 members. The number of female athletes is one more than twice the number of male athletes. How many members of the team are female?

PRACTICE EXERCISES • For Extra Practice, see page 611.

Evaluate each determinant.

12. $\begin{vmatrix} 0.5 & 1.2 \\ 4 & -3 \end{vmatrix}$
13. $\begin{vmatrix} 5 & -2 \\ -4 & 8 \end{vmatrix}$
14. $\begin{vmatrix} -10 & 9 \\ -8 & 6 \end{vmatrix}$
15. $\begin{vmatrix} 12 & 11 \\ 9 & -6 \end{vmatrix}$

Solve each system of equations using the method of determinants.

16. $-2x + 6y = 2$
$-3x + 5y = 11$

17. $4x + 5y = 34$
$-3x + 5y = 6$

18. $4x + 6y = 68$
$2x - 5y = 10$

19. $7x + 5y = -12$
$-3x + 6y = 25.5$

20. $-8x + 3y = -17$
$-3x + 5y = 44$

21. $3x - 2y = 19$
$-2x + 3y = -16$

22. $x + 8y = 9$
$-3x + 7y = 4$

23. $-23x + 25y = 42$
$-3x + 5y = 2$

24. $-4x + 4y = 8$
$-3x + 5y = 2$

25. $-7x + 8y = 10$
$-9x + 7y = 3$

26. $-6x + 5y = 13$
$3x + 4y = 13$

27. $-3x + y = 11$
$-4x - 5y = 21$

28. YOU MAKE THE CALL Colin and Angelo solved the system of equations consisting of $3x - 6y = 3$ and $-11x + 13y = -2$. Colin's solution is $(-1, -1)$. Angelo's solution is $(-1, 1)$. Who is correct? Explain.

29. CONSTRUCTION Fleetwood Construction Co. is building a walkway in the shape of an isosceles triangle. The combined length of the two congruent sides of the walkway is 5 m longer than the length of the third side. The perimeter of the walkway is 35 m. How long is each side of the walkway?

30. SPORTS In a recent basketball game, Kenji successfully made 11 baskets for a total of 25 points. The 25 points were a combination of 2-point goals and 3-point goals. How many 2-point goals did Kenji make? How many 3-point goals did he make?

31. DATA FILE Refer to the data on calories used by body weight on page 568. Nakita and Mary played tennis and then went swimming. Nakita weighs 150 lb, and Mary weighs 120 lb. Nakita used 514 calories, and Mary used 408 calories. How long did they spend at each activity?

◤ EXTENDED PRACTICE EXERCISES

32. WRITING MATH Describe the advantages and disadvantages of the four methods in this chapter used to solve a system of equations. Which methods are more difficult with more unknown variables? Which methods are more useful? Include clues that help you determine which method to use.

33. CHAPTER INVESTIGATION Choose both systems of equations in Lessons 8-3 and 8-4. Use the method of determinants to solve each system. Verify that your answers in Lessons 8-3 and 8-4 are correct.

◤ MIXED REVIEW EXERCISES

Graph the figure containing points P, Q, R, and S on a coordinate plane. Graph the image under the given rotation about (0, 0). (Lesson 7-3)

34. 180° clockwise

35. 90° counterclockwise

36. 320° counterclockwise

Solve and graph each inequality. (Lesson 3-7)

37. $4 + c > -2$

38. $2b \geq \dfrac{3}{2}$

39. $g - 3 \leq -2$

40. $9 + k < 4$

41. $4d - 8 > 16$

42. $3(a + 2) \geq 14$

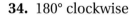

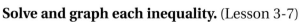

8-6 Problem Solving Skills: Directed Graphs

An effective problem solving strategy for many students is to **use a picture, diagram or model.** The visual aid shows the connection among the data and leads you to a solution.

If you look at a road map, you will see a collection of locations connected in different ways by roads. A **directed graph** is a geometrical representation of a map. It shows locations as points and roads as lines. Connections may be possible in only one direction on some paths and in both directions on other paths. You can analyze directed graphs by using matrices.

Problem Solving Strategies

Guess and check

Look for a pattern

Solve a simpler problem

Make a table, chart or list

✔ Use a picture, diagram or model

Act it out

Work backwards

Eliminate possibilities

Use an equation or formula

Problem

BUSINESS SummerSun Company has five plants in five different cities. The directed graph shown represents how supplies are transferred between plants.

a. Use a matrix to find the total number of ways in which supplies can be transferred between two plants without being routed through another plant.

b. What is the sum of the elements in the first row of the matrix? What does this number represent?

c. How many ways can supplies be transferred directly to Boston from another city?

Solve the Problem

a. Create a 5 × 5 matrix using the first letter of each location. Label the rows *From* and the columns *To*.

Use the directed graph to complete the matrix. Write *0* if there are no ways in which supplies can be directly transferred from one plant to another. Write *1* if there is one way. Then add all the number ones. There are 13 ways to transfer supplies between 2 plants without being routed through another plant.

b. The sum of the elements in the first row is 2. This represents the number of ways to transfer supplies directly from Fitchburg.

c. Find the sum of the elements in the column labeled B.

$$1 + 1 + 0 + 1 + 0 = 3$$

There are three ways to transfer supplies directly to Boston.

		To			
	F	L	B	P	W
F	–	–	–	–	–
L	–	–	–	–	–
B	–	–	–	–	–
P	–	–	–	–	–
W	–	–	–	–	–

		To			
	F	L	B	P	W
F	0	0	1	0	1
L	1	0	1	0	0
B	1	1	0	1	1
P	1	0	1	0	1
W	1	0	0	1	0

TRY THESE EXERCISES

TRANSPORTATION The directed graph shown represents the connections between subway stations.

1. Create a matrix for the subway stations that shows the number of direct connections from one station to another.

2. How many ways can you get directly to Station B?

3. Create a matrix for the subway stations that shows the number of ways someone can travel between two stations with one stop.

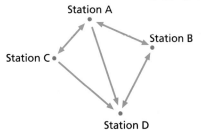

Subway System

PRACTICE EXERCISES

Use the situation from the problem on page 358.

4. Create a matrix to show the number of ways in which supplies can be shipped between any two plants if there is one stop between the two plants.

5. How many ways can supplies be transferred between two plants if there is one stop in between?

6. **WRITING MATH** Explain how the manager at SummerSun Company might use the information in Exercises 4 and 5.

ENTERTAINMENT Janeil and Rachel are visiting a theme park. The directed graph shows five locations in the park and the one-way roads connecting them.

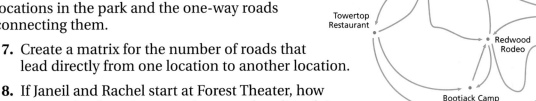

7. Create a matrix for the number of roads that lead directly from one location to another location.

8. If Janeil and Rachel start at Forest Theater, how many other locations can they travel to directly?

9. Create a matrix for the number of ways in which Janeil and Rachel can travel between two locations with one stop on the way.

10. **MODELING** The matrix shown represents the number of nonstop flights between cities A, B, C, D and E. Draw a directed graph that corresponds to the matrix.

$$\begin{array}{c c c c c c} & A & B & C & D & E \\ A & 0 & 1 & 1 & 1 & 0 \\ B & 1 & 0 & 0 & 1 & 0 \\ C & 1 & 0 & 0 & 1 & 0 \\ D & 1 & 1 & 1 & 0 & 1 \\ E & 0 & 0 & 0 & 1 & 0 \end{array}$$

MIXED REVIEW EXERCISES

Find the value of _y_ in each inverse variation. (Lesson 6-9)

11. Assume that _y_ varies inversely as _x_. When $x = 2$, $y = 150$. Find _y_ when $x = 4$.

12. Assume that _y_ varies inversely as _x_. When $x = 35$, $y = 21$. Find _y_ when $x = 7$.

Quadrilateral _ABCD_ has vertices $A(-1, 3)$, $B(4, 1)$, $C(4, -1)$ and $D(-3, -2)$. Graph the quadrilateral and its image under the given translations. (Lesson 7-1)

13. 4 units left
14. 3 units down
15. 2 units right
16. 6 units up
17. 3 units left and 2 units down
18. 2 units up and 5 units right

Review and Practice Your Skills

Evaluate each determinant.

1. $\begin{vmatrix} 3 & 5 \\ 2 & 4 \end{vmatrix}$

2. $\begin{vmatrix} 4 & -1 \\ -2 & 5 \end{vmatrix}$

3. $\begin{vmatrix} -4 & -3 \\ -5 & 4 \end{vmatrix}$

4. $\begin{vmatrix} 1 & 2 \\ 3 & 4 \end{vmatrix}$

5. $\begin{vmatrix} 0 & 11 \\ 4 & -6 \end{vmatrix}$

6. $\begin{vmatrix} -3 & 12 \\ 7 & -10 \end{vmatrix}$

7. $\begin{vmatrix} 5 & -6 \\ 2 & 7 \end{vmatrix}$

8. $\begin{vmatrix} 6 & -8 \\ 3 & -4 \end{vmatrix}$

Solve each system of equations using the method of determinants.

9. $3x - 4y = 13$
   $2x + y = 5$

10. $6x + 5y = 8$
    $4x - 3y = -1$

11. $4x - 5y = -19$
    $3x + 7y = 18$

12. $2x - 9y = 14$
    $6x - y = 42$

13. $2a - 4b = 18$
    $3a - b = 22$

14. $x - y = -1$
    $x + y = 9$

15. $8x - 2y = 6$
    $3x - 4y = -1$

16. $2y - 3x = 0$
    $3y - 2x = 10$

17. $5x - 2y = 9$
    $x - 4y = 9$

18. $x + 3y = -9$
    $5x + 2y = 7$

19. $-4x - 5y = -2$
    $x + 6y = 10$

20. $2x + 3y = 12$
    $x - 2y = -8$

TRANSPORTATION A bus company offers service between five towns A, B, C, D, and E. The directed graph represents the connections between the towns.

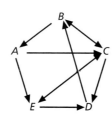

21. Create a matrix that shows the number of direct bus trips from one town to another.

22. How many ways can you get directly to town C?

23. Create a matrix that shows the number of ways someone can travel between two towns with one stop on the way.

TRAVEL The directed graph shows direct airline service between Denver, Omaha, Philadelphia, Atlanta, and Miami.

24. Create a matrix that shows the number of direct flights between each city.

25. According to the directed graph, which cities can you fly to directly from Omaha?

26. Create a matrix that shows the number of ways to get to each city with one stop on the way.

For each line identified by two points, state the slope of a line parallel and a slope of a line perpendicular to it. (Lesson 8-1)

27. $A(-3, 7)$ and $B(9, 1)$ **28.** $C(2, 5)$ and $D(-1, -4)$ **29.** $E(3, 5)$ and $F(-1, 8)$

Solve each system of equations graphically. Check the solution. (Lesson 8-2)

30. $y = 3x - 1$
$\quad y = -x + 3$

31. $y = 3x + 2$
$\quad y = 4x + 4$

32. $y = -2x + 5$
$\quad x = 3y + 6$

Use addition, subtraction or multiplication to solve each system of equations. Check the solution. (Lesson 8-4)

33. $x + y = 5$
$\quad -x + y = 1$

34. $6x + 3y = -15$
$\quad -2x + 4y = 0$

35. $x + y = 1$
$\quad x - y = 15$

MathWorks Career – Coach
Workplace Knowhow

While an athlete is the person ultimately responsible for performing in an athletic competition, the coach is crucial for directing the athlete's training before and during the event. To adequately prepare an athlete for competition, a coach considers the individual characteristics of each team member. One characteristic might be how the athlete responds to various types of foods eaten 24 h before a competition. In particular, how does the athlete's carbohydrate consumption affect performance? Below are two equations relating carbohydrate consumption (c) in grams and the time (t) in minutes required to complete the 400 m freestyle event.

Swimmer A: $t = -0.001737c + 6.2107$

Swimmer B: $t = 0.001520c + 4.5983$

1. Solve the system of equations using substitution.

2. Solve the system of equations by subtracting the second equation from the first.

3. Using the method of determinants, find the amount of carbohydrates consumed which will cause each swimmer to swim the 400 m freestyle in the same amount of time.

4. Draw a graph of each of the lines above on the same axes.

5. Explain the significance of each line's slope.

6. Complete the statement by filling in the blanks. Swimmer __?__ swims faster than swimmer __?__ before the "crossover point" of (__?__, __?__). Swimmer __?__ swims faster than swimmer __?__ after this "crossover point."

8-7 Systems of Inequalities

Goals
- Write a system of linear inequalities for a given graph.
- Graph the solution set of a system of linear inequalities.

Applications Sports, Entertainment, Retail, Finance

SPORTS The Tigers scored 20 points, but still lost the football game to the Eagles. The Eagles scored a combination of field goals (3 points each) and touchdowns that were each followed by the extra-point conversion (7 points each).

1. Write an inequality that represents the combination of field goals and touchdowns the Eagles could have scored.

2. Graph the inequality.

3. Based on your graph, name three possible combinations of field goals and touchdowns the Eagles could have scored.

◣ BUILD UNDERSTANDING

The slope-intercept form of an inequality is useful for writing a system of inequalities for a given graph. It is also useful for graphing a system of inequalities. The table shown can be used when graphing systems of inequalities.

Form of inequality	Boundary	Shaded region
$y > mx + b$	dashed	above the line
$y \geq mx + b$	solid	above the line
$y < mx + b$	dashed	below the line
$y \leq mx + b$	solid	below the line

Example 1

Write a system of linear inequalities for the graph shown.

Solution

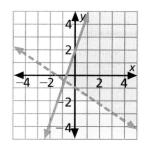

	Blue	**Yellow**
Determine m.	$\dfrac{\text{rise}}{\text{run}} = \dfrac{3}{1} = 3$	$\dfrac{\text{rise}}{\text{run}} = \dfrac{-2}{3} = -\dfrac{2}{3}$
Determine b.	$b = 2$	$b = -1$
Determine the inequality symbol.	Below $\leq$ Solid	Above $>$ Dashed

The system of linear inequalities for the graph is as follows:

$$y \leq 3x + 2$$
$$y > -\frac{2}{3}x - 1$$

COncepts in MOtion

Animation

mathmatters2.com

The solution set of a system of linear inequalities is the intersection of the graphs of the inequalities. For the system of inequalities graphed in Example 1, the solution set includes all points in the doubly shaded (green) region as well as points on the solid boundary line that are not on or below the dashed line.

Example 2

nline Personal Tutor at mathmatters2.com

Graph the solution set of the system of linear inequalities.

$$2x + y > 1$$
$$3x - y \geq 2$$

Problem Solving Tip

When graphing, a quick way to check your shading is to check the point (0, 0) in the original form of the inequality. If it is a solution, it should be in the shaded section.

Solution

First write each inequality in slope-intercept form. Then make a chart to use for graphing.

$$2x + y > 1 \qquad\qquad 3x - y \geq 2$$
$$2x - 2x + y > -2x + 1 \qquad 3x - 3x - y \geq -3x + 2$$
$$y > -2x + 1 \qquad\qquad -y \geq -3x + 2$$
$$\frac{-y}{-1} \leq \frac{-3x + 2}{-1} \qquad \text{Reverse the order of the inequality.}$$
$$y \leq 3x - 2$$

	$y > -2x + 1$	$y \leq 3x - 2$
Boundary	$y = -2x + 1$	$y = 3x - 2$
Shading	Above	Below
Line	Dashed	Solid

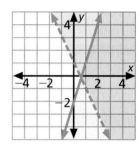

The solution set consists of all the points in the doubly shaded (green) region as well as all points on the solid boundary $y = 3x - 2$ that are not on or below the dashed boundary.

Example 3

ENTERTAINMENT Olympic Video offers a discount if you spend more than $12 and rent more than 4 DVDs.

a. Write an inequality for spending more than $12 and an inequality for renting more than 4 DVDs.

b. Graph the inequalities and shade the solution.

c. Give the coordinates of a point that satisfies the inequalities and the coordinates of a point that does not satisfy the inequalities. Explain the coordinates in terms of renting a DVD.

Solution

a. Let x = the amount of money spent, and let y = the number of DVDs rented.

$$x > 12 \qquad\qquad y > 4$$

b. Draw a dashed line at $x = 12$, and shade to the right of the line. Draw a dashed line at $y = 4$, and shade above the line.

c. The point (15, 5) satisfies both inequalities. This means that 5 DVDs were rented and $15 was spent. A discount was offered.

The point (10, 2) does not satisfy both inequalities. This means that 2 DVDs were rented and $10 was spent. A discount was not offered.

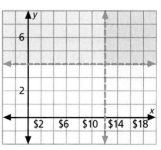

Write a system of linear inequalities for each graph.

1.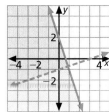

2.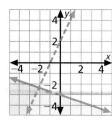

3.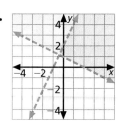

Graph the solution set of each system of inequalities.

4. $3x - y < -1$
$-4x - y < 2$

5. $2x + y > 3$
$3x - y \geq -2$

6. $-2x + y > 3$
$5x - 5y \leq -10$

7. **WRITING MATH** Write a brief paragraph explaining how to determine whether to shade above or below a line.

PRACTICE EXERCISES • For Extra Practice, see page 612.

Write a system of linear inequalities for each graph.

8.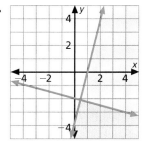

9.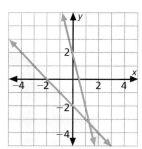

10.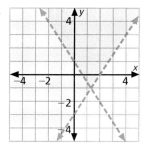

Graph the solution set of each system of inequalities.

11. $5x - y > -1$
$-2x + y \leq 3$

12. $-x + y \leq 4$
$3x - y < -2$

13. $-3x + y \geq 2$
$-2x - y > -1$

14. $4x - y > 1$
$-2x - y \leq -3$

15. $-x + y < -4$
$2x - y \geq 1$

16. $-x + y < -4$
$-x + 3y < 4$

17. $x - y > -1$
$-2x + y \leq 3$

18. $-2x + y \leq 4$
$3x - y < -2$

19. $-3x + 2y \leq 9$
$-3x - y < 9$

20. $5x - 3y > -1$
$-x + y \leq 3$

21. $-5x + 5y \leq 5$
$2x - y < -3$

22. $-x + 2y \geq 12$
$-2x - y < -1$

RETAIL Some CDs cost $10 and some cost $15. How many CDs can Keandre buy if he must spend less than $200 and buy more than 10 CDs?

23. Write two inequalities to represent the situation.

24. Graph the inequalities.

25. Give the coordinates of a point that satisfies the inequalities and the coordinates of a point that does not satisfy the inequalities. Explain the coordinates in terms of the problem.

26. Refer to Exercises 23–25. Can Keandre buy five of the $10 CDs and seven of the $15 CDs? Explain your reasoning.

FINANCE Akule plans on investing $2000 or less in two different accounts. The low-risk account pays 5% interest, and the high-risk account pays 10% interest. Akule wants to make more than $150 interest for the year.

27. Write two inequalities to show how Akule might split the total between the two accounts, assuming simple interest.

28. Graph the inequalities.

29. Give the coordinates of a point that satisfies the inequalities and the coordinates of a point that does not satisfy the inequalities.

30. What is the minimum amount that he can invest in the higher risk account?

◣ EXTENDED PRACTICE EXERCISES

31. CRITICAL THINKING Write a system of inequalities whose solution is the shaded region between $x = -2$ and $x = 5$.

Write a system of linear inequalities for each graph.

32.

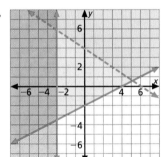

33.

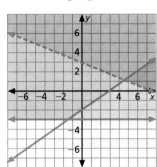

34.

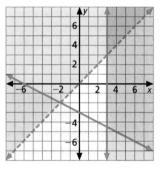

Graph the solution of each system of inequalities.

35. $-x - 2y > 4$
$3x + 4y \leq -12$
$y \leq -2$

36. $2x - 3y \geq -6$
$-x + 2y < -4$
$-x \geq -1$

◣ MIXED REVIEW EXERCISES

37. Shawna surveyed 700 adults about their favorite hobbies. Of the adults surveyed, 196 chose sports, 105 chose computers, 91 chose crafts, 78 chose camping, 63 chose dancing, 56 chose gardening, 48 chose traveling, 35 chose boating and 28 chose reading. Make a circle graph representing the favorite hobbies of the adults surveyed. (Lesson 5-9)

38. The ratio of girls to boys at Riverdale High School is 3:4. What is the probability that the first student to arrive at school one day is a boy? a girl? (Lesson 4-1)

39. DATA FILE Refer to the data on solid waste creation on page 567. If 56 T of solid waste is collected from a neighborhood, how much of it is metals? (Lesson 4-1)

Chapter 8 Review

VOCABULARY ◣

Choose the word from the list that best completes each statement.

1. If two nonvertical lines are __?__ they have the same slope.

2. A rectangular array of values is called a __?__.

3. The __?__ of equations is the set of ordered pairs that makes both equations true.

4. If two nonvertical lines are __?__ their slopes have a product of -1.

5. The method called __?__ uses determinants to solve a system of equations.

6. Two rational numbers that have a product of -1 are called __?__.

7. A __?__ is formed by two (or more) linear equations with the same two variables.

8. The first step in using __?__ to solve a system of equations is to solve one of the equations for a variable.

9. A __?__ has the same number of rows and columns.

10. A __?__ is a geometrical representation of a map.

a. Cramer's rule
b. determinant
c. directed graph
d. matrix
e. negative reciprocals
f. parallel
g. perpendicular
h. solution to a system
i. square matrix
j. substitution
k. system of equations
l. system of inequalities

LESSON 8-1 ◣ Parallel and Perpendicular Lines, p. 334

▶ If two nonvertical lines are parallel, they have the same slope. If two distinct lines have the same slope, they are parallel.

▶ If two nonvertical lines are perpendicular, the product of their slopes is -1. The slopes are **negative reciprocals** of each other.

11. Determine if the graphs $y = -\frac{1}{4}x + 2$ and $y - 4x = 1$ are parallel, perpendicular, or neither.

12. Identify the slope of a line parallel to $-3x + 4y = 24$. Then identify the slope of a line perpendicular to it.

13. Write an equation in slope-intercept form of the line passing through $(1, -3)$ and parallel to the line $y = 2x + 1$.

LESSON 8-2 ◣ Solve Systems of Equations Graphically, p. 338

▶ To solve a **system of equations** using graphs, graph both equations on the same coordinate plane.

▶ The coordinates of all points of intersection are **solutions of the system**.

Solve each system of equations graphically. Check the solution.

14. $x + y = 4$
    $y = x - 6$

15. $2x - y = 3$
    $y = -x + 6$

16. $x + 3y = 6$
    $x - y = 2$

LESSON 8-3 ◣ Solve Systems by Substitution, p. 344

▶ **Substitution** can be used to solve a system of equations algebraically. Solve one of the equations for one variable. Substitute that expression in the other equation and solve.

▶ When the solution of a system results in a true statement, the lines coincide and there are infinite solutions.

▶ When the solution of a system results in a false statement, the lines are parallel and there are no solutions.

Solve each system of equations. Check the solution.

17. $x + y = -4$
    $x - y = -2$

18. $3x - y = 1$
    $y = \frac{1}{2}x + 4$

19. $x - 2y = -10$
    $7x + 3y = 15$

20. The Chess Club wants to order T-shirts for its members. Shirt World will make the shirts for a $30 set-up fee and then $12 per shirt. T-Mania will make the shirts for $70 set-up fee and then $8 per shirt. For how many T-shirts will the cost be the same? What will be the cost?

LESSON 8-4 ◣ Solve Systems by Adding, Subtracting and Multiplying, p 348

▶ Solving a system of equations by an algebraic method can also be done by eliminating one of the variables using addition, subtraction, or multiplication and addition.

Solve each system of equations. Check the solution.

21. $3x - 2y = 4$
    $-3x + 4y = -2$

22. $3x - 5y = -16$
    $2x + 5y = 31$

23. $2m - 5n = -6$
    $2m - 7n = -14$

24. $4x - 2y = -10$
    $3x - 2y = -10$

25. $3x + 8y = -2$
    $5x + 3y = 7$

26. $4x - 7y = 10$
    $3x + 2y = -7$

LESSON 8-5 ◣ Matrices and Determinants, p. 354

▶ A **square matrix** has the same number of rows and columns.

▶ The value of a 2×2 **determinant** is the difference of the products of the diagonal entries in a matrix.

▶ To solve a system of equations using determinants, first write both equations in standard form, $Ax + By = C$.

Evaluate each determinant.

27. $\begin{vmatrix} 1 & 4 \\ 3 & 6 \end{vmatrix}$

28. $\begin{vmatrix} -1 & 0 \\ -2 & 4 \end{vmatrix}$

29. $\begin{vmatrix} 4 & -5 \\ -3 & 2 \end{vmatrix}$

30. Solve the systems of equations using the method of determinants.
    $3x - 5y = -1$ $8x + 8y = 3$
    $-2x + 4y = -2$ $32x - 12y = 1$

31. Play It Again sells used CDs and videos. In its first week, the store sold 40 used CDs and videos, at $4 per CD and $6 per video. The sales for both CDs and videos totalled $180. Use the method of determinants to find the number of CDs and videos the store sold in the first week.

LESSON 8-6 ◣ Problem Solving Skills: Directed Graphs, p. 358

▶ A **directed graph** is a geometrical representation of a map. It shows locations as points and roads as lines.

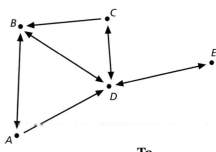

The directed graph represents the number of commuter train runs between cities A, B, C, and D.

32. Create a matrix for the number of runs from one city to another.

33. How many ways can you travel by commuter train directly to City C from another city?

34. The matrix shown represents the number of nonstop bus routes between streets A, B, C and D. Draw a directed graph that corresponds to the matrix.

$$\text{From} \quad \begin{array}{c} \\ A \\ B \\ C \\ D \end{array} \overset{\displaystyle \overset{\text{To}}{\begin{array}{cccc} A & B & C & D \end{array}}}{\begin{bmatrix} 0 & 1 & 1 & 1 \\ 0 & 0 & 0 & 1 \\ 1 & 1 & 0 & 1 \\ 1 & 0 & 0 & 0 \end{bmatrix}}$$

LESSON 8-7 ◣ Systems of Inequalities, p. 362

▶ The solution set of a system of linear inequalities is the intersection of the graphs of the inequalities.

▶ The slope-intercept form of an inequality indicates the boundary line, whether the boundary line should be solid or dashed, and which part of the plane to shade.

35. Write a system of linear inequalities for the graph shown.

Graph the solution set of each system of linear inequalities.

36. $-x + y < -2$

  $-2x - 3y \le -6$

37. $2y + x < 6$

  $3x - y > 4$

Kendra exercises every day by walking and jogging at least 3 mi. Kendra walks at a rate of 4 mi/h and Jogs at a rate of 8 mi/h. Suppose she can exercise no more than a half-hour today.

38. Write two inequalities to represent the situation.

39. Graph the inequalities.

40. Give the coordinates of a point that satisfies the inequalities. Explain the coordinates in terms of the problem.

CHAPTER INVESTIGATION

EXTENSION Obtain a football schedule for your school. The schedule can be for this year, a past year or the upcoming year. Be sure the schedule indicates whom all members of the league play each week. Work with a partner to create a simulation of the football season. Determine who wins each game by rolling a number cube. Whoever rolls the higher number chooses the winning team of each game. Determine which teams will be the top four to advance to the tournament. Use the point values you found in the Chapter Investigation of this chapter.

Chapter 8 Assessment

For each line identified by two points, state the slope of a line parallel and the slope of a line perpendicular to it.

1. $M(3, -4)$ and $N(8, -3)$ **2.** $A(2, -5)$ and $B(-2, 3)$ **3.** $S(3, 4)$ and $T(3, -5)$

Write an equation in slope-intercept form of a line passing through the given point and parallel to the given line.

4. $(3, 4); y = -4x + 5$ **5.** $(-1, 0); y = -\frac{3}{4}x + 2$ **6.** $(0, 6); x + y = 0$

7. Write an equation in slope-intercept form of a line passing through the given point and perpendicular to the given line in Exercises 4–6.

Use a graph to solve each system.

8. $3x - y = 2$
    $-x + 2y = -4$

9. $x + 2y = 6$
    $4x + 8y = -8$

10. $x + 3y \geq -3$
    $x - 2y < -6$

Solve each system of equations. Check the solution.

11. $x - y = 7$
    $2x - 3y = 11$

12. $3x - 4y = 5$
    $-x + 4y = -7$

13. $2x - 3y = 4$
    $-3x + 4y = -6$

Evaluate each determinate.

14. $\begin{vmatrix} 5 & 7 \\ -3 & 0 \end{vmatrix}$ **15.** $\begin{vmatrix} 3.1 & -1 \\ 0.5 & -1.2 \end{vmatrix}$ **16.** $\begin{vmatrix} 2 & -7 \\ 5 & -1 \end{vmatrix}$ **17.** $\begin{vmatrix} -1 & -2 \\ -1 & -1 \end{vmatrix}$

Solve each system of equations using the method of determinants.

18. $x - y = 5$
    $-x + 2y = -2$

19. $5x + 2y = 6$
    $x + 8y = -14$

20. $2x + 3y = -3$
    $6x - 7y = 11$

The directed map represents the connections between school buildings in one district.

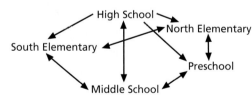

21. Create a matrix for the school district that shows the number of direct connections from one school building to another.

22. How many ways can you get directly to North Elementary?

23. Create a matrix for the school district that shows the number of ways a bus could travel between two schools with one stop.

24. Jack's and Matt's ages together total 19. Matt is five years older than Jack. How old is each boy?

25. Rashida has 16 coins in her pocket that total $2.95. They are dimes and quarters. How many of each coin does she have?

Standardized Test Practice

Part 1 Multiple Choice

Record your answers on the answer sheet provided by your teacher or on a sheet of paper.

1. Simplify $3x - 4 + (2x - 1) - (3x - 1)$. (Lesson 2-4)
   - (A) $2x - 4$
   - (B) $2x - 6$
   - (C) $8x - 6$
   - (D) $8x - 2$

2. What is the solution of $\frac{3x - 4}{2} = \frac{7x + 1}{4}$? (Lesson 3-5)
   - (A) -10
   - (B) -9
   - (C) 9
   - (D) 10

3. If the spinner is spun once, find P(gray or odd). (Lesson 4-4)
   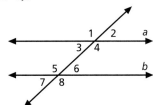
   - (A) $\frac{3}{16}$
   - (B) $\frac{3}{8}$
   - (C) $\frac{5}{8}$
   - (D) $\frac{7}{8}$

4. Luca remembered that the four digits of his locker combination were 4, 9, 15, and 22, but not in that order. What is the maximum number of different attempts Luca can make before his locker opens? (Lesson 4-6)
   - (A) 4
   - (B) 16
   - (C) 24
   - (D) 256

5. In the figure, $a \parallel b$. Find $m\angle 1$ if $m\angle 7$ is 42°. (Lesson 5-3).

   - (A) 42°
   - (B) 48°
   - (C) 58°
   - (D) 138°

6. What is the equation of the line? (Lesson 6-3)
   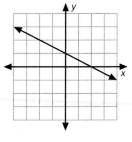
   - (A) $y = -\frac{1}{2}x + 2$
   - (B) $y = -\frac{1}{2}x + 1$
   - (C) $y = \frac{1}{2}x + 1$
   - (D) $y = 2x + 1$

7. Which of the following best describes the equation $y = 2x^3 + 1$? (Lesson 6-5)
   - (A) not a function
   - (B) a linear function
   - (C) a nonlinear function
   - (D) a quadratic function

8. The coordinates of the vertices of quadrilateral $ABCD$ are $A(-2, 4)$, $B(3, 7)$, $C(4, -2)$, and $D(-5, -3)$. If the quadrilateral is moved up 3 units and left 1 unit, which of these coordinates names a vertex of quadrilateral $A'B'C'D'$? (Lesson 7-1)
   - (A) $(1, 5)$
   - (B) $(-8, -2)$
   - (C) $(0, 6)$
   - (D) $(7, -3)$

9. Which equation represents a line parallel to the line given by $y - 3x = 6$? (Lesson 8-1)
   - (A) $y = -3x + 4$
   - (B) $y = 3x - 2$
   - (C) $y = \frac{1}{3}x + 6$
   - (D) $y = \frac{1}{3}x + 4$

10. The perimeter of a rectangular garden is 68 ft. The length of the garden is 4 more than twice the width. Which system of equations will determine the length ℓ and the width w of the garden? (Lesson 8-3)
    - (A) $2\ell + 2w = 68$
      $\ell = 2w + 4$
    - (B) $2\ell + 2w = 68$
      $w = 2\ell + 4$
    - (C) $2 + 2w = 68$
      $2\ell - w = 4$
    - (D) $2\ell + 2w = 68$
      $\ell = 4 - 2w$

Preparing for the Standardized Tests
For test-taking strategies and more
practice, see pages 627–644.

Part 2 Short Response/Grid In

Record your answers on the answer sheet provided by your teacher or on a sheet of paper.

11. On a standardized test, Jenny scored higher than 75 of the 95 people who took the test. In what percentile did she score? (Lesson 1-6)

12. Simplify $2(2x + 3) - 4(x + 2)$. (Lesson 2-6)

13. Solve $7 = \sqrt{5x + 2}$ for x. (Lesson 3-8)

14. A pair of adjacent, congruent angles are supplementary to each other. What is the measure of each angle? (Lesson 5-2)

15. Triangle ABC is inscribed in a circle. If $m\,\widehat{AB} = 150°$ and $m\,\widehat{AC} = 120°$, what is $m\angle BAC$? (Lesson 5-8)

16. Copy the figure. Then shade two squares so that the figure has rotational symmetry. (Lesson 7-4)

17. The sum of two numbers is 15. The difference between the numbers is 11. Find the value of the greater of the two numbers. (Lesson 8-4)

Test-Taking Tip

Questions 11, 14–16
Knowing mathematical terms is critical to your success on standardized tests. In preparation, make flash cards of the key terms using the glossary of your textbook.

Part 3 Extended Response

Record your answers on a sheet of paper. Show your work.

18. The capital letter A looks the same after a reflection over a vertical line. It does not look the same after a reflection over a horizontal line.

a. Name the other capital letters that look the same after a reflection over a vertical line. (Lesson 7-2)

b. Name the capital letters that look the same after a reflection over a horizontal line. (Lesson 7-2)

c. Which capital letters have line symmetry? (Lesson 7-4)

d. Which capital letters have rotational symmetry? (Lesson 7-4)

19. The manager of a movie theater found that Saturday's sales were \$3675. He knew that a total of 650 tickets were sold Saturday. Adult tickets cost \$7.50, and children's tickets cost \$4.50. (Lesson 8-3)

a. Write an equation to represent the number of tickets sold.

b. Write an equation to represent the amount of money collected.

c. How many of each kind of ticket were sold?

20. Consider the following system of equations. (Lesson 8-5)

$$2x - 3y = -4$$
$$-3x + 2y = 6$$

a. Write determinants A, A_x, and A_y for this system.

b. Find the value of each determinant and use these values to find the solution of the system.

Polynomials

THEME: Geography

Polynomials are often used in cartography (map making) and other areas of geography.

Whether you are walking to the store, riding your bike to a park, driving to visit relatives, or flying across the country, geography skills are used.

- **Truck drivers** (page 385) drive across the country, delivering goods to manufacturers and retail businesses for consumers to purchase. They must prepare for many unknown possibilities such as changes in weather, changes in road conditions, and the demands of delivery schedules.

- The geographic features of the land play a crucial role in decisions that **air traffic controllers** (page 403) must make on a routine basis. A flight path may vary, depending on the distance of the trip, noise restrictions, weather conditions, and the flight paths of other airplanes.

Math Online

mathmatters2.com/chapter_theme

Extreme Points of the United States (50 States)

Location	Latitude	Longitude	Distance[1]	
			miles	kilometers
Geographic center: Butte County, SD	44°58'N	103°46'W	0	0
Northernmost point: Point Barrow, AK	71°23'N	156°29'W	2507	4034
Easternmost point: West Quoddy Head, ME	44°49'N	66°57'W	1788	2997
Southernmost point: Ka Lae (South Cape), HI	18°55'N	155°41'W	3463	5573
Westernmost point: Cape Wrangell, AK (Attu Island)	52°55'N	172°27'E	3625	5833

[1] Distance is from geographic center of United States, Butte County, SD.

Data Activity: Extreme Points in the U.S.

Use the table for Questions 1–4.

1. Find the mean distance in miles from the geographic center of the 50 U.S. states to the northernmost point, easternmost point, southernmost point, and westernmost point.

2. What extreme location is farthest from Butte County, SD?

3. Why is the distance from the geographic center stated as 0?

4. Why do you think all of the longitudes are presented with a west location except Cape Wrangell, AK?

CHAPTER INVESTIGATION

Orienteering is an outdoor sport in which participants (orienteers) use an accurate, detailed map and a compass to find locations on a course, usually in the wilderness. The map's different colors show hills, valleys, streams, trails, fields and other landmarks.

A standard course consists of a start, a series of control sites and a finish. The control sites are marked with circles connected by dashed lines and numbered in the order they are to be visited. On the ground, a flag indicates a control site. To verify a visit to a control site, the orienteer marks the event's control card. To win an orienteering event, you must complete the course in the shortest amount of time.

Working Together

Use the orienteering map shown to find the approximate length of the course. Use the Chapter Investigation icons to guide your group.

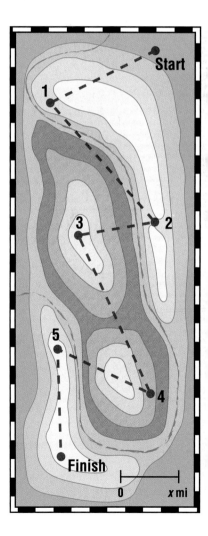

Are You Ready?

Refresh Your Math Skills for Chapter 9

The skills on these two pages are ones you have already learned. Use the examples to refresh your memory and complete the exercises. For additional practice on these and more prerequisite skills, see pages 576–584.

USING EXPONENTS

Remember the rules of exponents when multiplying and dividing terms with exponents.

Examples

$x^3 \cdot x^8 = x^{11}$ Add exponents when terms with like bases are multiplied.

$\dfrac{x^8}{x^3} = x^5$ Subtract exponents when terms with like bases are divided.

$(x^3)^8 = x^{24}$ Mulitply exponents when a power is raised to a power.

Simplify

1. $y^4 \cdot y^8$

2. $s^6 \cdot 2s^3$

3. $\dfrac{p^9}{p^3}$

4. $\dfrac{4k^4}{k^3}$

5. $(m^2)^3$

6. $(2g^2)^4$

7. $4x^3 \cdot 3x^5$

8. $\dfrac{16d^5}{6d^8}$

9. $\dfrac{(5w^5)^2}{10w^{10}}$

10. $\dfrac{(4x^3 + 2x^3)}{2x}$

11. $\dfrac{(9x \cdot 2x^3)}{3x^5}$

12. $\dfrac{4(y^4)^3}{2y^3}$

GREATEST COMMON FACTOR (GCF)

When you compare the factors of one or more numbers, the greatest common factor (GCF) is the greatest factor that is a factor of every number.

Examples

The factors of 36 are 1, 2, 3, 4, 6, 9, 12, 18 and 36.

The factors of 16 are 1, 2, 4, 8 and 16.

The GCF of 36 and 16 is 4.

When working with variables, look for the GCF of the constants and each variable.

The factors of $8xy^2$ are 1, 2, 4, 8, x, y and y^2.

The factors of $28y^3z$ are 1, 2, 4, 7, 14, 28, y, y^2, y^3 and z.

The GCF of $8xy^2$ and $28y^3z$ is $4y^2$.

Find the factors of each number or expression.

13. 35

14. 21

15. 81

16. 125

17. x^4

18. $3y^2$

19. $9z^3$

20. $14x^2y^3$

Find the GCF of each set of numbers.

21. 35 and 21

22. 21 and 81

23. 125 and 200

24. x^4 and $14x^2y^3$

25. $3y^2$ and $14x^2y^3$

26. $14x^2y^3$ and $12y^2z^3$

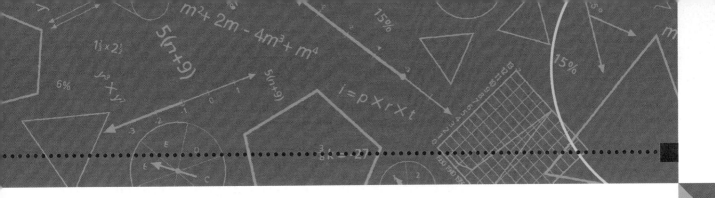

DISTRIBUTIVE PROPERTY

You can use the distributive property to simplify an expression. Multiply the expressions inside the parentheses by the factor outside the parentheses.

Example $4w^2(6w^2 - w + 6) = 4w^2 \cdot 6w^2 - 4w^2 \cdot w + 4w^2 \cdot 6$
$$= 24w^4 \qquad - 4w^3 \qquad + 24w^2$$

Simplify each expression using the distributive property.

27. $9t(3t^2 + 7)$

28. $-4(7y - 15)$

29. $-8x(x^5 + 3x^2)$

30. $4y^3(6y^2 + 7y - 12)$

31. $-3w(14w - 12w^2 - 5)$

32. $4y^2z(5y + 6z - 2yz)$

PERFECT SQUARES

It can be easier to recognize factoring patterns if you can recognize perfect squares. A perfect square is the product of a whole number multiplied by itself.

Examples

Base Number	Perfect Square	Base Number	Perfect Square
4	16	y	y^2
9	81	y^2	y^4
12	144	$3y^3$	$9y^6$

Determine if each quantity is a perfect square. If so, name the base number.

33. 25

34. 116

35. 256

36. 40

37. 169

38. 44

39. x^3

40. $2x^2$

41. x^2y^2

42. $25x^9$

43. $100y^{10}$

44. $36x^6y^4$

LIKE TERMS

When simplifying an expression, you combine like terms. Remember that the variable and exponent must be the same in order to be considered like terms.

Examples $4x^2y$, $16x^2y$, and $-13x^2y$ are like terms.

$12x^2z$, $12x^2$, and $32x^2z^2$ are not like terms.

Find each pair of like terms.

45. $3vw^3$, $-4vw$, $-9vw^3$

46. $-6x^3y$, $-4x^2y$, $19x^2y$

47. $21cs$, $7cs^2$, cs

48. $-\frac{1}{2}x^2y$, $\frac{1}{2}xy^2$, $\frac{3}{4}x^2y$

49. $16p^3q^4$, $17p^4q^3$, $18p^4q^3$

50. $0.8x^2yz^4$, $0.5x^2yz^4$, $1.5xy^2z^4$

9-1 Add and Subtract Polynomials

Goals
- Write polynomials in standard form.
- Add and subtract polynomials.

Applications Part-time Job, Travel, Geography, Modeling

Algeblocks can be used to model variable expressions.

1. What expression is represented on the Basic Mat shown?

2. Use Algeblocks and a Basic Mat to model the expression $3x^2 + x + 2$.

3. Add three more x-blocks to your model. What expression do the Algeblocks now represent?

4. Remove two x^2-blocks from your second model. What expression do the remaining Algeblocks represent?

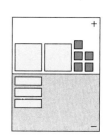

COncepts
in MOtion
Animation
mathmatters2.com

◤ BUILD UNDERSTANDING

A **monomial** is an expression that is a number, variable, or product of a number and one or more variables with whole number exponents. Each of these expressions is a monomial.

$$-11 \qquad k \qquad \frac{3}{4}mx^4 \qquad 7p^2f$$

The numerical factor of a monomial is the **coefficient**. For example, the coefficient of $-3x^2$ is -3. The coefficient of x is 1. A monomial with no variables, such as -11, is a **constant**.

A sum or difference of monomials is a **polynomial**. Each monomial is a **term** of the polynomial. A polynomial with two terms is a **binomial**. A polynomial with three terms is a **trinomial**. A monomial is a polynomial with one term.

binomial: $\underbrace{8n}_{\text{term}} + \underbrace{(-n^2)}_{\text{term}}$ trinomial: $\underbrace{a^2}_{\text{term}} + \underbrace{(-4ab)}_{\text{term}} + \underbrace{4b^2}_{\text{term}}$

A polynomial is written in **standard form** when its terms are ordered from the greatest to the least power of one of the variables. This order is referred to as *descending order*.

Example 1

Write each polynomial in standard form for the variable x.

a. $3x + 5 + 7x^3 - 2x^2$ **b.** $-8x^2y^3 + 4xy^4 + x^3$

Solution

Order the terms from greatest to least powers of x.

a. $3x + 5 + 7x^3 - 2x^2 = 7x^3 - 2x^2 + 3x + 5$ **b.** $-8x^2y^3 + 4xy^4 + x^3 = x^3 - 8x^2y^3 + 4xy^4$

Problem Solving Tip

Each term in a polynomial is separated by a plus or minus sign.

Terms such as $3xy^2$ and $-7xy^2$ that differ only in their coefficients are called **like terms**. To simplify a polynomial, combine like terms by adding their coefficients. A polynomial is simplified when only unlike terms remain.

Check Understanding

Sometimes polynomials are written with subtraction signs. How could you write the polynomials below to show that each is really a *sum* of monomials?

1. $6b - b^2$ **2.** $c^2 - 2cd + 3e^2$ **3.** $4p - 2q + 5$

Example 2

Simplify.

a. $4k^2 - 3k + 5k^2$

b. $2x^2y + 4x^2 + 3xy^2 - 7x^2y$

Solution

a. $4k^2 - 3k + 5k^2 = 4k^2 + 5k^2 - 3k$ Use the commutative property to rearrange like terms.

$\qquad\qquad\quad = (4 + 5)k^2 - 3k$ Use the distributive property to combine like terms.

$\qquad\qquad\quad = 9k^2 - 3k$

b. $2x^2y + 4x^2 + 3xy^2 - 7x^2y = 2x^2y - 7x^2y + 4x^2 + 3xy^2$

$\qquad\qquad\qquad\qquad\qquad = (2 - 7)x^2y + 4x^2 + 3xy^2$

$\qquad\qquad\qquad\qquad\qquad = -5x^2y + 4x^2 + 3xy^2$

To add polynomials, write the sum and simplify by combining like terms.

Example 3   Online **Personal Tutor at mathmatters2.com**

Simplify.

a. $7n + (5 - 3n)$

b. $(3x^2y - 7x + 2y) + (5x^2y + 2x - 3y)$

Solution

To simplify, use the associative and commutative properties as necessary to group like terms. Then use the distributive property to combine like terms.

a. $7n + (5 - 3n) = (7n - 3n) + 5$ Use the commutative and associative properties.

$\qquad\qquad\quad = (7 - 3)n + 5$ Use the distributive property to combine like terms.

$\qquad\qquad\quad = 4n + 5$

b. $(3x^2y - 7x + 2y) + (5x^2y + 2x - 3y)$

$\qquad\qquad = 3x^2y + (-7x) + 2y + 5x^2y + 2x + (-3y)$

$\qquad\qquad = (3x^2y + 5x^2y) + (-7x + 2x) + [2y + (-3y)]$

$\qquad\qquad = (3 + 5)x^2y + (-7 + 2)x + (2 - 3)y$

$\qquad\qquad = 8x^2y + (-5x) + (-y)$

$\qquad\qquad = 8x^2y - 5x - y$

To subtract one polynomial from another, add the opposite of the polynomial being subtracted.

Example 4

Simplify.

a. $8y - (5y + 3)$ **b.** $(2xy^2 - 5xy + 6) - (xy^2 + 3xy)$

Problem Solving Tip

You may find it easier to add or subtract polynomials by lining up like terms vertically.

$$3x^2y - 7x + 2y$$
$$+\ 5x^2y + 2x - 3y$$
$$\overline{8x^2y - 5x - y}$$

Solution

Change each term of the polynomial being subtracted to its opposite. Then follow the same procedure for adding polynomials.

a. $8y - (5y + 3) = 8y + [-5y + (-3)] = [8y + (-5y)] + (-3)$

$$= [8 + (-5)]y + (-3)$$

$$= 3y - 3$$

b. $(2xy^2 - 5xy + 6) - (xy^2 + 3xy) = (2xy^2 - 5xy + 6) + [-xy^2 + (-3xy)]$

$$= [2xy^2 + (-xy^2)] + [(-5xy) + (-3xy)] + 6$$

$$= (2 - 1)xy^2 + (-5 - 3)xy + 6$$

$$= 1xy^2 + (-8xy) + 6$$

$$= xy^2 - 8xy + 6$$

◼ TRY THESE EXERCISES

Write each polynomial in standard form for the variable x.

1. $9x^2 - 2x + 4 + 3x^3$ **2.** $-3 - 3x - 3x^2$ **3.** $x^2y^2 - y^3x$

Simplify.

4. $9p - 4p$ **5.** $k^2 + 2k - 3k - k^3$ **6.** $5x - 2xy + 3x + 4xy$

7. $(8h^2 - 2h) + (3h^2 + 5h)$ **8.** $(4a^2b + 8a) - (7a^2b - 3a)$ **9.** $(-2jk^2 + 6jk) - (5jk^2 + 3jk)$

10. WRITING MATH Will the sum and difference of two binomials always result in another binomial? Explain.

11. Write and simplify an expression for the perimeter of the figure shown.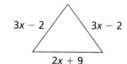

◼ PRACTICE EXERCISES • For Extra Practice, see page 612.

Write each polynomial in standard form for the variable x.

12. $2 + x^2 + 3x^3$ **13.** $3x^3 + 2x^4 + x + x^2$ **14.** $-x - x^5 + x^3 - 6$

15. $y^3 + 2xy^2 - 3x^2y - 5x^3$ **16.** $7x - 1 + 3x^2 - 4x^3$ **17.** $xy^4 + x^2y^2 + x^3y$

Simplify.

18. $3n - 5n$ **19.** $\frac{1}{4}x^2 + \frac{3}{7}x^2$ **20.** $7w + 5w - 4w - 8w$

21. $8y^3 - 4y^2 - 2y^3$ **22.** $-5c + 7d - 3d + 6$ **23.** $2x^3 + 4x + 7x$

24. $e^2 - 4e - 2e^2$ **25.** $-4n^2 + 2n + 6n + 8n^2$ **26.** $14a + 9b - 11a - 3b$

27. $(1.5x + 2.7) + (3.1x - 4.9)$ **28.** $(12p + 7) - (8p + 2)$ **29.** $-8hk + 10k + 3hk + 11hk$

Simplify.

30. $(5r + 3s + t) - (r + 9s + 7t)$

31. $(3x^2y + 2xy - 6) + (9x^2y + 8xy)$

32. $(6k^2 + 7k - 4) - (9k^2 - 11k + 7)$

33. $2.7m + 3.9n - 4.6m + 7.7n + 6.2 - 8.5$

DATA FILE For Exercises 34–36, refer to the data on the size of bird eggs on page 560. Let x represent the length of a grey heron egg. Express the length of the following bird eggs in terms of x.

34. Arctic tern   **35.** Partridge   **36.** Chicken (extra large)

37. **PART-TIME JOB** Yesterday Jason earned $(6r + 28t)$ dollars and today he earns $(9r + 18t)$ dollars. Write and simplify an expression for the total amount Jason earns in the two days.

38. **TRAVEL** The distance from Alpha City to Betaville on the highway is $(18x^2 + 15x + 13)$ mi. The distance by a shortcut on country roads is $(13x^2 - 11x - 1)$ mi. Write and simplify an expression to show the distance saved by the shortcut.

39. **GEOGRAPHY** The highest point on Earth above sea level is Mt. Everest, which is $(3x^2 - 9y - 72)$ ft above sea level. The lowest point above water is at the shores of the Dead Sea, which is $(13y - 1)$ ft below sea level. What is the difference in feet between these two points?

40. **MODELING** Show the sum of $(2x^2 + 4x) + (x^2 - 2x + 3)$ using Algeblocks.

41. Write and simplify an expression for the perimeter of the figure shown.

$5x - 7$

$3x + 2$ $3x + 2$

$5x - 7$

▰ EXTENDED PRACTICE EXERCISES

Simplify.

42. $\dfrac{3}{4}x^2 - \dfrac{1}{2}x + \dfrac{3}{8} - \dfrac{5}{8}x^2 + \dfrac{1}{4}x - \dfrac{1}{2}$

43. $7a^2b^2 + b^3 - b^2 - 4 - 3a^2b^2 + b^2$

44. $(10a^3 + 8a^2b - 4ab^2 - 7b^3) - (12a^3 - 3a^2b + 5ab^2 + 13b) + (-15a^3 - 9a^2b + ab^2 + 7b^3)$

45. Write two polynomials whose sum is $8p^2 - 6p + 4$.

46. Write two polynomials whose difference is $3x^2 - 5x + 7$.

47. **CRITICAL THINKING** The perimeter of the end of any box mailed by U.S. parcel post must not exceed 102 in. Find the maximum allowable value of x for the perimeter of the shaded face of this box.

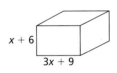

$x + 6$

$3x + 9$

▰ MIXED REVIEW EXERCISES

Write each equation in slope-intercept form. (Lesson 8-3)

48. $4x - 3y = -21$

49. $-6x + 2y = -5$

50. $3x = -5y + 12$

51. $9x + 3y = 27$

52. $-15 = 20x - 5y$

53. $\dfrac{1}{3}y - 2x = 7$

54. The formula for finding the area of a trapezoid is $\dfrac{1}{2}h(b_1 + b_2)$. Find the area of a trapezoid with height $h = 5$ in., base $b_1 = 7$ in. and base $b_2 = 3$ in. (Lesson 3-1)

Multiply Monomials

Goals ■ Use the rules for exponents to multiply monomials.

Applications Manufacturing, Sports, Photography, Modeling

Draw a rectangle. Label the length 5 cm and label the width 3 cm.

1. What is the area of the rectangle?

2. If you double both the length and width, what is the area?

3. If you triple both the length and width, what is the area?

4. Draw a new rectangle. Label the length *l* and label the width *w*.

5. What is the area of the new rectangle?

6. If you double both the length and the width, write an expression for the area of the new rectangle.

■ BUILD UNDERSTANDING

The commutative and associative properties of multiplication can be used to find a product of two monomials.

Example 1

Simplify.

a. $(5x)(7y)$ **b.** $(-2h)(4k)$ **c.** $\left(-\dfrac{1}{3}mn\right)(-12x)$

Solution

a. $(5x)(7y) = (5)(7)(x)(y)$
$$= 35xy$$

b. $(-2h)(4k) = (-2)(4)(h)(k)$
$$= -8hk$$

c. $\left(-\dfrac{1}{3}mn\right)(-12x) = \left(-\dfrac{1}{3}\right)(-12)(mn)(x)$ $-\dfrac{1}{3} \cdot -\dfrac{12}{1} = 4$
$$= 4mnx$$

Recall the *product rule for exponents*: $a^m \cdot a^n = a^{m+n}$

The product rule for exponents, together with the commutative and associative properties of multiplication, can be used to find a product of two monomials.

Example 2

MODELING Use Algeblocks and a Quadrant Mat to find the product $3x(-2x)$.

Solution

Step 1 Place three x-blocks in the positive part of the horizontal axis.

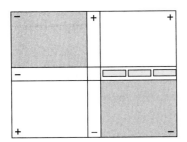

Step 2 Place two x-blocks in the negative part of the vertical axis.

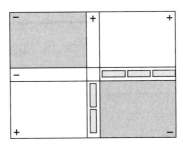

Step 3 Use x^2-blocks to form the area in the quadrant bounded by the x-blocks.

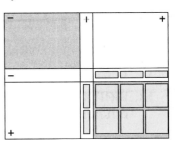

Read the answer from the mat: $-6x^2$.

Example 3

Simplify.

a. $(5e^2)(-6e^3)$

b. $(4m^2n^3)(-3mn^4p)$

Solution

a. $(5e^2)(-6e^3) = (5)(-6)(e^2 \cdot e^3)$

$\quad = -30e^{2+3}$

$\quad = -30e^5$

b. $(4m^2n^3)(-3mn^4p) = (4)(-3)(m^2 \cdot m)(n^3 \cdot n^4)p$

$\quad = -12(m^{2+1})(n^{3+4})p$

$\quad = -12m^3n^7p$

Recall the following rules for exponents.

$\quad$ *Power rule* : $(a^m)^n = a^{mn}$

$\quad$ *Power of a product rule:* $(ab)^m = a^m b^m$

Use these rules, with the commutative and associative properties of multiplication, to simplify monomials involving powers.

Check Understanding

In the solution of Example 3, part b, explain why you cannot add the exponents in the expression $-12m^3n^7$.

Example 4

Simplify.

a. $(5k^3)^2$

b. $(-3w^5y^6)^4$

Solution

a. $(5k^3)^2 = (5)^2(k^3)^2$

$\quad = 25k^{3 \cdot 2}$

$\quad = 25k^6$

b. $(-3w^5y^6)^4 = (-3)^4(w^5)^4(y^6)^4$

$\quad = (81w^{5 \cdot 4})(y^{6 \cdot 4})$

$\quad = 81w^{20}y^{24}$

Simplify.

1. $(9e)(3f)$
2. $(-5x)(8y)$
3. $\left(\frac{1}{2}a\right)(-16bc)$

4. $(12n^4)(4n^{12})$
5. $\left(\frac{2}{3}y^2\right)\left(-\frac{2}{5}y^5\right)$
6. $(-2c^5de^2)(-3c^2d^4)$

7. $(-2p^5)^4$
8. $(0.6x^3y)^2$
9. $(4a^3b^2c^5)^3$

10. Find the area of a rectangle that is $4ab^2c$ in. by $3a^3b$ in.

11. **MODELING** What product is modeled on the Algeblock Mat?

12. Use Algeblocks to find the product in Exercise 11.

13. **WRITING MATH** Write a short paragraph explaining the difference between $a^m \cdot a^n$ and $(a^m)^n$.

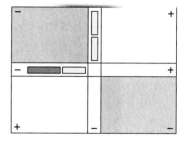

■ **PRACTICE EXERCISES** • For Extra Practice, see page 613.

Simplify.

14. $(3p)(7q)$
15. $(-2a)(4b)$
16. $(6m)(-2n)$

17. $(-xy)(-5hk)$
18. $9(-2a)$
19. $(-16w)\left(-\frac{3}{4}t\right)$

20. $(3.6k)(-2.5e)$
21. $(2a^5)(5b^6)$
22. $(-5m^2)(4m)$

23. $(8xy)(-3xy)$
24. $\left(\frac{2}{3}h^3\right)(-9h^2)$
25. $(4.3xy^2)(0.6y^4)$

26. $(a^5b^2)(a^3b^4)$
27. $(-20x^5y)\left(-\frac{3}{4}x^2y^2\right)$
28. $8e(-22e^3f)$

29. $(-abc)(abc)$
30. $(14hkn)(3.5hm)$
31. $\left(-\frac{2}{5}pq^2\right)\left(\frac{15}{16}p^2q\right)$

32. $(-3x)^2$
33. $(2h^3)^3$
34. $(2ab)^5$

35. $(-5x^2y)^2$
36. $(m^2n^5)^4$
37. $(-abc^2)^6$

38. $(m^3)(m^2)(m^6)$
39. $(-2a^5)(2a^4)(-2a^7)$
40. $(2y)(3y)^2$

41. $(-5x^2)(2x^3)^3$
42. $(4hk)^3(-h^3k^2)$
43. $(-a^2b)^5(a^2b)^4$

44. $(-x^3y^2)(3x^4yz)(-5xy^4z^3)$
45. $(4h^2)^2(2h^3)^3$
46. $(-4e^3f)(-e^2f^2)^2\left(\frac{1}{2}ef^4\right)^2$

Write and simplify an expression for the area of each figure.

47.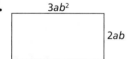
48.
49.

50. **ERROR ALERT** Tia says that $10^3 \cdot 10^6 = 100^9$. Explain Tia's mistake.

51. **GEOGRAPHY** On the Earth's surface, there is 3 times as much water as land. Write and simplify an expression to represent the surface area of Earth.

52. **MODELING** Use Algeblocks to sketch or show the product $-3x\,(y)$.

53. **SPORTS** The length of each side of a baseball diamond is $10x^2$ ft. Find the area of a baseball diamond in terms of x.

MANUFACTURING The diagram shows the net of a box that is being manufactured for a new product.

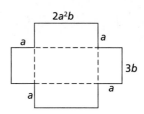

54. Find the area of the base of the box.

55. Find the area of each side of the box.

56. Find the total surface area of the box.

57. PHOTOGRAPHY Kyree frames a picture from his vacation. Write and simplify an expression to find the area of the frame.

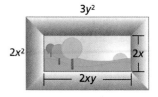

Write and simplify an expression for the volume of each prism.

58.

59.

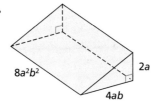

EXTENDED PRACTICE EXERCISES

Find the value of n in each equation.

60. $p^8 \cdot p^n = p^{11}$

61. $(k^2)^n = k^6$

62. $x^{12} = x^3 x^n$

63. A rectangle has an area that can be expressed as $64a^6b^9$. When its width is expressed as $4a^2b$, what is its length?

64. CRITICAL THINKING For the product $12x^2$, find nine pairs of monomial factors that have positive integer coefficients. (One possible pair is $3x \cdot 4x$.)

65. CHAPTER INVESTIGATION Write an expression in terms of x for the approximate distance between the following points.

    a. start and 1   **b.** 1 and 2   **c.** 2 and 3

    d. 3 and 4   **e.** 4 and 5   **f.** 5 and finish

MIXED REVIEW EXERCISES

Use data sets A and B to answer the following questions. Round to the nearest tenth when necessary.

    A: 6, 12, 7, 9, 6, 5, 4, 10, 12, 11, 12

    B: 35, 36, 40, 41, 46, 35, 32, 60, 54, 34, 48

66. Find the mean, median and mode for data set A. (Lesson 1-2)

67. Find the mean, median and mode for data set B. (Lesson 1-2)

68. Describe the outliers, clusters and gaps for data set B. (Lesson 1-3)

69. Create a box-and-whisker plot for data set A. (Lesson 1-6)

70. What is the greatest number in the first quartile for data set B? (Lesson 1-4)

Review and Practice Your Skills

Write each polynomial in standard form for the variable x.

1. $4 + 2x^2 + 5x^3$

2. $5x^2 + 2x^3 + x^4 + x$

3. $y^3 - xy^2 + 7x^2y - 3x^3$

4. $-2x - 4x^5 + 8 - x^3$

5. $9x - 10 - 4x^2 + 6x^3$

6. $x^2y + x^3y^3 + 3xy^3$

Simplify.

7. $8c - 3c$

8. $\frac{1}{3}x^2 + \frac{2}{5}x^2$

9. $9m - 3m - 7m + 5m$

10. $5y^2 - 3y^3 - 8y^3$

11. $3b - 6d + 2b + 10d$

12. $7x^3 - 3x - x$

13. $2k + k^2 + 2k - 2k^2$

14. $7x + 8y - 3x - 2y$

15. $(a^2 + ab) + (a^2 - ab)$

16. $(15p + 11) - (p - 1)$

17. $(18.5x - 3.9) - (13.2x + 5.1)$

18. $(7fg^2 + 3fg) - (-5fg^2 + 6fg)$

19. $5x - (3x + 7)$

20. $9m - 4n + 4m - 8n$

21. $(3ab^2 - 4ab + 6) - (ab^2 + 3ab)$

22. $14x^2y^3 - 6x^3y^2 - 11x^3y^2 + 3x^2y^3$

Simplify.

23. $(4a)(9b)$

24. $(5z)(-6x)$

25. $(-15c)\left(\frac{3}{5}b\right)$

26. $(xy)(-3xy)$

27. $(2c^6)(4b^4c^2)$

28. $5e(-11e^2g)$

29. $(-stu)(2stu)$

30. $7(-3x)$

31. $\left(\frac{2}{5}g^2\right)(25g^5)$

32. $(-4x)^3$

33. $(5g^2)^3$

34. $(2x^2y)^5$

35. $(x^2)(x^3)(x^5)$

36. $(q^2p^3)^3$

37. $(-ef^2g)^4$

38. $(-a^2b^3)(4a^4bc^2)(-3ab^5c)$

39. $(5h^4)^2(3h^2)^3$

40. $(-3xy)^2(-x^2y^2)(7xy^3)$

41. $(-3p^3)^4$

42. $(-4x^2)(9x^3)^2$

43. $(-xy^2z)^4$

44. $(8h^3)(-2h^2)^4$

45. $(-a^2b)(9a^3b^2)(2ab^3)$

46. $\left(-\frac{1}{3}x^3y\right)(9xy)^2$

47. Find the area of a rectangle that is $12n^2$ ft by $6n^3$ ft.

48. Find the area of a square that has a side measuring $2x^3$ cm.

49. Find the area of a triangle that has a base of $9x$ m and a height of $3x^2$ m.

50. Find the area of a parallelogram that has a base of $2.7xy^2$ in. and a height of $1.3x^2y$ in.

Simplify. (Lessons 9-1 and 9-2)

51. $(-2t - 7) + (6t - 3)$

52. $(-7x^4)^2$

53. $(5a - 8b) - (4a + h)$

54. $(3xy^3)^2$

55. $(-3xy^3)^3$

56. $(a^3b)^3(-ab^2)^2$

57. $(3p^5)(5p^2) + (4p^3)(4p^4)$

58. $(xy^2)^2 + (xy^3)$

59. $(-6a^3)\left(\frac{1}{6}a^3\right)$

Write and simplify an expression for the area of each figure. (Lesson 9-2)

60.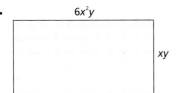

61.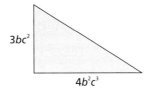

62.

MathWorks Career – Truck Driver
Workplace Knowhow

When truck drivers drive through the Mojave Desert in California, they drive through a very hot, arid, deserted region. Travel from Flagstaff, AZ, to Bakersfield, CA is 300 mi. Because there are few places along this route to refuel, many truck drivers attempt to drive this distance without stopping. They must constantly be aware of their speed, as this is directly proportional to the amount of fuel their vehicles consume. The equation gives the relationship between the fuel consumed in gallons per mile (f), average uphill speed in miles per hour (v_u) and average downhill speed in miles per hour (v_d).

$$f = 0.00001v_u{}^2 + 0.0002v_d{}^2$$

1. If the driver averages 40 mi/h uphill and 60 mi/h downhill, how much fuel will be required to travel the 300 mi between Flagstaff and Bakersfield?

2. Assuming that the driver has filled his 100-gal tank upon leaving Flagstaff, will the truck need to refuel at some point before arriving in Bakersfield? If so, how many gallons short of completing the trip on one tank of fuel is it? If the trip can be made without stopping, how many gallons of fuel are left in the tank?

3. A truck averages 30 mi/h uphill and has a 100-gal tank. What average speed must the truck maintain downhill in order to complete the trip without stopping?

4. What other factors besides average speed might affect fuel mileage?

Divide by a Monomial

Goals ■ Divide monomials and polynomials by monomials.

Applications Landscaping, Retail, Interior design, Geography, Modeling

---

For Questions 1–5, use the formula for the area of a rectangle.

1. The area of a rectangle is 45 ft^2 and the length is 9 ft. What is the width?

2. What operation did you use in Question 1?

3. The area of a rectangle is $12a^3$ ft^2 and the width is $3a$ ft. Explain how to find the length.

4. What do you think the length is in Question 3?

5. The area of a rectangle is $6x^2y + 12xy^2$ ft^2 and the length is $3xy$ ft. Explain how to find the width.

◤ BUILD UNDERSTANDING

Recall the quotient rule for exponents.

$$\text{Quotient rule: } \frac{a^m}{a^n} = a^{m-n}$$

You can use the quotient rule to divide one monomial by another. Since division by zero is undefined, throughout this lesson it is assumed that no denominator has a value of zero.

COncepts in MOtion
BrainPOP®
mathmatters2.com

Example 1

Simplify.

a. $\dfrac{12mn}{9m}$

b. $\dfrac{-14h^7k^5}{18hk^3}$

Solution

a. $\dfrac{12mn}{9m} = \dfrac{2^2 3mn}{3^2 m}$ *Write the prime factorization of each coefficient.*

$= \dfrac{4n}{3}$ *To simplify, divide the numerator and denominator by any common factors.*

b. $\dfrac{-14h^7k^5}{18hk^3} = \dfrac{(-1)(2)(7)h^7k^5}{(2)(3^2)hk^3}$ h^{7-1}, k^{5-3}

$= \dfrac{(-1)(7)h^6k^2}{3^2}$ *Simplify.*

$= \dfrac{-7h^6k^2}{9}$

> **Check Understanding**
>
> In Example 1, part a, what is the GCF of $12mn$ and $9m$? How is dividing a monomial by a monomial like simplifying a fraction?

The distributive property is used to multiply each term of a polynomial by a monomial.

$$3n(4n + 7) = 3n(4n) + 3n(7) = 12n^2 + 21n$$

Since division is the inverse of multiplication, you can reverse this process to divide a polynomial by a monomial.

$$3n(4n + 7) = 12n^2 + 21n, \text{ so } \frac{(12n^2 + 21n)}{3n} = 4n + 7.$$

To divide a polynomial by a monomial, divide each term of the polynomial by the monomial. You can use the greatest common factor (GCF) and the quotient rule for exponents to simplify the quotient $(12n^2 + 21n) \div 3n$.

$$\frac{12n^2 + 21n}{3n} = \frac{12n^2}{3n} + \frac{21n}{3n}$$

$$= \frac{12}{3} \cdot \frac{n^2}{n} + \frac{21}{3} \cdot \frac{n}{n}$$

Divide each pair of coefficients by their GCF, 3.
Divide each pair of variable parts by their GCF, n.

$$= 4 \cdot n^{2-1} + 7 \cdot 1$$

$$= 4n + 7$$

Example 2

Simplify.

a. $\dfrac{8n + 12}{4}$ b. $\dfrac{9x^4 - 12x^3 + 15x^2}{3x^2}$ c. $\dfrac{3a^2b + 9ab^2}{3ab}$

Solution

a. $\dfrac{8n + 12}{4} = \dfrac{8n}{4} + \dfrac{12}{4} = 2n + 3$

b. $\dfrac{9x^4 - 12x^3 + 15x^2}{3x^2} = \dfrac{9x^4}{3x^2} - \dfrac{12x^3}{3x^2} + \dfrac{15x^2}{3x^2} = 3x^2 - 4x + 5$

c. $\dfrac{3a^2b + 9ab^2}{3ab} = \dfrac{3a^2b}{3ab} + \dfrac{9ab^2}{3ab} = a + 3b$

> **Think Back**
>
> For any real numbers, a, b and c, where c is not equal to zero,
> $\dfrac{a + b}{c} = \dfrac{a}{c} + \dfrac{b}{c}$.

Example 3

The area of a rectangle is $(6x^3 + 2x^2 + x)$ cm². The width of the rectangle is $2x^2$ cm. Find the length of the rectangle.

Solution

Use the formula for the area of a rectangle, $A = l \cdot w$. So, $l = \dfrac{A}{w}$.

$$l = \frac{A}{w} = \frac{6x^3 + 2x^2 + x}{2x^2}$$

$$= \frac{6x^3}{2x^2} + \frac{2x^2}{2x^2} + \frac{x}{2x^2}$$

$$= 3x + 1 + \frac{1}{2x}$$

The area of the rectangle is $\left(3x + 1 + \dfrac{1}{2x}\right)$ cm.

Simplify.

1. $\dfrac{-36ab}{12b}$

2. $\dfrac{24m^5n^6}{30m^2n^6}$

3. $\dfrac{-15x^3y^2z}{20x^3y}$

4. $\dfrac{64x^2yz^3}{8x^2z^2}$

5. $\dfrac{6p-12r}{6}$

6. $\dfrac{16a^2+4ab}{4a}$

7. $\dfrac{14k^3m^2 \ \ 7km}{7km}$

8. $\dfrac{6h^2-2h^4+10h^3}{2h^3}$

9. $\dfrac{15c^3d^2-5cd+10c}{-5c}$

10. The area of a rectangle is $72mn$ yd^2, and the length is $9n$ yd. Find the width.

MODELING Use the Quadrant Mat shown for Exercises 11 and 12.

11. What division problem is represented on the Quadrant Mat?

12. What expression should be placed on the vertical axis to complete the Mat?

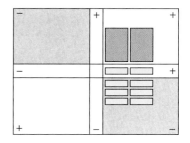

Simplify.

13. $\dfrac{8x^2}{2x}$

14. $\dfrac{12a^6}{12a^4}$

15. $\dfrac{-20y^3}{14y^3}$

16. $\dfrac{66uv^2}{-11v}$

17. $\dfrac{6ab}{12ac}$

18. $\dfrac{-16p^2q}{4p}$

19. $\dfrac{-150x^3y^2}{-10x^2y}$

20. $\dfrac{60x^4yz^2}{-12z}$

21. $\dfrac{110a^2bc^5}{10ac^2}$

22. $\dfrac{-30m^3n}{-6mn}$

23. $\dfrac{42a^2bc^3}{14a^2bc^3}$

24. $\dfrac{-24m^7n^9}{16m^3n^6}$

25. $\dfrac{-42a^4b^5c}{7a^4bc}$

26. $\dfrac{-81m^3n^3p^3}{-9mn^3p^3}$

27. $\dfrac{4a+20}{4}$

28. $\dfrac{9h-21}{3}$

29. $\dfrac{15a^2-10a}{5a}$

30. $\dfrac{21n^3+35n^2}{-7n}$

31. $\dfrac{14x^2-21x^3}{7x^2}$

32. $\dfrac{4xyz^2-4yz^2}{yz^2}$

33. $\dfrac{a-3a^3+5a^5}{a}$

34. $\dfrac{36c^6+30c^5-54c^4}{6c^3}$

35. $\dfrac{25w^3-15w^2-30w}{5w}$

36. $\dfrac{-24x^5+8x^4-64x^3}{-8x^3}$

37. $\dfrac{-45a^2b^3+18a^3b^4-9ab^2}{9ab}$

38. $\dfrac{12m^3n^4-20m^4n^3+32mn^2}{4mn^2}$

39. $\dfrac{3x^2yz-4xy^2z+2xyz^2-xyz}{xyz}$

40. $\dfrac{12ab-21a^2b+27ab^2-6a^2b^2}{-3ab}$

41. **LANDSCAPING** A rectangular garden has an area of $48pq$ square units. The width is $8q$ units. Write an expression for the length.

42. **RETAIL** How many CD's can you buy for $\$48xy^2$ if each CD sells for $\$3y$?

43. **MODELING** Sketch or show $\dfrac{-6x^2+9x}{-3x}$ using Algeblocks.

Write an expression for the unknown dimension of each rectangle.

44. area = $36ab$

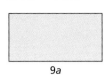

9a

45. area = $64mnp$

16mn

46. INTERIOR DESIGN Carmen is putting a wooden border 3 ft from the floor around the walls of her dining room. The perimeter of the room is $36a^2$ ft. The window and two doorways have a total width of $21a$ ft. To find the total number of yards needed, she can divide the total number of feet needed by 3 ft/yd. Write and simplify an expression for the total number of yards needed for the border.

47. GEOGRAPHY The Great Salt Lake Desert in Utah is rectangular in shape. The area of the desert is approximately $(8x^4 + 4x^3)$ mi^2. The width is about $2x^2$ mi. Find the approximate length of the Great Salt Lake Desert.

48. The volume of a rectangular prism is $(135mxy + 45my^2)$ cubic units. The length is $5y$ units and the height is $9m$ units. Write an expression for the width.

Great Salt Lake Desert

◤ EXTENDED PRACTICE EXERCISES

49. The area of a rectangle is $8x^2y^2$. The width of the rectangle is xy. Write an expression for the perimeter of the rectangle.

50. CRITICAL THINKING A square has a perimeter of $28x + 40y$. Express the area using the formula for the area of a square.

To divide a polynomial by a polynomial, find two factors of the numerator. Then divide by common factors. Divide the following.

51. $\dfrac{4n + 12}{n + 3}$

52. $\dfrac{2x - 12}{x - 6}$

53. $\dfrac{5a + xa}{5 + x}$

54. $\dfrac{3k^2 - 12k}{k - 4}$

◤ MIXED REVIEW EXERCISES

Use the figure for Exercises 55–60. (Lesson 5-1)

55. How many lines go through point M?

56. Name three points that lie on line CR.

57. Name all segments shown for which point I is an endpoint.

58. Name all the points that are collinear with points B and Q.

59. Identify the intersection of lines DF and CR.

60. Line FS bisects line QU. Name two congruent line segments made from this bisection.

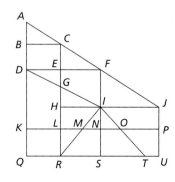

9-4 Multiply a Polynomial by a Monomial

Goals ■ Multiply polynomials by monomials.

Applications Travel, Part-time job, Sports, Finance, Geography, Modeling

Use Algeblocks for Questions 1–4.

1. What polynomial expressions are being multiplied on the Quadrant Mat shown?

2. What is the product represented by the Algeblocks?

3. Sketch or show $3x(x + 2)$ using Algeblocks.

4. Use Algeblocks to find the product $3x(x + 2)$.

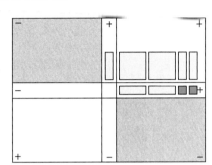

◣ BUILD UNDERSTANDING

The distributive property gives a method to multiply a polynomial by a monomial.

distributive property: $a(b + c) = ab + ac$

monomial polynomial

$$a(b - c) = ab - ac$$

> **Reading Math**
>
> The distributive property states that when a sum or difference of terms is multiplied by a factor, the factor must be *distributed*, through multiplication, over each term.
>
> $a(b + c) = ab + ac$
>
> Factor *a* is *distributed* over both terms, *b* and *c*.

Example 1

Simplify.

a. $5n(2n - 3)$

b. $-6k(-k^2 + 2k + 5)$

Solution

a. $5n(2n - 3) = 5n(2n) - 5n(3)$ Use the distributive property.

$= 10n^2 - 15n$ Multiply each pair of monomials.

b. $-6k(-k^2 + 2k + 5) = -6k(-k^2) + (-6k)(2k) + (-6k)(5)$

$= 6k^3 - 12k^2 - 30k$

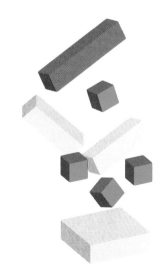

Example 2

Show the product $2x(x-3)$ using Algeblocks.

Solution

Step 1 Use a Quadrant Mat. Place two x-blocks on the positive horizontal axis.

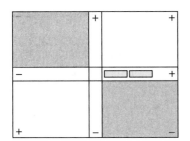

Step 2 Place one x-block on the positive vertical axis, and three unit blocks on the negative vertical axis.

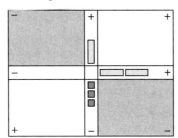

Step 3 Use x^2-blocks and x-blocks to form rectangular areas in all quadrants bounded by the pieces.

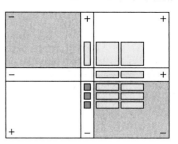

Example 3

Simplify.

a. $4(x^2 + xy) - 3(x^2 - xy)$

b. $7m(2m - 4) + 2m(2m - 4)$

Solution

a. $4(x^2 + xy) - 3(x^2 - xy)$

$= (4x^2 + 4xy) - (3x^2 - 3xy)$

$= 4x^2 + 4xy - 3x^2 + 3xy$

$= (4x^2 - 3x^2) + (4xy + 3xy)$

$= x^2 + 7xy$

b. $7m(2m - 4) + 2m(2m - 4)$

$= [7m(2m) - 7m(4)] + [2m(2m) - 2m(4)]$

$= (14m^2 - 28m) + (4m^2 - 8m)$

$= 14m^2 - 28m + 4m^2 - 8m$

$= (14m^2 + 4m^2) + (-28m - 8m)$

$= 18m^2 - 36m$

Example 4 **Personal Tutor at mathmatters2.com**

TRAVEL Renting a four-wheel-drive vehicle at Vacation Rentals costs \$43/day plus \$0.08/mi. The cost C is expressed by the formula $C = 43 + 0.08m$, where m is the number of miles the vehicle is driven.

Wilderness Camp rented a fleet of v vehicles for one day. Write a formula for the total cost of renting all the vehicles. Assume all vehicles travel the same number of miles.

Solution

total cost = number of vehicles · cost per vehicle

$TC = v(43 + 0.08m)$

$TC = 43v + 0.08mv$

> ### Problem Solving Tip
>
> In Example 3, part b, the term $(2m - 4)$ appears twice. Think of the binomial $(2m - 4)$ as a "chunk," and use the distributive property to simplify.
>
> $7m(2m - 4) + 2m(2m - 4)$
>
> $= (7m + 2m)(2m - 4)$
>
> $= 9m(2m - 4)$
>
> $= 18m^2 - 36m$

Simplify.

1. $6x(x + 4)$

2. $2n(8 - n^2)$

3. $12k^3(k^2 + k^4 + 3)$

4. $-5c^5(7c^2 - 2c + 11)$

5. $-4p(p^2 - 2p + 5)$

6. $4xy(x^2 - 3y)$

7. $-a^2b(3b + 4a^4)$

8. $5(2x^2 + xy) + 4(x^2 - 3xy)$

9. $14a(7a - 6) - 11a(7a - 6)$

10. **WRITING MATH** Explain how you could check your answer for Exercise 6.

11. **TRAVEL** Marcus drives 5 h through the Rocky Mountains at an average rate of 56 mi/h. Write a variable expression in simplest form that represents how far he would drive in the same amount of time if he drives m miles per hour slower.

12. **MODELING** What multiplication problem and product are represented on the Quadrant Mat shown?

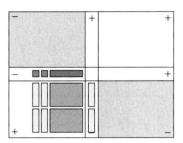

PRACTICE EXERCISES • For Extra Practice, see page 614.

Simplify.

13. $3m(2m + n)$

14. $11x(2x - y)$

15. $2a(3a + 4)$

16. $4k(5k - 7)$

17. $-8p(-3p^2 + 11p)$

18. $-\frac{2}{3}h^2\left(\frac{7}{8}h + \frac{2}{5}\right)$

19. $5x(3x + 2x^2)$

20. $-d^3(4d - 6d^2)$

21. $11n^6(-3n^4 - 12n^2)$

22. $2x(10x^2 - 7x + 13)$

23. $9c(3c + c^2 + 6c^3)$

24. $-4e(3e^2 + 5e - 3)$

25. $7x(1 + 2x + x^2)$

26. $-3y(y^3 - y^2 - y)$

27. $4r(-7r^3 + 2r - 10)$

28. $2a^2b(3a + b)$

29. $-12xy^3(2x^2 - y)$

30. $\frac{3}{7}d^3\left(\frac{1}{2}d^2 - \frac{2}{3}d + \frac{6}{7}\right)$

31. $-k^4(-12k^3 + 3k^2 - k)$

32. $10z^5(-6z^3 + 9z^2 - 5)$

33. $1.3c^4(2.7c^3 + 0.9c^2 - 3.4c)$

34. $-6s(4s^5 - 12s^3 + 9s)$

35. $-8t^3(12 + 5t - t^6)$

36. $4x^2y(3xy + 7xy^2 + 4x^2y^2)$

37. $4m(6m^2 + 3mn + n^2)$

38. $11y(-y^3 + 2y^2 - 8y + 7)$

39. $2(h^2 + 5) + 3(h^2 - 2)$

40. $-3(r^3 - 2) + 9(r^3 + 7)$

41. $8(2x - 3) - 3(2x - 3)$

42. $5x(4x^2 - 12) + 2x(7x^2 - 11)$

43. $14(w^2 - 4) - 9(w^2 - 4)$

44. $6(a - 2b) - 3(a - 2b)$

45. $7xy(2x - 3y) + 3xy(4x + 9y)$

46. **GRAPHING** Use a graphing calculator to graph $y = x(3x - 5)$ and $y = 6x^2 - 10x$ on the same screen. What do you notice about the two graphs?

47. **PART-TIME JOB** Rose works 16 h/wk and earns \$8.50/h. Write a variable expression in simplest form to represent her weekly earnings if her hourly wage is increased by d dollars per hour.

48. **TRAVEL** The fare for a cab is \$2 for the first mile and \$0.75 for each additional mile. The fare is expressed by the formula $F = 2 + 0.75m$, where m is the number of miles driven after the first mile. The Smith for President Committee engaged c cabs for a motorcade through the city. Write and simplify a formula for the total fare.

49. **SPORTS** The length of a soccer field is 40 m less than twice the width of the field. Write and simplify an expression for the area of the field.

FINANCE On Monday a share of XYZ stock sold for x dollars. On Tuesday the price rose \$3/share. On Wednesday the price fell \$5/share.

50. Akule buys 20 shares of XYZ on Monday, 30 shares on Tuesday, and 40 shares on Wednesday. Write a variable expression in simplest form that represents the total cost of Akule's purchases.

51. It cost Maria \$225 to buy 9 shares of XYZ on Tuesday and 6 shares on Wednesday. Find the cost of a share of XYZ on Monday.

Write and simplify an expression for the area of each figure.

52.
4m

6m + 3

53.
4x

12x² − 5x

54.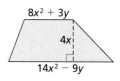
8x² + 3y

4x

14x² − 9y

MODELING Use Algeblocks to find each product.

55. $2x(2x + 2)$

56. $-2y(-y - 2)$

57. $3x(y + 1)$

EXTENDED PRACTICE EXERCISES

Simplify.

58. $\frac{1}{2}e\left(\frac{2}{3}e^4 - \frac{3}{8}e^3 + \frac{1}{4}e\right)$

59. $(x^3 - 9.2x^2 - 4.6x)2.5x^2$

60. $2y(y^3 + 3y^2 - 5y) - 5y(-3y^3 - 2y^2 + 3y)$

61. $5x^3y^2(-x^3y^4 - 2x^6y^2 + y^4 - 3xy)$

62. Write an expression in simplest form that represents the surface area of the prism.

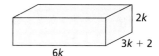
2k

3k + 2

6k

63. A picture frame measures 16 in. by 12 in. Write and simplify an expression for the area of the frame if the longer side is increased by n inches.

64. **GEOGRAPHY** Lake Ontario is located in both the U.S. and Canada. It has a width of about $0.5x$ mi, a length of about $(2x - 7)$ mi and a depth of about $8x$ mi. Write and simplify an expression for the volume of Lake Ontario.

65. **CHAPTER INVESTIGATION** Write an expression in terms of x for the length of the entire course, start to finish.

MIXED REVIEW EXERCISES

Solve each equation. Check the solution. (Lesson 3-2)

66. $x + 18 = 23$

67. $d - 64 = 128$

68. $s + (-17) = -56$

69. $15x = 5$

70. $\frac{g}{12} = -4$

71. $-\frac{2}{3}w = -6$

72. $y - 0.5 = 3.7$

73. $\frac{3}{4}x = 27$

74. $3.2 = 0.4n$

75. Kazuo is thinking of a whole number. If you multiply his number by 5 and add 15, the sum is greater than 100. What is the least number Kazuo could be thinking of? (Lesson 3-7)

Review and Practice Your Skills

Simplify.

1. $\dfrac{12h^2}{3b}$

2. $\dfrac{5f^7}{5f^2}$

3. $\dfrac{21x^4}{12x^4}$

4. $\dfrac{25a^4b}{-15ab}$

5. $\dfrac{6x - 9x^2 + 12x^3}{x}$

6. $\dfrac{15mn - 25m^2 - 5m^3n^2}{5m}$

7. $\dfrac{144x^7y^3z^2}{12x^2y^2}$

8. $\dfrac{18y^2 - 3y^4 + 15y^3}{3y^3}$

9. $\dfrac{-49a + 28}{7}$

10. $\dfrac{8abc^2 - 8bc^2}{bc^2}$

11. $\dfrac{9x^2y^2}{xy}$

12. $\dfrac{16x^2 - 24x^5}{4x^2}$

13. $\dfrac{8e^2 - 2e^4 + 12e^3}{2e^3}$

14. $\dfrac{30cd - 50c^2d^3 + 10c}{-10c}$

15. $\dfrac{(w^4)^3}{(w^5)^2}$

16. The area of a rectangle is $16x^2y$ ft^2, and the length is $4x$ ft. Find the width.

17. The area of a rectangle is $36pq^2$, and the width is $9pq$. Find the length.

18. How many CD's can you buy for $\$36ab$ if each CD sells for $\$3a$?

Simplify.

19. $6a(3a + b)$

20. $5m(2m - n)$

21. $-7b(4b - 5d^2)$

22. $-\dfrac{3}{4}k^2\left(\dfrac{5}{6}k + \dfrac{1}{2}\right)$

23. $-v^3(-10v^2 + 3v^2 - 4v)$

24. $-2x(x^3 - 3x - 6x^2)$

25. $3xy(y^2 - 3x + 1)$

26. $-4xy(x^3 - y^2 - xy)$

27. $2m(5m^2 + mn - 4n^2)$

28. $3a(5a - 4) + 6a(5a - 4)$

29. $\dfrac{1}{2}d^2\left(\dfrac{4}{3}d^2 + \dfrac{1}{5}d + 3\right)$

30. $8a(-a^3 + 2a^2 - 11a + 2)$

31. $-4(2b + 7) + 9(2b + 7)$

32. $7x(4x^3 - 10) + 8x(5x^3 + 5)$

33. $5rs(3r - 2s) - 2rs(r - 3s)$

34. $5x(d - 2) + 7x(d - 2)$

35. $12x(-x^3 + 2x^2 - 8x + 7)$

36. $5y(y^3 - 2y) + 5y(y - 2y^3)$

37. $-9(n^3 - 4) + 9(n^3 + 7n)$

38. $-8x^3(6x^2 - 8x + 11)$

39. $14b(7b - 6) - 11b(7b - 6)$

40. $3xy(-2x + 4y) + xy(3x - y)$

Write and simplify an expression for the area of each figure.

41. rectangle: $6x$ by $(3x + 9)$

42. rectangle: $3d^2$ by $(4d^3 - 3)$

43. triangle: base $= 4y$ and height $= 6y^2 + y$

44. triangle: base $= 12xy$ and height $= xy^2 + 3x^2y$

Write an expression for the unknown dimension of each rectangle. (Lesson 9-3)

45. area = $14a^2b^3$

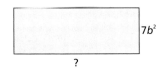

46. area = $54xyz$

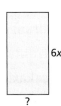

Write and simplify an expression for the area of each figure. (Lesson 9-4)

47.

48.

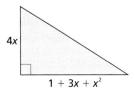

Simplify. (Lessons 9-1 and 9-2)

49. $(3xy^3)^3$

50. $(-4a^2b)^2$

51. $5b + b + b^2 - 6b^2$

52. $-(4rs^2)^3$

53. $4h - (h + 7)$

54. $x + x + x^3 + x^3$

Mid-Chapter Quiz

Write each polynomial in standard form for the variable x. (Lesson 9-1)

1. $x^2 + 7 + 2x^4 - 9x - 24x^3$

2. $2x - x^2y^2 + 3x^3y - 4$

3. $4y^3x + 6y^2x^2 + yx^3 + x$

Simplify. (Lessons 9-1 through 9-4)

4. $10w + 3w$

5. $y^2 - 2y^2 + 4y^2$

6. $\frac{2}{3}x^4 - \frac{1}{6}x^4$

7. $11n + n - 9n - 3n$

8. $7a^2 + 17a^3 - 4a^2$

9. $9p - q + 41q + 11p$

10. $6g^2 - 7g^2 - 4g + 3g^2$

11. $(3h^2 + 2) + (h^2 + 6h)$

12. $(7x^2 + 4x - 3) - (5x^2 - 2x)$

13. $(-7h)(-8j)$

14. $\left(\frac{1}{4}x^2\right)(-24x^4)$

15. $(3mn)(2pq)$

16. $(-8a^2b)(4b^2a)$

17. $(5x^3)(5x^4)$

18. $(9c^4d^2)(-6c^4d)$

19. $(-2k^4)^3$

20. $(3s^5ty^3)^2$

21. $(4y^2)^3(3yx)^2(x^2)^2$

22. $\frac{9vw}{15v}$

23. $\frac{30st^2(2y^3)}{25s^3tu^4}$

24. $\frac{12m - 15}{3}$

25. $\frac{6h^5 + 2h^3 - 8h^2}{2h^2}$

26. $\frac{-21bc^4 + 14b^3c}{7bc}$

27. $\frac{12b^4 + 2b^2 - 3b}{3b^2}$

28. $5x(4x - 2)$

29. $-2n(3n^2 + 4n - 5)$

30. $\frac{1}{2}y^3\left(\frac{3}{4}y^8 + \frac{5}{8}y^6\right)$

31. $2mn(7m - 3n) + 5mn(m + 6n)$

32. $4v(v^2 - 1) + 2v(-v^2 + 1)$

9-5 Multiply Binomials

Goals ■ Multiply binomials.

Applications Finance, Geography, Recreation, Photography

You can use Algeblocks to find the product of two binomials. Use the following steps to sketch or show the product $(2x + 1)(x - 3)$ on a Quadrant Mat.

Step 1 Place $(2x + 1)$ in the horizontal axis.

Step 2 Place $(x - 3)$ in the vertical axis. (Be sure to put the 3 unit blocks in the negative part of the axis.)

Step 3 Use x^2-blocks, x-blocks and unit blocks to form rectangular areas in all quadrants bounded by the binomials.

Step 4 Delete any zero pairs if possible. (Recall, a *zero pair* is any pair of terms whose sum is zero. For example, x^2 and $-x^2$ are zero pairs, or opposites.)

Step 5 Read the answer from the mat.

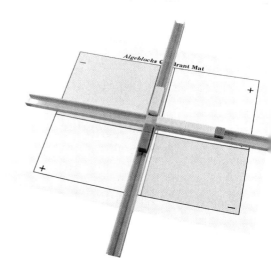

■ BUILD UNDERSTANDING

The product of two binomials can be found by applying the distributive property. The product is a polynomial of four terms that can usually be simplified to a trinomial or binomial.

Example 1

Find the product $(x + 3)(x + 7)$.

Solution

$$(x + 3)(x + 7) = (x)(x + 7) + (3)(x + 7)$$ Distribute $(x + 3)$ over the binomial $(x + 7)$.

$$= x(x) + x(7) + 3(x) + 3(7)$$ Use the distributive property twice.

$$= x^2 + 7x + 3x + 21$$ Combine like terms.

$$= x^2 + 10x + 21$$

In Example 1, notice that each term in the second step is the product of a different pair of monomials from the original expression.

This pattern provides a systematic method for multiplying two binomials. Multiply the *first* term of each binomial, then the *outer* terms, then the *inner* terms, and finally the *last* terms. Write these four products as a sum and simplify.

$$\text{(x + 6)(x + 4)} = x^2 + 4x + 6x + 24$$
$$= x^2 + 10x + 24$$

first — last
inner
outer

Many people use the acronym FOIL as a memory device to remember the order in which to multiply.

F   O   I   L
↑   ↑   ↑   ↑
first   outer   inner   last

Example 2

Simplify.

a. $(x + 5)(x - 8)$ **b.** $(m + 7)(m - 7)$ **c.** $(k - 9)^2$

Solution

Multiply binomials and use the distributive property.

a. $(x + 5)(x - 8) = (x + 5)[x + (-8)]$ Subtracting 8 is the same as adding -8.

$$= x(x) + x(-8) + 5(x) + 5(-8)$$
$$= x^2 + (-8x) + 5x + (-40)$$
$$= x^2 - 3x - 40$$

b. $(m + 7)(m - 7) = (m + 7)[m + (-7)]$

$$= m(m) + (-7m) + 7m + 7(-7)$$
$$= m^2 + 0m + (-49)$$
$$= m^2 - 49$$

c. $(k - 9)^2 = (k - 9)(k - 9)$

$$= [k + (-9)][k + (-9)]$$
$$= k(k) + (-9k) + (-9k) + (-9)(-9)$$
$$= k^2 - 18k + 81$$

COncepts in MOtion
Animation
mathmatters2.com

Check Understanding

In Example 2, part b, the sum of the "outer" and "inner" terms is zero, resulting in a product with only two terms m^2 and -49.

Describe a pattern in the original binomials that will always produce this result.

Example 3

Simplify.

$$5(n + 3)(n - 4) - 2(n - 6)(n + 5)$$

Solution

$5(n + 3)(n - 4) - 2(n - 6)(n + 5)$ Use FOIL twice.

$= 5[n^2 - 4n + 3n + 3(-4)] + (-2)[n^2 + 5n - 6n - 6(5)]$ Simplify inside parentheses.

$= 5(n^2 - n - 12) + (-2)(n^2 - n - 30)$ Distribute 5 and -2.

$= 5n^2 - 5n - 60 + (-2n^2) + 2n + 60$ Simplify.

$= 3n^2 - 3n$

Math Online   mathmatters2.com/extra_examples

Example 4

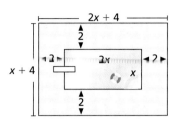 Personal Tutor at mathmatters2.com

A rectangular swimming pool is twice as long as it is wide. A walkway surrounding the pool is 2 yd wide and has an area of 196 yd². Find the dimensions of the pool.

Solution

Make a drawing. Let x represent the width of the pool. Label the other dimensions in terms of x.

area of walkway = area of pool and walkway − area of pool

$$196 = (2x + 4)(x + 4) - (2x)(x)$$
$$196 = 2x^2 + 8x + 4x + 16 - 2x^2 \quad \text{Combine like terms.}$$
$$196 = 12x + 16$$
$$180 = 12x$$
$$15 = x$$

The width of the pool is 15 yd. The length is twice the width, or 30 yd.

TRY THESE EXERCISES

Find the product.

1. $(x + 2)(x + 5)$ **2.** $(n - 7)(n + 3)$ **3.** $(p - 6)(p + 6)$ **4.** $(h - 4)^2$

5. $(y + 12)(y + 11)$ **6.** $(b + 4)^2$ **7.** $(p + 14)(p - 3)$ **8.** $(n - 9)^2$

Simplify.

9. $4(y + 3)(y + 2) + 3(y + 4)(y + 1)$ **10.** $-2(t - 4)(t + 7) + 9(t + 3)(t - 8)$

11. MODELING What expressions are being multiplied on the Quadrant Mat shown? Find the product of the binomials.

12. RECREATION A rectangular playground is 4 times as long as it is wide. The area of a 3-ft wide sidewalk around the playground is 1236 ft². Sketch the playground and then find its dimensions.

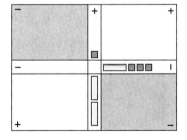

PRACTICE EXERCISES • For Extra Practice, see page 614.

Find the product.

13. $(a + 3)(a + 2)$ **14.** $(c + 4)(c - 5)$ **15.** $(p - 3)(p + 1)$ **16.** $(k + 6)^2$

17. $(h - 3)(h + 7)$ **18.** $(s - 8)(s + 2)$ **19.** $(x - 5)(x - 3)$ **20.** $(e + 2)^2$

21. $(c + 5)(c - 5)$ **22.** $(n + 12)(n - 6)$ **23.** $(k - 9)^2$ **24.** $(11 + r)(8 - r)$

25. $(6 + x)(6 - x)$ **26.** $(m - 10)(m + 8)$ **27.** $(7 + t)(12 + t)$ **28.** $(b + 12)^2$

29. $(w - 9)(w - 8)$ **30.** $(6b + 1)(7b + 3)$ **31.** $(2p + 10)(5p - 6)$ **32.** $(4k - 3)(6k - 7)$

33. $(20 + x)(-4 + x)$ **34.** $(2x + 3)(3x + 2)$ **35.** $(4k + 4)^2$ **36.** $(5c + 1)(5c - 1)$

37. $(7y + 3)(7y - 4)$ **38.** $(2x - 3)^2$ **39.** $(6x + 5)(5x + 6)$ **40.** $(3v + 5)(5v - 4)$

41. YOU MAKE THE CALL Ana used FOIL to find the product of $(2y - 7)(5y + 4)$. Her work is shown. Is she correct? If not, explain her error.

$$(2y - 7)(5y + 4) = (2y)(5y) + (2y)(4) + (-7)(5y) + (-7)(4)$$
$$= 10y^2 + 8y - 35y - 28$$
$$= 10y^2 - 27y - 28$$

Simplify.

42. $(y - 4)(y + 6) + (y - 9)(y + 3)$

43. $x(x + 1)(x - 3)$

44. $3(x + 4)(x + 2) - 2(x + 1)(x + 3)$

45. $n(3n + 3)(5n - 4) + 5n(n - 6)(2n + 3)$

46. WRITING MATH Explain why $(y + 4)^2 \neq y^2 + 16$ for $y \neq 0$.

47. Write an expression for the area of a square if the measure of each side is $5 - v$.

48. GEOGRAPHY The shape of Colorado is approximately a rectangle. The length is about $(2x^2 + 80)$ mi and the width is about $(3x^2 + 80)$ mi. Find the approximate area of Colorado in terms of x.

49. PHOTOGRAPHY A photo is 6 in. longer than it is wide. A $1\frac{1}{2}$ in. frame surrounds the photo. If the area of the frame is 99 in.^{2}, what are the dimensions of the photo?

50. MODELING Sketch or show $(x + 3)(2x - 1)$ using Algeblocks.

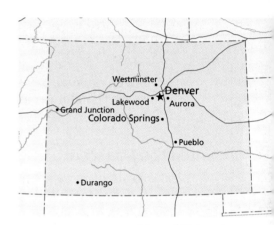

Map of Colorado

EXTENDED PRACTICE EXERCISES

Simplify.

51. $(3k - 1)(2k + 1) - (k - 1)^2$

52. $(5h - 7)^2 - (7 - 5h)^2$

53. $(x + 1)^3$

54. $(2c - 3)^3$

FINANCE Compound interest is the interest that is paid on money invested and previously earned interest. The formula for the amount of money A in an account that earns compound interest is $A = p(1 + r)^t$, where p is the amount invested, r is the rate of interest per time period, and t is the number of time periods.

55. Find A when $p = \$1000$, $r = 0.06$ and $t = 2$.

56. Find A when $p = \$4500$, $r = 0.03$ and $t = 3$.

MIXED REVIEW EXERCISES

Solve each system of equations using the substitution method. (Lesson 8-3)

57. $4x + y = 11$
$2x - 4y = -17$

58. $x - 3y = 6$
$6x - 6y = 12$

59. $3x + 2y = 15$
$x + y = -9$

60. Find the equation of the line perpendicular to the line $5x + 2y = 12$ and passing through the point $(-5, 4)$. (Lesson 8-1)

9-6

Problem Solving Skills: Work Backwards

When a problem involves a series of steps, you often solve it by working forward from the beginning of the problem to the end. Sometimes, however, you are told what happened at the end and asked to find what happened at the beginning. You can use the strategy **work backwards** to solve this type of problem.

Problem Solving Strategies

Guess and check

Look for a pattern

Solve a simpler problem

Use a picture, diagram or model

Make a table, chart or list

Act it out

✔ Work backwards

Eliminate possibilities

Use an equation or formula

Problem

Mr. Bogen drove to a gas station, where he spent $16 on gas. He spent half of his remaining cash for lunch, and then bought a magazine for $2.50. He has $9.50 left. How much cash did he have at the beginning?

Solve the Problem

The graphic organizer illustrates the steps described in the problem.

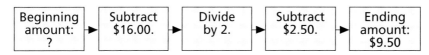

| Beginning amount: ? | → | Subtract $16.00. | → | Divide by 2. | → | Subtract $2.50. | → | Ending amount: $9.50 |

To find the beginning amount, work backwards from the end by reversing each step.

Ending amount: $9.50

Add $2.50:	$9.50 + $2.50 = $12.00	Addition is the opposite of subtraction.
Multiply by 2:	$12.00 · 2 = $24.00	Multiplication is the opposite of division.
Add $16.00:	$24.00 + $16.00 = $40.00	

So Mr. Bogen had $40 to begin with.

Check: $40 − $16.00 = $24.00

$24.00 ÷ 2 = $12.00

$12.00 − $2.50 = $9.50

◥ TRY THESE EXERCISES

1. A number is multiplied by 2 and then 11 is added to the result. The final number is 37. What is the original number?

2. A number is divided by 6, decreased by 10 and then multiplied by 5. The final number is 25. What is the original number?

3. Kayla has $5.40. One-third of the money she originally had went to lunch. Then she lent Marcus $3.00. How much money did Kayla originally have?

4. If you multiply Keshawn's age by 8, add 8, divide by 8 and then subtract 8, the result is 8. How old is Keshawn?

Five-step Plan

1 Read
2 Plan
3 Solve
4 Answer
5 Check

5. **GEOGRAPHY** The smallest U.S. state is Rhode Island. If you double the area of Rhode Island and then subtract 601 mi², the result will be 2489 mi², the number of square miles in Connecticut, the second smallest U.S. state. Find the area of Rhode Island in square miles.

6. **TRAVEL** After leaving her home in the morning, Gwen drove 25 mi at an average speed of 50 mi/h, then stopped 40 min for breakfast. Resuming her trip, she drove 90 mi at an average speed of 60 mi/h, arriving at 11 A.M. What time did she leave home?

7. Melvern was deciding what time to set his alarm. He has to be at a job interview by 8:30 A.M. He wants to be there 15 min early. It takes him 25 min to drive there. He needs 50 min in the morning to get ready. For what time should Melvern set his alarm?

8. **MONEY** The cash-in receipts in Parker's cash drawer total $823.37, and his cash-out receipts total $734.87. If he currently has $338.40 in his drawer, what was his opening balance?

9. Use the table of the longest highway tunnels in the U.S. Latoya drove at an average rate of 60 mi/h through a tunnel and then 18.85 mi farther to a rest stop. After a 10-min rest she noted that it had been exactly 30 min since she had started through the tunnel. Which tunnel did Latoya drive through?

Longest Highway Tunnels in the U.S.

Name	Length (miles)
A. Anderson Memorial	2.52
E. Johnson Memorial	1.70
Eisenhower Memorial	1.69
Allegheny	1.15
Liberty Tubes	1.12
Zion National Park	1.09
East River Mountain	1.03
Tuscarora	1.02

10. Pete is reading a book with 480 pages. When he has read three times as many pages as he already has read, he will be 144 pages from the end. How many pages has he read?

11. **PHYSICS** Each time a dropped ball bounces it returns to a height $\frac{2}{3}$ the height of the previous bounce. After the third bounce the ball returns to a height of 4 ft. From what height was it dropped?

12. **WRITING MATH** Write a problem that can be solved by working backwards. Include the answer to your problem.

Work backwards to find the factors of the given product. Use the distributive property in reverse order.

13. $x^2 + 5x = x(\blacksquare + \blacksquare)$

14. $m^2n - m^3 = m^2(\blacksquare - \blacksquare)$

15. $7c^2 + 21c - 42 = 7(\blacksquare + \blacksquare - \blacksquare)$

16. $9a^3 + 15a^2 - 6a = 3a(\blacksquare + \blacksquare - \blacksquare)$

17. $8w^4y - 10w^3k^2 + 2w^2 = 2w^2(\blacksquare - \blacksquare + \blacksquare)$

■ MIXED REVIEW EXERCISES

Simplify. (Lessons 2-4 and 2-5)

18. $4 + (-7) + (-9) + 12$

19. $4x + (-17x) + 21x + (-x)$

20. $-41 + 5x + (-11x) + 48$

21. $-8 \cdot 6 \cdot \left(-\frac{3}{4}\right) \cdot 5$

22. $6x \cdot 3y \cdot (-2z) \cdot (-x)$

23. $6z \cdot (-2) \cdot (-x) \cdot 4xz \cdot (-7x)$

24. $\dfrac{-5 \cdot (-4)}{-2}$

25. $\dfrac{-4x^2yz^2}{xyz}$

26. $\dfrac{7x \cdot (-8z)}{4z}$

Review and Practice Your Skills

Find the product.

1. $(x + 5)(x + 3)$ **2.** $(a - 2)(a - 4)$ **3.** $(m + 3)(m - 7)$

4. $(c - 2)^2$ **5.** $(3 + n)(2 + n)$ **6.** $(e + 6)(e - 8)$

7. $(8x + 7)(x - 2)$ **8.** $(y - 3)(y + 3)$ **9.** $(5x - 3)(4x - 2)$

10. $(4d - 8)(4d + 8)$ **11.** $(3x + 3)^2$ **12.** $(4x - 3)(2x + 1)$

13. $(5y - 12)^2$ **14.** $5(c - d)(a - b)$ **15.** $a(6a - 4)(5a - 3)$

Simplify.

16. $(x - 1)(x + 3) + (x + 1)^2$ **17.** $5(x - 2)(3x - 1) - (2x + 3)^2$

18. $(y - 3)(y + 2) + (y - 1)(y + 6)$ **19.** $n(n + 2)(n - 5)$

20. $6(a + 3)(a + 5) - (2a + 6)$ **21.** $b^2(b - 3)(b + 10)$

22. $(x - 9)(x + 11) + (x + 4)(x - 11)$ **23.** $-4(y + 7)(y - 1) + 8(y + 13)(y - 6)$

Use the work backwards strategy to solve.

24. A number is multiplied by 3 and then increased by 12. The final number is 36. What is the original number?

25. Grace is reading a book with 640 pages. When she has read two times as many pages as she has already read, she will be 340 pages from the end. How many pages has she read?

26. Carlos rode his bike to his sister's house. He biked 21 mi at an average speed of 7 mi/h, then stopped 30 min for lunch. He finished the trip to his sister's after biking 20 more miles averaging 10 mi/h. What time did he leave home to get to his sister's at 3 P.M.?

27. A number is divided by 3, decreased by 103 and then multiplied by 6. The final number is 48. What is the original number?

Work backwards to find the factors of the given product. Use the distributive property in reverse order.

28. $x^2 + 10x = x(\blacksquare + \blacksquare)$

29. $a^2b - a^3 = a^2(\blacksquare - \blacksquare)$

30. $2x^2 + 6x - 12 = 2(\blacksquare + \blacksquare - \blacksquare)$

31. $3c^3 + 9c^2 + c = c(\blacksquare + \blacksquare + \blacksquare)$

32. $4b^2 + 8b + 16 = 4(\blacksquare + \blacksquare + \blacksquare)$

33. $12x^2y^3 - 9x^2y^2 + 3xy = 3xy(\blacksquare - \blacksquare + \blacksquare)$

Simplify. (Lessons 9-2 and 9-3)

34. $\dfrac{15x^2y^4}{3xy}$ **35.** $(5a^2b)^3$ **36.** $(-x^4y^3z)^2$ **37.** $\dfrac{27x^3 - 3x}{3xy}$

Write and simplify an expression for the area of each figure. (Lesson 9-4)

38.

4x

4x

39.

6m

5m − 3

Find the product (Lesson 9-5)

40. $(x - 4)(x - 5)$ **41.** $(y - 8)^2$ **42.** $(a + 3)(a - 3)$

43. $(3c + 2)(c - 8)$ **44.** $(2a + 5)(3a + 2)$ **45.** $\left(2x + \dfrac{1}{2}\right)^2$

MathWorks Career – Air Traffic Controller

Workplace Knowhow

The landscape and geographical features of the earth partially determine what path an airplane will fly from one city to another. In order for a pilot to fly an airplane safely, an air traffic controller must plan the flight path. In addition, government noise restrictions require airplanes to fly above a minimum distance from the surface of the earth. The distance from New York City to Los Angeles, is 3000 mi. To safely fly this route, a flight path must consider geographical features such as the Rocky Mountains and the Great Salt Lake Desert as well as government restrictions. In the equation, x represents the distance in miles between New York City and Los Angeles, and y is the altitude above sea level in feet at which the airplane will fly at each point along the flight path.

$$y = -\left(\frac{1}{75}\right)x^2 + 40x$$

1. Graph the polynomial above. Among the x-values for the graph, use 0, 500, 1000, 1500, 2000, 2500 and 3000 mi.

2. At what distance from New York City is the airplane located at its highest altitude? What is the plane's altitude at this point?

3. The highest point in Utah is at Kings Peak in the Uinta Mountains, at an elevation of 13,528 ft. Assuming that the airplane flies directly over Kings Peak, how high above the Peak is the airplane at this point? (Kings Peak is approximately 2300 mi from New York City.)

4. At what two distances from New York City will the airplane have an altitude of 10,000 ft? Give your answer to the nearest 100 mi.

5. In constructing this equation, two assumptions were made about the elevation of New York City and Los Angeles. What are the assumed elevations of these two cities? Is this accurate or not? Explain why.

Factor Using Greatest Common Factor (GCF)

Goals ■ Factor polynomials using the Greatest Common Factor.

Applications Finance, Geography, Physics, Modeling

The formula for the surface area (*SA*) of a cylinder is usually given as $SA = 2\pi r^2 + 2\pi rh$, where r is the radius of the cylinder and h is the height. The formula can also be written $SA = 2\pi r(r + h)$.

1. Calculate the surface area of a cylinder with a radius of 3.6 cm and a height of 8.4 cm using each formula. Use 3.14 for π.

2. Which formula do you like better? Explain.

3. What advantage is there in writing a polynomial in a different form?

4. The formula for the surface area of a cone is $SA = \pi rs + \pi r^2$, where s is the slant height. Can this formula be written differently? Explain.

◤ BUILD UNDERSTANDING

COncepts in MOtion

Animation
mathmatters2.com

In earlier lessons you used the distributive property to simplify expressions.

$$4a(2a + 3) = 4a(2a) + 4a(3) = 8a^2 + 12a$$

In this process you begin with the factors $4a$ and $2a + 3$, then multiply to obtain the polynomial $8a^2 + 12a$.

This process can be reversed. **Factoring** a polynomial means to express it as a product of polynomials. To factor a polynomial like $8a^2 + 12a$, work backwards to find the factors. Begin by finding the greatest common factor (GCF) of its monomial terms.

Check Understanding

Find the greatest common factor (GCF) of the following.

1. 20, 12
2. 84, 120
3. $12x^3$, $18x^2$
4. $24a^3b^2$, $36a^2b$

Example 1

Factor each polynomial.

a. $10n + 6$

b. $4c^2 - 12c^5$

c. $9h^2k^2 - 12hk^2 + 24h^3k$

Solution

a. Find the GCF of $10n$ and 6.

$10n = 2 \cdot 5 \cdot n$ $\quad$ $6 = 2 \cdot 3$ $\quad$ The GCF is 2. $\qquad$ Write the prime factorization of each term.

Use the GCF and the distributive property to rewrite the polynomial.

$10n + 6 = 2 \cdot 5n + 2 \cdot 3$ $\qquad$ Write each term with 2 as a factor.

$\qquad\quad = 2(5n + 3)$

So, $10n + 6 = 2(5n + 3)$.

Check by multiplying. $\qquad$ $2(5n + 3) = 2(5n) + 2(3) = 10n + 6$

b. Find the GCF of $4c^2$ and $12c^5$.

$$4c^2 = 2^2c^2 \qquad\qquad 12c^5 = 2^2 \cdot 3 \cdot c^5 \qquad \text{Write the prime factorization of each term.}$$

The GCF is 2^2c^2, or $4c^2$.

So, $4c^2 - 12c^5 = 4c^2(1 - 3c^3)$

Check: $4c^2(1 - 3c^3) = 4c^2(1) + 4c^2(-3c^3) = 4c^2 - 12c^5$

c. Find the GCF for each term.

$$9h^2k^2 = 3^2h^2k^2 \qquad 12hk^2 = 2^2 \cdot 3hk^2 \qquad 24h^3k = 2^3 \cdot 3h^3k$$

The GCF is $3hk$.

So, $9h^2k^2 - 12hk^2 + 24h^3k = 3hk(3hk - 4k + 8h^2)$.

Check: $3hk(3hk - 4k + 8h^2) = 9h^2k^2 - 12hk^2 + 24h^3k$

Example 2

 Personal Tutor at mathmatters2.com

The formula for the surface area (*SA*) of a rectangular prism with length *l*, width *w* and height *h* is $SA = 2lw + 2wh + 2lh$. Rewrite the formula by factoring.

Solution

The GCF of $2lw$, $2wh$, and $2lh$ is 2.

$2lw + 2wh + 2lh = 2(lw + wh + lh)$

So, $SA = 2(lw + wh + lh)$.

Example 3

MODELING Use Algeblocks to factor $4xy - 2x^2$.

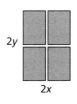

Solution

Step 1 Form rectangular areas of $4xy$ and $2x^2$ having one side as $2x$, the GCF.

Step 2 Place the GCF, $2x$, on the horizontal axis.

Step 3 Put the rectangular areas in the proper quadrant with the GCF as a boundary.

Step 4 Divide by making another boundary. The two boundaries are the factors.

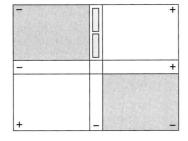
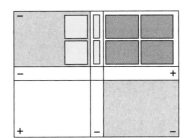
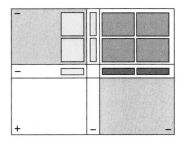

So, $4xy - 2x^2 = 2x(2y - x)$.

Factor each polynomial.

1. $7w - 21$

2. $4c^2 + 6c^3$

3. $9a^2 - 6a$

4. $16xy + 12x$

5. $45a^2b - 27ab^2$

6. $25mn^3 - 15mn^2 + 5m$

7. FINANCE The amount (A) in dollars returned after 1 year on a principal of P dollars invested at an annual rate of (r) percent is given by the formula $A = P + Pr$. Rewrite the formula by factoring.

8. MODELING Use Algeblocks to factor $-2y^2 + y$.

9. The formula for the perimeter (P) of a rectangle with length l and width w is $P = 2l + 2w$. Rewrite the formula by factoring.

■ **PRACTICE EXERCISES** • For Extra Practice, see page 615.

Factor each polynomial.

10. $8mn - 8mp$

11. $12k + 15$

12. $4x - 20$

13. $7e^2 + 21e$

14. $9p^3 + 27p^2$

15. $x^2 - xy$

16. $12k^2 - 42k$

17. $5y^4 - 20y^3$

18. $w^5 - w^4$

19. $100c - 200$

20. $3xy^2 + 6x$

21. $7xy - 56xz$

22. $v^5 - 6v^4 + 3v^3$

23. $5n^3 - 30m^2 - 15$

24. $3a^6 - 5a^3 + 2a^2$

25. $xy + xz + 2x$

26. $12x^4y^4 + 3x^3y^2 - 6x^2y^2$

27. $2xya - 4xyb + 6xyc$

28. $x^2 + 6xy - x$

29. $8h^2 - 16h + 24$

30. $50m^2 + 125mn + 25n^2$

31. $15a^3b + 20a^2b - 10ab$

32. $64x^6 - 48x^4 + 24x^2$

33. $6m^3n^3 + 3m^3n^2 + m^2n^2$

Evaluate each expression. Let $x = 2$ and $y = -3$.

34. $3xy^2 - 3x^2y$

35. $3xy(y - x)$

36. WRITING MATH What do you notice about your answers in Exercises 34 and 35? Explain.

37. Did Exercise 34 or Exercise 35 require fewer steps?

Mount McKinley, Alaska

38. The formula for the surface area (SA) of a cone with radius r and slant height s is $SA = \pi r^2 + \pi rs$. Rewrite the formula by factoring.

39. The formula for the number of diagonals (D) that can be drawn in a polygon with n sides is $D = \frac{1}{2}n^2 - \frac{3}{2}n$. Rewrite the formula by factoring.

40. GEOGRAPHY The highest mountain in the U.S. is Mount McKinley, located in Alaska. The height can be expressed as $(2x^4 + 4x^2 - 8x)$ ft. Factor this expression.

41. PHYSICS An expression used in connection with certain atomic particles is $\frac{1}{2}Z - \frac{1}{2}N$, where Z is the number of protons and N is the number of neutrons in the nucleus. Factor this expression.

MODELING Use Algeblocks to factor each polynomial.

42. $2x^2 - 4xy$ **43.** $xy - 3y$ **44.** $-x^2 + xy - x$

Factor each polynomial.

45. $18m^3n^2 + 45m^2n^3 + 27m^4n - 54m^2n^2$ **46.** $35x^5y - 40x^3y^2 + 10x^2 + 45x^4y^3$

47. $48ab - 40a^3b^2 + 24a^2b^3 + 28a^2b^2$ **48.** $x^2y^3z - x^2y^2z^2 + xy^4z - xy^3z^2$

Write an expression for the perimeter of each figure. Then factor the expression.

49.

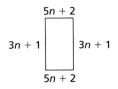

5n + 2
3n + 1   3n + 1
5n + 2

50.

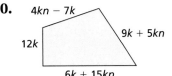

4kn − 7k
9k + 5kn
12k
6k + 15kn

51.

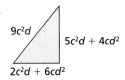

9c²d
5c²d + 4cd²
2c²d + 6cd²

If two polynomials have 1 as their greatest common factor, they are *relatively prime*. Tell whether the polynomials are relatively prime.

52. $2, 3x$ **53.** mn, n^2 **54.** $4ab, 3ab$

55. $c + c^2, 5c^4$ **56.** $k, k + 5$ **57.** $2m + 2, km^2 + km$

◤ EXTENDED PRACTICE EXERCISES

58. CRITICAL THINKING The perimeter of a square is $(8t^2 + 36t)$ ft. Find the area of the square.

Find each product.

59. $(x + 5)(x - 5)$ **60.** $(n + 7)(n - 7)$ **61.** $(p + 4)(p - 4)$

62. $(k + 9)(k - 9)$ **63.** $(y - 11)(y + 11)$ **64.** $(a + 1.5)(a - 1.5)$

Study the pattern in the products for Exercise 59–64. Write each polynomial as a product of two binomials.

65. $v^2 - 9$ **66.** $w^2 - 64$ **67.** $m^2 - 36$

68. $c^2 - 100$ **69.** $4x^2 - 25$ **70.** $36y^2 - 121z^6$

◤ MIXED REVIEW EXERCISES

Solve each equation. Check the solution. (Lesson 3-4)

71. $5x - 10 = 0$ **72.** $\frac{3}{4}y + 4 = 10$ **73.** $4a + 10 = 6$

74. $0.4b - 3.1 = 0.5$ **75.** $4(x - 6) = -20$ **76.** $\frac{w}{5} - 3 = 7$

Translate each sentence into an equation. Then solve the equation. (Lesson 3-4)

77. Three even consecutive numbers have a sum of -54. What are the three numbers?

78. What is the area of a swimming pool with a length that is 3 times the width and a perimeter of 96 ft.

Perfect Squares and Difference of Squares

Goals ■ Factor perfect square trinomials.
■ Factor difference of perfect squares.

Applications Travel, Number sense, Modeling, Geography

Use the Quadrant Mat shown for Questions 1–4.

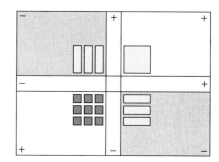

1. What polynomial expression is represented by the Algeblocks?

2. What factor should be placed along the horizontal axis?

3. What factor should be placed along the vertical axis?

4. Write an equation to represent the polynomial expression and its factors shown on the Quadrant Mat.

◣ BUILD UNDERSTANDING

Just as a number can be squared, such as $2^2 = 4$, a binomial can be squared. The result of squaring a binomial is a trinomial, called a **perfect square trinomial**.

$$(w + 6)^2 = (w + 6)(w + 6) = w^2 + 6w + 6w + 36$$
$$= w^2 + 12w + 36$$

Notice the pattern of a perfect square trinomial:

The first term is a perfect square. $w^2 = w \cdot w$

The last term is a perfect square. $36 = 6 \cdot 6$

The middle term is twice the product of $12w = 2\,(6 \cdot w)$
the square roots of the first and last terms.

So, $w^2 + 12w + 36$ fits the pattern. It is a perfect square trinomial.

Example 1

Tell whether the trinomial is a perfect square trinomial.

a. $n^2 + 10n + 25$

b. $k^2 + 7k + 49$

Think Back

Recall the following *perfect squares.*

$1 = 1 \cdot 1$	$81 = 9 \cdot 9$
$4 = 2 \cdot 2$	$100 = 10 \cdot 10$
$9 = 3 \cdot 3$	$121 = 11 \cdot 11$
$16 = 4 \cdot 4$	$144 = 12 \cdot 12$
$25 = 5 \cdot 5$	$169 = 13 \cdot 13$
$36 = 6 \cdot 6$	$196 = 14 \cdot 14$
$49 = 7 \cdot 7$	$225 = 15 \cdot 15$
$64 = 8 \cdot 8$	$256 = 16 \cdot 16$

Solution

a. The first term is a perfect square. $n^2 = n \cdot n$

The last term is a perfect square. $25 = 5 \cdot 5$

The middle term is twice the product of $10n = 2(5 \cdot n)$
the square roots of the first and last terms.

The trinomial is a perfect square trinomial.

b. First term: $k^2 = k \cdot k$ Last term: $49 = 7 \cdot 7$ Middle term: $7k \neq 2(7 \cdot k)$

The trinomial is not a perfect square trinomial.

You can use the patterns in a perfect square trinomial to factor it.

Example 2

Factor each polynomial.

a. $n^2 + 10n + 25$ 　　　　　　　　**b.** $y^2 - 20y + 100$

Solution

a. Determine if the trinomial is a perfect square trinomial.

First term: $n^2 = n \cdot n$ Last term: $25 = 5 \cdot 5$ Middle term: $10n = 2(n \cdot 5)$

Then, use the square roots to write the factors.

$$n^2 + 10n + 25 = (n + 5)(n + 5) = (n + 5)^2$$

b. Determine if the trinomial is a perfect square trinomial.

First term: $y^2 = y \cdot y$ Last term: $100 = 10 \cdot 10$ Middle term: $20y = 2(y \cdot 10)$

Since the sign of the middle term in the trinomial is negative, use a negative sign in each factor.

$$y^2 - 20y + 100 = (y - 10)(y - 10) = (y - 10)^2$$

A different pattern develops in the product of two binomials that are similar, where one binomial is a sum and the other is a difference.

$$
\underset{\substack{\text{sum of}\\ y \text{ and } 9\\ \downarrow}}{\,}\underset{\substack{\text{difference}\\ \text{of } y \text{ and } 9\\ \downarrow}}{\,}
$$
$$(y + 9)(y - 9) = y^2 - 9y + 9y - 81 = y^2 - 81$$

A polynomial such as $y^2 - 81$ is called the **difference of two squares**. If you recognize that a polynomial is a difference of two perfect squares, you can work backwards to find the factors.

> **Check Understanding**
>
> Verify the following:
>
> $k^2 + 9 \neq (k + 3)(k - 3)$
>
> $k^2 + 9 \neq (k + 3)(k + 3)$
>
> Can $k^2 + 9$ be written as a product of two binomials?

Example 3

Factor $k^2 - 9$.

Solution

$$k^2 - 9$$
$$\nearrow \qquad \nwarrow$$
$$k \cdot k \qquad 3 \cdot 3$$

To factor a difference of two squares, write the two binomials using the square roots of the terms. Make one binomial a sum and the other a difference.

$$k^2 - 9 = (k + 3)(k - 3)$$

> **COncepts in MOtion**
> **Animation**
> mathmatters2.com

Tell whether the trinomial is a perfect square trinomial.

1. $c^2 - 5c + 25$ **2.** $h^2 + 8h + 16$ **3.** $s^2 + 6s + 9$

4. WRITING MATH Explain why there is a multiple of 2 in the middle term of a perfect square trinomial.

Factor each polynomial if possible.

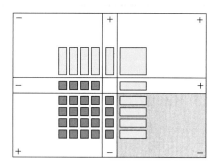

5. $m^2 + 18m + 81$ **6.** $f^2 - 2f + 1$

7. $p^2 + 121$ **8.** $q^2 - 25$

9. $h^2 + 4h + 8$ **10.** $s^2 - 196$

11. MODELING For the Quadrant Mat shown, write the trinomial and its factors.

Tell whether the trinomial is a perfect square trinomial.

12. $x^2 + 6x + 9$ **13.** $y^2 + 14y + 49$

14. $n^2 + 5n + 25$ **15.** $p^2 - 20p + 40$

16. $x^2 + 24x + 144$ **17.** $m^2 - 40m + 100$

Factor each polynomial if possible.

18. $c^2 - 4c + 4$ **19.** $k^2 - 18k + 81$ **20.** $n^2 - 100$

21. $x^2 - 64$ **22.** $y^2 + 22y + 121$ **23.** $h^2 - 225$

24. $v^2 - 20v + 40$ **25.** $y^2 - 1$ **26.** $c^2 - 24c + 144$

27. $e^2 - 6e + 9$ **28.** $f^2 + 12f + 36$ **29.** $p^2 - 256$

30. $d^2 + 2d + 1$ **31.** $c^2 + 50$ **32.** $w^2 + 49$

33. $h^2 + 24h + 144$ **34.** $p^2 - 20p + 100$ **35.** $p^2 - 144$

36. $k^2 - 4$ **37.** $x^2 - 7x + 49$ **38.** $m^2 + 9$

39. $c^2 - 8c + 16$ **40.** $n^2 - 16$ **41.** $q^2 - q + 1$

42. $t^2 - 400$ **43.** $b^2 + 26b + 169$ **44.** $z^2 - 81$

GRAPHING Use a graphing calculator to graph the two curves. Determine whether each equation represents the same graph. If they do not, factor the first polynomial.

45. $y = x^2 + 6x + 9$
$y = (x + 2)^2$

46. $y = x^2 - 25$
$y = (x - 5)(x + 5)$

Factor each polynomial.

47. $3m^2 - 27$ **48.** $n^2p - 25p$ **49.** $2m^2 + 32m + 128$

50. $x^2m^2 - 14x^2m + 49x^2$ **51.** $3ak^2 - 300a$ **52.** $4y^2 + 72y + 324$

53. MODELING Use Algeblocks to find the factors of $y^2 - 6y + 9$.

54. TRAVEL Use the table of longest highway tunnels on page 401. The length of the Schelde Tunnel is the square root of the length of the Eisenhower Tunnel in the U. S. How long is the Schelde Tunnel?

55. YOU MAKE THE CALL Eric says the binomial factors of $a^2 - 10a + 25$ are $(a - 5)$ and $(a + 5)$. Sumi says that the binomial factors are $(a - 5)(a - 5)$. Who is correct and why?

NUMBER SENSE Write a polynomial expression to represent each number trick. Let x equal the original number. Use an equation to check your work.

56. Think of a number. Subtract 7. Multiply by 3. Add 30. Divide by 3. Subtract the original number. The result is always 3.

57. Think of any non-zero number. Subtract 6. Multiply by 5. Add 30. Divide by the original number. The result is always 5.

■ EXTENDED PRACTICE EXERCISES

Factor each polynomial.

58. $9k^2 - 49$

59. $4n^2 - 25$

60. $4x^2 + 12x + 9$

61. $9m^2 + 30m + 25$

62. $49p^2 - 100$

63. $25x^2 - 40x + 16$

64. CRITICAL THINKING Suppose the area measures of these squares are both perfect squares. If the difference in the areas is 32 square units, find the values of x and y.

65. CHAPTER INVESTIGATION If the average speed of the orienteers is 2.5 mi/h, and the average time it takes to complete the course is 3 h, calculate the total length of the course in miles. ($d = r \cdot t$)

■ MIXED REVIEW EXERCISES

66. GEOGRAPHY The deepest part of the Pacific Ocean is the Marianna Trench, which is 35,840 ft deep. How many meters is this? How many miles is this? (Basic math skills)

67. Karen ran in a 10 km marathon. How many miles did she run? (Basic math skills)

68. If Karen can run an average of 10 ft/sec, how long did it take her to complete the marathon? (Basic math skills)

69. A living room measures 15 ft wide by 22 ft long. How many square yards of carpeting is needed to cover the floor? (Basic math skills)

Moonrise over the Pacific Ocean

Simplify. (Lesson 9-4)

70. $4y(-3y + 6)$

71. $-9n(11n^2 - 3n)$

72. $\frac{1}{3}x(-12x^3 + 6x^2 - 9)$

73. $4xy(3x^2 - 4y)$

74. $7m(m^2 + 3mn - 4n^2)$

75. $12(a - b) + 3(5a - 9b)$

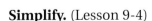

Chapter 9 Review

VOCABULARY ◣

Choose the word from the list that best completes each statement.

1. The __?__ of two or more monomials is the greatest factor that is a factor of every term.

2. In a polynomial, terms that are exactly alike, or that are alike except for their numerical coefficients, are called __?__.

3. To __?__ a polynomial means to express it as a product of polynomials.

4. In a __?__ the first and last terms are perfect squares and the middle term is twice the product of the square roots of the first and last terms.

5. A __?__ is a polynomial with two terms.

6. A __?__ is the numerical part of a monomial.

7. According to the __?__, $3(a + b)$ is equivalent to $3a + 3b$.

8. A __?__ is a monomial with no variables.

9. A binomial where one perfect square is subtracted from another perfect square is called a __?__.

10. A monomial or the sum or difference of two or more monomials is called a __?__.

a. binomial

b. coefficient

c. constant

d. difference of two squares

e. distributive property

f. factor

g. GCF

h. like terms

i. monomial

j. perfect square trinomial

k. polynomial

l. trinomial

LESSON 9-1 ◣ Add and Subtract Polynomials, p. 376

▶ The sum or difference of monomials is a **polynomial**. To simplify a polynomial, combine **like terms** by adding their **coefficients**.

▶ A polynomial is written in **standard form** when its **terms** are arranged in order from greatest to least powers of one of the variables.

Write each polynomial in standard form for the variable x.

11. $2x^2 + x^4 - 6$　　　**12.** $8xy^2 - 7x^2y + 5$　　　**13.** $6w^3x + 3wx^2 - 10x^6 + 5x^7$

Simplify.

14. $7a - 4a$　　　**15.** $8k^2 - 5k + 8k - 4k^2$　　　**16.** $4x + 3y - 6x + 7x - 10y$

LESSON 9-2 ◣ Multiply Monomials, p. 380

▶ Use the product rule for exponents to find a product of two monomials that have the same base.

▶ Use the power rule for exponents and the power of a product rule for exponents to simplify monomials that involve powers.

Simplify.

17. $(-3m)(2m^2n)$　　　**18.** $4x(2x^2)^3$　　　**19.** $(3a^2b^3)(-2ab^4c)$

20. $5r^3(4r^4)$　　　**21.** $(10x^3y)(-2xy^2)$　　　**22.** $3a^3b(5a^4c^2)$

LESSON 9-3 ◣ Divide by a Monomial, p. 386

▶ To simplify the quotient of two monomials, divide both monomials by their GCF.

Simplify.

23. $\dfrac{6xy^3}{8xy}$

24. $\dfrac{-9c + 21c^3}{3c}$

25. $\dfrac{3p^4q + 12p^3q^2 + 15p^2q}{3pq}$

26. $\dfrac{15a^3 - 9ab^3}{3a}$

27. $\dfrac{20x^2y^4 + 16x^3y^3 - 4x^4yz}{4x^2y}$

28. $\dfrac{2d^3 - 6d^6f^2 + 2d^7}{-2d^3}$

Write an expression for the unknown dimension of each rectangle.

29. area = $24x^5y^3$

30. area = $100a^3b$

31. area = $6x^4 + 12x^2$

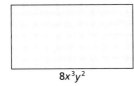

$8x^3y^2$

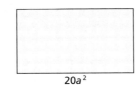

$20a^2$

$3x^2$

LESSON 9-4 ◣ Multiply a Polynomial by a Monomial, p. 390

▶ Use the distributive property and the rules for exponents to multiply a polynomial by a monomial.

Simplify.

32. $8m(2m^2 + 3m - 4)$

33. $3x(2x^2y + xy + 5y^2)$

34. $2(h^3 + 6h) + 7(4h^3 - h)$

35. $d(-2d + 4) + 15d$

36. $3w(6w - 4) + 2(5 - 3w)$

37. $-3c^2(2c + 7) + 2(c^2 - 4)$

38. A local business hires employees to stuff envelopes. Employees earn $50 per day plus $0.02 per envelope. Write a variable expression in simplest form that represents the cost to the business if x people each stuff y envelopes.

39. Spencer has $4000 to invest. He puts x dollars of this money into a savings account that earns 2% per year. He uses the rest of the money to buy a certificate of deposit that earns 4% per year. Write an equation for the total amount of money Spencer will earn after one year.

LESSON 9-5 ◣ Multiply Binomials, p. 396

▶ To multiply two binomials, write the products of the *first* terms, the *outer* terms, the *inner* terms and the *last* terms (FOIL), then simplify.

Find the product.

40. $(p + 5)(p - 7)$

41. $(2t - 4)(t - 1)$

42. $(5y - 3)(5y + 3)$

43. $(e + 3)(e + 7)$

44. $3(w - 6)(w + 6)$

45. $-2(y - 9)(y + 7) + 2(y + 3)(y - 3)$

46. A rectangular flower bed is 8 ft longer than it is wide. A walkway surrounding the flower bed is 3 ft wide and has an area of 192 ft². Find the dimensions of the flower bed.

LESSON 9-6 ◣ Problem Solving Skills: Work Backwards, p. 400

▶ You can use the strategy **work backwards** to solve some types of problems.

47. A number is multiplied by 3 and then 8 is added to the result. The final number is 56. What is the original number?

48. A bacteria population triples in number every day. If there are 2,187,000 bacteria on the seventh day, how many bacteria were there on the first day?

49. Kyle and Devin volunteer at the zoo at 9:00 A.M. on Saturdays. It takes 20 min to get from Devin's house to the zoo. Kyle picks up Devin, but it takes him 15 min to get to Devin's house. If it takes Kyle 45 min to get ready in the morning, what is the latest Kyle should get out of bed?

50. A store tripled the price it paid for a pair of sandals. After a month, the sandals were marked down $5. Two weeks later, the price was divided in half. Finally, the price was reduced by $3 and the sandals sold for $14.99. How much did the store pay for the pair of sandals?

LESSON 9-7 ◣ Factor Using Greatest Common Factor (GCF), p. 404

▶ **Factoring** a polynomial means to express it as a product of polynomials.

▶ To **factor** a polynomial, work backwards to find the factors. Begin by finding the greatest common factor (GCF) of its monomial terms.

Factor each polynomial.

51. $12x - 16$

52. $xy^2 - x^2y + xy$

53. $24m^2n - 3m^3n + 15m^4n^3$

54. $a + a^2b^2 + a^3b^3$

55. $3p^3q - 9pq^2 + 36pq$

56. $5x^3y^2 + 10x^2y + 25x$

Evaluate each expression. Let $x = -2$ and $y = 3$.

57. $4x^2y + xy$

58. $xy(4x + 1)$

59. $2x^2(y - xy)$

60. $y^3(10x - x^2)$

61. $8x^3 + x^2y - x$

62. $3x^2(x^2y + y)$

LESSON 9-8 ◣ Perfect Squares and Difference of Squares, p. 408

▶ Use these patterns to factor **perfect square trinomials** and polynomials that are the **differences of squares**.

$$a^2 + 2ab + b^2 = (a + b)^2 \qquad a^2 - 2ab + b^2 = (a - b)^2 \qquad a^2 - b^2 = (a + b)(a - b)$$

Factor each polynomial.

63. $k^2 + 12k - 36$

64. $c^2 - 20c + 100$

65. $x^2 - 64$

66. $r^2 - 49$

67. $m^2 - 10m + 25$

68. $a^2 + 22a + 121$

CHAPTER INVESTIGATION

EXTENSION Another type of orienteering is score orienteering, in which the course can be completed in any order. For the map on page 375, how many different ways can the course be completed? Which route is the shortest? Express this distance in terms of x.

Chapter 9 Assessment

Simplify.

1. $12x^2y^3 - 4xy^2 + 8xy^2 - 24x^2y^3$

2. $(3a^2b - 2a + 3b) + (4a^2b) + (4a^2b - 3a + 2b)$

3. $(6r^2s + 2rs - 3) - (4r^2s - 2rs)$

4. $(3a^3)(24a^4)$

5. $(24a^3b^4)^2$

6. $-2b(-4b^2 - b + 3)$

7. $4(x^2 + xy) - 3(x^2 - xy)$

8. $2n(4n - 1) + 3n(4n - 1)$

9. $(b + 4)(b + 2)$

10. $(p - 4)(p + 6)$

11. $(a + 8)(a - 8)$

12. $(k - 5)^2$

13. $3(x + 3)(x - 2) - 2(x - 1)(x + 4)$

Use the figure for Exercises 14 and 15.

14. Write an expression for the perimeter of the rectangle.

15. Write an expression for the area of the rectangle.

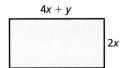

$4x + y$

$2x$

16. If you divide Ana's age by 6, add 5, multiply by 3, and subtract 7, the result is 29. How old is Ana?

17. Felix charges customers $25 plus $1.25/h to rent a snow blower. Write a variable expression in simplest form that represents how much Felix will take in if x people rent a snow blower for y hours.

Factor each polynomial.

18. $6m + 9$

19. $4x^2y - 8x^3y^2$

20. $12h^2k^3 + 6hk^2 - 30h^4k$

Evaluate each expression. Let $x = -1$, and $y = 3$.

21. $14x^2 - xy$

22. $-2x^3(y + 24x)$

23. $(x - 3)(x + 3)$

24. $4y^2(x + 3y)$

Simplify.

25. $\dfrac{10a^2b^3c}{12a^2b}$

26. $\dfrac{-15p + 20p^2}{25p}$

27. $\dfrac{x + x^2 + x^3}{x}$

28. $\dfrac{4a^2b + 8a^3b^2}{4a^2}$

29. $\dfrac{12x^3y + 3x^2y - 9xy}{3xy}$

30. $\dfrac{-14a^2b^4 - 21a^3b^5 + 7a^2b^2}{-7a^2b}$

Factor each polynomial.

31. $x^2 - 4x + 4$

32. $f^2 + 16f + 64$

33. $y^2 - 100$

34. $a^2b - 36b$

35. $t^2 + 25$

36. $x^2 - 14x + 49$

Standardized Test Practice

Part 1 Multiple Choice

Record your answers on the answer sheet provided by your teacher or on a sheet of paper.

1. A basketball team scored 70, 65, 75, 70, and 80 points in the first five games of the season. In the sixth game, they scored only 30 points. Which of these measures changed the most as a result of the sixth game? (Lesson 1-2)

   (A) mean   (B) median   (C) mode

   (D) They all change the same amount.

2. If 0.00023 is expressed as $2.3 \cdot 10^n$, what is the value of n? (Lesson 2-8)

   (A) -5 (B) -4

   (C) 4   (D) 5

3. Marcus and Antonio went shopping and spent $122 altogether. Marcus spent $25 less than twice as much as Antonio. How much did Antonio spend? (Lesson 3-4)

   (A) $49.00   (B) $73.00

   (C) $73.50   (D) $98.00

4. A company is producing padlocks that operate using a sequence of three different numbers. How many different sequences are there if the numbers can range from 0 through 99? (Lesson 4-6)

   (A) 100   (B) 941,094

   (C) 970,200   (D) 1,000,000

5. Which symbol represents the figure? (Lesson 5-2)

   (A) $\overleftrightarrow{AB}$ (B) $\overleftrightarrow{BA}$

   (C) $\overrightarrow{AB}$ (D) $\overrightarrow{BA}$

6. Find the distance between $R(-3, 2)$ and $S(2, 14)$. (Lesson 6-1)

   (A) $\sqrt{13}$ (B) $\sqrt{17}$

   (C) 13   (D) 17

7. Which figure has rotational symmetry? (Lesson 7-4)

   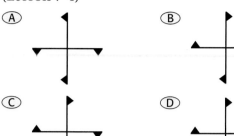

8. What is the slope of a line perpendicular to the line $2x + 3y = 4$? (Lesson 8-1)

   (A) $-\dfrac{3}{2}$ (B) $-\dfrac{2}{3}$

   (C) $\dfrac{2}{3}$ (D) $\dfrac{3}{2}$

9. Which graph represents the system of inequalities $y > x + 1$ and $y < -2x - 1$? (Lesson 8-7)

   (A) 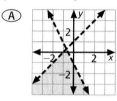   (B)

   (C) 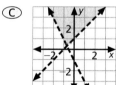   (D)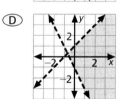

10. What is the product of $(3x^4y^2)^2$ and $4xy^3$? (Lesson 9-2)

    (A) $36x^9y^7$ (B) $36x^5y^5$

    (C) $12x^9y^7$ (D) $12x^5y^5$

11. Simplify $\dfrac{y^3z^9}{yz^2}$. (Lesson 9-3)

    (A) y^4z^7 (B) y^4z^{11}

    (C) y^2z^{11} (D) y^2z^7

12. Factor $m^2 - 16$. (Lesson 9-8)

    (A) $(m - 4)^2$ (B) $(m + 4)^2$

    (C) $(m + 4)(m - 4)$ (D) $4(m - 4)$

Preparing for the Standardized Tests
For test-taking strategies and more
practice, see pages 627–644.

Part 2 Short Response/Grid In

**Record your answers on the answer sheet
provided by your teacher or on a sheet of paper.**

13. What is the value of $-18 - (-4) \div 2 + 1$?
    (Lesson 2-2)

14. The length of a side of a square is $6x - 3$.
    What is the perimeter of the square?
    (Lesson 2-5)

15. What is the least integer that satisfies the
    inequality $2d + 3 > 2$? (Lesson 3-7)

16. What is the least integer that satisfies the
    equation $x^2 + 3 = 7$? (Lesson 3-8)

17. Caroline has 6 nickels, 4 pennies, and 3 dimes
    in her pocket. She takes one coin from her
    pocket at random. What is the probability it is
    a penny or a nickel? (Lesson 4-4)

18. You are required to read 5 books from a list of
    12 great American novels. How many different
    groups of books can you select? (Lesson 4-7)

Use circle X for Questions 19–21. (Lesson 5-8)

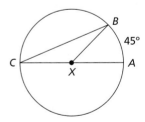

19. Find $m\angle AXB$.

20. Find $m\angle ACB$.

21. Find $m\widehat{ACB}$

22. What is the y-intercept of the line
    $5x + 2y = -4$? (Lesson 6-3)

23. Assume that y varies directly with x.
    When $x = 6$, $y = 27$. Find y when $x = 10$.
    (Lesson 6-8)

24. The current in an electrical circuit varies
    inversely with the resistance in the circuit.
    If the current is 1.5 amperes when the
    resistance is 4 ohms, what is the current
    when the resistance is 1.2 ohms? (Lesson 6-9)

25. How many lines of symmetry does a square
    have? (Lesson 7-4)

26. Simplify $\dfrac{54a^3b^4c^5}{9a^2b^3c^5}$. (Lesson 9-3)

27. In a game show, each question is worth twice
    as much as the question before it. The fifth
    question is worth $12,000. How much is the
    first question worth? (Lesson 9-6)

Part 3 Extended Response

**Record your answers on a sheet of paper.
Show your work.**

28. Tape is placed around a rectangular prism
    as shown.

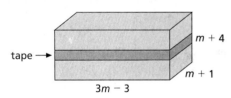

    a. Write an expression in simplest form
       that represents the length of the tape.
       (Lesson 9-1)
    b. Write an expression in simplest form that
       represents the area of the top of the prism.
       (Lesson 9-5)

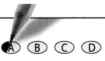

Test-Taking Tip

Question 28
When answering open-ended items on standardized tests,
follow these steps.
1. Read the item carefully.
2. Show a!! of your work. You may receive points for items
   that are only partially correct.
3. Check your work.

Three-Dimensional Geometry

THEME: History

History is a record of significant facts, events, and people. The study of history includes architecture, politics, music, population, language, religion, and mathematics. Specifically, the history of mathematics stretches to every part of the world and spans over thousands of years.

Three-dimensional geometry has been studied since ancient times. Consequently, people from every culture have required three-dimensional geometry to design shelters, clothing, and storage containers.

- **Urban planners** (page 431) supervise the construction of buildings, roads, bridges, and other structures within a city. Because it is important to maintain historically significant elements, while providing for the needs of a city, history plays its part in a city's development.

- **Exhibit designers** (page 451) create images and displays that show a product, historical event, or structure that is pleasing to one's eye.

Math
Online

mathmatters2.com/chapter_theme

Mathematicians throughout History

Mathematician	Life span	Accomplishment
Archimedes	c.298 –212 B.C.	Greatest mathematician of ancient times. Made contributions in geometry.
Easley, Annie	1933–	Developed computer code used to identify energy conversion systems for NASA.
Euclid	c.325 –265 B.C.	Wrote *The Elements*, 13 books covering geometry and number theory.
Gauss, Carl Friedrich	1777–1855	Published treatise on Number Theory. Discovered the asteroid Ceres.
Hypatia	c.370–415	She is considered the first woman of mathematicians.
Khayyam, Omar	c.1048–1131	Persian mathematician, astronomer and poet. Revised the Arabic calendar. Solved cubic equations through geometry.
Lovelace, Ada Byron	1815–1852	Suggested that Babbage's first "computer" calculate and play music.
Pascal, Blaise	1623–1662	Invented and sold the first adding machine in 1645. Developed probability theory.
Pythagoras of Samos	c.560 –c.480 B.C.	His theorem showed the existence of irrational numbers.

(c. stands for *circa*, meaning *about* or *approximately*)

Data Activity: Mathematicians throughout History

Use the table for Questions 1–4.

1. How long did Pythagoras live?

2. Who is considered the greatest mathematician of ancient times?

3. How many of these mathematicians did work with number theory? Who were they?

4. List the names of mathematicians in the table with whom you are familiar. Discuss with a partner one or two other accomplishments credited to each person on your list.

Gauss

CHAPTER INVESTIGATION

Many people have contributed to the modern understanding of geometry. By studying the work of early mathematicians, you can develop a better understanding of mathematical concepts.

Working Together

Design a timeline of significant mathematicians who made contributions to geometry. Use the timeline to understand how mathematics influenced life throughout history. Use the Chapter Investigation icons to guide you to a complete timeline.

Pascal

10 Are You Ready?

Refresh Your Math Skills for Chapter 10 · · · · · · · · · · ·

The skills on these two pages are ones you have already learned. Use the examples to refresh your memory and complete the exercises. For additional practice on these and more prerequisite skills, see pages 576–584.

PERIMETER FORMULAS

In this chapter you will work with three-dimensional solids. Formulas that apply to two-dimensional shapes can be a foundation for understanding solid geometry.

Perimeter of a square	Perimeter of a rectangle	Perimeter of any polygon
$P = 4s$, where s = side length	$P = 2l + 2w$, where l = length and w = width	P = sum of the lengths of the sides

Examples

Find the perimeter of each figure.

$$P = 2(8) + 2(5)$$
$$P = 16 + 10$$
$$P = 26 \text{ cm}$$

5 cm
8 cm

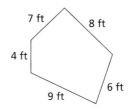
7 ft 8 ft
4 ft
9 ft 6 ft

$$P = 4 + 7 + 8 + 6 + 9$$
$$P = 34 \text{ ft}$$

Find the perimeter of each figure.

1.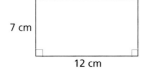
7 cm
12 cm

2.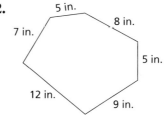
5 in.
8 in.
7 in.
5 in.
12 in.
9 in.

3.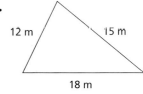
12 m 15 m
18 m

4.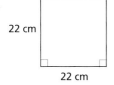
22 cm
22 cm

5.
6.5 m 6.5 m
6.5 m 6.5 m
6.5 m

6.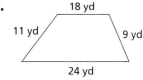
18 yd
11 yd 9 yd
24 yd

AREA FORMULAS

Area of a rectangle or square	Area of a parallelogram	Area of a triangle
$A = l \cdot w$, where l = length and w = width	$A = b \cdot h$, where b = length of the base and h = height	$A = \frac{1}{2}bh$, where b = length of the base and h = height

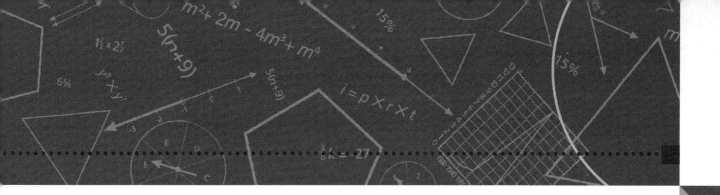

Examples

Find the area of each figure.

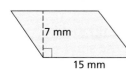
7 mm
15 mm

$A = 15 \cdot 7$

$A = 105 \text{ mm}^2$

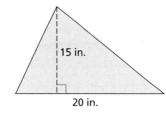
15 in.
20 in.

$A = \frac{1}{2}(20)(15)$

$A = \frac{1}{2}(300)$

$A = 150 \text{ in.}^2$

Find the area of each figure.

7.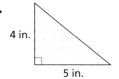
4 in.
5 in.

8.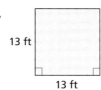
13 ft
13 ft

9.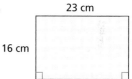
23 cm
16 cm

10.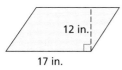
12 in.
17 in.

11.
32 m
45 m

12.
15 cm
9 cm

VOLUME FORMULAS

Volume of a rectangular prism	**Volume of a triangular prism**
$V = B \cdot h$, where B = area of the base and h = height	$V = B \cdot h$, where B = area of the base and h = height

Examples

Find the area of the base and substitute that measure to find the volume.

8 cm
3 cm
4 cm

$B = 3 \cdot 4$

$B = 12$

$V = 12 \cdot 8$

$V = 96 \text{ cm}^3$

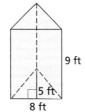
9 ft
5 ft
8 ft

$B = \frac{1}{2}(5)(8)$

$B = \frac{1}{2}(40)$

$B = 20$

$V = 20 \cdot 9 = 180 \text{ ft}^3$

Find the volume of each figure.

13.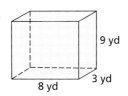
9 yd
8 yd
3 yd

14.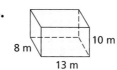
8 m
10 m
13 m

15.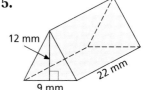
12 mm
9 mm
22 mm

10-1 Visualize and Represent Solids

Goals
- Identify properties of three-dimensional figures.
- Visualize three-dimensional geometric figures.

Applications   Packaging, History, Sports, Machinery, Recreation

Use a pencil, paper, tape and scissors.

1. Cut a small rectangle, a small right triangle and a small semicircle out of the paper.

2. Tape the rectangle near the top of the pencil. Spin the pencil on the tip to visualize what solid is created by spinning the rectangle. Describe the solid.

3. Repeat Question 2 for the right triangle.

4. Repeat Question 2 for the semicircle.

BUILD UNDERSTANDING

Some three-dimensional figures have both flat and curved surfaces.

A **cylinder** has a curved surface and two congruent circular bases that lie in parallel planes. The *axis* is a segment that joins the centers of the bases. If the axis forms a right angle with the bases, it is a *right* cylinder. If not, it is an *oblique* cylinder.

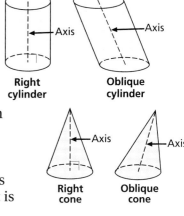

A **cone** has a curved surface and one circular base. The *axis* is a segment that joins the vertex to the center of the base. If the axis forms a right angle with the base, it is a *right* cone. Otherwise, it is an *oblique* cone.

A **sphere** is the set of all points that are a given distance from a given point, called the **center** of the sphere.

A **polyhedron** (plural: *polyhedra*) is a closed, three-dimensional figure made up of polygonal surfaces. The polygonal surfaces are **faces**. Two faces meet, or intersect, at an **edge**. A point at which three or more edges intersect is a **vertex**.

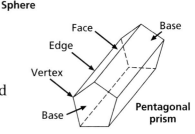

A **prism** is a polyhedron with two identical parallel faces called *bases*. The other faces are parallelograms. A prism is named according to the shape of its base.

A **pyramid** is a polyhedron with only one base. The other faces are triangles that meet at a **vertex**. A pyramid is named by the shape of its base.

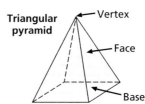

In prisms and pyramids, the faces that are not bases are *lateral faces*. The edges of these faces are *lateral edges*. Lateral faces can be either parallel or intersecting. The lateral edges can be intersecting or parallel. *Skew* edges are noncoplanar.

Check Understanding

Why is a cube also a rectangular prism?

Right rectangular prism

Oblique rectangular prism

Right square pyramid

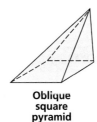

Oblique square pyramid

Example 1

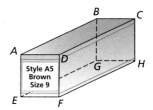

PACKAGING Identify each feature of the shoe box.

a. the shape of the shoe box

b. a pair of bases

c. a pair of parallel edges

d. a pair of intersecting edges

e. a pair of skew edges

f. a pair of intersecting faces

Solution

a. The shoe box is a right rectangular prism.

b. Two bases are *ADFE* and *BCHG*.

c. $\overline{DC}$ and $\overline{FH}$ are parallel edges.

d. $\overline{GH}$ and $\overline{FH}$ are intersecting edges.

e. $\overline{AD}$ and $\overline{CH}$ are skew edges.

f. *ABCD* and *CDFH* are intersecting faces.

Example 2

Draw a right triangular prism.

Solution

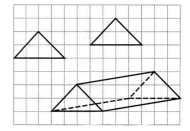

Step 1 Draw two congruent triangles.

Step 2 Draw segments that connect the corresponding vertices of the triangles. Use dashed segments to show the edges that cannot be seen.

◤ TRY THESE EXERCISES

Identify each figure and name its base(s).

1.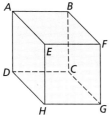

2.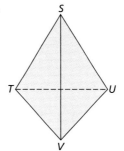

If possible, identify a pair of parallel faces, parallel edges, intersecting faces, and intersecting edges for each figure.

3. the figure in Exercise 1

4. the figure in Exercise 2

Draw each figure.

5. right circular cone

6. oblique rectangular prism

7. right hexagonal prism

8. right square pyramid

■ **PRACTICE EXERCISES** • **For Extra Practice, see page 616.**

Identify each figure and name its base(s).

9.
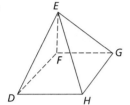

10.
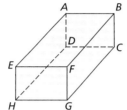

11.
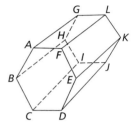

Identify a pair of parallel faces, parallel edges, intersecting faces, and intersecting edges for each figure.

12. the figure in Exercise 9

13. the figure in Exercise 10

14. the figure in Exercise 11

Draw each figure.

15. right rectangular prism

16. oblique square pyramid

17. oblique cone

18. right cylinder

19. sphere

20. right triangular prism

21. WRITING MATH Describe ways in which prisms and pyramids are similar. Describe differences between prisms and pyramids.

RECREATION If possible, identify the geometric figure represented and give the number of faces, vertices and edges.

22. tennis ball can

23. six-sided number cube

24. basketball

25. ice cream cone

26. rectangular gym

27. one-person tent

28. Which of the figures in Exercises 22–27 are polyhedra? Explain.

MACHINERY A cutting drill uses different bits to cut designs into furniture. Describe the geometric figure cut when using each of the following bits.

29.

30.

31.

32. YOU MAKE THE CALL Niki says that a right square pyramid is a polyhedron but an oblique square pyramid is not a polyhedron. Daniella says both figures are polyhedra since their faces are polygonal. Who is correct and why?

Decide whether each statement is *true* or *false*. Explain.

33. A cube is a polyhedron.

34. A polyhedron can have exactly three faces.

35. A polyhedron can have a circular face.

36. A polyhedron can have exactly four faces.

State whether each solid is a polyhedron. Explain.

37.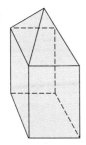

38.

39.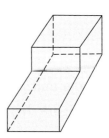

40. **HISTORY** The pyramids of Giza in Egypt are one of the seven wonders of the ancient world. Use the photo on page 433 to sketch the pyramid on paper. Then label the vertices.

EXTENDED PRACTICE EXERCISES

SPREADSHEET Create a spreadsheet for the following figures. List figures in Column A, the number of faces (*F*) in Column B, the number of vertices (*V*) in Column C, and the number of edges (*E*) in Column D. In Column E, write a formula to calculate the quantity $F + V - E$.

41. rectangular prism

42. hexagonal prism

43. pentagonal pyramid

44. square pyramid

45. Compare your results for Exercises 41–44. Make a generalization about the relationship between the numbers of faces, edges and vertices in a polyhedron.

46. **CHAPTER INVESTIGATION** Research ten mathematicians. Record the name, accomplishment, dates of birth and death, date of accomplishment and country of birth for each mathematician.

MIXED REVIEW EXERCISES

Find the slope of the line containing the given points. Identify any vertical or horizontal lines. (Lesson 6-2)

47. (4, 2), (6, 1)

48. (9, 3), (−2, 4)

49. (−1, 3), (2, 8)

50. (5, −4), (7, 7)

51. (−3, −5), (−3, −2)

52. (4, −5), (−2, 3)

53. (−2, 6), (4, 6)

54. (−1, −1), (−7, −7)

In the figure, $\overleftrightarrow{RS} \parallel \overleftrightarrow{MN}$ and $m\angle 4 = 80°$. Find each measure. (Lesson 5-3)

55. $m\angle 1$

56. $m\angle 2$

57. $m\angle 3$

58. $m\angle 5$

59. $m\angle 6$

60. $m\angle 7$

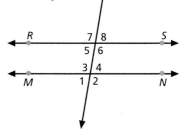

Nets and Surface Area

Goals
- Draw nets for three-dimensional figures.
- Use nets to find the surface area of polyhedra.

Applications Retail, History, Machinery, Architecture

Use grid paper and scissors.

1. Copy the figure shown on grid paper, cut it out and fold along the dotted lines.

2. What type of polyhedron is formed?

3. Draw a different arrangement of the six squares so when you cut and fold the figure it makes the same solid as the one created in Question 2.

4. Test the figure drawn in Question 3.

5. Discuss with your classmates those figures that did create the same solid and those that did not.

◤ BUILD UNDERSTANDING

A **net** is a two-dimensional pattern that can be folded to form a three-dimensional figure. A three-dimensional figure can have more than one net.

Example 1

What three-dimensional figure is represented by each net?

a. 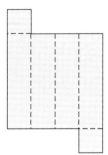 b. 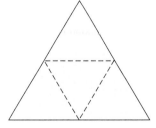 c.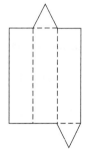

Solution

Visualize each net being folded to form a three-dimensional figure.

a. This figure will have four rectangular sides and two parallel square bases. It is a rectangular prism.

b. This figure will have three triangular sides and a triangular base. It is a triangular pyramid.

c. This figure will have three rectangular sides and two parallel triangular bases. It is a triangular prism.

Example 2

Draw a net for each three-dimensional figure.

a. right square pyramid

b. right cylinder

Problem Solving Tip

If you are unable to figure out the three-dimensional figure a net represents, copy the net, cut it out and fold it to find the answer.

Solution

a. 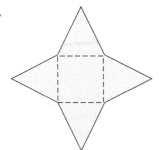   **b.**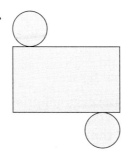

Nets can be used to calculate the surface area of certain three-dimensional figures. The **surface area** of a figure is the sum of the areas of all its bases and faces.

Example 3   ⊕nline **Personal Tutor at** mathmatters2.com

RETAIL Find the minimum amount of gift wrapping paper needed to cover a box with dimensions 1 ft by 2 ft by 3 in.

Solution

Draw the net for the rectangular prism, and calculate the areas of the six sides. Use the formula $A = l \cdot w$.

The top and bottom of the box are rectangles with dimensions 1 ft by 2 ft.

$$l \cdot w + l \cdot w$$
$$2 \cdot 1 + 2 \cdot 1 = 4 \text{ ft}^2$$ Find the combined surface area.

The left and right sides of the box are rectangles with dimensions 2 ft by 3 in.

$$3 \text{ in.} \cdot \frac{1 \text{ ft}}{12 \text{ in.}} = 0.25 \text{ ft}$$ Convert 3 in. to feet.

$$l \cdot w + l \cdot w$$ Find the combined surface area.
$$2 \cdot 0.25 + 2 \cdot 0.25 = 1 \text{ ft}^2$$

The front and back of the box are rectangles with dimensions 1 ft by 3 in. Use 0.25 ft for 3 in.

$$l \cdot w + l \cdot w$$
$$1 \cdot 0.25 + 1 \cdot 0.25 = 0.5 \text{ ft}^2$$

The total surface area is the sum of the areas of the sides.

$$4 + 1 + 0.5 = 5.5 \text{ ft}^2$$

So the box will require at least 5.5 ft² of gift wrapping paper.

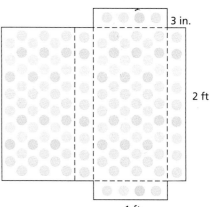

3 in.

2 ft

1 ft

Check Understanding

In Example 3, why are you asked to find the minimum amount of wrapping paper needed for the box?

Identify the three-dimensional figure for each net.

1.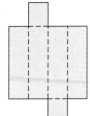

2.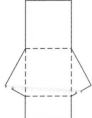

3.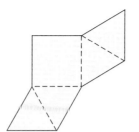

4.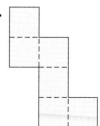

Draw a net for each three-dimensional figure.

5. cube

6. pentagonal prism

7. rectangular prism

Draw a net for each figure. Then find the surface area. Round each answer to the nearest tenth. Use 3.14 for π.

8.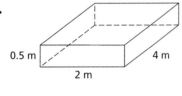

9.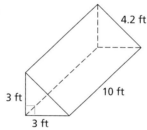

10.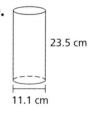

Identify the three-dimensional figure for each net.

11.

12.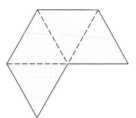

13.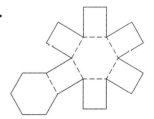

Draw a net for each three-dimensional figure.

14. octagonal prism

15. triangular prism

16. right cone

17. HISTORY Design and draw a Native American tepee. Then draw a net for the tepee.

18. Draw a net of a cube that has a side length of 6 m. Then find the surface area of the cube.

19. ARCHITECTURE What is the total surface area of the glass (including the roof) used to build the greenhouse? Draw a net of the walls and the roof to solve the problem.

Find the area of each net. Use 3.14 for π. Round to the nearest tenth if necessary.

20.

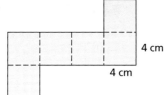

4 cm

4 cm

21.

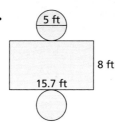

5 ft

8 ft

15.7 ft

22. The height of each triangle is $\frac{5\sqrt{3}}{3}$.

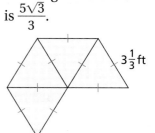

$3\frac{1}{3}$ ft

23. WRITING MATH How can nets help you find the surface area of three-dimensional figures? What kinds of mistakes do you make when calculating surface area?

◢ EXTENDED PRACTICE EXERCISES

Use nets to find the surface area of each prism. Use 3.14 for π.

24.

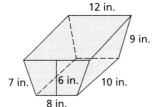

12 in.

9 in.

7 in.   6 in.   10 in.

8 in.

25.

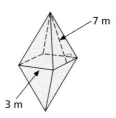

7 m

3 m

26.

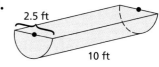

2.5 ft

10 ft

27. MACHINERY The wheel of a steamroller is a cylinder. If the wheel's diameter is 5 ft and its width is 7.2 ft, approximately how much area is covered by one complete wheel revolution? Use a net to visualize and solve the problem. Round the answer to the nearest whole number.

28. CRITICAL THINKING There are 11 different nets for a cube. Draw as many of them as you can.

29. Refer to exercise 28. To construct 20 cubes out of construction paper, which of the 11 nets would you use? Explain.

◢ MIXED REVIEW EXERCISES

Simplify each numerical expression. (Lesson 2-2)

30. $4 + 3 - 6 \cdot 4$

31. $8 \cdot 6 + 4^2 - 3$

32. $2 + (5 - 2)^2 \div 4$

33. $15 \div 3 + (7 - 3) \cdot 2$

34. $4 + 8(3 + 4) \div 2$

35. $(6 + 1)^2 - 15 \cdot 3 + 4$

Evaluate each expression when $c = 11$. (Lesson 2-2)

36. $18 - c$

37. $3(c + 8)$

38. $(c - 8)^2 + 4$

39. $9 + c(3c - 2)$

Simplify. (Lesson 9-2)

40. $(4r)(9p)$

41. $(6r^2)(4r^3g^2)$

42. $(-3a^2)(5ab^2)^2$

43. $(5c^3d)(2cd^4)$

44. $(-4k^4f^2)(-6kf^6)$

45. $(3s^3t^2)(-6st^3q)$

Math Online mathmatters2.com/self_check_quiz

Review and Practice Your Skills

Choose a word from the list to complete each statement.

1. A ___?___ has one square base and triangular faces.

2. A ___?___ has a curved surface and one circular base.

3. A ___?___ has two congruent circular bases that lie in parallel planes between a curved surface.

4. A ___?___ is a figure consisting of the set of all points in space that are a given distance from a given point.

5. A ___?___ has a triangular base and triangular faces.

6. A ___?___ has two triangular bases parallel to each other with the faces that are parallelograms.

> **a.** cone
>
> **b.** cylinder
>
> **c.** sphere
>
> **d.** square pyramid
>
> **e.** triangular prism
>
> **f.** triangular pyramid

Identify each figure. State the number of faces, vertices and edges of each.

7.

8.

9.

Identify the three-dimensional figure for each net.

10.

11.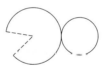

12.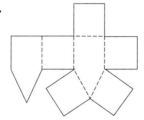

Draw a net for each three-dimensional figure.

13.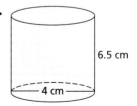
6.5 cm
4 cm

14.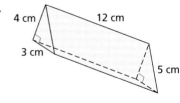
4 cm   12 cm
3 cm   5 cm

15.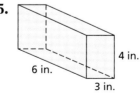
4 in.
6 in.
3 in.

Find the surface area of each figure.

16. the figure in Exercise 13

17. the figure in Exercise 14

18. the figure in Exercise 15

Draw a net for each three-dimensional figure. (Lesson 10-2)

19.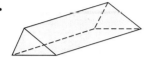

20.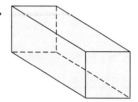

21.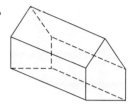

22. Draw a triangular pyramid.

23. How many vertices, faces and edges does a triangular pyramid have?

Match each pair of edges of the cube with an item in the box.
(Lesson 10-1)

24. $\overline{AD}$ and $\overline{CG}$

25. $\overline{BC}$ and $\overline{EH}$

26. $\overline{CG}$ and $\overline{GH}$

a. parallel edges

b. skew edges

c. perpendicular edges

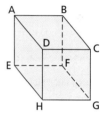

MathWorks
Workplace Knowhow

Career – Urban Planner

An urban planner designs and supervises the construction of roads, buildings, tunnels, bridges, and water supply systems. The city of Berlin, Germany, changed rapidly following the removal of the Berlin Wall. New construction was needed to reconstruct the city and bring its buildings to modern standards. An urban planner in Berlin must decide what types of buildings are most appropriate for the needs of the city while maintaining the historical look and structure. Suppose an urban planner is beginning the design process for a new government building in Berlin.

1. The shape of this new building is a right hexagonal prism. Make a sketch of the new structure, drawing the regular hexagons at the base and the top of the building.

2. Draw a net (to scale) of the building. The sides of the regular hexagonal bases should have a length of 30 m, and the height of the building should be 60 m.

3. What is the total surface area of the walls?

4. How many marble tiles will be required to cover the outside walls if the tiles are 60 cm wide and 10 cm tall?

10-3 Surface Area of Three-Dimensional Figures

Goals ■ Find the surface area of three-dimensional figures.

Applications Geography, Food service, Packaging, History

Use a box such as a cereal box or a shoe box for the activity.

1. Carefully tear the box apart at the edges. Do not keep the tabs that were used to glue the faces of the box together.

2. How many polygons do you now have? Describe them.

3. Use a ruler to measure. Then calculate the area of each of the polygons.

4. Add the area of all the polygons. What does the total area represent?

■ BUILD UNDERSTANDING

In the previous lesson, you learned that the surface area of a polyhedron is the sum of the areas of all its bases and faces. You saw how the net of a three-dimensional figure can be used to find the total surface area. In this lesson, you will use formulas to find the surface area of three-dimensional figures.

Example 1

Find the surface area of the rectangular prism.

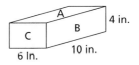

Solution

A rectangular prism has three pairs of congruent faces. The surface area is the sum of the areas of all the faces. Use the formula $SA = 2(\text{area } A) + 2(\text{area } B) + 2(\text{area } C)$, where each area can be found using the formula for the area of a rectangle, $A = lw$.

area of A $A = 10 \cdot 6$
$A = 60$

area of B $A = 10 \cdot 4$
$A = 40$

area of C $A = 6 \cdot 4$
$A = 24$

$SA = 2 \cdot (\text{area } A) + 2 \cdot (\text{area } B) + 2 \cdot (\text{area } C)$

$SA = 2 \cdot 60 + 2 \cdot 40 + 2 \cdot 24$

$SA = 248$

The surface area is 248 in.^{2}.

> **Check Understanding**
>
> How is using the surface area formula similar to using nets to find the surface area of a three-dimensional figure?

A square pyramid with four congruent triangular faces is called a **regular square pyramid**. If you know the dimensions of just one triangular face, you can calculate the surface area of the pyramid.

Example 2

GEOGRAPHY The Great Pyramid at Giza in Egypt has a square base approximately 756 ft long. The height of each triangular face, called the slant height, is approximately 612 ft. What is the approximate surface area of the faces of the Great Pyramid?

Solution

The Great Pyramid is a regular square pyramid, so its faces are four congruent triangles. Calculate the surface area of one of the faces and multiply by 4.

$A = \frac{1}{2}bh$ Use the formula for the area of a triangle.

$A = \frac{1}{2} \cdot 756 \cdot 612 = 231{,}336$

$SA = 4 \cdot 231{,}336 = 925{,}344$

The surface area of the faces of the Great Pyramid at Giza is approximately 925,344 ft².

A cylinder is a figure with two congruent circular bases and a curved surface. To find the surface area, add the areas of the bases to the area of the curved surface.

Surface Area of a Cylinder	$SA = 2\pi rh + 2\pi r^2$ where r is the radius of a base and h is the height of the cylinder.

Example 3 $\bigcirc$nline **Personal Tutor at** mathmatters2.com

Find the surface area of the cylinder.

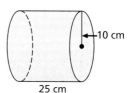

Solution

$SA = 2\pi rh + 2\pi r^2$

$SA \approx 2 \cdot 3.14 \cdot 10 \cdot 25 + 2 \cdot 3.14 \cdot 10^2$ $\pi \approx 3.14$

$SA \approx 1570 + 628 \approx 2198$

The surface area of the cylinder is approximately 2198 cm².

A cone is a three-dimensional figure with a curved surface and one circular base. The height of a cone is the length of a perpendicular segment from its vertex to its base. The **slant height** (s) of a cone is the length of a segment from its vertex to its base along the side of the cone.

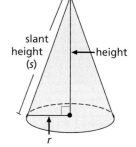

Surface Area of a Cone	$SA = \pi rs + \pi r^2$ where r is the radius of the base and s is the slant height.

Example 4

online Personal Tutor at mathmatters2.com

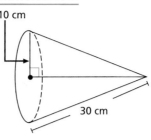

Find the surface area of the cone.

Solution

$SA = \pi rs + \pi r^2$

$SA \approx 3.14 \cdot 10 \cdot 30 + 3.14 \cdot 10^2 \qquad \pi \approx 3.14$

$SA \approx 942 + 314$

$SA \approx 1256$

The surface area of the cone is approximately 1256 cm^2.

◢ TRY THESE EXERCISES

Find the surface area of each figure. Round to the nearest hundredth.

1.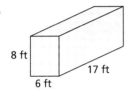

2.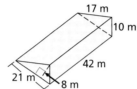

3.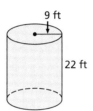

4.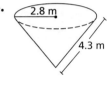

5.

6.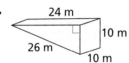

7.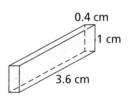

8.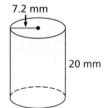

◢ PRACTICE EXERCISES • For Extra Practice, see page 617.

Find the surface area of each figure. Round to the nearest hundredth.

9.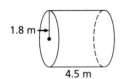

10.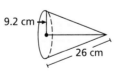

11.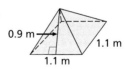

12.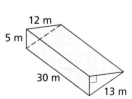

13.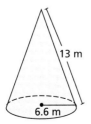

14.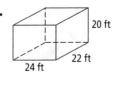

15.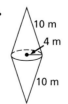

16.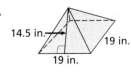

17. The base of each triangular face of a regular square pyramid is 9 in., and the height of each triangular face is 11 in. Find the surface area of the pyramid.

18. **PACKAGING** Tamara is wrapping a gift. The box is 14 in. by 9 in. by 8 in. What is the minimum amount of wrapping paper she will need to cover the box?

19. **WRITING MATH** Explain why the word "minimum" is used in Exercise 18.

20. **FOOD SERVICE** Find the surface area of a can of cranberry sauce that has a height of 4.2 in. and a radius of 1.5 in.

21. **SPREADSHEET** Use a spreadsheet to calculate the surface area of cylinders. Enter the formula for the surface area of a cylinder into the first cell. Then evaluate the formula for a cylinder that is 12 cm long with a radius of 3 cm, 6 cm, 12 cm, and 24 cm.

22. **HISTORY** An ancient Mayan pyramid has a square base. Each side of the base measures 230 m, and the slant height is 179 m tall. Find the area of the faces of the Mayan pyramid.

◤ EXTENDED PRACTICE EXERCISES

Find the surface area of each figure.

23.
24.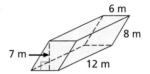
25.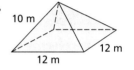

26. **CRITICAL THINKING** A cone has a height h and a radius r. Find an expression for the slant height. Then write a formula for the surface area of the cone using this expression.

27. A hole is drilled through a solid cube that has edges of 4 cm. The hole is drilled perpendicular to the top face of the cube. The diameter of the hole is 2 cm. To the nearest whole number, what is the total surface area of the resulting solid figure?

28. **DATA FILE** Refer to the data on U.S. shopping centers on page 570. What are possible dimensions of a rectangular prism whose surface area is equal to the area of Tysons Corner Center?

◤ MIXED REVIEW EXERCISES

Write the equation of each line using the given information. (Lesson 6-3)

29. $m = 3$, $(1, 4)$

30. $m = 2$, $(-2, -3)$

31. $(2, 7)$, $(3, 9)$

32. $(0, -2)$, $(2, -4)$

Find each product. (Lesson 9-5)

33. $(r + 3)(r - 4)$

34. $(z - 1)(z + 4)$

35. $(b - 4)^2$

36. $(f - 2)(f - 5)$

37. $(v + 4)(v - 8)$

38. $(w - 1)(w - 7)$

39. $(3k - 1)(2k + 4)$

40. $(2c + 3)^2$

10-4 Perspective Drawings

Goals
- Make one- and two-point perspective drawings.
- Locate the vanishing points of perspective drawings.

Applications Art, Interior design, History, Architecture

Use the three photographs for Questions 1–3.

1. Can you see depth in each of the photographs?

2. Do you think the photographer was located above, below or at eye level with the subject?

3. Does your eye travel to one point or two points in the background of each photograph?

◣ BUILD UNDERSTANDING

A **perspective drawing** is a way of drawing objects on a flat surface so that they look the same as they appear in real life. The eye and the camera are both constructed so that objects appear progressively smaller the farther away they are. Parallel lines drawn in perspective appear to come together in the distance.

Perspective drawings use vanishing points. A **vanishing point** is a point that lies on the horizon line. The horizon line is a line in the distance where parallel lines appear to come together. Perspective drawings can be made in either **one-point perspective** or **two-point perspective**.

Example 1

Draw a cube in one-point perspective.

Solution

Step 1 Draw a square to show the front surface of the cube. Draw a horizon line *j* and a vanishing point *A* on line *j*.

Step 2 Connect the vertices of the square to the vanishing point.

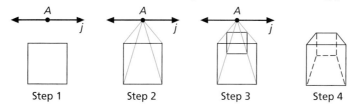

Step 1   Step 2   Step 3   Step 4

Step 3 Draw a smaller square whose vertices touch the four line segments.

Step 4 Connect the vertices of the two squares. Use a dashed segment to indicate the edges of the cube hidden from view. Remove line *j* and point *A*.

A **two-point perspective** drawing has two vanishing points. If you are looking at the corner of an object in a perspective drawing, then it is probably a two-point perspective drawing.

Math: Who, Where, When

Ancient Greeks and Romans developed techniques for drawing objects in perspective. Renaissance artists in the early 1400s used perspective drawings to create lifelike works.

Example 2

ARCHITECTURE Draw a building shaped like a rectangular prism in two-point perspective.

Solution

Step 1 Use a straightedge to draw a vertical line segment to represent the height of the building. Draw a vanishing point (VP) on each side of the segment. Sketch depth lines from the top and bottom of the segment to each point.

Step 2 Draw two segments parallel to the original segment to represent the height. Draw two segments to complete the top of the building.

Step 3 Draw two more depth lines from the top corners to the vanishing points.

Step 4 Erase the unused portions of the depth lines to complete the two-point perspective drawing.

Example 3

Locate the vanishing point of the perspective drawing.

Solution

To locate the vanishing point, use a straightedge to draw the depth lines from the top edges of the figure. The point of their intersection is the vanishing point of the perspective drawing.

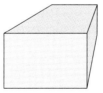
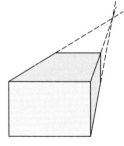

◥ TRY THESE EXERCISES

Sketch each object in one-point perspective.

1. cube 2. triangular prism 3. rectangular prism 4. cylinder

5. your first name written in capital block-style letters

Sketch each object in two-point perspective.

6. cube 7. shoe box 8. table

Trace each perspective drawing to locate the vanishing point(s).

9.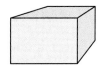

10.

11.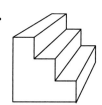

PRACTICE EXERCISES • For Extra Practice, see page 617.

Sketch each object in one-point perspective.

12. window

13. wall of a room

14. couch

Sketch each object in two-point perspective.

15. television

16. street sign

17. school bus

Trace each perspective drawing to locate the vanishing point(s).

18.

19.

20.

21. **WRITING MATH** Describe the location of the vanishing point when a viewer can see three sides of a box drawn in one-point perspective.

22. **HISTORY** The Sears Tower in Chicago is the tallest building in the U.S. Use a picture of the Sears Tower to make the one-point perspective drawing.

23. **INTERIOR DESIGN** Make a one-point perspective drawing of the square tile pattern of a floor.

A square is cut out of cardboard, and a flashlight shines behind it. Decide if it is possible to obtain the indicated shadow.

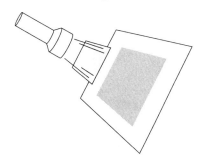

24. square

25. segment

26. triangle

27. point

28. trapezoid

29. kite

30. pentagon

31. parallelogram

32. ARCHITECTURE Describe the perspective used by the architect to create the illusion of a three-dimensional house in her plans.

33. Trace the house on a piece of paper. Locate a vanishing point.

34. GEOMETRY SOFTWARE Use geometry software to make a one- or two-point perspective drawing. Begin by drawing a figure such as a rectangle and locating one or two vanishing points. Sketch the depth lines, and use them to complete the perspective drawing.

■ EXTENDED PRACTICE EXERCISES

Determine the number of vanishing points used to draw each cube.

35.

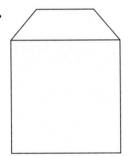

36.

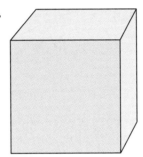

37.

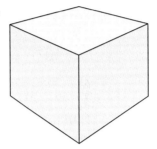

38. RESEARCH Use your school's library or another resource to find an example of a painting that uses one- and two-point perspectives. Identify the vanishing points used by the artists.

39. CRITICAL THINKING Make a two-point perspective drawing of a rectangular box with a square window cut out of one of the faces. Show that the box has depth or thickness.

40. CHAPTER INVESTIGATION Draw a line segment about 10 in. long across the center of a piece of paper with an arrowhead at each end. Place on the timeline in chronological order the name and accomplishment of each mathematician you researched. Label appropriately. Be sure the distances between the years recorded on your timeline are in proportion.

■ MIXED REVIEW EXERCISES

State if it is possible to have a triangle with sides of the given lengths. (Lesson 5-4)

41. 8, 5, 4   **42.** 10, 6, 4   **43.** 12, 7, 6   **44.** 9, 5, 5

45. 16, 14, 12   **46.** 11, 8, 5   **47.** 7, 7, 2   **48.** 12, 6, 5

Factor each polynomial. (Lesson 9-7)

49. $15z^2 + 18z$

50. $35r^2 + 60r$

51. $8y^5 - 24y^4 + 28$

52. $6r^3s + 9r^3s^2 - 18r^2s^2$

53. $28b^3c - 84b^2c^5$

54. $24wx^3 - 12w^2x^2 + 20w^3x^3$

55. $24g^4h - 40g^2h^2 - 16g^5$

56. $30a^3b^2 + 45a^2b^3 - 60a^3b^3$

Review and Practice Your Skills

Find the surface area of each figure.

1.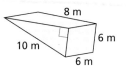

2.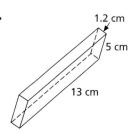

3.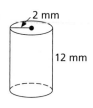

4.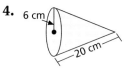

5.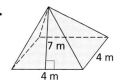

6.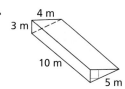

7. A regular square pyramid has a base area of 100 cm² and a height of 12 cm. What is the surface area? (Hint: Find the slant height *s*.)

8. A can of peaches has a height of 5 in. and a diameter of 3 in. How much metal is needed to make the can?

Sketch each object in one-point perspective.

9. railroad track   **10.** house   **11.** door

Sketch each object in two-point perspective.

12. stereo speaker   **13.** grandfather clock   **14.** building

Trace each perspective drawing to locate the vanishing point(s).

15.

16.

17.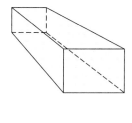

18. Explain the difference between a one-point perspective drawing and a two-point perspective drawing.

Draw a net for each three-dimensional figure. (Lesson 10-2)

19.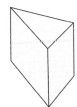

20.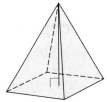

21.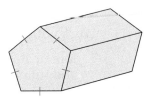

Refer to the figure. (Lesson 10-1)

22. How many vertices does the figure have?

23. How many faces does the figure have?

24. Name the figure.

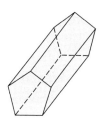

Mid-Chapter Quiz

Identify each figure and name its base(s). (Lesson 10-1)

1.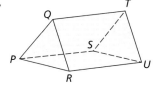

2.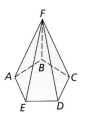

3.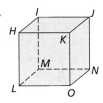

Identify the three-dimensional figure for each net. (Lesson 10-2)

4.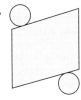

5.

6.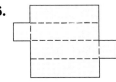

Find the surface area of each figure. (Lesson 10-3)

7.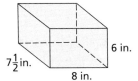

8.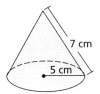

9.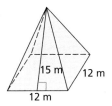

Sketch each object in one-point perspective. (Lesson 10-4)

10. pentagonal prism

11. building

12. highway

10-5 Isometric Drawings

Goals ■ Visualize and represent objects with isometric drawings.

Applications Art, Recreation, History, Architecture

Work with a partner. Use several cubes, preferably the kind that attach to each other.

1. Join or stack the cubes together to build a structure. Hold the structure in your hand, and rotate it to obtain different views.

2. Describe the structure to your partner. Which single view is easiest to describe?

3. Make a new structure. Describe it to your partner so that it is clear how to build the same structure.

4. Have your partner build the new structure based on your description.

◤ BUILD UNDERSTANDING

In the last lesson, you learned how to reproduce three-dimensional figures on paper using perspective drawings. Another way to show a three-dimensional object is with an isometric drawing. An **isometric drawing** shows an object from a corner view so that three sides of the object can be seen in a single drawing.

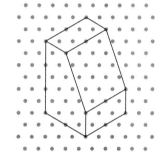

The figure shows an isometric drawing of a structure. In an isometric drawing, all parallel edges of the structure are shown as parallel line segments in the drawing. Perpendicular line segments do not necessarily appear perpendicular.

Example 1

Make an isometric drawing of a cube.

Solution

Step 1 Begin by drawing a vertical line, which will be the front edge of the cube. Then draw the right and left edges of the cube.

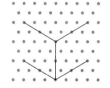

Step 2 Draw the vertical sides of the cube parallel to the front edge. Complete the isometric drawing by sketching the rear edges of the cube.

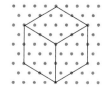

> **Problem Solving Tip**
>
> The length of all edges in an isometric drawing are either the true length or a scaled length. Angle measurements are not preserved in isometric drawings.

Example 2

Make an isometric drawing of a figure containing three cubes that form an L shape.

Solution

Step 1 Begin by drawing the left edges of the figure. Then draw the segments for the right edges of the figure. These segments are parallel to the edges of the left side and shown here in blue.

Step 2 Complete the isometric drawing by sketching the five remaining segments that outline the cubes.

Example 3

Use the isometric drawing to answer the questions.

a. How many cubes are used in the drawing?

b. How many cube faces are in the figure?

c. If each cube face represents 3.5 ft², what is the total surface area of the figure?

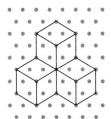

Solution

a. Even though there are only three visible cubes, the figure contains four cubes. The top cube is resting on a cube beneath it.

b. To count the total number of cube faces, proceed in a logical manner so that you do not miss any sides.

   Cube faces pointing up = 3

   Cube faces pointing to the front = 6

   Cube faces pointing to the back = 6

   Cube faces on the bottom = 3

   There are a total of 3 + 6 + 6 + 3 = 18 cube faces on the figure.

c. Since there are 18 cube faces and each face represents 3.5 ft², the total surface area of the figure is 18 · 3.5 ft² = 63 ft².

TRY THESE EXERCISES

Create an isometric drawing of each figure.

1. cube

2. rectangular prism

3. triangular prism

4. staircase composed of cubes

5. **ARCHITECTURE** Using isometric grid paper, design a building using a rectangular prism for the base and a triangular prism for the roof.

6. Use isometric grid paper to draw three different rectangular solids using a total of eight cubes each.

Use the isometric drawing for Exercises 7–9.

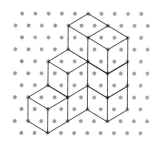

7. How many cubes are used in the drawing? How many are hidden?

8. How many cube faces are exposed in the figure?

9. If each face represents 1 yd², what is the total surface area of the figure?

◥ PRACTICE EXERCISES • For Extra Practice, see page 618.

Create an isometric drawing of each figure.

10. hexagonal prism

11. figure having 14 faces

12. pentagonal prism

13. figure composed of 11 cubes

14. capital block-style letter M

15. cube on top of a rectangular prism

16. Use isometric grid paper to make a sketch of the letters of your first name.

17. Use isometric grid paper to make a sketch of a rectangular prism with 18 faces.

Use the isometric drawing for Exercises 18–20. Assume that no cubes are hidden from view.

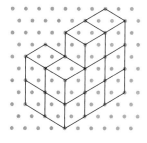

18. How many cubes are used in the drawing?

19. How many cube faces are exposed in the figure?

20. If the length of an edge of one of the cubes is 3.4 m, what is the total surface area of the figure?

Give the number of cubes used to make each figure and the number of cube faces exposed. Be sure to count the cubes that are hidden from view.

21.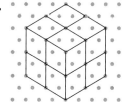

22.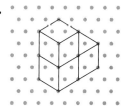

23.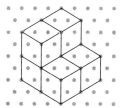

24.

25.

26.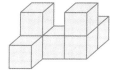

27. RECREATION Make an isometric drawing of a tent. Use a rectangular prism for the base and a triangular prism for the top of the tent.

28. WRITING MATH Explain the similarities and differences between regular square grid paper and isometric dot paper. Which do you prefer to use for drawing?

29. HISTORY Make an isometric drawing of the ancient Greek Parthenon.

30. YOU MAKE THE CALL Ken says that it is impossible to make an isometric drawing of a rectangular prism containing 6 cubes that has exactly 24 cube faces showing. Brandon says that you can do this by making one long stack of cubes. Who is correct and why?

The Parthenon; Athens, Greece

◣ EXTENDED PRACTICE EXERCISES

How many cubes are used to make each isometric figure? Be sure to count the cubes that are hidden from view.

31.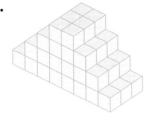

32.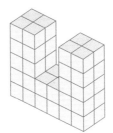

33.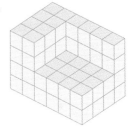

34. ART Isometric grid paper can be used to make artistic patterns and designs similar to the ones shown. Use isometric grid paper to create two different artistic designs. Color the designs.

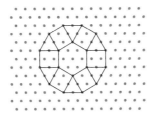   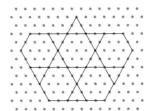

◣ MIXED REVIEW EXERCISES

Solve each system of equations graphically. (Lesson 8-2)

35. $3x - y = 10$
$x + y = 2$

36. $2x + y = 8$
$x - 3y = -10$

37. $x + 2y = 7$
$4x - 2y = -22$

38. $2x + 2y = -8$
$2x - 2y = 0$

39. $3x + 2y = 12$
$x - 3y = -7$

40. $x + 3y = 11$
$x - 2y = -7$

Show each sample space using ordered pairs. (Lesson 4-3)

41. A quarter and a dime are tossed.

42. A six-sided number cube is rolled and a dime is tossed.

43. A six-sided number cube is rolled and a spinner with five different colors is spun. (Use the numbers 1–5 for the colors on the spinner.)

44. Two spinners, one numbered 1–4 and the other lettered A–F, are spun.

10-6 Orthogonal Drawings

Goals
- ■ Sketch orthogonal drawings of figures.
- ■ Sketch and use foundation drawings.

Applications Engineering, Interior design, History, Safety

Use 11 cubes, preferably interlocking ones.

1. Label the four corners of a sheet of paper *A*, *B*, *C*, and *D*.

2. Build a structure by stacking the given number of cubes in each position as shown in the figure.

3. Match each figure below with corner views *A*, *B*, *C*, or *D*. Try to visualize the answer in your mind; then check the actual structure to verify.

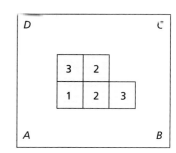

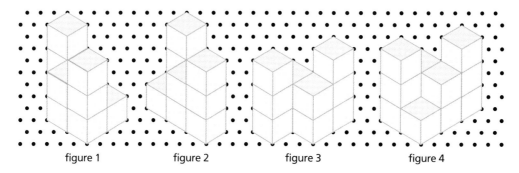

figure 1   figure 2   figure 3   figure 4

◣ BUILD UNDERSTANDING

While perspective drawings and isometric drawings are useful to help visualize a three-dimensional figure, they may not show all the details of an object that you need. An **orthogonal drawing**, or **orthographic drawing**, gives top, front, and side views of a three-dimensional figure as seen from a "straight on" viewpoint. In an orthogonal drawing, solid lines represent any edges that show.

Example 1

Make an orthogonal drawing of the figure.

Solution

Draw the front view first.

Draw the top view above it. Make sure it has the same width as the front view.

Draw the right-side view. Make sure it has the same height as the front view and the same depth as the top view.

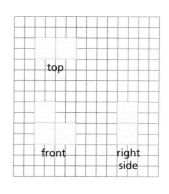

Example 2

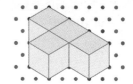

ENGINEERING The isometric drawing shows part of an air-conditioning duct.

a. Make an orthogonal drawing of the duct showing the front, top, and right-side views.

b. The length of one cube side is 1.6 m. The building engineer needs the surface area of the top to be between 10 m^2 and 10.5 m^2. Is the duct within the necessary requirements?

Solution

a. Think of the duct as a combination of four cubes. Then make each view of the orthogonal drawing.

b. Calculate the surface area of the top of the figure. It is composed of four cube faces.

$1.6 \cdot 1.6 = 2.56$ m^2 Calculate the area of one cube.

$2.56 \cdot 4 = 10.24$ m^2 Find the total surface area of the top.

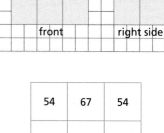

The duct is within the given requirements.

A **foundation drawing** shows the base of a structure and the height of each part. They are often used by architects and design engineers. The figure shows a foundation drawing of the Sears Tower in Chicago. Each number represents how many stories are in each section.

54	67	54
67	98	67
41	98	41

Example 3

Online **Personal Tutor at** mathmatters2.com

Create a foundation drawing for the isometric drawing. Assume the drawing is viewed from the lower left-hand corner.

Solution

First draw the orthogonal top view of the figure.

Then determine how many cubes belong in each section, and write the number to complete the foundation drawing.

TRY THESE EXERCISES

Make an orthogonal drawing labeling the front, top, and right-side views.

1.

2.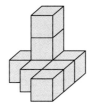

3.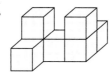

4.

The figure is a cube with a "half cube" on top of it. Tell which view of the figure is shown.

5.

6.

7.

Create a foundation drawing for each figure. Assume the drawing is viewed from the lower left-hand corner.

8.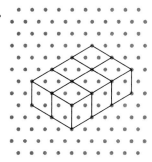

9.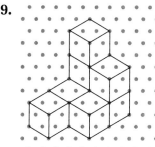

10.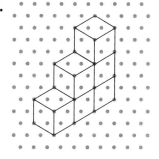

■ **PRACTICE EXERCISES** • For Extra Practice, see page 618.

Make an orthogonal drawing labeling the front, top, and right-side views.

11.

12.

13.

For each foundation drawing, sketch the front and right orthogonal views.

14.

5	5
3	1

front

15.

1	2	3
	2	1

front

16.

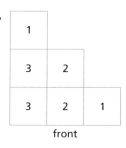

 17. **WRITING MATH** Choose an object and describe how the orthogonal drawings of the front, top, and right-side views are different or the same.

18. Make an orthogonal drawing of a regular six-sided number cube showing a view of each of the six sides.

19. **SAFETY** Most skyscrapers are built using steel girders shaped like the capital letter I. The shape avoids buckling and can support heavy loads. Make an orthogonal drawing of the girder showing the front, top and right-side views.

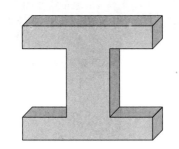

20. **WRITING MATH** Investigate the meaning of the word "orthogonal." Why do you think it is used to describe the "straight on" view of a figure?

21. **HISTORY** Make an orthogonal drawing of the Sears Tower in Chicago. (Refer to the foundation drawing of the Sears Tower found earlier in this lesson.)

◾ EXTENDED PRACTICE EXERCISES

Create an orthogonal drawing for each figure.

22.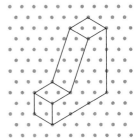

23.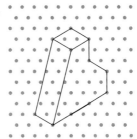

24.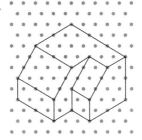

25. **CRITICAL THINKING** The drawing shows top and front views of a building. Draw three possible right-side views of the building.

26. Make an orthogonal drawing showing the front, top and right-side views of a staircase that is one cube wide and reaches a height of ten cubes. Then make a foundation drawing of the staircase.

27. **INTERIOR DESIGN** If the height of each cube in Exercise 26 is 8 in., how many square inches of carpeting are needed to completely cover the front and top edges of the staircase?

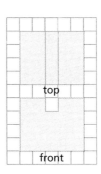

◾ MIXED REVIEW EXERCISES

Use grid paper. Graph each function for the domain of real numbers. (Lesson 6-6)

28. $y = x^2 - 1$

29. $y = x^2 - x + 2$

30. $y = x^2 + x - 4$

31. $y = 2x^2 + 3x + 1$

32. $y = 3x^2 - 4$

33. $y = 2x^2 - x - 1$

34. $y = 2x^2 + 2$

35. $y = x^2 - 3x + 4$

36. $y = 2x^2 + x - 5$

Write each phrase as a variable expression. (Lesson 2-3)

37. six times a number divided by three

38. the product of seven and a number

39. two more than 12 times a number

40. the quotient of negative six and a number

41. a number less nine

42. negative three times a number

Review and Practice Your Skills

Use the isometric drawing for Exercises 1–3. Assume that no cubes are hidden from view.

1. How many cubes are used in the drawing?

2. How many cube faces are exposed in the figure?

3. If the length of one of the cubes is 2.8 in., what is the total surface area of the figure to the nearest tenth of an inch?

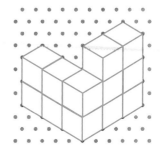

Create an isometric drawing of each figure.

4. pentagonal prism

5. figure composed of 10 cubes

6. figure with 7 faces

7. triangular prism

Create a foundation drawing for each isometric drawing. Assume the drawing is viewed from the lower left-hand corner.

8.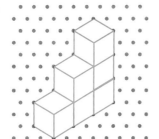

9.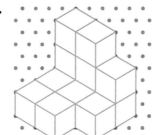

10.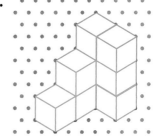

Make an orthogonal drawing labeling the front, top and right-side views.

11.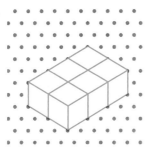

12.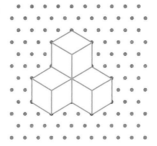

13.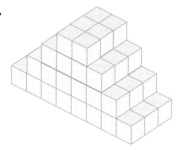

For each foundation drawing, sketch the front and right orthogonal views.

14.

3	2
2	1

front

15.

3	3
2	2
1	1

front

16.
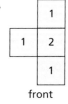

	1
1	2
	1

front

Draw a rectangular pyramid. (Lesson 10-1)

17. How many vertices does it have?

18. How many faces does it have?

19. How many bases does it have?

20. What is the surface area of a rectangular prism with dimensions of 2 in. by 6 in. by 4 in.? (Lesson 10-3)

21. Draw a net for a pentagonal prism. (Lesson 10-2)

For each foundation drawing, sketch the front and right orthogonal views. (Lesson 10-6)

22.

3	3	1
2		
1		

front

23.

2	1
2	1

front

24.

1	2	1
	1	

front

MathWorks
Workplace Knowhow
Career – Exhibit Designer

Exhibit designers organize and design products and materials so they serve the intended purpose and are visually pleasing. Some exhibit designers work in museums to create, design, install, disassemble and store historical displays. A museum exhibit designer is designing models of notable skyscrapers for a historical exhibit. Accurately communicating the design ideas to the model creator is important. The figure shown is a two-point perspective drawing of the Empire State Building (without the lightning rod) in New York City, NY.

Refer to the drawing and the table.

1. Create an orthogonal drawing of the Empire State Building.

2. Create an isometric drawing of the Empire State Building.

3. Create a one-point perspective drawing of the Empire State Building.

Empire State Building

Section	Width* (feet)	Depth* (feet)	Height* (feet)
1	833	375	125
2	667	292	250
3	500	208	750
4	250	125	125

* Dimensions are not actual measurements of the Empire State Building

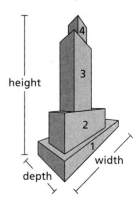

10-7 Volume of Prisms and Pyramids

Goals
- Use a formula to find the volume of prisms.
- Use a formula to find the volume of pyramids.

Applications Earth science, Hobbies, Packaging, History, Machinery

Use sheets of heavy paper, a ruler, tape and dried beans.

1. Use the patterns to make a cube and a square pyramid. Both shapes should be missing one side.

2. Completely fill the cube with dried beans. Record the number of beans used.

3. Pour the beans directly from the cube to the pyramid. Approximately how many times can you fill the pyramid?

4. What relationship do you see?

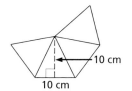

◥ BUILD UNDERSTANDING

Recall that **volume** is a measure of the number of cubic units needed to fill a region of space. To find the volume of any prism, multiply the area of the base by the height of the prism.

Volume of a Prism	$V = Bh$
	where B is the area of the base and h is the height of the prism.

Example 1

Find the volume of each prism.

a.

b.

Problem Solving Tip

Volume is always measured in cubic units since it is found by multiplying an area (which is measured in square units) by a length.

Solution

a. The base is a rectangle.

$$B = 14 \cdot 12 \qquad {\scriptstyle B = lw}$$

$$B = 168$$

Use the volume formula.

$$V = 168 \cdot 25 \qquad {\scriptstyle V = Bh}$$

$$V = 4200$$

The volume is 4200 m³.

b. The base is a square.

$$B = 7^2 \qquad {\scriptstyle B = s^2}$$

$$B = 49$$

Use the volume formula.

$$V = 49 \cdot 7 \qquad {\scriptstyle V = Bh}$$

$$V = 343$$

The volume is 343 ft³.

Example 2

Find the volume of the triangular prism.

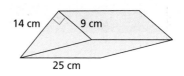

Solution

The base is a right triangle.

$$B = \frac{1}{2} \cdot 9 \cdot 14 \qquad B = \frac{1}{2}bh$$

$$B = 63$$

Use the volume formula.

$$V = 63 \cdot 25 \qquad V = Bh$$

$$V = 1575$$

The volume is 1575 cm^3.

> ### Problem Solving Tip
>
> To find the volume of a rectangular prism, use the formula $V = lwh$.
>
> To find the volume of a cube, use the formula $V = s^3$.

Example 3 nline **Personal Tutor at mathmatters2.com**

RECREATION A rectangular swimming pool is filled with 1750 ft^3 of water. The pool width is 14 ft and the pool depth is 5 ft. Find the length of the pool.

Solution

Use the volume formula for a rectangular prism.

$$1750 = l \cdot 14 \cdot 5 \qquad V = lwh$$

$$1750 = 70l \qquad \text{Solve for } l.$$

$$\frac{1750}{70} = \frac{70l}{70}$$

$$25 = l$$

The length of the pool is 25 ft.

The volume of a prism is 3 times the volume of a pyramid. To find the volume of a pyramid, take $\frac{1}{3}$ the product of the area of the base and the height.

Volume of a Pyramid	$V = \frac{1}{3}Bh$ where B is the area of the base and h is the height.

Example 4

Find the volume of the rectangular pyramid.

Solution

Find the area of the rectangle base.

$$B = 24 \cdot 16 \qquad B = lw$$

$$B = 384 \text{ m}^2$$

Use the volume formula.

$$V = \frac{1}{3} \cdot 384 \cdot 30 \qquad V = \frac{1}{3}Bh$$

$$V = 3840$$

The volume of the pyramid is 3840 m^3.

Math nline mathmatters2.com/extra_examples

Find the volume of each figure.

1.

2.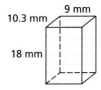

3.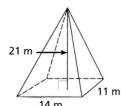

4.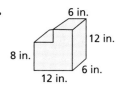

5. PACKAGING A gift box that is 15 in. high is a prism with a square base. It has a volume of 540 in.³. What is the length of the sides of its base?

6. Find the volume of a square pyramid if the length of a side of its base is 23 ft and its height is 65 ft. Round the answer to the nearest tenth.

7. A triangular prism has a volume of 54 cm³. The triangular base has a height of 4 cm and a base of 6 cm. What is the height of the prism?

8. WRITING MATH Given the volume of a pyramid and the area of its base, describe how you can find its height.

■ **PRACTICE EXERCISES** • **For Extra Practice, see page 619.**

Find the volume of each figure.

9.

10.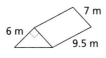

11.

12.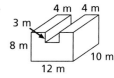

13.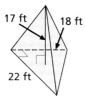

14.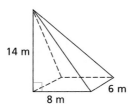

15.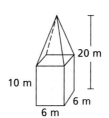

16.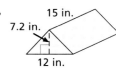

17. WRITING MATH Explain the difference between the surface area and the volume of a three-dimensional figure.

18. A cereal box is 8 in. long and 4 in. wide. Its volume is 368 in.³. What is the height of the box?

19. The perimeter of the base of a square prism measures 60 cm. The height of the prism is 25 cm. What is the volume of the prism?

20. The base of a prism is an isosceles right triangle with sides measuring $5\sqrt{2}$ cm, $5\sqrt{2}$ cm and 10 cm. The height of the prism is 20 cm. What is the volume of the prism?

21. MACHINERY A dump truck has a bed that is 12 ft long and 8 ft wide. The walls of the bed are 4.5 ft high. When the truck is loaded, no material can be higher than the walls of the bed. If topsoil costs $18/yd³, what is the cost of a full truckload of topsoil?

22. EARTH SCIENCE When water freezes, its volume increases by about 10%. A tin container that measures 1 ft by 10 in. by 9 in. is exactly half full of water. It is left outside on a winter day, and all of the water freezes. What is the approximate volume of ice in the tin?

23. HISTORY The square base of an Egyptian pyramid is 62 m long. If the height of the pyramid is 78 m, what is its volume?

■ EXTENDED PRACTICE EXERCISES

Find the volume of each figure.

24.

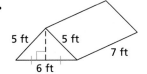

25.

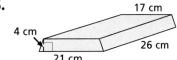

26.

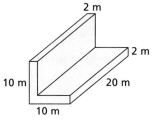

27. HOBBIES An aquarium has a length of 15 in. and a width of 11 in. A rock put into the aquarium causes the water level to rise by 2 in. The rock is completely submerged. What is the volume of the rock?

28. Suppose you wish to make a cardboard box with a volume of 1000 cm³. What dimensions would you give to the box in order to use the least amount of cardboard?

29. CRITICAL THINKING The sides of a cube each measure 1 ft. If each side is increased by 1 in., by how many cubic inches would the volume increase?

30. CHAPTER INVESTIGATION Research a significant historical event that was happening at each date on your timeline.

■ MIXED REVIEW EXERCISES

Simplify. (Lesson 9-4)

31. $2z(4z + w)$

32. $3d(2d - 3c)$

33. $5rs(3r - 4s)$

34. $-6xy(2x - 5y)$

35. $2p(4p + 3r) + 5p(7p - 8r)$

36. $3st(3s - 3t) + 2st(2s - 5t)$

37. $4yz(y - 3z) + 2yz(6y + 4z)$

38. $-5cd(2c + 3d) + 2cd(c - 5d)$

Copy quadrilateral *KLMN* on grid paper. Draw each dilation. (Lesson 7-5)

39. scale factor 2, center (0, 0)

40. scale factor $\frac{1}{2}$, center (0, 0)

41. scale factor 3, center *L*

42. scale factor $\frac{2}{3}$, center *N*

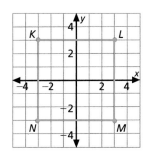

Volume of Cylinders, Cones, and Spheres

Goals ■ Find the volume of cylinders, cones, and spheres.

Applications Sports, Horticulture, Astronomy, History, Chemistry

FITNESS A person's **vital capacity** is the measure of the volume of air held in his or her lungs. Take a deep breath, and blow into a balloon as much air as possible. Trap the air by tying off the balloon.

1. Push on the end of the balloon so that it forms a sphere. Then use a tape measure to find the circumference of the balloon. Let this measure be C.

2. Find your vital capacity (VC) by using the formula $VC = \dfrac{C^3}{6\pi^2}$.

■ BUILD UNDERSTANDING

Recall that the general formula for finding the volume of a prism is $V = Bh$, where B is the area of the base. Since the base of a cylinder is circular, replace B in this formula with πr^2, the formula for the area of a circle.

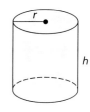

Volume of a Cylinder	$V = \pi r^2 h$ where r is the radius of the base and h is the cylinder's height.

Example 1

Find the volume of the cylinder.

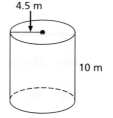

4.5 m

10 m

Solution

$V = \pi r^2 h$ Use the formula for the volume of a cylinder.

$V \approx 3.14 \cdot (4.5)^2 \cdot 10$ $\pi \approx 3.14$

$V \approx 635.85$

The volume of the cylinder is approximately 635.85 m³.

Check Understanding

How is the formula for the volume of a cone similar to the formula for the volume of a pyramid?

The volume of a cylinder is 3 times the volume of a cone that has the same radius and height. To find the volume of a cone, take $\frac{1}{3}$ the product of the area of the base and the height of the cone.

Volume of a Cone	$V = \dfrac{1}{3}\pi r^2 h$ where r is the radius of the base and h is the cone's height.

Example 2

 Personal Tutor at **mathmatters2.com**

Find the volume of the cone.

Solution

$V = \dfrac{1}{3}\pi r^2 h$ Use the formula for the volume of a cone.

$V \approx \dfrac{1}{3} \cdot 3.14 \cdot 4^2 \cdot 6$ $\pi \approx 3.14$

$V \approx 100.48$

The volume of the cone is approximately 100.48 cm³.

There is also a formula for the volume of a sphere.

Volume of a Sphere	$V = \dfrac{4}{3}\pi r^3$ where r is the radius of the sphere.

Example 3

ASTRONOMY The diameter of the planet Mars is approximately 6800 km. What is the volume of Mars?

Solution

Assume that Mars is a sphere. Mars' radius is half of 6800 km, or 3400 km.

$V = \dfrac{4}{3}\pi r^3$ Use the formula for the volume of a sphere.

$V \approx \dfrac{4}{3} \cdot 3.14 \cdot (3400)^3$ $\pi \approx 3.14$

$V \approx 1.6455 \cdot 10^{11}$ km³

The volume of Mars is approximately 165,000,000,000 km³.

Mars

Example 4

Find the volume of the figure.

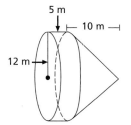

Solution

The total volume of the figure is the sum of the cylinder's volume and the cone's volume.

Volume of the cylinder	Volume of the cone
$V = \pi r^2 h$	$V = \dfrac{1}{3}\pi r^2 h$
$V \approx 3.14 \cdot 12^2 \cdot 5$	$V \approx \dfrac{1}{3} \cdot 3.14 \cdot 12^2 \cdot 10$
$V \approx 2260.8$	$V \approx 1507.2$

Total volume $\approx 2260.8 + 1507.2 \approx 3768$

The volume of the figure is approximately 3768 m³.

Find the volume of each figure. Round to the nearest whole number.

1.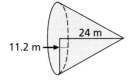

2.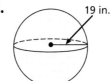

3.

4.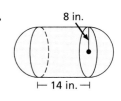

5. A cone with a radius of 12 cm has a volume of 753.6 cm³. What is the height of the cone?

6. What is the volume of a sphere with a radius of 35 m?

7. A cubic foot of sand has a mass of 75 lb. How many pounds of sand would fit into a cylindrical container with a radius of 10 ft and a height of 5 ft?

8. What is the volume of a hemisphere with a radius *r*?

◥ **PRACTICE EXERCISES** • **For Extra Practice, see page 619.**

Find the volume of each figure. Round to the nearest whole number.

9.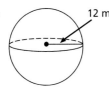

10.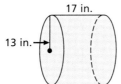

11.

12.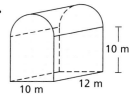

13. A gasoline storage tank is a cylinder with a radius of 10 ft and a height of 6 ft. How many cubic feet of gasoline will the tank hold?

14. A sphere has a volume of $523\frac{1}{3}$ in.³. What is the radius of the sphere?

15. **SPORTS** Tennis balls are sold in cylindrical cans. Each can holds three tennis balls. If the volume of the can is 150.72 in.³, what is the approximate radius of a tennis ball?

16. **HORTICULTURE** The figure at the right is a sketch of a proposed greenhouse that is to be shaped like a hemisphere. To the nearest cubic foot, what is the amount of space inside this greenhouse?

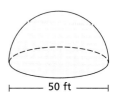

Find the volume of each figure. Round to the nearest tenth.

17.

18.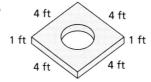

The radius of the hole is 1 ft.

19.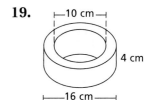

20. **WRITING MATH** Explain how to find the volume of a cylinder, a cone, and a sphere. Describe the similarities among the three different volume formulas.

21. HISTORY Early astronomers estimated the equatorial radius of the earth to be approximately 520 km. The actual radius is about 6378 km. What is the approximate volume of the earth? How inaccurate was the astronomers' estimate of the earth's volume?

22. A cylinder has a circumference of 12π m and a height of 4.7 m. What is the volume of the cylinder? Round the answer to the nearest whole number.

23. CHEMISTRY A chemist pours 188.4 cm^3 of a liquid into a glass cylinder that has a radius of 2 cm. How many centimeters deep is the liquid?

DATA FILE For Exercises 24–27, refer to the data on various balls used in sports on page 571. Find the volume of each sports ball. Round to the nearest tenth.

24. baseball   **25.** golf ball   **26.** volleyball   **27.** croquet ball

EXTENDED PRACTICE EXERCISES

CRITICAL THINKING Each of the cans shown in the diagram has a volume of 1000 cm^3. (Recall that 1000 cm^3 = 1 L.)

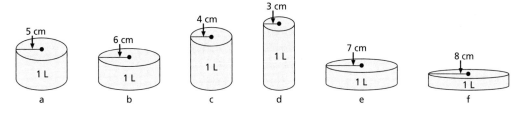

28. Find the height of each can rounded to the nearest tenth.

29. Which two cans have the least surface area?

30. If you manufacture one liter cans, what radius would you use so that you save the most material?

31. SPREADSHEET Write a spreadsheet program to calculate the volume of a sphere. Input the formula in the first cell, and calculate the volumes of spheres with radii 2 m, 4 m, 8 m and 16 m. Do you notice a pattern? Explain.

MIXED REVIEW EXERCISES

Find the value of the determinant of each matrix. (Lesson 8-5)

32. $\begin{vmatrix} 4 & 5 \\ 3 & 6 \end{vmatrix}$ **33.** $\begin{vmatrix} 1 & 8 \\ -2 & 5 \end{vmatrix}$ **34.** $\begin{vmatrix} -6 & -2 \\ 4 & 7 \end{vmatrix}$ **35.** $\begin{vmatrix} 5 & 6 \\ -12 & 8 \end{vmatrix}$ **36.** $\begin{vmatrix} 0 & -1 \\ 2 & 6 \end{vmatrix}$

37. $\begin{vmatrix} -3 & 3 \\ 4 & -6 \end{vmatrix}$ **38.** $\begin{vmatrix} 8 & 13 \\ -10 & -9 \end{vmatrix}$ **39.** $\begin{vmatrix} -4 & 7 \\ 2 & -8 \end{vmatrix}$ **40.** $\begin{vmatrix} 4 & 9 \\ 2 & 5 \end{vmatrix}$ **41.** $\begin{vmatrix} 11 & 4 \\ -9 & -3 \end{vmatrix}$

In the figure, $m\angle MLN = 35°$ and $m\angle NLO = 55°$.
Find each measure. (Lesson 5-8)

42. $m\widehat{NO}$ **43.** $m\widehat{MN}$

44. $m\widehat{MNO}$ **45.** $m\widehat{MPO}$

Review and Practice Your Skills

PRACTICE ◤ LESSON 10-7

Find the volume of each figure. Round to the nearest tenth.

1.
4 m, 15 m, 5 m

2.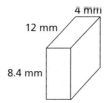
4 mm, 12 mm, 8.4 mm

3.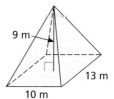
9 m, 13 m, 10 m

4.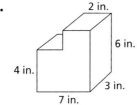
2 in., 6 in., 4 in., 3 in., 7 in.

5.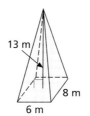
13 m, 8 m, 6 m

6.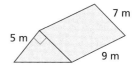
7 m, 5 m, 9 m

7.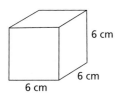
6 cm, 6 cm, 6 cm

8.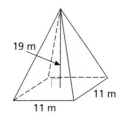
19 m, 11 m, 11 m

9.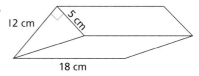
12 cm, 5 cm, 18 cm

10. A triangular prism has a volume of 108 m³. The triangular base has a height of 9 m and a base of 6 m. What is the height of the prism?

PRACTICE ◤ LESSON 10-8

Find the volume of each figure. Round to the nearest whole number.

11.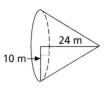
24 m, 10 m

12.
7 in.

13.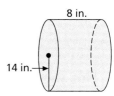
8 in., 14 in.

14.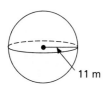
11 m

15.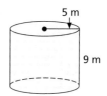
5 m, 9 m

16.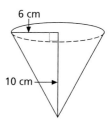
6 cm, 10 cm

17. The volume of a cone with a 14-mm diameter is 820 mm³. Find the height.

18. The radius of an inflated beach ball is 15 in. What is the amount of air inside?

19. Find the minimum amount of gift wrap needed to cover a box that measures 2 ft by 3 ft by 6 in. (Lesson 10-2)

20. Sketch a net for the three-dimensional figure. (Lesson 10-2)

21. Create an isometric drawing of a triangular prism. (Lesson 10-5)

22. Create an isometric drawing of a figure composed of six cubes. (Lesson 10-5)

Find the surface area and volume of each figure. Round to the nearest tenth.
(Lessons 10-3, 10-7 and 10-8)

23.

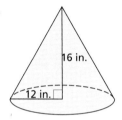

24.

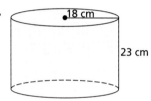

25.

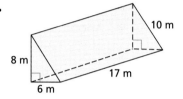

26.

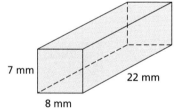

27.

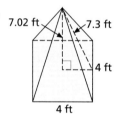

28.

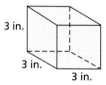

Identify each figure and state the number of faces, vertices and edges.
(Lesson 10-1)

29.

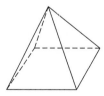

30.

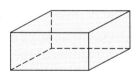

31.

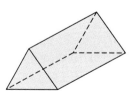

Trace each perspective drawing to locate the vanishing point(s). (Lesson 10-4)

32.

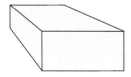

33.

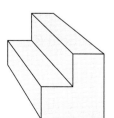

34.

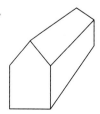

Problem Solving Skills: Length, Area, and Volume

Formulas are mathematical tools that help guide you to a solution. The strategy, **use an equation or formula**, is appropriate when you have data that can be substituted into a formula or equation. If you do not know the formula, write an equation that states the relationships in the problem. In either case, solve the equation to find the solution.

Problem Solving Strategies

Guess and check

Look for a pattern

Solve a simpler problem

Make a table, chart or list

Use a picture, diagram or model

Act it out

Work backwards

Eliminate possibilities

✔ Use an equation or formula

Problem

CONSTRUCTION The tank of a raised cylindrical water tower is 26 ft high with a radius of 11 ft. There is a 2-ft-wide walkway around the base of the tank.

a. About how many gallons of paint are needed to paint the exterior of the tank if 1 gal of paint will cover 300 ft²?

b. There is a railing around the walkway. How long is it?

c. How many gallons of water does the tank hold? (The volume of 1 gal is approximately 0.1337 ft³.)

Solve the Problem

a. Each gallon of paint covers an area of 300 ft².

$SA = 2\pi r^2 + 2\pi rh$ Use the formula for surface area of a cylinder.

$SA \approx (2 \cdot 3.14 \cdot 11^2) + (2 \cdot 3.14 \cdot 11 \cdot 26)$ $\pi \approx 3.14$

$SA \approx 760 + 1796 \approx 2556$

The surface area is approximately 2556 ft². Since each gallon of paint covers 300 ft², divide the surface area by 300 ft.

$2556 \div 300 \approx 8.5$

It will take approximately 8.5 gal of paint.

b. The length of the railing is the circumference of a circle with a radius 2 ft greater than the radius of the tank.

$C = 2\pi r$ Use the formula for the circumference of a circle.

$C \approx 2 \cdot 3.14 \cdot (11 + 2)$ $\pi \approx 3.14$

$C \approx 81.6$

The railing is approximately 81.6 ft long.

c. The number of gallons the tank will hold is the volume of the tank.

$V = \pi r^2 h$ Use the formula for the volume of a cylinder.

$V \approx 3.14 \cdot 11^2 \cdot 26 \approx 9878.4$ $\pi \approx 3.14$

The volume is approximately 9878.4 ft³. Since 1 gal is approximately 0.1337 ft³, the tank holds approximately $9878.4 \div 0.1337 \approx 73{,}884.8$ gal.

Five-step Plan

1 Read
2 Plan
3 Solve
4 Answer
5 Check

Round answers to the nearest tenth.

1. A cypress tree has a radius of 18.75 ft. If it is fenced in so that there is a border 8 ft wide around the trunk of the tree, how many feet of fencing are needed?

2. South African ironwood is the world's heaviest wood, weighing about 90 lb/ft^3. How much would an 8-in. cube of ironwood weigh?

3. **PACKAGING** Find the minimum amount of paper needed for the label of a cylindrical soup can that is 7 in. high with a radius of 2.25 in.

4. The windows of a building have a total area of 32.4 m^2. Each window is a rectangle measuring 1.2 m by 1.8 m. How many windows are in the building?

▧ PRACTICE EXERCISES

Round answers to the nearest tenth.

5. **LANDSCAPING** A pound of grass seed covers an area of 250 ft^2. How many pounds of seed would you need for a rectangular lawn measuring 35 ft by 50 ft?

6. The roof of a shed is a square pyramid with sides of 2.4 m. The height of each triangular face is 2 m. How many square meters of tar paper would it take to cover the roof?

7. A circular swimming pool has a diameter of 17 ft and a depth of 5 ft. If the pool is considered to be full when the water level is 1 ft below the rim of the pool, how many cubic feet of water does it take to fill the pool?

8. A cardboard hat is made of a cone with radius 5 in. and slant height 14 in. The circular brim is 4-in. wide. Find the minimum amount of cardboard needed to make the hat.

9. **WRITING MATH** Write a problem that can be solved by applying the formula for length, area, or volume. Provide an answer for your problem.

10. **CRITICAL THINKING** The names and sizes of wooden boards specify the dimensions before they are dried and planed. For example, a one-by-ten board is actually $\frac{3}{4}$ in. by $9\frac{1}{4}$ in. If a patio is built using 15 one-by-ten boards that are each 12 ft long, what is the area of the patio?

▧ MIXED REVIEW EXERCISES

Simplify. (Lesson 9-3)

11. $\dfrac{14b^2}{2b}$

12. $\dfrac{51a^2c^3}{3ac}$

13. $\dfrac{27g^3h}{9gh}$

14. $\dfrac{7m^3n^4}{m^2n}$

15. $\dfrac{49k^3l^2}{14k^3l^2}$

Calculate each permutation. (Lesson 4-6)

16. $_8P_2$

17. $_5P_2$

18. $_{12}P_4$

19. $_{10}P_6$

20. $_9P_4$

21. $_{12}P_8$

22. $_9P_6$

23. $_{10}P_3$

24. $_{14}P_5$

25. $_{11}P_3$

Chapter 10 Review

VOCABULARY

Choose the word from the list that best completes each statement.

1. A(n) __?__ is a polyhedron with only one base.

2. A(n) __?__ is a polyhedron with two identical parallel bases.

3. The __?__ of a cone is the length of a segment drawn from its vertex to its base along the side of the cone.

4. A(n) __?__ is the set of all points in space that are a given distance from a given point.

5. A one-point perspective drawing has one __?__.

6. A(n) __?__ is a three-dimensional figure with two congruent circular bases that lie in parallel planes.

7. The sum of the areas of all bases and faces of a three-dimensional figure is called its __?__.

8. The number of cubic units needed to fill a three-dimensional figure is called its __?__.

9. A(n) __?__ drawing shows three sides of a three-dimensional figure.

10. A(n) __?__ drawing shows the individual sides of a three-dimensional figure.

a. cone

b. cylinder

c. face

d. isometric

e. orthogonal

f. prism

g. pyramid

h. slant height

i. sphere

j. surface area

k. vanishing point

l. volume

LESSON 10-1 ■ Visualize and Represent Solids, p. 422

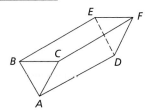

▶ A **polyhedron** is a closed, three-dimensional figure made up of polygonal surfaces called **faces**. A segment that is the intersection of two faces is an **edge**. The point at which three or more edges intersect is a **vertex**.

Use the figure at the right.

11. Identify the figure.

12. Identify its base(s).

13. Identify its lateral face(s).

14. State the number of vertices.

15. State the number of edges.

16. Identify all edges that are skew to $\overline{CF}$.

LESSON 10-2 ■ Nets and Surface Area, p. 426

▶ A net is a two-dimensional pattern that can be folded to form a three-dimensional figure. A net can be used to find surface area.

Draw a net for each three-dimensional figure. Then find the surface area.

17.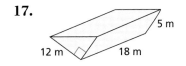

18.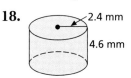

19.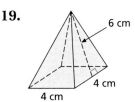

LESSON 10-3 ◣ Surface Area of Three-Dimensional Figures, p. 432

▶ **Surface areas** can be found using these formulas.

rectangular prism:
$SA = 2(lw + lh + wh)$

cylinder:
$SA = 2\pi r^2 + 2\pi rh$

cone:
$SA = \pi rs + \pi r^2$

Find the surface area of each figure. Round to the nearest hundredth.

20.
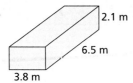
2.1 m
6.5 m
3.8 m

21.

8 m
20.2 m

22.

5 ft
11 ft

23. A soup can has a height of 10 cm and a diameter of 6.5 cm. Find the amount of steel needed to make the can.

24. A pet carrier is in the shape of a rectangular prism. It is 2.5 ft long, 1 ft high and 1.25 ft wide. What is the surface area of the carrier?

LESSON 10-4 ◣ Perspective Drawings, p. 436

▶ A **perspective drawing** is a way of drawing objects on a flat surface so that they look the same as they appear in real life.

25. Sketch a rectangular tissue box in one-point perspective.

26. Sketch a rectangular tissue box in two-point perspective.

27. Locate the vanishing point in the perspective drawing.

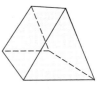

LESSON 10-5 ◣ Isometric Drawings, p. 442

▶ An **isometric drawing** shows an object from a corner view so that three sides of the object can be seen in a single drawing.

Give the number of cubes used to make each figure and the number of cube faces exposed. Be sure to count the cubes that are hidden from view.

28.

29.

30.

LESSON 10-6 ◣ Orthogonal Drawings, p. 446

▶ An **orthogonal drawing** gives top, front and side views of a three-dimensional figure as seen from a "straight on" viewpoint.

Make an orthogonal drawing labeling the front, top and right-side views.

31.

32.

33.
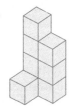

For each foundation drawing, sketch the front and right orthogonal views.

34.

2	1	1
1	2	3

35.

	1
2	1
2	4

36.

			3
			2
3	4	1	1

LESSON 10-7 ◣ Volume of Prisms and Pyramids, p. 452

▶ **Volume** can be found using these formulas.

$$\text{prism: } V = Bh \qquad\qquad \text{pyramid: } V = \tfrac{1}{3}Bh$$

37. Find the volume of a square pyramid if the length of a side of its base is 15 cm and its height is 23 cm.

38. A rectangular prism has a base that measures 5 ft by 3 ft. Its volume is 300 ft^3. What is its height?

39. A rectangular cake pan is 2 in.-by-13 in.-by-9 in. A round cake pan has a diameter of 8 in. and a height of 2 in. Which will hold more cake batter, the rectangular pan or two round pans?

LESSON 10-8 ◣ Volume of Cylinders, Cones and Spheres, p. 456

▶ **Volume** can be found using these formulas.

$$\text{cylinder: } V = \pi r^2 h \qquad \text{cone: } V = \tfrac{1}{3}\pi r^2 h \qquad \text{sphere: } V = \tfrac{4}{3}\pi r^3$$

Find the volume of each figure. Round to the nearest tenth.

40. cylinder: $r = 6.1$ mm, $h = 3.8$ mm

41. cone: $r = 4.2$ cm, $h = 11$ cm

42. sphere: $r = 6$ mm

43. cone: $r = 5$ in., $h = 15$ in.

LESSON 10-9 ◣ Problem Solving Skills: Length, Area and Volume, p. 462

▶ The strategy, **use an equation or formula**, is appropriate when you have data that can be substituted into a formula or equation.

44. A rectangular room measures 12 yd by 22 ft. How many square yards of carpet will it take to carpet the entire room?

45. A rectangular room is 12 ft-by-21 ft. The walls are 8 ft tall. Paint is sold in 1-gal containers. If a gallon of paint covers 450 ft^2, how many gallons of paint should be purchased to paint the walls of the room?

CHAPTER INVESTIGATION

EXTENSION Write a report about how the study of mathematics influences history. Your report could answer some of the following questions. How did cultural and historical events affect mathematics? How does geography affect the spread of information and new discoveries? How did mathematics help trigger major cultural events like the Industrial Revolution and the Information Age? Have all mathematicians been formally educated? How did computers change the way mathematicians work?

Chapter 10 Assessment

Use the figure to name the following.

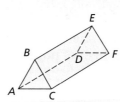

1. a pair of intersecting edges
2. a pair of parallel edges
3. a pair of parallel faces
4. a pair of edges that are skew
5. the bases

Draw a net for each figure. Then identify the figure and find its surface area.

6.
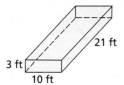

7.

8.

9. Find the surface area of a cone with a radius of 3.1 cm and a slant height of 12.4 cm.

10. Find the surface area of the figure.

11. Locate the vanishing point of the figure.

12. Create an isometric drawing of a figure composed of 5 cubes.

13. Make an orthogonal drawing of the figure you drew in Exercise 12. Show the front, top and right-side views.

Find the volume of each figure. Round to the nearest tenth if necessary.

14.

15.

16.
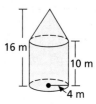

17. To find the number of square feet of wrapping paper needed to cover a box shaped like a rectangular prism, which formula should you use?

   a. $P = 2l + 2s$ **b.** $V = lwh$ **c.** $SA = 2(lw + lh + wh)$

18. Write a problem that can be solved by applying the formula for the volume of a sphere. Provide an answer for your problem.

Standardized Test Practice

Part 1 Multiple Choice

Record your answers on the answer sheet provided by your teacher or on a sheet of paper.

1. Desiree's test scores are 83, 75, 86, and 82. If her teacher uses the mean, what score does she need on the fifth test in order to have an average of 85? (Lesson 1-2)
   - Ⓐ 85
   - Ⓑ 90
   - Ⓒ 99
   - Ⓓ 105

2. Of the people who buy raffle tickets, 500 win nothing, 1 wins $25, and 1 wins $100,000. If you were promoting the raffle and wanted to give a misleading statistic about the average winning, which measure of central tendency would you use? (Lesson 1-7)
   - Ⓐ mean
   - Ⓑ median
   - Ⓒ mode
   - Ⓓ range

3. Which expression is *not* represented by $2x - 1$? (Lesson 2-3)
   - Ⓐ twice a number minus one
   - Ⓑ twice a number less than one
   - Ⓒ two times a number decreased by one
   - Ⓓ two times a number minus one

4. Which graph represents the solution of $3t - 5 \geq -2$? (Lesson 3-7)
   - Ⓐ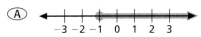
   - Ⓑ
   - Ⓒ
   - Ⓓ

5. What is the measure of each angle of a regular polygon that has 9 sides? (Lesson 5-7)
   - Ⓐ 126°
   - Ⓑ 140°
   - Ⓒ 150°
   - Ⓓ 180°

6. What is the slope of the line that passes through $A(-3, 2)$ and $B(5, -4)$? (Lesson 6-2)
   - Ⓐ $-\frac{4}{3}$
   - Ⓑ $-\frac{3}{4}$
   - Ⓒ $\frac{3}{4}$
   - Ⓓ $\frac{4}{3}$

7. Which line is *not* parallel to $2x - 3y = 5$? (Lesson 8-1)
   - Ⓐ $-2x + 3y = 1$
   - Ⓑ $2x - 3y = 2$
   - Ⓒ $4x - 6y = 5$
   - Ⓓ $3x + 2y = 5$

8. If $2x + y = 3$ and $x + y = 1$, what is the value of y? (Lesson 8-4)
   - Ⓐ -2
   - Ⓑ -1
   - Ⓒ 1
   - Ⓓ 2

9. Factor $x^2 - 4x + 4$. (Lesson 9-8)
   - Ⓐ $(x - 2)^2$
   - Ⓑ $(x + 2)^2$
   - Ⓒ $x(x - 4)$
   - Ⓓ $(x + 2)(x - 2)$

10. How many faces does a pentagonal pyramid have? (Lesson 10-1)
    - Ⓐ 5
    - Ⓑ 6
    - Ⓒ 7
    - Ⓓ 8

11. What is the volume of the rectangular pyramid? (Lesson 10-7)
    - Ⓐ 176 ft³
    - Ⓑ 404 ft³
    - Ⓒ 528 ft³
    - Ⓓ 576 ft³

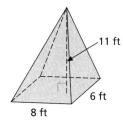

Test-Taking Tip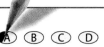

Question 11

Most standardized tests include any commonly used formulas at the front of the test booklet, but it will save you time to memorize many of these formulas. For example, you should memorize that the volume of a pyramid is one-third the area of the base times the height of the pyramid.

Preparing for the Standardized Tests
For test-taking strategies and more
practice, see pages 627–644.

Part 2 | Short Response/Grid In

**Record your answers on the answer sheet
provided by your teacher or on a sheet of paper.**

12. What is the value of t^{-2} for $t = 5$?
(Lesson 2-8)

13. The area of a triangle can be determined by
using the formula $A = \frac{1}{2}bh$. Solve this formula
for h. (Lesson 3-2)

14. Solve $2(3t - 6) = t + 8$. (Lesson 3-4)

15. A bag contains 4 red marbles, 3 blue marbles,
and 2 white marbles. One marble is chosen
without replacement. Then another marble is
chosen. What is the probability that the first
marble is red and the second marble is blue?
(Lesson 4-5)

16. Find the value of x in circle Q. (Lesson 5-8)

17. The center of a circle is located at the origin
of a coordinate plane. If $A(5, 12)$ is on the
circle, what is the radius of the circle?
(Lesson 6-1)

18. The distance a vehicle travels at a given speed
is a direct variation of the time it travels. If a
vehicle travels 30 mi in 45 min, how far can it
travel in 2 h? (Lesson 6-8)

19. Triangle RST with vertices $R(5, 4)$, $S(3, -1)$,
and $T(0, 2)$ is translated so that R' is at $(3, 1)$.
What are the coordinates of S'?
(Lesson 7-1)

20. Solve the system of equations.
(Lesson 8-3)
$$y = 3x$$
$$x + 2y = -21$$

21. The perimeter of the rectangle is $16a + 2b$.
Write an expression for the length of the
rectangle. (Lessons 9-3 and 9-4)

22. A bowling league has n teams. You can use
the expression $\frac{1}{2}n^2 - \frac{1}{2}n$ to find the total
number of games that will be played if each
team plays each other team exactly once.
Factor this expression. (Lessons 9-7)

23. What is the surface area of a cube with sides
of 7 in.? (Lesson 10-3)

24. A can of soup is 12 cm high and has a diameter
of 8 cm. A rectangular label is being designed
for this can of soup. If the label will cover the
surface of the can except for its top and
bottom, what is the width and length of the
label, to the nearest centimeter? (Lesson 10-9)

Part 3 | Extended Response

**Record your answers on a sheet of paper.
Show your work.**

25. Describe the three-dimensional figure.
Include the number of cubes needed to
make the figure and the number of cube
faces exposed. Draw an orthogonal drawing
labeling the front, top, and right-side views.
Then, draw a foundation drawing.
(Lessons 10-5 and 10-6)

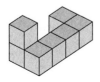

26. Draw a cylinder and a cone that have the
same volume. Explain your response and
calculate the volume and surface area of each
figure. (Lesson 10-8)

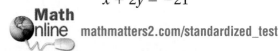

Right Triangle Trigonometry

THEME: Photography

Photography is a popular way to record events, capture memories, and create art. The technical process of photography involves light, chemistry, and right triangle trigonometry. Right triangle trigonometry has many other practical applications, including architecture and engineering.

- **Aerial photographers** (page 483) use trigonometry to take images from an elevated location. Real estate developers, cartographers, and transportation engineers use the services of aerial photographers.

- The work of **camera designers** (page 503) is complicated and intense. They use angles, planes, lengths, and proportions to create cameras with unique capabilities.

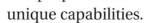
**Math
Online**
mathmatters2.com/chapter_theme

Events in the History of Photography

Year	Historical Event
1558	Giovanni Battista Porta is the first to describe the concept of the Camera Obscura, or dark chamber. Leonardo da Vinci is also attributed with developing the Camera Obscura.
1826	During an 8 h process, Joseph Nicephore Niepce produces the first successful picture.
1829	Louis Daguerre discovers a way to develop film that took 4 h. Deguerre also discovers that an image could be made permanent by immersing it in salt.
1835	William Henry Fox Talbot develops a paper negative.
1845	Friederich von Martens introduces a panoramic camera with a lens that moves along an arc of over 150°.
1851	Frederick Scott Archer introduces a high-quality negative to produce numerous prints.
1861	Physicist J.C. Maxwell produces the first color photograph.
1873	Photographs are printed in newspapers for the first time in America.
1925	The flashbulb is invented.
1963	Kodak introduces the Instamatic.
1987	Both Kodak and Fuji introduce the "Quick Snap," the first 35-mm disposable camera.
1991	The Kodak Professional Digital Camera System is introduced.
1992	Eastman Kodak Company introduces the Photo CD for the photo-retail market.
1994	Apple Quick Take 100 was the first mass-market color digital camera.
1996	Advantix Camera is introduced.

Data Activity: Events in the History of Photography

Use the table for Questions 1–4.

1. Who invented the first 35-mm disposable camera?

2. How many years passed between the first color photograph and the photo CD?

3. Were photographs first printed in newspapers before or after the Civil War?

4. Make a timeline using the information in the table.

CHAPTER INVESTIGATION

The photograph shown at the right was achieved by the photographer being at a different level than the subject of the image. In other words, the picture was taken "at an angle." This technique of having the camera above or below the subject can be used to emphasize a specific characteristic, create a visual effect or produce an abstract view. Even amateur photographers want a variety of photograph styles.

Working Together

The distance from the subject and the angle of the camera are critical for photographers to get their desired results. Determine the distance a photographer needs to be from the subject when the angle of the camera is known. Use the Chapter Investigation icons to guide your group to find the distances for angles whose measures are 15°, 28° and 35°.

The skills on these two pages are ones you have already learned. Use the examples to refresh your memory and complete the exercises. For additional practice on these and more prerequisite skills, see pages 576–584.

MEASURING TRIANGLES

In this chapter you will study trigonometry, which deals with the ratios of sides and angles of triangles. It is helpful to be able to measure the angles of any triangle.

Example   Measure each angle of $\triangle ABC$.

To measure $\angle A$, line up the protractor's midpoint on point A and the base along $\overline{AB}$. Read the angle measure on the outer edge of the protractor. Measure $\angle B$ the same way. Once you know $m\angle A$ and $m\angle B$, you can calculate $m\angle C$.

$\angle A = 50°$ and $\angle B = 70°$ $50° + 70° + \angle C = 180°$
$\angle C = 60°$

Find the measure of each angle.

1.

2.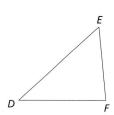

3.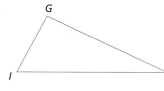

EQUIVALENT RATIOS

Relationships between sides and angles of triangles can be expressed as ratios. It will be helpful to be able to recognize and find equivalent ratios.

Example   Are the ratios equivalent? $\dfrac{2}{3}$ and $\dfrac{8}{12}$ $\dfrac{3}{5}$ and $\dfrac{12}{25}$

Cross-multiply to check. If the products are equal, the ratios are equivalent.

$\dfrac{2}{3} \overset{?}{=} \dfrac{8}{12}$ $\dfrac{3}{5} \overset{?}{=} \dfrac{12}{25}$

$2 \cdot 12 \overset{?}{=} 3 \cdot 8$ $3 \cdot 25 \overset{?}{=} 5 \cdot 12$

$24 = 24$ The ratios are equivalent. $75 \neq 60$ The ratios are not equivalent.

Example   Find three equivalent ratios. $\dfrac{20}{25}$

Multiply or divide both numbers in the ratio by the same number.

$\dfrac{20}{25} \div \dfrac{5}{5} = \dfrac{4}{5}$ $\dfrac{20}{25} \cdot \dfrac{3}{3} = \dfrac{60}{75}$ $\dfrac{20}{25} \cdot \dfrac{4}{4} = \dfrac{80}{100}$

$\dfrac{20}{25}$ is equivalent to $\dfrac{4}{5}, \dfrac{60}{75}$ and $\dfrac{80}{100}$.

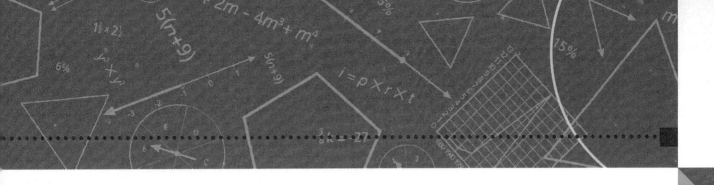

Are the ratios equivalent? Write *yes* or *no*.

4. $\dfrac{44}{112}$ and $\dfrac{11}{28}$

5. $\dfrac{9}{13}$ and $\dfrac{27}{39}$

6. $\dfrac{1}{4}$ and $\dfrac{17}{69}$

7. $\dfrac{18}{24}$ and $\dfrac{3}{4}$

8. $\dfrac{11}{25}$ and $\dfrac{66}{150}$

9. $\dfrac{6}{15}$ and $\dfrac{42}{105}$

10. $\dfrac{4}{7}$ and $\dfrac{48}{85}$

11. $\dfrac{38}{120}$ and $\dfrac{19}{60}$

12. $\dfrac{7}{11}$ and $\dfrac{84}{130}$

13. $\dfrac{8}{72}$ and $\dfrac{1}{9}$

14. $\dfrac{5}{12}$ and $\dfrac{22}{48}$

15. $\dfrac{51}{60}$ and $\dfrac{17}{20}$

16. $\dfrac{15}{69}$ and $\dfrac{5}{23}$

17. $\dfrac{18}{27}$ and $\dfrac{72}{105}$

18. $\dfrac{14}{17}$ and $\dfrac{70}{85}$

19. $\dfrac{31}{84}$ and $\dfrac{62}{166}$

Find three equivalent ratios.

20. $\dfrac{90}{100}$

21. $\dfrac{5}{6}$

22. $\dfrac{12}{16}$

23. $\dfrac{3}{5}$

24. $\dfrac{22}{44}$

25. $\dfrac{75}{100}$

26. $\dfrac{9}{81}$

27. $\dfrac{1}{2}$

SOLVING RADICAL EQUATIONS

Trigonometry frequently involves solving equations that contain radicals.

Example

Use a calculator to find a decimal approximation for each variable. If necessary, round your answer to the nearest thousandth.

$$y = \sqrt{\dfrac{37}{4}}$$
$$y \approx 3.041$$

$$7r = \sqrt{\dfrac{10}{6}}$$
$$\dfrac{7r}{7} \approx \dfrac{1.2909}{7}$$
$$r \approx 0.184$$

Use a calculator to find the decimal approximation for each variable. If necessary, round your answer to the nearest thousandth.

28. $b = 7\sqrt{16}$

29. $r = 4\sqrt{22}$

30. $\dfrac{6}{\sqrt{5}} = c$

31. $\dfrac{k}{\sqrt{12}} = 5$

32. $\sqrt{\dfrac{28}{4}} = g$

33. $5a = 3\sqrt{42}$

34. $3z = \sqrt{80}$

35. $c = 6\sqrt{15}$

36. $7m = \dfrac{18}{\sqrt{29}}$

37. $4n = 9\sqrt{314}$

38. $y = \dfrac{(2\sqrt{58})}{6}$

39. $6d = 2\sqrt{190}$

40. $p = \dfrac{(4\sqrt{92})}{3}$

41. $w = 5\sqrt{34}$

42. $3g = \dfrac{\sqrt{55}}{2}$

43. $4x = \sqrt{\dfrac{11}{4}}$

44. $12y = 6\sqrt{21}$

45. $\dfrac{\sqrt{17}}{3} = 9a$

11-1 Similar Polygons

Goals
- Identify similar polygons.
- Find measures of similar polygons.

Applications   Architecture, Sports, Art, Photography

Refer to the two triangles.

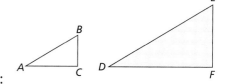

1. Triangle *ABC* and triangle *DEF* appear to have the same shape, but they are different sizes. Measure each angle of the two triangles. What do you notice?

2. Find the following ratios of the side lengths of each triangle: *AB* : *DE*, *BC* : *EF* and *AC* : *DF*. What do you notice?

3. Describe how the two triangles are similar.

4. Describe how the two triangles are different.

■ BUILD UNDERSTANDING

COncepts in MOtion
Animation
mathmatters2.com

Similar figures have the same shape, but not necessarily the same size. To indicate that two figures are similar, the symbol ~ is used.

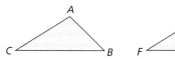   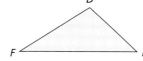   △*ABC* ~ △*DEF*

In similar polygons, vertices can be matched up so that all pairs of corresponding angles are congruent and all pairs of corresponding sides are in proportion.

Similarity	Congruent angles	Corresponding sides
△*ABC* ~ △*DEF*	$\angle A \cong \angle D$	*AB* : *DE*
	$\angle B \cong \angle E$	*BC* : *EF*
	$\angle C \cong \angle F$	*AC* : *DF*

When identifying two polygons as similar, name their corresponding vertices in the same order. You can show that two triangles are similar if the corresponding angles are congruent or the corresponding sides are in proportion.

Example 1

Determine if each pair of triangles is similar.

a.

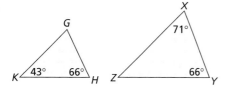

b.

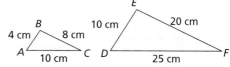

Solution

a. Find the missing angle measures.

$$m\angle G = 180° - (66° + 43°)$$
$$= 180° - 109°$$
$$= 71°$$

$$m\angle Z = 180° - (66° + 71°)$$
$$= 180° - 137°$$
$$= 43°$$

So $\angle G \cong \angle X$ and $\angle Z \cong \angle K$. Since corresponding pairs of angles are congruent, $\triangle GHK \sim \triangle XYZ$.

b. Find the ratios of corresponding sides.

$$\frac{AB}{DE} = \frac{4}{10} = \frac{2}{5}$$
$$\frac{BC}{EF} = \frac{8}{20} = \frac{2}{5}$$
$$\frac{AC}{DF} = \frac{10}{25} = \frac{2}{5}$$

The ratios are equivalent, so corresponding sides are in proportion and $\triangle ABC \sim \triangle DEF$.

To determine if two polygons that are not triangles are similar, compare *both* their angles and the lengths of their sides.

Example 2

Determine if *ABCD* is similar to *PQRS*.

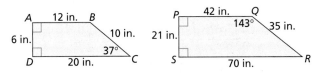

Solution

Find the missing measures of the angles.

$$m\angle B = 360° - (90° + 90° + 37°)$$
$$= 360° - 217°$$
$$= 143°$$

$$m\angle R = 360° - (90° + 90° + 143°)$$
$$= 360° - 323°$$
$$= 37°$$

Find the ratios of the lengths of corresponding sides.

$$\frac{AB}{PQ} = \frac{12}{42} = \frac{2}{7} \qquad \frac{BC}{QR} = \frac{10}{35} = \frac{2}{7} \qquad \frac{CD}{RS} = \frac{20}{70} = \frac{2}{7} \qquad \frac{DA}{SP} = \frac{6}{21} = \frac{2}{7}$$

Pairs of corresponding angles are congruent and pairs of corresponding sides are in proportion, so $ABCD \sim PQRS$.

> **Check Understanding**
>
> If $\triangle URL \sim \triangle JKT$, what are the congruent angles and corresponding sides?

Example 3   **Personal Tutor at** mathmatters2.com

SPORTS Miguel is making a model of a basketball court. The dimensions of a basketball court are 84 ft by 50 ft. If the length of Miguel's model is 3.5 ft, how wide is the model?

Solution

Since Miguel's model is similar to an actual basketball court, the corresponding sides are in proportion. Let w represent the width of Miguel's model.

$$\frac{84}{3.5} = \frac{50}{w}$$
$$84 \cdot w = 3.5 \cdot 50 \qquad \text{Find the cross-products.}$$
$$w = \frac{175}{84} = 2\frac{7}{84} = 2\frac{1}{12} \qquad \text{Solve for } w.$$

So the width of Miguel's model is $2\frac{1}{12}$ ft, or 2 ft 1 in.

Determine if each pair of polygons is similar.

1.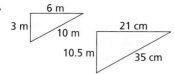

2.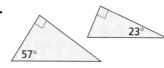

3.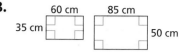

Find the length of $\overline{AB}$ in each pair of similar figures.

4.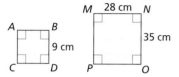

5.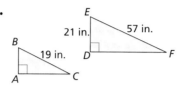

6.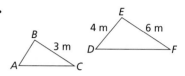

7. WRITING MATH Write a short paragraph to explain the similarities and differences between the terms "similar" and "congruent." Include examples to illustrate these similarities and differences.

8. If $\square ABCD \sim \square RSTU$ and if $AB = 10$, $RS = 8$, and $ST = 4$, what is the length of $\overline{BC}$?

◥ **PRACTICE EXERCISES** • For Extra Practice, see page 620.

Determine if each pair of polygons is similar.

9.

10.

11.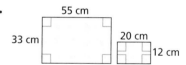

Find the length of $\overline{AB}$ in each pair of similar figures.

12.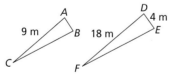

13.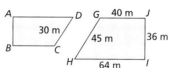

14.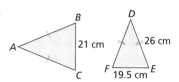

Find $m\angle A$ in each pair of similar polygons.

15.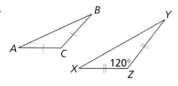

16.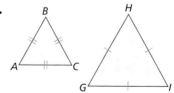

17.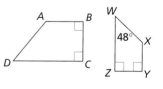

18. ARCHITECTURE An architect draws a blueprint to show the floor plan for a new home. If the scale ratio is 1 cm : 4 m and the master bedroom has dimensions 2.4 cm by 1.7 cm in the blueprint, what are the actual dimensions of the master bedroom?

19. ART *ABCD* and *EFGH* are similar rectangular picture frames. The length of $\overline{AB}$ is 12 in., and the length of $\overline{BC}$ is 16 in. If $\overline{EF}$ has a length of 21 in., what is the length of $\overline{FG}$?

20. WRITING MATH Consider this statement: *All squares are similar.* Is this statement true or false? Write a paragraph to support your answer.

21. Triangle *ABC* is similar to triangle *DEF*. List three equivalent ratios for the sides of the triangles.

22. PHOTOGRAPHY Photographers often take a film negative and enlarge the image. Is this an example of similar figures? Explain why or why not.

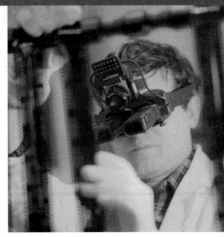

23. YOU MAKE THE CALL Tyra claims that in the figure shown, it cannot be determined that the two triangles are similar. Matt says that since $\overline{AC} \parallel \overline{ED}$, $\angle A \cong \angle D$ and $\angle C \cong \angle E$ because they are two pairs of alternate interior angles. Therefore, Matt says that the two triangles can be shown to be similar. Who is correct and why? If you agree with Matt, name the triangle similarity.

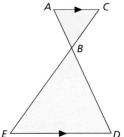

EXTENDED PRACTICE EXERCISES

Write *sometimes, always* or *never* for each statement.

24. A rhombus is similar to another rhombus.   **25.** Circles are similar to each other.

26. Congruent figures are similar.   **27.** Similar figures are congruent.

28. Figures with different numbers of sides can be similar.

29. CRITICAL THINKING *ABCD* and *MNOP* are similar rectangles. The length of $\overline{AB}$ is 4 cm, and the length of $\overline{MN}$ is 15 cm. If the area of *ABCD* is 48 cm^2, what is the area of *MNOP*?

30. GEOMETRY SOFTWARE Use geometry software to create several different types of triangles. Copy each triangle, and increase or decrease its size to create pairs of similar triangles.

MIXED REVIEW EXERCISES

Solve. Round answers to the nearest tenth when necessary. (Lesson 6-9)

31. Assume that *y* varies inversely as *x*. When $x = 8$, $y = 16$. Find *y* when $x = 2$.

32. Assume that *y* varies inversely as *x*. When $x = 5$, $y = 20$. Find *y* when $x = 7$.

33. Assume that *y* varies inversely as *x*. When $x = 6$, $y = 14$. Find *y* when $x = 2$.

State whether each pair of triangles is congruent by SAS, ASA, or SSS. If the pair is not necessarily congruent, write *not congruent*. (Lesson 5-5)

34. 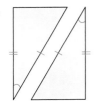   **35.**    **36.**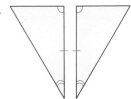

11-2 Indirect Measurement

Goals ■ Use similar triangles to find indirect measurements.

Applications Nature, Recreation, Advertising, Earth Science, Photography

HEALTH Carla is hiking in the woods and wishes to know approximately how far she walks. So she takes a tape measure and marks a distance of 20 m. She walks the 20-m distance several times and finds that she makes on average 26 steps in approximately 14 sec.

1. If Carla takes 5400 steps during her hike walking at a fairly constant rate, approximately how far does she walk?

2. If Carla's hike takes her approximately 50 min to complete, about how far did she hike?

3. How can you explain the difference in the approximations? Which do you think is more accurate?

■ BUILD UNDERSTANDING

In some situations, it is impossible or impractical to find a length by measuring the actual distance. An **indirect measurement** is one in which you take other measurements that allow you to calculate the required measurement.

Example 1

NATURE A tree casts a shadow that is 12 m long. At the same time, a forest ranger 170 cm tall notices that her shadow is 300 cm long. What is the height of the tree?

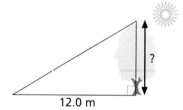

12.0 m

Solution

The angle at which the sun's rays meet the ground is the same in both right triangles. Since the triangles have corresponding pairs of congruent angles, they are similar. Use this similarity to write and solve a proportion.

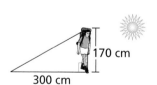

170 cm

300 cm

$$\frac{170}{h} = \frac{300}{12}$$

$170 \cdot 12 = h \cdot 300$ Find the cross-products.

$2040 = 300h$

$$\frac{2040}{300} = \frac{300h}{300}$$ Divide both sides by 300.

$6.8 = h$

The height of the tree is 6.8 m.

Example 2

The wire supporting a radio transmitting tower touches the top of a 3.6-m pole. What is the height of the tower?

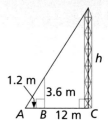

Solution

The right triangles share a common acute angle, so they are similar. Use this similarity to write and solve a proportion.

$$\frac{1.2}{1.2 + 12} = \frac{3.6}{h}$$

$1.2 \cdot h = (1.2 + 12)\, 3.6$ Find the cross-products.

$1.2h = 47.52$

$h = 39.6$

The height of the tower is 39.6 m.

The **mirror method** is another means of indirectly measuring the approximate height of an object. Place a mirror on the ground between you and the object you wish to measure. Position yourself so that you can see the top of the object in the mirror. Then the right triangles shown in the diagram are similar since their acute angles that have the mirror as a vertex are congruent.

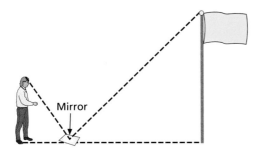

Mirror

The mirror method is useful on a cloudy day when there are no shadows.

Problem Solving Tip

When you write a proportion, you do not always have to convert measurements to the same units. But the corresponding units of the ratios must be the same.

For example, the proportion $\frac{6\ \text{ft}}{34\ \text{in.}} = \frac{7\ \text{ft}}{x\ \text{in.}}$ is suitable to solve for x in inches.

Example 3

Marty uses the mirror method to find the height of a telephone pole. He places the mirror 33 ft from the pole's base and steps back 6 ft, which allows him to see the top of the telephone pole in the mirror. Marty is 5.5 ft tall. Find the height of the telephone pole.

Solution

Marty has established a pair of similar triangles. Marty's distance from the mirror corresponds to the pole's distance. His height corresponds to the pole's height.

$$\frac{6}{33} = \frac{5.5}{h}$$

$6 \cdot h = 33 \cdot 5.5$

$6h = 181.5$

$h = 30.25$ The telephone pole is approximately 30.25 ft tall.

Check Understanding

Explain why the mirror method produces an approximate height of the object being measured.

Use indirect measurement to find the unknown length of x.

1.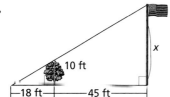

2.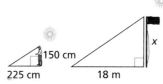

3.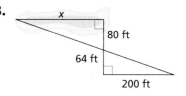

4. Find the height of the building in the figure to the right.

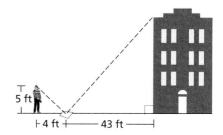

5. **RECREATION** Darnell places a mirror on the ground 48 ft from the base of a rollercoaster hill. He then stands 4 ft from the mirror so that he can see the top of the hill. If Darnell is 5.9 ft tall, how high is the hill?

6. Lian is 5 ft tall. To measure a flagpole that is 60 ft tall, she places a mirror on the ground 4 ft away from her. How far from the flagpole is Lian standing?

7. A building casts a shadow 72 ft long. At the same time, a 5-ft tall boy casts a shadow 12 ft long. What is the height of the building?

8. **WRITING MATH** Write a paragraph describing the difference between direct measurement and indirect measurement. Include examples of each type.

■ **PRACTICE EXERCISES** • **For Extra Practice, see page 620.**

Use indirect measurement to find the unknown length of x.

9.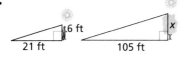

10.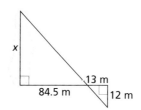

11.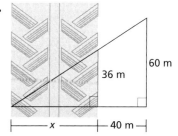

12. Mykaela stands 7.5 ft from a mirror she has placed on the ground to measure the height of her house. If the mirror is 23 ft from the house and Mykaela is 5.3 ft tall, how tall is the house?

13. **ADVERTISING** To measure the height of a billboard, Vijay places a mirror on the ground 13 m from the base of the billboard. He then stands 3.7 m away from the mirror so that he can see the reflection of the top of the billboard. If Vijay is 1.9 m tall, how high off the ground is the top of the billboard?

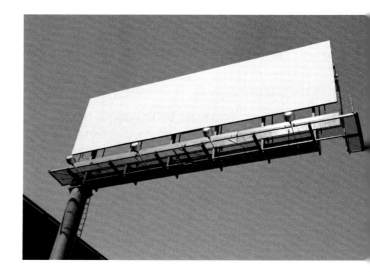

14. A tree casts a shadow 26 m long. At the same time, a flagpole 25 m high casts a shadow 32.5 m long. How tall is the tree?

15. A 45-ft tall building casts a shadow 10 yd long. At the same time, Simon casts a shadow that is 3 ft long. How tall is Simon?

16. A lumberjack places a mirror 50 ft from a 100-ft tall tree. If the lumberjack is standing 53 ft from the base of the tree and can see the top of the tree through the mirror, how tall is the lumberjack?

17. EARTH SCIENCE Each time Old Faithful in Yellowstone National Park erupts, rangers can estimate the height of the geyser by comparing it to the height of a tree. The rangers locate a tree of about the same height as the geyser. If the shadow of the tree is 93 ft when the shadow of a 6-ft ranger is 4 ft, what is the height of the tree and the approximate height of the geyser?

18. A 25-ft ladder is leaning against a wall. A rung that is 5 ft from the bottom of the ladder is 4 ft above the ground. How far above the ground does the top of the ladder touch the wall?

◤ EXTENDED PRACTICE EXERCISES

19. MONUMENTS A monument is 96 ft tall and a tree next to it is 84 ft tall. What is the ratio of the lengths of their shadows? Does the ratio depend on the time of day? Explain.

20. CRITICAL THINKING Suppose you know the height of a flagpole on the beach of Chesapeake Bay and that it casts a shadow 4 ft long at 4:00 P.M. (EST). You also know the height of a flagpole on the shoreline of Lake Michigan whose shadow is hard to measure at 3:00 P.M. (CST). Since 4:00 (EST) = 3:00 (CST), you propose the following proportion of heights and lengths to find the length of the shadow of the Michigan flagpole. Explain whether this proportion will give an accurate measure.

$$\frac{\text{height of Chesapeake flagpole}}{\text{shadow of Chesapeake flagpole}} = \frac{\text{height of Michigan flagpole}}{\text{shadow of Michigan flagpole}}$$

21. PHOTOGRAPHY Annie takes a picture of a pine tree measuring 12.19 m tall. She is 180 cm tall and casts a shadow of 300 cm. What is the greatest distance that Annie can be from the tree if she does not want the lens exposed to direct sunlight?

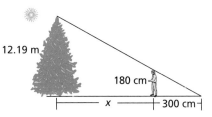

22. CHAPTER INVESTIGATION The photographer, subject and ground are the vertices of a right triangle. Draw a diagram of the triangle formed by a photographer who is shooting at an upward angle at the subject of the picture. Use the distance between the subject and the ground as one leg of the right triangle.

◤ MIXED REVIEW EXERCISES

Give the coordinates of the image of each point under a reflection across the given axis. (Lesson 7-2)

23. (4, 6); *y*-axis   **24.** (2, −4); *x*-axis   **25.** (−3, 7); *x*-axis   **26.** (−5, 1); *y*-axis

27. Two number cubes are rolled. Find the probability that the difference of the numbers rolled is 1 or 2. (Lesson 4-4)

28. One card is drawn from a standard deck of cards. Find the probability that the card drawn is a heart or a queen. (Lesson 4-4)

Review and Practice Your Skills

PRACTICE ◼ LESSON 11-1

Determine if each pair of triangles is similar.

1.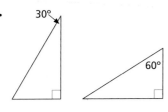

2.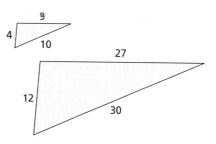

3.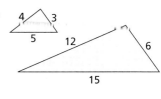

Find the length of $\overline{AB}$ in each pair of similar figures.

4.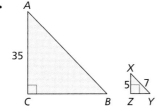

5.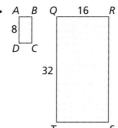

6.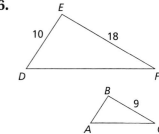

Find $m\angle B$ in each pair of similar polygons.

7.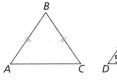

8.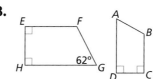

9.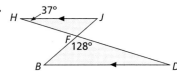

PRACTICE ◼ LESSON 11-2

Use indirect measurement to find the unknown length x.

10.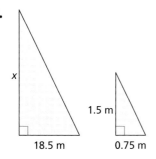

11.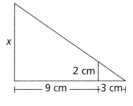

12.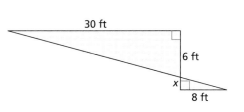

13. A tree casts a 20 m shadow while at the same time a 1.5 m person casts a shadow that is 1.2 m long. How tall is the tree?

14. If a person who is 2 m tall casts a 5 m shadow at 7 P.M., how long should the shadow of the 445 m tall Sears Tower be at the same time?

MathWorks Career – Aerial Photographer
Workplace Knowhow

Aerial photographs are useful in many professions such as real estate development, transportation planning, construction and cartography. These photographs are used to design a new neighborhood, determine the necessity and size of new roads, as well as the development of new maps. Three different techniques are used to take aerial photographs.

High Oblique Technique

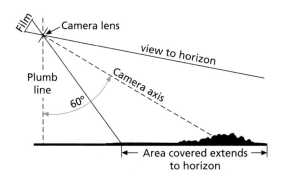

Vertical Technique

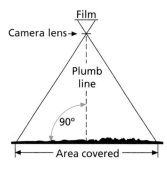

Low Oblique Technique

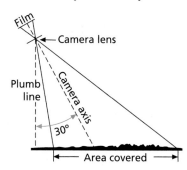

Use the three techniques, Figure A, and Figure B for Exercises 1–3.

1. Figure A is a sketch of an aerial view of a city street and a skyscraper ($\overline{FD}$) that is positioned along a street. The height of the skyscraper is 933 ft. The camera lens is located at point A. A portion of the road between points C and D must be resurfaced. Using similar triangles, determine the length of the section of the road that must be repaired.

2. The vertical technique, in Figure B, uses two similar triangles. Name the triangles and explain why they are similar.

3. If the width of the film ($\overline{MN}$) in Figure B is 35 mm and the distance from the center of the film to the camera lens is 100 mm, determine the width of the area covered in the photograph. Assume that the camera lens is 500 m above the ground.

Figure A

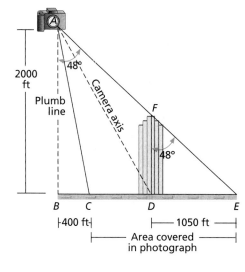

Figure B

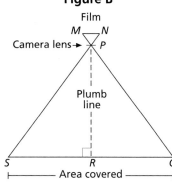

11-3 The Pythagorean Theorem

Goals ■ Use the Pythagorean Theorem to find unknown lengths.

Applications Fitness, Recreation, Sports, Hobbies, Photography

HISTORY An ancient Chinese mathematical manuscript called the Zhoubi is between 2000 and 3000 years old. The manuscript includes an illustration similar to Figure 1. The figure is a square that encloses four congruent triangles and a smaller square.

Figure 1

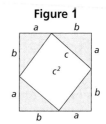

1. The triangles can be rearranged as shown in Figure 2. How do the areas of Figure 1 and Figure 2 compare?

2. How do the areas of the unshaded regions of the two figures compare?

3. How do the unshaded regions relate to the length of the triangle sides?

Figure 2

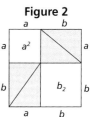

COncepts in MOtion

Interactive Lab

mathmatters2.com

◣ BUILD UNDERSTANDING

In a right triangle, the side opposite the right angle is the **hypotenuse**. The other two sides are the **legs**. About 2500 years ago, a Greek mathematician named Pythagoras proved a property about the hypotenuse and legs of a right triangle. This property is called the **Pythagorean Theorem**.

Pythagorean Theorem	$c^2 = a^2 + b^2$ In any right triangle, the square of the hypotenuse (c^2) is equal to the sum of the squares of the legs ($a^2 + b^2$).

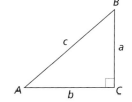

Example 1

Find the length of the hypotenuse of triangle *ABC*.

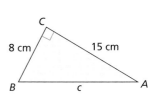

Solution

$$c^2 = a^2 + b^2 \qquad \text{Use the Pythagorean Theorem.}$$

$$c^2 = 8^2 + 15^2 \qquad \text{Substitute 8 for } a \text{ and 15 for } b.$$

$$c^2 = 64 + 225$$

$$c^2 = 289$$

$$\sqrt{c^2} = \sqrt{289} \qquad \text{Take the square root of each side.}$$

$$c = 17$$

The length of the hypotenuse is 17 cm.

Math: Who, Where, When

A 1940's collection of 370 different proofs of the Pythagorean Theorem includes work by the 12th century Hindu Bhaskara, the 15th century Italian Da Vinci, and the 19th century American James Garfield, the 20th U.S. president.

Example 2

Find the length of $\overline{MO}$ to the nearest tenth.

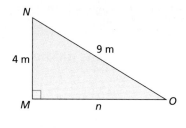

Solution

$$9^2 = 4^2 + n^2 \qquad c^2 = a^2 + b^2$$

$$81 = 16 + n^2$$

$$65 = n^2 \qquad \text{Subtract 16 from both sides.}$$

$$\sqrt{65} = \sqrt{n^2} \qquad \text{Take the square root of both sides.}$$

$$8.062 \approx n$$

The length of $\overline{MO}$ is approximately 8.1 m.

Example 3   Online Personal Tutor at mathmatters2.com

RECREATION The longest side of a right triangular sail measures 6 m, and the base of the sail measures 3 m. What is the minimum height of the mast to the nearest tenth?

Solution

Draw a diagram of the problem. The triangle is a right triangle, and the longest side is the hypotenuse.

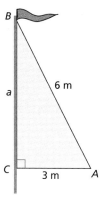

$$c^2 = a^2 + b^2$$

$$6^2 = a^2 + 3^2 \qquad \text{Substitute 6 for } c \text{ and 3 for } b.$$

$$36 = a^2 + 9$$

$$27 = a^2 \qquad \text{Subtract 9 from both sides.}$$

$$\sqrt{27} = \sqrt{a^2}$$

$$5.196 \approx a$$

The height of the mast must be at least 5.2 m.

Example 4

What is the length of the diagonal of rectangle *ABCD*?

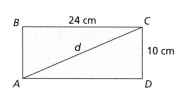

Solution

Since the diagonal forms the hypotenuse of a right triangle, the Pythagorean Theorem can be used to solve the problem.

$$d^2 = 10^2 + 24^2$$

$$d^2 = 100 + 576$$

$$\sqrt{d^2} = \sqrt{676}$$

$$d = 26$$

The length of the diagonal of rectangle *ABCD* is 26 cm.

> **Check Understanding**
>
> Use the Pythagorean Theorem to write expressions for a^2 and b^2.

Math Online mathmatters2.com/extra_examples

Find the unknown length. Round to the nearest tenth.

1.
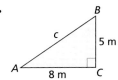

2.

3.

4.
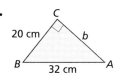

5. A flagpole is supported by a wire cable connected to its highest point. The cable is 40 ft long and is attached to the ground 15 ft from the base of the pole. How tall is the flagpole? Round to the nearest foot.

6. How long is the diagonal of a square with sides of 4 cm? Round to the nearest tenth.

7. What is the length of a rectangle that has a diagonal of 16 ft and width 7 ft?

8. **WRITING MATH** Explain how you could find the perimeter of a rhombus whose diagonals measure 12 m and 16 m. Find the perimeter.

■ PRACTICE EXERCISES • For Extra Practice, see page 621.

Find the length of each hypotenuse. Round to the nearest tenth.

9.
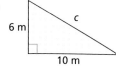

10.
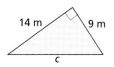

11.
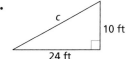

Find the unknown length. Round to the nearest tenth.

12.
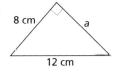

13.
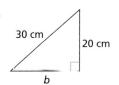

14.
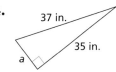

15. What is the length of the diagonal of a rectangle with sides 40 ft and 9 ft?

16. How wide is a rectangle that is 20 in. long and has a diagonal of length 25 in.?

17. A string of pennants is 65 ft long. It is stretched from the top of a tree to a point on the ground 39 ft from the base of the tree. How tall is the tree?

18. **FITNESS** A rectangular swimming pool is 48 ft long and 14 ft wide. A life preserver must be located nearby and have a rope at least as long as the diagonal of the pool. What is the minimum length of the rope?

19. **SPORTS** A baseball diamond is a square with 90-ft sides. To the nearest foot, what is the approximate distance a catcher must throw from home plate to second base? (Hint: Find the length of a diagonal of the square.)

20. A rectangular park is 1384 ft long and 1293 ft wide. To the nearest foot, how long is the walk from one corner of the park to the opposite corner? If Leya walks at an average rate of 264 ft/min, approximately how long will it take her to make this walk? Round to the nearest minute.

21. HOBBIES Sarah is embroidering a square design in an embroidery hoop with a 6-in. diameter. How long are the sides of the largest square that will fit in the hoop? Round to the nearest tenth.

Find the unknown length in each figure. Round to the nearest tenth.

22.

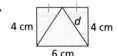

4 cm   4 cm d

6 cm

23.

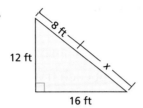

8 ft

12 ft x

16 ft

24.

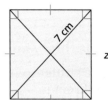

7 cm z

▶ EXTENDED PRACTICE EXERCISES

25. The sloping sides of the triangular front of a tent are 7 ft long from the peak to the ground. If the base of the tent is 6 ft wide, how tall is the tent? Round to the nearest tenth.

26. DATA FILE Refer to the data on the planet's heart on page 565. If the earth is cut in half at the equator and the cross section is divided into 4 equal sectors of 90°, how long would the chord of a sector be? Round to the nearest mile.

27. A lumber mill is cutting lengths of wood with square crosssections that have 12-in. sides. What is the minimum diameter of the tree trunks that the lumber mill can use? Round to the nearest tenth.

28. CRITICAL THINKING Evaluate the expressions $2n$, $n^2 - 1$ and $n^2 + 1$ for the values $n = 2, 3, 4$ and 8. Study the three numbers you obtained for each value of n. What do you notice? How can each set of three numbers be related to the Pythagorean Theorem?

29. PHOTOGRAPHY A photographer takes a picture of the Washington Monument for a magazine. He stands 1000 ft away from the base of the monument, and the distance from this point to the peak is 1143.7 ft. How tall is the Washington Monument? Round to the nearest foot.

▶ MIXED REVIEW EXERCISES

Solve each system of equations. Check the solution. (Lesson 8-4)

30. $3x - y = 3$
$5x - y = 7$

31. $2x + 2y = 10$
$2x - y = 7$

32. $3x + 4y = 5$
$x + 4y = -1$

33. $2x - 3y = -14$
$5x - 3y = -17$

34. $2x + 2y = -8$
$6x + 2y = -16$

35. $4x - 2y = -16$
$4x + 6y = 0$

36. $x + 3y = 16$
$2x - y = -3$

37. $2x - 4y = 20$
$3x + 2y = -2$

Determine whether the graph of the inequality has a dashed or solid boundary. Tell whether to shade above or below the boundary. (Lesson 6-4)

38. $y > 3x - 4$

39. $y < 2x + 8$

40. $y \leq 5x - 7$

41. $y > 2x - 5$

42. $y \geq x + 5$

43. $y \leq -3x - 6$

44. $y < 4x + 2$

45. $y \geq x + 13$

Sine, Cosine, and Tangent Ratios

Goals
■ Identify the sine, cosine, and tangent ratios in a right triangle.
■ Compute the sine, cosine, and tangent ratios for different angles.

Applications Travel, Photography, Navigation

Use the three similar triangles shown to answer the questions.

1. Use a ruler to measure the sides of each triangle to the nearest millimeter.

2. Find each ratio to the nearest hundredth.

   a. $\dfrac{BC}{AB}$, $\dfrac{EF}{DE}$, and $\dfrac{HI}{GH}$ **b.** $\dfrac{AC}{AB}$, $\dfrac{DF}{DE}$, and $\dfrac{GI}{GH}$ **c.** $\dfrac{BC}{AC}$, $\dfrac{EF}{DF}$, and $\dfrac{HI}{GI}$

3. What can you conclude about the ratio of the length of one leg of a right triangle to its hypotenuse as compared to the ratio of the length of the corresponding leg and hypotenuse of a similar triangle?

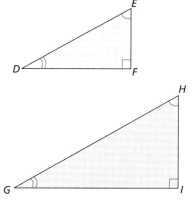

4. What can you conclude about the ratio of the lengths of the legs of one right triangle as compared to the ratio of the lengths of the corresponding legs of a similar right triangle?

◥ BUILD UNDERSTANDING

The legs of a right triangle are often described by relating them to one of the acute angles of the triangle. In relation to $\angle R$, $\overline{RT}$ is the leg *adjacent* to $\angle R$ and $\overline{ST}$ is the leg *opposite* $\angle R$.

For any right triangle, there are three **trigonometric ratios** of the lengths of the sides of the triangle. These ratios are the same for all congruent angles in right triangles even though the side lengths of the triangles may be different.

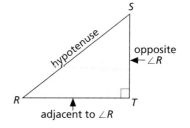

The trigonometric ratios for right triangle ABC are summarized in the chart.

Trigonometric Ratios

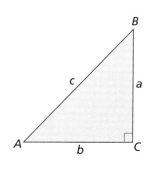

Name of ratio	Abbreviation	Ratio
sine of $\angle A$	sin A	$\dfrac{\text{length of leg opposite } \angle A}{\text{hypotenuse}}$ or $\dfrac{a}{c}$
cosine of $\angle A$	cos A	$\dfrac{\text{length of leg adjacent to } \angle A}{\text{hypotenuse}}$ or $\dfrac{b}{c}$
tangent of $\angle A$	tan A	$\dfrac{\text{length of leg opposite } \angle A}{\text{length of leg adjacent to } \angle A}$ or $\dfrac{a}{b}$

Example 1

Find sin *D*, cos *D*, and tan *D*.

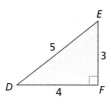

Solution

$$\sin D = \frac{\text{length of leg opposite } \angle D}{\text{hypotenuse}} = \frac{3}{5}$$

$$\cos D = \frac{\text{length of leg adjacent to } \angle D}{\text{hypotenuse}} = \frac{4}{5}$$

$$\tan D = \frac{\text{length of leg opposite } \angle D}{\text{length of leg adjacent to } \angle D} = \frac{3}{4}$$

Problem Solving Tip

To remember the trigonometric ratios, many people use the memory device SOH CAH TOA.

SOH Sine is Opposite over Hypotenuse.

CAH Cosine is Adjacent over Hypotenuse.

TOA Tangent is Opposite over Adjacent.

Example 2

In right triangle *PQR*, if sin *P* = $\frac{12}{13}$, find cos *P* and tan *P*.

Solution

Make a diagram of $\triangle PQR$.

$$\sin P = \frac{\text{length of leg opposite } \angle P}{\text{hypotenuse}} = \frac{12}{13}$$

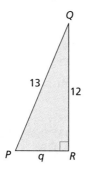

Find the length of *q*, the side adjacent to $\angle P$.

$$12^2 + q^2 = 13^2 \qquad \text{Use the Pythagorean Theorem.}$$

$$144 + q^2 = 169$$

$$q^2 = 25 \qquad \text{Subtract 144 from both sides.}$$

$$\sqrt{q^2} = \sqrt{25}$$

$$q = 5$$

So the length of the side adjacent to $\angle P$ is 5. Therefore,

$$\cos P = \frac{\text{length of leg adjacent to } \angle P}{\text{hypotenuse}} = \frac{5}{13}$$

$$\tan P = \frac{\text{length of leg opposite } \angle P}{\text{length of leg adjacent to } \angle P} = \frac{12}{5}$$

When right triangles are similar, corresponding angles are congruent and the ratios of the lengths of corresponding sides are equal. This means that the trigonometric ratios for corresponding angles are the same. Because these ratios are the same, scientific calculators have sine, cosine and tangent keys, which store the values of these ratios for every possible angle measure.

Example 3

CALCULATOR Use a calculator to find sin 35°, cos 35°, and tan 35° to four decimal places.

Solution

Use the key sequence required by your calculator. Be sure the calculator is set in degree mode.

$$\sin 35° \approx 0.5736 \qquad \cos 35° \approx 0.8192 \qquad \tan 35° \approx 0.7002$$

◼ TRY THESE EXERCISES

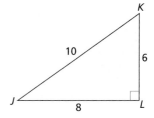

In △JKL, find each trigonometric ratio.

1. sin J
2. sin K
3. cos K
4. tan J
5. tan K
6. cos J

In △XYZ, ∠Z is a right angle. If cos Y = $\frac{8}{10}$, find these ratios.

7. sin Y
8. tan Y
9. cos X

CALCULATOR Use a calculator to find each ratio to four decimal places.

10. sin 60°
11. cos 42°
12. tan 65°

◼ PRACTICE EXERCISES • For Extra Practice, see page 621.

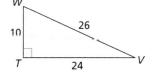

In △TVW, find each trigonometric ratio in lowest terms.

13. sin V
14. cos V
15. cos W
16. sin W
17. tan W
18. tan V

CALCULATOR Use a calculator to find each ratio to four decimal places.

19. cos 52°
20. tan 17°
21. tan 85°
22. sin 45°
23. sin 81°
24. tan 58°
25. cos 33°
26. cos 76°

27. **PHOTOGRAPHY** For his photography class, Mike takes a picture of an oak tree during autumn. Mike stands 75 ft from the base of the tree and 139 ft from the top of the tree. What is the cosine of the angle formed by the base of the tree, Mike's position and the top of the tree?

28. **WRITING MATH** Restate in your own words the definition of the sine ratio, the cosine ratio and the tangent ratio.

29. In right triangle *EFG*, if sin *E* is greater than sin *F*, which side of the triangle is longer, $\overline{EF}$ or $\overline{FG}$? Drawing a picture and selecting some values for the lengths of the sides may help you solve the problem.

For each right triangle, find the trigonometric ratio in lowest terms.

30. In $\triangle ABC$, $m\angle C = 90°$, $BC = 6$ and $AC = 8$. Find $\cos A$.

31. In $\triangle DEF$, $m\angle F = 90°$, $EF = 15$ and $DE = 25$. Find $\sin E$.

32. In $\triangle MAT$, $m\angle A = 90°$, $MT = 10$ and $AT = 8$. Find $\tan T$.

33. **NAVIGATION** A ship returning from sea is 295 m from the base of a lighthouse on the beach. If the lighthouse is 82 m tall, what are the sine, cosine, and tangent of the angle formed by the base of the lighthouse, the ship, and the top of the lighthouse?

34. **ERROR ALERT** In $\triangle JKL$, $\angle L$ is a right angle, $\tan J = \frac{9}{40}$ and $\sin J = \frac{9}{41}$. Jim says this means that $KL = 9$, $JL = 40$ and $JK = 41$. Do you agree with Jim's reasoning? Explain why or why not.

35. **TRAVEL** An airplane travels down a 2200-ft runway to takeoff. By the time the plane reaches the end of the runway, it is 310 ft above the ground. What is the tangent of the angle formed by the end of the runway, the plane's starting position and the plane's current position? What is the cosine of the angle?

EXTENDED PRACTICE EXERCISES

36. **CRITICAL THINKING** If the tangent of an angle is greater than 1, which side of the triangle is longer, the leg adjacent to the angle or the leg opposite the angle?

CALCULATOR Determine how each ratio changes as an angle increases from 0° to 90°.

37. sine

38. cosine

39. tangent

Tell whether each statement is *true* or *false* for $\triangle ABC$.

40. $\sin A = \cos B$

41. $\cos A = \frac{1}{\sin A}$

42. $\tan A = (\sin A)(\cos A)$

43. $\tan A = \frac{\sin A}{\cos A}$

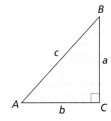

MIXED REVIEW EXERCISES

Simplify. (Lesson 9-4)

44. $4k(6k + 2p)$

45. $3r(6r^2 + 8)$

46. $6st(4s + 7t)$

47. $6p(4p^2 - 3) + 7p(3p^2 - 5)$

48. $2a^2(a - b) + 5b^2(a + b)$

49. $4g^2h(6h - 3g + 5h^2)$

Solve each equation. Check the solution. (Lesson 3-4)

50. $7z + 4 = 25$

51. $\frac{x}{4} - 8 = 6$

52. $6b - 18 = 0$

53. $5 = \frac{2}{3}k + 3$

54. $4y - 3 = 7.8$

55. $\frac{1}{3}(g + 12) = 9$

56. $-6b + 5 = -8$

57. $3x + 2.8 = -3.2$

58. $4b - 27 - 2b = 14$

Review and Practice Your Skills

PRACTICE ◣ LESSON 11-3

Find the unknown length. Round to the nearest tenth.

1.
19
x
41

2.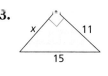
x
53 45

3.
x 11
15

Find the length of each hypotenuse. Round to the nearest tenth.

4.
27 x
36

5.
12 7
x

6.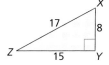
x 17
13

7. How long is the diagonal of a square with 5-in. sides?

PRACTICE ◣ LESSON 11-4

In △XYZ, find each trigonometric ratio in lowest terms.

8. sin X

9. tan Z

10. cos X

11. cos Z

12. sin Z

13. tan X

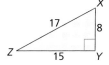
X
17 8
Z 15 Y

In △ABC, ∠B is a right angle. If sin $C = \frac{12}{15}$, find these ratios.

14. cos A

15. cos C

16. tan C

 CALCULATOR Use a calculator to find each ratio to four decimal places.

17. cos 30°

18. sin 60°

19. sin 30°

20. cos 60°

21. sin 81°

22. cos 9°

23. cos 45°

24. sin 45°

25. What do you notice about the sine and cosine of two pairs of complementary angles in Exercises 17–24?

In △MNO, ∠N is a right angle. If sin $M = \frac{3}{5}$, find these ratios.

26. cos M

27. sin O

28. tan O

Determine if the figures are similar. (Lesson 11-1)

29.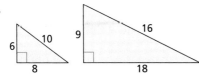

30.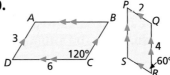

31.

Find the unknown length. Round to the nearest tenth. (Lesson 11-3)

32.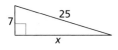

33.

34.

Mid-Chapter Quiz

Determine if each pair of polygons is similar. (Lesson 11-1)

1.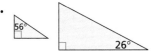

2.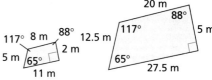

Use indirect measurement to find the length of _x_. (Lesson 11-2)

3.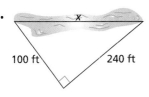

4.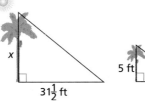

5.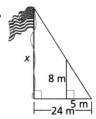

Find the unknown length. Round to the nearest tenth. (Lesson 11-3)

6.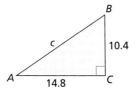

7.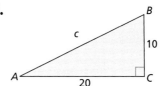

8.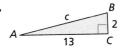

In △_FGH_, find each trigonometric ratio. (Lesson 11-4)

9. sin _F_

10. sin _G_

11. cos _G_

12. tan _G_

13. cos _F_

14. tan _F_

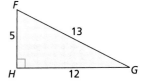

Chapter 11 **Review and Practice Your Skills** | **493**

11-5 Find Lengths of Sides in Right Triangles

Goals ■ Use trigonometric ratios to find the lengths of sides in right triangles.

Applications Engineering, Construction, Photography, Archaeology

For each equation, decide if you can substitute two values using a calculator and the figure shown. If so, show the substitution.

1. $\sin 40° = \dfrac{\text{opposite}}{\text{hypotenuse}}$ **2.** $\cos 40° = \dfrac{\text{adjacent}}{\text{hypotenuse}}$ **3.** $\tan 40° = \dfrac{\text{opposite}}{\text{adjacent}}$

4. Solve the equations for which you are able to find two substitutions in Questions 1 through 3. Find the unknown lengths of the triangle. Round to the nearest tenth.

5. Use the Pythagorean Theorem to verify your answers.

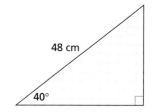

◥ BUILD UNDERSTANDING

Recall that the trigonometric ratios for any given acute angle are the same for any right triangle. If you know the measures of one acute angle and one side of a right triangle, you can find the lengths of the triangle's other sides.

Example 1

In △*ABC*, find *AB* to the nearest tenth.

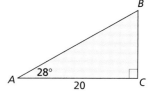

Solution

Decide which trigonometric ratio relates the unknown side to the known angle and the known side. $\overline{AC}$ is *adjacent* to ∠*A*. $\overline{AB}$ is the *hypotenuse*. The ratio that relates an adjacent leg to the hypotenuse is the cosine ratio.

Write and solve an equation involving the trigonometric ratio and the known values. Use a calculator to approximate cos 28°.

$$\cos 28° = \frac{AC}{AB} \qquad \text{\small adjacent leg} \atop \text{\small hypotenuse}$$

$$0.8829 \approx \frac{20}{AB}$$

$$0.8829 \cdot AB \approx 20 \qquad \text{\small Find the cross-products.}$$

$$\frac{0.8829AB}{0.8829} \approx \frac{20}{0.8829} \qquad \text{\small Divide both sides by 0.8829.}$$

$$AB \approx 22.7$$

The length of $\overline{AB}$ is approximately 22.7.

> **Check Understanding**
>
> What trigonometric ratio would you use to find the length of $\overline{BC}$ in △*ABC* of Example 1?

Example 2

ENGINEERING A bridge engineer surveys a completed bridge from a distance of 106.5 m from the center of the bridge's base. She looks up at an angle of 32° to see the top of the bridge. How tall is the bridge?

Solution

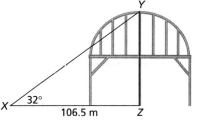

Draw a picture of the bridge, and label the picture with the given information to decide which trigonometric ratio to use.

$\overline{YZ}$ is opposite of $\angle X$. $\overline{XZ}$ is adjacent of $\angle X$. The ratio that relates an *opposite* leg with an *adjacent* leg is the tangent ratio.

Write and solve an equation involving the trigonometric ratio and the known values.

$$\tan 32° = \frac{YZ}{106.5} \quad \text{opposite leg} \atop \text{adjacent leg}$$

$$0.6249 \approx \frac{YZ}{106.5}$$

$$0.6249 \cdot 106.5 \approx YZ \quad \text{Find the cross-products.}$$

$$66.6 \approx YZ$$

The height of the bridge is approximately 66.6 m.

Example 3

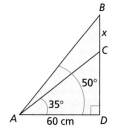

Find the value of *x* in the figure shown.

Solution

Use trigonometric ratios to find *BD* and *CD*.

$$\tan 50° = \frac{BD}{60} \qquad\qquad \tan 35° = \frac{CD}{60}$$

$$1.1918 \approx \frac{BD}{60} \qquad\qquad 0.7002 \approx \frac{CD}{60}$$

$$1.1918 \cdot 60 \approx BD \qquad\qquad 0.7002 \cdot 60 \approx CD$$

$$71.508 \approx BD \qquad\qquad 42.012 \approx CD$$

Subtract *CD* from *BD* to find *x*, the length of *BC*.

$$71.508 - 42.012 = 29.496$$

So *x* is approximately 29.5 cm.

> **Math: Who, Where, When**
>
> The Hindu mathematician Aryabhata first tabulated sines of angles around the year 500 A.D.

▧ TRY THESE EXERCISES

Find each length to the nearest tenth.

1. *DF*   **2.** *EF*

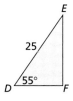

3. *AB*   **4.** *BC*

5. CONSTRUCTION A guy wire is secured near the top of a television transmitting tower. The guy wire meets the ground at an angle of 48°. If the height of the tower is 32 m, how far from the base of the tower is the guy wire secured? Round to the nearest tenth.

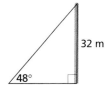

6. An angle of a right triangle measures 75°. If the length of the side opposite this angle is 27 ft, what are the lengths of the two other sides?

7. Find the value of x in the figure shown.

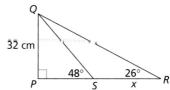

8. WRITING MATH Explain how you would find the length of $\overline{AC}$ in isosceles triangle ABC, where $m\angle C = 90°$ and $\overline{AB}$ measures 10 cm.

◼ PRACTICE EXERCISES • For Extra Practice, see page 622.

Find each length to the nearest tenth.

9. PQ **10.** PR **11.** ML **12.** LN

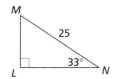

13. RECREATION A parasailor is being towed behind a boat on a 250-ft rope. If the parasailor is being towed at an angle of elevation of 32°, approximately how high is he above the surface of the water?

14. ARCHAEOLOGY All but two of the Egyptian pyramids have faces that are inclined at 52° angles. Suppose an archaeologist discovers the ruins of a pyramid. Most of the pyramid has eroded, but he is able to determine that the length of a side of the square base is 82 m. How tall was the pyramid assuming its faces were inclined at 52°? Round your answer to the nearest meter.

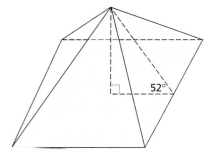

15. An angle of a right triangle measures 39°. If the length of the leg adjacent to this angle is 6 m, what are the lengths of the other two sides?

16. An angle of a right triangle measures 58°. If the length of the hypotenuse is 40 cm, what are the lengths of the other two sides of the triangle?

17. The tailgate of a truck is 1.12 m from the ground. How long should a ramp be so that the incline from the ground up to the tailgate is 9.5°? Round your answer to the nearest tenth.

Find the area of each figure. Round to the nearest tenth.

18.

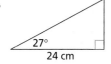

19.

20.

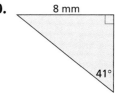

21. PHOTOGRAPHY An aerial photographer takes pictures of two remote islands in the Pacific Ocean. The islands are 6 mi apart. When the plane is directly above Island 1, the line of sight of the plane from Island 1 to Island 2 forms a 54° angle. How high above Island 1 is the plane? Round to the nearest tenth of a mile.

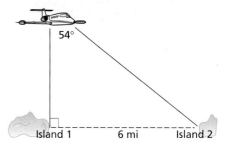
54°
Island 1   6 mi   Island 2

22. WRITING MATH If the measure of an acute angle of a right triangle and the length of its opposite side are given, explain how to find the measures of the other two sides.

23. A 15-ft ladder leaning against a wall makes a 53° angle between the ground and the ladder. To the nearest foot, how far up the wall does the ladder reach?

24. DATA FILE Refer to the data on the size and weight of balls used in sports on page 572. If the sun's rays make an angle of 28° with the ground, how long is the shadow of a soccer ball? a large softball? a croquet ball? Round to the nearest tenth.

◢ EXTENDED PRACTICE EXERCISES

25. GEOMETRY SOFTWARE Use geometry software to draw an equilateral triangle that has sides of length 5 cm. What is the area of this triangle? (Hint: Use the software to draw an altitude of the triangle, which is a segment perpendicular to a side from one of the vertices.) Round to the nearest centimeter.

26. CRITICAL THINKING Jamaal is facing north when he sees an airplane. The angle that his line of sight makes with the ground is 62° when he looks up. At the same time, Jill is standing 10 mi away and also facing north. Her line of sight makes an angle of 26° with the ground when she looks up. What is the altitude of the airplane in feet?

27. MODELING On grid paper, draw three different right triangles for which the tangent of one of the acute angles is 0.8. What is the tangent of the other acute angle of each triangle? Explain how you arrived at the three right triangles you have drawn.

28. CHAPTER INVESTIGATION In your diagram, label the right angle, the legs of the right triangle and the angle of the camera. What trigonometric ratio will you use to find the distance the photographer is from the subject?

◢ MIXED REVIEW EXERCISES

Trace each figure. Draw all the lines of symmetry, or write *none*. (Lesson 7-4)

29.   **30.**   **31.** G   **32.**

Evaluate each expression when $a = -2$ and $b = 3$. (Lessons 2-7 and 2-8)

33. a^2 **34.** b^2 **35.** a^3 **36.** b^3

37. $(a^2 + 5)^2$ **38.** $(a^2 - b)^2$ **39.** $(b^2)^2$ **40.** $(a - b^2)^3$

Find Measures of Angles in Right Triangles

Goals ■ Use trigonometric ratios to find the measures of angles in a right triangle.

Applications Travel, Safety, Photography, Astronomy

Use △EFG for Questions 1–3.

1. Given that the tan $E \approx 1.3764$ and $EG = 8$, find all measures of the triangle that you can determine.

2. Use your measures from Question 1 to verify that tan $E \approx 1.3764$, then find the sine of $\angle E$ and the cosine of $\angle E$.

3. Using the information you found in Questions 1 and 2, can you find $m\angle E$? If so, state the $m\angle E$ and explain how you found it. If not, what information is missing to determine the $m\angle E$?

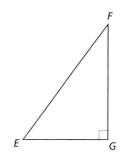

■ BUILD UNDERSTANDING

If you know the sine, cosine, or tangent of an angle in a right triangle, you can find the measure of the angle.

In addition to having keys for the sine, cosine, and tangent functions, calculators also have keys for the inverses of each of these functions. They are symbolized $\sin^{-1}$, $\cos^{-1}$, and $\tan^{-1}$. These inverse functions "undo" the sine, cosine, and tangent functions. So if you know that the sine of an angle is 0.7071067812, the inverse sine key on a calculator shows that the angle is about 45°.

Example 1

Reading Math

Read "$\sin^{-1}(0.75)$" as "the angle whose sine is 0.75."

CALCULATOR Use a calculator to find what angles have the given trigonometric ratios. Round to the nearest degree.

a. $\sin A = 0.866025$ **b.** $\cos B = 0.087156$ **c.** $\tan C = 0.344327$

Solution

Use the key sequence required by your calculator. Be sure the calculator is set in degree mode.

a. $\sin^{-1}(0.866025) \approx 60$

$m\angle A \approx 60°$

Press [2nd] [sin⁻¹] 0.866025 [)] [ENTER].

b. $\cos^{-1}(0.087156) \approx 85$

$m\angle B \approx 85°$

Press [2nd] [cos⁻¹] 0.087156 [)] [ENTER].

c. $\tan^{-1}(0.344327) \approx 19$

$m\angle C \approx 19°$

Press [2nd] [tan⁻¹] 0.344327 [)] [ENTER].

Sometimes instead of being given one of the trigonometric ratios of an angle, you may be given the lengths of two sides of a right triangle. You can use this information to write a trigonometric ratio and, in turn, use the ratio to find the measure of the angle.

Example 2   **Personal Tutor at** mathmatters2.com

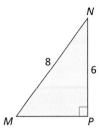

Find $m\angle M$ **in the right triangle.**

Solution

Decide which trigonometric ratio relates the angle whose measure you want to find and the sides whose lengths are known. $\overline{NP}$ is the leg opposite $\angle M$. $\overline{MN}$ is the hypotenuse of the triangle. The ratio that relates an opposite leg with the hypotenuse is the sine ratio.

Write and solve an equation involving the sine ratio and the given values.

$$\sin M = \frac{NP}{MN} \qquad \frac{\text{opposite leg}}{\text{hypotenuse}}$$

$$\sin M = \frac{6}{8} \qquad \text{Substitute the known values.}$$

$$\sin M = 0.75$$

Use a calculator to find the inverse sine of 0.75.

In the triangle, $m\angle M \approx 49°$.

Sometimes you need to combine your knowledge of trigonometric ratios with other geometric concepts to find the measure of an angle.

Example 3

In the figure $\overleftrightarrow{AD} \parallel \overleftrightarrow{BC}$. **Find** $m\angle DAC.$

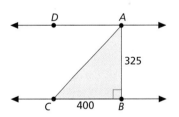

Solution

Note that $\angle DAC$ and $\angle ACB$ are alternate interior angles of two parallel lines cut by a transversal, so they are congruent. Use the inverse tangent function.

$$\tan \angle ACB = \frac{325}{400} \qquad \frac{\text{opposite leg}}{\text{adjacent leg}}$$

$$\tan \angle ACB = 0.8125$$

Using a calculator, $m\angle ACB \approx 39°$. Since $\angle DAC \cong \angle ACB$, $m\angle DAC \approx 39°$.

◣ TRY THESE EXERCISES

CALCULATOR Use a calculator to find what angles have the given trigonometric ratios. Round to the nearest degree.

1. $\sin A = 0.345$ 2. $\tan Q = 1.15036$ 3. $\cos P = 0.93358$ 4. $\cos U = 0.43837$

5. $\tan H = 0.4592$ 6. $\sin I = 0.5$ 7. $\cos Y = 0.1$ 8. $\tan E = 0.265$

9. Find $m\angle A$ in the right triangle shown.

10. In right triangle XYZ, $\cos Z = 0.4848$. Find $m\angle Z$.

11. In the figure, $\overrightarrow{PQ} \parallel \overrightarrow{SR}$. Find $m\angle QPR$.

12. **WRITING MATH** Explain how you might find the angles in a right triangle whose sides measure 5 m, 12 m, and 13 m.

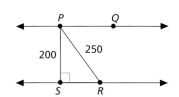

CALCULATOR Use a calculator to find what angles have the given trigonometric ratios. Round to the nearest degree.

13. $\tan N = 0.895$ 14. $\cos W = 0.69753$ 15. $\cos Q = 0.2323$ 16. $\sin K = 0.5612$

17. $\sin T = 0.54321$ 18. $\cos D = 0.9$ 19. $\sin B = 0.3816$ 20. $\tan G = 1.9$

Find each measure to the nearest whole degree.

21. $m\angle A$

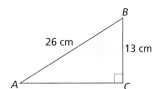

22. $m\angle S$

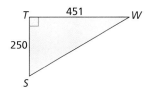

23. $m\angle G$

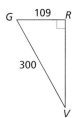

24. An airplane takes off from an airport and flies due south. When it has reached a ground distance of 1500 m from its starting point, it is 970 m above the ground. What angle does the plane's path make with the ground?

25. **TRAVEL** San Francisco's Filbert Street is the steepest street in the world. It rises 1 ft for every 3.17 ft of horizontal distance. Find the angle at which Filbert Street rises.

26. The Chamonix Line of the French National Railroad is the steepest climbing train track in the world. It ascends 1 ft for every 11 ft of horizontal distance. Find the angle at which the Chamonix Line rises.

Filbert Street, San Francisco

SAFETY In order to prevent ladders from accidentally slipping off of a wall, many construction companies follow the rule that the measure of the angle that an unsecured ladder makes with the ground should not be less than 75°. Determine if each ladder is safe to use.

27. The base of an 8-m ladder is 1.5 m away from the wall.

28. A ladder's base is 6.9 yd away from the wall and touches 4.7 yd up the wall.

29. The base of a 3.6-m ladder is 0.8 m away from the wall.

30. PHOTOGRAPHY A reporter photographs a firefighter fighting a fire on the fourth floor of a burning building. The photographer is 60 ft away from the bottom of the building, and the second floor window is 48 ft above the ground. At what angle does the photographer hold her camera?

31. Without using a table of trigonometric values or a calculator, find the angles of a right triangle with two sides measuring 17.5 m. (Hint: Drawing a picture may help solve the problem.)

32. When is the cosine of an angle greater than the sine of the angle as the angle varies from 0° to 90°? Use a calculator to experiment with several different angles to solve the problem.

33. ASTRONOMY A large telescope is being used to track an object circling the earth. If the distance between the telescope and the object is 1400 mi and the object is 610 mi above the surface of the earth, at what angle is the telescope raised?

 34. WRITING MATH Explain why your answer in Exercise 33 is an approximation and cannot be precise.

EXTENDED PRACTICE EXERCISES

35. One leg of a right triangle is twice as long as the other leg. To the nearest whole degree, what are the measures of the acute angles of the triangle?

36. TRAVEL A commercial jet is flying at a height of 2.8 km, and the ground distance between the plane and the point of takeoff is 6 km. To the nearest degree, what is the angle of the plane's path with the ground?

37. CRITICAL THINKING A circle has a radius of 22.4 mm. What is the measure of the central angle that determines a chord of length 16.8 mm? Round to the nearest degree.

 38. CHAPTER INVESTIGATION Suppose the subject is 5.5 ft from the ground. Use a trigonometric ratio to write an equation for each degree: 15°, 28° and 35°. For each angle measure, write the distance to the nearest foot that the photographer needs to be from the subject.

MIXED REVIEW EXERCISES

Find the surface area of each figure. Use 3.14 for π. Round to the nearest tenth. (Lesson 10-3)

39.

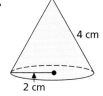

4 cm

2 cm

40.

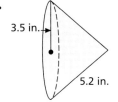

3.5 in.

5.2 in.

41.

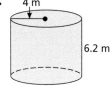

4 m

6.2 m

Factor each polynomial. (Lesson 9-7)

42. $12p^2 + 8p^3$

43. $12f^4 + 6f^3 + 18f^2$

44. $24x^3 - 32x^4 + 40x^3$

Review and Practice Your Skills

Find the value of _y_ rounded to the nearest hundredth. State whether you used sine, cosine, or tangent to solve each exercise.

1.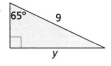

2.

3.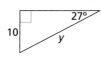

Find the value of _y_ to the nearest hundredth.

4.

5.

6.

7.

8.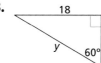

9.

Find _m∠A_ rounded to the nearest degree.

10. $\sin A = 0.05463$

11. $\cos A = 0.2318$

12. $\cos A = \dfrac{5}{9}$

13. $\sin A = 0.16$

14. $\tan A = \dfrac{5}{3}$

15. $\sin A = \dfrac{1}{2}$

Find _m∠B_ rounded to the nearest degree.

16.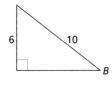

17.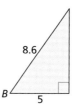

18.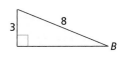

19.

20.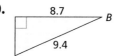

21.

22. In the rectangle shown, what is the measure of ∠_RPE_?

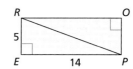

Find the unknown length. Round to the nearest tenth. (Lesson 11-3)

23.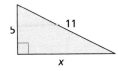

24.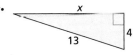

25.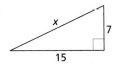

In △CDE, ∠D is a right angle and sin $C = \frac{7}{25}$. Find these ratios. (Lesson 11-4)

26. tan E

27. cos C

28. tan C

29. After taking off, an airplane climbs at a steady 18° angle for 5000 ft of air distance. How far above the ground is the plane? (Lesson 11-5)

30. Find the m∠B to the nearest degree. (Lesson 11-6)

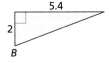

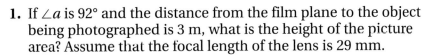

Career – Camera Designer

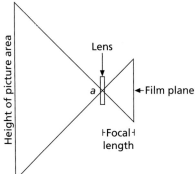

Although cameras are very complicated and intricate, the basic functioning can be explained in general terms. Light from an object is focused on a film plane through a lens. The size of the image produced on a photograph is affected by several parameters, one of which is the focal length. The focal length is the distance from the front of the lens to the film plane. In general, as the focal length increases, the size of the picture area decreases.

1. If ∠a is 92° and the distance from the film plane to the object being photographed is 3 m, what is the height of the picture area? Assume that the focal length of the lens is 29 mm.

2. If the focal length of the lens doubles from 40 mm to 80 mm, what do you predict will happen to the size of ∠a? What will happen to the height of the picture area? Assume that the distance from the film plane to the object remains the same.

3. Comment on the validity of your conjecture about ∠a in Exercise 2 after calculating m∠a for the following focal lengths. Assume that the film plane is 6 cm high.

    a. 40 mm   **b.** 80 mm

4. Using the two values of ∠a calculated in Exercise 3, determine the heights of the picture areas that may be photographed for the following focal lengths. Assume that the object remains 3 m from the film. Does this confirm your prediction in Exercise 3 concerning the heights of the picture areas?

    a. 40 mm   **b.** 80 mm

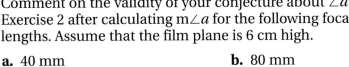

11-7

Special Right Triangles

Goals
- Explore the relationships in 30°-60°-90° right triangles.
- Explore the relationships in 45°-45°-90° right triangles.

Applications Architecture, Entertainment, Photography, Sports, Nature

Use a ruler and a protractor.

1. Draw a 2-in. square and its diagonal. Cut along the diagonal.

2. Measure each angle and the legs of one of the triangles formed by the halves of the square. Record these on the triangle.

3. Use the Pythagorean Theorem to find the measure of the hypotenuse of the triangle. Express this measure as a square root, and record it on the triangle.

4. How do the measures of the legs of a right triangle with a 45° angle compare?

5. How does the measure of the hypotenuse of a right triangle with a 45° angle compare to the measure of its legs?

◥ BUILD UNDERSTANDING

Many engineers, draftspeople, architects and designers use right triangles to maintain drawing proportions. These include two special forms of right triangles. The triangles are a **30°-60°-90° right triangle** and a **45°-45°-90° right triangle**.

These special right triangles can be used to solve problems in a manner similar to using trigonometric ratios. The special geometric properties of 30°-60°-90° right triangles and 45°-45°-90° right triangles are summarized in the table.

30°-60°-90° right triangles	45°-45°-90° right triangles
Length of side opposite 30° angle: x	Length of each leg: x
Length of side opposite 60° angle: $x\sqrt{3}$	Length of hypotenuse: $x\sqrt{2}$
Length of hypotenuse: $2x$	

Problem Solving Tip

Use this memory device to help you remember which triangle uses $\sqrt{2}$ and which uses $\sqrt{3}$.

There are two different side lengths in a 45°-45°-90° right triangle, so it uses $\sqrt{2}$.

There are three different side lengths in the 30°-60°-90° right triangle, so it uses $\sqrt{3}$.

Example 1

Find *MP* and *NP* in the triangle.

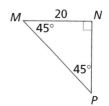

Solution

Triangle *MNP* is a 45°-45°-90° triangle, and $\overline{MP}$ is the hypotenuse.

$$MP = MN\sqrt{2} = 20\sqrt{2}$$

Since the legs are congruent, $NP = 20$.

Example 2

Find *BC* and *AC* in the triangle.

Solution

Triangle *ABC* is a 30°-60°-90° triangle, and $\overline{AB}$ is the hypotenuse.

Leg $\overline{BC}$ is the side opposite the 30° angle.

$$AB = 2 \cdot BC$$

$$16 = 2 \cdot BC \qquad \text{Divide both sides by 2.}$$

$$8 = BC$$

Leg $\overline{AC}$ is the side opposite the 60° angle.

$$AC = BC \cdot \sqrt{3}$$

$$AC = 8\sqrt{3}$$

Example 3

 Personal Tutor at <u>mathmatters2.com</u>

ARCHITECTURE The angle formed by the roof of a monument measures 60°. Since this is such a steep angle, an architect places a support beam at a slant distance of 8 ft along the roof from the corner. How tall is the support beam to the nearest tenth?

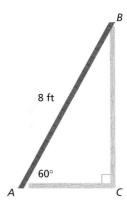

Solution

Let the distance from the corner to the support beam equal *x*. Triangle *ABC* is a 30°-60°-90° right triangle. Since $\overline{AB}$ is the hypotenuse, 8 ft = 2*x*. Solve for *x*.

$$2x = 8$$

$$x = \frac{8}{2} = 4$$

Leg $\overline{BC}$, the height of the support beam, is the side opposite the 60° angle.

$$BC = x\sqrt{3}$$

$$BC = 4\sqrt{3}$$

$$BC \approx 4 \cdot 1.7321$$

$$BC \approx 6.9284$$

So the height of the support beam is approximately 6.9 ft.

> **Check Understanding**
>
> When is it necessary to convert square roots into their approximate decimal equivalents?

◤ TRY THESE EXERCISES

Find each length. Approximate any square roots with decimals.

1. *RS*   **2.** *QR*

3. *GF*   **4.** *GH*

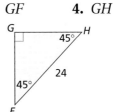

5. ENTERTAINMENT Jeff is flying his kite in an open, flat field. He lets out 840 ft of taut string, which forms an angle of 60° with the ground. How high above the ground is the kite? Round to the nearest tenth of a foot.

6. An isosceles triangle has a right angle and a hypotenuse of length 20.8 cm. Find the lengths of the other two sides of the triangle. Round to the nearest tenth.

7. The shortest side of a right triangle with a 60° angle is 10 m. Find the lengths of the other two sides of the triangle. Leave your answers in square root form.

8. The legs of a 45°-45°-90° triangle each measure 2 in. How long is the hypotenuse of the triangle? Round to the nearest tenth.

◣ PRACTICE EXERCISES • For Extra Practice, see page 623.

Find each length. Leave your answers in square root form.

9. *BC*   **10.** *AC*   **11.** *NP*   **12.** *MN*

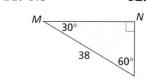

13. *GH*   **14.** *GK*   **15.** *VW*   **16.** *SW*

17. PHOTOGRAPHY A reconnaissance aircraft flies at an elevation of 30,000 ft and takes pictures of the terrain below. The camera mounted on the bottom of the jet points downward to form an angle of 60° with the bottom of the plane. How far in front of the jet does the reconnaissance camera point? Round to the nearest foot.

18. A 4.2-m ladder leaning against a wall makes a 75° angle with the ground. How far from the wall is the foot of the ladder? Round to the nearest tenth.

19. How far up the wall is the top of the ladder in Exercise 18? Find a decimal value to the nearest tenth.

20. When the sun is at an angle of 30°, a tree casts a 90-ft shadow. Find a decimal value for the height of the tree. Round to the nearest tenth.

21. NATURE Pearl and Joey use a directional compass to find the distance across a river. They select an object directly across the river and take a compass reading to that object. Pearl then walks along the river in a direction perpendicular to her original line of sight until the compass reading has changed by 45°. Joey measures the distance Pearl walked as 30 m. What is the width of the river?

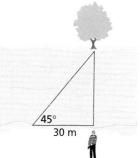

22. SPORTS A professional baseball diamond is a square with sides that are 90 ft long. Find the distance from first base to third base to the nearest foot.

23. The figure is a regular octagon divided into several non-overlapping regions. Explain how you can use these regions to find the area of the octagon.

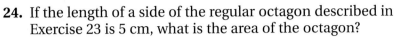

24. If the length of a side of the regular octagon described in Exercise 23 is 5 cm, what is the area of the octagon?

25. A swimmer tries to swim across a river from point *K* to point *M*. Because of the current, the swimmer reaches point *L* instead. How far does the swimmer travel? Find a decimal value to the nearest tenth.

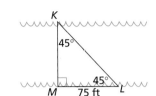

26. WRITING MATH Are all right isosceles triangles similar? Explain.

■ EXTENDED PRACTICE EXERCISES

27. Find the measures of the other sides of the triangles shown. Use these measures to complete the table. Leave your answers in square root form.

	Sine	Cosine	Tangent
30°			
45°			
60°			

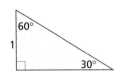

28. CALCULATOR Use a calculator to rewrite the table in Exercise 27 with decimal values rounded to four places.

Write an expression that represents each of the following.

29. The area of a right triangle with a 30° angle and shortest side of length *x*.

30. The area of a right triangle with a 45° angle and legs of length *x*.

31. The area of a rectangle whose diagonal forms two triangles, each having a 60° angle and whose shorter legs measure *x*.

32. DATA FILE Refer to the data on road congestion at ten major U.S. cities on page 575. Use an atlas, a ruler and a protractor to find three cities that approximately form a 45°-45°-90° or 30°-60°-90° right triangle.

■ MIXED REVIEW EXERCISES

Find the volume of each figure. Use 3.14 for π. Round to the nearest tenth. (Lesson 10-7)

33.

34.

35.

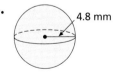

On grid paper, graph each figure and its image under the given rotation. (Lesson 7-3)

36. *A*(2, 3), *B*(8, 3), *C*(8, 7); 90° clockwise around (0, 0)

37. *A*(−9, 5), *B*(−3, 5), *C*(6, −5), *D*(1, −5); 180° clockwise around (0, 0)

38. *A*(2, −2), *B*((3, −5), *C*(8, −5); 360° clockwise around (0, 0)

When you solve a multistep problem or a problem that requires logical reasoning, you need to examine all possibilities and combinations. One strategy, **eliminate possibilities**, is often used to search for reasonable solutions. With this strategy, all possible solutions to a problem are considered. Then they are tested and either dismissed as impossible or accepted as reasonable.

Problem Solving Strategies

Guess and check

Look for a pattern

Solve a simpler problem

Make a table, chart or list

Use a picture, diagram or model

Act it out

Work backwards

✓ Eliminate possibilities

Use an equation or formula

Problem nline Personal Tutor at mathmatters2.com

FITNESS Sam, Queisha and Carlos climb Mt. Everest, which has an elevation of 29,028 ft. They take a break at 25,740 ft. Sam tells the group that they have approximately 3200 ft left to climb to the top. Queisha disagrees, saying that they have approximately 9000 ft left to climb. Carlos thinks that they have approximately 7500 ft left to climb. They use surveying instruments to find that the angle of elevation to the peak is 20.5°. Use trigonometry to determine which hiker has the most reasonable answer.

Solve the Problem

Draw a diagram of the situation.

First find a, the difference between the top of Mt. Everest and where the group is now.

$$29,028 - 25,740 = 3288$$

Sam's estimate of 3200 ft can be eliminated because the distance left to climb is the hypotenuse shown in the diagram, not a.

Use a trigonometric ratio to find x, the remaining distance to the top.

$$\sin 20.5° = \frac{3288}{x}$$
$$0.3502 \approx \frac{3288}{x}$$
$$0.3502 \cdot x \approx 3288$$
$$\frac{0.3502x}{0.3502} \approx \frac{3288}{0.3502}$$
$$x \approx 9388.9$$

The distance left to climb is approximately 9388.9 ft, so Carlos' estimate can also be eliminated. Queisha's estimate of approximately 9000 ft is close to the actual answer, so her answer is the most reasonable.

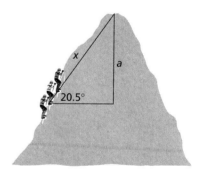

Use trigonometry to find the most reasonable answer.

Five-step Plan
1 Read
2 Plan
3 Solve
4 Answer
5 Check

1. If $m\angle A = 40°$ and $AC = 12.4$, is $BC \approx 9.4$, 10.4, or 11.4?

2. If $m\angle B = 42°$ and $AC = 9.3$, is $AB \approx 8.7$, 6.9, or 13.9?

3. **YOU MAKE THE CALL** Emma knows that $m\angle A = 54°$ and $AC = 13$ cm in right triangle ABC. She reasons that $m\angle B = 36°$ and, using trigonometric ratios, that $AB \approx 17.9$ and $BC \approx 10.5$. Are her answers reasonable? Explain.

◤ PRACTICE EXERCISES

Use trigonometry and the figure for Exercises 1 and 2 to find the most reasonable answer.

4. If $m\angle A = 12°$ and $AC = 8.5$, is $AB \approx 7.9$, 10.1, or 8.7?

5. If $m\angle B = 63°$ and $BC = 6.9$, is $AB \approx 12.6$, 15.2, or 19.3?

6. **BUSINESS** A business developer plans to construct an office near the intersection of Howe Rd. and Schlomer Ave. The two streets will be connected. Is it reasonable for the new road to be 200 yd, 750 yd, or 950 yd?

GEOGRAPHY Ricki is certain that $\triangle ABC$, whose vertices are an office building, a school, and a stadium, is a right triangle. Her best estimate for AB is between 6.0 km and 6.2 km and between 35° and 40° inclusive for $\angle A$.

7. If Ricki uses 35° for $m\angle A$ and 6.0 for AB, is it reasonable to expect that she will get the smallest estimate of BC?

8. If she uses 40° for $m\angle A$ and 6.2 for AB, is it reasonable to expect that she will get the largest estimate of BC?

9. Estimate BC using each set of measurements in Exercises 7 and 8.

10. If Ricki estimates BC using $m\angle A = 37.58$ and $AB = 6.1$, is it reasonable to assume her answer will be between the two estimates in Exercise 9? Explain. Estimate BC using these measurements to verify your answer.

◤ MIXED REVIEW EXERCISES

Factor each polynomial, if possible. (Lesson 9-8)

11. $d^2 - 12d + 36$

12. $r^2 + 8r + 16$

13. $p^2 - 36$

Find the sum of the interior angles of each convex polygon. (Lesson 5-7)

14.

15.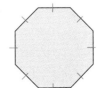

16.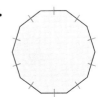

Chapter 11 Review

VOCABULARY

Choose the word from the list that best completes each statement.

1. In $\triangle ABC$, $m\angle C = 90°$. The __?__ of $\angle A$ is the ratio of the length of the leg adjacent to $\angle A$ to the length of the hypotenuse.

2. The __?__ is a method of indirect measuring that is useful on a cloudy day when there are no shadows.

3. In $\triangle ABC$, $m\angle C = 90°$. The __?__ of $\angle A$ is the ratio of the length of the leg opposite to $\angle A$ to the length of the leg adjacent to $\angle A$.

4. In a right triangle, the side opposite the right angle is the __?__.

5. __?__ have the same shape, but not necessarily the same size.

6. In a right triangle, the sides that form the right angle are called __?__.

7. The __?__ relates the lengths of all three sides of right triangles.

8. In $\triangle ABC$, $m\angle C = 90°$. The __?__ of $\angle A$ is the ratio of the length of the side opposite to $\angle A$ to the length of the hypotenuse.

9. Three __?__ are sine, cosine, and tangent.

10. Determining a length by using similar triangles is an example of __?__ measurement.

a. congruent figures

b. cosine

c. direct

d. hypotenuse

e. indirect

f. legs

g. mirror method

h. Pythagorean Theorem

i. similar figures

j. sine

k. tangent

l. trigonometric ratios

LESSON 11-1 ◢ Similar Polygons, p. 474

▶ In **similar polygons**, vertices can be matched so that pairs of corresponding angles are congruent and all pairs of corresponding sides are in proportion.

Determine if each pair of polygons is similar.

11.

55 m   40 m

22 m   16 m

24 m   60 m

12.

7 ft

40°

7 ft

LESSON 11-2 ◢ Indirect Measurement, p. 478

▶ An indirect measurement is one in which you take other measurements that allow you to calculate the required measurement.

Use indirect measurement to find the unknown length x.

13.
140 ft 200 ft ⊢320 ft⊣
x

14.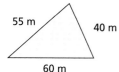
48 m 64 m 17 m
x

15.
x
1.5 m
◄3 m►
30 m

LESSON 11-3 ◥ The Pythagorean Theorem, p. 484

▶ The **Pythagorean Theorem** states that in any right triangle, the square of the **hypotenuse** is equal to the sum of the squares of the **legs**. In right triangle *ABC*, this property can be stated as $c^2 = a^2 + b^2$.

Find the unknown length. Round to the nearest tenth.

16.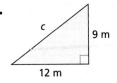

17.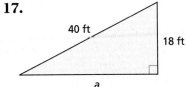

18.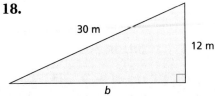

19. What is the length of the hypotenuse of a right triangle whose legs measure 13 cm and 6 cm? Round to the nearest tenth.

LESSON 11-4 ◥ Sine, Cosine and Tangent Ratios, p. 488

▶ For each acute angle in a right triangle, the lengths of the sides can be used to form **trigonometric ratios: sine, cosine,** and **tangent**.

In △*KLM*, find each trigonometric ratio.

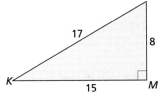

20. sin *K*   21. cos *L*   22. tan *L*

23. cos *K*   24. sin *L*   25. tan *K*

CALCULATOR Use a calculator to find each ratio to four decimal places.

26. sin 42°   27. cos 53°   28. tan 22°

LESSON 11-5 ◥ Find Lengths of Sides in Right Triangles, p. 494

▶ If you know the measure of one acute angle and the measure of one side of a right triangle, you can use trigonometric ratios to find the lengths of the other two sides.

Find each length to the nearest tenth.

29. *ST*   30. *SU*   31. *AB*   32. *AC*

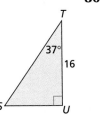

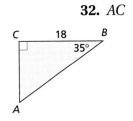

33. To guard against a fall, a ladder should make an angle of 75° or less with the ground. What is the maximum height that a 20-ft ladder can reach safely?

34. The angle that a wheelchair ramp forms with the ground is 6°. What is the height of the ramp if it is 20 ft long?

LESSON 11-6 ◣ Find Measures of Angles in Right Triangles, p. 498

▶ If you know the sine, cosine or tangent of an angle in a right triangle, you can find the measure of the angle.

35. In right triangle *DEF*, cos *D* = 0.25. Find *m∠D*.

36. In right triangle *RST*, tan *S* = 0.7265. Find *m∠S*.

37. In △*WXY*, *m∠W* = 90°, leg *WX* is 14 in., and hypotenuse *XY* is 20 in. Find *m∠X*.

38. In △*LMN*, *m∠L* = 90°, leg *LM* is 5 ft, and hypotenuse *MN* is 8 ft. Find *m∠N*.

39. In △*HIJ*, *m∠H* = 90°, leg *HI* is 22 cm, and leg *HJ* is 9 cm. Find *m∠I*.

LESSON 11-7 ◣ Special Right Triangles, p. 504

▶ Sometimes it is easier to use the special geometric properties of a **30°-60°-90° right triangle** or a **45°-45°-90° right triangle** to solve problems than to use trigonometric ratios.

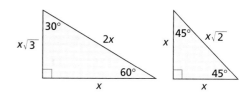

Find each length. Leave your answer in square root form.

40. *MP* **41.** *MN* **42.** *EF* **43.** *DF*

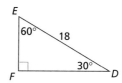

44. The length of the hypotenuse of a 30°-60°-90° right triangle is 7.5 m. Find the length of the side opposite the 30° angle.

45. The length of one of the legs of a 45°-45°-90° right triangle is 6.5 in. Find the lengths of the other sides.

LESSON 11-8 ◣ Problem Solving Skills: Reasonable Solutions, p. 508

▶ One strategy for solving problems is to **eliminate possibilities**.

Use trigonometry to find the most reasonable answer.

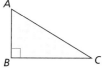

46. If *m∠A* = 20° and *AB* = 5.6, is *BC* ≈ 2.04 , 5.9, or 8.7?

47. If *m∠C* = 36° and *AB* = 15.9, is *AC* ≈ 11.2 , 21.9, or 27.1?

48. If *m∠A* = 55° and *AC* = 7.5, is *AB* ≈ 4.3, 8.0, or 13.1?

49. If *m∠C* = 28° and *BC* = 12.3, is *AC* ≈ 5.8, 10.9, or 13.9?

50. If *m∠C* = 26° and *BC* = 26.2, is *AB* ≈ 12.8, 25.4, or 53.7?

CHAPTER INVESTIGATION

EXTENSION Since most amateur photographers do not carry a tape measure to mark distances, estimating is necessary. Use a yardstick to measure in feet the length of your stride. Take this measurement four times and find the average length of your stride to the nearest quarter foot. Once you know the length of your stride, estimate the number of steps you would take to shoot pictures at the angles of 15°, 28°, and 35°.

Chapter 11 Assessment

Find the unknown measure in each pair of similar polygons.

1.

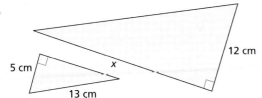

2.

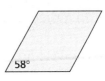 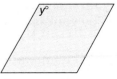

3. A flagpole casts a shadow 49.2 m long, while Joe, who is 1.8 m tall, casts a shadow 2.4 m long. How tall is the flagpole?

Find each unknown measure. Round to the nearest tenth.

4.

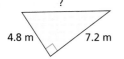

5.

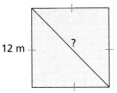

6.

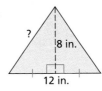

7.

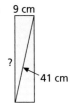

In △QRS, find each trigonometric ratio.

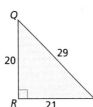

8. cos S

9. tan Q

10. sin Q

Use a calculator to find each value to four decimal places.

11. *KL*

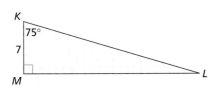

12. *m∠C*

Find each length. Leave your answer in square root form.

13. *AB* and *CB*

14. *YZ* and *XZ*

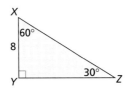

15. The length of a kite string fastened to the ground is 86 m. The vertical height of the kite is 52 m. Find the angle that the string makes with the ground. Round to the nearest degree.

16. A guy wire is anchored to the ground 95.6 m from the base of a transmitting tower. If it forms a 62.5° angle with the ground, how long is the wire? Round to the nearest tenth.

Standardized Test Practice

Part 1 Multiple Choice

Record your answers on the answer sheet provided by your teacher or on a sheet of paper.

1. The high temperature of twelve cities one day in March were 40°F, 72°F, 74°F, 35°F, 58°F, 64°F, 40°F, 67°F, 40°F, 75°F, 68°F, and 51°F. What is the range of the data? (Lesson 1-2)

    (A) 75°F
    (B) 51°F
    (C) 40°F
    (D) 11°F

2. Which angle is *not* congruent to ∠1? (Lesson 5-3)

    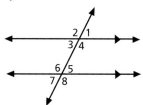

    (A) ∠3
    (B) ∠5
    (C) ∠6
    (D) ∠7

3. Which graph is the solution of $y > 2x + 1$? (Lesson 6-4)

    (A)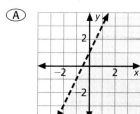
    (B)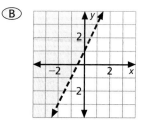
    (C)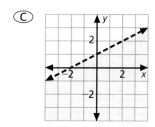
    (D)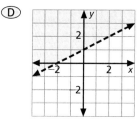

4. If $-5xy + 2x^2y - xy^2 + 4$ is multiplied by $3x^2y$, what is the coefficient of the x^2y term? (Lesson 9-1)

    (A) -15
    (B) -3
    (C) 6
    (D) 12

For Exercises 5 and 6, use the prism at the right.

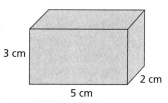

3 cm
2 cm
5 cm

5. What is the surface area of the prism? (Lesson 10-3)

    (A) 25 cm²
    (B) 31 cm²
    (C) 50 cm²
    (D) 62 cm²

6. What is the volume of the prism? (Lesson 10-7)

    (A) 15 cm³
    (B) 30 cm³
    (C) 50 cm³
    (D) 60 cm³

7. In △PQR, what is the value of tan R? (Lesson 11-4)

    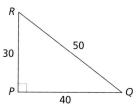

    (A) $\dfrac{3}{4}$
    (B) $\dfrac{4}{5}$
    (C) $\dfrac{5}{4}$
    (D) $\dfrac{4}{3}$

8. A pole is supported with a guy wire. The wire is secured at the ground to form an angle of 75°. It is attached to the pole 15 ft above the ground. What is the distance between the pole and the place where the guy wire is secured to the ground? (Lesson 11-8)

    (A) 4.0 ft
    (B) 5.6 ft
    (C) 6.1 ft
    (D) 15.5 ft

Test-Taking Tip

Question 8
You can often eliminate incorrect answers. In Question 8, you know that the side opposite the 75° angle must be longer than the side opposite the 15° angle. So you can eliminate answer choice D from consideration since 15.5 ft > 15 ft.

Part 2 Short Response/Grid In

Record your answers on the answer sheet provided by your teacher or on a sheet of paper.

9. What is the value of $12(9 + 5) - 6 \cdot 3$? (Lesson 2-2)

The formula $T = 40 + \frac{c}{4}$ shows the relationship between T, the temperature in degrees Fahrenheit, and c, the number of cricket chirps per minute. Use the formula for Questions 10 and 11.

10. If the number of cricket chirps in 1 min is 88, what is the approximate temperature? (Lesson 3-1)

11. If the temperature is 75°F, how many cricket chirps would you expect to hear in 1 min? (Lesson 3-4)

In a lottery, the jackpot is won by choosing four numbers between 1 and 50 in the correct order. Numbers cannot be repeated. Use this information for Questions 12 and 13.

12. How many permutations of the numbers can be chosen? (Lesson 4-6)

13. A smaller prize can be won if you choose the 4 correct numbers in any order. How many combinations can be chosen? (Lesson 4-7)

14. The circumference of a circle varies directly as the length of the diameter. What is the constant of variation? (Lesson 6-8)

15. What is the slope of a line perpendicular to the graph of $4x - 2y = 2$? (Lesson 8-1)

16. The sum of two numbers is 2. Twice the first number minus three times the second is -11. What are the numbers? (Lesson 8-4)

17. Evaluate the determinant of $\begin{bmatrix} -2 & 3 \\ -1 & 5 \end{bmatrix}$. (Lesson 8-5)

18. In a blueprint, 1 in. represents an actual length of 16 ft. If the dimensions of the living room are $1\frac{1}{4}$ in.-by-$1\frac{1}{2}$ in. on the blueprint, what are the actual dimensions of the room? (Lesson 11-1)

19. Find the height of the lamppost. (Lesson 11-2)

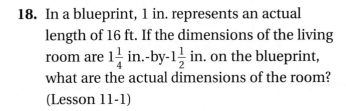

20. If the hypotenuse of a 30°-60°-90° right triangle is 60 cm long, what is the length of the shorter leg of the triangle? (Lesson 11-7)

Part 3 Extended Response

Record your answers on a sheet of paper. Show your work.

21. Two cars leave at the same time and both drive to Nashville. The cars' distance from Knoxville, in miles, can be represented by the two equations below, where t respresents time in hours. (Lessons 6-2 and 9-1)
Car A: $A = 65t + 10$ Car B: $B = 55t + 20$

   a. Which car is faster? Explain.

   b. How far did Car B travel after 2 h?

   c. Find an expression that models the distance between the two cars.

22. Diego hikes 4 mi north, 5 mi west, and then 6 mi north again. Draw a diagram showing the direction and distance of each segment of Diego's hike. At the end of his hike, how far is Diego from his starting point? Explain how you determined this distance. (Lesson 11-3)

23. Explain how you could find the area of the parallelogram. (Lesson 11-5)

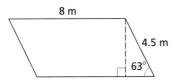

Logic and Sets

THEME: Music

In everyday life, you encounter many types of groups that are arranged in
sets. For example, the nutrition that you receive each day comes from food
groups consumed in meals and snacks. The music you listen to each day
brings together groups of instruments and recording artists to create a
pleasing sound.

You can group almost all musical components into sets and arrangements.
Notes must be arranged using properties of sets and logic to create a song.
The voice ranges in a choir are grouped to produce the perfect sounds.
Instruments in a band are arranged so that the sounds complement those
instruments in the same area.

- **Symphony orchestra conductors** (page 529) combine subsets of range,
  sound, and intensity to produce harmonious music.

- A **professor of music history** (page 547) uses sets
  and logic to teach the timeline of the development of
  musical instruments.

Math Online

mathmatters2.com/chapter_theme

Range of Frequencies for Selected Instruments and Voices

Instrument	Frequency range (Hz)	Voice	Frequency range (Hz)
Flute	256-2304	Soprano	240-1365
Clarinet	160-1536	Alto	171-683
French horn	106-853	Tenor	128-683
Trombone	80-840	Baritone	96-384
Tuba	43-341	Bass	80-341
Violin	192-3072		
Cello	64-683		
Double bass	40-240		

Data Activity: Range of Frequencies for Selected Instruments and Voices

Use the table for Questions 1–5.

1. Which instruments or types of voice are able to produce sounds with a frequency of 950 Hz?

2. Which instrument has the widest frequency range? Which has the narrowest range?

3. Which voice has the widest frequency range? Which has the narrowest range?

4. Name an instrument and voice that have the same highest frequency. Which instrument and voice have the same lowest frequency?

5. How could you use frequencies to create new groupings of instruments?

CHAPTER INVESTIGATION

Western music is divided into time periods or eras. Each musical era has specific characteristics of style, theme and genre. The eras include the Middle Ages, the Renaissance, the Baroque Age, the Classical Period, the Romantic Era and the Twentieth Century. Each of these eras and their composers constitute a set. Musical styles gradually change from one era to the next. Cross-over artists are musicians that show characteristics of more than one era in their compositions.

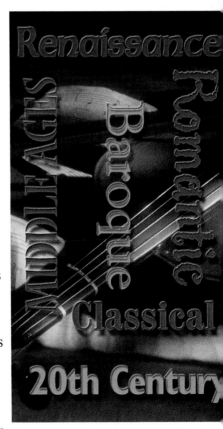

Working Together

Make a Venn diagram of composers from the Baroque Age, Classical Period and Romantic Era, including crossover composers from each time period. Use the Chapter Investigation icons to guide your group's progress.

12 Are You Ready?

Refresh Your Math Skills for Chapter 12

The skills on these two pages are ones you have already learned. Use the examples to refresh your memory and complete the exercises. For additional practice on these and more prerequisite skills, see pages 576–584.

VENN DIAGRAMS

Some problems are more easily solved if you can see a picture of the problem. Venn diagrams provide a clear organized picture of information.

Example Of the 30 students on the field trip, 14 ordered pasta salad with their lunch and 22 ordered potato salad. Four students ordered neither pasta salad nor potato salad. How many students ordered both pasta salad and potato salad?

Draw a Venn diagram to help solve the problem.

The diagram shows that 10 students ordered both pasta salad and potato salad.

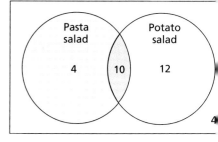

Students' Fruit Preference

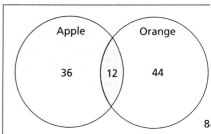

Use the Venn diagram for Exercises 1–4.

1. How many students prefer apples over oranges?
2. How many students prefer both apples and oranges?
3. How many students do not prefer apples or oranges?
4. How many students are represented in the Venn diagram?

Use the Venn diagram for Exercises 5–9.

5. How many people have driven a dozer and a forklift, but not a loader?
6. How many people have driven a loader and a dozer, but not a forklift?
7. How many people have driven only a dozer?
8. How many people have driven all three machines?
9. How many people have driven none of machines?
10. Make a Venn diagram to show the following information. Of 100 people surveyed, 10 like only country music, 8 like only opera and 26 like only rock. Six like country and opera, but not rock. Twelve like rock and country, but not opera. Eighteen like rock and opera, but not country. Twenty people like all three kinds of music.

Experience on Heavy Equipment (Zoo Construction Workers Surveye

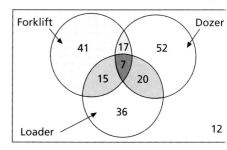

LOGICAL REASONING

Logical reasoning involves examining facts and drawing conclusions from the facts.

Example   Three cars entered the race: Car 1, Car 2 and Car 3. The drivers in no special order were Karl, Taro and Melinda. At the 100-mi mark, Taro was ahead of Car 3 and behind Melinda. Karl could see Car 2 and Melinda ahead. Who drove each car?

Fact	**Conclusion**
Taro was ahead of Car 3.	Taro was not driving Car 3.
Taro was behind Melinda.	Melinda was not driving Car 3.
	Therefore, Karl was driving Car 3.
Karl could see Car 2.	Karl was not driving Car 2.
Karl could see Melinda ahead.	Melinda was not driving Car 2.
	Therefore, Taro was driving Car 2.
	Therefore, Melinda was driving Car 1.

Draw a conclusion from the following sets of facts.

11. Miguel has either gym class or chemistry class fourth period. He does not need to change his clothes for his fourth period class.

12. Natalie will watch one of two movies tonight: Creature From the Slime Pit or Poodles on Parade. She does not like scary movies.

13. We have a Dalmatian, a long-haired cat, and a short-haired cat. Pealer does not have long hair, but he barks. Spats has white feet. Trophy sheds long hair. What is the name of each animal?

14. The Blue Sox baseball team has three men on base. Willie batted first of the three. Gerry's walk on balls forced Juan to move up one base. Which man is on which base?

NON-ROUTINE PROBLEM SOLVING

Some problems require you to think logically to determine the correct answer.

Solve.

15. Mrs. Miller wants to put up a clothes line 250-ft long. The line requires a pole to support it every 5 ft. How many poles will she need?

16. A hat and matching scarf and gloves cost $49 altogether. The hat costs $3 more than the scarf and $2 more than the gloves. What is the cost of each item?

17. A target has areas worth 13, 15, 17, 19, 21 and 26 points. Raymond scored exactly 100 points with 5 darts. Which areas did he hit?

18. Charlene's piggy bank contains $12.95 in quarters and dimes. There are 95 coins in all. How many of each coin are in the bank?

12-1 Properties of Sets

Goals ■ Define sets using different notations.

■ Explore and use properties of sets.

Applications History, Food service, Music

Work in groups of four or five students.

1. Have each member of the group write the names of two classroom objects, such as eraser and pen, on slips of paper.

2. Put the slips of paper in a bag, and mix them well. Then have each member draw two slips from the bag and describe how the objects are related.

3. After you have finished, put all the slips back into the bag and begin again. This time have each student draw three slips.

◣ BUILD UNDERSTANDING

A **set** is a well-defined collection of items. Each item is called an **element**, or a **member**, of the set. A set is usually named with a capital letter and may be defined in three ways.

Description notation describes the set.

W = the set of whole numbers

S = the set of even whole numbers less than 20

Roster notation lists the elements of the set.

$W = \{0, 1, 2, 3, \ldots\}$

$S = \{0, 2, 4, 6, 8, 10, 12, 14, 16, 18\}$

Set-builder notation gives the rule that defines each element.

$W = \{x \mid x$ is a whole number$\}$

$S = \{x \mid x$ is an even whole number less than 20$\}$

A set whose elements cannot be counted or listed is called an **infinite set**. If all of the elements of a set can be counted or listed, the set is called a **finite set**.

> **Reading Math**
>
> Braces are used to enclose the elements of a set.
>
> Three dots, or an ellipsis, indicate that the pattern of a set continues.
>
> In set-builder notation, {$x|x$ is a whole number} is read as "the set of all numbers x such that x is a whole number."

Example 1

Define each set in roster notation and in set-builder notation. Then determine whether the set is finite or infinite.

a. Z, the set of integers

b. P, the set of odd whole numbers less than 10

Solution

a. Roster notation: $Z = \{\ldots, -3, -2, -1, 0, 1, 2, 3, \ldots\}$

Set-builder notation: $Z = \{x \mid x \text{ is an integer}\}$

The ellipsis indicates that all the elements cannot be listed, so Z is infinite.

b. Roster notation: $P = \{1, 3, 5, 7, 9\}$

Set-builder notation: $P = \{x \mid x \text{ is an odd whole number less than } 10\}$

All the elements in the set are listed, so P is a finite set.

To show that 5 is an element of the set $\{1, 3, 5\}$, write $5 \in \{1, 3, 5\}$.

To show that 7 is not an element of the set $\{1, 3, 5\}$, write $7 \notin \{1, 3, 5\}$.

Example 2

Use set notation to write the following.

a. 5 is an element of $\{0, 1, 2, 3, 4, 5\}$.

b. The letter e is not an element of the letters in "banana."

Math: Who, Where, When

Georg Cantor (1845-1918) is credited with being the first to develop set theory as a separate branch of mathematical logic.

Solution

a. $5 \in \{0, 1, 2, 3, 4, 5\}$ **b.** $e \notin \{b, a, n\}$

Two sets A and B are **equal sets** (written $A = B$) if they contain the same members. The members of the sets do not necessarily have to be in the same order. Two sets are **equivalent sets** if they contain the same number of elements.

Example 3

Determine whether the following sets are *equal* or *equivalent*.

$S = \{E, G, B, D, F\}$, $T = \{F, B, D, G, E\}$, $Q = \{3, 7, 11, 5, 9\}$

Solution

Sets S, T, and Q each contain five elements, so they are equivalent sets. Also, each set is equivalent to itself. Since S and T have exactly the same elements, $S = T$.

If every element of set A is also an element of set B, then A is called a **subset** of B. Read $A \subseteq B$ as "A is a subset of B." Consider the sets $X = \{1, 2\}$, $Y = \{1, 2, 3\}$ and $Z = \{1, 3, 5\}$. All of the elements of X are also elements of Y, so X is a subset of Y, or $X \subseteq Y$. Not all of the elements of X are elements of Z, so X is not a subset of Z, or $X \nsubseteq Z$.

Any set is equal to itself. For set Y, $\{1, 2, 3\} = \{1, 2, 3\}$, or $Y = Y$. For this reason, every set is a subset of itself. So $Y \subseteq Y$.

Consider W, weeks containing eight days. Since no week contains eight days, W has no elements. A set having no elements is called an **empty set**, or the **null set**. To indicate that W is an empty set, write either $W = \{\ \}$ or $W = \varnothing$. The null set is a subset of every set.

Example 4

HISTORY List all of the subsets of the set {World War II, Independence Day}.

Solution

Each single-element set that uses an element of a set is a subset of that set. Also, the set itself is a subset, and the null set is a subset. So there are four subsets: {World War II}, {Independence Day}, {World War II, Independence Day}, ∅.

◥ TRY THESE EXERCISES

Define each set in roster notation and in set-builder notation. Then determine whether the set is finite or infinite.

1. *L*, the set of whole numbers less than 7

2. *W*, the set of whole numbers

3. *P*, the set of positive integers less than 22

4. *R*, the set of negative integers

Use set notation to write the following.

5. 3 is an element of {3, 6, 9, 12, . . . }.

6. *m* is not an element of {*a, e, i, o, u*}.

7. Determine whether the sets are *equal* or *equivalent*.
 P = {1, 3, 5, 7}, *Q* = {3, 7, 1, 4}, *R* = {1, 7, 3, 4}

8. List all of the subsets of {*a, c, t*}.

◥ PRACTICE EXERCISES • For Extra Practice, see page 623.

Define each set in roster notation and in set-builder notation. Then determine whether the set is finite or infinite.

9. *U*, the set of whole numbers less than 15

10. *GL*, the set of the Great Lakes

11. *M*, the set of months having 32 days

12. *V*, the set of vowels in *Figueroa*

13. *K*, the set of integers greater than −4

14. *J*, the set of whole numbers greater than 3

Use set notation to write the following.

15. 4 is an element of {4, 6, 8, 10}.

16. Set *N* is not a subset of set *M*.

17. The null set is a subset of set *B*.

18. Set *R* has no elements.

Determine if the following sets are *equal* or *not equal*.

19. {*r, o, v, e*} and {*o, v, e, r*}

20. {*a, b, c, d, e*} and {*c, d, e*, 1, 2}

21. Write a set that is equivalent to the pair of sets in Exercise 19.

22. Write a set that is equivalent to the pair of sets in Exercise 20.

23. MUSIC List all subsets of the set of musical instruments {cello, harp, flute}.

24. WRITING MATH Explain why the null set is a subset of every set.

Determine if each statement is *true* or *false*. If *false*, explain why.

25. $2 \in \{x \mid x$ is a whole number$\}$

26. $8 \in \{1, 3, 5, 7, \ldots\}$

27. $\{8, 12, 16\} \subseteq \{4, 16, 12, 8\}$

28. $\{x \mid x$ is a square$\} \subseteq \{x \mid x$ is a rectangle$\}$

29. If two sets are equal, then they are equivalent.

30. The empty set is a subset of itself.

31. If two sets are equivalent, then they are equal.

32. The set $\{1\}$ is a subset of $\varnothing$.

33. **ERROR ALERT** To indicate that 5 is an element of $\{5, 10, 15, 20\}$, Marcellus writes $\{5\} \in \{5, 10, 15, 20\}$. What mistake has Marcellus made? How should he correct it?

34. **FOOD SERVICE** Determine if the sets of lunch specials offered in the cafeteria are *equal* or *equivalent*. $A = \{$pizza, meatloaf, lasagna$\}$; $B = \{$meatloaf, lasagna, pizza$\}$; $C = \{$lasagna, hamburger, pizza$\}$

Write all the subsets of each set.

35. $\{1\}$

36. $\{8, 9\}$

37. $\{m, a, t\}$

38. $\{\}$

EXTENDED PRACTICE EXERCISES

Determine the number of subsets for each set.

39. $\{1\}$

40. $\{1, 2\}$

41. $\{1, 2, 3\}$

42. $\{1, 2, 3, 4\}$

43. **CRITICAL THINKING** How many subsets does a set of five elements have? How many subsets does a set of six elements have? How many subsets does the null set have?

44. **WRITING MATH** Write a rule or definition that expresses the relationship between the number of elements in a set and its number of subsets.

45. **CHAPTER INVESTIGATION** Find and list five composers from the Baroque Age.

MIXED REVIEW EXERCISES

Find the number of possible outcomes. (Lesson 4-6)

46. Nick has 6 pairs of jeans, 2 pairs of sneakers and 5 T-shirts. From these clothes, how many different outfits can he make?

47. For lunch, the cafeteria offers 3 entrees, 6 side dishes, 2 salads and 5 drinks. How many different meals are possible?

48. A new car comes in 7 colors, 4 body styles, with 3 engines and 4 kinds of tires. How many different cars can be ordered?

Use the distance formula to find the distance between the points. Round answers to the nearest tenth. (Lesson 6-1)

49. $A = (5, 3), B = (4, -2)$

50. $C = (6, -2), D = (4, 3)$

51. $E = (-4, 3), F = (2, -5)$

52. $G = (6, 4), H = (-3, 2)$

53. $I = (-1, 3), J = (2, -5)$

54. $K = (-5, 4), L = (4, -5)$

12-2 Union and Intersection of Sets

Goals
■ Find the complement of a set.
■ Find the union and the intersection of two sets.

Applications Biology, Machinery, Cooking, Music

HOBBIES Suppose there are 32 students in your homeroom. Of the 32 students, 8 play in the band, 12 sing in the choir and 4 are in both band and choir.

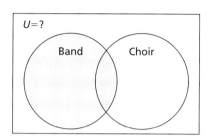

1. Copy and complete the Venn diagram.

2. How many students are in band only?

3. How many students are in choir only?

4. How many of the 32 students are in neither group?

◤ BUILD UNDERSTANDING

In working with sets, you must define the general set of elements being discussed. The set of all such elements is the **universal set**. The universal set (U) can be an infinite set, such as the set of whole numbers or the set of real numbers. It can be a finite set, such as $\{a, e, i, o, u\}$.

From the universal set, several subsets can be formed.

> If $U = \{0, 1, 2, 3, 4, 5, 6, 7, 8, 9\}$, then one possible subset is $A = \{2, 4, 6, 8\}$.

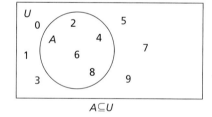

Another subset contains all elements of U that are *not* elements of A. This subset is the **complement** of A, symbolized as A'.

> In the universal set above, $A' = \{0, 1, 3, 5, 7, 9\}$.

In set-builder notation, $A' = \{x \mid x \in U \text{ and } x \notin A\}$.

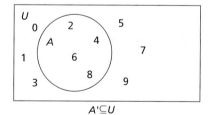

Example 1

Use roster notation to represent each complement.

$U = \{0, 1, 2, 3, 4, 5, 6, 7, 8\}$ $A = \{0, 2, 4, 6, 8\}$ $B = \{1, 3, 5\}$

a. A'

b. B'

Solution

a. A' is the set of elements in U that are not in A.
   $A' = \{1, 3, 5, 7\}$

b. B' is the set of elements in U that are not in B.
   $B' = \{0, 2, 4, 6, 7, 8\}$

Two or more sets can be combined to form new sets. The **union** of any two sets, *A* and *B*, is symbolized as $A \cup B$ and read as "*A* union *B*."

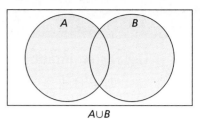

$A\cup B$

The set $A \cup B$ contains all the elements that are in *A*, in *B*, or in both, as shown in the Venn diagram of $A \cup B$.

$$A \cup B = \{x \mid x \in A \text{ or } x \in B\}$$

The **intersection** of two sets, *A* and *B*, is symbolized by $A \cap B$ and read as "*A* intersect *B*."

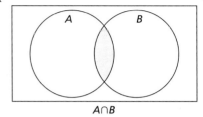

$A\cap B$

The set $A \cap B$ contains the elements that are common to both *A* and *B*, as shown in the Venn diagram of $A \cap B$.

$$A \cap B = \{x \mid x \in A \text{ and } x \in B\}$$

Example 2

List the members of each set.

a. $C \cup D$

b. $C \cap D$

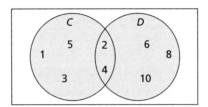

Solution

a. $C \cup D$ is the set of those elements that are in *C*, in *D*, or in both.

$$C = \{1, 2, 3, 4, 5\} \text{ and } D = \{2, 4, 6, 8, 10\}$$

$$C \cup D = \{1, 2, 3, 4, 5\} \cup \{2, 4, 6, 8, 10\}$$

$$C \cup D = \{1, 2, 3, 4, 5, 6, 8, 10\}$$

b. $C \cap D$ is the set of elements common to both *C* and *D*.

$$C \cap D = \{1, 2, 3, 4, 5\} \cap \{2, 4, 6, 8, 10\}$$

$$C \cap D = \{2, 4\}$$

> ### Problem Solving Tip
>
> The beginning sound of "union" and "intersection" can help you remember what symbol to use.
>
> "U"nion − ∪
>
> "In"tersection − ∩

If two sets have no elements in common, their intersection will be the empty set, $\varnothing$. Two sets whose intersection is the empty set are called **disjoint sets**.

Example 3

BIOLOGY On a hiking trip, Team A collected the objects indicated by the set *A* = {maple leaves, soil, tree sap}. Team B collected the objects indicated by the set *B* = {river water, flower pollen, oak leaves}. Find $A \cap B$.

Solution

No item was collected by both teams, so the two sets are disjoint. Therefore, $A \cap B = \varnothing$.

Math
Online   mathmatters2.com/extra_examples

Use roster notation to represent each set.

$U = \{1, 2, 4, 6, 8, 9\}$ $A = \{1, 2, 4\}$ $B = \{4, 6, 9\}$ $C = \{1, 2\}$

1. A' **2.** B' **3.** C' **4.** $A \cup B$

5. $B \cup C$ **6.** $A \cap C$ **7.** $B \cap C$ **8.** $A \cup C$

Find each set by listing the members.

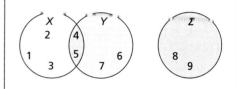

9. $X \cup Y$ **10.** $Y \cup Z$ **11.** $X \cap Y$

12. $Y \cap Z$ **13.** $X \cap Z$ **14.** $X \cup Z$

15. WRITING MATH Explain how drawing a diagram to represent data can be helpful in naming unions, intersections and complements of sets.

Use roster notation to represent each set.

$U = \{-3, -2, -1, 0, 2, 4, 8\}$ $A = \{0, 2, 8\}$ $B = \{-3, -2, 2\}$ $C = \{-2, -1, 4, 8\}$

16. A' **17.** B' **18.** C' **19.** $A \cup B$

20. $B \cup C$ **21.** $A \cap C$ **22.** $B \cap C$ **23.** $A \cup C$

MACHINERY Find each set by listing the members. Making a diagram may help.
$A = \{\text{chain saw, tractor, lawn mower}\}$
$B = \{\text{tractor, table saw, edger, backhoe, chain saw}\}$
$C = \{\text{hedge trimmer, forklift}\}$

24. $A \cup B$ **25.** $A \cap B$ **26.** $A \cup C$

27. $A \cap C$ **28.** $B \cap C$ **29.** $B \cup C$

30. Let $A = \{f, l, o, a, t\}$ and $B = \{b, r, o, t, h\}$. Find $A \cap B$.

31. Let $C = \{2, 4, 8, 16\}$ and $D = \{3, 9, 27\}$. Find $C \cap D$.

Use roster notation to represent each set.

$U = \{1, 2, 3, 4, 5, 6, 7, 8, 9\}$ $E = \{3, 5, 7\}$ $F = \{4, 6\}$ $G = \{2, 4, 6, 8\}$

32. $E' \cap F'$ **33.** $E \cap F$ **34.** $E' \cap F$

35. $E \cup F'$ **36.** $G \cap E'$ **37.** $(G \cup F)'$

COOKING U is the set of eight baking ingredients shown in the diagram. Name each set.

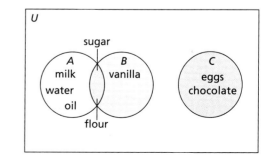

38. $A \cup B$ **39.** $A \cap C$

40. $B' \cap A$ **41.** $C \cup A'$

42. $(C \cup A)'$ **43.** $(C \cap A)'$

44. $A' \cap B$ **45.** $(C' \cup A')'$

Use roster notation to represent each set.

$U = \{1, 2, 3, \ldots, 19, 20\}$ $A = \{2, 3, 5, 7, 11, 13, 17, 19\}$ $B = \{2, 4, 6, 8, \ldots, 18\}$

$C = \{1, 2, 3, 4, 5\}$ $D = \varnothing$

46. A' **47.** C' **48.** B' **49.** D'

50. $A \cap B$ **51.** $C \cup D$ **52.** $B \cap C$ **53.** $A \cup C$

Let $A = \{x \mid x$ is a real number and $x > -1\}$ and $B = \{x \mid x$ is a real number and $x < 3\}$.

54. On two separate number lines, graph A and B.

55. Use set-builder notation to describe $A \cup B$.

56. Use set-builder notation to describe $A \cap B$.

57. WRITING MATH Describe the complement of the complement of a set. Given a set A, describe $(A')'$.

◼ EXTENDED PRACTICE EXERCISES

MUSIC Let A = {flute, drums, cymbals, french horn, tuba, saxophone, oboe}; B = {flute, guitar, piccolo, violin, tuba, keyboard}; C = {flute, guitar, tuba, violin, tamborine}; $D = \varnothing$. List the members of each set.

58. $B \cap (C \cap D)$ **59.** $B \cup (C \cap D)$

60. $A \cap (B \cup C)$ **61.** $C \cup (B \cap A)$

CRITICAL THINKING Determine whether each statement is *true* or *false*. If *false*, provide the correct answer.

U = {whole numbers} $S = \{0, 2, 4, 6, 8, \ldots\}$ $R = \{1, 3, 5, 7, \ldots\}$ $T = \{1, 2, 3, 4, \ldots\}$

62. $R' \cup S = S$ **63.** $R \cap S = \varnothing$

64. $S \cup T = T$ **65.** $R \cup S = S$

66. CHAPTER INVESTIGATION Find and list five composers from the Classical Period.

◼ MIXED REVIEW EXERCISES

State whether it is possible to have a triangle with sides of the given lengths. (Lesson 5-4)

67. 12, 8, 6   **68.** 14, 7, 7   **69.** 15, 8, 9   **70.** 9, 5, 5

71. 3, 4, 9   **72.** 5, 8, 17   **73.** 3, 4, 5   **74.** 2, 9, 10

Find the unknown angle measures in each figure. (Lesson 5-4)

75.
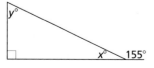

76.
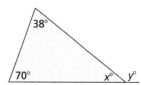

Review and Practice Your Skills

PRACTICE ◣ LESSON 12-1

Define each set in roster notation and in set-builder notation. Then determine whether the set is finite or infinite.

1. *M*, the set of months having 30 days
2. *C*, the set of consonants in *achieve*
3. *S*, the set of integers greater than −10
4. *T*, the set of whole numbers greater than 3
5. *P*, the set of positive integers less than 18
6. *O*, the set of odd one-digit numbers

Use set notation to write the following.

7. *C* is a subset of *E*.
8. *e* is not an element of {*a*, *b*, *c*, *d*}.
9. Set *M* has no elements.
10. 5 is an element of {5, 10, 15, 20, . . .}.
11. List all of the subsets of {*e*, *f*, *g*}.

PRACTICE ◣ LESSON 12-2

Use roster notation to represent each set.

$U = \{-5, -3, -1, 0, 1, 3, 5\}, A = \{1, 3, 5\}, B = \{-5, -3, -1\}, C = \{0, 1\}$

12. A'
13. B'
14. C'
15. $A \cup B$
16. $B \cup C$
17. $A \cap C$
18. $B \cap C$
19. $A \cup C$

Find each set by listing the members.

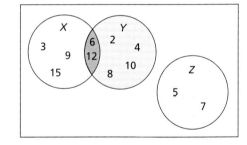

20. $X \cup Y$
21. $Y \cup Z$
22. $X \cap Y$
23. $Y \cap Z$
24. $X \cap Z$
25. $X \cup Z$

26. Let $A = \{f, l, o, w, e, r\}$ and $B = \{r, o, s, e\}$. Find $A \cap B$.
27. Let $C = \{0, 3, 6, 9\}$ and $T = \{2, 4, 6, 8\}$. Find $C \cup T$.

PRACTICE ◣ LESSON 12-1–LESSON 12-2

Use set notation to write the following. (Lesson 12-1)

28. 6 is an element of {2, 4, 6, 8, . . .}
29. Set *D* is a subset of set *E*
30. Set *M* has no elements.

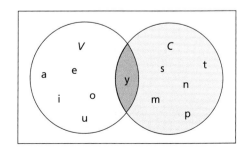

Find each set by listing the members. (Lesson 12-2)

31. V
32. V'
33. $V \cup C$
34. $V \cap C$

MathWorks
Workplace Knowhow

Career – Symphony Orchestra Conductor

A symphony orchestra is composed of instruments grouped into four subsets: woodwinds, brass, percussion and strings. A conductor directs the tempo of a musical piece and cues the musicians. They must also understand the range, sound, and intensity of each instrument in order to produce harmony among them. An orchestra conductor must sometimes arrange subsets of the full orchestra to perform for private or community events.

Subsets of a Full Symphony Orchestra

Woodwinds	Brass	Percussion	Strings
Clarinet	Trumpet	Timpani	Violin
Bass Clarinet	Trombone	Snare drum	Viola
Bassoon	Tuba	Glockenspiel	Cello
Contrabassoon	Horn	Bass drum	Bass
Flute		Triangle	Harp
Oboe		Cymbals	
Piccolo		Celesta	
English horn		Piano	

Use the information in the table for Exercises 1 and 2.

1. A couple would like a chamber group to perform during a wedding ceremony. They would like the group to consist of two different brass instruments and three different string instruments. List two possible configurations for the chamber group.

2. For the opening of the new civic center, the city would like a quartet of four woodwind players to perform. Two of the four instruments must be the same and the remaining two should be different. List three possible groups using all of the instruments from the woodwind subset of the orchestra.

During the summer months, a Pops orchestra is formed using the same instruments as a full symphony, but composed of performers from a variety of backgrounds. Some instrumentalists perform only during the summer with the Pops. Others participate in the Pops as well as the symphony.

3. If the "universe" (in mathematical terms) is considered all the performers in the Pops or the Symphony Orchestra, draw a Venn diagram representing these two groups.

4. When both the Pops and the Symphony Orchestra perform together, this is considered the __?__ of these two sets. When only the members belonging to both groups perform together, this is considered the __?__ of the two sets.

12-3 Problem Solving Skills: Conditional Statements

An *if-then* statement is called a **conditional statement**. Conditional statements have two parts, a *hypothesis* and a *conclusion*. Statements that are not in the form of conditional statements can sometimes be rephrased as *if-then* statements.

hypothesis   conclusion

If **two lines are perpendicular**, then **the lines form four right angles**.

Another conditional statement can be formed by interchanging the hypothesis and conclusion of the original statement. The second statement is called the **converse** of the original.

hypothesis   conclusion

If **two lines form four right angles**, then **the lines are perpendicular**.

A conditional statement may be either true or false. To show that a conditional statement is false, find a **counterexample**. This is an instance that satisfies the hypothesis but not the conclusion. Just one counterexample proves that the conditional statement is false.

Finding a counterexample is a way of guessing an answer and then checking it. Determining the truth of a conditional statement is a form of the **guess and check** problem solving strategy.

Problem Solving Strategies

✔ Guess and check

Look for a pattern

Solve a simpler problem

Make a table, chart or list

Use a picture, diagram or model

Act it out

Work backwards

Eliminate possibilities

Use an equation or formula

Problem

Refer to the following statement.

> Two right angles are congruent.

a. Rephrase the statement as an *if-then* statement.

b. Write the converse.

c. Determine if the statement and its converse are *true* or *false*. If *false*, give a counterexample.

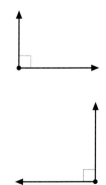

Solve the Problem

a. If two angles are right angles, then the angles are congruent.

b. *Converse:* If two angles are congruent, then the angles are right angles.

c. The original statement is true since all right angles measure 90°. The converse, however, is false. A counterexample would be two angles that both measure 30°. They are two congruent angles, but they are not right angles.

Five-step Plan

1 Read
2 Plan
3 Solve
4 Answer
5 Check

Write the converse of each statement. Then determine whether the statement and its converse are *true* or *false*. If *false*, give a counterexample.

1. If two numbers are negative, then their product is positive.

2. If it is raining, then there is a cloud in the sky.

3. If a triangle has two congruent sides, then it is isosceles.

Rephrase each statement as an *if-then* statement.

4. NUMBER SENSE Two numbers whose sum is even are even.

5. PHYSICS All objects thrown into the air will fall back to the ground.

■ **PRACTICE EXERCISES**

Write the converse of each statement. Then determine whether the statement and its converse are *true* or *false*. If *false*, give a counterexample.

6. If a is a negative number, then a does not equal zero.

7. If $3b + 7 = 25$, then $b = 6$.

8. If the only factors of a positive number are 1 and itself, then the number is prime.

Rephrase each statement as an *if-then* statement.

9. All squares are rectangles.

10. Paintings by Monet are masterpieces.

11. To climb the wall you need a ladder.

12. Toothbrushing prevents cavities.

13. WRITING MATH Describe the relationship between a conditional statement and its converse. If a statement is true, is the converse necessarily true?

DATA FILE For Exercises 14 and 15, refer to the data on audible frequency ranges on page 561. Use this information to complete the conditional statements.

14. If a dog is able to hear a noise, then the noise is at least ___?___ Hz.

15. If a noise is emitted at 235 Hz, then it will sound as loud as ___?___.

■ **MIXED REVIEW EXERCISES**

16. There are 7 horses in the parade. If they march single file, how many possible ways can they be ordered? (Lesson 4-6)

17. Of the 14 students trying out for volleyball, only 5 will be selected to join the team. How many groups of students can be selected? (Lesson 4-6)

Solve each system of equations. (Lesson 8-2–Lesson 8-4)

18. $y = 3x - 9$
$y = -x - 1$

19. $y = 2x - 5$
$y = 3x - 9$

20. $y = -x - 3$
$y = 2x + 12$

Converse, Inverse, and Contrapositive

Goals
- Compare the converse, inverse, and contrapositive of conditional statements.
- Determine if conditional statements are true or false.

Applications Advertising, Weather, Music, Geography

Read each pair of statements.

a. If the instrument is a guitar, then it has strings.
If the instrument has strings, then it is a guitar.

b. If the instrument is a guitar, then it has strings.
If the instrument is not a guitar, then it does not have strings.

c. If the instrument is a guitar, then it has strings.
If the instrument does not have strings, then it is not a guitar.

1. In which pair(s) are both statements true?

2. In which pair(s) is only one statement true?

◣ BUILD UNDERSTANDING

Any statement can be negated. To write the **negation** of a statement, add "not" to the statement or add the words "It is not the case that..."

> *Statement* The instrument is a guitar.
> *Negation* The instrument is not a guitar.
> *Negation* It is not the case that the instrument is a guitar.

In logic, the letters p and q are often used to symbolize statements. The negation can then be symbolized as $\sim p$ or $\sim q$, read "not p" or "not q." If a statement is true, its negation is false. If a statement is false, then its negation is true.

> **Problem Solving Tip**
>
> The negation of a negative statement is a positive statement.
>
> For example, the negation of "It is not raining," is "It is raining."

Example 1

Write the negation of the statement two ways.

> *Statement* The instrument has strings.

Solution

> *Negation* The instrument does **not** have strings.
> *Negation* **It is not the case that** the instrument has strings.

Three related conditional statements can be made from a given statement.
Consider the statement: *If you are in Chicago, then you are in Illinois.*
Let p stand for the hypothesis, *you are in Chicago.*
Let q stand for the conclusion, *you are in Illinois.*
Use the symbol $\rightarrow$ to represent the *if-then* relationship.

Statement If you are in Chicago, then you are in Illinois.
  If p, then q. $p \rightarrow q$

Converse If you are in Illinois, then you are in Chicago.
  If q, then p. $q \rightarrow p$

Inverse If you are not in Chicago, then you are not in Illinois.
  If $\sim p$, then $\sim q$. $\sim p \rightarrow \sim q$

Contrapositive If you are not in Illinois, then you are not in Chicago.
  If $\sim q$, then $\sim p$. $\sim q \rightarrow \sim p$

Recall that the converse of a true conditional statement is not necessarily true.

**COncepts
in MOtion**
Animation
mathmatters2.com

Example 2

WEATHER Write the inverse of the statement, and tell whether it is *true* or *false*. Explain.

Statement If it is raining, then there is a cloud in the sky.

Solution

Inverse If it is not raining, then there is not a cloud in the sky.

The original statement is true, but the inverse is false. This is because it is possible to not have rain even though there is a cloud in the sky.

If a conditional statement is true, its contrapositive is also true. If a conditional statement is false, its contrapositive is also false.

Example 3

Find the contrapositive of each statement. Tell whether each statement and its contrapositive are *true* or *false*. Explain.

a. *Statement* If a figure is a rectangle, then the figure is a polygon.

b. *Statement* If a figure is a rectangle, then it is a square.

Solution

a. *Contrapositive* If a figure is not a polygon, then the figure is not a rectangle.

The statement is true. A rectangle is a type of polygon. The contrapositive is also true. If a figure is not a polygon, then it cannot be a special polygon such as a rectangle.

b. *Contrapositive* If a figure is not a square, then the figure is not a rectangle.

The statement is false. A figure may be a rectangle but not necessarily a square. Its contrapositive is also false. A figure may not be a square, but it might be a rectangle.

Write the negation of each statement two ways.

1. The trumpet is a brass instrument.

2. The number 14 is divisible by 3.

3. All cars have four wheels.

4. Baseball is America's favorite pastime.

Write the converse, inverse, and contrapositive of each statement. Determine if each is *true* or *false*. If *false*, give a counterexample to explain why.

5. If a plant is a tree, then it has leaves.

6. If a number is greater than 25, then that number is greater than 15.

7. If you play the violin, then you play a stringed instrument.

8. **GEOGRAPHY** If you live in Colorado, then you live in Denver.

PRACTICE EXERCISES • For Extra Practice, see page 624.

Write the negation of each statement two ways.

9. History is Jasmine's favorite subject.

10. Tomorrow is Wednesday.

11. The factory recycles unused paper.

12. The house is made out of wood.

Write the converse, inverse, and contrapositive of each statement. Determine if each is *true* or *false*. If *false*, give a counterexample to explain why.

13. If the object is an airplane, then it has wings.

14. If you are studying geometry, then you are studying mathematics.

15. If two lines intersect, then they are not parallel to each other.

16. If a triangle is acute, then all its angles are less than 90°.

17. **WRITING MATH** Choose a conditional statement, and draw the corresponding Venn diagram. Describe how the Venn Diagram would be different for the contrapositive of the statement.

ADVERTISING For Exercises 18–22, refer to the following statement: *Those who wear Swift shoes run fast.*

18. Rewrite the statement as a conditional statement.

19. Identify the hypothesis and conclusion of the statement.

20. Write the converse of the statement. Is it necessarily true? Explain.

21. Write the inverse of the statement. Is it necessarily true? Explain.

22. Write the contrapositive of the statement. Is it necessarily true? Explain.

23. **DATA FILE** Refer to the data on calories used per minute by people of different body weights on page 568 . Using the chart, write an if-then statement. Then write its converse, inverse, and contrapositive.

MUSIC For Exercises 24–28, refer to the following statement:
Those who play the french horn play a brass instrument.

24. Rewrite the statement as a conditional statement.

25. Identify the hypothesis and conclusion of the conditional statement.

26. Write the converse of the statement. Is it necessarily true? Explain.

27. Write the inverse of the statement. Is it necessarily true? Explain.

28. Write the contrapositive of the statement. Is it necessarily true? Explain.

Write the negation of each statement.

29. My car is not a convertible.

30. There is no snow on the ground.

31. The bananas are not ripe.

32. The tent is not torn.

Write the inverse and contrapositive of each statement.

33. If the bill is not paid, then the phone service will be turned off.

34. If the bus arrives on time, then we will not miss our plane.

35. If Akiko does not study, then he will not pass the test.

◢ EXTENDED PRACTICE EXERCISES

Complete each statement with *converse*, *inverse*, or *contrapositive*.

36. The inverse of the converse of a conditional statement is the ___?___.

37. The contrapositive of the converse of a conditional statement is the ___?___.

38. The converse of the inverse of a conditional statement is the ___?___.

39. The converse of the contrapositive of a conditional statement is the ___?___.

40. **CRITICAL THINKING** Write a conditional statement for which the converse, inverse and contrapositive are all true.

41. **CHAPTER INVESTIGATION** List five composers from the Romantic Era.

◢ MIXED REVIEW EXERCISES

State whether each pair of triangles is congruent by SAS, ASA, or SSS. (Lesson 5-5)

42. 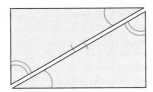　**43.** 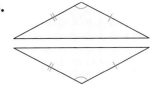　**44.**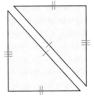

Find the range of each function using the given domain. (Lesson 6-5)

45. $y = 5x - 3$, when the domain is $\{-3, -2, -1, 0, 1, 2, 3\}$

46. $y = x + 6$, when the domain is $\{-5, -4, -3, -2, -1, 0\}$

47. $y = 3x - 7$, when the domain is $\{-3, -2, -1, 0, 1, 2, 3\}$

48. $y = x^2 - 4$, when the domain is $\{1, 2, 3, 4, 5, 6\}$

Review and Practice Your Skills

PRACTICE ■ LESSON 12-3

Write the converse of each statement. Then determine whether the statement and its converse are *true* or *false*. If *false*, give a counterexample.

1. If Mahala lives in Alabama, then he lives south of Canada.

2. If $6x = 24$, then $x = 4$.

3. If a triangle has 3 congruent angles, then it has 3 congruent sides.

4. If $x = 6$, then $x^2 = 36$.

Rephrase each statement as an *if-then* statement.

5. All squares are rhombuses.

6. To drive a car you need a license.

7. Music composed by Mozart is classical.

8. In bad weather your mom drives you to school.

9. To get an A for the semester, Adam must get a 95% on the final exam.

10. Plastic bottles are recyclable.

PRACTICE ■ LESSON 12-4

Write the negation of each statement.

11. The house is made of brick.

12. The flute is wooden.

13. Soccer is a popular sport in Brazil.

14. The car is not red.

Refer to the following statement for Exercises 15–20.

Those who play the oboe play a wind instrument.

15. Rewrite the statement as a conditional statement.

16. Identify the hypothesis of the statement.

17. Identify the conclusion of the statement.

18. Write the converse of the statement. Is it *true* or *false*? If *false*, give a counterexample.

19. Write the inverse of the statement. Is it *true* or *false*? If *false*, give a counterexample.

20. Write the contrapositive of the statement. Is it *true* or *false*? If *false*, give a counterexample.

Use the set $M = \{g, l, o, v, e\}$ for Exercises 21 and 22. (Lessons 12-1)

21. Write a set that is equal to set M.

22. Write a set that is equivalent to set M.

23. Rephrase the following statement as an if-then statement: *All parallelograms have interior angles whose measures have a sum of 360°.* (Lesson 12-3)

24. Write the converse of the statement in Exercise 23. (Lesson 12-4)

Tell whether each statement is *true* or *false*. Then tell if the converse is *true* or *false*. (Lessons 12-3 and 12-4)

25. If today is Monday, then tomorrow is Tuesday.

26. If $y > 0$, then $y^2 > 0$.

Mid-Chapter Quiz

Define each set in roster notation and set-builder notation. Then determine whether the set is finite or infinite. (Lesson 12-1)

1. D, the set of odd integers less than 100

2. N, the set of natural numbers

3. S, the set of seasons in a year

4. T, the set of positive three-digit numbers

Use set notation to write the following. (Lesson 12-1)

5. B is the set of blue apples.

6. -5 is not an element of $\{-6, -4, -2\}$.

7. The null set is a subset of the universal set.

8. Set Q has no elements.

Use roster notation to represent each set. (Lesson 12-2)

$U = \{10, 11, 12, \ldots, 18\}, A = \{10, 12, 14, 16, 18\}, B = \{10, 11, 12\}, C = \{15, 17\}$

9. $A \cup B$

10. A'

11. $A' \cap B$

12. $B \cup C$

13. $B \cap C$

14. $A \cap B$

15. $A \cup C$

16. $B' \cap C'$

Write the converse, inverse and contrapositive of each statement. Determine if each is *true* or *false*. If *false*, give a counterexample. (Lesson 12-4)

17. If you are in Tokyo, then you are in Japan.

18. If n is a whole number, then n is a natural number.

19. If two lines are parallel, then they do not intersect.

20. If the weather is foggy, then driving visibility is low.

12-5 Inductive and Deductive Reasoning

Goals ■ Identify and use inductive reasoning.
■ Identify and use deductive reasoning.

Applications Number theory, Biology, Music, Food service, Travel

Suppose you are drawing marbles from a bag. There are exactly 15 marbles in the bag. The first marble you draw is green, as are the next four marbles.

1. If you know the bag contains an equal number of red, green and white marbles. What can you conclude with certainty about the next marble you will draw?

2. Suppose you have been told nothing about the color of the marbles in the bag. You draw five more marbles, and they are all green. What might you conclude about the color of the next marble you will draw? Can you make this conclusion with certainty?

■ BUILD UNDERSTANDING

When you examine individual instances of an event and then make a conjecture that is supposed to apply to all such events, you use **inductive reasoning**. The greater the number of instances that support a conclusion, the greater the experimental probability that the conclusion is correct.

You cannot be sure, however, since you cannot test all possible instances. So you cannot know that an inductive conclusion is true beyond a doubt.

Example 1

Find a pattern for the sum of the first n odd integers.

$$1 = 1 = 1^2$$
$$1 + 3 = 4 = 2^2$$
$$1 + 3 + 5 = 9 = 3^2$$

$$1 + 3 + 5 + 7 = 16 = 4^2$$
$$1 + 3 + 5 + 7 + 9 = 25 = 5^2$$
$$1 + 3 + 5 + 7 + 9 + 11 = 36 = 6^2$$

Solution

Let n = the number of odd numbers added together. The series of statements suggests the following conjecture.

Conjecture The sum of the first n odd whole numbers equals n^2.

Test the conjecture. Shown are the next two sums in the series.

$$1 + 3 + 5 + 7 + 9 + 11 + 13 = 49 = 7^2$$

$$1 + 3 + 5 + 7 + 9 + 11 + 13 + 15 = 64 = 8^2$$

The conjecture that the sum of the first n odd whole numbers probably equals n^2 seems to be true.

Inductive reasoning can lead to conclusions that are true, but since every instance cannot be tested, inductive reasoning alone cannot make the conclusions certain.

Deductive reasoning begins with a set of statements, called **premises**, that are accepted as true. The **conclusion** of a deductive argument follows from the premises and is implied by them. Together the premises and conclusion make up an **argument**.

In a deductive argument, the premises themselves lead logically to the conclusion. If the logic used to reach a conclusion is sound, it can be accepted as true, since the premises are true.

Example 2

MUSIC Complete the deductive argument by writing the conclusion that follows from the premises.

Premise 1 If a student is a member of the school orchestra, then the student plays a musical instrument.

Premise 2 Francine is a member of the school orchestra.

Solution

Francine is a member of the school orchestra, so she fits the condition stated in Premise 1. Given these premises, the following conclusion is certain.

Conclusion Francine plays a musical instrument.

COncepts in MOtion
Interactive Lab
mathmatters2.com

Example 3

Tell whether the reasoning is *inductive* or *deductive*.

a. Frank finds that on five different Friday evenings before the basketball game, a coffee and dessert cart is outside the gymnasium. He concludes that the cart is outside the gymnasium every night before a basketball game.

b. Kolene uses the definition of a square (a rectangle with four congruent sides) and of perimeter (the sum of the lengths of the sides of a figure) to conclude that the perimeter of any square is equal to 4 times the length of a side.

Check Understanding

Which type of reasoning involves drawing a conclusion based on a pattern of examples?

Which type of reasoning involves drawing conclusions based on a set of definitions or premises?

Solution

a. Frank draws a conclusion about the coffee and dessert cart from evidence about five Friday evenings. He reasons from evidence of a few instances to a conclusion about every instance. The reasoning is inductive.

b. Kolene reasons from the definition of a square and the definition of perimeter to a conclusion about the perimeter of a square. The conclusion follows logically from the premises. The reasoning is deductive.

Predict the next number in each pattern.

1. $1 \cdot 1 = 1$; $11 \cdot 11 = 121$; $111 \cdot 111 = 12{,}321$; $1111 \cdot 1111 = $ __?__

2. 4, 12, 36, 108, __?__   **3.** 1, 1, 2, 3, 5, 8, 13, __?__   **4.** 18, 6, −6, −18, __?__

Complete each argument by drawing a conclusion.

5. *Premise 1* If you are 16 years old or older, you may apply for a driver's license.
Premise 2 Ramon is 17 years old.

6. *Premise 1* If the animal is a spaniel, then it is a canine.
Premise 2 Bowzer is a spaniel.

Tell whether the reasoning is *inductive* or *deductive*.

7. Whenever Damica enters a room where there is a vase of roses, her eyes water and she begins to sneeze. So she must be allergic to roses.

8. Tomás knows that to be in the school choir, you must be able to read music. Mark is a member of the choir. Tomás concludes that Mark can read music.

◤ **PRACTICE EXERCISES** • For Extra Practice, see page 625.

Predict the next number in each pattern.

9. 3, 6, 12, 24, __?__   **10.** 2, 4, 7, 11, 16, __?__

11. 10, 1, 0.1, 0.01, __?__   **12.** $12, 3, \frac{3}{4},$ __?__

Complete each argument by drawing a conclusion.

13. *Premise 1* If Alita finishes her homework, then she will join us for dinner.
Premise 2 Alita has finished her homework.

14. *Premise 1* All parallelograms have two pairs of parallel sides.
Premise 2 A rectangle is a parallelogram.

15. *Premise 1* If the electricity is off, then our appliances do not work.
Premise 2 There is a storm and our electricity is off.

Tell whether the reasoning is *inductive* or *deductive*.

16. FOOD SERVICE Marcia notices that pizza has been on the cafeteria menu every Friday for the past six weeks. She reasons that the cafeteria always serves pizza on Fridays.

17. BIOLOGY Albert knows that the Earth's nitrogen, oxygen and water are all necessary for life. He reasons that evidence of nitrogen, oxygen and water on a planet means that life must be present.

18. If x is greater than y and y is greater than z, what conclusion can be drawn about the relationship between x and z? What type of reasoning did you use to draw your conclusion?

TRAVEL When Reshad turned 24 years old, he bought a new red sports car. He had never gotten a speeding ticket before that time. During the first six months after buying the car, Reshad got four speeding tickets.

19. What conjecture might Reshad make about red sports cars?

20. Are the officers who issued the tickets likely to make the same conjecture? Why or why not?

21. What observations might Reshad make to test his conjecture?

22. Could Reshad's behavior have changed after he bought the car? Explain.

23. **WRITING MATH** Explain the similarities and differences between inductive and deductive reasoning. Give an example of each type.

Draw a conclusion if possible. If not, write *no conclusion*.

24. Kelly is older than Franchesca. Franchesca is older than Megan.

25. Kendall is taller than Doug. Hoshiko is taller than Doug.

◢ EXTENDED PRACTICE EXERCISES

26. **YOU MAKE THE CALL** Ron notices a pattern in the whole numbers shown. He conjectures that every whole number can be written as the sum of consecutive whole numbers. Is he correct? Test his conjecture to find a counterexample, if possible.

$$5 = 2 + 3 \qquad 6 = 1 + 2 + 3 \qquad 9 = 2 + 3 + 4$$
$$15 = 1 + 2 + 3 + 4 + 5 \qquad 42 = 9 + 10 + 11 + 12 \qquad 51 = 16 + 17 + 18$$

27. **CRITICAL THINKING** If Chloe is taking French III, then she will be able to translate the passage. Chloe is able to translate the passage. Can a conclusion be drawn from these premises? Explain why or why not.

28. Suppose that you are taking a quiz. You know that if the first answer is true, then the next answer is false. The last answer is the same as the first answer. The second answer is true. How can you show that the last answer on the quiz is false? (Hint: Assume that the last answer on the quiz is true.)

◢ MIXED REVIEW EXERCISES

Jolly Trolley Riders

Use the table for Exercises 29–31. (Lesson 4-1)

Age group	21-30 yr	31-40 yr	41-50 yr	51-60 yr
Monday	123	256	304	193
Tuesday	204	372	302	188
Wednesday	251	400	255	194
Thursday	294	415	278	253
Friday	309	420	391	266
Saturday	422	518	476	283

29. What is the probability of a Thursday rider being in the 31–40 age range?

30. What is the probability of a rider in the 41–50 age range riding on a Monday?

31. What is the probability of any rider riding on a Tuesday?

Write the polynomial in standard form. (Lesson 9-1)

32. $8x + 4x^2 - 3$

33. $3y - 2y^2 + 6 + 2y^4$

34. $3x^4 + 2x - 3x^2 + 4$

35. $5x^2 + x^5 - 3x^2 + 4x + x^4$

Patterns of Deductive Reasoning

Goals ■ Identify arguments as valid or invalid.
 ■ Identify arguments as sound or unsound.

Applications   Number sense, Music, Horticulture, Retail, Advertising

Read the arguments and answer the questions.

Argument 1 If you live in San Francisco, then you live in California. You live in San Francisco. Therefore, you live in California.

Argument 2 If you live in San Francisco, then you live in California. You do not live in California. Therefore, you do not live in San Francisco.

Argument 3 If you live in San Francisco, then you live in California. You live in California. Therefore, you live in San Francisco.

Argument 4 If you live in San Francisco, then you live in California. You do not live in San Francisco. Therefore, you do not live in California.

San Francisco, California

1. In which arguments does the conclusion follow logically from the premises?

2. In which arguments does the conclusion not follow from the premises?

◥ BUILD UNDERSTANDING

An argument may be valid or invalid depending on the relationship between the premises and the conclusion. In a **valid argument**, if the premises are true, then the conclusion must be true.

One valid argument form is the **Law of Detachment,** or *modus ponens.* This says that given a premise $p \rightarrow q$, if p is affirmed, then q logically follows.

> *Premise* 1 If an animal is a dog, then it has four legs. ($p \rightarrow q$)
> *Premise* 2 Clarence is a dog. (p)
> *Conclusion* Therefore, Clarence has four legs. (q)

Another valid argument form is the **Law of the Contrapositive,** or *modus tollens.* In this form, when given a premise $p \rightarrow q$, if q is denied, then the negation of p logically follows.

> *Premise* 1 If an animal is a dog, then it has four legs. ($p \rightarrow q$)
> *Premise* 2 Polly does not have four legs. ($\sim q$)
> *Conclusion* Therefore, Polly is not a dog. ($\sim p$)

Valid Argument Forms

Law of Detachment
Premise 1 $p \rightarrow q$
Premise 2 p
Conclusion q

Law of the Contrapositive
Premise 1 $p \rightarrow q$
Premise 2 $\sim q$
Conclusion $\sim p$

Reading Math

Latin phrases are sometimes used in mathematics. They are understood even though most people don't know the Latin language.

In an **invalid argument**, even if the premises are true, the conclusion does not logically follow. Given a premise $p \rightarrow q$, an invalid argument occurs when q is affirmed or when p is denied.

> *Premise* 1 If an animal is a dog, then it has four legs. ($p \rightarrow q$)
> *Premise* 2 Fifi has four legs. (q)
> *Conclusion* Therefore, Fifi is a dog. (p)

> *Premise* 1 If an animal is a dog, then it has four legs. ($p \rightarrow q$)
> *Premise* 2 Sam is not a dog. ($\sim p$)
> *Conclusion* Therefore, Sam does not have four legs. ($\sim q$)

Example 1

Determine by form whether the arguments are *valid* or *invalid*.

a. If $2x = 4$, then $x = 2$. $2x = 4$. Therefore, $x = 2$.

b. If a polygon is a square, then the measures of its angles total $360°$. This polygon is not a square. Therefore, the measures of its angles do not total $360°$.

Solution

a. Let p represent "$2x = 4$" and q represent "$x = 2$." The argument has premises $p \rightarrow q$ and p and conclusion q. This is the valid form of the Law of Detachment.

b. Let p represent "this polygon is a square" and q represent "the measures of its angles total $360°$." The argument has premises $p \rightarrow q$ and $\sim p$ and conclusion $\sim q$. This argument form is invalid. Both premises can be true without the conclusion necessarily being true. The polygon could be any nonsquare quadrilateral.

The validity of an argument does not depend on the truth or falsity of the statements in it. Validity is concerned only with the form of an argument.

A valid argument form guarantees that *if* the premises are true, then the conclusion *must* be true. A deductive argument whose form is valid and whose premises are true is called a **sound argument**. A deductive argument whose form is valid but contains at least one false premise is an **unsound argument**.

Example 2  **Online** Personal Tutor at mathmatters2.com

HORTICULTURE Determine the validity and soundness of the argument.

If a flower has red petals, then it is a rose.
This flower has red petals.
Therefore, this flower is a rose.

Solution

Let p represent "a flower has red petals" and q represent "it is a rose." The argument has premises $p \rightarrow q$ and p and conclusion q. So the argument form is Law of Detachment, which is valid. However, the first premise is not true. There are many different types of flowers with red petals, such as a carnation or a tulip.

So the argument is unsound. It is valid because of its form, but it is unsound because its first premise is false.

Determine by form whether the arguments are *valid* or *invalid*. If *valid*, name the argument form.

1. If $3b = 12$, then $b = 4$. $3b = 12$. Therefore, $b = 4$.

2. If it rains, then our picnic will be canceled. Our picnic will be canceled. Therefore, it must have rained.

3. If you like to eat tacos, then you will like Taco Joe's. You like Taco Joe's. Therefore, you like to eat tacos.

Determine the validity of each argument. If valid, name the argument form and determine its soundness.

4. NUMBER SENSE If a number is not a whole number, then it is not greater than zero. The number 1.2 is greater than zero, so it must be a whole number.

5. If the bulb is burned out, then the lamp will not light. The lamp does not light. Therefore, the bulb is burned out.

6. WRITING MATH If the conclusion of an argument is untrue, must the argument necessarily be invalid? Explain.

Determine by form whether the arguments are *valid* or *invalid*. If *valid*, name the argument form.

7. MUSIC If you dedicate years to learning how to play the guitar, then you will be a talented player. Natasha has not dedicated years to learning how to play the guitar, so Natasha will not be a talented guitar player.

8. If the animal is a whale, then it is a mammal. The animal is a whale. Therefore, the animal is a mammal.

9. If two coplanar lines intersect, then they intersect at one point. These two coplanar lines intersect, so their intersection is a point.

Determine the validity of each argument. If valid, name the argument form and determine its soundness.

10. If the food is an apple, then it is a vegetable. This food is not an apple. So this food is not a vegetable.

11. RETAIL If a customer buys two sweaters at regular price, then the third sweater is 50% off. Suppose the customer does not receive 50% off the price of a third sweater. Therefore, the customer has not purchased two sweaters at regular price.

12. If a number is greater than zero, then it is an integer. The number 23.49 is greater than zero, so it must be an integer.

13. If a number is prime, then its only factors are 1 and itself. The number 4 has factors 1, 2 and 4. Therefore, the number 4 is not prime.

14. **DATA FILE** Refer to the data on music media and the year they were introduced on page 563. Use the information to complete the argument. Then determine its validity and soundness.

  If ___?___ was introduced before 1982, then it does not play CDs.
  The ___?___ was invented in 1979.
  Therefore, ___?___.

Determine the validity. If valid name the argument form and determine its soundness.

15. If a basketball player is fouled while taking a three-point shot, then he shoots three free throws. John shoots three free throws, so he must have been fouled while taking a three-point shot.

16. All humans are mammals. Terrance's dog is a mammal. Therefore, Terrance's dog is a human.

17. If dogs were cats, pigs would whistle. Pigs whistle. Dogs are cats.

18. If a number is a whole number, it is a perfect square. The number 14.5 is not a perfect square. So the number 14.5 is not a whole number.

EXTENDED PRACTICE EXERCISES

ADVERTISING Many advertisements are actually arguments in which only a premise and a conclusion are stated. This often leads to invalid arguments. Determine the unstated key premise in each argument.

19. Feeling tired? Eat Pep Crackles for breakfast.

20. Don't be a fool. Buy the best—Springsole running shoes!

21. Don't like rich coffee flavor? Then don't buy Golden Bean coffee.

22. Without life insurance, you can't feel secure about your future.

23. **WRITING MATH** Write an example of an invalid argument in advertising. Explain why a certain conclusion cannot be drawn from such an argument.

24. **CRITICAL THINKING** Show that if the square of a number is odd, then the number is odd. (Hint: Show that the contrapositive is true. Show that if a number is not odd, then the square of the number is not odd. Use $2x$ to represent any even number.)

MIXED REVIEW EXERCISES

Find the surface area of each figure. (Lesson 10-3)

25.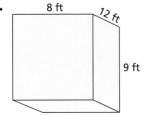
8 ft 12 ft 9 ft

26.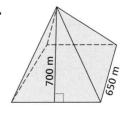
700 m 650 m

Find the trigonometric ratios in lowest terms. (Lesson 11-4)

27. sin R 28. tan R 29. cos R

30. cos S 31. sin S 32. tan S

R 16 20 T 12 S

Review and Practice Your Skills

Predict the next number in each pattern.

1. 1, 4, 16, 64, ___?___

2. 10, 5, 0, ___?___

3. 3, 30, 15, 150, 75, ___?___

Complete each argument by drawing a conclusion.

4. *Premise 1* All rhombuses have diagonals that are perpendicular.
Premise 2 A square is a rhombus.

5. *Premise 1* Courtney is older than Lamar.
Premise 2 Lamar is older than Al.

6. *Premise 1* If the electricity is out, our cordless phone does not work.
Premise 2 The flood caused the power to go out.

7. *Premise 1* If you are over 12 years old, you can babysit.
Premise 2 Sakima is 13 years old.

Determine by form whether the arguments are *valid* or *invalid*. If *valid*, name the argument form.

8. If Shantel gets home early, then she will play tennis.
Shantel got home early. Therefore, she will play tennis.

9. If the animal is a kangaroo, it is a mammal. The animal is not a mammal.
Therefore, the animal is not a kangaroo.

10. If an animal is a horse, it has a mane.
Shadow has a mane. Therefore, Shadow is a horse.

11. If the temperature is 28°F, then water freezes.
The water is frozen. Therefore, the temperature is 28°F.

12. If the figure is a triangle, then the figure is a polygon.
The figure is not a polygon. Therefore, the figure is not a triangle.

Write each statement in if-then form. Then write its converse. Determine if the converse is *true* or *false*. (Lessons 12-3 and 12-4)

13. The angles in a linear pair are supplementary.

14. The secretary is in a bad mood when it rains.

Predict the next two numbers in the pattern. (Lesson 12-5)

15. 7, 9, 11, 13, ___?___, ___?___

16. $\frac{1}{32}, \frac{1}{16}, \frac{1}{8}$, ___?___, ___?___

Let A = {f, o, t, b, a, l} and B = {b, a, s, l}. (Lesson 12-2)

17. Find $A \cup B$.

18. Find $A \cap B$.

In the process of studying the development of music in any culture, a professor of music history examines a broad spectrum of topics relating to music, from the basic construction of musical instruments to the role it plays in cultural development. Professors teach courses, maintain grades, advise students, conduct research and write for scholarly journals.

1. A *box zither*, native to the Middle East, is a rectangular- or trapezoidal-shaped box with strings that lie parallel to and stretch the length of the resonating box. The strings are either struck with light hammers or plucked. Construct an argument using deductive reasoning that explains why the box zither is not a percussion instrument. Begin by assuming that the box zither is a percussion instrument.

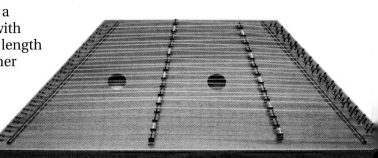

box zither

2. The *English horn* is a woodwind instrument utilizing two thin reeds that are connected laterally and vibrate jointly. However, the following argument could be made regarding the English horn. **If an instrument is a type of horn, then it must be a brass instrument. The English horn is a type of horn. Therefore, it must be a brass instrument.** Evaluate the validity and soundness of this argument.

3. The *alphorn*, a long horn used by Alpine herdsman and villagers, is carved or bored in wood and covered with bark from a birch tree. On a two-week visit to Switzerland, a music professor observed that the instrument was played in a village at noon and 1:00 P.M. each day. She also noticed that within five minutes after the horn was sounded at noon, all of the village farmers gathered in the town hall for a meal. She reasoned that the sound of the alphorn signaled the beginning and end of the daily "lunch hour." What type of reasoning was the professor applying? Justify your answer.

4. The *jew's harp*, a percussion instrument played in many Southeast Asian countries, is commonly used during courtship rituals. While observing these rituals in Bali, a professor determined that if the jew's harp were played by a man three consecutive nights in front of the home of an unwed woman, then he was seeking her hand in marriage. Hayak had stated that he was not yet prepared to enter into a marriage relationship with Maya. Therefore, one can conclude that Hayak had not played the jew's harp for three consecutive nights in front of Maya's home. What valid argument form is used to draw this conclusion?

Logical Reasoning and Proof

Goals
- Use logical reasoning to prove algebraic statements.
- Use logical reasoning to prove geometric statements.

Applications Travel, Number theory, Landscaping, Safety, Music

Refer to line segment AD.

1. Suppose $\overline{AB}$ and $\overline{CD}$ have the same length. What conclusion can be drawn about the lengths of $\overline{AC}$ and $\overline{BD}$?

2. Describe how you can be certain about your conclusion from Question 1.

3. Do you need to know the length of $\overline{BC}$ to draw this conclusion? Explain.

4. Suppose you know that $\overline{AC}$ and $\overline{BD}$ have the same measure. Could you conclude from this information that $\overline{AB} \cong \overline{CD}$? Explain.

◢ BUILD UNDERSTANDING

Deductive reasoning is important in mathematics. In geometry you use true premises, including definitions and postulates, to reach true conclusions. In algebra you use properties and laws to simplify algebraic expressions and to solve equations. Knowing which facts to use and how to organize your argument logically is an important part of solving problems successfully.

Example 1

Show that $m\angle a = m\angle b$ in the figure shown. Present your reasons in a logical order.

Solution

Since the triangle is isosceles, its base angles are congruent, or So, $m\angle 2 = m\angle 3$. Also, angles 1 and 2 and angles 3 and 4 are supplementary.

Therefore, $m\angle 1 + m\angle 2 = 180°$ and $m\angle 3 + m\angle 4 = 180°$. Use these facts from geometry and the laws of algebra to reach the desired conclusion.

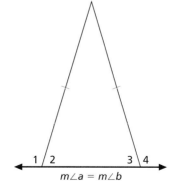

$m\angle 2 = 180° - m\angle 1$
$m\angle 3 = 180° - m\angle 4$
$180° - m\angle 1 = 180° - m\angle 4$
$m\angle 1 = m\angle 4$

Example 1 is a deductive proof in which definitions and rules are applied in order to arrive at a new conclusion.

Example 2

Prove that the statement below is true.

Statement The sum of an odd integer and an even interger is an odd integer.

Solution

Use a deductive argument to show that the statement is true.

Let n be any integer. Then any odd integer can be represented by the expression $2n + 1$. Then any even integer can be represented by the expression $2m$, where m is an integer.

Use these expressions to represent the sum of any odd integer and any even integer.

$$(2n + 1) + 2m = (2n + 2m) + 1$$
$$= 2(n + m) + 1$$

The expression $2(n + m) + 1$ is in the form of an odd integer. So the statement is true for any odd and any even integer. ∎

Example 2 considers the general case of odd and even integers. Any value can be selected for n and m and the result will still hold true. Therefore, we can be certain that the statement is true for *all* odd and even integers.

Check Understanding

Why do the expressions $2n + 1$ and $2m$ represent any odd and even number, respectively?

Find the values of n and m that result in the numbers 23, −56, 1, 120 and 45.

Example 3

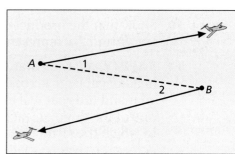

TRAVEL A plane leaves airport A and flies 240 mi northeast. Another plane leaves airport B and flies 240 mi southwest. Show that $m\angle 1 = m\angle 2$.

Solution

Since the two airplanes are flying in opposite directions, their flight routes are parallel to each other. This means that $\overline{AB}$ is a transversal cutting the two parallel lines formed by the airplanes. Since $\angle 1$ and $\angle 2$ are alternate interior angles of the two parallel lines cut by a transversal, they are congruent to each other. ∎

◼ TRY THESE EXERCISES

Show that the conclusion given is true. Present your argument in a logical order.

1. $a° + b° = 180°$

2. $a° = b° = 45°$

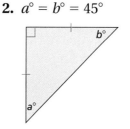

3. **NUMBER THEORY** Show that the sum of any two odd numbers is even. Let $2n + 1$ and $2m + 1$ represent any two odd numbers.

4. Show that the product of two odd numbers is odd. Let $2n + 1$ and $2m + 1$ represent any two odd numbers.

5. Show that the supplements of two congruent angles are congruent to each other.

6. **LANDSCAPING** Find the measure of each angle in the four-sided garden shown in the figure.

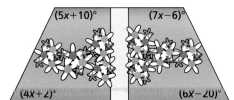

$(5x+10)°$ $(7x-6)°$ $(4x+2)°$ $(6x-20)°$

PRACTICE EXERCISES • For Extra Practice, see page 626.

Show that the conclusion given is true. Present your argument in a logical order.

7. $m\angle 2 = 32°$

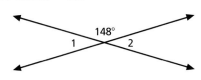

148° 1 2

8. $m\angle 1 < 66°$

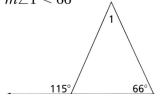

1 115° 66°

9. Show that the difference of two even numbers is another even number. Let $2n$ and $2m$ represent any two even numbers.

10. Show that the product of two even numbers is an even number. Let $2n$ and $2m$ represent any two even numbers.

11. **SAFETY** A farmer wants to safely transport a wolf, a goat and a head of cabbage across a river. If left alone, the wolf will eat the goat and the goat will eat the cabbage. The farmer can take only one of the three on each trip. Use logical reasoning to list the order in which the farmer needs to escort the wolf, goat and cabbage across the river so that they all arrive safely.

12. Use the diagram of the river and its banks to find the width of the river. Give geometric reasons in a logical order to support your answer.

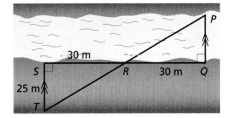

P
30 m
S R 30 m Q
25 m
T

13. **MODELING** Draw a Venn diagram to justify the following statement: *If A is a subset of B and B is a subset of C, then A is a subset of C.*

14. Show that the difference of two odd numbers is an even number. Let $2n + 1$ and $2m + 1$ represent any two odd numbers.

15. The angles of a hexagon have measures $10x°$, $(7x + 25)°$, $(12x - 5)°$, $(11x - 20)°$, $(12x + 5)°$, and $(9x + 65)°$. Find the measure of each angle of the hexagon.

16. MUSIC Use deductive reasoning to arrange the statements in a logical order.
   a. Pam sits closer to the conductor than Smitty.
   b. Smitty plays either a woodwind or a percussion instrument.
   c. Pam does not play a percussion instrument.
   d. All woodwind players are seated closer to the conductor than the percussion players.

17. Write the conclusion that follows from the statements in Exercise 16.

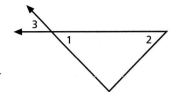

18. In the figure, show that if $\angle 1 \cong \angle 2$, then $\angle 2 \cong \angle 3$.

19. GEOMETRY SOFTWARE Use geometry software to show that the sum of the angles of a triangle is 180°. Draw several different types and sizes of triangles, and measure the angles. Is this an example of inductive or deductive reasoning? Explain.

EXTENDED PRACTICE EXERCISES

20. Show that the product of two rational numbers is a rational number. (Hint: Let $\frac{a}{b}$ and $\frac{c}{d}$ represent any two rational numbers, where a, b, c and d are integers and b and d are not zero.)

21. CRITICAL THINKING Do you think the quotient of two rational numbers is a rational number? Use algebraic facts to prove this statement, or find a counterexample that disproves it.

22. WRITING MATH Explain the different ways that a rational number can be expressed.

23. CHAPTER INVESTIGATION Which composer(s) is a crossover from Baroque to Classical? Which composer(s) is a crossover from Classical to Romantic? Create a Venn diagram of the composers from these three eras.

MIXED REVIEW EXERCISES

Find each product. (Lesson 9-5)

24. $(x + 4)(x + 5)$

25. $(x + 1)(x + 3)$

26. $(x + 2)(x - 4)$

27. $(x + 3)^2$

28. $(x - 4)^2$

29. $(x - 5)(x + 5)$

30. $(2x + 5)(x - 3)$

31. $(3x - 4)(2x + 1)$

32. $(5x - 7)(2x - 3)$

33. $(4x + 2)^2$

34. $(3x - 8)^2$

35. $(3x + 3)^2$

Find the volume of each solid. (Lesson 10-7)

36.

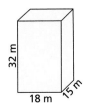

37.

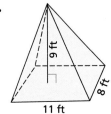

38.

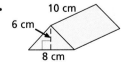

Chapter 12 Review

VOCABULARY ◣

Choose the word from the list that best completes each statement.

1. Two sets *A* and *B* are __?__ if they contain the same members.
2. Two sets are __?__ if they contain the same number of elements.
3. Two sets whose intersection is the empty set are called __?__.
4. To write the __?__ of a statement, add "not" to the statement.
5. In a deductive argument, the __?__ themselves lead logically to the conclusion.
6. If a set of statements is accepted as true, you are using __?__.
7. A set that contains no elements is called a(n) __?__
8. Given the statement $p \rightarrow q$, if *p* is true then *q* is also true. This is an example of the __?__.
9. The set of integers is a __?__ of the set of whole numbers.
10. A set that contains elements common to two other sets is the __?__ of the two sets.

> a. deductive reasoning
> b. disjoint sets
> c. equal sets
> d. equivalent sets
> e. inductive reasoning
> f. intersection
> g. Law of Detachment
> h. Law of the Contrapositive
> i. negation
> j. null set
> k. premises
> l. subset

LESSON 12-1 ◣ Properties of Sets, p. 520

▶ **Description notation** describes the set; **roster notation** lists the elements of the set; and **set-builder notation** gives the rule that defines each element.

Define each set in roster notation and in set-builder notation. Then determine whether the set is finite or infinite.

11. *A*, the set of whole numbers less than 5
12. *B*, the set of negative integers less than −3.
13. *C*, the set of whole numbers greater than −8
14. *D*, the set of positive integers
15. *R*, the set of negative integers greater than −100
16. *S*, the set of natural numbers less than or equal to 100

Use set notation to write the following.

17. 5 is an element of {5, 10, 15, 20, 25, …}
18. 8 is an element of {2, 4, 6, 8, …}

LESSON 12-2 ◣ Union and Intersection of Sets, p. 524

▶ If *A* is a subset of *U* then the **complement** of *A*, *A′*, is the subset containing all elements of *U* that are not elements of *A*.

▶ The **union** of two sets *A* and *B* is the set of all elements that are in *A*, in *B*, or in both *A* and *B*. The **intersection** of two sets *A* and *B* is the set of all elements that are elements of *A* and elements of *B*.

Let $U = \{1, 2, 3, 4, \ldots, 12\}, A = \{1, 2, 3, 4\}$ and $B = \{3, 6, 9, 12\}$. Find the following.

19. A' **20.** $A \cup B$ **21.** $A \cap B$

22. B' **23.** $A' \cap B$ **24.** $A' \cup B$

List the members of each set.

25. $R \cup S$ **26.** $R \cap S$

27. $S \cap T$ **28.** $R \cup T$

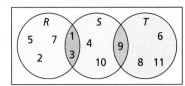

LESSON 12-3 ◣ Problem Solving: Conditional Statements, p. 530

▶ An *if-then* statement is called a **conditional statement**.

▶ Interchanging the hypothesis and conclusion of a conditional statement results in the **converse** of the original statement.

Refer to the following statement: *All rectangles are quadrilaterals.*

29. Rephrase the statement as an *if-then* statement.   **30.** Write the converse.

31. Determine if the statement and its converse are *true* or *false*. If *false*, give a counterexample.

Refer to the following statement: *When it's raining people use umbrellas.*

32. Rephrase the statement as an *if-then* statement.   **33.** Write the converse.

34. Determine if the statement and its converse are *true* or *false*. If false, give a counterexample.

Refer to the following statement: *All figures that are rhombi have perpendicular diagonals.*

35. Rephrase the statement as an *if-then* statement.   **36.** Write the converse.

37. Determine if the statement and its converse are *true* or *false*. If *false*, give a counterexample.

LESSON 12-4 ◣ Converse, Inverse and Contrapositive, p. 532

▶ The **converse** and **inverse** of a true conditional statement are not necessarily true.

▶ A conditional statement and its **contrapositive** are either both true or both false.

Refer to the following statement: *If you are studying Russian, then you are studying a foreign language.*

38. Write the converse of the statement.

39. Write the inverse of the statement.

40. Write the contrapositive of the statement.

Refer to the following statement: *If you like vegetables then you like broccoli.*

41. Write the converse of the statement.

42. Write the inverse of the statement.

43. Write the contrapositive of the statement.

LESSON 12-5 ◣ Inductive and Deductive Reasoning, p. 538

▶ In **inductive reasoning**, a limited number of individual instances provide evidence for a conjecture about all instances of something.

▶ In **deductive reasoning**, a conclusion is drawn from premises that are accepted as true.

Complete each argument by drawing a conclusion.

44. *Premise 1* If the animal is a lion, then it has a mane.
Premise 2 Elsa is a lion.

45. *Premise 1* If you live in Chicago, then you live in Illinois.
Premise 2 Jacquie lives in Chicago.

46. *Premise 1* Green leafy vegetables are a good source of calcium.
Premise 2 Spinach is a green leafy vegetable.

LESSON 12-6 ◣ Patterns of Deductive Reasoning, p. 542

▶ In a **valid argument**, if the premises are true then the conclusion must be true.

Determine by form whether the argument is *valid* or *invalid*. If *valid*, name the argument form.

47. If the bird is a canary, then the bird is not a macaw.
My bird Chatter is a macaw.
Therefore, my bird is not a canary.

48. If you earn an A on the test, then you studied.
Abby earned an A on the test.
Therefore she studied for the test.

49. If your pet swims, then it is a fish.
Fluffy doesn't swim.
Therefore, Fluffy is not a fish.

LESSON 12-7 ◣ Logical Reasoning and Proof, p. 548

▶ Knowing which facts to use and how to organize your argument logically is an important part of solving problems successfully.

Show that the conclusion given is true. Present your argument in a logical order.

50. $m\angle A = 51°$

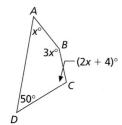

51. $a° = b° = c° = 60°$

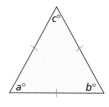

CHAPTER INVESTIGATION

EXTENSION Find composers from the Renaissance, the time period before the Baroque Age, and the Twentieth Century. Add these two time periods to your Venn Diagram.

Chapter 12 Assessment

1. Determine which sets are *equal* and which are *equivalent*.

   $A = \{9, 3, 1, 7, 11\}$ $B = \{11, 7, 1, 9, 3\}$ $C = \{7, 1, 5, 9, 4\}$

Let $U = \{t, i, m, e, s\}$, $P = \{t, i, m, e\}$ and $Q = \{s, i, t, e\}$. Find the following.

2. $P' \cup Q'$ 3. $(P \cup Q)'$ 4. $P' \cap Q'$

Write the converse, inverse, and contrapositive of each statement. Determine whether each statement is *true* or *false*. If *false*, give a counterexample.

5. If a triangle is equilateral, then each angle of the triangle has a measure of 60°.

6. If a number is a member of the set $\{1, 3, 5, 7, 11\}$, then the number is prime.

7. If a plant is a tree, then it has roots.

8. Write the following statement in if-then form. Our collie Max barks whenever he hears a knock at the door.

Write the negation of each statement two ways.

9. The young man is entering college in the fall.

10. Mia is building her science project out of mylar.

Tell whether the reasoning used is inductive or deductive.

11. Ravi measured 12 pairs of alternate interior angles formed by parallel lines and transversals. He then concluded that alternate interior angles have the same measure.

12. Harry used the definitions of a square and of the area of a rectangle to conclude that the area of a square with sides of length x is equal to x^2.

Determine by form whether the arguments are *valid* or *invalid*. If *valid*, name the argument form.

13. If a number is odd, then the square of the number is odd. The number 7 is odd. Therefore, the square of 7 is odd.

14. If the figure is a pentagon, then the figure is a polygon. The figure is not a pentagon. Therefore, the figure is not a polygon.

Show that the conclusion given is true. Present your argument in a logical order.

15. $AB = 5$ cm

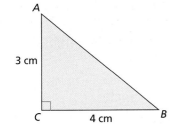

16. The slope of $\overline{AB} = \frac{4}{5}$.

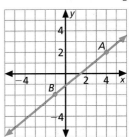

Standardized Test Practice

Part 1 Multiple Choice

Record your answers on the answer sheet provided by your teacher or on a sheet of paper.

1. Simplify: $\dfrac{x^2 y^{-3}}{x^{-2} y^2}$. (Lesson 2-8)

   Ⓐ $\dfrac{1}{y^5}$ Ⓑ x^4 Ⓒ $\dfrac{x^2}{y^6}$ Ⓓ $\dfrac{x^4}{y^5}$

2. A die is rolled, and the spinner shown is spun. What is the probability of rolling a 3 and spinning the letter D? (Lesson 4-4)

   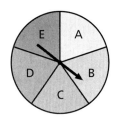

   Ⓐ $\dfrac{1}{5}$ Ⓑ $\dfrac{1}{6}$ Ⓒ $\dfrac{1}{11}$ Ⓓ $\dfrac{1}{30}$

3. The coordinates of the vertices of $\triangle RST$ are $R(3, 3)$, $S(5, 4)$, and $T(4, -3)$. If the triangle is reflected over the x-axis, what are the coordinates of S'? (Lesson 7-2)

   Ⓐ $(-5, -4)$ Ⓑ $(5, -4)$
   Ⓒ $(-5, 4)$ Ⓓ $(4, 5)$

4. When solving the system of equations, which expression could be substituted for x? (Lesson 8-3)

   $$x + 4y = 1$$
   $$2x - 3y = -9$$

   Ⓐ $4y - 1$ Ⓑ $3y - 9$
   Ⓒ $1 - 4y$ Ⓓ $-9 - 3y$

5. What is the determinant of matrix M? (Lesson 8-5)

   $$M = \begin{bmatrix} -3 & 2 \\ 0 & 8 \end{bmatrix}$$

   Ⓐ -26 Ⓑ -24 Ⓒ 0 Ⓓ 22

Test-Taking Tip

Question 1
When answering a multiple-choice question, first find an answer on your own. Then, compare your answer to the answer choices given in the item. If your answer does not match any of the answer choices, check your calculations.

---

6. What is the distance from one corner of the rectangular garden to the opposite corner? (Lesson 11-3)

   Ⓐ 13 yd
   Ⓑ 14 yd
   Ⓒ 15 yd
   Ⓓ 17 yd

7. Which of the following represents the union of X and Y? (Lesson 12-2)

   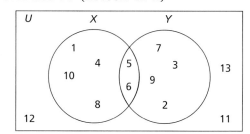

   Ⓐ {1, 2, 3, 4, 5, 6, 7, 8, 9, 10}
   Ⓑ {1, 2, 3, 4, 5, 6, 7, 8, 9, 10, 11, 12, 13}
   Ⓒ {5, 6}
   Ⓓ {11, 12, 13}

8. Which of the following is a counterexample to the statement: *If Eric did not eat lunch, then he must not feel well?* (Lesson 12-3)
   Ⓐ Eric was not hungry.
   Ⓑ Eric ate lunch so he must feel fine.
   Ⓒ Eric ate at a deli.
   Ⓓ Eric does not feel well.

9. Which of the following is the converse of the statement: *All right angles measure 90°?* (Lesson 12-4)
   Ⓐ If an angle is a right angle, then its measure is 90°.
   Ⓑ If an angle is not a right angle, then its measure is not 90°.
   Ⓒ If an angle has a measure of 90°, then it is a right angle.
   Ⓓ If an angle does not have a measure of 90°, then it is not a right angle.

Part 2 Short Response/Grid In

Record your answers on the answer sheet provided by your teacher or on a sheet of paper.

10. Solve the system of equations. (Lesson 8-3)

$$x + 2y = 13$$
$$-2x - 3y = -18$$

11. What polynomial expression describes the area of the rectangle shown? (Lesson 9-5)

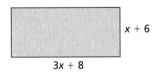

$x + 6$

$3x + 8$

12. Shelly bought a triangular prism at the science museum. The bases of the prism are equilateral triangles with side lengths of 2 cm. The height of the prism is 4 cm. What is the surface area of Shelly's prism? Round to the nearest square centimeter? (Lesson 10-3)

13. Allison is 5 ft tall. She is standing next to a tree with a shadow that is 22.5 ft long. If her shadow is 11 ft long, how tall in feet is the tree? Round to the nearest tenth of a foot. (Lesson 11-2)

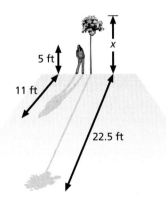

5 ft

11 ft

x

22.5 ft

14. A 17-ft ladder is propped against a window ledge. If the top of the ladder is 8 ft above ground, how far away from the building is the base of the ladder? (Lesson 11-3)

15. What is the length of $\overline{RT}$? Round to the nearest tenth. (Lesson 11-7)

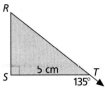

R

S   5 cm   T   135°

Part 3 Extended Response

Record your answers on a sheet of paper. Show your work.

16. A movie theatre concession is ordering cones for popcorn. There are two sizes available. (Lesson 10-8)

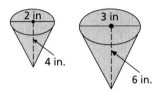

2 in   3 in

4 in.   6 in.

a. What is the volume of each cone? Use 3.14 for π and round to the nearest tenth.

b. What is the ratio of the height of the larger cone to the height of the smaller cone?

c. What is the ratio of the volume of the larger cone to the volume of the smaller cone?

17. Miguel hikes 3 mi north, 7 mi east, and then 6 mi north again. (Lesson 11-3)

a. Draw a diagram showing the direction and distance of each segment of Miguel's hike. Label the starting point, ending point, and the distance in miles of each segment of his hike.

b. To the nearest tenth of a mile, how far is Miguel from his starting point?

c. How did your diagram help you find Miguel's distance from his starting point?

Student Handbook

Data File

Data File

Animals

American Pet Ownership

Pet	Millions of U.S. Households
Dogs	34.6
Cats	29.2
Birds	5.4
Fish	2.7
Rabbits	2.3
Horses	1.9
Rodents	1.4
Reptiles	0.9

How Big is a Bird Egg?

Type of Bird Egg	Length (centimeters)	Mass (grams)
Arctic Tern	4.0	19
Barn Owl	3.9	20.7
Chicken (extra large)	6.3	63.8
Chicken (small)	5.3	42.5
Grey Heron	6.0	60
Hummingbird	Less than 1.25	Less than 0.5
Louisiana Egret	4.5	27.5
Partridge	3.6	14
Swallow	1.9	2
Swift	2.5	3.6
Turtledove	3.1	9

Cricket Chirps Per Minute in Relation to Temperature

Temperature (°F)	Chirps/Minute
50	40
52	48
55	60
56	64
58	76
60	80
64	98
68	116
72	125
73	135
75	140
80	160

Percent of Americans Owning a Pet for Particular Reasons

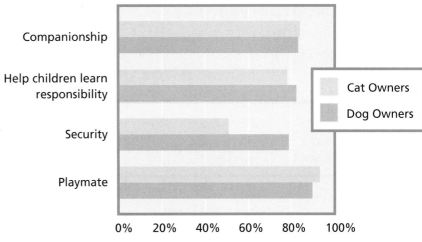

Companionship

Help children learn responsibility

Security

Playmate

☐ Cat Owners
▨ Dog Owners

0% 20% 40% 60% 80% 100%

Data File

Audible Frequency Ranges

Animal	Emission (Hertz: vibrations/second)		Reception (Hertz: vibrations/second)	
	Minimum	Maximum	Minimum	Maximum
Bats	10,000	120,000	1,000	120,000
Cats	760	1,520	60	65,000
Dogs	452	1,080	15	50,000
Dolphins	7,000	120,000	150	150,000
Grasshoppers	7,000	100,000	100	15,000
Humans	85	1,100	20	20,000

Sleep Times

Creature	Average hours/day
Armadillo	19
Cow	7
Elephant	3
Giraffe	4
Horse	5
Human adult	8
Human child	10-12
Jaguar	11
Mole	8
Mountain Beaver	14
Pig	13
Rabbit	10
Sheep	6
Shrew	Less than 1
Two-toed Sloth	20

Arts & Entertainment

Teen Attendance at Various Events

Event	Percent of U.S. Teens Who Attended
Pro sports	44
Art museum	31
Rock concert	28
Other museum	26
Symphony concert	11
Ballet	7
Opera	4

Popular Symphony Orchestras

Instrument	Orchestra Members		
	Boston Symphony Orchestra	Japan Philharmonic Symphony Orchestra	Tenerife Symphony Orchestra
	Boston, Massachusetts	Tokyo, Japan	Canary Islands, Spain
Violins	25	24	26
Violas	11	9	10
Cellos	11	9	9
Basses	9	6	7
Flutes	3	4	3
Piccolo	1	0	2
Oboes	3	4	3
English horn	1	0	2
Clarinets	4	4	6
Bassoons	3	4	3
Contrabassoon	1	0	1
Horns	6	6	5
Trumpets	3	4	2
Trombones	3	4	2
Tuba	1	1	1
Timpani & Percussion	5	4	3
Harps	1	1	1

Music Media and Year Introduced

Music Type	Year Introduced
Record player	1877
33-RPM records	1948
45-RPM records	1949
Transistor radio	1957
Audio cassette	1965
Walkman	1979
CD's	1982
CD walkman	1986
MP3 player	1997

Top Concert Tours

Rank	Performer	Year	Gross revenues
1	The Rolling Stones	1994	$121,200,000
2	U2	2001	$109,700,000
3	Pink Floyd	1994	$103,500,000
4	Paul McCartney	2002	$103,300,000
5	The Rolling Stones	1989	$98,000,000
6	The Rolling Stones	1997	$89,300,000
7	The Rolling Stones	2002	$87,900,000
8	'N Sync	2001	$86,800,000
9	Backstreet Boys	2001	$82,100,000
10	Tina Turner	2000	$80,200,000
11	U2	1997	$79,900,000
12	The Eagles	1994	$79,400,000
13	'N Sync	2000	$76,400,000
14	The New Kids on the Block	1990	$74,100,000
15	Cher	2002	$73,600,000
16	Dave Matthews Band	2000	$68,200,000
17	U2	1992	$67,000,000
18	Billy Joel/Elton John	2002	$65,500,000
19	The Rolling Stones	1999	$64,700,000
20	The Eagles	1995	$63,300,000

Earth Science

Recent Notable Earthquakes

Date	Place	Magnitude (Richter Scale)
September 19, 1985	Michoacan, Mexico	8.1
December 7, 1988	Armenia	7.0
October 17, 1989	San Francisco Bay Area, CA	7.1
January 16, 1995	Kobe, Japan	6.9
May 10, 1997	Northern Iran	7.5
May 30, 1998	Northeastern Afghanistan	6.9
August 17, 1999	Western Turkey	7.4
January 26, 2001	Gujarat, India	7.9
June 23, 2001	Arequipa, Peru	8.1
March 25–26, 2002	Nahrin, Afghanistan	6.1
January 22, 2003	Colima, Mexico	7.6
May 21, 2003	Northern Algeria	6.8

The Planet's Heart

Crust: average depth is 19-22 mi (30-35 km)

Mantle: total depth is 1780 mi (2870 km)

Outer core: 1300 mi (2100 km) of molten iron

Inner core: solid, mostly iron ball, 1700 mi (2740 km) thick

HOW BIG IT IS
Diameter at poles: 7,900 mi (12,713 km).
Diameter at equator: 7,926 mi (12,756 km).
Surface area: 196,885,000 square mi (510,066,000 km²).
Land area: 57,294,000 square mi (148,429,000 km²).
Sea area: 139,591,000 square mi (361,637,000 km²).

Measuring Earthquakes

Richter Scale	Description
2.5	Generally not felt, but recorded on seismometers.
3.5	Felt by many people.
4.5	Some local damage may occur.
6.0	A destructive earthquake.
7.0	A major earthquake.
8.0 and above	Great earthquakes.

The energy of an earthquake is generally reported using the Richter scale, a system developed by American geologist Charles Richter in 1935, based on measuring the heights of wave measurements on a seismograph. On the Richter scale, each single-integer increase represents 10 times more ground movement and 30 times more energy released. The change in magnitude between numbers on the scale can be represented by 10^x and 30^x, where x represents the change in the Richter scale measure. Therefore, a 3.0 earthquake has 100 times more ground movement and 900 times more energy released than a 1.0 earthquake.

How Often Quakes Occur

Richter Scale	World-wide Occurrence
8 and higher	1 per year
7.0-7.9	18 per year
6.0-6.9	120 per year
5.0-5.9	800 per year
4.9 or less	9150 per year

Data File

10 Largest Lakes of the World

Name and Location	Area		Length		Maximum Depth	
	square miles	square kilometers	miles	kilometers	feet	meters
Caspian Sea, Azerbaijan-Russian-Kazakhstan-Turkmenistan-Iran	152,239	394,299	745	1199	3104	946
Superior, U.S.- Canada	31,820	82,414	383	616	1333	406
Victoria, Tanzania-Uganda	26,828	69,485	200	322	270	82
Aral, Kazakhstan-Uzbekistan	25,659	66,457	266	428	223	68
Huron, U.S.-Canada	23,010	59,596	247	397	750	229
Michigan, U.S.	22,400	58,016	321	517	923	281
Tanganyika, Tanzania-Congo	12,700	32,893	420	676	4708	1435
Baikal, Russia	12,162	31,500	395	636	5712	1741
Great Bear, Canada	12,000	31,080	232	373	270	82
Nyasa, Malawi-Mozambique-Tanzania	11,600	30,044	360	579	2316	706

Lake Michigan

Wind Speeds

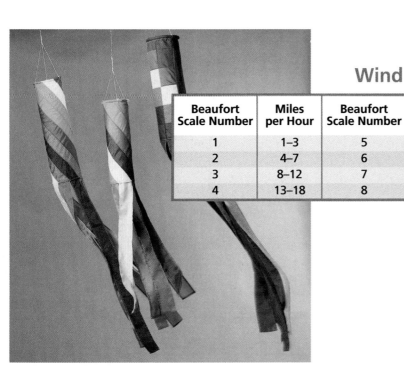

Beaufort Scale Number	Miles per Hour	Beaufort Scale Number	Miles per Hour	Beaufort Scale Number	Miles per Hour
1	1–3	5	19–24	9	47–54
2	4–7	6	25–31	10	55–63
3	8–12	7	32–38	11	64–75
4	13–18	8	39–46	12	greater than 75

Environment

Largest National Parks, in Acreage

National park	Location	Acreage (millions)
Wrangell-St. Elias	Alaska	7.6
Gates of the Arctic	Alaska	7.0
Denali	Alaska	4.7
Katmai	Alaska	3.5
Glacier Bay	Alaska	3.2
Lake Clark	Alaska	2.5
Yellowstone	Montana/Wyoming	2.2
Koouk Valley	Alaska	1.7
Everglades	Florida	1.4
Grand Canyon	Arizona	1.2

Data File

Yellowstone National Park
Montana/Wyoming

Most Visited Sites in the National Park System, 2002

Site (Location)	Recreation Visits
Blue Ridge Parkway (NC & VA)	21,538,760
Golden Gate National Recreation Area (CA)	13,961,267
Great Smoky Mountains National Park (NC & TN)	9,316,420
Gateway National Recreation Area (NJ & NY)	9,014,438
Lake Mead National Recreation Area (AZ & NV)	7,550,284
George Washington Memorial Parkway (VA, MD, and D.C.)	7,419,375
Natchez Trace Parkway (MS, AL, & TN)	5,643,170
Delaware Water Gap National Recreation Area (NJ & PA)	5,165,415
Gulf Islands National Seashore (FL & MS)	4,561,862
Cape Cod National Seashore (MA)	4,455,931

Cape Cod

Threatened and Endangered Species

Group	Endangered[1]		Threatened[2]	
	U.S.	Foreign	U.S.	Foreign
Mammals	56	252	7	16
Birds	75	178	15	6
Reptiles	14	65	18	14
Amphibians	9	8	7	1
Fishes	65	11	40	0
Snails	15	1	7	0
Clams	56	2	6	0
Crustaceans	15	0	3	0
Insects	24	4	9	0
Arachnids	5	0	0	0
Flowering plants	500	1	111	0
Conifers	2	0	0	2
Ferns & others	26	0	2	0
Total	**862**	**522**	**225**	**39**

1. Endangered species are those in danger of extinction.
2. Threatened species are those likely to become an endangered species within the foreseeable future.

Where the Water Goes for a Family of Four in the U.S.

Usage	Gallons/Day
Toilet flushing	100
Shower and bathing	80
Laundry	35
Dishwashing	15
Other	13

What Creates Solid Waste?

- 37.5% — Paper and paperboard
- 17.9% — Yard wastes
- 14.6% — Rubber, leather, textile, wood, other
- 8.3% — Metals
- 8.3% — Plastics
- 6.7% — Food wastes
- 6.7% — Glass

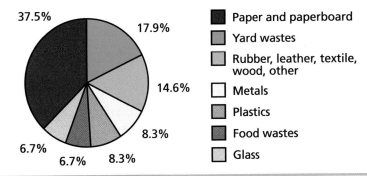

Health & Fitness

Physicians' Office Visits

Specialist Visited	Percent of all Visits
General & Family Practice	29.8
Pediatrics	12.6
Internal Medicine	11.4
Obstetrics & Gynecology	8.4
Opthalmology	5.6
Orthopedic Surgery	5.1
Dermatology	3.8
General Surgery	3.7
Psychiatry	2.4
Otolaryngology	2.3
Cardiovascular Disease	1.6
Urological Survey	1.5
Neurology	0.9
All other specialists	11.0

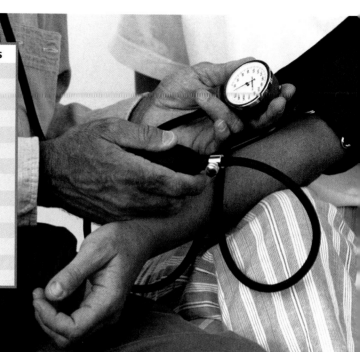

Pizza Toppings

Topping	Calories	Fat Grams
Bacon	110	9
Black olives	56	5
Extra cheese	168	8
Green peppers	5	0
Ham	41	2
Mushrooms	5	0
Onions	11	Less than 1
Pepperoni	80	7
Sausage	97	8

Calories Used Per Minute by People of Different Body Weights

Activity	100 lb	120 lb	150 lb	200 lb
Volleyball	2.3	2.7	3.4	4.6
Walking (3 mi/h)	2.7	3.2	4.0	5.4
Tennis	4.5	5.4	6.8	9.1
Swimming (crawl)	5.8	6.9	8.7	11.6
Skiing (cross-country)	7.2	8.7	10.8	14.5

Target Heart Rates

Age (years)	Target HR Zone, 50–75% (beats per minute)	Average Maximum Heart Rate, 100% (beats per minute)
20	100–150	200
25	98–146	195
30	95–142	190
35	93–138	185
40	90–135	180
45	88–131	175
50	85–127	170
55	83–123	165
60	80–120	160
65	78–116	155
70	75–113	150

Leading Causes of Death for Men and Women in the U.S., 2000

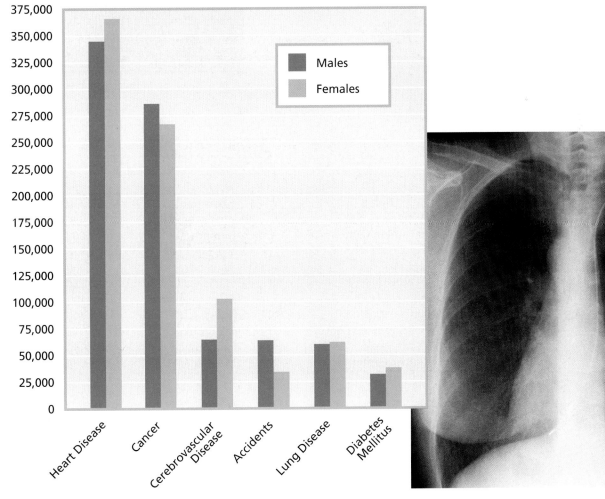

Leisure

Favorite Pastimes

Activity	All Adults (percent)	Men (percent)	Women (percent)
Reading	28	15	41
TV watching	19	19	20
Fishing	12	21	4
Spending time with family/kids	12	13	12
Gardening	11	8	13
Team Sports	9	15	4
Golf	8	14	2
Walking	8	6	10
Going to movies	7	4	9
Swimming	6	6	6
Renting movies	5	6	5
Traveling	5	4	5
Sewing/crocheting	4	-	8
Exercise	4	4	4
Hunting	4	7	-
Church/church activities	4	3	5

Ten Largest U.S. Shopping Centers
(Gross Leasable Area)

Name	Location	Size (square feet)
Mall of America	Bloomington, MN	4,200,000
Del Amo Fashion Center	Torrance, CA	3,000,000
South Coast Plaza/Crystal Court	Costa Mesa, CA	2,918,236
Woodfield Mall	Schaumburg, IL	2,700,000
Sawgrass Mills	Sunrise, FL	2,350,000
Roosevelt Field Mall	Garden City, NY	2,100,000
The Galleria	Houston, TX	2,100,000
Oak Brook Shopping Center	Oak Brook, IL	2,013,000
Garden State Plaza	Paramus, NJ	1,909,804
Tysons Corner Center	McLean, VA	1,900,000

Sizes and Weights of Balls Used in Various Sports

Type	Diameter (centimeters)	Average Weight (grams)
Baseball	7.6	145
Basketball	24.0	596
Croquet ball	8.6	340
Field hockey ball	7.6	160
Golf ball	4.3	46
Handball	4.8	65
Soccer ball	22.0	425
Softball (large)	13.0	279
Softball (small)	9.8	187
Table tennis ball	3.7	2
Tennis ball	6.5	57
Volleyball	21.9	256

Shopping Day Preference

Day	Percent U.S. Shoppers
Sunday	7
Monday	4
Tuesday	5
Wednesday	12
Thursday	13
Friday	17
Saturday	29
No preference	13

Age of Internet Users

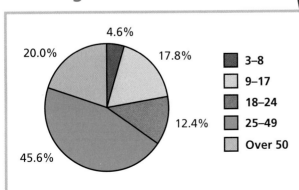

4.6%
17.8%
20.0%
12.4%
45.6%

- 3–8
- 9–17
- 18–24
- 25–49
- Over 50

Data File

Travel & Transportation

Trips to Foreign Destinations from the U.S.

Destinations	Number (thousands)	Percent
All trips to foreign destinations	41,295	100.0
Canada	11,534	27.9
Mexico	9,579	23.2
Central America	857	2.1
Caribbean	4,470	10.8
South America	1,016	2.5
Europe	7,305	17.7
Africa	508	1.2
Asia	3,312	8.0
Pacific	443	1.1
All other foreign destinations	2,271	5.5

Weekly Time Spent Traveling by Trip Purpose of Persons Age 16 and Older

Trip Purpose	Time Spent (hours)
To or from work	1.76
Work-related business	0.16
Shopping	0.85
Other family of personal business	1.30
School/church	0.36
Doctor/dentist	0.08
Vacation	0.12
Visiting friends/relatives	0.70
Pleasure driving	0.05
Other social and recreational	1.04
Other	0.05
Total time per week	6.47

Ways People Get to Work
(workers 16 years and older)

Method	Amount	Percent
Car, truck, or van-drive alone	97,102,050	75.7
Car, truck, or van-carpool	15,634,051	12.2
Public transportation (including taxicab)	6,067,703	4.7
Walked	3,758,982	2.9
Other means	1,532,219	1.2
Worked at home	4,184,223	3.3

Record Transportation Speeds

Type of Transportation	Record Speed (mi/h)
Steam locomotive	126
Magnetic levitation train (with passengers)	321.2
Supersonic transport (SST) (Concorde)	1,037.5
Combat jet fighter	2,110
Unmanned monorail	3,090
Fixed wing aircraft	4,534
Unmanned rocket sled (railed)	6,121
Apollo command module	24,791

Road Congestion at 10 Major U.S. Cities

	Daily Vehicle Miles (thousands)	Vehicle Hours of Delay (thousands)	Delay and Fuel Cost (millions of dollars)
New York, NY	101,295	400,115	7660
Boston, MA	22,890	84,845	1595
Cincinnati, OH	15,745	25,385	505
St. Louis, MO	25,740	41,690	805
Atlanta, GA	42,940	97,245	1885
Miami, FL	13,585	74,850	1365
Dallas, TX	48,700	141,125	2640
Phoenix, AZ	19,425	72,590	1360
Los Angeles, CA	126,495	791,970	14,635
Seattle, WA	22,455	67,550	1315

United States of America

U.S. Population

Year	Population (thousands)	Year	Population (thousands)	Year	Population (thousands)	Year	Population (thousands)
1950	152,271	1964	191,889	1978	222,585	1992	256,894
1951	154,878	1965	194,303	1979	225,055	1993	260,255
1952	157,553	1966	196,560	1980	227,726	1994	263,436
1953	160,184	1967	198,712	1981	229,966	1995	266,557
1954	163,026	1968	200,706	1982	232,188	1996	269,667
1955	165,931	1969	202,677	1983	234,307	1997	272,912
1956	168,903	1970	205,052	1984	236,348	1998	276,115
1957	171,984	1971	207,661	1985	238,466	1999	279,295
1958	174,882	1972	209,896	1986	240,651	2000	282,434
1959	177,830	1973	211,909	1987	242,804	2001	285,545
1960	180,671	1974	213,854	1988	245,021	2002	288,600
1961	183,691	1975	215,973	1989	247,342	2003	290,810
1962	186,538	1976	218,035	1990	250,132		
1963	189,242	1977	220,239	1991	253,493		

Top 10 Coldest U.S. Temperatures on Record

Location	Date	Temperature (°F)
Prospect Creek Camp, AK	January 23, 1971	–80
Rogers Pass, MT	January 20, 1954	–70
Peter's Sink, UT	February 1, 1985	–69
Moran, WY	February 9, 1933	–63
Maybell, CO	February 1, 1985	–61
Island Park Dam, ID	January 18, 1943	–60
Parshall, ND	February 15, 1936	–60
Pokegama Dam, MN	February 16, 1903	–59
McIntosh, SD	February 17, 1936	–58
Seneca, OR	February 10, 1933	–54

Birth Years of the States

State	Date	State	Date	State	Date
AL	1819	LA	1812	OH	1803
AK	1959	ME	1820	OK	1907
AZ	1912	MD	1788	OR	1859
AR	1836	MA	1788	PA	1787
CA	1850	MI	1837	RI	1790
CO	1876	MN	1858	SC	1788
CT	1788	MS	1817	SD	1889
DE	1787	MO	1821	TN	1796
FL	1845	MT	1889	TX	1845
GA	1788	NE	1867	UT	1896
HI	1959	NV	1864	VT	1791
ID	1890	NH	1788	VA	1788
IL	1818	NJ	1787	WA	1889
IN	1816	NM	1912	WV	1863
IA	1846	NY	1788	WI	1848
KS	1861	NC	1789	WY	1890
KY	1792	ND	1889		

10 Windiest U.S. Cities

City	Mean Speed (mi/h)
Blue Hill, MA	15.4
Dodge City, KS	13.9
Amarillo, TX	13.5
Rochester, MN	13.1
Cheyenne, WY	12.9
Casper, WY	12.8
Great Falls, MT	12.6
Goodland, KS	12.5
Boston, MA	12.5
Lubbock, TX	12.4

Top 10 Hottest U.S. Temperatures on Record

Location	Date	Temperature (°F)
Greenland Ranch, CA	July 10, 1913	134
Parker, AZ	July 7, 1905	127
Laughlin, NV	June 26, 1990	122
Alton, KS	July 24, 1936	121
Steele, ND	July 6, 1936	121
Gannvalley, SD	July 5, 1936	120
Ozark, AK	August 10, 1936	120
Seymour, TX	August 12, 1936	120
Tishmomingo, OK	July 26, 1943	120
Pendleton, OR	August 10, 1898	119

Prerequisite Skills

❶ Place Value and Order

Example 1

Write 2,345,678.9123 in words.

Solution

The place value chart shows the value of each digit. The value of each place is ten times the place to the right.

millions	hundred thousands	ten thousands	thousands	hundreds	tens	ones	.	tenths	hundredths	thousandths	ten thousandths
2	3	4	5	6	7	8	.	9	1	2	3

The number shown is *two million, three hundred forty-five thousand, six hundred seventy-eight and nine thousand one hundred twenty-three ten-thousandths.*

Example 2

Use < or > to make this sentence true.   6 ▇ 2

Solution

Remember, < means "less than" and > means "greater than." So, 6 > 2.

◣ EXTRA PRACTICE EXERCISES

Write each number in words.

1. 3647

2. 6,004,300.002

3. 0.9001

Write each of the following as a number.

4. two million, one hundred fifty thousand, four hundred seventeen

5. five thousand, one hundred twenty and five hundred two thousandths

6. nine million, ninety thousand, nine hundred and ninety-nine ten-thousandths

Use < or > to make each sentence true.

7. 9 ▇ 8

8. 164 ▇ 246

9. 63,475 ▇ 6,435

10. 52 ▇ 50

11. 5.39 ▇ 9.02

12. 43.94 ▇ 53.69

❷ Add and Subtract Whole Numbers and Decimals

To add or subtract whole numbers and decimals, write the digits so the place values line up. Add from right to left, renaming when necessary. When adding or subtracting decimals, be sure to place the decimal point directly below the aligned decimals in the problem.

Example 1

Add 0.058, 25.39, 6346, and 1.57. The answer is called the *sum*.

Solution

$$
\begin{array}{r}
0.058 \\
25.39 \\
6346. \\
+1.57 \\
\hline
6373.018
\end{array}
$$

The zero to the left of the decimal point is used to show there are no ones.

The decimal point is at the end of whole numbers.

Add from right to left.

Example 2

Subtract 6.37 from 27. The smaller number is subtracted from the larger number. The answer is called the *difference*.

Solution

$$
\begin{array}{r}
27.00 \\
-6.37 \\
\hline
20.63
\end{array}
$$

Add zeros if that helps you complete the subtraction.

◤ EXTRA PRACTICE EXERCISES

Add or subtract.

1. $23.146 - 17.215$

2. $46.48 - 6.57$

3. $52 - 1.95$

4. $0.86 + 0.75$

5. $83 - 82.743$

6. $9.45 + 13.2$

7. $14.5 - 9.684$

8. $913.03 - 79$

9. $0.8523 - 0.794$

10. $30,000 - 1237.64$

11. $45.3 + 160.09$

12. $6000 - 4362$

13. $462.09 + 32.7$

14. $78,850 - 56,176.8$

15. $3160.915 + 1920.03$

16. $17,347.85 - 12,516.90$

17. $107,285 - 61,500.25$

18. $6.4 + 54.2 + 938.05 + 3.7 + 47.3$

19. $1765.36 + 1587.50 + 1400$

20. $51,876.36 + 48,156.95 + 1417.86$

21. $76.2 + 80 + 56 + 9.321$

22. $567.1 + 6 + 13.452 + 100$

23. $5.93 + 0.06 + 96.021 + 0.0378$

24. $51.5 + 87 + 68.2$

25. $6532.03 + 861.006 + 3170.95$

➌ Multiply Whole Numbers and Decimals

To multiply whole numbers, find each partial product and then add.

When multiplying decimals, locate the decimal point in the product so that there are as many decimal places in the product as the total number of decimal places in the factors.

Example 1

Multiply 2.6394 by 3000.

Solution

$$
\begin{array}{r}
2.6394 \\
\times \quad 3000 \\
\hline
7918.2000 \text{ or } 7918.2
\end{array}
$$

Zeros after the decimal point can be dropped because they are not significant digits.

Example 2

Multiply 3.92 by 0.023.

Solution

$$
\begin{array}{r}
3.92 \\
\times \quad 0.023 \\
\hline
1176 \\
+ \quad 7840 \\
\hline
0.09016
\end{array}
$$

2 decimal places
+ 3 decimal places

5 decimal places

The zero is added before the nine, so that the product will have five decimal places.

◢ EXTRA PRACTICE EXERCISES

Multiply.

1. 36×45	**2.** 500×30	**3.** $17{,}000 \times 230$
4. 6.2×8	**5.** 950×1.6	**6.** 3.652×20
7. 179×83	**8.** 257×320	**9.** 8560×275
10. 467×0.3	**11.** 2.63×183	**12.** 0.758×321.8
13. 49.3×1.6	**14.** 6.859×7.9	**15.** 794.4×321.8
16. 0.08×4	**17.** 0.062×0.5	**18.** 0.0135×0.003
19. 21.6×3.1	**20.** 8.76×0.005	**21.** 5.521×3.642
22. 5.749×3.008	**23.** 8.09×0.18	**24.** $89{,}946 \times 2.85$
25. 6.31×908	**26.** 391.05×25	**27.** $35{,}021 \times 76.34$

Prerequisite Skills

④ Divide Whole Numbers and Decimals

Dividing whole numbers and decimals involves a repetitive process of estimating a quotient, multiplying and subtracting.

$$
\begin{array}{r}
34 \quad \leftarrow \text{quotient} \\
\text{divisor} \rightarrow 7\overline{)239} \quad \leftarrow \text{dividend} \\
\underline{21{\downarrow}} \quad \leftarrow 3 \times 7 \\
29 \quad \leftarrow \text{Subtract. Bring down the 9.} \\
\underline{28} \quad \leftarrow 4 \times 7 \\
1 \quad \leftarrow \text{remainder}
\end{array}
$$

Example 1

Find: 283.86 ÷ 5.7

Solution

When dividing decimals, move the decimal point in the divisor to the right until it is a whole number. Move the decimal point the same number of places in the dividend. Then place the decimal point in the answer directly above the new location of the decimal point in the dividend.

If answers do not have a remainder of 0, you can add 0's after the last digit of the dividend and continue dividing.

$$
5.7.\overline{)283.8.6} \rightarrow
\begin{array}{r}
49.8 \\
57\overline{)2838.6} \\
\underline{228} \\
558 \\
\underline{513} \\
45\ 6 \\
\underline{45\ 6} \\
0
\end{array}
$$

EXTRA PRACTICE EXERCISES

Divide.

1. $72 \div 6$
2. $6000 \div 20$
3. $26{,}568 \div 8$
4. $5.6 \div 7$
5. $120 \div 0.4$
6. $936 \div 12$
7. $3.28 \div 4$
8. $0.1960 \div 5$
9. $1968 \div 0.08$
10. $16 \div 0.04$
11. $1525 \div 0.05$
12. $109.94 \div 0.23$
13. $0.6 \div 24$
14. $7.924 \div 0.28$
15. $32.6417 \div 9.1$
16. $24 \div 0.6$
17. $1784.75 \div 29.5$
18. $0.01998 \div 0.37$
19. $7.8 \div 0.3$
20. $12{,}000 \div 0.04$
21. $820.94 \div 0.02$
22. $89{,}946 \div 28.5$
23. $15 \div 0.75$
24. $7.56 \div 2.25$
25. $0.19176 \div 68$
26. $0.168 \div 0.48$
27. $5.1 \div 0.006$
28. $55{,}673 \div 0.05$
29. $84.536 \div 4$
30. $261.18 \div 10$
31. $134{,}554 \div 0.14$
32. $90{,}294 \div 7.85$
33. $59{,}368 \div 47.3$
34. $11{,}633.5 \div 439$
35. $28.098 \div 14$
36. $16.309 \div 0.09$
37. $55.26 \div 1.8$
38. $8276 \div 0.627$
39. $10{,}693 \div 92.8$
40. $48.8 \div 1.6$
41. $27{,}268 \div 34$
42. $546.702 \div 0.078$

❺ Multiply and Divide Fractions

To multiply fractions, multiply the numerators and then multiply the denominators. Write the answer in simplest form.

Example 1

Multiply $\frac{2}{5}$ and $\frac{7}{8}$.

Solution

$$\frac{2}{5} \times \frac{7}{8} = \frac{2 \times 7}{5 \times 8} = \frac{14}{40} = \frac{7}{20}$$

To divide by a fraction, multiply by the reciprocal of that fraction. To find the reciprocal of a fraction, invert the fraction (turn upside down). The product of a fraction and its reciprocal is 1. Since $\frac{2}{3} \times \frac{3}{2} = \frac{6}{6}$ or 1, $\frac{2}{3}$ and $\frac{3}{2}$ are reciprocals of each other.

Example 2

Divide $1\frac{1}{5}$ by $\frac{2}{3}$.

Solution

$$1\frac{1}{5} \div \frac{2}{3} = \frac{6}{5} \div \frac{2}{3} = \frac{6}{5} \times \frac{3}{2} = \frac{6 \times 3}{5 \times 2} = \frac{18}{10}, \text{ or } 1\frac{4}{5}$$

◤ Extra Practice Exercises

Multiply or divide. Write each answer in simplest form.

1. $\frac{2}{3} \div \frac{5}{6}$

2. $\frac{3}{5} \times \frac{10}{12}$

3. $\frac{5}{8} \div \frac{1}{4}$

4. $\frac{1}{2} \times \frac{2}{3}$

5. $\frac{2}{3} \times \frac{1}{2}$

6. $\frac{3}{4} \times \frac{5}{8}$

7. $\frac{1}{2} \div \frac{2}{3}$

8. $\frac{2}{3} \div \frac{1}{2}$

9. $\frac{3}{4} \div \frac{5}{8}$

10. $2\frac{2}{3} \div 1\frac{3}{5}$

11. $1\frac{1}{5} \times 2\frac{1}{4}$

12. $3\frac{1}{3} \times 1\frac{1}{10}$

13. $5\frac{2}{5} \div 2\frac{4}{7}$

14. $2\frac{4}{7} \div 5\frac{2}{5}$

15. $2\frac{4}{7} \times 5\frac{2}{5}$

16. $1\frac{7}{8} \div 1\frac{7}{8}$

17. $\frac{3}{4} \times \frac{2}{3} \times 1\frac{5}{8} \times 2\frac{2}{3}$

18. $7\frac{1}{2} \div 2\frac{1}{4}$

19. $6\frac{2}{3} \times 4\frac{1}{2} \times 5\frac{3}{8}$

20. $11\frac{5}{9} \times 6\frac{1}{12}$

21. $\frac{25}{42} \div \frac{5}{21}$

22. $\frac{13}{18} \div \frac{8}{9}$

23. $\frac{3}{8} \times \frac{11}{12} \times \frac{16}{33}$

24. $\frac{51}{56} \div \frac{17}{24}$

❻ Add and Subtract Fractions

To add and subtract fractions, you need to find a common denominator and then add or subtract, renaming as necessary.

Example 1

Add $\frac{3}{4}$ and $\frac{5}{6}$.

Solution

$$\frac{3}{4} = \frac{3}{4} \times \frac{3}{3} = \frac{9}{12}$$
$$+\frac{5}{5} = \frac{5}{6} \times \frac{2}{2} = +\frac{10}{12}$$
$$\frac{19}{12}$$

Add the numerators and use the common denominator.

Then simplify. $\frac{19}{12} = 1\frac{7}{12}$

Example 2

Subtract $1\frac{3}{5}$ from $5\frac{1}{2}$.

Solution

$$5\frac{1}{2} = \quad 5\frac{5}{10} = \quad 4\frac{15}{10}$$
$$-1\frac{3}{5} = -1\frac{6}{10} = -1\frac{6}{10}$$
$$3\frac{9}{10}$$

You cannot subtract $\frac{6}{10}$ from $\frac{5}{10}$, so rename again.

◥ EXTRA PRACTICE EXERCISES

Add or subtract.

1. $\frac{1}{5} + \frac{1}{10}$

2. $\frac{2}{3} + \frac{1}{3}$

3. $\frac{5}{8} + \frac{3}{4}$

4. $\frac{6}{7} - \frac{2}{7}$

5. $\frac{3}{4} - \frac{1}{3}$

6. $\frac{5}{8} - \frac{1}{4}$

7. $2\frac{1}{2} + 3\frac{1}{2}$

8. $6\frac{5}{8} + 3\frac{7}{8}$

9. $3\frac{2}{3} + 4\frac{1}{2}$

10. $2\frac{3}{4} - 1\frac{1}{4}$

11. $5\frac{1}{8} - 3\frac{7}{8}$

12. $1\frac{1}{3} - \frac{2}{3}$

13. $6\frac{1}{2} + 5\frac{7}{9}$

14. $9\frac{2}{5} - 1\frac{1}{8}$

15. $7\frac{2}{3} + 6\frac{1}{5}$

16. $8\frac{1}{10} - 5\frac{2}{3}$

17. $6\frac{1}{2} - 5\frac{3}{5}$

18. $10\frac{5}{8} - 9\frac{3}{4}$

19. $1\frac{1}{5} + 2\frac{1}{3} + 5\frac{1}{4}$

20. $9\frac{2}{3} + 4\frac{3}{5} + 6\frac{1}{2}$

21. $10\frac{7}{8} + 3\frac{3}{4} + 6\frac{1}{2} + 2\frac{5}{8}$

❼ Fractions, Decimals and Percents

Percent means per hundred. Therefore, 35%
means 35 out of 100. Percents can be written as
equivalent decimals and fractions.

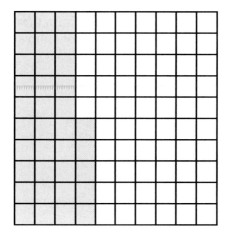

$$35\% = 0.35$$ Move the decimal point
two places to the left.

$$= \frac{35}{100}$$ Write the fraction with
a denominator of 100.

$$= \frac{7}{20}$$ Then simplify.

Example 1

Write $\frac{3}{8}$ as a decimal and as a percent.

Solution

$$\frac{3}{8} = 0.375$$ Divide to change a fraction to a decimal.

$$0.375 = 37.5\%$$ To change a decimal to a percent move the decimal point two
places to the right and insert the percent symbol.

Percents greater than 100% represent whole numbers or mixed numbers.

$$200\% = 2 \text{ or } 2.00 \qquad 350\% = 3.5 \text{ or } 3\frac{1}{2}$$

▶ EXTRA PRACTICE EXERCISES

Write each fraction as a decimal and as a percent.

1. $\frac{1}{2}$
2. $\frac{1}{4}$
3. $\frac{3}{4}$
4. $\frac{9}{10}$
5. $\frac{3}{10}$
6. $\frac{1}{25}$
7. $3\frac{7}{8}$
8. $1\frac{1}{5}$
9. $\frac{13}{25}$

Write each decimal as a fraction and as a percent.

10. 0.63
11. 0.15
12. 0.4
13. 2.35
14. 10.125
15. 0.625
16. 0.05
17. 0.125
18. 0.3125

Write each percent as a decimal and as a fraction.

19. 10%
20. 12%
21. 100%
22. 150%
23. 160%
24. 75%
25. 8%
26. 87.5%
27. 0.35%

Prerequisite Skills

⑧ Multiply and Divide by Powers of Ten

To multiply a number by a power of 10, move the decimal point to the right. To multiply by 100 means to multiply by 10 two times. Each multiplication by 10 moves the decimal point one place to the right.

To divide a number by a power of 10, move the decimal point to the left. To divide by 1000 means to divide by 10 three times. Each division by 10 moves the decimal point one place to the right.

Example 1

Multiply 21 by 10,000.

Solution

$21 \times 10,000 = 210,000$ The decimal point moves four places to the right.

Example 2

Find $145 \div 500$.

Solution

$$145 \div 500 = 145 \div 5 \div 100$$
$$= 29 \div 100$$
$$= 0.29$$ The decimal point moves two places to the left.

◤ EXTRA PRACTICE EXERCISES

Multiply or divide.

1. 15×100

2. $96 \times 10,000$

3. $1296 \div 100$

4. $9687.03 \div 1000$

5. $36 \times 20,000$

6. $7500 \div 3000$

7. 9×30

8. 94×6000

9. $561 \div 30$

10. $1505 \div 500$

11. $71 \times 90,000$

12. $9 \times 120,000$

13. $3159 \div 10,000$

14. $1,000,000 \times 0.79$

15. $601 \times 30,000$

16. $75 \div 300$

17. 4000×12

18. $14 \times 7,000,000$

19. $49,000 \div 7000$

20. $980 \div 10,000$

21. $216 \div 2000$

22. $108,000 \div 900$

23. $72 \times 10,000,000$

24. $953.16 \div 10,000$

25. $1472 \div 8000$

26. $490,000 \div 700$

27. $80 \times 90,000$

28. $8001 \div 90$

29. 50×6000

30. $950,000 \div 50,000$

31. $81,000 \times 5$

32. $1458 \times 30,000$

33. $452.3 \div 10$

34. $986,856.008 \div 10,000$

35. $316 \times 70,000$

36. $60 \div 1200$

⑨ Round and Order Decimals

To round a number, follow these rules:

1. Underline the digit in the specified place. This is the place digit. The digit to the immediate right of the place digit is the test digit.

2. If the test digit is 5 or larger, add 1 to the place digit and substitute zeros for all digits to its right.

3. If the test digit is 4 or smaller, substitute zeros for it and all digits to the right.

Example 1

Round 4826 to the nearest hundred.

Solution

4<u>8</u>26	Underline the place digit.
4800	Since the test digit is 2, and 2 is less than 5, substitute zeros for 2 and all digits to the right.

To place decimals in ascending order, write them in order from least to greatest.

Example 2

Place in ascending order: 0.34, 0.33, 0.39.

Solution

Compare the first decimal place, then compare the second decimal place.

0.33 (least), 0.34, 0.39 (greatest)

◼ EXTRA PRACTICE EXERCISES

Round each number to the place indicated.

1. 367 to the nearest ten
2. 961 to the nearest ten
3. 7200 to the nearest thousand
4. 3070 to the nearest hundred
5. 41,440 to the nearest hundred
6. 34,254 to the nearest thousand
7. 208,395 to the nearest thousand
8. 654,837 to the nearest ten thousand

Write the decimals in ascending order.

9. 0.29, 0.82, 0.35
10. 1.8, 1.4, 1.5
11. 0.567, 0.579, 0.505, 0.542
12. 0.54, 0.45, 4.5, 5.4
13. 0.0802, 0.0822, 0.00222
14. 6.204, 6.206, 6.205, 6.203
15. 88.2, 88.1, 8.80, 8.82
16. 0.007, 7.0, 0.7, 0.07

Extra Practice

Chapter 1

Extra Practice 1–1 • Surveys and Sampling Methods • pages 6–9

RETAIL The owners of a new music store want to survey local residents to find out what kinds of music they prefer. What kind of sampling method is represented by each of these possibilities that the owners have considered using?

1. Ask every tenth customer who enters the store on a certain day.

2. Ask all people who enter the store between 2:00 P.M. and 5:00 P.M.

3. Send a survey form to people listed in the telephone directory whose telephone numbers end with the digit 7.

4. Ask all customers who were checked out at a certain cash register chosen at random.

NUTRITION Heather is preparing a report on nutrition for health class and she wants to find out the favorite vegetable among teenagers in her school. Name the sampling method represented by each of the following. Then give one reason why the findings that would result from each method could be biased.

5. Ask every tenth teenager who enters a fast-food restaurant.

6. Ask ten teenagers whose names are drawn at random from the names of all students in your school.

7. Ask ten students in another health class.

8. Ask the first ten students who arrive at a health fair.

Extra Practice 1–2 • Measures of Central Tendency and Range • pages 10–13

For Exercises 1–5, use the following ages of students in a theater club. If necessary, round your answers to the nearest tenth.

13 18 14 15 14 17 17 16 13 16 17 17 18 5 13

1. Find the mean.
2. Find the median.
3. Find the mode.
4. Find the range.
5. Which measure of central tendency is the best indicator of the age of a member of the club?

Use the table for Exercises 6–10. If necessary, round your answers to the nearest tenth.

6. Find the mean.
7. Find the median.
8. Find the mode.
9. Find the range.
10. Which measure of central tendency is the best indicator of the number of hours per week spent engaging in aerobic exercise?

Hours of Aerobic Exercise Per Week

Number of hours	Frequency
0	4
1	5
2	5
3	19
4	27
5	16
6	14
7	10

For Exercises 1–6, use the following number of push-ups by students in 1 min.

18	18	0	15	1	3	4
2	25	29	5	6	7	15
12	10	9	8	10	15	7
0	1G	10	5	G	12	10
9	14	14	11	20	21	14
13	22	24	12	14	13	13

1. Use a frequency table to make a histogram of the data.

2. How many students are there in the data?

3. How many students did from 10 to 14 push-ups?

4. How many students were unable to do as many as 15 push-ups?

5. How many students did 20 or more push-ups?

6. Which number of push-ups were performed least frequently?

For Exercises 7–10, use the following number of book covers sold each day at a school store.

104	121	182	170	185	93	251	177	93	57	93	115
184	92	117	110	174	118	183	116	171	129	174	

7. Organize the data into a stem-and-leaf plot.

8. On how many days were at least 170 book covers sold?

9. What is the median number of book covers sold?

10. Find the mode of the set of data.

Use the scatter plot for Exercises 1–9.

1. How many students were surveyed if no two students answered in exactly the same way?

2. What is the range of hours spent watching television?

3. What is the range of hours spent exercising?

4. Find the mode(s) of the hours spent watching television.

5. Find the mode(s) of the number of hours spent exercising.

6. Predict the weekly number of hours of television watching of students who exercise 8 h per week.

7. One student watches television 12 h per week. How many more hours is this than the median number of hours of television watching?

8. Which point lies farthest from the trend line? What could account for this?

9. Is there a positive or negative correlation between watching television and exercising?

TV Watching And Exercise

Hours of TV watching (vertical axis: 0, 4, 8, 12, 16)
Weekly hours of exercise (horizontal axis: 0, 4, 8, 12, 16)

For each set of data, find the first quartile, the median, and the third quartile.

1. 47   70   47   63   52   55   62   59   62

2. 13   10   15   19   12   17   10   14   11   13   18   15

3. Make a box-and-whisker plot for the following scores on a math achievement test.

   8   6   7   9   5   4   8   3   6   3   5   8   6

This box-and-whisker plot shows the results of a survey on the amount of money families budgeted for vacations. Use the plot for Exercises 6–8.

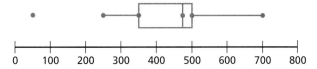

4. What are the least and greatest amounts reported?

5. What percent of those surveyed budgeted $450 or more?

6. In which interval are the data most closely clustered? The most spread out?

7. On a test, Erica has the 8th highest test score. If 28 students took the test, what is Erica's percentile rank?

8. Nick scored in the 80th percentile on a test, how many people scored above him out of 40 students?

REAL ESTATE A real-estate broker sold eight houses last month for $90,000, $75,000, $150,000, $80,000, $140,000, $80,000, $100,000 and $95,000.

1. Which measure of central tendency might the broker use to tell homeowners the average selling price in order to encourage them to want to sell their houses? Explain.

2. Which measure of central tendency should the broker use to describe the average selling price to people looking for a house? Explain.

FOOD SERVICE A school cafeteria charges $3 for the daily lunch special. The manager is considering raising the price and asked 25 students to name the greatest amount they would be willing to pay for the lunch special. The results are shown in the table.

3. In order to justify a large increase in the lunch price, which measure of central tendency might the manager use? Explain.

4. If students that buy their lunch saw this survey, which measures of central tendency might they point out to the manager to keep the price increase low? Explain.

Greatest Amount People Would Pay For The Daily Lunch Special

Amount (dollars)	Frequency
6.00	3
5.00	4
4.50	3
4.00	5
3.75	7
3.50	1
3.00	1
2.00	1

Extra Practice

Write each set of data as a matrix. Name its dimensions.

1.

	A	B	C	D
1	Age			
2	0-3	8	3	9	2
3	4-7	6	5	0	4
4	8-11	8	7	6	5

2.

	A	B	C
1		Male	Female
2	Monday	200	321
3	Tuesday	218	282
4	Wednesday	135	129

3.

	A	B	C	D	E	F
1		Jan	Feb	Mar	Apr	May
2	Paid	6	0	5	9	2
3	Unpaid	3	9	4	0	7

Use matrices _A–D_ for Exercises 4–11.

4. Give the elements of _A_.

5. Name the dimensions of _D_.

6. Find $A + C$.

7. Find $C + A$.

8. Find $A - C$.

9. Find $C - A$.

10. Find $A + C + B$.

11. Find $A + B - C$.

$$A = \begin{bmatrix} 3 & 4 & 3 \\ 1 & 1 & 2 \\ 2 & 7 & 6 \end{bmatrix} \quad B = \begin{bmatrix} 4 & -1 & 3 \\ 3 & -2 & 0 \\ -5 & 8 & 3 \end{bmatrix}$$

$$C = \begin{bmatrix} 6 & 1 & 3 \\ 1 & 4 & -2 \\ 5 & 3 & -1 \end{bmatrix} \quad D = \begin{bmatrix} 4 & -5 \\ 2 & 0 \\ -7 & 6 \end{bmatrix}$$

12. Use a graphing calculator to find the sum and difference of matrices _G_ and _H_.

$$G = \begin{bmatrix} 82 & 23 \\ 25 & 66 \\ 13 & -47 \end{bmatrix} \quad H = \begin{bmatrix} -34 & 83 \\ 27 & 53 \\ 97 & -18 \end{bmatrix}$$

Chapter 2

Graph each set of numbers on a number line.

1. $\left\{ -3, -2.5, -\dfrac{1}{2}, 2, 3.6 \right\}$

2. $\{-3, -2, 0, 2.9, 5\}$

3. the integers from -5 to 3

4. the integers from -5 to -12

5. all real numbers less than 4

6. all real numbers less than or equal to -4

7. all real numbers greater than -3

8. all real numbers less than or equal to 5

9. all real numbers greater than or equal to -4

10. integers from -3 to -8

Evaluate each expression.

11. $-x$, when $x = -2.4$

12. $|n|$, when $n = -3$

13. $-|-n|$, when $n = 2$

14. $-|x|$, when $x = -4$

15. $-|f|$, when $f = -5.3$

16. $-b$, when $b = 4.78$

17. $-\left(\dfrac{b}{c} \right)$, when $b = 2$ and $c = -4$

18. Which depth is closer to sea level, a depth of -1689 ft or a depth of -3792 ft?

Simplify each numerical expression.

1. $3 \cdot 8 - 2 + 6$

2. $(30 + 5) \div (5 \cdot 3)$

3. $4 \cdot 9 - 7$

4. $3^2 + 8$

5. $(31 - 7) \div 6 \cdot 9$

6. $9.4 - 1.2 \cdot 18 \div 9$

7. $\frac{1}{4}\left(\frac{1}{4} + \frac{3}{8}\right)$

8. $\frac{1}{3}\left(\frac{1}{2} + \frac{4}{5}\right)$

9. $(16 - 5) \cdot 2 + 2^3$

Evaluate each variable expression when $n = 12$.

10. $20 - n$

11. $\frac{1}{2}n$

12. $n + 6$

13. $n \div 4$

14. $4n - 18$

15. $\frac{60}{n}$

16. $n + 3n$

17. $\frac{2}{3}n + 7$

18. $5n + 36 \div 3^2$

19. Write a variable expression for the phrase: multiply a number n by 3 and then add 2.

20. Ben charges \$4/h to babysit. He charges an extra \$3 if he babysits on a Friday night. Write and simplify a numerical expression for the amount Ben charges to babysit for 5 h on a Friday night.

Write each phrase as a variable expression.

1. the sum of five and a number

2. the quotient of a number and 30

3. negative two times a number

4. twice a number decreased by 11

5. a number decreased by eight

6. negative eight times a number

7. twice a number divided by negative four

8. the quotient of one and twice a number

9. nine less than the sum of five and a number

10. the quotient of twice a number and three

11. negative one-third times a number

12. the difference of nine and a number

13. a number less 18

14. twice a number multiplied by three

15. the difference of one and a number

16. the sum of one and twice a number

Translate each variable expression into a word phrase.

17. $8 + t$

18. $\frac{x}{2}$

19. $-4p$

20. $3n - 2$

21. $\frac{1}{2}g$

22. $-\frac{14}{2y}$

23. $12s - 4$

24. $6x + 10$

25. $-\frac{y}{2}$

26. $h^2 - 12$

27. $0.4 - 7j$

28. $22 + \frac{1}{3}n$

29. When water freezes, its volume increases. The volume of ice equals the sum of the volume of water and the product of one-eleventh and the volume of water. If x cm^3 of water is frozen, write an expression for the volume of the ice that is formed.

Simplify.

1. $4x + 9x$
2. $6n - 9n$
3. $-5ab + 6ab$
4. $4s + 4s - 3m$
5. $6j + (-j) + 9j$
6. $3jk + (-jk) + 7j$
7. $\frac{1}{3}a + \frac{1}{4}a$
8. $-3w + 8w + (-w)$
9. $-5d - (2d + d)$
10. $2g - (-g) + 4 - 3g$
11. $-6e - 4 + 3e - e$
12. $3r + 9pq + 6r - 3pq$
13. $3x + (-9x)$
14. $6n + 9n + 3mn - 8mn$
15. $\frac{1}{4}ab + \frac{3}{4}ab - \frac{5}{6}b + \frac{1}{6}a$
16. $4mn + 4m - 3mn$
17. $-j + (-4j) - 6j$
18. $0.13jk + (-0.5jk) + 1.2jk$
19. $\frac{1}{8}xy + \frac{1}{4}xy + \frac{1}{2}xy$
20. $-\frac{3}{7}w - \frac{6}{7}w + \left(-\frac{1}{7}w\right)$
21. $-5de + 2d + de - de - 4de - 2$
22. $2g + (-g) - 4 + 3g$
23. $-6e + 4 - 3e + e$
24. $-3rs + (-2rs) - 6r + 6s$

Evaluate each expression when $a = -3$ and $b = 10$.

25. $a - 3a$
26. $-b + 4b$
27. $a - b + 3a - 9b$
28. $\frac{1}{3}a - \frac{1}{5}b$
29. $-2a - (-3a)$
30. $3ab - ab + 2ab$
31. $4a + 4b - 6a - 6b$
32. $\frac{3}{5}ab + \frac{2}{3}ab - b$
33. $3a + 4b + 2b - 5a$

34. Theresa sold $20x$ sweaters on Tuesday. On Wednesday she sold $8x$ sweaters fewer than on Tuesday. Write and simplify an expression for the number of sweaters Theresa sold on Wednesday.

Simplify.

1. $5(3a + 2a)$
2. $-(3z + x)$
3. $-3(4b + 9b)$
4. $-8(x - y)$
5. $\frac{1}{3}(6a + 12ab - 21a)$
6. $4(5 - z)$
7. $-2(e + 3)$
8. $\frac{1}{5}(10f - 5)$
9. $-3(-n + 9)$
10. $\frac{-6p + 4}{-2}$
11. $\frac{5g - 10}{5}$
12. $\frac{-14v - 7}{-2}$
13. $\frac{12 - 36d}{4}$
14. $\frac{3.9 + 4.5h}{0.3}$
15. $\frac{225r + 30}{-15}$
16. $\frac{6t + 12st - 21s}{3}$
17. $\frac{10n + 5}{-5}$
18. $\frac{3a + (-6)}{-3}$

Evaluate each expression when $a = \frac{1}{2}$, $b = -2$, and $c = 10$.

19. $-4(a - 22)$
20. $6(-b - 8)$
21. $\frac{c - 24}{7}$
22. $c(8 + b)$
23. $\frac{-10a + 12b}{-2}$
24. $\frac{3c - 2b}{17}$

25. A piece of wood measures $(6a + 3b)$ in. long. Write and simplify an expression for the length of a piece of wood 6 times as long.

Extra Practice

Simplify.

1. $4(n + 3s) - 6s$

2. $-8(b - 5c) + 10c$

3. $(9a - 3b) - (4a + 6b)$

4. $-\frac{1}{2}(4m - 5n) - \frac{1}{2}n$

5. $4(5f - 2g) - (2f + 3g)$

6. $1.5(3b + 4c) - 3(2.5b + 5c)$

7. $5(3b - 4c) - 3(5b - 5c)$

8. $(3x + 2y) - 3(2x - y)$

9. $-2(3m + n) - 2(2m + 7n)$

10. $6(-2b - 3c) + 3(2b + 3c)$

11. $\frac{1}{2}(4p + 8q) - \frac{3}{4}(8p + 4q)$

12. $15(3b + 3) - 19$

13. $-(2x - 4y) - 6(x - 5y)$

14. $7\left(\frac{1}{7}b + \frac{1}{7}c\right) + 6\left(\frac{1}{3}b + \frac{2}{3}c\right)$

15. $1.2(5f - 20g) - 5.2(5f + 30g)$

16. $13(3s + 4t) - 3(15t)$

17. $17(3m + 4n) - 17(3m + 4n)$

18. $9(-a + b) - 3(-a - 5c)$

19. Heta wrapped 6 gifts for a birthday party. Some of the gifts were wrapped in green paper and the rest in blue paper. It cost $1.25 for each sheet of green paper and $0.75 for each sheet of blue paper. Write and simplify a variable expression for the total amount Heta spent on x sheets of wrapping paper.

20. Michael bought 8 paperback books. Some of the books were mysteries and the rest were science fiction. The mystery books cost $4.95 and the science fiction books were on sale for $2.95. Write and simplify a variable expression for the total amount Michael spent on x books.

Evaluate each expression when $m = 5$ and $n = -2$.

1. n^2

2. $m^2 + n^3$

3. $(m + n)^2$

4. $3mn^2$

5. $4(n + 3)^2$

6. $(m - n)^2$

7. $6(2 + n)^2$

8. $\dfrac{m^3}{m^2}$

9. $(1 - m)^2$

10. $(n^2 - 5)^2$

11. $8(m - 1)^2$

12. $\left(\dfrac{n}{4}\right)^2$

13. $(n^2 - 3)^2$

14. $\left(\dfrac{m}{2}\right)^2$

15. $\dfrac{n^8}{n^7}$

16. $4n^2 + m$

Simplify.

17. $c \cdot c^3$

18. $(m^2)^3$

19. $4(b^2)^3$

20. $x^9 \cdot x^3$

21. $\dfrac{y^3}{y}, y \neq 0$

22. $\left(\dfrac{1}{z}\right)^3, z \neq 0$

23. $(x^2)^2$

24. $4(a^3)^3$

25. $\left(\dfrac{y}{3}\right)^3$

26. $\dfrac{18n^8}{3n^3}, n \neq 0$

27. $m^2 \cdot m \cdot m^3$

28. $3(x^2)^4$

29. $t^{15} \cdot t^8$

30. $\left(\dfrac{1}{d}\right)^{11}, d \neq 0$

31. $(2yx^3)^3$

32. $(r^2)(r^5)(r^3)$

33. A piece of wood x^4 in. long is to be cut into pieces x^2 in. long. How many pieces will there be?

34. If the value of x in Exercise 33 is 6, how long would each piece be?

35. A piece of string x^5 in. long is to be cut into x^2 pieces. How long will each piece be?

Simplify.

1. $a^2 \cdot a^{-4}$ 2. $b^3 \div b^{-2}$ 3. $3a^2 \cdot 5a^{-2}$ 4. $x^2 \div x^5$

5. $(m^2)^{-1}$ 6. $x^{-4} \cdot x^{-3}$ 7. $a^{-3} \div a^3$ 8. $(x^3)^{-2}$

Evaluate each expression when $x = -2$ and $y = 3$.

9. x^3 10. $x^3 \cdot x^{-2}$ 11. $(xy)^{-3}$ 12. $x^2 \cdot y^{-2}$

13. $(y - x)^{-2}$ 14. $(x^3 - y)^2$ 15. $y^3 \div y^2$ 16. $(xy)^{-3}$

Write each number in scientific notation.

17. 9560   18. 0.0598   19. 3450

20. 34,900   21. 0.0096   22. 0.0000007

Write each number in standard form.

23. $3.75 \cdot 10^{-2}$ 24. $1.9 \cdot 10^5$ 25. $1.23 \cdot 10^{-3}$

26. $6.3 \cdot 10^{-8}$ 27. $2.590 \cdot 10^{-4}$ 28. $1.095 \cdot 10^6$

29. $4.5 \cdot 10^{-4}$ 30. $5.890 \cdot 10^3$ 31. $9.65 \cdot 10^{-5}$

32. A mid-western state spends $1,900,000 annually on books for its libraries. At that rate, how much would it spend in eight years? Write the number in scientific notation.

Chapter 3

Which of the given values is a solution of the equation?

1. $a + 5 = 3$; 8, −8, −2   2. $-4n - 7 = 1$; 2, −2, −3

3. $12 + w = 9$; −3, −9, 9   4. $15 + 3n = 3$; 3, 4, −4

Use mental math to solve each equation.

5. $m - 12 = 32$ 6. $-4 + x = -8$ 7. $21 = 3 + t$

8. $\frac{1}{3}x = 15$ 9. $7t = 4.9$ 10. $n - 7.6 = 0$

11. $-4 = \frac{1}{2}x$ 12. $-5 - x = 1$ 13. $y + 6 = 21$

Find the unknown measure of each triangle. Use the formula $A = \frac{1}{2}bh$.

14. The area of the triangle is 32 ft^2 and the height is 8 ft. Find the base.

15. The area of the triangle is 50 m^2 and the base is 25 m. Find the height.

16. The height of the triangle is 46 cm and the base is 11 cm. Find the area.

Solve each equation. Check the solution.

1. $3x = 48$
2. $1.5 = n - 3$
3. $11 - m = -8.7$
4. $\frac{2}{7}p = 4$
5. $5 - b = 6$
6. $-3w = 1$
7. $\frac{5}{6}a = 5$
8. $-4p = 80$
9. $-\frac{n}{5} = -4$
10. $-85 = 17r$
11. $1.4x = 8.4$
12. $5.6 = 7 + n$
13. $\frac{n}{6} = 5$
14. $\frac{7}{8} = \frac{1}{4} + t$
15. $50b = -20$
16. $j - 4.7 = 12$
17. $n + 13 = 7$
18. $2.4g = 6$
19. $\frac{5}{6}s = -10$
20. $-6d = 40$
21. $18x = -12$

22. During a 4-wk period Makota ran a total of 196 mi. To show the situation, he let n equal "the average number of miles run per week" and wrote the equation as $4n = 196$. What was the average number of miles Makota ran each week?

23. Twelve times a certain number equals -4. To model the situation, Elaine let x equal "a certain number." Then she wrote the equation $12x = -4$. What was the original number?

Solve each equation. Check the solution.

1. $4x - 2x + 3 = 27$
2. $4b - 7 = 0$
3. $16 = -2 - 3w$
4. $\frac{4}{5}d + 7 = 19$
5. $\frac{n}{4} + 3 = 15$
6. $2d + 4d - 2 = -14$
7. $-y - y + 7 = -1$
8. $3x - 2.7 = 5.1$
9. $3(a + 4) = 21$
10. $-6a + 4 + 2a + 1 = 13$
11. $5.1x - 4 = 31.7$
12. $7(t - 2) = 21$
13. $5z + 4 = z + \frac{16}{5}$
14. $4(2x + 3) = 52$
15. $6y - 5 + y + 2 = 11.7$
16. $3(a + 2) = -2(a - 13)$
17. $7(12 - x) = -14$
18. $\frac{1}{4}(6r - 2) = \frac{1}{4}$

19. Find $m\angle YXZ$ and $m\angle XZY$ in the figure below.

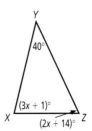

20. Find $m\angle CDB$ and $m\angle CDA$ in the figure below.

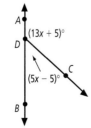

21. Find $m\angle AYD$ and $m\angle CYB$ in the figure below.

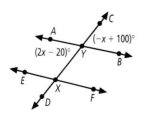

22. Twice a number decreased by 7 is the same as four times the number decreased by 23. Find the number.

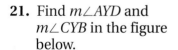

Solve each proportion. Check the solution.

1. $\dfrac{5}{3} = \dfrac{a}{9}$

2. $\dfrac{33}{z} = \dfrac{15}{10}$

3. $\dfrac{-36}{b} = \dfrac{24}{16}$

4. $18 : x = 20 : 60$

5. $14 : 16 = 35 : a$

6. 15 is to z as 21 is to 14

7. $\dfrac{e + 3}{5} = 8$

8. $\dfrac{8}{f - 5} = \dfrac{1}{2}$

9. $\dfrac{n - 9}{6} = 4$

10. $\dfrac{p + 9}{p} = \dfrac{2}{11}$

11. $\dfrac{g - 7}{-3} = \dfrac{g}{4}$

12. $\dfrac{3v - 7}{5} = v + 3$

13. $\dfrac{2 - d}{7} = \dfrac{2 + d}{3}$

14. $\dfrac{9 + h}{h} = \dfrac{4}{5}$

15. $\dfrac{5r + 3}{9} = \dfrac{r - 1}{2}$

16. $\dfrac{6t + 1}{2} = \dfrac{t - 1}{5}$

17. $\dfrac{3n + 5}{7} = \dfrac{1 - 3n}{5}$

18. $\dfrac{3}{7} = \dfrac{a + 6}{3a + 2}$

19. A company produces 1450 tires. Of these, 12% are for recreational vehicles. How many tires are for recreational vehicles?

20. For one year, a company's income was \$620,980. The company's expenses for that year were \$496,784. What percent of the company's income was profit? (Hint: income − expenses = profit)

21. Of all the employees at a company, 15% have reported in sick in the last month. There are 560 employees at this company. How many reported in sick last month?

Graph the solutions of each inequality on a number line.

1. $b > 4$

2. $x < 5$

3. $4 > f$

4. $b \geq 4$

5. $n > 0$

6. $x \leq 5$

7. $-4 \geq c$

8. $4 \leq f$

9. $x \geq 5$

10. $x > 3$

11. $1.5 > f$

12. $-1 \leq h$

13. $n < 2$

14. $m < -2$

15. $b > -4$

16. $j \leq 18$

17. $0 > k$

18. $2.5 \geq t$

State the inequality that is represented on each number line.

19.

20.

22.

23.

Extra Practice

Solve and graph each inequality.

1. $4x - 2 < 22$
2. $4 - r > 12$
3. $4 \le \frac{1}{5}y$
4. $5m + 2 \ge -3$
5. $3r + 2 > -7$
6. $7 + 2a < 19$
7. $15 < 3n + 21$
8. $8 - 2m > 14$
9. $12 \ge -3n + 5$
10. $\frac{1}{4}d - 2 \le 1$
11. $20 \le -3n + 5$
12. $0.4p - 3 \le -1$
13. $8 - 4a > 12$
14. $5 - \frac{1}{2}w \ge 6$
15. $18 \le 3 - 5x$
16. $0.3x - 4 > 0.2$

17. Pablo wants to get a grade of at least 95 in math this semester. So far he has test scores of 89, 98, 97 and 92. He will take his last test on Friday. What is the lowest grade he can get on the test and still get a 95 this semester?

18. Ellie can lift a maximum of 225 lb. The bar already has 105 lb of weights on it. She wants to add more weight. Each disk weighs 20 lb. How many disks can she add to the bar so that she can lift her maximum weight?

Solve each equation. Check the solutions.

1. $x^2 = 121$
2. $b^2 = \frac{25}{64}$
3. $n^2 = 81$
4. $x^2 = 1.44$
5. $\sqrt{f} = \frac{1}{10}$
6. $\sqrt{x + 3} = 1$
7. $\sqrt{x} = 3$
8. $\sqrt{y} = 0.5$
9. $\sqrt{v - 5} = 8$
10. $\sqrt{4 + x} = 4$
11. $3 = \sqrt{x} - 7$
12. $8\sqrt{x} = 24$
13. $\sqrt{3r} - 2 = 7$
14. $7x^2 = 210$
15. $8 + 3n^2 = 83$
16. $6 = \sqrt{2(x + 4)}$
17. $\sqrt{5h} = \frac{1}{5}$
18. $\sqrt{6y - 7} = 2$
19. $4t^2 = 9$
20. $r^2 + 8 = 908$
21. $12 = \sqrt{x + 3} - 8$

22. The formula $d = 16t^2$ gives the distance d in feet traveled by an object dropped from a resting position and allowed to fall freely. The variable t represents the time of the fall in seconds. If you dropped an object from a height of 1024 ft above the ground, how long would it take to land on the ground?

23. Payton will place 4 square tiles over a floorboard with a total area of 576 cm^2. Each of the tiles has the same dimensions and when placed together fit exactly over the floorboard. Write and solve an equation to find the dimensions of each tile.

Chapter 4

Extra Practice 4–1 • Experiments and Probabilities • pages 150–153

Use the table for Exercises 1–5.

Science Fair Attendance

	10th graders	11th graders	12th graders	Teachers	Students' family
Wednesday	162	129	135	27	87
Thursday	336	180	144	24	36
Friday	288	240	179	144	109

1. Find the probability that a person at the fair on Wednesday is a 10th grader.

2. Find the probability that a person at the fair on Wednesday is a teacher.

3. Find the probability that a person at the fair on Thursday is an 11th grader.

4. Find the probability that a person at the fair on Thursday is a 12th grader.

5. Find the probability that a person at the fair on Friday is a family member.

SPORTS On a particular day at a ski resort, 19 people skied on the beginner slope, 123 people skied on the intermediate slope and 58 skied on the advanced slope. Find the probability that a skier used the following slope.

6. intermediate slope 7. beginner slope 8. advanced slope

Extra Practice 4–3 • Sample Spaces and Theoretical Probability • pages 158–161

1. A coin is tossed and a marble drawn from a bag that contains 1 red marble, 1 blue marble, 1 green marble, and 1 yellow marble. Show the sample space using ordered pairs. Then show the sample space using a tree diagram.

2. A spinner with six equal sectors labeled A through F is spun and a coin is tossed. Show the sample space using ordered pairs. Then show the sample space using a tree diagram.

3. **SPORTS** At a baseball tryout there were 3 candidates for pitcher, 5 candidates for catcher and 7 candidates for shortstop. How many different ways can these positions be filled?

The spinners shown are both spun for Exercises 4–10.

4. How many possible outcomes are there?

5. Find *P*(letter before F in the alphabet, odd number)

6. Find *P*(C, 4)

7. Find *P*(D, multiple of 3)

8. Find *P*(F, odd number)

9. Find *P*(vowel, number less than 3)

10. Find *P*(consonant, prime number)

1. Two number cubes are rolled. Find the probability that the sum of the numbers is 3 or 4.

2. A card is drawn at random from a standard deck of 52 cards. Find the probability that the card is a 7 or a queen.

3. Two number cubes are rolled. Find the probability that the sum of the numbers is even and greater than 9.

4. A card is drawn at random from a standard deck of 52 cards. Find the probability that the card is a club or a king.

The spinner shown is spun one time. Use the diagram of the spinner to solve Exercises 5–16. Find each probability.

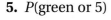

5. P(green or 5)
6. P(green or even)

7. P(blue or yellow)
8. P(green or less than 5)

9. P(blue or prime)
10. P(yellow or less than 4)

11. P(blue and odd)
12. P(green and even)

13. P(green and prime)
14. P(green and a multiple of 3)

15. P(green or yellow)
16. P(yellow and even)

17. A card is drawn at random from a standard deck of 52 cards. Find the probability that the card is a diamond or a multiple of 3.

18. Four coins are tossed. Find the probability that the coins show 2 heads or 2 tails.

A group of numbered cards contains one 1, two 2s, four 3s, five 4s and six 5s. Cards are picked, one at a time, then replaced. Find each probability.

1. P(1, then 2)
2. P(1, then 3)
3. P(1, then 4)

4. P(2, then 3)
5. P(2, then 4)
6. P(2, then 2)

7. P(1, then 5)
8. P(3, then 5)
9. P(4, then 5)

A bag contains 4 red, 5 blue, and 6 green marbles. Marbles are taken at random and not replaced. Find each probability.

10. P(red, then blue)
11. P(red, then green)
12. P(red, then red)

13. P(blue, then red)
14. P(green, then green)
15. P(blue, then green)

As a result of a storm, a sign reading NO SWIMMING lost three of its letters.

16. What is the probability that the letters were all vowels?

17. What is the probability that the letters were not vowels?

18. What is the probability that the letters were lost in this order: O, S, M?

1. In how many different ways can a teacher arrange 6 students in a row of seats?

2. From 8 candidates, a committee of judges must choose 5 candidates and rank them. In how many ways can this be done?

3. A radio disc jockey has a set of 12 songs to play. In how many different ways can the disc jockey play the next 3 songs?

4. A video store owner wants to display 7 different video boxes in a row on a shelf. In how many ways can this be done?

5. How many different 5-digit numbers can be formed from the digits 1, 2, 3, 4 and 5 if each digit can be used only once?

6. Find the number of 5-letter "words" that can be formed from the letters of the word "chair"?

7. The class offices of president, vice president and secretary are to be filled. There are 20 candidates. In how many ways can the positions be filled?

Calculate each of the following permutations.

8. $_4P_0$　　　　**9.** $_6P_3$　　　　**10.** $_3P_3$　　　　**11.** $_{10}P_6$　　　　**12.** $_7P_2$

1. There are 8 one-liter cans each containing a different fruit juice. How many different types of fruit juice can be obtained by mixing 3 full containers of juice together?

2. A deck of 26 cards, marked A through Z, is shuffled and 3 cards are dealt. What is the probability that the cards are A, B and C in any order?

3. Alana, Gina and Lenisa are among 20 candidates competing for 3 open positions on the track team. What is the probability that they will be the 3 who are selected?

4. A taco restaurant offers 10 different fillings for tacos. If you were to choose 3 fillings, how many different tacos could you order?

5. How many "words" can be made with the letters, *a, c, e, g, i, k* and *m* if only three different letters are used in each word?

6. A random drawing is held to determine which 2 of the 8 members of the chorus will be sent to a district competition. How many different pairs of 2 could be sent to the competition?

7. How many different ways can a 4-person committee be chosen from 10 people if there are no restrictions?

Calculate each of the following combinations.

8. $_4C_0$　　　　**9.** $_6C_3$　　　　**10.** $_3C_3$　　　　**11.** $_{10}C_6$　　　　**12.** $_7C_2$

598 |

Chapter 5

Extra Practice 5–1 • Elements of Geometry • pages 192–195

State whether each statements is *true* or *false*.

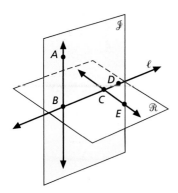

1. *B*, *C* and *D* lie in plane 𝒥.

2. *A*, *B* and *C* line in plane ℛ.

3. Planes 𝒥 and ℛ intersect in line ℓ.

4. Exactly one line contains *C* and *E*.

5. *D* does not lie in plane ℛ.

6. $\overleftrightarrow{AB}$ lies on plane 𝒥.

7. $\overleftrightarrow{CE}$ lies on plane 𝒥 and plane ℛ.

Identify the points in each figure as *collinear* or *noncollinear* and *coplanar* or *noncoplanar*.

8.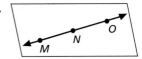

9.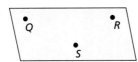

10.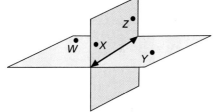

Extra Practice 5–2 • Angles and Perpendicular Lines • pages 196–199

Find the measure of a complement and supplement of each angle.

1. $m\angle X = 18°$

2. $m\angle Y = 60°$

3. $m\angle Z = 20°$

4. $m\angle M = 8°$

5. $m\angle N = 85°$

6. $m\angle O = 50°$

In the figure shown, $\overline{AD} \perp \overline{XC}$.

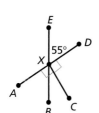

7. Name all right angles.

8. Name a pair of vertical angles.

9. Name two adjacent complementary angles.

10. Find $m\angle BXC$.

11. Find $m\angle BXD$.

For Exercises 12 and 13, use the figure shown.

12. Name all pairs of perpendicular lines.

13. Name two angles that are congruent to $\angle ADB$.

14. Write a rule for finding the supplement of any angle.

Refer to the figure to classify each pair of angles.

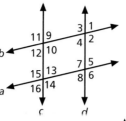

1. ∠8 and ∠13
2. ∠11 and ∠15
3. ∠6 and ∠15
4. ∠4 and ∠9
5. ∠12 and ∠16
6. ∠4 and ∠10

In the figure, $\overrightarrow{AB} \parallel \overrightarrow{CD}$. Find each measure.

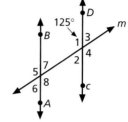

7. $m\angle 5$
8. $m\angle 6$
9. $m\angle 7$
10. $m\angle 8$
11. $m\angle 2$
12. $m\angle 3$

State whether the lines cut by the transversal are parallel or not.

13.

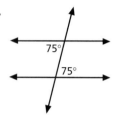

14.

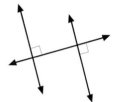

15.

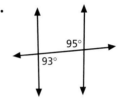

State whether it is possible to have a triangle with sides of the given lengths.

1. 11, 8, 3
2. 15, 8, 8
3. 4, 5, 6
4. 100, 150, 160

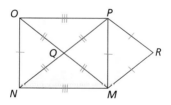

Use the figure to classify each triangle by its sides.

5. △OQN
6. △MPQ
7. △MPR

In the figure $\overrightarrow{WX} \parallel \overrightarrow{YZ}$ and $\overrightarrow{WZ} \perp \overrightarrow{VX}$ find each measure.

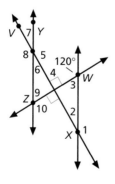

8. $m\angle 1$
9. $m\angle 2$
10. $m\angle 3$
11. $m\angle 4$
12. $m\angle 5$
13. $m\angle 6$
14. $m\angle 7$
15. $m\angle 10$

Find the unknown angle measures in each figure.

16.

17.

18.

For Exercises 1–5 use △*XYZ*.

1. Which angle is included between $\overline{XY}$ and $\overline{YZ}$?

2. Which angle is included between $\overline{YX}$ and $\overline{XZ}$?

3. Which side is included between $\angle X$ and $\angle Y$?

4. Which side is included between $\angle Y$ and $\angle Z$?

5. Which side is included between $\angle X$ and $\angle Z$?

State whether the pair of triangles is congruent by SAS, ASA, or SSS.

6.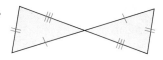

7.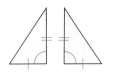

8.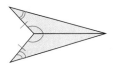

9.

10.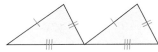

11.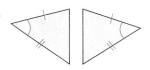

12. Can you show that two triangles are congruent if two of their corresponding parts are congruent? Why or why not?

Match the property named with all the quadrilaterals listed that have that property: *parallelogram, rectangle, rhombus, square*.

1. Consecutive angles are supplementary.

2. The sum of the angle measure is 360°.

3. Four sides are congruent.

4. All four angles are right angles.

5. The diagonals are perpendicular.

6. The diagonals are congruent.

Find the unknown angle measure in each parallelogram.

7. $m\angle Q$

8. $m\angle R$

9. $m\angle S$

10. $m\angle Y$

11. $m\angle X$

12. $m\angle W$

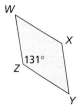

Find each unknown measure in each parallelogram.

13. $m\angle P$

14. NO

15. $m\angle M$

16. OP

17. $m\angle O$

18. PS

19. RS

20. QS

21. RT

22. PR

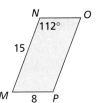

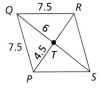

Classify each polygon by its number of sides. Tell whether it is *convex* or *concave*, *regular* or *not regular*.

1.

2.

3.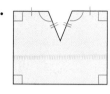

Find the sum of the interior angles of each convex polygon.

4. heptagon   5. nonagon   6. decagon   7. quadrilateral

8. 16-gon   9. 18-gon   10. pentagon   11. octagon

Find the measure of an interior angle of each regular polygon.

12. pentagon   13. hexagon   14. octagon

15. What is the number of sides of a regular polygon in which each interior angle measures 168°?

16. What is the measure of each interior angle of a convex 90-gon?

In ⊙T, $m\angle BXC = 90°$ and $m\angle AXB = 45°$. Find each measure.

1. $m\angle \overset{\frown}{DC}$ 2. $m\angle \overset{\frown}{DA}$

3. $m\angle \overset{\frown}{ABC}$ 4. $m\angle \overset{\frown}{ADC}$

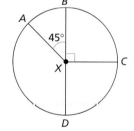

In ⊙T, $\overline{NQ}$, $\overline{PS}$, and $\overline{OR}$ are diameters. Use ⊙T for Exercises 5–10.

5. Name four semicircles.

6. Name two inscribed angles.

7. Find the measure of $\overset{\frown}{SR}$.

8. Find the measure of $\overset{\frown}{SNO}$.

9. If $\overline{NT}$ has a length of 5, what is the length of $\overline{TQ}$?

10. Name two chords that are twice the length of $\overline{RT}$.

State whether each of the following statements is *true* or *false*.

11. A central angle is an angle whose vertex lies on the circle.

12. A radius is a chord.

13. A diameter is a chord.

14. The measure of a minor arc is always less than 180°.

Extra Practice

Chapter 6

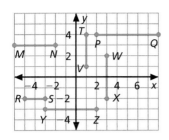

Use the graph to calculate the length of each segment.

1. $\overline{RS}$
2. $\overline{TV}$
3. $\overline{WX}$
4. $\overline{YZ}$
5. $\overline{MN}$
6. $\overline{PQ}$

Find the distance between the points. Round to the nearest tenth.

7. $A(2, 3), B(-4, 3)$
8. $A(2, 3), D(2, -2)$
9. $B(-4, 3), E(-4, 0)$
10. $D(2, -2), F(-1, -2)$
11. $E(-4, 0), G(0, 0)$
12. $H(5, 0), E(-4, 0)$

Use the given endpoints of each circle's diameter. Find each circle's center and the lengths of its diameter and radius. Round to the nearest tenth, if necessary.

13. $A(-1, 2), B(-3, -4)$
14. $A(3, 4), D(6, 9)$
15. $B(5, 1), E(6, -2)$
16. $D(-5, 9), F(-2, 0)$
17. $A(6, 1), B(-4, 5)$
18. $A(2, 3), B(-4, 5)$

Find the slope of each line segment shown.

1. $\overline{AB}$
2. $\overline{CD}$
3. $\overline{EF}$
4. $\overline{GH}$
5. $\overline{IJ}$
6. $\overline{KL}$
7. $\overline{MN}$
8. $\overline{OP}$
9. $\overline{QR}$
10. $\overline{ST}$
11. $\overline{UV}$
12. $\overline{WX}$

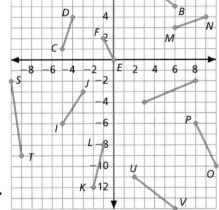

Graph the line that passes through the given point and has the given slope.

13. $(-1, 2), m = \dfrac{1}{5}$
14. $(0, -3), m = -\dfrac{1}{2}$
15. $(1, 4), m = 3$
16. $(3, -2), m = -\dfrac{5}{2}$
17. $(-2, -2), m = \dfrac{1}{4}$
18. $(4, 0), m = -2$
19. $(-1, 2), m$ is undefined
20. $(-4, -5), m = 0$

Find the slope of the line containing the given points. Name any vertical or horizontal lines.

21. $(1, 1), (0, 4)$
22. $(-5, 2), (9, 0)$
23. $(5, -3), (5, 8)$
24. $(-2, 0), (0, -4)$
25. $(5, 8), (3, -2)$
26. $(4, -3), (8, -21)$
27. $(8, 8), (2, 5)$
28. $(3, 7), (-3, -8)$

Identify the slope and the *y*-intercept for each line.

1. $y = \frac{1}{2}x + 5$ **2.** $y = 3x - 4$ **3.** $y = -7$ **4.** $-2x - 4y = 8$

Graph each line.

5. $y = -x + 3$ **6.** $y = 4x - 7$ **7.** $y = -\frac{1}{2}x - 5$

8. $-3x - y = 6$ **9.** $5x - 2y = 10$ **10.** $y = 3x - 4$

Write an equation of each line using the given information.

11. $m = 2, b = -1$ **12.** $(1, -1), (-2, -4)$ **13.** $m = -3, (5, -2)$

14. $m = -2, (3, -1)$ **15.** $m = -\frac{3}{2}, b = 0$ **16.** $m = \frac{1}{2}, (-8, 4)$

17. $(1, 3), (2, -2)$ **18.** $(-1, 4), (-3, -2)$ **19.** $m = 10, b = \frac{1}{2}$

Write an equation of each line whose graph is shown below.

20. 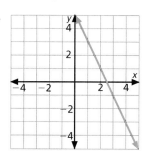   **21.** 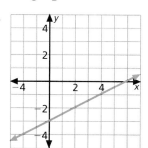   **22.**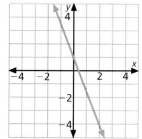

State whether the graph of each inequality has a *dashed* or *solid* line.

1. $y > x + 3$ **2.** $y \leq 2x - 5$ **3.** $y \geq 4$ **4.** $x - 2y < 4$

Determine whether solutions of the inequality are *above* or *below* the boundary. State if the boundary line is included.

5. $y + 2x > 5$ **6.** $y - 2x > 5$ **7.** $y \leq x - 4$ **8.** $y < -3$

9. $-y < 3$ **10.** $x - y > 2$ **11.** $-2y > 8$ **12.** $4x + y \leq 6$

Graph each inequality.

13. $y < 2$ **14.** $y \geq 3$ **15.** $x \geq 2$ **16.** $x < -1$

17. $y > -x$ **18.** $y \leq -x + 1$ **19.** $y > 2 - x$ **20.** $y \leq 3 - 2x$

21. Write and graph an inequality that describes the set of points for which the *y*-coordinate is greater than or equal to the *x*-coordinate decreased by 2.

22. Write and graph an inequality that describes the set of points for which the sum of the *y*-coordinate and twice the *x*-coordinate is less than 5.

Determine if each graph represents a function. Explain.

1.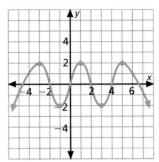

2.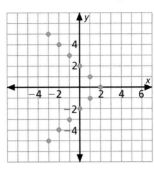

3.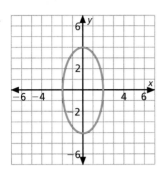

Graph each function for the given domain.

4. $y = 4x + 1$, $\{-2, -1, 0, \frac{1}{2}, 1\}$

5. $y = 3x - 2$, $\{-3, -1, 0, 2, 4\}$

6. $y = 3x + 1$, {all real numbers}

7. $y = -2x$, {all real numbers}

Graph each function when the domain is $\{-2, -1, 0, 1, 2\}$.

8. $y = 4x - 1$

9. $y = -x - 1$

10. $y = 3$

11. If a car is moving at a constant speed of 40 mi/h, then the distance it covers is a function of the amount of time it is moving. Let y be the distance in miles and let x be the time in hours. Then the distance function is $y = 40x$. Graph the function when the domain is $\left\{\frac{1}{4}, \frac{1}{2}, 1, 1\frac{1}{2}, 1\frac{3}{4}\right\}$.

Graph each function for the domain of real numbers.

1. $y = x^2 - 1$

2. $y = 3x^2$

3. $y = 2x^2 + 1$

4. $y = 2x^2 - 4$

PHYSICS The equation for the height of an object shot straight up into the air from ground level at a velocity of 64 ft/sec is $y = -16t^2 + 64t$, where y is the height of the object in feet and t is the time in seconds.

5. Graph the function for the domain of real numbers from $t = 0$ through $t = 4$.

6. What is the maximum height the object reaches?

7. How long does it take the object to reach the maximum height?

8. At what time(s) is the object 48 ft above the ground?

9. At what time does the object return to the ground?

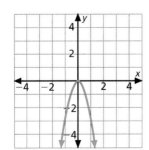

Complete each ordered pair so that it corresponds to a point on the graph.

10. $(0, ?)$

11. $(?, -3)$

12. $(-1, ?)$

1. Assume that *y* varies directly as *x*. When *x* = 15, *y* = 20. Find *y* when *x* = 36.

2. Assume that *y* varies directly as *x*. When *x* = 10, *y* = 4. Find *x* when *y* = 8.

3. Assume that *y* varies directly as *x*. When *x* = 4, *y* = 3. Find *y* when *x* = 5.

4. Assume that *y* varies directly as *x*. When *x* = 3, *y* = 5. Find *y* when *x* = 6.

5. The annual simple interest on a loan varies directly as the amount of the loan. The interest on a $1000 loan is $65. Find the interest on an $800 loan.

6. If you drop an object out of a window, the distance that it falls varies directly as the square of the amount of time it is falling. If an object falls 64 ft in 2 sec, how far will it fall in 5 sec?

7. The distance a spring stretches varies directly as the weight attached to it. If a 15-kg weight stretches the spring 24 cm, how many centimeters will a 20-kg weight stretch it?

8. The distance an object rolls down a slope varies directly as the square of the time it rolls. A ball rolls 80 ft down the slope in 5 sec. How far will it roll in 10 sec?

1. Assume that *y* varies inversely as *x*. When *x* = 8, *y* = 12. Find *y* when *x* = 2.

2. Assume that *y* varies inversely as *x*. When *x* = 3, *y* = 7. Find *y* when *x* = 5.

3. Assume that *y* varies inversely as *x*. When *x* = 45, *y* = 20. Find *y* when *x* = 60.

4. Assume that *y* varies inversely as *x*. When *x* = 9, *y* = 30. Find *y* when *x* = 10.

5. Assume that *y* varies inversely as *x*. When *x* = 3, *y* = 4. Find *y* when *x* = 2.

6. Assume that *y* varies inversely as *x*. When *x* = 12, *y* = 3. Find *y* when *x* = 4.

7. The time it takes to fill a water tank varies inversely as the square of the diameter of the pipe used to fill it. A pipe with a diameter of 3 cm takes 8 min to fill the tank. How long would it take to fill the tank with a pipe having a diameter of 5 cm?

8. The amount paid by each member of a group chartering a bus varies inversely as the number of people in the group. When there are 24 people in the group, the cost is $35. What is the cost per person when there are 32 people in the group?

Identify each relationship as a *direct variation* or an *inverse variation*.

9. the surface area of a sphere and its radius

10. the speed of a train and the time it takes to travel 10 mi

11. the weight of a steel rod and the length of the rod

12. the intensity of light and the distance from the light source

Chapter 7

Extra Practice 7–1 • Translations in the Coordinate Plane • pages 296–299

On a coordinate plane, graph $\triangle MNP$ with vertices $M(2, 2)$, $N(1, -2)$ and $P(-2, -1)$. Then graph its image under each translation from the original position.

1. 2 units up

2. 3 units right

3. 4 units left and 2 units down

On a coordinate plane, graph trapezoid $RSTV$ with vertices $R(-5, 1)$, $S(-2, 4)$, $T(-1, 2)$ and $U(-3, 0)$. Then graph its image under each translation from the original position.

4. 5 units down

5. 3 units right

6. 1 unit left and 3 units up

Write the rule that describes each translation.

7.
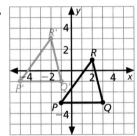

8.
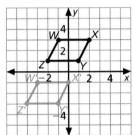

9.
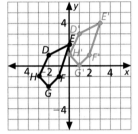

10. The coordinates of a point after a translation of 7 units left and 7 units down are $(4, -2)$. Find the coordinates of the point in its original position.

Extra Practice 7–2 • Reflections in the Coordinate Plane • pages 300–303

Give the coordinates of the image of each point under a reflection across the given line.

1. $(-4, 3)$; x-axis

2. $(6, 2)$; y-axis

3. $(-2, 0)$; y-axis

4. $(-5, 7)$; $y = x$

5. $(-4, 4)$; $y = -x$

6. $(4, -6)$; $y = -x$

7. $(4, 0)$; $x = 1$

8. $(3, 5)$; $y = 1$

Copy each figure on a coordinate plane. Then graph its image under a reflection across the given line.

9. $y = x$
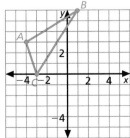

10. $y = -x$
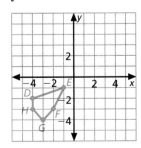

11. $y = 1$
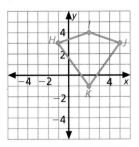

Copy each figure on a coordinate plane. Then graph its image after each rotation about the origin.

1. 90° clockwise

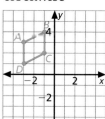

2. 180° clockwise

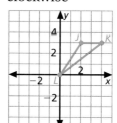

3. 90° counterclockwise

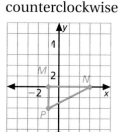

4. 180° counterclockwise

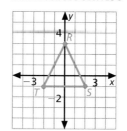

Trace each figure and its rotated image. Identify the center of rotation, the angle of rotation and the direction of rotation.

5.

6.

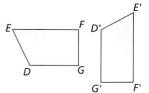

7.

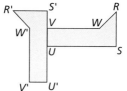

Give the coordinates of the image of *A*(5, 2) after each rotation about the origin.

8. 90° counterclockwise

9. 90° clockwise

10. 180° counterclockwise

11. 180° clockwise

Tell whether each dashed line is a line of symmetry. If not, trace the line and one side of the figure. Complete the drawing so that it has line symmetry.

1.

2.

3.

4.

Give the order of rotational symmetry for each figure.

5.

6.

7.

8.

Draw a figure with the given number of lines of symmetry.

9. 0

10. 1

11. 2

12. 3

13. infinite

Extra Practice

Copy △*PQR* on grid paper. Draw each dilation.

1. scale factor: 2, center (0, 0)

2. scale factor: $\frac{1}{2}$, center (0, 0)

3. scale factor: 1.5, center (0, 0)

Copy rectangle *ABCD* on grid paper. Draw each dilation.

4. scale factor: 3, center (0, 0)

5. scale factor: $\frac{1}{3}$, center (0, 0)

Give the scale factor and center for each dilation.

6. Triangle *ABC* has vertices *A*(3, 1), *B*(6, −1) and *C*(7, 6). Its dilated image is △*A'B'C'* with vertices *A'*(6, 2), *B'*(12, −2) and *C'*(14, 12).

7. Quadrilateral *LMNO* has vertices *L*(6, 0), *M*(−8, −2), *N*(−4, 4) and *O*(8, 6). Its dilated image is quadrilateral *L'M'N'O'* with vertices *L'*(3, 0), *M'*(−4, −1), *N'*(−2, 2) and *O'*(4, 3).

Chapter 8

For each line identified by two points, state the slope of a line parallel and the slope of a line perpendicular to it.

1. *A*(3, 1) and *B*(6, 10)

2. *C*(−1, −4) and *D*(−5, −5)

3. *E*(0, −2) and *F*(−5, 6)

4. *G*(6, 0) and *H*(0, 6)

5. *J*(−2, 3) and *K*(−5, 7)

6. *L*(−3, −5) and *M*(2, −2)

Determine if the graphs will show parallel or perpendicular lines, or neither.

7. $y = -3x + 4$
   $y = \frac{1}{3}x - 1$

8. $y = 2x + 6$
   $y = -2x$

9. $y = -\frac{1}{4}x - 1$
   $y = 4x + 10$

10. $x + 2y = 12$
    $3x + 6y = 3$

11. $2x + y = -7$
    $2x + y = 0$

12. $3x - y = 5$
    $3x + y = -10$

Write an equation in slope-intercept form of a line passing through the given point and parallel to the given line.

13. (3, 4); $y = -3x + 4$

14. (4, −2); $2x + y = 7$

15. (3, −2); $y = -2x + 5$

Write an equation in slope-intercept form of a line passing through the given point and perpendicular to the given line.

16. (4, −2); $y = -\frac{1}{2}x + 5$

17. (3, −2); $y = -\frac{2}{3}x + 5$

18. (−3, −2); $y = 2x + 5$

Extra Practice

Determine if the given ordered pair is a solution of the system of equations.

1. $(3, 1)$; $3x + y = 10$
$x - y = -1$

2. $(-2, -5)$; $6x - 2y = -2$
$2x + y = -9$

3. $(6, 0)$; $x + 5y = 12$
$2x + 2y = 6$

4. $(-1, 2)$; $-x + 2y = 5$
$4x - 3y = -10$

5. $(2, 1)$; $-2x + 3y = -1$
$5x + y = 11$

6. $(0, 4)$; $4x - 2y = -8$
$-3x + y = -4$

Solve each system of equations graphically. Check the solution.

7. $y = 2x + 5$
$y = -\dfrac{2}{3}x - 3$

8. $y = \dfrac{1}{2}x - 2$
$y = 2x - 11$

9. $y = \dfrac{1}{3}x + 6$
$y = x + 8$

10. $y = x - 2$
$y = -2x - 8$

11. $y = \dfrac{3}{4}x - 7$
$y = -\dfrac{1}{4}x - 9$

12. $y = 2x + 1$
$2y = -\dfrac{1}{2}x + 6$

 GRAPHING Use a graphing calculator to solve each system of equations.

13. $-3x + y = 28$
$3x + y = -14$

14. $x + 6y = 42$
$-x + 2y = 6$

15. $5x + 6y = 0$
$-x + 3y = -21$

Solve each system of equations. Check the solution.

1. $2x + 3y = 18$
$-x + 4y = 13$

2. $5x - y = -10$
$5x + y = 0$

3. $4x - 2y = 14$
$x + 8y = 12$

4. $3x - 9y = 15$
$x - 3y = 5$

5. $-x - y = 2$
$3x + y = 2$

6. $5x - 4y = -19$
$x + 2y = -1$

7. $-x + 4y = -23$
$-2x - y = 17$

8. $7x - 3y = -20$
$-2x + y = 16$

9. $3x + 3y = -12$
$5x - 2y = 22$

10. $4x + y = 6$
$-x + 2y = -15$

11. $-3x + 2y = 8$
$-7x + 4y = 20$

12. $-x + 3y = -11$
$2x + 7y = 9$

13. $9x - 2y = -5$
$5x + y = 12$

14. $4x + 4y = 15$
$-x - y = 5$

15. $7x + y = -1$
$-10x - 2y = -2$

16. $3x - 4y = 5$
$-x + y = -10$

17. $-2x + 8y = -18$
$-x - 2y = -9$

18. $5x + 2y = -1$
$-3x - y = 2$

19. The schoolyard has 12 trees. The number of maple trees is 3 less than twice the number of oak trees. How many maple trees are there in the schoolyard?

20. At the farm market, Jed bought 4 lb of apples and 6 lb of peaches for $18. Maria bought 2 lb of apples and 10 lb of peaches for $23. Find the cost per pound of the apples and peaches.

Extra Practice

Solve each system of equations. Check the solutions.

1. $x + 2y = 5$
$-x + 2y = 3$

2. $2x - 3y = 9$
$3x + 3y = 6$

3. $2x - y = -3$
$5x + y = -4$

4. $4x - y = 5$
$-4x + 2y = -2$

5. $-x - 3y = 2$
$5x + 3y = 14$

6. $2x + y = -4$
$-2x - 6y = -6$

7. $3x + 4y = 1$
$3x + 2y = -7$

8. $6x - y = 2$
$5x - y = 1$

9. $x - 3y = -16$
$x + 2y = 9$

10. $-x + 4y = -6$
$2x - 6y = 10$

11. $2x - y = 2$
$3x + 3y = -33$

12. $4x - 8y = 8$
$x - y = 4$

13. $3x + 2y = -2$
$5x - y = -25$

14. $5x + y = 11$
$10x - 3y = -8$

15. $x + 7y = 18$
$-4x - 4y = 0$

16. $3x - 4y = 7$
$-2x + 2y = -6$

17. $-x + 8y = -17$
$3x + 9y = 18$

18. $5x - 8y = 3$
$-x + y = 0$

19. Taylor has 25 coins made up of quarters and dimes. The value of the coins is $3.55. How many of each coin does she have?

20. Evan has 50 coins made up nickels and dimes. The value of the coins is $4.20. How many of each coin does he have?

Evaluate each determinant.

1. $\begin{vmatrix} 0 & 3 \\ -1 & 4 \end{vmatrix}$

2. $\begin{vmatrix} -6 & 6 \\ 4 & -4 \end{vmatrix}$

3. $\begin{vmatrix} 3.1 & -2.4 \\ 0.2 & -2.5 \end{vmatrix}$

4. $\begin{vmatrix} 5 & -1 \\ 4 & 0 \end{vmatrix}$

5. $\begin{vmatrix} 3 & 0 \\ 0 & -3 \end{vmatrix}$

6. $\begin{vmatrix} 4 & -3 \\ 5 & 3 \end{vmatrix}$

7. $\begin{vmatrix} 6 & 6 \\ 6 & 6 \end{vmatrix}$

8. $\begin{vmatrix} 2 & 3 \\ 4 & 8 \end{vmatrix}$

Solve each system of equations using the method of determinants. Round to the nearest thousandth, if necessary.

9. $-3x + y = 14$
$x + 2y = -3$

10. $6x + y = 2$
$5x - y = 9$

11. $x - 3y = -7$
$x + 2y = -2$

12. $-5x + 4y = -5$
$2x - 7y = 2$

13. $2x - 4y = -6$
$3x + 3y = 18$

14. $5x - y = -17$
$x - 3y = -2$

15. $3x + 4y = 17$
$x + 2y = 17$

16. $3x - 5y = 12$
$5x + y = 42$

17. $x - 2y = 11$
$5x + y = 9$

18. $-x + 4y = 21$
$9x - 7y = 3$

19. $2x - 9y = 22$
$5x + y = 33$

20. $14x - 6y = 23$
$11x - 7y = 46$

Write a system of linear inequalities for each graph.

1.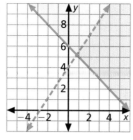

2.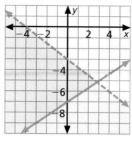

3.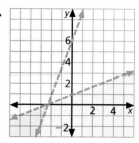

4.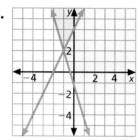

5.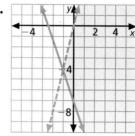

6.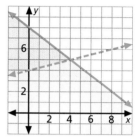

Graph the solution set of each system of inequalities.

7. $2x + y > 2$
$-3x + y \leq -7$

8. $7x - y \geq -1$
$2x + y < -8$

9. $4x + y \geq 4$
$-x + 2y \leq 7$

10. $-x + y < 7$
$-2x - y \leq 8$

Chapter 9

Extra Practice 9–1 • Add and Subtract Polynomials • pages 376–379

Write each polynomial in standard form for the variable x.

1. $5 + x^2 + 3x + x^3$

2. $x^2 + x^4 + 4x + 2x^3$

3. $2x^3 + 3x^5 + 5x - x^4$

4. $1 - 4x - x^2 - 3x^3$

5. $-2x^2 + 3 - x^5 + 6x$

6. $3x^3 - 2x^7 - 2 + 5x^4$

Simplify.

7. $6k - k$

8. $10m - 15m$

9. $3h - 4h + 8h - 5h$

10. $4t^2 + 5t^2 - 9t^2$

11. $x + 3y - 4x + 6y$

12. $5 - 2z + 4z^2 - 3z$

13. $6w - 3 + 4v + w + 6$

14. $10mn + 2m^2 - 4mn + 2n$

15. $(4a - 6) + (3a - 1)$

16. $(3b + 4c) - (8c - 5b)$

17. $(4r^2 - 2r + 3) + (2r^2 - 6r - 9)$

18. $(w^2 + 5w + 8) - (w^2 + 6w + 12)$

19. $(6s + 4st - t) + (-3s - 6st + 7t)$

20. $(4d^2 + de - e^2) - (d^2 + 3de + 8e^2)$

Extra Practice

Simplify.

1. $(2a)(6b)$

2. $(4xy)(-5y)$

3. $(-10c)(-3d)$

4. $(ab)(-4d)$

5. $-5(12x)$

6. $(3d)\left(\dfrac{1}{3}e\right)$

7. $\left(\dfrac{1}{?}w\right)(12v)$

8. $15(3z)$

9. $(-3x^2)(4y^3)$

10. $(5j^2)(4k^2)$

11. $(-m^4)(2n^2)$

12. $(c^3d^2)(e^5d^3)$

13. $(m^2n^3p^3)(m^2n^2p^3)$

14. $(-2de)(9de)$

15. $(8f)(-3f^2)$

16. $(5a^2b)(-b)$

17. $(2.5r)(0.3r^2)$

18. $\left(\dfrac{5}{6}gh^2\right)\left(-\dfrac{12}{25}h\right)$

19. $(4w)^2$

20. $(-3x)^2$

21. $(2k)^3$

22. $(3st^2)^3$

23. $(x^2y^3z)^4$

24. $(-de^2f^3)^3$

Write and simplify an expression for the area of each figure.

25.
$6xy^2$
$6xy^2$

26.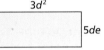
$3d^2$
$5de$

27.
$5m^2n$
$12mn^3$

Simplify.

1. $\dfrac{6n^3}{3n}$

2. $\dfrac{14x^5}{2x}$

3. $\dfrac{8h^2}{-8h}$

4. $\dfrac{18f}{2f}$

5. $\dfrac{3x}{9xz}$

6. $\dfrac{100a^2b^2}{10ab^2}$

7. $\dfrac{22e^2f}{11e^2}$

8. $\dfrac{-32j^4k^2}{-4j^3k^2}$

9. $\dfrac{49f^3gh^2}{7fh}$

10. $\dfrac{35a^8b^7}{21a^3b^5}$

11. $\dfrac{6x+18}{6}$

12. $\dfrac{10y-50}{2}$

13. $\dfrac{42j^3+28j^2}{7j}$

14. $\dfrac{22k^4+44k^5}{11k^2}$

15. $\dfrac{100s^5-50s^7}{25s^4}$

16. $\dfrac{d^5-5d^4+3d^3}{d}$

17. $\dfrac{3e^4+12e^3-9e}{3e}$

18. $\dfrac{4r^2st+9rs^2t-rst^2}{rst}$

19. $\dfrac{v^3w^2x^3+2v^2w^3x^4+6vw^4x^4}{vwx^2}$

20. $\dfrac{8cd-12c^2d+10cd^2-6c^2d^2}{-2cd}$

21. The area of a rectangle is $24x^2y^3$ ft^2 and the length is $6x^2y$ ft. Find the width.

22. The area of a rectangle is $150a^4b^2c^3$ yd^2 and the width is $10ab^2c$ yd. Find the length.

Simplify.

1. $x(3x + 4y)$

2. $2a(5a + 3b)$

3. $8m(2m - n)$

4. $4s(4s - 3)$

5. $6j(j + 9)$

6. $3k(-k + 7)$

7. $3a(4a + 5a^3)$

8. $w(w^4 - w^5)$

9. $-5d(2d - d^2)$

10. $2g(-g^3 + 4g^2 - 3)$

11. $-6e(4 + 3e^3 + e^5)$

12. $\frac{1}{3}f(9f^2 + 6f - 3)$

13. $4v\left(\frac{1}{2}v + \frac{1}{4}v^2\right)$

14. $7x(2x^2 - x^3 + 4x^5)$

15. $-3m(4m^2 + 6mn + 2n^2)$

16. $4(t^2 - 2) + 2(-t^2 + 1)$

17. $5(2z + 3) - 3(4z - 5)$

18. $7w(3w + 3) - 6w(4w + 5)$

19. $6v(2v - 2) + 5v(4v - 8)$

20. $\frac{1}{3}a(9a^3 - 12a^2 + 15a)$

21. $(c^4 - 3.5c^3 - 2.5c^2)1.6c$

22. $5z^2(3 - 6z + 8z^2)$

23. Jessica's weekly salary can be represented by the formula $S = 50 + 3n$ where n is the number of widgets she makes in one week. Write and simplify an expression for her salary for two weeks if she made d widgets the first week and $2d$ widgets the second week.

24. Last week Chiko mowed 12 lawns at $25.00 per lawn. This week, he mowed an additional k lawns. Write a variable expression in simplest form that represents the total amount Chiko was paid last week and this week.

Find each product.

1. $(x + 2)(x + 1)$

2. $(b - 3)(b + 2)$

3. $(y - 5)(y - 2)$

4. $(c + 3)(c - 1)$

5. $(4 - a)(3 - a)$

6. $(5 - z)(1 + z)$

7. $(e + 3)^2$

8. $(f - 5)^2$

9. $(n + 9)(n - 3)$

10. $(p + 4)(p - 6)$

11. $(g + 10)^2$

12. $(v - 7)(v - 6)$

13. $(12 - d)(-1 + d)$

14. $(3 + h)(9 - h)$

15. $(r + 3)(r - 3)$

Simplify.

16. $2(n + 2)(n - 4)$

17. $3(a + 3)(a + 8)$

18. $k(k - 2)(k + 1)$

19. $j(j - 3)(j - 5)$

20. $(v + 4)(v - 5) - (v + 6)(v + 8)$

21. $(d + 6)(d - 3) + (d + 5)(d - 8)$

22. $5(x + 3)(x - 3) + 3(x - 1)(x + 7)$

23. $2(y + 3)(y + 1) - 3(y + 2)(y + 1)$

24. Write an expression for the area of a square if the measure of each side is $(s + 7)$ in.

Extra Practice

Factor each polynomial.

1. $8x - 4$
2. $6xy - 6xz$
3. $15m + 5$
4. $14c - 21$
5. $-3d + 18$
6. $16k - 14$
7. $8f^3 + 24f$
8. $4g^4 - 16g^2$
9. $7h^3 + 35h$
10. $ab + b^2$
11. $s^2t^3 + st^4$
12. $wv^3 - w^2v^5$
13. $10m^5 - 15m^3$
14. $8n^3 + 6n$
15. $z^8 - z^5$
16. $120d - 80$
17. $20x^2y - 4xy^2$
18. $-3ef^2 + 24e^3g$
19. $x^4 - 2x^3 + 3x^2$
20. $5y^5 + y^4 - 2y^3$
21. $4n^2 - 8n + 20$
22. $12b^4 + 6b^3 - 9$
23. $3de + 5de^2 + 4e$
24. $r^3s^2t + r^2s^3v + r^2s^2w$
25. $35c^2 + 14cd + 21cd^2$
26. $15jk^3 - 15jk + 10j^3k$
27. $4x^2y^2 - x^4y^2 + 2xy^2z$
28. $g^3h^3 + 3g^4h^2 - 5g^5h$
29. $2m^3n^2 - m^3n^3 - 4m^4n^2$
30. $p^2q^3 + 9p^4 - 8p^3q^2$

Evaluate each expression. Let $x = 2$ and $y = -2$.

31. $4xy^3 + 2x^2y$
32. $2xy(2y^2 + x)$

Rewrite each formula by factoring.

33. $C = 3r + 6s + 3t$
34. $A = 10d + 35d^2$

Tell whether the trinomial is a perfect square trinomial.

1. $x^2 + 4x + 4$
2. $y^2 + 25y + 50$
3. $k^2 + 20k + 121$
4. $a^2 + 12a + 36$
5. $e^2 + 16e + 81$
6. $d^2 + 16d + 64$
7. $m^2 + 30m + 150$
8. $t^2 + 26t + 169$

Factor each polynomial if possible.

9. $s^2 + 14s + 49$
10. $p^2 - 60p + 90$
11. $n^2 - 16n + 64$
12. $j^2 + 8j + 16$
13. $b^2 + 14b + 144$
14. $f^2 - 2f + 1$
15. $t^2 + 60t + 900$
16. $a^2 + 12a + 38$
17. $k^2 - 36$
18. $c^2 - 64$
19. $x^2 - 10$
20. $z^2 - 49$
21. $w^2 - 81$
22. $d^2 - 96$
23. $t^2 - 49$
24. $v^2 - 75$
25. $g^2 - 25$
26. $a^2 - 16$
27. $c^2 - 1$
28. $h^2 - 9$
29. $q^2 - 40$
30. $r^2 - 100$
31. $n^2 - 225$
32. $x^2 + 12x + 36$
33. $y^2 - 256$
34. $b^2 - 7b + 49$
35. $k^2 - 100$
36. $2x^2 + 32x + 128$
37. $y^2 + 16$
38. $6b^2 - 600b$

Chapter 10

Extra Practice 10–1 • Visualize and Represent Solids • pages 422–425

Identify each figure and name its base(s).

1.

2.

3.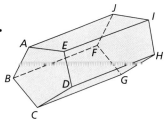

Identify a pair of parallel edges, intersecting faces, and intersecting edges for each figure.

4. the figure in Exercise 1

5. the figure in Exercise 2

6. the figure in Exercise 3

Identify each figure.

7.

8.

9.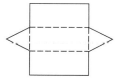

Draw each figure.

10. right triangular prism

11. cube

12. right cylinder

13. oblique cone

14. triangular pyramid

15. right cone

Extra Practice 10–2 • Nets and Surface Area • pages 426–429

Identify the three-dimensional figure for each net.

1.

2.

3.

4.

5.

6.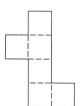

7.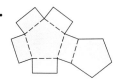

8.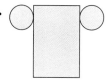

Draw a net for each figure. Then find the surface area.

9.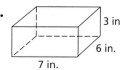

10.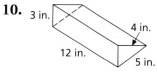

11.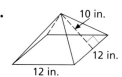

12.

Extra Practice

Find the surface area of each figure. Use 3.14 for π. If necessary, round to the nearest tenth.

1.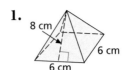
8 cm
6 cm
6 cm

2.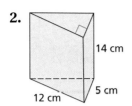
14 cm
5 cm
12 cm

3.
4 ft
2.6 ft

4.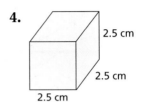
2.5 cm
2.5 cm
2.5 cm

5.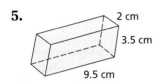
2 cm
3.5 cm
9.5 cm

6.
6 cm

7.
6 cm
14 cm

8.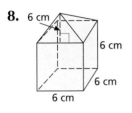
6 cm
6 cm
6 cm
6 cm

9.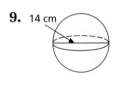
14 cm

10.
2 m
5 m

11.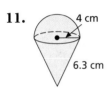
4 cm
6.3 cm

12.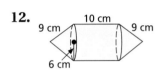
10 cm
9 cm
9 cm
6 cm

13. A metal cylinder has a radius of $\frac{1}{2}$ ft and a surface area of 30 ft². What is the height of the cylinder to the nearest whole number?

Sketch each object in one-point perspective.

1. a three-dimensional letter F

2. a pentagonal prism

3. a flat envelope

Sketch each object in two-point perspective.

4. a triangular prism

5. an open book

6. a computer screen

Use each perspective drawing to locate the vanishing point(s).

7.

8.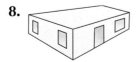

9.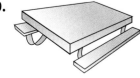

Create an isometric drawing of each figure.

1. a rectangular pyramid

2. a right triangular prism

3. a 2-unit cube

4. a 2-unit by 4-unit rectangular prism

Use the isometric drawing for Exercises 5–7.

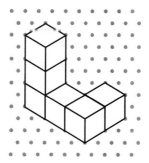

5. How many cubes are used in the drawing?

6. How many cube faces are exposed in the figure?

7. If the length of an edge of one of the cubes is 5 yd, what is the total surface area of the figure?

Give the number of cubes used to make each figure and the number of cube faces exposed. Be sure to count the cubes that are hidden from view.

8.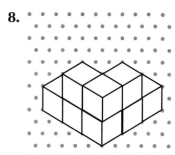

9.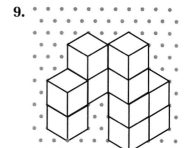

10.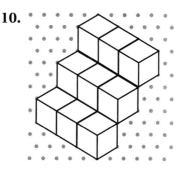

Make an orthogonal drawing labeling the front, top and right-side views.

1.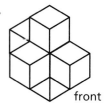
front

2.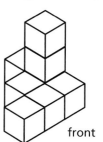
front

3.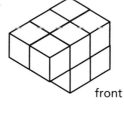
front

For each foundation drawing, sketch a figure composed of cubes.

4.

3	4
1	1

front

5.

4	3	2
2	1	

front

6.

2	1	2
1		1

front

Extra Practice

Create a foundation drawing for each figure.

7.

8.
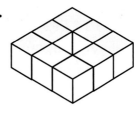

9.
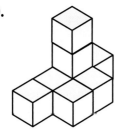

Extra Practice 10–7 • Volume of Prisms and Pyramids • pages 452–455

Find the volume of each figure.

1.
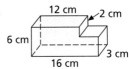

2.
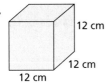

3.
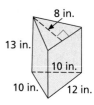

4.
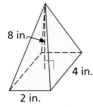

5.
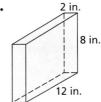

6.
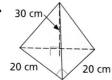

7.
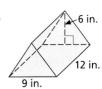

8.
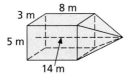

9. A crate is 1.5 ft long, 2.5 ft wide and 15 in. high. What is the volume of the crate? Round your answer to the nearest tenth.

Extra Practice 10–8 • Volume of Cylinders, Cones and Spheres • pages 456–459

Find the volume of each figure. Use 3.14 for π. Round to the nearest whole number.

1.

2.

3.

4.

5.
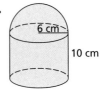

6.

7.

8.
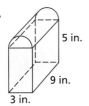

9. A water tank in the shape of a sphere has a diameter of 50 ft. How many cubic feet of water, to the nearest whole number, can the tank hold? Use 3.14 for π.

Extra Practice

Chapter 11

Extra Practice 11–1 • Similar Polygons • pages 474–477

Determine if each pair of polygons is similar.

1.

2.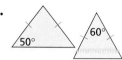

3.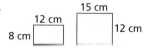

Find the length of $\overline{AB}$ in each pair of similar figures.

4.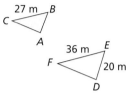

5.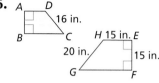

6.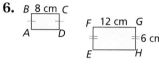

Find the $m\angle A$ in each pair of similar polygons.

7.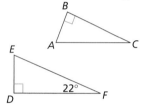

8.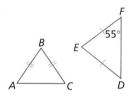

9.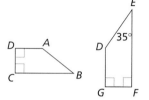

10. Triangle ABC and $\triangle DEF$ are similar triangles. If $m\angle A = 30°$ and $m\angle F = 50°$, what are the measures of $\angle D$ and $\angle E$?

Extra Practice 11–2 • Indirect Measurement • pages 478–481

Use indirect measurement to find the unknown length x.

1.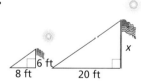

2.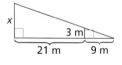

3.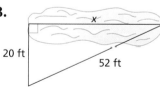

4.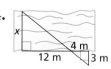

5.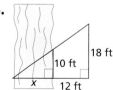

6.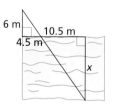

7. Brian uses the mirror method to find the height of a tree. He places the mirror 30 ft from the tree's base and he stands 5 ft from the mirror. Brian is 4.5-ft tall. Find the height of the tree.

8. A 1-m stick casts a 0.8-m shadow. At the same time, a flagpole casts a 4.8-m shadow. How tall is the flagpole?

Extra Practice

Find the length of each hypotenuse. Round to the nearest tenth.

1.

2.

3.

4.

Find the unknown length. Round to the nearest tenth.

5.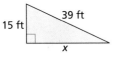

6.

7.

8.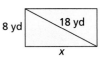

Find each unknown measurement. Round to the nearest tenth.

9.

10.

11.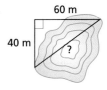

12. If you traveled 36 mi south and then 48 mi east, what is the shortest distance you could travel to return to your starting point?

In △RST, find each ratio in lowest terms.

1. sin T

2. tan S

3. cos T

4. tan T

5. cos S

6. sin S

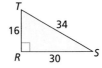

For each triangle, use the information given to find the trigonometric ratio indicated. Write the answer in lowest terms.

7. In △ABC, $m\angle B = 90°$, $AB = 12$, and $BC = 9$. Find cos C.

8. In △DEF, $m\angle F = 90°$, $FE = 24$, and $DE = 26$. Find cos E.

9. In △JKL, $m\angle L = 90°$, $JL = 10$, and $JK = 26$. Find cos J.

10. In △PQV, $m\angle Q = 90°$, $PV = 34$, and $QV = 30$. Find cos P.

11. In △TAG, $m\angle G = 90°$, $TA = 39$, and $TG = 15$. Find cos T.

12. In △HOP, $m\angle O = 90°$, $OP = 36$, and $HO = 15$. Find cos P.

CALCULATOR Use a calculator to find each ratio to four decimal places.

13. cos 50°

14. tan 20°

15. cos 37°

16. sin 74°

17. tan 51°

18. sin 15°

19. tan 8°

20. tan 82°

21. sin 44°

22. cos 45°

23. sin 50°

24. cos 25°

Extra Practice

Find each length to the nearest tenth.

1. *AB*   **2.** *BC*   **3.** *RP*   **4.** *MP*

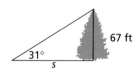

5. Find the height of the flagpole, *p*.   **6.** Find the length of the tree's shadow, *s*.

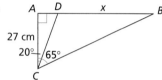

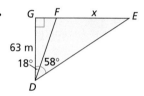

In the figure shown, find the value of *x* to the nearest tenth.

7.

8.

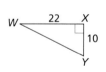

Find each measure to the nearest whole degree.

1. $m\angle B$ **2.** $m\angle W$ **3.** $m\angle B$ **4.** $m\angle G$

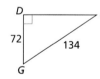

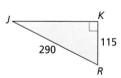

5. $m\angle R$ **6.** $m\angle T$ **7.** $m\angle H$ **8.** $m\angle R$

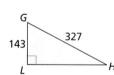

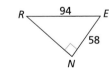

9. When a certain airplane has reached a ground distance of 1800 ft from its lift-off point, it is 1050 ft above the ground. To the nearest whole degree, what angle does the plane's path make with the ground?

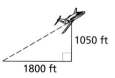

10. The base of a ladder is 32 in. from a wall. The top of the ladder is touching the wall at a place that is 132 in. from the ground. To the nearest degree, what is the measure of the angle formed by the ladder and the wall?

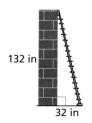

Extra Practice

Use △ABC for Exercises 1–6. Leave your answers in square root form.

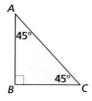

1. If $AB = 9$, find BC and AC.
2. If $AB = 15$, find BC and AC.

3. If $BC = 4$, find AB and AC.
4. If $BC = 12$, find AB and AC.

5. If $AC = 27$, find AB and BC.
6. If $AC = 2$, find AB and BC.

Use △XYZ for Exercises 7–12. Leave your answers in square root form.

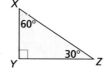

7. If $XY = 4$, find YZ and XZ.
8. If $XY = 7$, find YZ and XZ.

9. If $YZ = 18$, find XY and XZ.
10. If $YZ = 2$, find XY and XZ.

11. If $XZ = 24$, find XY and YZ.
12. If $XZ = 1$, find XY and YZ.

A 300-ft guy wire that supports an antenna makes a 60° angle with the ground.

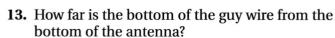

13. How far is the bottom of the guy wire from the bottom of the antenna?

14. How tall is the antenna? Round to the nearest tenth.

Chapter 12

Define each set in roster notation and in set-builder notation. Then determine whether the set is finite or infinite.

1. W, the set of whole numbers less than 6
2. D, the set of letters in *dinosaur*

3. N, the set of the odd whole numbers
4. V, the set of vowels

5. F, the set of whole number factors of 12
6. M, the set of multiples of 5 less than 30

Write a set equivalent to each set.

7. $\left\{\frac{1}{2}, 1, \frac{3}{2}, 2\right\}$
8. $\{d, z, m, h, v, q\}$
9. $\{4, 12, 24, 48\}$

Determine if the sets are *equal or not equal*.

10. $\{c, x, k, r, g\}$ and $\{k, x, g, r, c\}$
11. $\{5, 31, 7, 27, 9, 53\}$ and $\{7, 9, 37, 53, 5, 29\}$

12. $\{t, j, a, w, n, f\}$ and $\{a, n, f, v, t, w\}$
13. $\{18, 33, 79, 57, 82\}$ and $\{57, 18, 79, 82, 33\}$

Determine if each statement is *true* or *false*.

14. $62 \in \{0, 2, 4, 6, \ldots\}$
15. $34 \in \{4, 8, 12, 16, \ldots, 48\}$

16. $\{30, 40\} \subseteq \{3, 6, 9, 12, \ldots\}$
17. $\{d, m\} \subseteq \{a, s, j, m, y, c, n, d, f\}$

18. $\{x|x$ is a parallelogram$\} \subseteq \{x|x$ is a quadrilateral$\}$

Write the subsets of each set.

19. $\{3, 6, 9\}$
20. $\{a, b\}$
21. $\{15, 30\}$
22. $\{d, e, n\}$

Extra Practice

Find each set by listing the members.

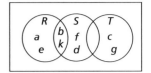

1. $R \cup S$
2. $R \cup T$
3. $R \cap S$
4. $S \cap T$

Let $U = \{s, t, r, i, p, e\}$, $B = \{t, r, i, p\}$ and $C = \{p, i, t\}$. Find each union.

5. $B' \cup C'$
6. $B' \cup C$
7. $B \cup C'$

Let $U = \{q, t, f, j, b, w\}$, $R = \{t, j, w\}$ and $S = \{q, j, b\}$. Find each union.

8. $R \cup S'$
9. $R' \cup S$
10. $R' \cup S'$

11. Let $M = \{t, r, u, c, k\}$ and $N = \{c, h, a, r, t\}$. Find $M \cap N$.
12. Let $F = \{6, 12, 24, 48\}$ and $G = \{8, 12, 48\}$. Find $F \cap G$.

Use roster notation to represent each set.

$U = \{2, 4, 6, 8, 10, 12, 14, 16, 18\}$ $A = \{6, 10, 16\}$ $B = \{4, 12, 18\}$ $C = \{4, 8, 12, 14\}$

13 $A' \cap B$
14. $A \cap C'$
15. $A \cap B$
16. $B \cap C$
17. $B' \cap C$
18. $A' \cap C'$

Write the converse, inverse and contrapositive of each statement. Determine if each is *true* or *false*. If *false*, give a counterexample to explain why.

1. If an animal is a cat, then it has paws.

2. If a figure is a square, then it has four sides.

3. If an angle is obtuse, then its measure is greater than 90°.

4. If a vehicle is a station wagon, then it has four wheels.

5. If you have a dog, then you have a pet.

6. If a figure is a circle, then it has a diameter.

7. If a number is greater than 100, then the number is greater than 50.

8. If you vacation in Paris, then you vacation in France.

9. If an object is a baseball, then it is round.

10. If a whole number is even, then it is divisible by two.

11. If you are studying Italian, then you are studying a foreign language.

12. If a set contains all the factors of 48, then one element in that set is 8.

13. If two sides of a triangle are congruent, then the angles opposites those sides are congruent.

Extra Practice

Predict the next number in each pattern.

1. 5, 15, 35, 65, 105, __?__

2. 1, 2, 4, 8, 16, 32, __?__

3. 50, 10, 2, $\frac{2}{5}$, __?__

4. 30, 25, 21, 18, 16, __?__

5. 3, 6, 18, 72, __?__

6. 24, 12, 6, 3, __?__

Complete each argument by drawing a conclusion.

7. *Premise 1* All A+ students are on the High Honor Roll.
Premise 2 Kim is an A+ student.

8. *Premise 1* If the Rams win the basketball game, they are the conference champions.
Premise 2 The Rams win the basketball game.

9. *Premise 1* All four-year olds may register for story-time at the Library.
Premise 2 Mike is four years old.

Tell whether the reasoning is *inductive* or *deductive*.

10. The football team has been undefeated for the past three years. Since the coach is the same this year the team should be undefeated this year as well.

11. The swim club requires lifeguards to be at least 18 years old. Sam is working as a lifeguard at the swim club, so he must be at least 18 years old.

12. If *x* is one more than *y* and *y* is two more than *z*, what conclusion can be drawn about the relationship between *x* and *z*? What type of reasoning did you use to draw your conclusion?

Determine by form whether the arguments are *valid* or *invalid*. If *valid*, name the argument form.

1. If an animal is a cat, then it has a tail. Fluffy is a cat. Therefore Fluffy has a tail.

2. If my bicycle is broken, then I cannot exercise. I cannot exercise. Therefore, my bicycle is broken.

3. If a number is a multiple of ten, then it is an even number. The number is not an even number. Therefore, the number is not a multiple of ten.

4. If Julie becomes a cheerleader, then she will go to all the soccer games. Julie does not become a cheerleader. Therefore, Julie does not go to all the soccer games.

5. If you are a member of Lake City Swim Club, then you are a resident of Lake City. You are not a resident of Lake City. Therefore, you are not a member of Lake City Swim Club.

6. If a rectangle has a width of *m* and a length of *n*, then its area is *mn*. The area of a rectangle is *mn*. Therefore, the rectangle has a width of *m* and a length of *n*.

Show that the conclusion given is true. Present your argument in a logical order.

1. $a° + b° = 100°$

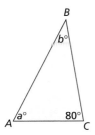

2. $m\angle 1 + m\angle 2 = 180°$

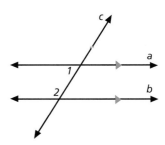

3. $m\angle 1 = 60°$

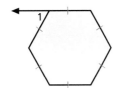

4. $w = 50$ in.

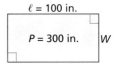

5. Find the measure of each central angle in circle E.

6. The exterior angles of a pentagon have measures $5x°$, $3x°$, $(2x + 7)°$, $(3x − 1)°$, and $(5x − 6)°$. Find the measure of each interior angle.

7. Show that when the square of a number is cubed, the result is the same as that number to the sixth power. Let n represent the number.

8. Show that 2 less than any even integer is an even integer. Let n be any integer.

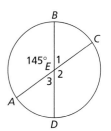

Extra Practice

Preparing for Standardized Tests

Becoming a Better Test-Taker

At some time in your life, you will probably have to take a standardized test. Sometimes this test may determine if you go on to the next grade or course, or even if you will graduate from high school. This section of your textbook is dedicated to making you a better test-taker.

TYPES OF TEST QUESTIONS   In the following pages, you will see examples of four types of questions commonly seen on standardized tests. A description of each type of question is shown in the table below.

Type of Question	Description	See Pages
multiple choice	4 or 5 possible answer choices are given from which you choose the best answer.	628–631
gridded response	You solve the problem. Then you enter the answer in a special grid and shade in the corresponding circles.	632–635
short response	You solve the problem, showing your work and/or explaining your reasoning.	636–639
extended response	You solve a multi-part problem, showing your work and/or explaining your reasoning.	640–644

PRACTICE   After being introduced to each type of question, you can practice that type of question. Each set of practice questions is divided into five sections that represent the concepts most commonly assessed on standardized tests.

- Number and Operations
- Algebra
- Geometry
- Measurement
- Data Analysis and Probability

USING A CALCULATOR   On some tests, you are permitted to use a calculator. You should check with your teacher to determine if calculator use is permitted on the test you will be taking, and if so, what type of calculator can be used.

TEST-TAKING TIPS   In addition to Test-Taking Tips like the one shown at the right, here are some additional thoughts that might help you.

- Get a good night's rest before the test. Cramming the night before does not improve your results.

- Budget your time when taking a test. Don't dwell on problems that you cannot solve. Just make sure to leave that question blank on your answer sheet.

- Watch for key words like NOT and EXCEPT. Also look for order words like LEAST, GREATEST, FIRST, and LAST.

> **Test-Taking Tip**
>
> If you are allowed to use a calculator, make sure you are familiar with how it works so that you won't waste time trying to figure out the calculator when taking the test.

Multiple-Choice Questions

Multiple-choice questions are the most common type of questions on standardized tests. These questions are sometimes called *selected-response questions*. You are asked to choose the best answer from four or five possible answers.

To record a multiple-choice answer, you may be asked to shade in a bubble that is a circle or an oval, or to just write the letter of your choice. Always make sure that your shading is dark enough and completely covers the bubble.

The answer to a multiple-choice question is usually not immediately obvious from the choices, but you may be able to eliminate some of the possibilities by using your knowledge of mathematics. Another answer choice might be that the correct answer is not given.

Incomplete Shading
ⓐ ⓑ Ⓒ ⓓ
Too light shading
ⓐ ⓑ Ⓒ ⓓ
Correct shading
ⓐ ⓑ Ⓒ ⓓ

Example 1

A storm signal flag is used to warn small craft of wind speeds that are greater than 38 miles per hour. The length of the square flag is always three times the length of the side of the black square. If y is the area of the black square and x is the length of the side of the flag, which equation describes the relationship between x and y?

Ⓐ $y = \dfrac{1}{3}x^2$

Ⓑ $y = \dfrac{1}{9}x^2$

Ⓒ $y = x^2 - 1$

Ⓓ $y = 3x$

Ⓔ $y = 9x$

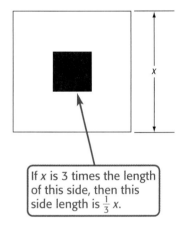

If x is 3 times the length of this side, then this side length is $\frac{1}{3}x$.

> For the area of a square, $A = s^2$. So, $A = x \cdot x$ or x^2.

The area of the black square is part of the area of the flag, which is x^2. Eliminate choices D and E because they do not include x^2.

$A = \left(\dfrac{1}{3}x\right)^2$ or $\dfrac{1}{9}x^2$ square units

So, $y = \dfrac{1}{9}x^2$. This is choice B.

Use some random numbers to check your choice.

> Multiples of 3 make calculations easier.

Length of Flag (x)	Length of Black Square	Area of Black Square	Area $= \frac{1}{9}x^2$
12	4	16	$16 \stackrel{?}{=} \frac{1}{9}(12^2)$ ✓
27	9	81	$81 \stackrel{?}{=} \frac{1}{9}(27^2)$ ✓
60	20	400	$400 \stackrel{?}{=} \frac{1}{9}(60^2)$ ✓

Many multiple-choice questions are actually two- or three-step problems. If you do not read the question carefully, you may select a choice that is an intermediate step instead of the correct final answer.

Example 2

Barrington can skateboard down a hill five times as fast as he can walk up the hill. If it takes 9 min to walk up the hill and skateboard back down, how many minutes does it take him to walk up the hill?

   (F) 1.5 min   (G) 4.5 min   (H) 7.2 min   (J) 7.5 min

Before involving any algebra, let's think about the problem using random numbers.

Skating is five times as fast as walking, so walking time equals 5 times the skate time. Use a table to find a pattern

Skate Time	Skate Time × 5 = Walk Time
6 min	$6 \cdot 5 = 30$ min
3 min	$3 \cdot 5 = 15$ min
2 min	$2 \cdot 5 = 10$ min
x min	$x \cdot 5 = 5x$ min

> Use the pattern to find a general expression for walk time given any skate time.

The problem states that the walk time and the skate time total 9 min.

> Use the expression to write an equation for the problem.

$x + 5x = 9$	skate time + walk time = 9 min
$6x = 9$	Add like terms.
$x = 1.5$	Divide each side by 6.

Looking at the choices, you might think that choice F is the correct answer. But what does x represent, and what is the problem asking?

The problem asks for the time it takes to walk up the hill, but the value of x is the time it takes to skateboard. So, the actual answer is found using $5x$ or $5(1.5)$, which is 7.5 min.

The correct choice is J.

Example 3

The Band Boosters are making ice cream to sell at an Open House. Each batch of ice cream calls for 5 c of milk. They plan to make 20 batches. How many gallons of milk do they need?

   (A) 800   (B) 100   (C) 25   (D) 12.5   (E) 6.25

The Band Boosters need 5×20 or 100 c of milk. However, choice B is not the correct answer. The question asks for *gallons* of milk.

4 c = 1 qt and 4 qt = 1 gal, so 1 gal = 4×4 or 16 c.

$100 \cancel{c} \times \dfrac{1 \text{ gal}}{16 \cancel{c}} = 6.25$ gal, which is choice E.

Multiple Choice Practice

Choose the best answer.

Number and Operations

1. One mile on land is 5280 ft, while one nautical mile is 6076 ft. What is the ratio of the length of a nautical mile to the length of a land mile as a decimal rounded to the nearest hundredth?

   (A) 0.87   (B) 1.01   (C) 1.15   (D) 5.68

2. The star Proxima Centauri is 24,792,500 million mi from Earth. The star Epsilon Eridani is 6.345×10^{13} mi from Earth. In scientific notation, how much farther from Earth is Epsilon Eridani than Proxima Centauri?

   (A) 0.697×10^{14} mi   (B) 3.866×10^{13} mi

   (C) 6.097×10^{13} mi   (D) 38.658×10^{12} mi

3. In 1976, the cost per gallon for regular unleaded gasoline was 61 cents. In 2002, the cost was \$1.29 per gal. To the nearest percent, what was the percent of increase in the cost per gallon of gas from 1976 to 2002?

   (A) 1%   (B) 53%   (C) 95%   (D) 111%

4. The serial numbers on a particular model of personal data assistant (PDA) consist of two letters followed by five digits. How many serial numbers are possible if any letter of the alphabet and any digit 0–9 can be used in any position in the serial number?

   (A) 676,000,000   (B) 67,600,000

   (C) 6,760,000   (D) 676,000

Algebra

5. The graph shows the approximate relationship between the latitude of a location in the Northern Hemisphere and its distance in miles from the equator. If y represents the distance of a location from the equator and x represents the measure of latitude, which equation describes the relationship between x and y?

   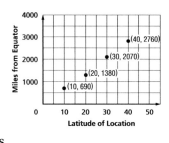

   (A) $y = x + 69$ (B) $y = x + 690$

   (C) $y = 69x$ (D) $y = 10x$

6. A particular prepaid phone card can be used from a pay phone. The charge is 30 cents to connect and then 4.5 cents per minute. If y is the total cost of a call in cents where x is the number of minutes, which equation describes the relation between x and y?

   (A) $y = 4.5x + 30$ (B) $y = 30x + 4.5$

   (C) $y = 0.45x + 0.30$ (D) $y = 0.30x + 0.45$

7. Katie drove to the lake for a weekend outing. The lake is 100 mi from her home. On the trip back, she drove for an hour, stopped for lunch for an hour, and then finished the trip home. Which graph best represents her trip home and the distance from her home at various times?

   (A)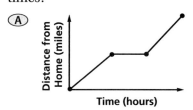

   (B)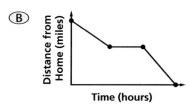

   (C)

   (D)

8. Temperature can be given in degrees Fahrenheit or degrees Celsius. The formula $F = \frac{9}{5}C + 32$ can be used to change any temperature given in degrees Celsius to degrees Fahrenheit. Solve the formula for C.

   (A) $C = \frac{5}{9}(F - 32)$ (B) $C = F + 32 - \frac{9}{5}$

   (C) $C = \frac{5}{9}F - 32$ (D) $C = \frac{9}{5}(F - 32)$

Geometry

9. Which of the following statements are true about the 4-in. quilt square?

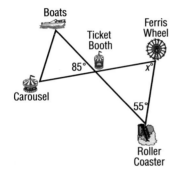

R 2 in.
2 in.
U   V   S
X   T   W

(A) *VSWT* is a square.

(B) *UVTX* ≅ *VSWT*

(C) Four right angles are formed at *V*.

(D) Only A and B are true.

(E) A, B, and C are true.

10. At the Daniels County Fair, the carnival rides are positioned as shown. What is the value of *x*?

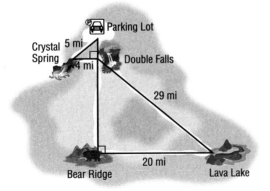

(A) 40   (B) 47.5

(C) 55   (D) 70

(E) 85

11. The diagram shows a map of the Clearwater Wilderness hiking area. To the nearest tenth of a mile, what is the distance from the Parking Lot to Bear Ridge using the most direct route?

(A) 24 mi   (B) 25.5 mi

(C) 26 mi   (D) 30.4 mi

Measurement

12. Laura expects about 60 people to attend a party. She estimates that she will need one quart of punch for every two people. How many gallons of punch should she prepare?

(A) 7.5

(B) 15

(C) 30

(D) 34

13. Stone Mountain Manufacturers are designing two sizes of cylindrical cans below. What is the ratio of the volume of can A to the volume of can B?

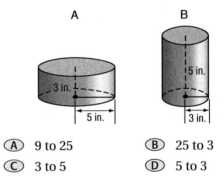

(A) 9 to 25   (B) 25 to 3

(C) 3 to 5   (D) 5 to 3

Data Analysis and Probability

14. The 2000 populations of the five least-populated U.S. states are shown in the table. Which statement is true about this set of data?

State	Population
Alaska	626,932
North Dakota	642,200
South Dakota	754,844
Vermont	608,827
Wyoming	493,782

(A) The mode of the data set is 642,200.

(B) The median of the data set is 626,932.

(C) The mean of the data set is 625,317.

(D) A and C are true.

(E) B and C are true.

Gridded-Response Questions

Preparing for Standardized Tests

Gridded-response questions are other types of questions on standardized tests. These questions are sometimes called *student-produced responses* or *grid-ins,* because you must create the answer yourself, not just choose from four or five possible answers.

For gridded response, you must mark your answer on a grid printed on an answer sheet. The grid contains a row of four or five boxes at the top, two rows of ovals or circles with decimal and fraction symbols, and four or five columns of ovals, numbered 0–9. Since there is no negative symbol on the grid, answers are never negative. At the right is an example of a grid from an answer sheet.

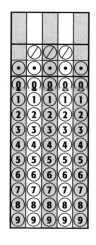

Example 1

Diego drove 174 mi to his grandmother's house. He made the drive in 3 h without any stops. At this rate, how far in miles can Diego drive in 5 h?

What value do you need to find?

You need to find the number of miles Diego can drive in 5 h.

Write a proportion for the problem. Let m represent the number of miles.

$$\text{miles} \longrightarrow \frac{174}{3} = \frac{m}{5} \longleftarrow \text{miles}$$
$$\text{hours} \longrightarrow \qquad\qquad \longleftarrow \text{hours}$$

Solve the proportion.

$\dfrac{174}{3} = \dfrac{m}{5}$ Original proportion

$870 = 3m$ Find the cross products.

$290 = m$ Divide each side by 3.

How do you fill in the grid for the answer?

- Write your answer in the answer boxes.

- Write only one digit or symbol in each answer box.

- Do not write any digits or symbols outside the answer boxes.

- You may write your answer with the first digit in the left answer box, or with the last digit in the right answer box. You may leave blank any boxes you do not need on the right or the left side of your answer.

- Fill in only one bubble for every answer box that you have written in. Be sure not to fill in a bubble under a blank answer box.

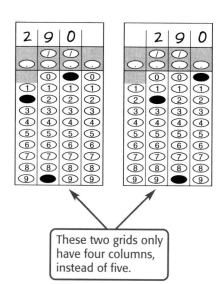

These two grids only have four columns, instead of five.

Many gridded response questions result in an answer that is a fraction or a decimal. These values can also be filled in on the grid.

Example 2

What is the slope of the line that passes through $(-2, 3)$ and $(2, 4)$?

Let $(-2, 3) = (x_1, y_1)$ and $(2, 4) = (x_2, y_2)$.

$m = \dfrac{y_2 - y_1}{x_2 - x_1}$ Slope formula

$= \dfrac{4 - 3}{2 - (-2)}$ or $\dfrac{1}{4}$ Substitute and simplify.

How do you grid the answer?

You can either grid the fraction $\dfrac{1}{4}$, or rewrite it as 0.25 and grid the decimal. Be sure to write the decimal point or fraction bar in the answer box. The following are acceptable answer responses that represent $\dfrac{1}{4}$ and 0.25.

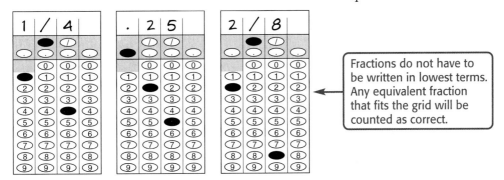

Fractions do not have to be written in lowest terms. Any equivalent fraction that fits the grid will be counted as correct.

Some problems may result in an answer that is a mixed number. Before filling in the grid, change the mixed number to an equivalent improper fraction or decimal. For example, if the answer is $1\dfrac{1}{2}$, do not enter 11/2 as this will be interpreted as $\dfrac{11}{2}$. Instead, either enter 3/2 or 1.5.

Example 3

Amber's cookie recipe calls for $1\dfrac{1}{3}$ c of coconut. If Amber plans to make 4 batches of cookies, how much coconut does she need?

Find the amount of coconut needed using a proportion.

coconut $\longrightarrow$ $\dfrac{1\frac{1}{3}}{1} = \dfrac{x}{4}$
batches $\longrightarrow$

$4(1\dfrac{1}{3}) = 1x$

$4(\dfrac{4}{3}) = x$

$\dfrac{16}{3} = x$

Do not leave a blank answer box in the middle of an answer.

Leave the answer as the improper fraction $\dfrac{16}{3}$, as you cannot correctly grid $5\dfrac{1}{3}$.

Gridded-Response Practice

Solve each problem. Then copy and complete a grid like the one shown on page 632.

Number and Operations

1. China has the most days of school per year for children with 251 days. If there are 365 days in a year, what percent of the days of the year do Chinese students spend in school? Round to the nearest tenth of a percent.

2. Charles is building a deck and wants to buy some long boards that he can cut into various lengths without wasting any lumber. He would like to cut any board into all lengths of 24 in., 48 in., or 60 in. In feet, what is the shortest length of boards that he can buy?

3. At a sale, an item was discounted 20%. After several weeks, the sale price was discounted an additional 25%. What was the total percent discount from the original price of the item?

4. The Andromeda Spiral galaxy is 2.2×10^6 light-years from Earth. The Ursa Minor dwarf is 2.5×10^5 light-years from Earth. How many times as far is the Andromeda Spiral as Ursa Minor dwarf from Earth?

5. Twenty students want to attend the World Language Convention. The school budget will only allow for four students to attend. In how many ways can four students be chosen from the twenty students to attend the convention?

Algebra

6. Find the y-intercept of the graph of the equation $3x + 4y - 5 = 0$.

7. Name the x-coordinate of the solution of the system of equations $2x - y = 7$ and $3x + 2y = 7$.

8. Solve $2b - 2(3b - 5) = 8(b - 7)$ for b.

9. Kersi read 36 pages of a novel in 2 h. Find the number of hours it will take him to read the remaining 135 pages if he reads them at the same rate?

10. The endpoints of a segment are $(-5, 7)$ and $(-2, 9)$. Find the length to the nearest tenth.

11. Ms. Blackwell needs to rent a car for her family vacation. She has found a company that offers the following two options.

Plan	Flat Rate	Cost per Mile
Option A	$40	$0.25
Option B	$30	$0.35

How many miles must the Blackwells drive for the plans to cost the same?

Geometry

12. Triangle MNP is reflected over the x-axis. What is the x-coordinate of the image of point N?

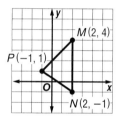

13. The pattern for the square tile shown in the diagram is to be enlarged so that it will measure 15 in. on a side. By what scale factor must the pattern be enlarged?

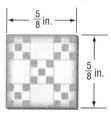

14. Find the measure of $\angle A$ to the nearest degree.

Test-Taking Tip

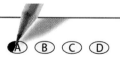

Question 14
Remember that the hypotenuse of a right triangle is always opposite the right angle.

15. ∠KLM and ∠XYZ are complementary. If $m\angle KLM = 3x - 1$ and $m\angle XYZ = x + 7$, find the measure of the larger angle.

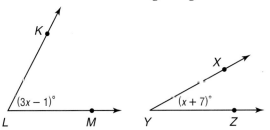

16. A triangle has a perimeter of 96 cm. The ratio of measures of its three sides is 6:8:10. Find the length of the longest side in centimeters.

17. The scale on a map of Texas is 0.75 in. = 5 mi. The distance on the map from San Antonio to Dallas is 8.25 in. What is the actual distance from San Antonio to Dallas in miles?

Measurement

18. Pluto is the farthest planet from the Sun in this solar system at 2756 million mi. If light travels at 186,000 mi/sec, how many minutes does it take for a particular ray of light to reach Pluto from the Sun? Round to the nearest minute.

19. Noah drove 342 mi and used 12 gal of gas. At this same rate, how many gallons of gas will he use on his entire trip of 1140 mi?

20. The jumping surface of a trampoline is shaped like a circle with a diameter of 14 ft. Find the area of the jumping surface. Use 3.14 for π and round to the nearest square foot.

21. A cone is drilled out of a cylinder of wood. If the cone and cylinder have the same base and height, find the volume of the remaining wood. Use 3.14 for π and round to the nearest cubic inch.

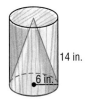

14 in.

6 in.

22. Chi-Yo wants to make a quilt pattern using similar triangles as shown. What is the length of the third side of the larger triangle?

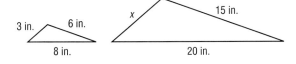

3 in.   6 in.   8 in.   x   15 in.   20 in.

Data Analysis and Probability

23. The table shows the average size in acres of farms in the six states with the largest farms. Find the median of the farm data in acres.

Average Size of Farms for 2001	
State	**Acres per Farm**
Alaska	1586
Arizona	3644
Montana	2124
Nevada	2267
New Mexico	2933
Wyoming	3761

24. The Lindley Park Pavilion is available to rent for parties. There is a fee to rent the pavilion and then a charge per hour. The graph shows the total amount you would pay to rent the pavilion for various numbers of hours. If a function is written to model the charge to rent the pavilion, where x is the number of hours and y is the total charge, what is the slope of the function?

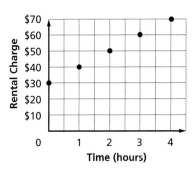

25. A particular game is played by rolling three tetrahedral (4-sided) dice. The faces of each die are numbered with the digits 1–4. How many outcomes are in the sample space for the event of rolling the three dice once?

26. In a carnival game, the blindfolded contestant draws two toy ducks from a pond without replacement. The pond contains 2 yellow ducks, 10 black ducks, 22 white ducks, and 8 red ducks. The best prize is won by drawing two yellow ducks. What is the probability of drawing two yellow ducks? Write your answer as a percent rounded to the nearest tenth of a percent.

Short-Response Questions

Short-response questions require you to provide a solution to the problem, as well as any method, explanation, and/or justification you used to arrive at the solution. These are sometimes called *constructed-response, open-response, open-ended, free-response,* or *student-produced questions.* The following is a sample rubric, or scoring guide, for scoring short-response questions.

Credit	Score	Criteria
Full	2	Full credit: The answer is correct and a full explanation is provided that shows each step in arriving at the final answer.
Partial	1	Partial credit: There are two different ways to receive partial credit. • The answer is correct, but the explanation provided is incomplete or incorrect. • The answer is incorrect, but the explanation and method of solving the problem is correct.
None	0	No credit: Either an answer is not provided or the answer does not make sense.

On some standardized tests, no credit is given for a correct answer if your work is not shown.

Example

Susana is painting two large rooms at her art studio. She has calculated that each room has 4000 ft² to be painted. It says on the can of paint that one gallon covers 300 ft² of smooth surface for one coat and that two coats should be applied for best results. What is the minimum number of 5-gal cans of paint Susana needs to buy to apply two coats in the two rooms of her studio?

FULL CREDIT SOLUTION

First find the total number of square feet to be painted.
$$4000 \times 2 = 8000 \text{ ft}^2$$
Since 1 gal covers 300 ft², multiply 8000 ft² by the unit rate $\frac{1 \text{ gal}}{300 \text{ ft}^2}$.
$$8000 \text{ ft}^2 \times \frac{1 \text{ gal}}{300 \text{ ft}^2} = \frac{8000}{300} \text{ gal}$$
$$= 26\frac{2}{3} \text{ gal}$$
Each can of paint contains 5 gal, so divide $26\frac{2}{3}$ gal by 5 gal.

$$26\frac{2}{3} \div 5 = \frac{80}{3} \div 5 = \frac{\overset{16}{\cancel{80}}}{3} \times \frac{1}{\cancel{5}_{1}} = \frac{16}{3} = 5\frac{1}{3}$$

Since Susana cannot buy a fraction of a can of paint, she needs to but 6 cans of paint.

The steps, calculations, and reasoning are clearly stated.

The solution of the problem is clearly stated.

PARTIAL CREDIT SOLUTION

In this sample solution, the answer is correct; however there is no justification for any of the calculations.

> There is no explanation of how $26\frac{2}{3}$ was obtained.

$$26\frac{2}{3} \div 5 = \frac{80}{3} \div 5$$

$$= \frac{\overset{16}{\cancel{80}}}{3} \times \frac{1}{\underset{1}{\cancel{5}}}$$

$$= \frac{16}{3}$$

$$= 5\frac{1}{3}$$

Susana will need to buy 6 cans of paint.

PARTIAL CREDIT SOLUTION

In this sample solution, the answer is incorrect. However, after the first statement, all of the calculations and reasoning are correct.

There are 4000 ft² to be painted and one gallon of paint covers 300 ft².

$$4000\text{ft}^2 \times \frac{1 \text{ gal}}{300 \text{ ft}^2} = \frac{4000}{300} \text{ gal}$$

$$= 13\frac{1}{3} \text{ gal}$$

> The first step of doubling the square footage for painting the second room was left out.

Each can of paint contains five gallons. So 2 cans would contain 10 gal, which is not enough. Three cans of paint would contain 15 gal which is enough.

Therefore, Susana will need to buy 3 cans of paint.

NO CREDIT SOLUTION

The wrong operations are used, so the answer is incorrect. Also, there are no units of measure given with any of the calculations.

$$300 \times 2 = 600$$
$$600 \div 5 = 120$$
$$4000 \div 120 = 33\frac{1}{3}$$

Susana will need 34 cans of paint.

Short-Response Practice

Solve each problem. Show all your work.

Number and Operations

1. The world's slowest fish is the sea horse. The average speed of a sea horse is 0.001 mi/h. What is the rate of speed of a sea horse in feet per minute?

2. Two buses arrive at the Central Avenue bus stop at 8 A.M. The route for the City Loop bus takes 35 min, while the route for the By-Pass bus takes 20 min. What is the next time that the two buses will both be at the Central Avenue bus stop?

3. Toya's Clothing World purchased some denim jackets for $35. The jackets are marked up 40%. Later in the season, the jackets are discounted 25%. How much does the store lose or gain on the sale of one jacket at the discounted price?

4. A femtosecond is 10^{-15} sec, and a millisecond is 10^{-3} sec. How many times faster is a millisecond than a femtosecond?

5. Find the next three terms in the sequence.

   $$1, 3, 9, 27, \ldots$$

Algebra

6. Find the slope of the graph of $5x - 2y + 1 = 0$.

7. Simplify $5 + x(1 - x) + 3x$. Write the result in the form $ax^2 + bx + c$.

8. Solve $17 - 3x \geq 23$.

9. The table shows what Gerardo charges in dollars for his consulting services for various numbers of hours. Write an equation that can be used to find the charge for any amount of time, where y is the total charge in dollars and x is the number of hours.

Hours	Charge	Hours	Charge
0	$25	2	$55
1	$40	3	$70

10. The population of Clark County, Nevada, was 1,375,765 in 2000 and 1,464,653 in 2001. Let x represent the years since 2000 and y represent the total population of Clark County. Suppose the county continues to increase at the same rate. Write an equation that represents the population of the county for any year after 2000.

Geometry

11. Triangle ABC is dilated with scale factor 2.5. Find the coordinates of dilated $\triangle A'B'C'$.

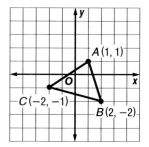

12. At a particular time in its flight, a plane is 10,000 ft above a lake. The distance from the lake to the airport is 5 mi. Find the distance in feet from the plane to the airport.

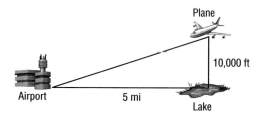

13. Refer to the diagram of the two similar triangles below. Find the value of a.

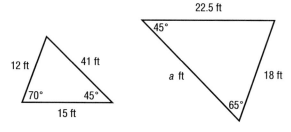

14. The vertices of two triangles are $P(3, 3)$, $Q(7, 3)$, $R(3, 10)$, and $S(-1, 4)$, $T(3, 4)$, $U(-1, 11)$. Which transformation moves $\triangle PQR$ to $\triangle STU$?

15. Find the coordinates of the vertices of quadrilateral $G'H'I'J'$ after a reflection of quadrilateral $GHIJ$ across the x-axis.

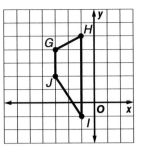

20. A child's portable swimming pool is 6 ft across and is filled to a depth of 8 in. One gallon of water is 231 in.³. What is the volume of water in the pool in gallons? Use 3.14 as an approximation for π and round to the nearest gallon.

6 ft

8 in.

Measurement

16. One inch is equivalent to approximately 2.54 cm. Nikki is 61 in. tall. What is her height in centimeters?

17. During the holidays, Evan works at Cheese Haus. He packages gift baskets containing a variety of cheeses and sausages. During one four-hour shift, he packaged 20 baskets. At this rate, how many baskets will he package if he works 26 h in one week?

18. Ms. Ortega built a box for her garden and placed a round barrel inside to be used for a fountain in the center of the box. The barrel touches the box at its sides as shown. She wants to put potting soil in the shaded corners of the box at a depth of 6 in. How many cubic feet of soil will she need? Use 3.14 for π and round to the nearest tenth of a cubic foot.

4 ft

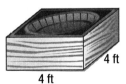
4 ft

4 ft

19. A line segment has its midpoint located at $(1, -5)$ and one endpoint at $(-2, -7)$. Find the length of the line to the nearest tenth.

Data Analysis and Probability

21. The table shows the five lowest recorded temperatures on Earth. Find the mean of the temperatures.

Location	Temperature (°F)
Vostok, Antarctica	-138.6
Plateau Station, Antarctica	-129.2
Oymyakon, Russia	-96.0
Verkhoyansk, Russia	-90.0
Northice, Greenland	-87.0

22. Two six-sided dice are rolled. The sum of the numbers of dots on the faces of the two dice is recorded. What is the probability that the sum is 10?

23. The table shows the amount of a particular chemical that is needed to treat various sizes of swimming pools. Write the equation for a line to model the data. Let x represent the capacity of the pool in gallons and y represent the amount of the chemical in ounces.

Pool Capacity (gal)	Amount of Chemical (oz)
5000	15
10,000	30
15,000	45
20,000	60
25,000	75

24. Fifty balls are placed in a bin. They are labeled from 1 through 50. Two balls are drawn without replacement. What is the probability that both balls show an even number?

Test-Taking Tip

Ⓐ Ⓑ Ⓒ Ⓓ

Question 20
Most standardized tests will include any commonly used formulas at the front of the test booklet. Quickly review the list before you begin so that you know what formulas are available.

Extended-Response Questions

Extended-response questions are often called *open-ended* or *constructed-response questions.* Most extended-response questions have multiple parts. You must answer all parts correctly to receive full credit.

Extended-response questions are similar to short-response questions in that you must show all of your work in solving the problem, and a rubric is used to determine whether you receive full, partial, or no credit. The following is a sample rubric for scoring extended-response questions.

Credit	Score	Criteria
Full	4	A correct solution is given that is supported by well-developed, accurate explanations.
Partial	3, 2, 1	A generally correct solution is given that may contain minor flaws in reasoning or computation or an incomplete solution. The more correct the solution, the greater the score.
None	0	An incorrect solution is given indicating no mathematical understanding of the concept, or no solution is given.

On some standardized tests, no credit is given for a correct answer if your work is not shown.

Make sure that when the problem says to *Show your work,* show every aspect of your solution including figures, sketches of graphing calculator screens, or reasoning behind computations.

Example 1

The table shows the population density in the United States on April 1 in each decade of the 20th century.

a. Make a scatter plot of the data.

b. Alaska and Hawaii became states in the same year. Between what two census dates do you think this happened. Why did you choose those years?

c. Use the data and your graph to predict the population density in 2010. Explain your reasoning.

U.S. Population Density	
Year	People Per Square Mile
1910	31.0
1920	35.6
1930	41.2
1940	44.2
1950	50.7
1960	50.6
1970	57.4
1980	64.0
1990	70.3
2000	79.6

FULL CREDIT SOLUTION

Part a A complete scatter plot includes a title for the graph, appropriate scales and labels for the axes, and correctly graphed points.

- The student should determine that the year data should go on the *x*-axis while the people per square mile data should go on the *y*-axis.

- On the *x*-axis, each square should represent 10 years.

- The *y*-axis could start at 0, or it could show data starting at 30 with a broken line to indicate that some of the scale is missing.

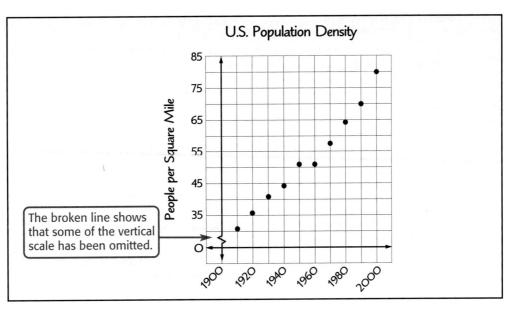

U.S. Population Density

The broken line shows that some of the vertical scale has been omitted.

You might know when Alaska became a state. So, another acceptable reason is that both Alaska and Hawaii became states in 1959.

Part b

1950–1960 because when Alaska became a state it added little population but a lot of land, which made the people per square mile ratio less.

Part c

About 85.0, because the population per square mile would probably get larger so I connected the first point and the last point. The rate of change for each year was $\frac{79.6 - 31.0}{2000 - 1910}$ or about 0.54. I added 10×0.54, or 5.4 to 79.6 to get the next 10-year point.

Actually, any estimate from 84 to 86 might be acceptable. You could also use different points to find the equation for a line of best fit for the data, and then find the corresponding y value for $x = 2010$.

PARTIAL CREDIT SOLUTION

Part a This sample answer includes no labels for the graph or the axes and one of the points is not graphed correctly.

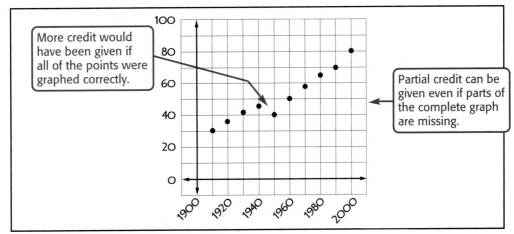

More credit would have been given if all of the points were graphed correctly.

Partial credit can be given even if parts of the complete graph are missing.

Part b Partial credit is given because the reasoning is correct, but the reasoning was based on the incorrect graph in Part a.

> 1940–1950, because when Alaska became a state it added little population but a lot of land, which made the people per square mile ratio less.

Part c Full credit is given for Part c

> Suppose I draw a line of best fit through points (1910, 31.0) and (1990, 70.3). The slope would be $\frac{70.3 - 31.0}{1990 - 1910}$ or about 0.49. Now use the slope and one of the points to find the y-intercept.
>
> $$y = mx + b$$
> $$70.3 = 0.49(1990) + b$$
> $$-904.8 = b$$
>
> So an equation of my line of best fit is
> $$y = 0.49x - 904.8.$$
>
> If $x = 2010$, then $y = 0.49(2010) - 904.8$ or about 80.1 people per square mile in the year 2010.

This sample answer might have received a score of 2 or 1. Had the student graphed all points correctly and gotten Part B correct, the score would probably have been a 3.

NO CREDIT SOLUTION
Part a

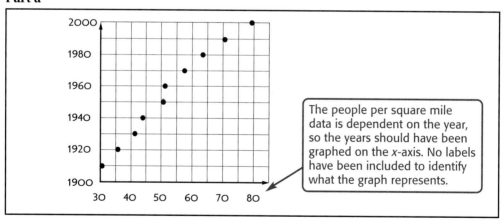

The people per square mile data is dependent on the year, so the years should have been graphed on the x-axis. No labels have been included to identify what the graph represents.

Part b

> I have no idea.

Part c

> 85, because it is the next grid line.

In this sample answer, the student does not understand how to represent data on a graph or how to interpret the data after the points are graphed.

Extended Response Practice

Solve each problem. Show all your work.

Number and Operations

1. The table shows what one dollar in U.S. money was worth in five countries in 1970 and in 2001.

Money Equivalent to One U.S. Dollar		
Country	1970 Value	2001 Value
France	5.5 francs	7 francs
Germany	3.6 marks	2 marks
Great Britain	0.4 pounds	0.67 pounds
Italy	623 lire	2040 lire
Japan	358 yen	117 yen

a. For which country was the percent of increase or decrease in the number of units of currency that was equivalent to $1 the greatest from 1970 to 2001?

b. Suppose a U.S. citizen traveled to Germany in 1970 and in 2001. In which year would the traveler receive a better value for their money? Explain.

c. In 2001, what was the value of one franc in yen?

2. The table shows some data about the planets and the Sun. The radius is given in miles and the volume, mass, and gravity quantities are related to the volume and mass of Earth, which has a value of 1.

	Volume	Mass	Density	Radius	Gravity
Sun	1,304,000	332,950	0.26	434,474	28
Mercury	0.056	0.0553	0.98	1516	0.38
Venus	0.857	0.815	0.95	3760	0.91
Moon	0.0203	0.0123	0.61	1079	0.17
Mars	0.151	0.107	0.71	2106	0.38
Jupiter	1321	317.83	0.24	43,441	2.36
Saturn	764	95.16	0.12	36,184	0.92
Uranus	63	14.54	0.23	15,759	0.89
Neptune	58	17.15	0.30	15,301	1.12
Pluto	0.007	0.0021	0.32	743	0.06

a. Make and test a conjecture relating volume, mass, and density.

b. Describe the relationship between radius and gravity.

c. Can you be sure that the relationship in part b holds true for all planets? Explain.

Algebra

3. The graph shows the altitude of a glider during various times of his flight after being released from a tow plane.

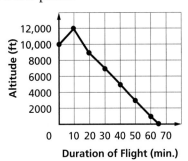

a. What point on the graph represents the moment the glider was released from the tow plane? Explain the meaning of this point in terms of altitude.

b. During which time period did the greatest rate of descent of the glider take place? Explain your reasoning.

c. How long did it take the glider to reach an altitude of 0 ft? Where is this point on the graph?

d. What is the equation of a line that represents the glider's altitude y as the time increased from 20 to 60 min?

e. Explain what the slope of the line in part d represents?

4. John has just received his learner's permit which allows him to practice driving with a licensed driver. His mother has agreed to take him driving every day for two weeks. On the first day, John will drive for 20 min. Each day after that, John's mother has agreed he can drive 15 min more than the day before.

a. Describe the pattern.

b. For how many minutes will John drive on the last day? Show how you found the number of minutes.

c. John's driver's education teacher requires that each student drive for 30 h with an adult outside of class. Will John fulfill this requirement? Explain.

Geometry

5. Polygon *QUAD* is shown on a coordinate plane.

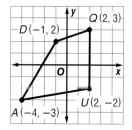

a. Find the coordinates of the vertices of *Q'U'A'D'*, which is the image of *QUAD* after a reflection over the *y*-axis. Explain.

b. Suppose point *M*(*a, b*) is reflected over the *y*-axis. What will be the coordinates of the image *M'*?

c. Describe a reflection that will make *QUAD* look "upside down." What will be the coordinates of the vertices of the image?

6. The diagram shows a sphere with a radius of *a* and a cylinder with a radius and a height of *a*.

a. What is the ratio of the volume of the sphere to the volume of the cylinder?

b. What is the ratio of the surface area of the sphere to the surface area of the cylinder?

c. The ratio of the volume of another cylinder is 3 times the volume of the sphere shown. Give one possible set of measures for the radius and height of the cyilinder in terms of *a*.

Measurement

7. Alexis is using a map of the province of Saskatchewan in Canada. The scale for the map shows that 2 cm on the map is 30 km in actual distance.

a. The distance on the map between two cities measures 7 cm. What is the actual distance between the two cities in kilometers? Show how you found the distance.

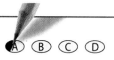

Test-Taking Tip

Question 5
In a reflection over the *x*-axis, the *x*-coordinate remains the same, and the *y*-coordinate changes its sign. In a reflection over the *y*-axis, the *y*-coordinate remains the same, and the *x*-coordinate changes its sign.

b. Alexis is more familiar with distances in miles. The distance between two other cities is 8 cm. If one kilometer is about 0.62 mi, what is the distance in miles?

c. Alexis' entire trip measures 54 cm. If her car averages 25 mi/gal of gasoline, how many gallons will she need to complete the trip? Round to the nearest gallon. Explain.

8. The diagram shows a pattern for a quilt square called Colorful Fan.

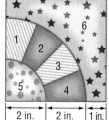

a. What is the area of region 6? Explain.

b. What is the area of region 1? Explain.

c. What is the ratio of the area of region 5 to the area of the entire square? Show how you found the ratio. Leave the ratio in terms of π.

Data Analysis and Probability

9. The table shows the Olympics winning times in the women's 1500-meter speed skating event. The times are to the nearest second.

Year	Time (s)	Year	Time (s)
1960	172	1984	124
1964	143	1988	121
1968	142	1992	126
1972	141	1994	122
1976	137	1998	118
1980	131	2002	114

a. Make a scatter plot of the data.

b. Use the data and your graph to predict the winning time in 2010.

10. There are 1320 ways for three students to win first, second, and third place during a debate.

a. How many students are on the debate team?

b. What is the probability that a student will come in one of the first three places if each student has an equal chance of succeeding?

c. If the teacher announces the third place winner, what is the probability that one particular other student will win first or second place?

Technology Reference Guide

Graphing Calculator Overview

This section summarizes some of the graphing calculator skills you might use in your mathematics classes using the TI-83 Plus or TI-84 Plus.

General Information

- Any yellow commands written above the calculator keys are accessed with the [2nd] key, which is also yellow. Similarly, any green characters or commands above the keys are accessed with the [ALPHA] key, which is also green. In this text, commands that are accessed by the [2nd] and [ALPHA] keys are shown in brackets. For example, [2nd] **[QUIT]** means to press the [2nd] key followed by the key below the yellow **[QUIT]** command.
- [2nd] **[ENTRY]** copies the previous calculation so it can be edited or reused.
- [2nd] **[ANS]** copies the previous answer so it can be used in another calculation.
- [2nd] **[QUIT]** will return you to the home (or text) screen.
- [2nd] **[A-LOCK]** allows you to use the green characters above the keys without pressing [ALPHA] before typing each letter.
- Negative numbers are entered using the [(-)] key, not the minus sign, [−].
- The variable x can be entered using the [X,T,θ,n] key, rather than using [ALPHA] **[X]**.
- [2nd] **[OFF]** turns the calculator off.

Key Skills

Use this section as a reference for further instruction. For additional features, consult the TI-83 Plus or TI-84 Plus user's manual.

ENTERING AND GRAPHING EQUATIONS
Press [Y=]. Use the [X,T,θ,n] key to enter *any* variable for your equation. To see a graph of the equation, press [GRAPH].

SETTING YOUR VIEWING WINDOW
Press [WINDOW]. Use the arrow or [ENTER] keys to move the cursor and edit the window settings. Xmin and Xmax represent the minimum and maximum values along the x-axis. Similarly, Ymin and Ymax represent the minimum and maximum values along the y-axis. Xscl and Yscl refer to the spacing between tick marks placed on the x- and y-axes. Suppose Xscl = 1. Then the numbers along the x-axis progress by 1 unit. Set Xres to 1.

THE STANDARD VIEWING WINDOW
A good window to start with to graph an equation is the **standard viewing window.** It appears in the [WINDOW] screen as follows.

To easily set the values for the standard viewing window, press [ZOOM] 6.

ZOOM FEATURES
To easily access a viewing window that shows only integer coordinates, press [ZOOM] 8 [ENTER].

To easily access a viewing window for statistical graphs of data you have entered, press [ZOOM] 9.

USING THE TRACE FEATURE

To trace a graph, press $\boxed{\text{TRACE}}$. A flashing cursor appears on a point of your graph. At the bottom of the screen, x- and y-coordinates for the point are shown. At the top left of the screen, the equation of the graph is shown. Use the left and right arrow keys to move the cursor along the graph. Notice how the coordinates change as the cursor moves from one point to the next. If more than one equation is graphed, use the up and down arrow keys to move from one graph to another.

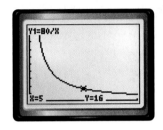

SETTING OR MAKING A TABLE

Press $\boxed{\text{2nd}}$ **[TBLSET]**. Use the arrow or $\boxed{\text{ENTER}}$ keys to move the cursor and edit the table settings. Indpnt represents the x-variable in your equation. Set Indpnt to *Ask* so that you may enter any value for x into your table. Depend represents the y-variable in your equation. Set Depend to *Auto* so that the calculator will find y for any value of x.

USING THE TABLE

Before using the table, you must enter at least one equation in the $\boxed{\text{Y=}}$ screen. Then press $\boxed{\text{2nd}}$ **[TABLE]**. Enter any value for x as shown at the bottom of the screen. The function entered as Y_1 will be evaluated at this value for x. In the two columns labeled X and Y_1, you will see the values for x that you entered and the resulting y-values.

PROGRAMMING ON THE TI–83 PLUS

When you press $\boxed{\text{PRGM}}$, you see three menus: EXEC, EDIT, and NEW. EXEC allows you to execute a stored program by selecting the name of the program from the menu. EDIT allows you to edit or change an existing program. NEW allows you to create a new program. For additional programming features, consult the TI–83 Plus or TI–84 Plus user's manual.

ENTERING INEQUALITIES

Press $\boxed{\text{2nd}}$ **[TEST]**. From this menu, you can enter the $=$, $\neq$, $>$, $\geq$, $<$, and $\leq$ symbols.

ENTERING AND DELETING LISTS

Press $\boxed{\text{STAT}}$ $\boxed{\text{ENTER}}$. Under L_1, enter your list of numerical data. To delete the data in the list, use your arrow keys to highlight L_1. Press $\boxed{\text{CLEAR}}$ $\boxed{\text{ENTER}}$. Remember to clear all lists before entering a new set of data.

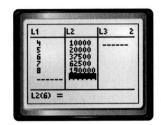

PLOTTING STATISTICAL DATA IN LISTS

Press $\boxed{\text{Y=}}$. If appropriate, clear equations. Use the arrow keys until Plot1 is highlighted. Plot1 represents a Stat Plot, which enables you to graph the numerical data in the lists. Press $\boxed{\text{ENTER}}$ to turn the Stat Plot on and off. You may need to display different types of statistical graphs. To see the details of a Stat Plot, press $\boxed{\text{2nd}}$ **[STAT PLOT]** $\boxed{\text{ENTER}}$. A screen like the one below appears.

At the top of the screen, you can choose from one of three plots to store settings. The second line allows you to turn a Stat Plot on and off. Then you may select the type of plot: scatter plot, line plot, histogram, two types of box-and-whisker plots, or a normal probability plot. Next, choose which lists of data you would like to display along the x- and y-axes. Finally, choose the symbol that will represent each data point.

Cabri Jr. Overview

Cabri Junior for the TI-83 Plus and TI-84 Plus is a geometry application that is designed to reproduce the look and feel of a computer on a handheld device.

General Information

Starting Cabri Jr. To start Cabri Jr., press APPS and choose Cabri Jr. Press any key to continue. If you have not run the program on your calculator before, the F1 menu will be displayed. To leave the menu and obtain a blank screen, press CLEAR. If you have run the program before, the last screen that was in the program before it was turned off will appear. See Quitting Cabri Jr. for instructions on clearing this screen to obtain a blank screen.

In Cabri Jr., the four arrow keys (◀, ▶, ▼, ▲), along with ENTER, operate as a mouse would on a computer. The arrows simulate moving a mouse, and ENTER simulates a left click on a mouse. For example, when you are to select an item, use the arrow keys to point to the selected item and then press ENTER. You will know you are accurately pointing to the selected item when the item, such as a point or line, is blinking.

Quitting Cabri Jr. To quit Cabri Jr., press 2nd [QUIT], or [OFF] to completely shut off the calculator. Leaving the calculator unattended for approximately 4 minutes will trigger the automatic power down. After the calculator has been turned off, pressing ON will result in the calculator turning on, but not Cabri Jr. You will need to press APPS and choose Cabri Jr. Cabri Jr. will then restart with the most current figure in its most recent state.

As Cabri Jr. resembles a computer, it also has dropdown menus that simulate the menus in many computer programs. There are five menus, F1 through F5.

Navigating Menus To navigate each menu, press the appropriate key for F1 through F5. The arrow keys will then allow you to navigate within each menu. The ▲ and ▼ keys allow you to move within the menu items. The ▶ and ◀ keys allow you to access a submenu of an item. If an item has an arrow to the right, this indicates there is a submenu. Although not displayed, the menu items are numbered. You can also select a menu item by pressing the number that corresponds to each item. For example, to select the fourth menu item in the list, press [4]. If you press a number greater than the number of items in the list, the last item will be selected. If you press [0], you will leave the menu without selecting an item. This is the same as pressing CLEAR.

Key Skills

Use this section as a reference for further instruction. For additional features, consult the TI-84 Plus user's manual.

[F1] MANAGING FIGURES

The menu below provides the basic operations when working in Cabri Jr. These are commands normally found in menus within computer applications.

[F2] CREATING OBJECTS

This menu provides the basic tools for creating geometric figures. You can create points in three different ways, a line and line segment by selecting two points, a circle by defining the center and radius, a triangle by finding three vertices, and a quadrilateral by finding four vertices.

[F3] CONSTRUCTING OBJECTS

This menu provides the tools to construct new objects from existing objects. You can construct perpendicular and parallel lines, a perpendicular bisector of a segment, an angle bisector, a midpoint of a line segment, a circle using the center and a point on the circle, and a locus.

[F4] TRANSFORMING OBJECTS

This menu provides the tools to transform geometric figures. Using figures that are already created, you can access this menu to create figures that are symmetrical to other figures, reflect figures over a line of reflection, translate figures using a line segment or two points that define the translation, rotate figures by defining the center of rotation and angle of rotation, and dilate figures using the center of the dilation and a scale factor.

[F5] COMPUTING OBJECTS

This menu provides the tools for displaying, labeling, measuring, and computing. You can make an object visible or invisible, label points on figures, alter the way objects are displayed, measure length, area, and angle measures, display coordinates of points and equations of lines, make calculations, and delete objects from the screen.

While a graphing calculator cannot do everything, it can make some tasks easier. To prepare for whatever lies ahead, you should try to learn as much as you can about the technology. The future will definitely involve technology and the people who are comfortable with it will be successful. Using a graphing calculator is a good start toward becoming familiar with technology.

Glossary/Glosario

A mathematics multilingual glossary is available at **www.math.glencoe.com/multilingual_glossary.** The glossary is available in the following languages.

Arabic	English	Korean	Tagalog
Bengali	Haitian Creole	Russian	Urdu
Cantonese	Hmong	Spanish	Vietnamese

English

Español

▪ A ▪

absolute value (p. 54) The distance a number is from zero on a number line. The absolute value of an integer, a, is written as $|a|$.

valor absoluto (p. 54) Valor que tiene una cifra por su figura, por ejemplo, en el número 592 el valor absoluto de 5 es 5.

acute angle (p. 196) An angle measuring less than 90°.

$0° < m\angle A < 90°$

ángulo agudo (p. 196) Ángulo que mide menos de 90°.

acute triangle (p. 206) A triangle with three acute angles.

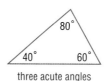

three acute angles
tres ángulos agudos

triángulo acutángulo (p. 206) Triángulo que sus tres ángulos son agudos.

adjacent angles (p. 197) Two angles that have a common vertex and a common side, but no interior points in common.

ángulos adyacentes (p. 197) Dos ángulos que tienen el mismo vértice y un lado común pero no comparten ningún punto interior.

alternate exterior angles (p. 202) In the figure, transversal t intersects lines l and m. $\angle 5$ and $\angle 3$, and $\angle 6$ and $\angle 4$ are alternate exterior angles.

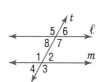

ángulos alternos-externos (p. 202) En la figura, la transversal t intersec las rectas l y m. $\angle 5$ y $\angle 3$, $\angle 6$ y $\angle 4$ son ángulos alternos enternos.

alternate interior angles (p. 202) In the figure, transversal t intersects lines l and m. $\angle 1$ and $\angle 7$, and $\angle 2$ and $\angle 8$ are alternate interior angles.

ángulos alternos-internos (p. 202) En a figura, la transversal t intersec las rectas l y m. $\angle 1$ y $\angle 7$, y $\angle 2$ y $\angle 8$ son ángulos alternos internos.

angle (pp. 196, 206) The figure formed by two rays that have a common endpoint.

ángulo (pp. 196, 206) Figura formada por dos rayos o líneas rectas que parten del mismo vértice.

angle of rotation (p. 306) The amount of turn of a rotation expressed as a fractional part of a whole turn or in degrees.

ángulo de rotación (p. 306) Cantidad de un movimiento de rotación que se expresa ya sea como parte fraccional de una rotación completa o en grados.

Angle-Side-Angle Postulate (ASA) (p. 213) If two angles and the included side of one triangle are congruent to two corresponding angles and the included side of another triangle, then the triangles are congruent.

Postulado de ángulo-lado-ángulo (ALA) (p. 213) Si dos ángulos y el lado incluido de un triángulo son congruentes a dos ángulos y al lado incluido de otro triángulo, entonces los triángulos son congruentes.

arc (p. 227) A section of the circumference of the circle.

arco (pp. 227) Porción de una circunferencia o de un círculo.

area (p. 418) The number of square units needed to fill a two-dimensional space.

área (p. 418) El número de unidades al cuadrado que se necesitan para llenar una figura bidimensional.

English

Español

■ B ■

bar graph (p. 2) A means of displaying statistical information in which horizontal or vertical bars are used to compare quantities.

gráfica de barra (p. 2) Representación gráfica de una serie de datos valiéndose de barras horizontales o verticales.

base (of an exponent) (p. 82) The factor being multiplied in a number written in exponential form. For example, in b^2, b is the base.

base (de un exponente) (p. 82) Factor que es afectado por el exponente en una potencia. Por ejemplo, en b^2 la base es b.

base angle (p. 207) For an isosceles triangle, the two congruent angles opposite the congruent sides.

base angular (p. 207) En un triángulo isósceles, el lado que se opone a los dos ángulos congruentes.

biased (p. 7) The characteristic of a survey whose findings are not truly representative of the entire population.

tendencioso (p. 7) Característica de una encuesta que no representa el total de una población.

binomial (p. 376) A polynomial with two terms.

binomio (p. 376) Polinomio formado por dos términos.

bisector (of an angle) (p. 197) A ray that divides the angle into two congruent adjacent angles.

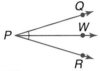

$\overrightarrow{PW}$ is the bisector of $\angle P$.
$\overrightarrow{PW}$ es la bisectriz del $\angle P$.

bisectriz (de un ángulo) (p. 197) Rayo o línea recta que divide un ángulo en dos ángulos iguales y adyacentes.

bisector (of a segment) (p. 193) Any line, segment, ray or plane that intersects the segment at its midpoint.

bisectriz (de un segmento o de una recta) (p. 193) Cualquier línea, segmento, rayo o plano que interseque a una recta en su punto medio.

box-and-whisker plot (p. 29) A method of displaying data that uses quartiles to draw a box to illustrate the interquartile range and to draw whiskers to illustrate data values outside the interquartile range.

diagrama de bloque (p. 29) Representación gráfica que identifica tendencias y síntesis informativas mostrando la distribución de datos dividida en cuatro partes iguales, de las cuales las dos interiores se representan con una caja y las dos exteriores con pelo.

■ C ■

center of dilation (p. 316) The point of which the distance between it and every point on the image is equal to the distance between it and every corresponding point on the preimage times the scale factor.

centro de homotecia (p. 316) Punto donde se intersecan las líneas que parten de los vértices de la primera figura y pasan por los vértices correspondientes de la figura o las figuras homólogas hechas a escala.

center of rotation (p. 306) The point about which the figure is rotated in a rotation.

centro de rotación (p. 306) Punto donde se apoya la rotación de una figura.

central angle (of a circle) (p. 227) An angle with its vertex at the center of a circle. The measure of a central angle is always less than 180°.

ángulo central (de un círculo) (p. 227) Ángulo cuyo vértice es el centro del círculo. La medida de un ángulo central es siempre menor de 180 grados.

chord (p. 226) A segment with both endpoints on the circle.

cuerda (p. 226) Recta o segmento de línea en la que ambos polos son puntos de la circunferencia.

circle (p. 226) In a plane, the set of all points that are a given distance for a fixed point. That fixed point is the center of the circle.

P is the center of the circle.
P es el centro del círculo.

círculo (p. 226) En un plano, el conjunto de todos los puntos a una distancia dada de un punto fijo. El punto fijo es el centro del círculo.

circle graph (p. 232) A means of displaying data where items are represented as parts of the whole circle. Each part, or percent of the data, is represented by a *sector*.

gráfica dentro del círculo o diagrama de sectores (p. 232) Representación gráfica en la cual los datos se muestran como porciones de un todo representado por un círculo. Cada porción o porcentaje representa un *sector*.

English

closed half-plane (p. 259) The region that is a solution of an inequality that includes the solid line.

cluster sampling (p. 6) A sampling method in which the members of the population are randomly selected from particular parts of the population and then surveyed in clusters.

coefficient (p. 376) The numerical part of a monomial.

coefficient of correlation (p. 26) A statistical measure, r, between 1 and -1, that tells if a correlation is positive or negative, strong or weak.

collinear points (p. 192) Points that lie on the same line.

P, Q, and R are collinear.
P, Q y R son colineales.

combination (p. 178) A set of items in which order is not important. For example, *acb* and *bac* are both combinations of the letters *a*, *b*, and *c*. The number of combinations of *n* items taken *r* items at a time is written

$$_nC_r = \frac{n!}{(n-r)!\,r!}.$$

complement (of an event) (p. 165) The complement of event *A* is the event *not A*.

complement (of a set) (pp. 147, 524) If *A* is a subset of *U*, then the subset of all elements of *U* that are not elements of *A* is called the complement of *A*, symbolized as A'.

complementary angles (p. 196) Two angles whose measures have a sum of 90°.

$m\angle ABC + m\angle DEF = 90°$

compound event (p. 162) An event made up of two or more simple events.

concave polygon (p. 222) A polygon is concave if a line that contains a side of the polygon also contains a point in its interior.

conclusion (p. 539) That which logically follows the premises of a deductive argument.

conditional statement (p. 530) An *if-then* statement having two parts, a hypothesis and a conclusion, symbolized $p \rightarrow q$.

cone (p. 422) A three-dimensional figure with a curved surface and one circular base. The *axis* is a segment that joins the vertex to the center of the base. If the axis forms a right angle with the base, it is a *right* cone. Otherwise it is an *oblique* cone.

vertex
vértice

base
base

congruent angles (pp. 197, 206) Angles that have the same measure.

congruent line segments (p. 193) Line segments that have the same measure.

Español

medio plano cerrado (p. 259) Región que es una solución de una desigualdad que incluye la línea sólida.

muestra por conglomerado (p. 6) De una parte de la población se escogen algunos miembros al azar y luego se les registra como conglomerado, no como individuos.

coeficiente (p. 376) Parte numérica de un monomio.

coeficiente de correlación (p. 26) Medida estadística, r, entre 1 y -1, que señala si una relación es positiva o negativa.

puntos colineales (p. 192) Dos o más puntos que forman parte de la misma línea.

combinación (p. 178) Conjuntos de artículos en que el orden no es importante. Por ejemplo, *acb* y *bac* son ambas combinaciones de las letras *a*, *b*, y *c*. El número de combinaciones de *n* artículos tomados *r* a la vez se escribe $_nC_r = \frac{n!}{(n-r)!\,r!}$.

suceso de complemento (p. 165) El complemento del evento *A* es el evento *no A*.

conjunto de complemento (pp. 147, 524) Si *A* es un subconjunto de *U*, entonces el subconjunto de todos los elementos que no forman parte de *A* sería el complemento de *A*, simbolizado como A'.

ángulos complementarios (p. 196) Dos ángulos cuyas medidas al sumarse dan como resultado 90°.

evento compuesto (p. 162) Evento que está compuesto por dos o más eventos simples.

polígono cóncavo (p. 222) Un polígono es cóncavo si una línea que contiene un lado del polígono además contiene un punto de su interior.

conclusión (p. 539) Lo que se desprende lógicamente de las premisas de un argumento deductivo.

declaración condicional (p. 530) Declaración compuesta por dos partes, *si* y *entonces*, llamadas respectivamente *premisa* o *antecedente*, y *conclusión* o *consecuente*. Se representa $p \rightarrow q$.

cono (p. 422) Figura tridimensional con una base circular y un vértice. La línea perpendicular a la base que inicia en el vértice viene siendo la *altura*. Si el punto extremo opuesto al vértice es el centro de la base, entonces será un *cono regular*. De otra manera será un *cono oblicuo*.

ángulos congruentes (pp. 197, 206) Ángulos que tienen la misma medida.

rectas o segmentos congruentes (p. 193) recta o segmento de línea que tienen la misma medida.

Glossary/Glosario

English

Español

congruent triangles (p. 212) Triangles whose vertices can be matched so that corresponding parts of the triangle are congruent.

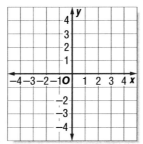

$\triangle ABC \cong \triangle EDF$

triángulos congruentes (p. 212) Triángulos cuyos tres vértices se corresponden.

constant (p. 376) A monomial that has no variables.

continuous (p. 265) A characteristic of a function in which the domain is all real numbers and the function is defined for all values of the domain.

contrapositive (p. 533) A conditional statement formed by interchanging and negating the hypothesis and conclusion of the original conditional statement, symbolized $\sim q \rightarrow \sim p$. A statement and its contrapositive are either both true or both false.

convenience sampling (p. 6) A sampling method is which members of a population are selected because they are readily available, and all are surveyed.

converse (p. 530) A conditional statement formed by interchanging the hypothesis and conclusion of the original conditional statement, symbolized $q \rightarrow p$. The converse of a true statement is not necessarily true.

convex polygon (p. 222) A polygon is convex if each line containing a side has no points in the interior of the polygon.

coordinate plane (p. 244) A mathematical system in which two number lines are drawn perpendicular to each other and form four quadrants. The horizontal number line is called the *x*-axis. The vertical number line is called the *y*-axis.

coplanar points (p. 192) Points that lie in the same plane.

corresponding angles (p. 202) Two angles in corresponding positions relative to two lines cut by a transversal. Also, angles in the same position in congruent or similar polygons.

cosine (p. 488) The trigonometric ratio of an $\angle A$, abbreviated cos *A*, of the length of the leg adjacent to $\angle A$ to the length of the hypotenuse.

counterexample (p. 530) An instance that satisfies the hypothesis, but not the conclusion of the conditional statement. A single counterexample proves that the conditional statement is false.

cross-products (p. 122) The cross-products of $\frac{a}{b} = \frac{c}{d}$ are *ad* and *bc*. In a proportion, the cross-product of the means is equal to the cross product of the extremes.

constante (p. 376) Monomio que no contiene variables.

continuo (p. 265) Característica de una función en la cual el dominio está compuesto por números reales y los valores de dominio definen la función.

contrapuesto (p. 533) Declaración condicional en la que se niegan tanto las premisas o hipótesis como la conclusión de una declaración condicional antes expuesta. Se simboliza $\sim q \Rightarrow \sim p$. Una declaración y su contrapuesto pueden ser ambas verdaderas o ambas falsas.

muestra por conveniencia (p. 6) Método de muestreo en el cual los miembros de una Población son escogidos sólo porque están disponible o porque ya han sido encuestados.

converso (p. 530) Declaración condicional en la que se intercambian la premisa o hipótesis por la conclusión. Se simboliza $q \Rightarrow p$. El opuesto de una declaración verdadera no es necesariamente verdadero.

polígono convexo (p. 222) Un polígono es convexo si cada línea que contiene un lado no tiene puntos al interior del polígono.

plano de coordenadas o plano cartesiano (p. 244) Dos rectas numéricas perpendiculares que forman una cuadrícula. La recta numérica horizontal se llama eje *x*, y la vertical se llama eje *y*.

puntos coplanares (p. 192) Puntos que forman parte del mismo plano.

ángulos correspondientes (p. 202) Ángulos que están en la misma posición con relación a la secante y a las líneas paralelas. También los ángulos que se encuentran en la misma posición en polígonos congruentes o iguales.

coseno (p. 488) Razón trigonométrica del $\angle A$, se abrevia cos *A*, y se define como el cociente del cateto adyacente del $\angle A$, sobre la hipotenusa.

contraejemplo (p. 530) Instancia que satisface la hipótesis o la premisa, pero no la conclusión de la declaración condicional. Un contraejemplo prueba que la conjetura no es válida.

productos cruzados (p. 122) Para $\frac{a}{b}$ y $\frac{c}{d}$, los productos cruzados son *ad* y *bc*. En una proporción los productos cruzados son iguales.

English / Español

English

cylinder (p. 422) A three-dimensional figure made up of a curved quadrilateral region and two congruent circular bases that lie in parallel planes. The *axis* is a segment that joins the centers of the bases. If the axis forms a right angle with the bases it is a *right* cylinder. Otherwise it is an *oblique* cylinder.

base
base
base
base

Español

cilindro (p. 422) Figura tridimensional que esta compuesta por una superficie curva, cuadrilátera y cerrada, además de dos bases circulares las cuales son paralelas. El *eje* es una recta o segmento que unifica los centros de cada base. Si el eje forma un ángulo recto con ambas bases, será un cilindro *regular*. De otra manera será un cilindro *oblicuo*.

■ D ■

deductive reasoning (p. 539) A process in which a conclusion is reasoned based on a set of premises. The conclusion necessarily follows from the premises.

razonamiento deductivo (p. 539) Razonamiento donde la conclusión se deduce de las premisas. La conclusión necesariamente se desprende de las premisas.

dependent events (p. 169) If the result of the second event is affected by the result of the first event, the second event is dependent on the first event. If event B is dependent on event A, then $P(A \text{ and } B) = P(A) \cdot P(B, \text{ given } A)$.

eventos dependientes (p. 169) Dos eventos en los que el resultado del primero afecta el resultado del segundo y viceversa. Si el evento B depende del evento A, entonces $P(A \text{ y } B) = P(A) \cdot P(B, \text{ dado a } A)$.

description notation (p.520) Set notation that describes the set.

notación descriptiva (p. 520) Notación del conjunto que describe al mismo conjunto.

determinant (p. 354) The difference of the products of the diagonal entries of a 2×2 square matrix. The determinant of a matrix A, named det A, is symbolized by using vertical bars in place of matrix brackets:

$$\det A = \begin{vmatrix} a & b \\ c & d \end{vmatrix} = ad - bc.$$

determinado (p. 354) Diferencia de los productos de las notaciones diagonales 2×2 matriz cuadrada. Lo determinado de la matriz A, llamado det A, se simboliza usando barras verticales en lugar de llaves:

$$A = \begin{vmatrix} a & b \\ c & d \end{vmatrix} = ad - bc.$$

diagonal (p. 223) A segment that joins two vertices of a polygon but is not a side. Diagonals separate the interior of the polygon into nonoverlapping triangular regions.

$\overline{SQ}$ is a diagonal.
$\overline{SQ}$ *es una diagonal.*

diagonal (p. 223) Recta o segmento de línea que conecta dos vértices de un polígono pero que no es un lado. Las diagonales dividen el interior del polígono en una región triangular no superpuesta.

diameter (p. 226) A chord that passes through the center of a circle and has endpoints on the circle. The diameter is equal to twice the measure of the radius.

diámetro (p. 226) Línea recta que pasa por el centro del círculo y que sus puntos extremos se localizan en la circunferencia. El diámetro está formado por dos radios.

dilation (p. 316) A transformation that produces an image that is the same shape as the original figure but a different size.

homotecia (p. 316) Transformación que da como resultado una imagen que tiene la misma forma de la figura original pero de diferente tamaño.

directed graph (p. 358) A geometrical representation of a map that shows locations as points and roads as lines. Connections may be possible in only one direction on some paths and in both directions on other paths. Directed graphs can be analyzed by using matrices.

gráfica directa (p. 358) representación geométrica de un mapa en el que las localidades se señalan con puntos y las caminos con líneas. Las conexiones serán posibles en un sentido en algunos senderos, pero en otros serán posibles en doble sentido. Las gráficas directas se pueden analizar usando matrices.

direct square variation (p. 277) A function that can be written in the form $y = kx^2$, where k is a nonzero constant.

variación cuadrada directa (p. 277) Función que se puede escribir en la forma $y = kx^2$, donde k es una constante que no es igual a cero.

direct variation (p. 276) A function that can be written in the form $y = kx$, where k is a nonzero constant.

variación directa (p. 276) Función que se puede escribir en la forma $y = kx$, donde k es una constante que no es igual a cero.

disjoint sets (p. 525) Two sets whose intersection is the empty set.

conjuntos disyuntivos (p. 525) Dos conjuntos en los que su intersección es el conjunto vacío.

Glossary/Glosario

English

distance formula (p. 245) For any points $P_1(x_1, y_1)$ and $P_2(x_2, y_2)$, the distance between P_1 and P_2 is given by the formula $P_1P_2 = \sqrt{(x_2 - x_1)^2 + (y_2 - y_1)^2}$.

distributive property (p. 390) Each factor outside parentheses can be used to multiply each term within the parentheses. $a(b + c) = ab + ac$.

division property of inequality (p. 133) If you divide each side of an inequality by the same positive number, the order of the inequality remains the same. If you divide each side of an inequality by the same negative number, the order of the inequality is reversed.

domain (p. 264) The set of all possible values of x for the function $y = f(x)$.

Español

fórmula de la distancia (p. 245) Para cualquier par de puntos $P_1(x_1, y_1)$ y $P_2(x_2, y_2)$, la distancia entre P_1 y P_2 es dada por la fórmula $P_1P_2 = \sqrt{(x_2 - x_1)^2 = (y_2 - y_1)^2}$.

propiedad distributiva (p. 390) Cada factor fuera del paréntesis puede multiplicarse con cada término dentro del paréntesis. $a(a + c) = ab + ac$.

propiedad de la división en la desigualdad (p. 133) Si se dividen ambos lados de una desigualdad entre el mismo número positivo, la desigualdad proporcional seguirá siendo la misma. Si se dividen ambos lados de una desigualdad entre el mismo número negativo, la desigualdad proporcional será la opuesta.

dominio (p. 264) Conjunto de todos los valores posibles de x para la función $y = f(x)$.

■ E ■

edge (p. 422) The intersection of the faces of a polygon.

element (pp. 38, 520) A member of a set or a number in a matrix.

enlargement (p. 316) A dilation that creates an image that is larger than the preimage.

equal sets (p. 521) Sets that contain the same members, written $A = B$.

equation (p. 104) A statement in which two numbers or expressions are equal.

equiangular triangle (p. 206) A triangle with all three angles congruent.

equilateral triangle (p. 206) A triangle with all three sides congruent.

$\overline{AB} \cong \overline{BC} \cong \overline{AC}$
$\angle A \cong \angle B \cong \angle C$

equivalent sets (p. 521) Sets that contain the same number of elements.

event (p. 158) Any one of the possible outcomes or combination of possible outcomes of an experiment.

experiment (p. 150) An activity that is used to produce data that can be observed and recorded.

experimental probability (p. 150) The probability of an event based on the results of an experiment in which the number of favorable observations is divided by the total number of observations.

exponent (pp. 57, 82) A number showing how many times the base is used as a factor. For example, in b^x the exponent is x.

exterior angle (of a triangle) (p. 207) The angle formed by extending one of the sides of a triangle and that is equal to the two remote interior angles of the triangle.

∠1 is an exterior angle.
∠1 *es un ángulo externo.*

arista (p. 422) La intersección de las caras de un *poliedro*.

elemento (pp. 38, 520) Miembro de un conjunto o un número en una matriz.

ampliación (p. 316) Dilatación que crea una figura más grande que la figura original.

conjuntos iguales (p. 521) Conjuntos que contienen los mismos elementos, se escribe $A = B$.

ecuación (p. 104) Planteamiento matemático donde dos números o expresiones algebraicas son iguales.

triángulo equiangular (p. 206) Triángulo que tiene sus tres ángulos iguales.

triángulo equilátero (p. 206) Triángulo que tiene sus tres lados iguales.

conjuntos equivalentes (p. 521) Conjuntos que contienen el mismo número de elementos.

evento (p. 158) Resultado o la combinación de varios resultados de un experimento.

experimento (p. 150) Actividad que se usa para producir datos que pueden ser observados y grabados.

probabilidad experimental (p. 150) La probabilidad de un evento basándose en los resultados de un experimento en el cual el número de observaciones favorables se divide entre el número total de observaciones.

exponente (pp. 57, 82) Número que muestra las veces que la base se usa como factor. Por ejemplo, en b^x el exponente es x.

ángulo exterior (de un triángulo) (p. 207) Ángulo que se forma al extender uno de los lados de un triángulo y su medida será igual a la suma de los dos ángulos interiores que no le sean adyacentes.

English

Español

exterior angles (of a transversal system)   (p. 202) The four angles formed by a transversal and the two lines it intersects that are outside the two lines.

ángulos exteriores (de un sistema transversal)   (p. 202) Los cuatro ángulos exteriores que se forman en los puntos de intersección de la secante o transversal con las dos líneas paralelas.

■ F ■

face   (p. 422) A polygonal surface of a polyhedron.

cara   (p. 422) Superficie poligonal de un poliedro.

factor   (p. 404) Any number or polynomial multiplied by another to produce a product.

factor   (p. 404) Cualquier número o polinomio multiplicado por otro número que da como resultado un producto.

factorial   (p. 172) The product of all whole numbers from n to 1. The factorial function is used to compute the number of permutations of n different items, $n(n-1)(n-2) \ldots (2)(1)$ and is written as $n!$.

factorial   (p. 172) Producto de todos los números enteros de n a 1. La función factorial se usa para ordenar el número de permutaciones de n cifras diferentes, $n(n-1)(n-2)...(2)(1)$ y se escribe como n.

factoring   (p. 404) To express a number or polynomial as a product of numbers or polynomials.

factorizar   (p. 404) Expresar un número o polinomio como un producto de otros números o polinomios.

Fibonacci sequence   (p. 93) A sequence in which each term is the sum of the previous two terms.

secuencia fibonacci   (p. 93) Secuencia en la cual cada término es la suma de los dos términos previos.

finite set   (p. 520) A set whose elements can all be counted or listed.

conjunto finito   (p. 520) Conjunto en el que todos sus elementos se pueden contar o enlistar.

FOIL   (p. 397) An acronym and memory device used to remember the order to multiply binomials (*first, outer, inner, last*).

PFDU   (p. 397) Siglas y ardid memorístico que se usa para recordar los pasos para multiplicar binomios (*Primero, Fuera, Dentro, Último*).

formula   (p. 105) An equation stating a relationship between two or more quantities.

fórmula   (p. 105) Ecuación que determina las relaciones entre dos o más cantidades o variables.

foundation drawing   (p. 447) A drawing that shows the base of a structure and the height of each part.

dibujo de cimientos   (p. 447) Dibujo que muestra la base de una estructura y la longitud de cada parte.

frequency table   (p. 16) A method of recording data that shows how often an item appears in a set of data.

tabla de frecuencia   (p. 16) Método de recopilación de datos que muestra la frecuencia con que un dato aparece en un conjunto de datos.

front-end estimation   (p. 3) A method of estimating in which only initial digits, or *front-end* digits, are used. This method results in a simpler computation but also in a less precise estimation.

estimación de digitos en los extremos delanteros (p. 3) Método de estimación en el que se usan sólo los dígitos iniciales o de los *extremos delanteros*. Este método da como resultado una manera más simple de ordenar aunque la estimación es menos precisa.

function   (p. 264) An equation in two variables that has a relationship where each x-coordinate is paired with exactly one y-coordinate.

función   (p. 264) Ecuación en dos variables en la que se da una relación donde cada coordenada de x se aparea con una de la coordenada y.

function notation   (p. 264) Writing a function as equal to $f(x)$ in contrast to writing it as equal to y.

notación de la función   (p. 264) Descripción de una función como igual a $f(x)$ en contraste con la descripción de que ésta es igual a y.

fundamental counting principle   (p. 159) If there are two or more stages of an activity, the total number of possible outcomes is the product of the number of possible outcomes for each stage of the activity.

principio fundamental de conteo   (p. 159) Si hay dos o más etapas en una actividad, el número total de resultados posibles es el producto del número de resultados posibles por cada etapa de la actividad.

■ G ■

geometry   (p. 192) The study of points in space (from the Greek *geo*, meaning "earth" and *metria* meaning "measurement").

geometría   (p. 192) Estudio de los puntos en el espacio (viene de las raíces griegas *geo*, que significa tierra; y *metro*, que significa "medida").

graph of an equation   (p. 254) The set of all points whose coordinates are solutions of an equation.

gráfica de una ecuación   (p. 254) Conjunto de puntos cuyas coordenadas son las soluciones de una ecuación.

English

graph of a number   (p. 52) The point that corresponds to a number and is indicated on the number line by a solid dot.

graph of an inequality   (p. 258) A graph on the coordinate plane that includes a boundary, either a solid line or dashed line, and a shaded region.

greatest common factor (GCF)   (p. 404) The greatest expression that is a factor of two or more expressions.

Español

gráfica de una desigualdad   (p. 52) Gráfica en el plano de coordenadas que incluye un límite representado ya sea por una línea cerrada o fragmentada.

gráfica de un número   (p. 258) Punto que corresponde a un número; se indica en la recta numérica con un punto cerrado.

máximo factor común (MFC)   (p. 404) La expresión mayor que es factor de dos o más expresiones.

■ H ■

histogram   (p. 16) A type of bar graph used to show frequencies in which no space is between the bars and the bars usually represent grouped intervals of numbers.

hypotenuse   (p. 484) The side opposite the right angle in a right triangle.

histograma   (p. 16) Tipo de gráfica de barras que se usa para mostrar frecuencias en las cuales no hay espacio entre las barras, y las barras por lo general representan intervalos agrupados de números.

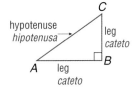

hipotenusa   (p. 484) Lado opuesto al ángulo recto en un triángulo rectángulo.

■ I ■

image   (p. 296) The new figure that results from a transformation of the original figure.

independent events   (p. 168) If the result of the second event is not affected by the result of the first event, the second event is independent of the first. If A and B are independent events, $P(A \text{ and } B) = P(A) \cdot P(B)$.

indirect measurement   (p. 478) A measurement that is made in situations where it is impossible or impractical to find a length by measuring the actual distance (direct measurement). Using similar triangles, by means of shadows or mirrors, is one method of indirect measurement.

inductive reasoning   (p. 538) A process in which a conclusion is reasoned based on the examination of a pattern of instances of an event. The conclusion is called a *conjecture* and cannot be known to be true beyond a doubt.

inequality   (p. 126) A mathematical sentence that contains one of the symbols $\neq$, $<$, $>$, $\leq$, or $\geq$.

infinite set   (p. 520) A set whose elements cannot be counted or listed.

inscribed angle   (p. 228) An angle whose vertex lies on the circle and whose sides contain chords of the circle. The measure of an inscribed angle is one-half the measure of the arc it intercepts.

integers   (p. 52) The set of whole numbers and their opposites.

imagen   (p. 296) Nueva figura que resulta de una transformación.

eventos independientes   (p. 168) Si el resultado de un evento no es afectado por el resultado del primero, el segundo evento es independiente del primero. Si A y B son eventos independientes, $P(A \text{ y } B) = P(A) \cdot P(B)$.

medida indirecta   (p. 478) Cálculo de una medida que se hace en situaciones donde es imposible o impráctico encontrar la longitud midiendo la distancia real (medida directa). El uso de triángulos similares por medio de sombras y espejos, es un método de medida indirecta.

razonamiento inductivo   (p. 538) Proceso lógico donde una conclusión se desprende del examen de un conjunto de eventos específicos. La conclusión es llamada conjetura y no puede ser aceptada si hay por lo menos una duda.

desigualdad   (p. 126) Declaración matemática que usa uno de los símbolos $\neq$, $<$, $>$, $\leq$, o $\geq$.

conjunto infinito   (p. 520) Conjunto en el cual sus elementos no se pueden contar o enlistar.

ángulo inscrito   (p. 228) Ángulo que su vértice recae en el círculo y cuyos lados vienen a ser cuerdas del mismo. La medida de un ángulo inscrito es la mitad de lo que mide el arco que lo intercepta.

enteros   (p. 52) Conjunto de los números enteros y sus opuestos.

English

Español

interior angles   (p. 202) The four angles formed by a transversal and the two lines it intersects that are inside the two lines. $\angle 1$, $\angle 2$, $\angle 7$, and $\angle 8$ are interior angles.

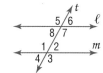

ángulos interiores   (p. 202) Los cuatro ángulos interiores formados por las intersecciones de la secante o transversal y las dos líneas paralelas. $\angle 1$, $\angle 2$, $\angle 7$, y $\angle 8$ son ángulos interiores.

interquartile range   (p. 20) The difference between the values of the first and third quartiles.

alcance intercuartil   (p. 28) Diferencia entre los valores del primero y del tercer cuartil.

intersection (of two sets)   (pp. 147, 525) The intersection of two sets, A and B, symbolized $A \cap B$, contains all the elements that are common in both A and B.

intersección (de dos conjuntos)   (pp. 147, 525) La intersección de dos conjuntos, A intersección B, simbolizado $A \cap B$, contiene todos los elementos que ambos conjuntos tienen en común.

invalid argument   (p. 543) A deductive argument such that even if the premises are true, the conclusion does not logically follow.

razonamiento inválido   (p. 543) Razonamiento deductivo que incluso si las premisas son verdaderas, la conclusión es ilógica.

inverse (of a statement)   (p. 533) A conditional statement formed by negating the hypothesis and conclusion of the original conditional statement, symbolized $\sim p \rightarrow \sim q$. The inverse of a true statement is not necessarily true.

opuesto (de una declaración)   (p. 533) Declaración condicional formada por la negación de la hipótesis o antecedente y de la conclusión o consecuente de la declaración condicional original, se simboliza $\sim p \Rightarrow \sim q$, el opuesto de una declaración verdadera no es necesariamente falso.

inverse square variation   (p. 283) A function that can be written in the form $y = \frac{k}{x^2}$ or $x^2 y = k$, where k is a nonzero constant.

variación cuadrada inversa   (p. 283) Función que puede escribirse en la forma $y = \frac{k}{x^2}$ o $x^2 y = k$, donde k no es igual a cero.

inverse trigonometric functions   (p. 498) Functions $\sin^{-1}$, $\cos^{-1}$ and $\tan^{-1}$ that "undo" the sin, cos and tan functions and are useful in finding measures of angles in right triangles when only the length of the sides is known.

función trigonométrica inversa   (p. 498) Las funciones sen^{-1}, $\cos^{-1}$ y $\tan^{-1}$ que "deshacen" las funciones sen, cos y tan, y son prácticas para encontrar medidas de ángulos en triángulos rectángulos cuando sólo las medidas de los lados se conocen.

inverse variation   (p. 282) A function that can be written in the form $y = \frac{k}{x}$, where k is a nonzero constant.

variación inversa   (p. 282) Función que puede escribirse en la forma $y = \frac{k}{x}$, donde k es una constante que no es igual a cero.

irrational numbers   (p. 52) Numbers that are non-terminating and non-repeating decimals, such as π and $\sqrt{2}$.

número irracional   (p. 52) Números con decimales como π y (raíz cuadrada de 2) que no terminan y que además no llegan a una cifra que se repita.

isometric drawing   (p. 442) A drawing of a three-dimensional object in such a way that all lines are drawn to scale, all parallel edges of the structure are drawn parallel, but all perpendicular lines are not necessarily drawn perpendicular.

dibujo isométrico   (p. 442) Dibujo de un cuerpo tridimensional que se hace de tal forma que todas las rectas se dibujan a escala, todas las aristas paralelas de la estructura se dibujan paralelas, pero no todas las rectas perpendiculares se dibujan de forma perpendicular necesariamente.

isosceles triangle   (p. 206) A triangle with at least two congruent sides and angles.

triángulo isósceles   (p. 206) Triángulo que tiene dos lados y dos ángulos iguales y un lado y un ángulo desigual.

legs   (p. 484) In a right triangle, the two sides that are not the hypotenuse.

catetos   (p. 484) En el triángulo rectángulo, los dos lados que no son la hipotenusa.

English   # Español

■ **L** ■

like terms (pp. 66, 377) Terms that have identical variable parts, or in other words, monomials that differ only in their coefficients.

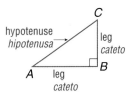

términos comunes (pp. 66, 377) Términos que tienen las mismas variables, o en otras palabras, términos que sólo difieren en sus coeficientes.

line (p. 192) A set of points that extends infinitely in two opposite directions.

línea (p. 192) Conjunto infinito de puntos que se extiende en dos direcciones opuestas.

linear equation (pp. 254, 257) An equation for which the graph is a line. The *standard form* of a linear equation is written $Ax + By = C$ where A, B and C are real numbers and A and B are not both zero.

ecuación lineal (pp. 254, 257) Ecuación que representa una función lineal. La forma estándar de una ecuación lineal se escribe $Ax + By = C$ donde A, B y C son números reales y A y B no son ambos igual a cero.

line graph (p. 3) A means of displaying data using points and line segments to show changes in data over periods of time.

gráfica de línea (p. 3) Representación gráfica de datos valiéndose de puntos y rectas o segmentos de línea para mostrar los cambios de datos en un periodo de tiempo.

line of best fit or trend line (p. 21) A line that can be drawn near most of the points on a scatter plot. A trend line that slopes upward to the right indicates a *positive correlation* between the sets of data. A trend line that slopes downward to the right indicates a *negative correlation.*

línea más apropiada o línea de tendencia (p. 21) Línea que se puede dibujar cerca de casi todos los puntos en un diagrama disperso. Una línea de tendencia que suba hacia la derecha indica una *correlación positiva* entre el conjunto de datos. Una línea de tendencia que baje hacia la derecha indica una *correlación negativa.*

line of symmetry (p. 310) A line on which a figure can be folded so that when one half is reflected over that line it matches the other part exactly.

línea simétrica (p. 310) Línea que divide a una figura en dos parte de tal manera que al doblarse una mitad es exactamente igual a la otra.

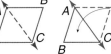

$\overleftrightarrow{AC}$ is a line of symmetry.
$\overleftrightarrow{AC}$ es un eje de simetría.

line segment (p. 193) Part of a line consisting of two endpoints and all points that lie between these two endpoints.

recta o segmento de línea (p. 193) Conjunto infinito de puntos que apuntan hacia la misma dirección contenidos entre dos puntos extremos.

■ **M** ■

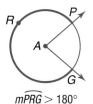

$m\overparen{PRG} > 180°$

major arc (p. 227) An arc that is larger than a semicircle.

arco mayor (p. 227) Arco que es más largo que un semicírculo.

matrix (pp. 38, 354) A rectangular array of data in rows and columns and enclosed by brackets. The number of rows and columns in a matrix determine its dimensions. When two matrices have equal dimensions, corresponding elements are elements in the same position of each matrix.

matriz (pp. 38, 354) Formación rectangular de datos en filas y en columnas enmarcadas por llaves. El número de filas y de columnas en una matriz determina sus dimensiones. Cuando dos matrices tienen las mismas dimensiones, Los elementos que se corresponden están en la misma posición en cada matriz.

mean, or *arithmetic average* (p. 10) The sum of the values in a data set divided by the number of data.

media o *promedio aritmético* (p. 10) La suma de un conjunto de números dividida entre la cantidad de números que forman el conjunto.

English | Español

measures of central tendency (p. 10) Statistical values mean, median and mode that represent a central, or middle, value of a data set.

median (p. 10) The middle value of a data set when the data are arranged in numerical order.

midpoint (p. 193) The point that divides the segment into two congruent segments.

S ———— M ———— T

$SM = MT$

midpoint formula (p. 245) For any points P_1 and P_2, the midpoint between P_1 and P_2 is given by the formula $\left(\dfrac{x_1 + x_2}{2}, \dfrac{y_1 + y_2}{2} \right)$.

minor arc (p. 227) An arc that is smaller than a semicircle.

misleading data (p. 34) Data that leads to a false perception.

mode (p. 10) The value that occurs most often in a set of data.

model (p. 114) A physical or numerical representation of a real-life situation.

monomial (p. 376) An expression that is a number, variable, or product of a number and one or more variables with whole number exponents.

multiplication property of equality (p. 108) When two expressions are equal, you can multiply each expression by the same number and the resulting products will be equal. If $a = b$, then $a \cdot c = b \cdot c$ and $c \cdot a = c \cdot b$.

multiplication property of inequality (p. 133) If you multiply each side of an inequality by the same positive number, the order of the inequality remains the same. If you multiply each side of an inequality by the same negative number, the order of the inequality is reversed.

mutually exclusive (p. 162) Events that cannot occur at the same time are mutually exclusive. If A and B are mutually exclusive events, then $P(A \text{ or } B) = P(A) + P(B)$. If A and B are not mutually exclusive, then $P(A \text{ or } B) = P(A) + P(B) - P(A \text{ and } B)$.

medidas de tendencia central (p. 10) Medidas que se usan en la estadística para analizar datos. Estas medidas son la *media*, la *mediana* y la *moda*.

mediana (p. 10) El valor que queda en el medio al ordenarse los datos de menor a mayor. Si quedaran dos valores, la mediana sería el promedio de ambos.

punto medio (p. 193) Punto que divide a una recta en dos rectas de igual medida.

fórmula del punto medio (p. 245) Para dos puntos P_1 y P_2, el punto medio entre P_1 y P_2 es dado por la fórmula $\left(\dfrac{x_1 + x_2}{2}, \dfrac{y_1 + y_2}{2} \right)$.

arco menor (p. 227) Arco que es más pequeño que el semicírculo.

datos engañosos (p. 34) Datos que te llevan a una falsa percepción.

moda (p. 10) Número o artículo que se da con más frecuencia en un conjunto de datos.

modelo (p. 114) Representación física o numérica de una situación de la vida real.

monomio (p. 376) Expresión que puede ser un número, una variable o el producto de un número y de una o más variables con números enteros como exponentes.

propiedad en la multiplicación de la igualdad (p. 108) Cuando dos expresiones son iguales, se puede multiplicar cada expresión por el mismo número y los resultados serán iguales. Si $a = b$, entonces $ac = bc$ y $ca = cb$.

propiedad en la multiplicación de la desigualdad (p. 133) Si multiplicas ambos lados de una desigualdad por el mismo número positivo, la proporción de la desigualdad seguirá siendo la misma. Si multiplicas ambos lados de una desigualdad por el mismo número negativo, la proporción seguirá siendo la misma pero opuestas.

mutuamente exclusivo (p. 162) Eventos que no pueden ocurrir a la vez son mutuamente exclusivos. Si A y B son eventos mutuamente exclusivos, entonces $P(A \text{ o } B) = P(A) + P(B)$. Si A y B no son mutuamente exclusivos, entonces $P(A \text{ o } B) = P(A) + P(B) - P(A \text{ y } B)$.

■ N ■

negation (p. 532) To write the negation of a statement, add "not" to the statement or add the words "it is not the case that." The symbol for negation is $\sim$.

negative correlation (p. 21) A relationship between the sets of data on a scatter plot such that the slope of the line of best fit is down and to the right, or in other words, as the horizontal axis value increases, the vertical axis value decreases.

negative reciprocals (p. 334) Two numbers that have a product of -1.

negación (p. 532) Para señalar la negación de una declaración se le añade "no" o la frase "no se da el caso de que". El símbolo de la negación es $\sim$.

correlación negativa (p. 21) Relación tal entre los conjuntos de datos en un diagrama disperso que la inclinación de la línea más apropiada es hacia abajo y a la derecha, o en otras palabras, cuando el valor del eje horizontal aumenta y el valor del eje vertical disminuye.

recíprocos negativos (p. 334) Dos números cuyo producto es igual a -1.

English

Español

net (p. 426) A two-dimensional pattern that can be folded to form a three-dimensional figure. A three-dimensional figure can have more than one net.

red (p. 426) Patrón dimensional que, cuando se dobla, forma una figura tridimensional. Una figura tridimensional puede tener más de una red.

noncollinear points (p. 192) Points that do not lie on the same line.

puntos no colineales (p. 192) Puntos que no forman parte de la misma línea.

noncoplanar points (p. 192) Points that do not lie in the same plane.

puntos no coplanares (p. 192) Puntos que no forman parte del mismo plano.

null set (p. 521) A set having no elements. The symbol for the null set is $\varnothing$ or { }.

conjunto nulo (p. 521) Conjunto que no tiene ningún elemento. El símbolo para el conjunto vacío es $\varnothing$ o { }.

numerical expression (p. 56) An expression that contains two or more numbers joined by operations.

expresión numérica (p. 56) Expresión que contiene dos o más números conectados por una operación.

■ O ■

obtuse angle (p. 196) An angle whose measure is greater than 90° but less than 180°.

obtuse triangle (p. 206) A triangle that has one obtuse angle.

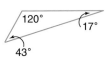

one obtuse angle
un ángulo obtuso

ángulo obtuso (p. 196) Cualquier triángulo que tiene un ángulo obtuso.

triángulo obtuso (p. 206) Triángulo que tiene un ángulo obtuso.

one-point perspective (p. 436) A perspective drawing that has one vanishing point.

punto de perspectiva (p. 436) Dibujo a perspectiva que tiene un punto de fuga.

open half-plane (p. 258) The region on either side of a line on a coordinate plane. The line separating the half-planes forms the boundary, or edge, of each half-plane.

medio plano abierto (p. 258) Región en cualquier lado de una línea en un plano de coordenadas. La línea que separa los medios planos forma un límite o borde de cada medio plano.

opposite rays (p. 196) Rays that extend in opposite directions.

rayos o líneas opuestas (p. 196) Rayos o líneas que se extienden en direcciones opuestas.

opposites (p. 52) Numbers that are the same distance from zero on a number line but in the opposite direction.

opuestos (p. 52) Números que están a la misma distancia del cero pero en direcciones opuestas.

ordered pair (p. 244) A point in the coordinate plane stated as (x, y).

par ordenado (p. 244) Punto en un plano de coordenadas establecido como (x, y).

order of operations (p. 56) A set of rules that specify which operations must precede other operations in order to properly evaluate expressions.

orden de las operaciones (p. 56) Conjunto de pasos que especifican cuales operaciones deben de preceder a otras operaciones para evaluar expresiones correctamente.

order of rotational symmetry (p. 311) The number of times a figure fits exactly over its original position when the figure is being turned about a point during a complete rotation.

orden de simetría en rotación (p. 311) Número de veces que una figura cabe exactamente sobre su posición original cuando la figura gira alrededor de un punto durante una rotación completa.

origin (p. 244) The point (0, 0) of intersection of the x-axis and y-axis in a coordinate plane.

origen (p. 244) El punto (0, 0) de intersección del eje x y del eje y en un plano coordenado o cartesiano.

orthogonal drawing or orthographic drawing (p. 446) A drawing of the top, front and side views of a three dimensional figure as seen from a "straight on" viewpoint.

dibujo ortogonal o proyección ortogonal (p. 446) Vista tridimensional de un objeto en el cual desde uno de los "vértices superiores" se muestran las aristas de una figura sin distorsionar las dimensiones de la misma.

outliers (pp. 17, 19, 29) Specifically, any data value that is less than the median of the lower half of data minus 1.5 times the interquartile range or greater than the median of the upper half of the data plus 1.5 times the interquartile range. Generally, see *stem-and-leaf plot*.

datos extremos (pp. 17, 19, 29) Específicamente, cualquier valor que sea menor que la mediana de la mitad más baja de los datos menores 1.5 veces al alcance intercuartil o mayor que la mediana más alta de los datos positivos mayores 1.5 veces al alcance intercuartil. Generalmente, véase *diagrama de tallo y hoja*.

English   # Español

■ P ■

palindrome (pp. 175, 313) A number or word that reads the same backward or forward.

parabola (p. 268) A curve that is the graph of a quadratic function. Not all parabolas represent functions.

parallel lines (pp. 202, 334) Coplanar lines that do not intersect. Parallel lines have equal slopes.

$$\overleftrightarrow{AB} \parallel \overleftrightarrow{CD}$$

parallelogram (p. 216) A quadrilateral with two pairs of parallel sides.

percentile (p. 30) A measure of a rank, or standing, within a group. A percentile divides a group of data into two parts, those at or below a certain score and those above. The percentile represents the percent of scores at or below a particular score by dividing the number of scores less than or equal to a given score by the total number of scores.

perfect number (p. 93) A number equal to the sum of its factors, excluding the number itself.

perfect square (p. 373) An integer whose square roots are integers.

perimeter (p. 418) The number of units needed to go around a two-dimensional space.

permutation (p. 172) An arrangement of items in a particular order. The number of permutations of n different items is $n!$. The number of permutations of n different items taken r items at a time and with no repetitions is $_nP_r = \dfrac{n!}{(n-r)!}$.

perpendicular lines (pp. 197, 334) Two lines that intersect to form right angles. The slopes of perpendicular lines have a product of -1.

line $m \perp$ line n
recta $m \perp$ recta n

perspective drawing (p. 436) A drawing of a three-dimensional object in such a way that parallel lines that extend into the distance are not actually drawn parallel but instead are drawn toward a *vanishing point*.

pictograph (p. 2) A means of displaying data that uses pictures or symbols to represent data. The key identifies the number of data items represented by each symbol.

plane (p. 192) A flat surface that exists without end in all directions.

point (p. 192) A location in space having no dimensions.

palíndromo (pp. 175, 313) Número o palabra que dice lo mismo al leerse desde el principio o desde el final.

parábola (p. 268) Curva que viene a ser la gráfica de una función cuadrática. No todas las parábolas representan funciones.

rectas paralelas (pp. 202, 334) rectas coplanares que no se intersecan. Las rectas paralelas tienen la misma inclinación.

paralelogramo (p. 216) Cuadrilátero con dos pares de lados paralelos.

porcentil (alcance) (p. 30) Medida de un alcance o categoría dentro de un grupo. Un porcentil divide un grupo de datos en dos partes, Aquellos que están por debajo de una cierta puntuación y aquellos que están por encima. El porcentil representa el por ciento de posiciones debajo de una posición particular y resulta al dividir el número de posiciones menores o iguales que la posición dada entre el total de posiciones.

números perfectos (p. 93) Número igual a la suma de sus factores, excluyendo el número mismo.

cuadrado perfecto (p. 373) Número cuya raíz cuadrada es un entero.

perímetro (p. 418) Distancia alrededor de una figura de dos dimensiones.

permutación (p. 172) Arreglo de objetos en un orden particular. El número de permutaciones de diferentes objetos n se denomina $n!$. El número de permutaciones de diferentes artículos n que toma un artículo r sólo una vez se lee $_nP_r = \dfrac{n!}{(n-r)!}$.

rectas perpendiculares (pp. 197, 334) Dos rectas que se intersecan para formar ángulos rectos. Las pendientes de las rectas perpendiculares tienen un producto de -1.

dibujo en perspectiva (p. 436) Dibujo de un objeto de tres dimensiones en el que las líneas paralelas no se dibujan en realidad en forma paralela sino que son dirigidas hacia un *punto* de fuga.

pictografía (p. 2) Gráfica que usa figuras o símbolos para representar datos. La clave identifica el número de datos representados por cada símbolo.

plano (p. 192) Superficie llana que se extiende sin fin en todas direcciones.

punto (p. 192) Lugar específico en el espacio que no tiene dimensiones.

Glossary/Glosario

English

polygon (p. 222) A closed plane figure formed by joining three or more line segments at their endpoints. Each segment, or *side* of the polygon, intersects exactly two other segments, one at each endpoint. The point at which the endpoints meet is called a *vertex*.

polyhedron (p. 422) A closed, three-dimensional figure made of polygons. The polygonal surfaces are called *faces*. Two faces meet, or intersect, to form an *edge*. The point at which three or more edges intersect is called a *vertex*.

polynomial (p. 376) The sum or difference of monomials. Each monomial is called a *term* of the polynomial. A polynomial is in *standard form* when its terms are ordered from the greatest to the least powers of one of the variables.

population (p. 6) The entire set or group of individuals from which a sample is taken.

postulate (p. 193) A hypothesis that is assumed to be true.

power of a product rule (p. 83) A property of exponents which states that when factors of a product are raised to an exponent, the power of each factor is that exponent. For all real numbers a and b, if m and n are integers, then $(ab)^m = a^m b^m$.

power of a quotient rule (p. 83) A property of exponents which states that when factors of a quotient are raised to an exponent, the power of each factor is that exponent. For all real numbers a and b, if m and n are integers, then $\left(\dfrac{a}{b}\right)^m = \dfrac{a^m}{b^m}$ if $b \neq 0$.

power rule (p. 83) A property of exponents which states that when a number in exponential form is raised to an exponent, the power of the base is the product of the exponents. For all real numbers a and b, if m and n are integers, then $(a^m)^n = a^{mn}$.

preimage (p. 296) The original figure of the image that resulted from a transformation.

premises (p. 539) A set of statements accepted as true and from which a conclusion is drawn in deductive arguments.

Español

polígono (p. 222) Figura plana y cerrada que se forma al conectarse tres o más líneas rectas en sus puntos extremos. Cada *lado* del polígono se interseca exactamente con otros dos, uno en cada extremo. El punto en el que unen los lados se llama *vértice*.

poliedro (p. 422) Figura cerrada tridimensional formada por polígonos. Cada polígono recibe el nombre de *cara*. Dos caras se unen, o se intersecan, para formar una línea de encuentro que se llama *arista*. El punto de encuentro de tres o más aristas se llama *vértice*.

polinomio (p. 376) Suma o resta de dos o más monomios. Cada monomio es un *término* del polinomio. Un polinomio se halla en su *forma estándar* cuando sus términos se encuentran ordenados de la mayor a la menor potencia de una de sus variables.

población (p. 6) Totalidad de un conjunto o grupo de individuos que son tomados de una muestra.

postulado (p. 193) Hipótesis que se asume como cierta.

regla de la potencia de un producto (p. 83) Propiedad de los exponentes que establece que cuando los factores de un producto son elevados, el poder de dicho factor es ese exponente. Para todos los números reales a y b, si m y n son números enteros, entonces $(ab)^m = a^m b^m$.

potencia de la regla del cociente (p. 83) Propiedad de los exponentes que establece que cuando los factores de un cociente elevados a un exponente, la potencia de cada factor es ese exponente. Para todos los números reales a y b, si m y n son números enteros, entonces $\left(\dfrac{a}{b}\right)^m = \dfrac{a^m}{b^m}$, si b es desigual a 0.

regla de la potencia (p. 83) Propiedad de los exponentes que establece que cuando un número en forma exponencial es elevado al valor de un exponente, la potencia de la base es el producto de los exponentes. Para todos los números reales a y b, si m y n son números enteros, entonces $(a^m)^n = a^{mn}$.

pre-imagen (p. 296) Figura original que resulta de una transformación.

premisas (p. 539) Conjunto de afirmaciones que se aceptan como ciertas y del que se llega a conclusiones por medio de argumentos deductivos.

---

prism (p. 422) A polyhedron with two identical parallel faces called *bases*. The other faces are parallelograms. A prism is named according to the shape of its bases.

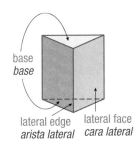

base
base

lateral edge
arista lateral

lateral face
cara lateral

triangular prism
prisma triangular

prisma (p. 422) Poliedro con dos caras paralelas idénticas llamadas *bases*. Las otras caras son paralelogramos. Los prismas reciben su nombre a partir de la forma de sus bases.

English | Español

probability   (p. 148) The chance or likelihood that an event will occur, written $P(E)$. An impossible event has a probability of 0. A certain event has a probability of 1.

property of negative exponents   (p. 86) For any nonzero real number a, if n is a positive integer, $a^{-n} = \frac{1}{a^n}$.

proportion   (p. 122) An equation stating that two ratios are equal. A proportion can be written as *a is to b as c is to d*, $a : b = c : d$, or $\frac{a}{b} = \frac{c}{d}$. In any of these forms, b and c are called the *means* of the proportion, and a and d are called the *extremes*.

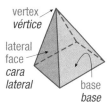

vertex / *vértice*
lateral face / *cara lateral*
base / *base*
rectangular pyramid / *pirámide rectangular*

pyramid   (p. 422) A polyhedron with only one base. The other faces are triangles that meet at a vertex. A pyramid is named by the shape of its base.

Pythagorean theorem   (p. 484) In any right triangle, the square of the length of the hypotenuse c is equal to the sum of the squares of the lengths of the legs a and b. The Pythagorean theorem is expressed as $c^2 = a^2 + b^2$.

probabilidad   (p. 148) Posibilidad de que ocurra un evento y que se lee $P(E)$. Un evento imposible tiene una probabilidad de 0. Un evento que es cierto tiene una probabilidad de 1.

propiedad de exponentes negativos (p. 86) Para cualquier número real que no sea cero a, si n es un número entero positivo, $a^{-n} = \frac{1}{a^n}$.

proporción   (p. 122) Ecuación que establece que dos razones son equivalentes. Una proporción puede ser escrita de la siguiente manera: *a es a b como c es a d*, $a : b = c : d$, o $\frac{a}{b} = \frac{c}{d}$. En cualquiera de estas formas, b y c se denominan *medios* de la proporción, y a y d se denominan *extremos*.

pirámide   (p. 422) Poliedro con una sola base. Las otras caras son triángulos. Una pirámide recibe el nombre de acuerdo a la forma de su base.

teorema de Pitágoras   (p. 484) En un triángulo rectángulo, la suma de los cuadrados de las longitudes de los catetos a y b es igual al cuadrado de la longitud de la hipotenusa c. El Teorema de Pitágoras se expresa así: $c^2 = a^2 + b^2$.

■ Q ■

quadrant   (p. 244) One of the four regions formed by the axes of the coordinate plane.

quadratic equation   (p. 268) An equation of the form $Ax^2 + Bx + C = 0$, where A, B and C are real numbers and A is not zero.

quadrilateral   (p. 216) A closed plane figure that has four sides.

quartiles   (p. 28) The three values which divide an ordered set of data into four equal parts. The *first quartile* is the median of the lower half of the data. The *third quartile* is the median of the upper half. The *second quartile* is another name for the median of the entire set of data.

cuadrante   (p. 244) Una de las cuatro regiones formadas por los ejes del plano de coordenadas.

ecuación cuadrática   (p. 268) Ecuación de la forma $Ax^2 + Bx + C = 0$, donde A, B y C son números reales y A no es cero.

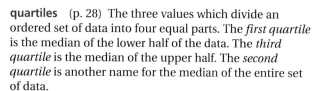

cuadrilátero   (p. 216) Polígono que tiene cuatro lados.

cuartiles   (p. 28) Se refiere a los tres valores que dividen un conjunto ordenado de datos en cuatro partes iguales. El *primer cuartil* es la mediana de la mitad inferior de los datos. El *tercer cuartil* es la mediana de la mitad superior. El *segundo cuartil* es otro nombre para la mediana del conjunto entero de datos.

■ R ■

radical   (p. 136) An expression in the form $\sqrt{a}$, where $\sqrt{}$ is called the radical symbol and a is called the radicand.

radius   (p. 226) A line segment that has one endpoint at the center of the circle and the other endpoint on the circle.

radical   (p. 136) Expresión en la forma $\sqrt{a}$ donde $\sqrt{}$ se llama símbolo radical y a es el radicando.

radio   (p. 226) Distancia que hay entre el centro del círculo y cualquier punto de la circunferencia.

Glossary/Glosario

English

Español

random sampling   (p. 6) A sampling method in which each member of the population has an equal chance of being selected.

muestra aleatoria   (p. 6) Cada miembro de una población tiene la misma posibilidad de ser seleccionado.

range   (p. 11) The difference between the greatest and least values in a set of data.

alcance   (p. 11) Diferencia entre el número mayor y el menor en un conjunto de datos.

range (of a function)   (p. 264) The set of all possible values of y for the function $y = f(x)$.

alcance (de una función)   (p. 264) Conjunto de todos los valores posibles de y para la función $y = f(x)$.

rate   (p. 101) A ratio that compares different units.

tasa   (p. 101) Razón que compara diferentes tipos de unidades.

ratio   (p. 100) A comparison of two quantities represented in one of the following ways: $a{:}b$, $\frac{a}{b}$, a to b.

razón   (p. 100) Comparación de un número con otro que se representa de las tres siguientes formas: $a{:}b$, $\frac{a}{b}$, a es a b.

rational number   (p. 52) A number that can be expressed as a ratio of two numbers a and b, where b is not equal to zero. This is written $\frac{a}{b}$, $b \neq 0$.

número racional   (p. 52) Cualquier número que pueda ser expresado en forma $\frac{a}{b}$, donde a es cualquier entero y b es también cualquier entero excepto 0. Se escribe $\frac{a}{b}$, b no es igual a b.

ray   (p. 196) Part of a line that starts at one endpoint and extends without end in one direction.

D   F

rayo o línea recta   (p. 196) Parte de una línea que tiene un punto extremo y se extiende sin fin en una dirección.

real numbers   (p. 52) The set of rational and irrational numbers.

números reales   (p. 52) Conjunto formado por los números racionales e irracionales.

reciprocals   (p. 334) Two numbers that have a product of 1. To find the reciprocal of a number, switch the numerator and the denominator.

recíprocos   (p. 334) Dos números son recíprocos cuando su producto es 1. Para encontrar el recíproco de un número intercambie el numerador por el denominador.

rectangle   (p. 216) A parallelogram that has four right angles.

rectángulo   (p. 216) Paralelogramo que tiene cuatro ángulo rectos.

reflection   (p. 300) A transformation in which a figure is reflected, or flipped, across a line of reflection. A reflection image is congruent to the preimage but oriented in the opposite direction.

reflexión   (p. 300) Transformación en la cual una figura se voltea o se refleja sobre una línea de reflejo. La reflexión de una imagen es congruente con la pre-imagen pero orientada en dirección opuesta.

regular polygon   (p. 222) A polygon that has all sides congruent and all angles congruent.

regular pentagon
pentágono regular

polígono regular   (p. 222) Polígono que tiene todos sus lados y sus ángulo congruentes.

relative frequency   (p. 150) A comparison of the number of times a particular outcome occurs to the total number of observations.

frecuencia relativa   (p. 150) Comparación del número de veces que un resultado particular ocurre con el número total de observaciones.

repeating decimal   (p. 52) A decimal in which a digit or group of digits repeats. A bar above the group of repeating digits is used to express the repeating decimal.

decimal periódico   (p. 52) Decimal en donde un dígito o grupo de dígitos se repiten. Se usa una barra encima del grupo de dígitos que se repite para expresar el decimal periódico.

rhombus   (p. 216) A parallelogram that has four congruent sides.

rombo   (p. 216) Paralelogramo que tiene sus cuatro lados iguales.

right angle   (p. 196) An angle measuring exactly 90°.

ángulo recto   (p. 196) Ángulo que mide exactamente 90°.

English

right triangle (p. 206) A triangle having one right angle.

roster notation (p. 520) Set notation that lists the elements of the set.

rotation (p. 306) A transformation in which a figure is rotated, or turned, about a point.

rotational symmetry (p. 311) A figure has rotational symmetry if it fits exactly over its original position at least once during a complete rotation about a point.

Español

triángulo rectángulo (p. 206) Triángulo que tiene un ángulo recto.

lista de anotaciones (p. 520) Conjunto de anotaciones que ponen en una lista de determinados elementos.

rotación (p. 306) Transformación en la que una figura gira o rota sobre un punto.

simetría rotacional (p. 311) Propiedad de una figura que puede quedar exactamente en su posición original después de haber rotado una vez.

■ S ■

sample (p. 6) A representative part of a population.

sample space (p. 158) The set of all possible outcomes of an experiment.

scale drawing (p. 122) A drawing that represents an object. All lengths in the drawing are proportional to actual lengths in the object. The *scale* of the drawing is the ratio of the size of the drawing to the actual size of the object.

scale factor (p. 316) The number that is multiplied by the length of each side of a figure to create an altered image in a dilation.

scalene triangle (p. 206) A triangle with no congruent sides and no congruent angles.

scatter plot (p. 20) A method of displaying the relationship between two sets of data in which the data are represented as ordered pairs and graphed as unconnected points.

scientific notation (p. 87) A notation for writing a number as the product of a factor that is greater than or equal to 1 and less than 10 and a second factor that is a power of 10.

sector (p. 232) A part of a circle graph formed by a central angle.

muestra (p. 6) Parte representativa de una población.

espacio de muestra (p. 158) El conjunto de todos los resultados posibles en un experimento.

dibujo a escala (p. 122) Dibujo que representa un objeto. Todas las longitudes en el dibujo son proporcionales a las longitudes reales del objeto. La *escala* del dibujo es la razón del tamaño del dibujo al tamaño del objeto real.

factor escala (p. 316) Número que se multiplica por la longitud de cada uno de los lados de una figura para crear una imagen alterada a través de una dilatación.

triángulo escaleno (p. 206) Triángulo que tiene tanto sus lados como sus ángulos desiguales.

diagrama disperso (p. 20) Representación gráfica que muestra la relación entre dos conjuntos de datos en los cuales los datos se representan como pares ordenados y también como puntos sin conexión.

notación científica (p. 87) Notación de un número como el producto de un primer factor que es igual o mayor que 1 y menor que 10 y un segundo factor que es una potencia de 10.

sector (p. 232) Parte de una gráfica dentro del círculo formada por un ángulo central.

The shaded region is a sector of ⊙*A*.
La región sombreada es un sector de ⊙A.

semicircle (p. 227) An arc of a circle with endpoints that are the endpoints of a diameter.

sequence (p. 92) A set of numbers that is arranged according to a pattern. Each number is a *term* of the sequence.

semicírculo (p. 227) Arco de un círculo cuyos puntos extremos son también los puntos extremos del diámetro.

secuencia (p. 92) Conjunto de números que se ordena de acuerdo a un patrón. Cada número es un *término* de la secuencia.

English

set (p. 520) A well-defined collection of items. Each item is called an *element*, or *member*, of the set. A set is usually named with a capital letter.

set-builder notation (p. 520) Set notation that gives a rule that defines each element in the set.

side (pp. 206, 222) A line segment that joins the vertices of a two-dimensional figure or a plane that joins the edges of a three-dimensional figure.

Side-Angle-Side Postulate (SAS) (p. 213) If two sides and the included angle of one triangle are congruent to two corresponding sides and the included angle of another triangle, then the triangles are congruent.

Side-Side-Side Postulate (SSS) (p. 213) If three sides of one triangle are congruent to three corresponding sides of another triangle, then the triangles are congruent.

similar figures (p. 474) Figures that have the same shape but not necessarily the same size, indicated by the symbol ~. In *similar polygons*, all pairs of corresponding angles are congruent and all pairs of corresponding sides are in proportion.

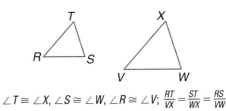

$\angle T \cong \angle X, \angle S \cong \angle W, \angle R \cong \angle V; \frac{RT}{VX} = \frac{ST}{WX} = \frac{RS}{VW}$

simplify (pp. 56, 66) To find the value of a numerical expression or to perform as many of the indicated operations as possible in a variable expression.

simulation (p. 154) A model of a situation in which you carry out trials, collect data and calculate probabilities. The model is used because it is easier to implement than the actual problem.

sine (p. 488) The trigonometric ratio of an $\angle A$, abbreviated sin A, of the length of the leg opposite $\angle A$ to the length of the hypotenuse.

skew lines (p. 202) Noncoplanar lines that do not intersect and are not parallel.

slant height (p. 433) The slant height of a cone is the length of a line drawn from its vertex to its base along the side of the cone. The slant height of a pyramid is the height of a face.

slope (p. 248) The ratio of a line segment's change in vertical distance compared to its change in horizontal distance, given by the formula $m = \frac{y_2 - y_1}{x_2 - x_1}$.

slope-intercept form (p. 254) A linear equation in the form $y = mx + b$ where m is the slope of the graph of the equation and b is the y-intercept.

space (p. 192) The set of all points.

Español

conjunto (p. 520) Colección bien definida de datos. Cada dato recibe el nombre de *elemento* o *miembro* de el conjunto. A un conjunto por lo general se le define con una letra mayúscula

notación de un conjunto construido (p. 520) Notación de un conjunto que da la pauta que define a cada elemento del conjunto.

lado (pp. 206, 222) Recta o segmento de línea que comparte los vértices ya sea de una figura bimensional o de un plano que comparte las aristas de una figura tridimensional.

Postulado de lado-ángulo-lado (LAL) (p. 213) Si dos lados y el ángulo contenido de un triángulo se corresponden con los lados y el ángulo de otro triángulo, entonces los triángulos son congruentes.

Postulado de lado-lado-lado (p. 213) Si tres lados de un triángulo son congruentes con los lados correspondientes de otro triángulo, entonces los triángulos son congruentes.

figuras similares (p. 474) Figuras que tienen la misma forma pero no necesariamente el mismo tamaño, indicado con el símbolo ~. En polígonos similares, todos los pares de ángulos correspondientes son congruentes y todos los pares de lados correspondientes son proporcionales.

simplificar (pp. 56, 66) Encontrar el valor de una expresión numérica o hacer todas las operaciones indicadas que sean posibles.

simulación (p. 154) Modelo de una situación en la cual realizas pruebas, recolectas datos y calculas probabilidades. El modelo se usa porque es más fácil de implementar que el problema real.

seno (p. 488) Razón geométrica de un $\angle A$, se abrevia sen A, que es el cociente del cateto opuesto al ángulo en cuestión, sobre la hipotenusa.

líneas oblicuas (p. 202) Líneas que no son coplanares, que no se intersecan y tampoco son paralelas.

altura diagonal (p. 433) La altura diagonal de un cono es la longitud de la recta que sale del vértice de su punta y baja por todo lo largo de su lado. La altura diagonal de una pirámide es la altura de una de sus caras.

pendiente (p. 248) La razón de cambio de una recta en distancia vertical comparada a su cambio en distancia horizontal, es dada por la fórmula $m = \frac{y_2 - y_1}{x_2 - x_1}$.

forma de pendiente e intersección (p. 254) Ecuación lineal en forma de $y = mx + b$ donde m es la pendiente de la gráfica de la ecuación y b es el interceptor en y.

espacio (p. 192) Conjunto de todos los puntos.

English   Español

sphere (p. 422) The set of all points in space that are a given distance from a given point, called the *center* of the sphere.

C is the center of the sphere.
C es el centro de la esfera.

esfera (p. 422) Figura tridimensional formada por puntos que están a la misma distancia de un punto dado llamado el centro de la esfera.

square (p. 136) A factor multiplied by itself, or in other words, a number raised to the second power.

cuadrado (p. 136) Factor multiplicado por sí mismo, o en otras palabras, un número elevado a la segunda potencia.

square (p. 216) A parallelogram that has four right angles and four congruent sides.

cuadrado (p. 216) Paralelogramo con cuatro ángulos rectos y todos los lados iguales.

square matrix (pp. 39, 354) A square matrix is a matrix that has the same number of rows and columns.

matriz cuadrada (pp. 39, 354) Es una matriz que tiene el mismo número de filas y de columnas.

square root (p. 136) A number, a, is a square root of another number, b, if $a^2 = b$. A square root is one of two equal factors of a number.

raíz cuadrada (p. 136) Un número, a, es la raíz cuadrada de otro número, b, si $a^2 = b$. Una raíz cuadrada es uno de los dos factores iguales de un número.

stem-and-leaf plot (p. 17) A method of displaying data in which certain digits are used as *stems* and the remaining digits are used as *leaves*. *Outliers* are data values that are much greater than or much less than most of the other values. *Clusters* are isolated groups of values. *Gaps* are large spaces between values.

diagrama de tallo y hoja (p. 17) Representación gráfica donde se organizan y se muestran datos donde algunos dígitos se usan como tallos, y otros como hojas. Los datos extremos son valores que son mucho más grandes o mucho más pequeños que los otros valores. Los datos extremos son grupos de valores aislados. Las separaciones son largos espacios entre los valores.

straight angle (p. 196) An angle measuring 180°.

ángulo recto (p. 196) Ángulo que mide 180°.

subset (p. 521) If every element of set *A* is also an element of set *B*, then *A* is called a subset of *B*, written $A \subseteq B$.

subconjunto (p. 493) Si cada elemento del conjunto *A* es también un elemento del conjunto *B*, entonces *A* es subconjunto de *B*, se escribe $A \subseteq B$.

substitution method (p. 345) A method of solving systems of equations algebraically. Solve one of the equations for one variable, and substitute that expression in the other equation and solve.

método de substitución (p. 345) Método para resolver sistemas de ecuaciones algebraicas. Se resuelve una ecuación por una variable, y se substituye esa expresión en la otra ecuación y se resuelve.

supplementary angles (p. 196) Two angles whose measures have a sum of 180°.

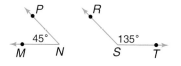

$m\angle MNP + m\angle RST = 180°$

ángulos suplementarios (p. 196) Dos ángulos cuya suma de sus medidas es igual a 180°.

surface area (p. 427) The sum of the areas of all the faces and bases of a three-dimensional figure.

superficie del área (p. 427) La suma de las superficies de todas las caras y bases de una figura tridimensional.

systematic sampling (p. 6) A method of sampling in which members of a population that have been ordered in some way are selected according to a pattern.

muestra sistemática (p. 6) Después de que una población se ordena de algún modo, los miembros de la población se eligen de acuerdo a un patrón.

system of linear equations (p. 338) Two or more linear equations with the same two variables. The *solution of a system* of linear equations is the ordered pair that makes both equations true.

sistema de ecuaciones lineales (p. 338) Dos o más ecuaciones con las mismas variables. La solución de un sistema de ecuaciones lineales es el par ordenado que hace a ambas ecuaciones verdaderas.

■ T ■

tangent (p. 488) The trigonometric ratio of an $\angle A$, abbreviated tan *A*, of the length of the leg opposite $\angle A$ to the length of the leg adjacent to $\angle A$.

tangente (p. 488) Razón trigonométrica de un $\angle A$, se abrevia tan *A*, que es el cociente del cateto opuesto sobre el cateto adyacente al ángulo en cuestión.

Glossary/Glosario

English

terminating decimal (p. 52) A decimal in which the only repeating digit is 0.

terms (pp. 66, 376) The parts of a variable expression or polynomial that are separated by addition or subtraction signs. Terms that have identical variable parts are called *like terms*. Terms that have different variable parts are called *unlike terms*.

tessellation (p. 320) A repeating pattern of figures that completely covers a plane without gaps or overlaps.

theoretical probability (p. 159) The probability of an event based on the value of a mathematical formula in which the number of favorable outcomes is divided by the total number of possible outcomes.

transformation (p. 296) A way of moving or changing the size of a geometric figure in the coordinate plane by translation, reflection, rotation or dilation. The new figure is referred to as the *image* of the original figure, and the original is referred to as the *preimage* of the new.

translation (p. 296) A transformation in which a figure is slid to produce a new figure exactly like the original. Imagine all the points of the figure sliding along a plane at once in the same direction and for the same distance. A translation image is congruent to the preimage.

transversal (p. 202) A line that intersects each of two other coplanar lines in different points to produce interior and exterior angles.

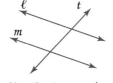

Line *t* is a transversal.
La recta t es una transversal.

trapezoid (p. 216) A quadrilateral that has exactly one pair of parallel sides.

tree diagram (p. 158) A diagram that shows all the possible outcomes in a sample space.

triangle (p. 206) The closed plane figure formed by three line segments joining three noncollinear points. Each point is called a *vertex*. Each vertex names an *angle* of the triangle. Each of the line segments that joins the vertices is called a *side*. The sum of the angles is 180°.

trinomial (p. 376) A polynomial with three terms.

two-point perspective (p. 436) A perspective drawing that has two vanishing points.

two-step equations (p. 116) Equations involving two operations. To solve, use the correct order of operations.

Español

decimal terminal (p. 52) Decimal en donde el único dígito que se repite es 0.

términos (pp. 66, 376) Partes de una expresión algebraica separadas por el signo de la suma o de la resta. Los términos que tienen variables idénticas se llaman *términos semejantes*. Los términos que tienen variables diferentes se llaman *términos diferentes*.

mosaico (p. 320) Patrón de una figura en donde copias idénticas caben sin dejar huecos y sin sobrepasar los bordes.

probabilidad teórica (p. 159) Probabilidad de un evento basado en el valor de una fórmula matemática en la cual el número de resultados favorabes se divide entre el número total de resultados posibles.

transformación (p. 296) Movimiento de una figura ya sea por traslación, reflexión o rotación. La nueva figura recibe el nombre de *imagen* del original, y al original se le llama *pre-imagen* de la nueva.

traslación (p. 296) Transformación en la cual todos los puntos de una figura se deslizan hacia la misma dirección produciendo una nueva figura idéntica a la original. Imagina los puntos de una figura deslizándose a lo largo de un plano a la vez en la misma dirección y a la misma distancia. Una imagen de traslación es congruente con la pre-imagen.

transversal (p. 202) Línea que interseca dos o más líneas en puntos distintos de un plano para producir ángulos interiores y exteriores.

trapezoide (p. 216) Cuadrilátero que tiene solamente un par de lados paralelos.

diagrama de árbol (p. 158) Diagrama que muestra todos los resultados posibles en un espacio de muestra.

triángulo (p. 206) Polígono que tiene tres lados y tres puntos no colineales. Cada punto se llama *vértice*. Cada vértice determina un *ángulo* del triángulo. Cada línea que une a los vértices se conoce como *lado*. La suma de los ángulos es igual a 180 grados.

trinomio (p. 376) Polinomio compuesto por tres términos.

perspectiva de dos puntos (p. 436) Dibujo en perspectiva que muestra dos puntos de fuga.

ecuaciones en dos pasos (p. 116) Ecuaciones que encierran dos operaciones. Para resolverlas, es necesario usar las operaciones en el orden correcto.

English

Español

union (of two sets) (pp. 147, 525) The union of two sets, *A* and *B*, symbolized $A \cup B$, contains all the elements that are in *A*, in *B*, or in both.

unión (de dos conjuntos) (pp. 147, 525) Unión de dos conjuntos, *A* y *B*, simbolizados como $A \cup B$. Esta unión contiene todos los elementos que hay en *A*, en *B* o en ambos.

unit rate (p. 101) A rate that is a comparison to 1 unit.

tasa de unidad (p. 101) Valor que tiene un denominador de una sola unidad.

universal set (p. 524) The general set of all elements being considered in a discussion. The universal set, *U*, can be an infinite set or a finite set.

conjunto universal (p. 524) Conjunto general de todos los elementos que se consideran en una discusión. El conjunto Universal se simboliza con *U* y se puede referir a conjuntos infinitos y finitos.

unlike terms (p. 66) See *terms*.

términos diferentes (p. 66) Véase *términos*.

unsound argument (p. 543) A deductive argument whose form is valid but contains at least one false premise.

razonamiento inestable (p. 543) Razonamiento deductivo cuya forma es válida pero contiene por lo menos una premisa falsa.

valid argument (p. 542) A deductive argument such that if the premises are true, then the conclusion must be true. In other words, the premises are assumed or known to be true and the conclusion logically follows.

razonamiento válido (p. 542) Razonamiento deductivo cuyas premisas son verdaderas y por lo tanto la conclusión también es verdadera. En otras palabras, si las premisas se asumen como verdaderas, la conclusión por lógica también lo es.

value (p. 56) The number represented by a numerical expression.

valor (p. 56) Número asignado a una expresión numérica.

vanishing point (p. 436) A point on the horizon line where parallel lines appear to come together.

punto de fuga (p. 436) Punto en el cual las líneas paralelas parecen hacer intersección pero que en un dibujo a perspectiva tiende a desaparecer.

variable (p. 54) A symbol used to represent a number.

variable (p. 54) Símbolo que se usa para representar un número.

variable expression (p. 57) An expression that contains one or more variables.

expresión variable (p. 57) Expresión que contiene por lo menos una variable.

Venn diagram (pp. 147, 516) A diagram that can help you see sets and their relationships more clearly. The diagram is composed of a rectangle that represents the universal set and one or more circles that represent sets within the universe.

diagrama de Venn (pp. 147, 516) Diagrama en el cual un conjunto de datos se representan. En un diagrama de Venn cada cosa se representa por una región circular dentro de un rectángulo. El rectángulo se llama el *conjunto universal*.

vertex (pp. 196, 206, 222, 422) The common endpoint of two rays that form an angle or of two sides of a polygon. The rays are called the *sides* of the angle or polygon.

vértice (pp. 196, 206, 222, 422) Punto donde se intersecan dos lados para formar un ángulo o donde se unen dos lados de un polígono. Los rayos se consideran lados de un ángulo o de un polígono.

vertical angles (p. 197) The angles that are not adjacent to each other when two lines intersect. Vertical angles are congruent.

∠1 and ∠3 are vertical angles.
∠2 and ∠4 are vertical angles.
∠1 *y* ∠3 *son ángulos opuestos por el vértice.*
∠2 *y* ∠4 *son ángulos opuestos por el vértice.*

ángulos verticales (p. 197) Ángulos de la misma medida formados por dos líneas que se intersecan. Los ángulos verticales son congruentes.

vertical line test (p. 265) If a vertical line drawn anywhere on the graph does not cross the graph more than one time, then the graph represents a function.

prueba de la línea vertical (p. 265) Si cualquier línea vertical pasa a través de la gráfica diagramada más de una vez, entonces la relación no es una función.

Glossary/Glosario

English

vital capacity (p. 456) The measure of the volume of air in a person's full lungs.

volume (p. 452) The number of cube units needed to fill a three-dimensional space.

Español

capacidad vital (p. 456) Medida del mayor volumen de aire en los pulmones de una persona.

volumen (p. 452) Número de unidades cúbicas que se necesita para llenar un objeto tridimensional.

W

weighted mean (p. 13) A mean in which certain data values are given different degrees of emphasis by multiplying them by a variable factor called the *weight* of the data value.

whiskers (p. 29) Lines that are drawn from the sides of the box in the box-and-whisker plot.

whole numbers (p. 52) The set of natural numbers and zero.

media cargada (p. 13) Media en que a los valores de ciertos datos se les dan diferentes grados de importancia al multiplicarlos por una variable llamada el *peso* de los datos.

pelos (p. 29) Líneas que se dibujan a los lados de un diagrama de bloque.

números enteros (p. 52) Conjunto de los números naturales que incluye al cero.

X

x-axis (p. 244) The horizontal number line of the coordinate plane.

eje x o de las abscisas (p. 244) Recta numérica horizontal en un plano cartesiano.

Y

y-axis (p. 244) The vertical number line of the coordinate plane.

y-intercept (p. 254) The y-coordinates of the point (or points) where the graph intersects the y-axis, represented by the variable b in the equation $y = mx + b$.

eje y o de las ordenadas (p. 244) Recta numérica vertical en un plano cartesiano.

intersección en y (p. 254) Véase *intersección*.

Z

zero pairs (p. 67) Of Algeblocks, when equal numbers of the same block are on opposite sides of the mat. To simplify an expression using Algeblocks, remove zero pairs from the mat.

zero property of exponents (p. 86) For any nonzero real number a, $a^0 = 1$.

pares de cero (p. 67) Es cuando los números iguales de un mismo bloque se encuentran en lados opuestos en los bloques algebraicos.

propiedad cero de los exponentes (p. 86) Se refiere a cualquier número real que no sea cero, a, $a^0 = 1$

Selected Answers

Selected Answers

Chapter 1: Sample and Display Data

Lesson 1-1, pages 6–9

1. systematic   3. cluster   5. convenience   7. Answers will vary, but a possible answer is to ask every fifth teenager in line at the cafeteria.   9. Systematic. People who attend a sporting event usually have a greater interest in that sport than another.   11. Convenience. Teenagers at the same party are most likely friends, and a common interest of friends can be the sport they participate in or enjoy watching.   13. Convenience. Only people with a strong opinion are likely to call in.   15. Random. The people may not be home or may not eat pizza.   17. Answers will vary.   19. Answers will vary.   21. convenience   23. A possible method is to survey fifty people at a recycling center.   25. Answers will vary.   27. A possible method is to survey all citizens whose phone number ends with a particular digit.   29. Answers will vary.   31. 5983   33. 656.23   35. 11,873   37. 1440.444   39. 1471   41. 5465   43. 451   45. 2108

Lesson 1-2, pages 10–13

1. mean = 15.3 mi; median = 13.15 mi; no mode; range = 15.7 mi   3. mean   5. Removing the $15.87 meal will affect the measures of central tendency except for the mode, since in this case there is no mode. The mean decreases by $1.09, the median decreases by $0.51, and the range decreases by $5.87.   7. 4.5   9. 4   11. Answers will vary. Have students experiment with their grades. Different students' grades will benefit from different measures.   13. 93   15. Mean. Some animals hibernate or are more dormant at certain times of the year than they are at other times.   17. 13   19. always   21. Sometimes; change 1 to 4 in 1, 1, 2, 4.   23. No. She needs to average all test scores, not just the previous average. Her average is 86.7.   25. 800   27. 70   29. 21,576   31. 40,194   33. 12,155   35. 14,990.112   37. 178.577   39. 122.133   41. 116.279   43. 39.484

Review and Practice Your Skills, pages 14–15

1. systematic   3. convenience   5. cluster   7. convenience   9. mean = $27,285.70; median = $27,000; mode = $22,000; range = $13,000   11. mean = $95.70; median = $89; mode = $105; range = $90   13. mean = $315; median = $305; no mode; range = $150   15. 57   17. 225   19. Sample answer: Send a survey to 50 random addresses of one neighborhood within the community.   21. Sample answer: Send a survey in every tenth water bill, with a total of 50 surveys sent in all.   23. mean = 6.6 mi; median = 6.2 mi; mode = 6; range = 4 mi

Lesson 1-3, pages 16–19

1.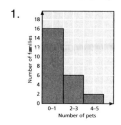

3. 2   5. median = 1; mode = 1; range = 5   7. outlier: 520; clusters: 146–189, 229–305, 412–427; gaps: 189–229, 305–366, 367–412, 427–520   9. 253

11.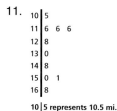

10 | 5 represents 10.5 mi.

13. 120–129   15. 100–109, 110–119, 120–129, 130–139, 140–149

17.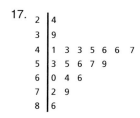

2 | 4 represents 24%.

19. 6   21. The mean and median are almost equal in this case.   23. Answers will vary.

25.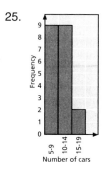

27. The data has no outliers.   29. $18   31. 80,238 people

Lesson 1-4, pages 20–23

1. 8   3. 7   5. Answers will vary.   7. 72   9. There are two points that are farthest from the trendline. One point reflects that the student only watched one hour of TV, but scored only a 50. This suggests that perhaps the student participates in other activities that take away from study time. The other point reflects that a student did not watch any TV, but only scored a 64. Answers will vary as to why for both points.   11. positive   13. Answers will vary.   15. Answers will vary.   17. 4000   19. 15   21. 2000   23. 2000   25. 30   27. 6

Review and Practice Your Skills, pages 24–25

1.
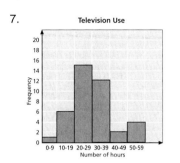

2	3 7 9
3	1 4 5 8
4	2

2|3 represents $23.

3.

4	8 9
5	0 1 4 7 7 7 7 8
6	1 1 4 5 8

4|8 represents 48 yr.

5. 57

7.
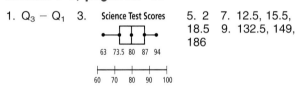

Television Use

9. 50–59 11. 10
13. positive 15. none
17. yes 19. 80
21. 6.9, 8.1 23. 7.2
25. 23.5

Lesson 1-5, pages 26–27

1. B; negative 3. C; no correlation 5. $r = -0.996$
7. Negative. As temperature rises, snow ski sales will fall.
9. Negative. In general, as a car experiences wear and tear, gas mileage will suffer. 11. $\frac{1}{4}, \frac{2}{8}, \frac{4}{16}$
13. $\frac{4}{5}, \frac{8}{10}, \frac{12}{15}$ 15. $\frac{14}{16}, \frac{21}{24}, \frac{28}{32}$ 17. 14 19. -12
21. 7 23. 10 25. 4

Lesson 1-6, pages 28–31

1. $Q_3 - Q_1$ 3.
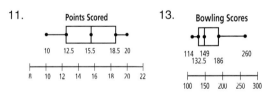

Science Test Scores

5. 2 7. 12.5, 15.5, 18.5 9. 132.5, 149, 186

11.

Points Scored

13.

Bowling Scores

15. The range for Mario's scores was less than the range for Shannon's scores. However, the interquartile range was greater for Mario than Shannon. 17. 95th 19. 55th
21. 45th 23. If Bly ranked 19th in the class then 18 students scored higher and 13 scored lower on the history exam. The percentile rank is the number of scores less than or equal to Bly's, 14, divided by the total number of students, 32. His percentile rank is 44. The information about his score of 27 out of 50 is not needed. 25. To be ranked in the 90th percentile means that 90% of the scores were equal or less than yours. To be ranked in the 10th percentile means that 90% of the scores were greater than yours. 27. 75th 29. Answers will vary. 31. -12
33. -91 35. -9 37. 48

Review and Practice Your Skills, pages 32–33

1. positive; $r = 0.991$ 3. negative: $r \approx -0.99$
5. negative 7. 20.5, 29.5, 31.5

9.
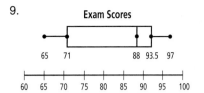

Exam Scores

10–13.
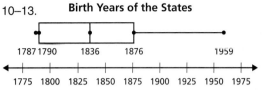

Birth Years of the States

11. $Q_1 = 1790$, $Q_2 = 1836$, 6 mo, $Q_3 = 1876$ 13. yes
15. 27th 17. 91st 19. mean = 85.1; median = 85; mode = 76, 90 21. Mr. Pascal's

Lesson 1-7, pages 34–37

1. No. By tripling the diameter, the area is increased 9 times. 3. median 5. The mean would be a little better (12); however, since the data varies greatly, only the summer month data should be used.
7.
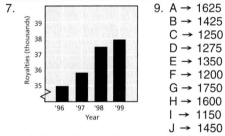

9. A → 1625
   B → 1425
   C → 1250
   D → 1275
   E → 1350
   F → 1200
   G → 1750
   H → 1600
   I → 1150
   J → 1450

11. Only positions of the complete bars are shown.
13. He should use the mean to justify the largest increase.
15. Yes. Answers will vary. 17. Answers will vary.
19. A marketer may use the first graph to advertise for a potential new trend to fad buyers. The second graph could be used to advertise the reliability and consistency of its product line. 21. 16 ft 23. 14 m 25. false 27. true
29. false 31. false

Lesson 1-8, pages 38–41

1. $\begin{bmatrix} 106 & 255 \\ 348 & 491 \\ 196 & 304 \end{bmatrix}$ 3. $\begin{bmatrix} 390 & 1 \\ 400 & 5 \\ 402 & 1 \\ 404 & 3 \\ 410 & 3 \\ 440 & 1 \end{bmatrix}$
3×2 6×2

7. 2×2 9. $\begin{bmatrix} 35 & 19 \\ 17 & 12 \\ 21 & 32 \end{bmatrix}$ 11. $\begin{bmatrix} 23 & 10 \\ 9 & 6 \\ 1 & 19 \end{bmatrix}$ 13. $\begin{bmatrix} 38 & 35 \\ 25 & 40 \end{bmatrix}$

15. $\begin{bmatrix} 142 & 114 \\ 110 & 125 \end{bmatrix}$ 17. 1916 19. $\begin{bmatrix} 992 & 1008 & 960 \\ 888 & 912 & 920 \\ 1184 & 1128 & 1152 \\ 816 & 872 & 824 \end{bmatrix}$

21. $\begin{bmatrix} 11 & 15 & 24 \\ 19 & 31 & 28 \end{bmatrix}$ 23. Yes. Addition of each element is commutative.

25. 1.96 ft² 27.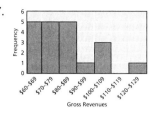

Chapter 1 Review, pages 42–44

1. h 3. c 5. d 7. f 9. l 11. convenience
13. mean = $39.69; median = $38; mode = $34; range = $24 15.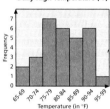

17. between 92 and 99 19. positive 21. 25 23. negative; $r = -0.83$ 25. negative 27. 12.5 29. mode 31. 3, 10, 5, 6, 11, 4, 8, 1, 0, 5

33. $\begin{bmatrix} 10 & 23 \\ 11 & 12 \\ 20 & 6 \\ 8 & 4 \\ 7 & 7 \end{bmatrix}$

Chapter 2: Foundations of Algebra

Lesson 2-1, pages 52–55

1.

3. 5. $<$

7. 9.

11. 0 13. -3 15. $\dfrac{3}{4}$

17.

19. 21.

23. Fairbanks; $-10°F > -13°F$ 25. $=$ 27. $<$ 29. $<$
31. $-2\dfrac{1}{2}$ 33. -4 35. 20 37. -80
39.

41.

43.

45. False; there are no integers between 3 and 4.
47. false; $-\pi$ 49. convenience sampling

Lesson 2-2, pages 56–59

1. 2 3. 3.3 5. 7 7. 29 9. 255 11. Answers will vary depending on calculator function. 13. 34 15. 11
17. 0.5 19. 61 21. 21 23. 263 25. 32
27. $15 + (15 - 6) = 24$ mi 29. $(x + y + z) \div 3$
31. $x^2 + 3; 7$ 33. $y - 5x; -6$ 35. 0.283
37. 53.04 39. $\dfrac{17}{18}$ 41. 15.6 43. Answers will vary.
45. ≈ 86

Review and Practice Your Skills, pages 60–61

1. 3.

5. 7. $>$ 9. $<$
11. $<$

13. 15.

17. 19. 17 21. 111 23. -4

25. 27 27. 4 29. 71 31. 6 33. 73 35. 10 37. 3
39. 21 41. 27 43. 40 45. 152 47. 36
49. $<$ 51. $>$ 53. $=$
55.

57. 59. 2 61. 54 63. 3

Lesson 2-3, pages 62–65

1. $7x$ 3. $-2 + x$ 5. $x - 22$ 7. the product of eight and a number 9. the quotient of -20 and a number 11. half of a number 13. 9 sec before blastoff
15. $x - 101$ 17. $\dfrac{-6}{x}$ 19. $32 + 2x$ 21. $16x$
23. $\dfrac{5}{10} + x$ 25. $-4 - x$ 27. $8 - x$ 29. $\dfrac{30x}{-62}$
31. six decreased by a number 33. the sum of one-half and a number 35. a number divided by 25 37. -21 divided by three times a number 39. 12 less than three times a number 41. the quotient of negative five times a number and negative 11 43. the product of a number and three 45. $n + 1,000,000$, a number increased by one million 47. $n + 200,000$, a number increased by 200 thousand 49. $\dfrac{1}{2}x + 2$ 51. $2x - 101,000$
53. $180(n - 2)$ For 55–57, answers will vary. 59. -12 multiplied by the sum of three times a number and nine

61.
```
2 | 2 6 6
3 | 3 5 9 9
4 | 1 2 3 5 8 8
5 | 2 6 8
6 | 4 5 7 8
```
3|5 represents 35.

63. 20

Lesson 2-4, pages 66–69

1. $7t$ 3. $-12t$ 5. $\dfrac{1}{2}x + \dfrac{1}{5}y$ 7. $8ab - 9ac$ 9. 20

Selected Answers

11. 43 13. 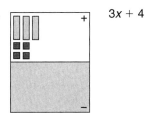 $3x + 4$

15. $14s$ 17. $-7k$ 19. $\frac{1}{9}m$ 21. $7c$ 23. $8xy + x$
25. $18a - 8b$ 27. $\frac{1}{6}d$ 29. $8.7st - 0.4t$ 31. $-6b$
33. $\frac{13}{8}xy - \frac{1}{4}x$ 35. $6wz + 15w$ 37. -30 39. 41
41. -16 43. 68 45. Answers will vary. 47. 160.847 cm
49. $-y - 3$ 51. $x - 5$ 53. $x - 2$
55. $4x + 9$ 57. $6a - 9b$

59.

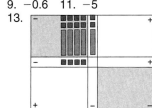

Review and Practice Your Skills, pages 70–71

1. $8 + x$ 3. $x - 5$ 5. $8y$ 7. $4x - 7$ 9. $\frac{1}{2}x + 6$
11. $\frac{2}{3}x - 14$ 13. the difference of a number and six
15. negative four decreased by a number 17. the product of negative seven and a number 19. the product of five and a number divided by ten 21. six times a number decreased by 11 23. the sum of a number and -15 25. twice a number decreased by seven 27. a product of five and two other numbers 29. $2x$ 31. $-2y$
33. $11c$ 35. $\frac{5}{4}t$ 37. $2xy + 4x$ 39. $2.1r + 7.5t$
41. $-8y$ 43. $3cd - 3d$ 45. $3n - 3m$ 47. $7\frac{2}{3}a - \frac{1}{3}b$
49. $1.6x + 2.2t$ 51. $2.4j - 9.1k$ 53. -25 55. -2
57. 37 59. 16 61. -6 63. 7 65. 20 67. $4\frac{1}{2}$
69. 2 71. $5 + 2x$ 73. $\frac{12}{y}$ 75. the difference of 8.7
and a number 77. the quotient of a number and three
79. $5x$ 81. 0 83. $15 + 13z$ 85. 29 87. 4 89. 43

Lesson 2-5, pages 72–75

1. $10 + 2s$ 3. $18.6 - 24b$ 5. $d - 3$ 7. $-6w - 5$
9. -0.6 11. -5
13. 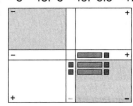 15. Answers will vary but may include to distribute -7 over the expression $(x + 4)$.
17. $-5m - 15$
19. $-9 - \frac{1}{4}x$
21. $0.32h - 0.24k$
23. $9q + 12r$
25. $-56q - 49r$

27. $14c - 35d$ 29. $-6b - 2$ 31. $7 - 8p$ 33. $4m - 10n$
35. $3q - 3r$ 37. $2d + g$ 39. $\frac{-169}{11}w + \frac{39}{11}v$
41. -8 43. 8 45. 8.5 47. 4 $-2y - 2$
49.

51. $-2x - 3$ 53. 627 ft^2
55. A2 · 21 57. $2667
59. No; he did not multiply -5 and $4xy$ and he combined unlike terms $-10x$ and $15y$.
61. $-369a + 1025b - 41c + 492d$ 63. 9184 m^3
65. Answers will vary. 67. 58–62

Lesson 2-6, pages 76–79

1. $-9x + 21$ 3. $-6n + 6m$ 5. $4c - 84$
7. $-0.5a - 1.5b - 2$ 9. $25x + 8y - 21$
11. $83x + 56(15 - x) = 27x + 840$ or $56x + 83(15 - x) = 1245 - 27x$ 13. $6m - 5$ 15. $-3x + 12$ 17. $4x + 24$
19. $3d + 4.3$ 21. $15x + 30$ 23. $3\frac{1}{2}y - 9$ 25. $32f - 148$
27. $-47t + 58$ 29. $20 - 11w$ 31. $-p + 14q$
33. $-9r - 8s$ 35. $-16c - 5cd - 12d$ 37. $18x + 21(200 - x) = 4200 - 3x$ or $21x + 18(200 - x) = 3x + 3600$
39. e 41. a 43. d 45. $14.9a + 2.5b$ 47. $-88a + 32b$
49. $-2ab + 65a - 83b$ 51. $-24x - 24$ 53. 41
55. positive 57. 61–70

Review and Practice Your Skills, pages 80–81

1. $-6c + d$ 3. $\frac{3}{2}y + 8$ 5. $10e - 30f$ 7. $x - 2$
9. $-a - 2b$ 11. $-18bc + 21$ 13. $-4s - 6t$
15. $11c - 5$ 17. 67 19. 9 21. 4 23. -31 25. 9
27. -5 29. $7x + 16y$ 31. $15a + 40$ 33. $9b - 4a + 35$
35. $-6x + 2y$ 37. $-1.3x - 8.4$ 39. $13x + 7$
41. $-2a + b$ 43. $\left(2\frac{1}{2}x + 11\frac{1}{3}z\right)$ cm 45. d 47. a
49. $-2x + 5$ 51. $3r - 11$ 53. 0 55. $-3x + 21$
57. $16t + 24$ 59. $10ab - 1$ 61. -7 63. $\frac{9}{2}$ 65. 30

Lesson 2-7, pages 82–85

1. 4 3. -50 5. c^{11} 7. d^{10} 9. v^4 11. z^4 13. It is impossible to divide by 0. 15. -7 17. $\frac{16}{49}$
19. 3 21. 16 23. x^{15} 25. g^6 27. j^{12} 29. 0
31. $5w^2$ 33. d^9 35. 819,200 bytes 37. 2^{10} or 1024
times 39. x^{18} 41. fg^2 43. $\frac{r^5}{s^{10}}$ 45. v^4
47. 271 million 49. 4 51. 12 53. true 55. 3
57. 38.4 59. 10

61.

Selected Answers

Lesson 2-8, pages 86–89

1. x^{-3} or $\dfrac{1}{x^3}$ 3. w^{-20} or $\dfrac{1}{w^{20}}$ 5. $\dfrac{1}{16}$ 7. $\dfrac{1}{25}$ 9. $2.9 \cdot 10^7$

11. $8.08 \cdot 10^{-5}$ 13. 0.0095 15. $2.7 \cdot 10^7$

17. y^{-10} or $\dfrac{1}{y^{10}}$ 19. $a^0 = 1$ 21. $b^0 = 1$ 23. $\dfrac{1}{a^{12}}$ or a^{-12}

25. $\dfrac{1}{1024}$ 27. 16 29. $\dfrac{1}{16}$ 31. -1 33. $4.658 \cdot 10^{-2}$

35. $1.5 \cdot 10^7$ 37. $3.658 \cdot 10^6$ 39. 530,000,000
41. 3907 43. 60,046,000 45. $3.46 \cdot 10^7$
47. 123,840,000 km $= 1.2384 \cdot 10^8$ km

49. $(6 \cdot 3)^0$, $2^3 \cdot 2^{-2}$, $\left(\dfrac{1}{3}\right)^{-1}$ 51. 4; a^n 53. $\dfrac{y^2}{x^2}$

55. $\dfrac{1}{m^7 n^3}$ 57. $-5v - 2w$ 59. $-11tw - 6w$

61. $-6r + 12$ 63. $-8.5v + 4$ 65. $2x - \dfrac{3}{2}y + 3$

Review and Practice Your Skills, pages 90–91

1. 144 3. $\dfrac{9}{16}$ 5. 73 7. 2 9. $\dfrac{8}{9}$ 11. 18 13. a^7

15. $c^8 d^2$ 17. $4x^6$ 19. $\dfrac{16}{y^4}$ 21. x^2 23. m^8

25. a^9 27. $64x^6$ 29. $-3d^6 e^3$ 31. y^{-8} or $\dfrac{1}{y^8}$

33. b^{-2} or $\dfrac{1}{b^2}$ 35. m^{-5} or $\dfrac{1}{m^5}$ 37. $\dfrac{1}{16}$ 39. 8

41. $\dfrac{1}{16}$ 43. $1 \cdot 10^6$ 45. $3.334 \cdot 10^3$ 47. $1.3 \cdot 10^{-4}$

49. 0.0072 51. 0.000012 53. 55,000

55. ![number line with points at -2/3, 1/2, 3/2] 57. ![number line]

59. ![number line] 61. 5 63. $11\dfrac{1}{2}$

65. 32 67. the sum of a number and seven
69. the product of negative three and a number
71. -0.5 decreased by a number 73. $2b$ 75. $2a^3 b^2$
77. $2x + 6$ 79. $-3a - 8$ 81. 32 83. $8a^6 b^3$
85. $-17x - 73$ 87. false 89. false 91. true 93. true
95. false 97. true 99. 31 101. 18 103. -11

105. 18 107. $\dfrac{1}{2}$ 109. 8.9 111. 72

Lesson 2-9, pages 92–93

1. 12, 19, 26, 33, 40 3. -20, $-\dfrac{20}{3}$, $-\dfrac{20}{9}$, $-\dfrac{20}{27}$, $-\dfrac{20}{81}$
5. $3x + 2$, $4x + 7$, $5x + 12$, $6x + 17$, $7x + 22$ 7. 2323,
2424, 2525, 2626, 2727 9. 11,109, 22,218, 33,327,
44,436, 55,545
11. ![dot grid patterns]

Increase the number of rows and columns of dots by 1.

13. add 2, 3, 4, . . . ; 21, 28, 36 15. divide by 3; 1, $\dfrac{1}{3}$, $\dfrac{1}{9}$

17. \$588 19. 1: 5%; 2: 10%; 3: 15%; 4: 20%;
5: 25%; 6: 30%; 7: 35%; 8: 40% 21. \$1687.50
23. yes; $28 = 1 + 2 + 4 + 7 + 14$ 25. 8, 13, 21, 34, 55,
89 27. > 29. $-25°$F 31. -18

Chapter 2 Review, pages 94–96

1. f 3. a 5. e 7. d 9. k 11. ![number line from -2 to 2 with open circle at 1] 13. < 15. >

17. 12 19. 17 21. 65 23. 144 25. $n - 9$ 27. $4 - 2n$
29. the quotient of a number and three 31. one less than
sixteen times a number 33. $2.2mn + 1$ 35. $-2xy - x$
37. -10 39. 17 41. $-c - 3d$ 43. $-6y + 12x$ 45. $3y - 3x$
47. $2x + \dfrac{1}{2}y$ 49. $14x - 3$ 51. $3x - 8$ 53. $6x + 18$

55. $-x + 16$ 57. $(6x - 8)$ in. 59. 33 61. 20 63. a^8
65. $9w^4$ 67. $\dfrac{1}{4}$ 69. $\dfrac{1}{8}$ 71. 0.00075 73. $5.46 \cdot 10^{-5}$

75. $\dfrac{1}{81}$, $\dfrac{1}{243}$, $\dfrac{1}{729}$ 77. 36, 34, 45

Chapter 3: Equations and Inequalities

Lesson 3-1, pages 104–107

1. -1 3. 1 5. 6 7. 0.6 m 9. 27 mi/h 11. 3 13. $\dfrac{3}{4}$

15. 35 17. 0 19. $-4, 4$ 21. 2.5 23. -6 25. 13.3°C
27. 17.8°C 29. 63 ft 31. 154 ft 33. 8, -8 35. $\varnothing$,
absolute value is positive. 37. 0.22 kWh 39. Answers
will vary. 41. irrational 43. rational 45. 58 47. 34
49. 4 51. 8.5

Lesson 3-2, pages 108–111

1. divide by 3 3. multiply by 8 5. subtract 3 7. 12
9. 4.6 11. $-\dfrac{17}{30}$ 13. $s = \dfrac{P}{4}$ 15. $r = \dfrac{I}{pt}$ 17. 21

19. -0.9 21. 0.12 23. -9 25. 13.5 27. -4
29. $-\dfrac{25}{2}$ 31. -2.5 33. $-6\dfrac{1}{2}$ 35. -2.5 37. 65

39. -20 41. $-\dfrac{14}{15}$ 45. No; He needs to multiply both

sides by $-\dfrac{5}{2}$. 47. 6.5%

49.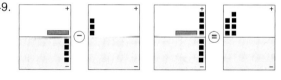

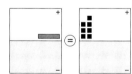

49. $y = 7$ 51. $x + 6 = 10$; 4
53. $\dfrac{1}{3}x = 24$; 72

55. 20 57. The value of x decreases. 59. -28
61. 16–25 and 36–45; 26–35 and 46–55 63. 4
65. 3.8 67. 85

Review and Practice Your Skills, pages 112–113

1. -2 3. $-6, 6$ 5. $11\dfrac{3}{4}$ 7. 0 9. 4 in. 11. 6 ft

13. 2 15. $\dfrac{1}{7}$ 17. 12 19. 9 21. -7 23. -32 25. 1

27. -11 29. $\dfrac{16}{25}$ 31. -90 33. 36 35. 32 37. $-\dfrac{1}{32}$

39. 2.2 41. $\dfrac{1}{2}$ 43. $-\dfrac{2}{3}$ 45. -7 47. -5 49. $t = \dfrac{I}{pr}$

51. $7x = -35$; -5 53. $10 + x = 27$; 17 55. 5 57. -5
59. -27 61. $r = \dfrac{C}{2\pi}$ 63. $C = A - B - D$ 65. 10

Lesson 3-3, pages 114–115

1. 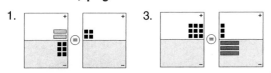 3.

5.

Weight on Earth	100	125	150	175	200
Weight on Pluto	4	5	6	7	8

7. $P = \dfrac{1}{25}E$

9.

Week	1	2	3	4	5	6	7
Distance of run (miles)	1	$1\frac{1}{2}$	2	$2\frac{1}{2}$	3	$3\frac{1}{2}$	4

$1 + \dfrac{1}{2}w(w - 1),\ d = 1 + \dfrac{1}{2}(w - 1)$

11.

No. of bi-weekly checks	0	1	2	3	4	5
Amount in savings account	$300	$350	$400	$450	$500	$550

$300 + 50w,\ A = 300 + 50w$

13.

Number of children at camp	1	2	3	4	5
Total cost	$20	$25	$30	$35	$40

15. $C = 15 + 5n$ 17. $45 + 5n$ 19. Answers will vary, but should include a discussion of the flat rate and cost per child.

21. $8a$ 23. $14k$ 25. $\dfrac{11}{24}b$ 27. $1.3g + 3.9h$ 29. $3t - s$

31. h^{12} 33. $2f^{21}$ 35. m^5

Lesson 3-4, pages 116–119

1. 4 3. 16 5. 7 7. 17 9. $x = \dfrac{y - b}{m}$ 11. $n + 16 +$

$n + 32 + 34 = 100;\ n = 9,\ n + 16 = 25$ 13. 4 15. 9

17. -5 19. -4 21. -1 23. 2.2 25. -11 27. -2

29. 1 31. $-\dfrac{6}{7}$ 33. 2 35. $r = \dfrac{8}{7}(p - 20)$

37. $C = \dfrac{5}{9}(F - 32)$ 39. 27.8°C

41.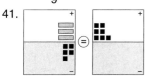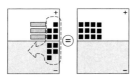

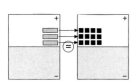 $x = 4$

43.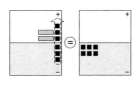

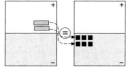 $x = -3$

45. $m\angle EFG = 51°,\ m\angle GFH = 39°$ 47. $l = 120$ yd,

$w = 53$ yd 49. $\dfrac{1}{2}x - 5 = -12,\ x = -14$ 51. 24°, 54°, 102°

53. 0 55. 5 57. 43, 45, 50, 129 59. clusters: high 80s/low 90s; gaps: 50–82, 108–129 61. 86

Review and Practice Your Skills, pages 120–121

1.

Week	1	2	3	4	5	6
Amount in bank	$120	$150	$180	$210	$240	$270

3. $A = 90 + 30w$

5.

Miles	100	200	300	400	500	600
Charge per day	$47	$59	$71	$83	$95	$107

7. $35 + 0.12m$ 9. $B = 35 + 0.12m$ 11. 3 13. $-\dfrac{19}{2}$

15. -4 17. 5 19. 12.5 21. 117 23. 4 25. -28

27. $\dfrac{3}{4}x + 2 = 8,\ x = 8$ 29. $5x - 10 = 0,\ x = 2$ 31. 30

33. -3.1 35. -8

37.

Week	1	2	3	4	5	6
Amount in savings account	$250	$300	$350	$400	$450	$500

$200 + 50w,\ A = 200 + 50w$ 39. 30 41. 4 43. 2
45. -4

Lesson 3-5, pages 122–125

1. 12 3. 33 5. -54 7. 3 9. 12 11. 39 13. 32

15. 17 17. 16 19. 44 21. 37.5 23. 51 25. 5

27. 25 29. -3 31. $\dfrac{1}{2}$ 33. 5 35. $\dfrac{34}{25}$ 37. Jenisa;

Erica added 8 to -18 and received an answer of 10.

Dividing 2 into 10 resulted in 5. 39. 8% 41. $\dfrac{\text{part}}{\text{whole}} =$

$\dfrac{x}{100}$ 43. about 47% 45. 3 m 47. 7 49. -33

51. 4.5%

53.

55.

57. 59.

Lesson 3-6, pages 126–129

1. $y < 2.8$ 3. $y = 2.8$

5. 7.

9. $x > 0$ 11. $p \ge 13$ 13. $d < -\dfrac{1}{3}$ 15. $d = -\dfrac{1}{3}$

17. substitution and number line

19. 21.

23. 25.

27. 29.

31. 33.

35. $x < 13$ 37. $x \le \dfrac{2}{3}$ 39. $x < 212$

41. 43. $x < 1199$

45. always 47. sometimes; true when $a = b$

49. terminating 51. terminating 53. repeating
55. terminating 57. z^{-3} 59. r^2 61. p 63. v^{-25}
65. 93 min

Review and Practice Your Skills, pages 130–131

1. 14 3. 14 5. 12 7. 81 9. 49 11. 11 13. 7
15. $15\frac{2}{3}$ 17. 31 19. 0 21. 3 23. 20%
25.   27.
29.   31.
33.   35.
37. $x > -8$ 39. 30 41. 4.5 43. $\frac{7}{20}$ 45. 10 47. 14
49.

Lesson 3-7, pages 132–135

1. $z \leq 2$
3. $y < -6$
5. $t \geq 12$
7. $n < -4$
9. $w \leq 18$
11. He can consume at most 675 calories.
13. $x \geq -4$
15. $x \geq 3$ 17. $h \leq -8$
19. $r \leq -\frac{7}{2}$ 21. $k \geq -2$
23. $y < 8$ 25. $q < -3$
27. $b \leq 6$ 29. $x \leq 9$
31. $a \leq -8$ 33. $x \leq -4$
35. $p < -5.2$ 37. $x \geq -1$
39. 28 41. $-6 < 18x, x > -\frac{1}{3}$ 43. $8 + 4x \leq -12, x \leq -5$
45. 12.5 yd 47. $a < 8\frac{1}{3}$ 49. Sometimes; for example if
$a = 0$, $b = 2$, $c = 4$, and $d = 1$, then $2 > 1$, or if $a = 1$,
$b = 0$, $c = 4$, and $d = 1$, then $-3c - 1$. 51. $7.382047 \cdot$
10^6 53. $3.84719 \cdot 10^7$ 55. 0.0005697 57. $10z + 40$
59. $f - 3g$ 61. $-2a - 6$

Lesson 3-8, pages 136–139

1. ± 5 3. $\pm \sqrt{7}$ 5. $\pm \frac{7}{10}$ 7. 1 9. ± 12 11. 0 13. In
Exercise 7 the entire expression $d + 3$ is under the radical
and both sides of the equation can be easily squared
immediately. In Exercise 12 add 3 to both sides before

squaring them. 15. 81 17. ± 8 19. 4
21. $\pm \sqrt{59}$ 23. $\varnothing$ 25. ± 4 27. ± 16 29. 36 31. 72
33. $\pm \sqrt{50}$ 35. 10 37. $\pm \sqrt{105}$ 39. 40 mi/h
41. $3x^2 = 588$, $x = 14$ in.; 14 in. by 14 in. 43. $\left(\frac{T}{2\pi}\right)^2 \cdot$
$981 = L$ 45. ± 0.12 47. ± 0.5 49. ± 9.055 51. ± 4.583
53. 13 55. 576 lb 57. No. No. An object's velocity
increases the further it falls. 59. $x + 7$ 61. $\frac{5x}{-12}$
63. 4 65. -6

Chapter 3 Review, pages 140–142

1. 1 3. g 5. f 7. c 9. d 11. 7 13. 12
15. 282.7 cm^3 17. -4 19. 9 21. $\frac{1}{4}x = -15$, $x = -60$
23.

Age of alligator (years)	0	1	2	3	4	5
Length of alligator (inches)	8	20	32	44	56	68

Let a = age in years; $8 + 12a$. Let l = length of alligator;
$l = 8 + 12a$.
25.

Days overdue	0	1	2	3	4	5
Amount owed	\$1.80	\$2.10	\$2.40	\$2.70	\$3.00	\$3.30

Let d = days overdue; $1.80 + 0.30d$. Let a = amount
owed; $a = 1.80 + 0.30d$.
27. 0.5 29. -56 31. 3 33. 10 35. 10 gal
37.   39.
41.   43. $z > 2$
45. $t > 5$ 47. $s < 11$
49. 625 51. 64 53. ± 6 55. 1.8 h

Chapter 4: Probability

Lesson 4-1, pages 150–153

1. ≈ 0.66 3. 0.4, 0.6 5. 0.71 7. 0.31 9. 0.25
11. Answers will vary. 13. 0.12 15. ≈ 0.06 17. 0.16
19. ≈ 0.41 21. ≈ 0.19 23. 0.004 25.

Wind Speeds

Miles per hour	Tally	Frequency
1-3	I	1
4-7	II	4
8-12	III	8
13-18	I	3
19-24	I	2
25-31	I	1
32-38	I	3
39-46		0
47-54	I	2
55-63		0
64-75		0
75+		0

Lesson 4-2, pages 154–155

1. Models will vary. Use a coin to model gender. Let
heads represent a girl and tails represent a boy. Since the
family is planning to have 4 children, flip 4 coins at a time.
3. Answers will vary. 5. Models and experimental
probabilities will vary. 5. Use a number cube to model the
chance for a cure. Let the outcomes 1, 3, and 5 represent
that the drug works. Let the outcomes 2 and 4 represent
that the drug does not work. Disregard the outcome
6. Toss 2 number cubes at a time to represent the 2
patients chosen randomly 7. Models and experimental
probabilities will vary. Use a spinner divided into three

Selected Answers

equal sections labeled 1, 2, and 3. Let 1 and 2 represent a made free throw and 3 represent a missed free throw. Spin 5 spinners at a time to model 5 free throws shot by Tia.   **9.** Jolon is correct. Any device, such as a number cube, that has an even number of outcomes can be used since the outcomes can be separated into two equal categories.   **11.** -8 **13.** $0.8\overline{8}$ **15.** -3 **17.** $-\dfrac{8}{3}$

19. 40,800   **21.** 0.0000917   **23.** 3,298,700
25. 0.000659

Review and Practice Your Skills, pages 156–157

1. 0.75   **3.** 0.84   **5.** ≈ 0.48 **7.** ≈ 0.46 **9.** ≈ 0.24
For 11–13, models and experimental probabilities will vary.
11. Use a number cube with each number representing a different baseball card. Roll 6 cubes at once to represent 6 boxes.   **13.** Use a coin to model gender. Let heads represent a girl and tails represent a boy. Since the family wants to have 3 children, flip 3 coins at a time.
15. ≈ 0.94 **17.** ≈ 0.79 **19.** 60   **21.** 0.90

Lesson 4-3, pages 158–161

1. 24
3.

5. $\dfrac{1}{24}$ **7.** $\dfrac{1}{4}$ **9.** 160   **11.** 64   **13.** $\dfrac{3}{64}$ **15.** $\dfrac{1}{16}$
17. 30   **19.** $\dfrac{5}{13}$ **21.** 72   **23.** $\dfrac{1}{216}$ **25.** $\dfrac{125}{216}$
27. $\dfrac{1}{9}$ **29.** 7   **31.** -6 **33.** 2   **35.** 10   **37.** 0.6   **39.** 2

Lesson 4-4, pages 162–165

1. $\dfrac{1}{2}$ **3.** $\dfrac{2}{9}$ **5.** $\dfrac{3}{26}$ **7.** $\dfrac{3}{4}$ **9.** $\dfrac{1}{4}$ **11.** $\dfrac{1}{6}$ **13.** $\dfrac{5}{8}$ **15.** $\dfrac{7}{8}$

17. $\dfrac{5}{6}$ **19.**

21. $\dfrac{4}{5}$ **23.** $\dfrac{1}{5}$ **25.** Mary is probably incorrect because the results of the survey are most likely not mutually exclusive. In other words, just because a voter lies about voting on one county position does not exclude them from lying about voting for the other position.   **27.** 40   **29.** Yes. They are mutually exclusive.   **31.** $\dfrac{4}{9}$ **33.** mean: 1.8; median: 2; mode: 2   **35.** mean: 8.7; median: 8; mode: 6, 12   **37.** 32   **39.** 7

41. 98   **43.** -4 **45.** $0.8\overline{8}$ **47.** 4.76

Review and Practice Your Skills, pages 166–167

1. 30   **3.** $\dfrac{1}{10}$ **5.** $\dfrac{1}{5}$ **7.** $\dfrac{1}{26}$
9.

11. $\dfrac{1}{36}$ **13.** $\dfrac{25}{36}$

15. $\dfrac{11}{26}$ **17.** $\dfrac{4}{5}$ **19.** $\dfrac{4}{15}$ **21.** $\dfrac{1}{12}$ **23.** $\dfrac{1}{9}$ **25.** 0.81
27. (dime, nickel) $\rightarrow$ (H, H), (H, T), (T, H), (T, T)   **29.** $\dfrac{4}{5}$

Lesson 4-5, pages 168–171

1. $\dfrac{1}{10}$ **3.** $\dfrac{1}{25}$ **5.** $\dfrac{1}{15}$ **7.** $\dfrac{5}{12}$ **9.** $\dfrac{1}{50}$ **11.** $\dfrac{4}{25}$
13. $\dfrac{868}{43,500} \approx 0.02$ **15.** $\dfrac{4}{91}$ **17.** 0.1024   **19.** 0.2105
21. Independent events are events whose outcomes do not affect one another. For example, when a card is drawn from a deck of cards it is then replaced and reshuffled before drawing the second card. Dependent events are events for which the outcome of one can affect the other. When a card is drawn from a deck of cards and not replaced, the probability of the next card drawn is different than if the card were replaced and the deck reshuffled.
23. 0.013   **25.** $\dfrac{25}{1352}$ **27.** 3; $\dfrac{8}{81}$ **29.** No; there are 12 ways to get a multiple of 3 and 9 ways to get a multiple of 4.   **31.** 9   **33.** 52   **35.** -8 **37.** 9   **39.** $\dfrac{81}{4}$
41. $-4ab - 4a$ **43.** $3k - 3m$ **45.** $-11c + b$

Lesson 4-6, pages 172–175

1. 20   **3.** 4   **5.** 720   **7.** 120   **9.** 30   **11.** 1   **13.** 60
15. 24   **17.** 5040   **19.** 35   **21.** 3125; 120   **23.** 5040
25. 72   **27.** 24   **29.** $\dfrac{24}{625}$ **31.** 48
33.

35.

37.

39.

41.

43.

45. $x \geq 4$ **47.** $r < 8$ **49.** $f < 2$ **51.** $r \leq 3$ **53.** $s \leq 50$
55. $a < 4$

Review and Practice Your Skills, pages 176–177

1. $\dfrac{5}{36}$ **3.** $\dfrac{5}{48}$ **5.** $\dfrac{25}{144}$ **7.** $\dfrac{3}{26}$ **9.** $\dfrac{1}{13}$ **11.** $\dfrac{1}{26}$ **13.** $\dfrac{5}{51}$
15. $\dfrac{2}{51}$ **17.** $\dfrac{28}{153}$ **19.** 210   **21.** 20   **23.** 504   **25.** 5

27. 332,640 29. 5040 31. 210 33. 48 35. $\frac{1}{8}$ 37. $\frac{5}{48}$
39. $\frac{11}{26}$ 41. $\frac{1}{24}$ 43. 120 45. 12

Lesson 4-7, pages 178–181

1. 10 3. 1 5. $\frac{1}{36}$ 7. 20 9. 165 11. 1 13. $\frac{1}{325}$
15. 252 17. 38,760 19. Answers will vary. An insightful answer will note that a combination is equal to the permutation divided by r! 21. 1140 23. $\frac{1}{20,825}$
25. 7315 27. $\frac{1}{120}$ 29. Answers will vary. 31. 20
33. 2.5 35. 7 37. $\pm5\sqrt{3}$ 39. 36 41. $\frac{33}{4}$
43. $t = 190p$

Chapter 4 Review, pages 182–184

1. g 3. k 5. d 7. e 9. h 11. ≈ 0.21 13. 0.23 15. Answers will vary. Sample answer: Use the numbers 1 and 2 on a number cube to represent whole wheat bread. Use heads on a coin to represent bread dated to be sold by the middle of the week. Roll the number cube and flip the coin. Record whether or not the number cube shows 1 or 2 and the coin shows heads. Repeat the simulation 30 times and determine the experimental probability. 17. Answers will vary. Sample answer: Let 1 and 2 represent one action figure, 3 and 4 represent another action figure, and 5 and 6 represent the third action figure. Roll a number cube five times and record the result. Determine whether all 3 action figures are represented. Repeat the simulation 30 times and determine the experimental probability. 19. 32 cars 21. $\frac{1}{4}$ 23. $\frac{2}{3}$
25. $\frac{1}{10}$ 27. $\frac{1}{3}$ 29. 24 31. 378 33. 70

Chapter 5: Logic and Geometry

Lesson 5-1, pages 192–195

1. any three of A, B, X and Y 3. 1 5. neither
7.   z • 9.

11. X and Y 13. line ℓ 15. true 17. true 19. true
21. Postulate 1 23. Postulate 2 25. Postulate 3
27. U 29. $\overline{LB}$, $\overline{LC}$, $\overline{LU}$, $\overline{LN}$, $\overline{LP}$ 31. It would most likely be a point, but it could be a line. 33. Atlanta, New York, Boston 35. 24 37. 4

Lesson 5-2, pages 196–199

1. 76°, 166° 3. 60°, 150° 5. ∠BGC, ∠CGE, ∠EGF, ∠FGB 7. 70° 9. 37° since the two angles are complementary and 90° − 53° = 37°. 11. 90° since the supplement of a right angle is a right angle. 13. $\overline{TO} \perp \overline{RO}$; $\overline{TO} \perp \overline{OS}$ 15. ∠ROP and ∠QOS, ∠ROQ and ∠POS 17. 87°

19.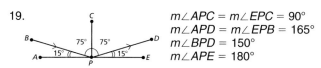

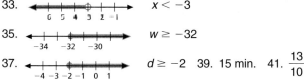

21. 72° 23. the bisector of the angle 25. 90° 27. lowered
29. $m\angle 4 = 45°$, $m\angle 1 = 45°$, $m\angle 3 = 45°$ 31. False; the complement of an acute angle is an acute angle.
33. 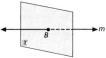 $x < -3$

35. ──── $w \geq -32$
    −34 −32 −30

37. ──── $d \geq -2$ 39. 15 min. 41. $\frac{13}{10}$
    −4 −3 −2 −1 0 1

Review and Practice Your Skills, pages 200–201

1. 3 3. F and C 5. $\overrightarrow{DE}$, $\overrightarrow{EF}$, $\overrightarrow{DF}$ or line ℓ 7. true
9. false 11.

13. 70° 15. none 17. 40° 19. 13°, 103° 21. 31°, 121° 23. 30°, 120° 25. 45°, 135° 27. 59°, 149°
29. 85°, 175° 31. b 33. c 35. 131°, supplementary
37. 30°, complementary

Lesson 5-3, pages 202–205

1. ∠6 and ∠3, ∠5 and ∠4 3. ∠5 and ∠3, ∠6 and ∠4
5. corresponding 7. alternate interior 9. 120° 11. 60°
13. corresponding 15. corresponding 17. alternate exterior 19. alternate exterior 21. 105° 23. 105°
25. 75° 27. parallel by alternate interior angles
29. parallel by alternate exterior angle 31. Answers will vary. 33. $m\angle 1 = 58°$; $m\angle 3 = 58°$; $(9x − 13)° = 122°$; $(7x + 17)° = 122°$ 35. Skew 37. They are parallel.
39. alternate interior angles 41. False; by definition, skew lines do not lie in the same plane. 43. Answers will vary, but one idea is parallel lines in poetry. 45. $\frac{1}{66}$
47. P(white or blue or green or red)

Lesson 5-4, pages 206–209

1. yes 3. yes 5. $x = 35$ 7. $x = 64$, $y = 58$, $z = 122$
9. no 11. yes 13. scalene 15. equilateral 17. $3p = 54$; $2p = 36$ 19. $y = 99$ 21. 123°; The sum of angles is 180°. 23. 40° 25. 90° 27. 40° 29. 140° 33. Yes. The angles opposite of congruent sides are congruent.
35. equilateral; interior angles: 60°; exterior angle: 120°
37. sometimes 41. $35t^5$ 43. $5s^2 − 25s$ 45. $−7st + 12$ 47. $−x^4 + 3x^2$

Review and Practice Your Skills, pages 210–211

1. vertical 3. alternate, interior 5. corresponding
7. supplementary, same-side interior 9. alternate, exterior
11. 60° 13. 60° 15. 120° 17. parallel by corresponding angles 19. not enough information 21. yes 23. no
25. no 27. 65° 29. 50° 31. 65° 33. 115°

Selected Answers

35.  37.

39. true 41. 49°, 139° 43. 76°, 166° 45. 8°, 98°
47. a = 70°, b = 70°, c = 110° 49. a = 80°, b = 80°

Lesson 5-5, pages 212–215

1. $\overline{AB} \cong \overline{DE}$, $\overline{AC} \cong \overline{DF}$, $\overline{BC} \cong \overline{EF}$, $\angle A \cong \angle D$, $\angle B \cong \angle E$, $\angle C \cong \angle F$ 3. $\triangle USR \cong \triangle UST$ by SSS 5. $\angle B$ 7. $\overline{AB}$
9. $\overline{RS} \cong \overline{XY}$, $\overline{ST} \cong \overline{YZ}$, $\overline{RT} \cong \overline{XZ}$, $\angle R \cong \angle X$, $\angle S \cong \angle Y$, $\angle T \cong \angle Z$ 11. SAS 13. SAS or ASA 15. $\angle R$ 17. $\overline{RU}$ 19. yes; by SSS 21. yes; by SAS, SSS and ASA
23. 6 25. Latravis is correct. Since two angles of one triangle are congruent to two angles of another triangle, their third angles must also be congruent. This is because the sum of the angles in every triangle is 180°. The triangles then are congruent by ASA. 27. No. The sides of the two triangles are not necessarily congruent.
29. Rochester, MN 31. −15,625, 78,125, −390,625
33. 8, 13, 21 35. 34, 43, 53 37. $\frac{1}{100}$

Lesson 5-6, pages 216–219

1. 138° 3. 138° 5. 10 7. 8 9. 6 11. 122°
13. 122° 15. 58° 17. 18 19. 106° 21. 106°
23. rectangle, square 25. parallelogram, rectangle, rhombus, square 27. rhombus, square 29. rectangle, square 31. 59°, 59°, 121° 33. 10 35. 111° 37. false
39. false 41. true 43. false 45. 360°; Turn U will be more difficult because it is a sharper turn. 47. a quadrilateral has two pairs of parallel sides 49. The legs represent the diagonals of a quadrilateral where $\overline{AB}$ and $\overline{DC}$ are opposite sides. Since $\overline{AO} \cong \overline{OC}$ and $\overline{BO} \cong \overline{OD}$, these diagonals bisect each other. Thus, the quadrilateral $ABCD$ is a parallelogram and $\overline{AB} \parallel \overline{DC}$. 51. $\frac{47}{4}$ 53. 25
55. 2 57. 10 59. 48 61. −3.2

Review and Practice Your Skills, pages 220–221

1. ASA 3. SSS or SAS 5. ASA 7. $\angle S$ 9. $\angle T$ 11. $\overline{TU}$ 13. 8 15. 81° 17. 128° 19. 52° 21. $\overline{DN}$ 23. $\angle NIF$ 25. ASA 27. not necessarily congruent 29. 82°
31. 82° 33. 82°

Lesson 5-7, pages 222–225

1. convex; regular pentagon 3. concave; not regular hexagon 5. 1440° 7. 1980° 9. 144° 11. 156°
13.  15. 360° 17. 3240° 19. 165°
21. convex, not regular, 12-gon
23. quadrilateral, yes
25. octagon, yes 27. false; the sum is 1620° 29. true

31. 33.

35. (sum of interior angles) + (sum of exterior angles) = n(180°) the sum of interior angles = (n − 2)180°. So, (n − 2)180° + (sum of exterior angles) = n(180°)

(180°n − 360°) + (sum of exterior angles) = 180°n sum of exterior angles = 360° 37. 30 39. 21 41. $\frac{1}{7}$

Lesson 5-8, pages 226–229

1. $\overline{LP}$, $\overline{LN}$, $\overline{LM}$ 3. $\overparen{PMN}$, $\overparen{PNM}$, $\overparen{NOP}$, $\overparen{MPO}$, $\overparen{PNO}$ 5. 180°
7. 263° 9. $\angle MNO$ 11. 60° 13. 105° 15. $\overparen{POQ}$, $\overparen{PNQ}$, $\overparen{NQO}$, $\overparen{OPN}$ 17. 110° 19. Both are radii. 21. Answers will vary. 23. 60° 25. True; the measure of a radius is one half the measure of a diameter. 29. 160° 31. 26.4 ft
33. $\frac{1}{16}$ 35. $\frac{1}{8}$ 37. $\frac{1}{12}$

Review and Practice Your Skills, pages 230–231

1. 3. 5. 1800°, 150° 7. 6840°, 171° 9. hexagon, concave, not regular 11. rectangle, convex, not regular 13. b
15. g 17. c 19. f 21. 300° 23. 120° 25. 180° 27. point, line 29. congruent
31. congruent 33. supplementary 35. neither 37. yes
39. yes 41. no 43. 30° 45. ASA 47. SSS 49. false 51. true 53. true

Lesson 5-9, pages 232–233

1. 3.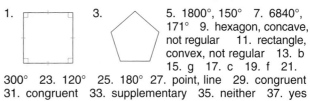

5. 7.

9. 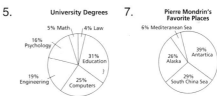 11. $\frac{3}{4}d + 15\frac{3}{4}$ 13. $-9n + 27$
15. $23 + 25.5a$

Chapter 5 Review, pages 234–236

1. h 3. 1 5. c 7. b 9. i 15. Answers will vary. Sample answer: the floor and the ceiling 17. 41°, 131°
19. 78°, 168° 21. $\angle 1$ and $\angle 5$; $\angle 2$ and $\angle 6$; $\angle 3$ and $\angle 7$; $\angle 4$ and $\angle 8$ 23. $\angle 1$ and $\angle 8$; $\angle 2$ and $\angle 7$ 25. The lines are parallel because the angles are alternate exterior angles. 27. no 29. x = 30 31. x = 60, y = 70, z = 50
33. cannot be determined 35. 10 37. 100° 39. false
41. 1620° 43. 135° 45. $\angle KML$ 47. $\overparen{KL}$, $\overparen{KM}$ 49. $\overparen{KL}$, $\overparen{KM}$

51.
Ocean Surface Areas
Arctic 4%
Indian 22%
Pacific 49%
Atlantic 25%

Chapter 6: Graphing Functions

Lesson 6-1, pages 244–247

1. 8 3. 13 5. (−2, −0.5) 7. 7.2; 3.6; (7, 9)
9. (3.5, 5); (6.5, 5); (5, 3) 11. 10 13. 7 15. (−2, 2)
17. (3, 3.5) 19. 3.2 21. 13 23. 3.2 25. 2.2
27. 1.0 29. $C(9, -5)$; $d \approx 6.3$; $r \approx 3.2$ 31. $C(4, 0.5)$;
$d = 15$; $r = 7.5$ 33. $C(6, -6)$; $d \approx 12.8$; $r \approx 6.4$
35. 136.5 mi^2 37. (−146.75, 43), (−103.5, 39),
(−60.25, 35) 39. (−8, 14)

41.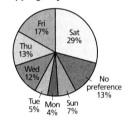
Shopping Day Preference
Fri 17% Sat 29%
Thu 13%
Wed 12%
Tue 5% Mon 4% Sun 7%
No preference 13%

43. −3 45. −1

Lesson 6-2, pages 248–251

1. $m_a = -\frac{5}{9}$; $m_b = \frac{1}{2}$ 3. $m_l = -\frac{1}{3}$; $m_m = \frac{1}{6}$

5.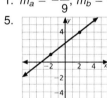
7.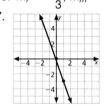

9.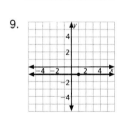

11. $-\frac{3}{4}$
13. $\frac{2}{5}$ 15. $\frac{2}{5}$ 17. −2
19. $\frac{2}{3}$

21.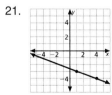
23.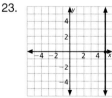

25.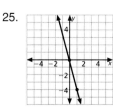
27.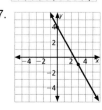

29.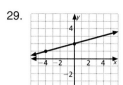
31. $x = 4$ 33. The coordinates have the same x- or y-value.
35. −2 37. 40 mph; 60 mph
39. returning home 41. 12 43. 6

Review and Practice Your Skills, pages 252–253

1. 13 3. 16.3 5. 15 7. $\left(1, -3\frac{1}{2}\right)$ 9. $\left(4, -1\frac{1}{2}\right)$
11. $\left(5\frac{1}{2}, -3\right)$ 13. 11.3 15. 8 17. (2, 0)
19. (−2, 0) 21. $C(3, 5)$; $d = 14$, $r = 7$ 23. $C(0, 2)$;
$d \approx 8.9$; $r \approx 4.5$ 25. $C\left(5, 1\frac{1}{2}\right)$; $d = 9$; $r = 4\frac{1}{2}$
27. $m_a = 0$; $m_b =$ undefined

29.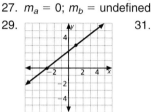
31.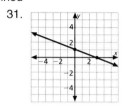

33.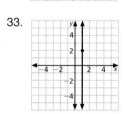
35. $\frac{6}{7}$ 37. $y = 4$ 39. $C\left(\frac{1}{4}, -\frac{1}{3}\right)$;
$d = \frac{5}{6}$; $r = \frac{5}{12}$ 41. $C\left(-1\frac{1}{2}, -1\frac{1}{2}\right)$;
$d \approx 11.0$; $r = 5.5$
43. $C\left(-7\frac{1}{2}, 4\right)$; $d \approx 2.2$; $r \approx 1.1$

45. 18.8 47. 2.8 49. 17.2

51.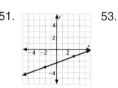
53.
55.

Lesson 6-3, pages 254–257

1. $y = \frac{1}{2}x - 2$ 3. $y = -\frac{1}{3}x + 5$
5. $y = -x$ 7.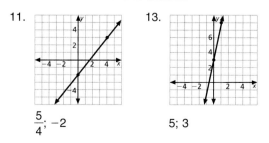
9. $y = -x - 3$

11.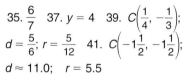
$\frac{5}{4}$; −2
13. 5; 3

15. First, plot the *y*-intercept. Then use the slope to locate one or more other points. Next draw the line. If $ax + by = c$ is the form, first solve for *y*.

17. $y = \frac{5}{11}x - \frac{12}{11}$ 19. $y = -\frac{1}{4}x + \frac{29}{4}$ 21. $x = 0$

23. $y = -x - 1$ 25. $y = -x + 6$ 27. $y = -4$

29. $y = -x - 3$ 31. $y = \frac{1}{2}x - 6$ 33. $y = \frac{5}{2}x - 3$

35. The change in barometric pressure.

Pressure decreases as altitude increases.   37. -13.5 mm

39. $y = -\frac{A}{B}x + \frac{C}{B}$; $m = -\frac{A}{B}$; *y*-int $= \frac{C}{B}$

41. 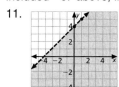   43.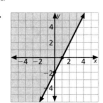

45. 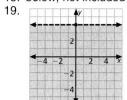   47.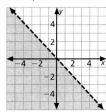

49. 72°, 162°   51. 7°, 97°

Lesson 6-4, pages 258–261

1. yes   3. no   5. below; not included   7. below; included   9. above; included

11. 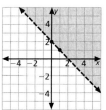   13.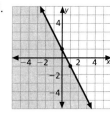

15. below; not included   17. above; not included

19. 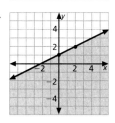   21.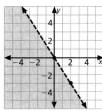

23. 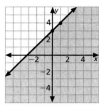   25.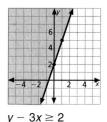

27. 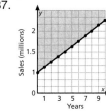   29.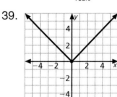

31. 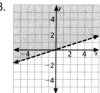   33.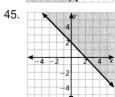

$y \le x + 3$ $y - 3x \ge 2$

35. $y < x + 1$

37. 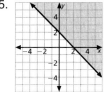 $y \ge 125{,}000x + 1{,}000{,}000$

39.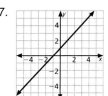

41. yes   43. yes

45. Answers will vary.

47. $\frac{1}{50}$ 49. p^9 51. m^4

53. f 55. $\frac{1}{r^8}$ or r^{-8}

Review and Practice Your Skills, pages 262–263

1. 1; 2   3. 2; 0   5. 5; -6 7. $y = \frac{2}{3}x + 4$

9. $y = -\frac{4}{3}x + \frac{2}{3}$ 11. $y = -\frac{1}{2}x + \frac{3}{2}$ 13. $y = x - 8$

15. $y = 6x - 1$ 17. $y = 5$ 19. $y = x + 4$

21. $y = \frac{1}{3}x - 4$ 23. $y = -x - 5$ 25. $y = -\frac{5}{3}x + 18$

27. $y = -\frac{1}{2}x - 1$ 29. no   31. yes   33. yes

35. above; included   37. above; included

39. below; included

41.   43.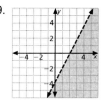

45.   47. $y \ge -x - 3$

49. 10.2   51. 13   53. $y = -2x + 2$ 55. $y = -5x - 6$

57.   59.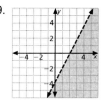

1; 1

Selected Answers

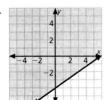

61. 63. 6.3   65. isosceles

Lesson 6-5, pages 264–267

1. no   3. no   5. yes

7. 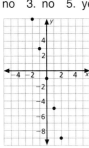   9.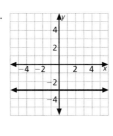

11. no; vertical line test fails
13. yes; vertical line test passes

15.   17.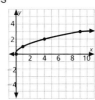

19.   21.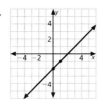

23. $c = 12t + 20$; linear function   25. 13 mL   27. -3
29. $-4, 4$ 31. $z < 6$ 33. $a \leq -6$ 35. $r < \dfrac{11}{2}$

Lesson 6-6, pages 268–271

1. 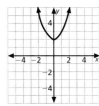   3.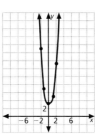

5.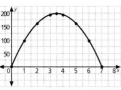

t	h
0	0
1	96
2	160
3	192
4	192
5	160
6	96
7	0

7. 3.5 sec   9. 2
11. 0

13. 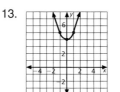   15.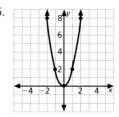

17.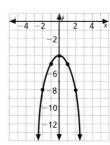

19. Answers will vary but may include 25 × 25, 30 × 20, 35 × 15, 40 × 10, etc.

21.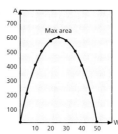

23. 625 ft²   25. -2
27. 1 or -1 29. 2.5 sec
33. $y = x^2 + 10$; $y = x^2 - 10$
35. Yes; changing the term shifts the graph horizontally.
39. Yes. If a is between 0 and 1 the parabola opens wider than the graph of $y = x^2$. It is narrower if a is greater than 1.

41. 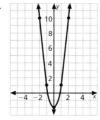   43.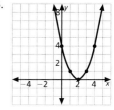

45. 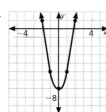   47.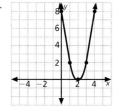

49. $\left(\dfrac{-b}{2a}, c - \dfrac{b^2}{4a}\right)$ 51. 128°   53. 128°   55. 113°
57. $27x$ 59. $13x$

Review and Practice Your Skills, pages 272–273

1. 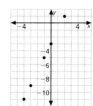   3.

5. yes   7. yes

Selected Answers

9. 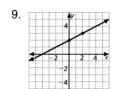 **11.**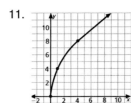

13.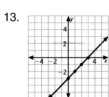

15. 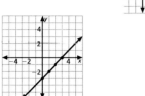 **17.**

19.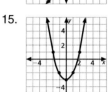

21. 1 **23.** −2 or 2 **25.** 2 sec
27. Above. Although h is negative at $t = 3$ sec, we know from experience that balls bounce.
29. $\left(2\frac{1}{16}, -2\frac{3}{4}\right)$ **31.** −1; 0
33. 2; −4 **35.** $y < \frac{3}{2}x - 2$

37. $y \leq -\frac{2}{3}x + 2$

Lesson 6-7, pages 274–275

1. c **3.** (4, 7), (−2, −5), (0, −1), (1, 1), (2, 3) **5.** $y = \sqrt{x}$
7.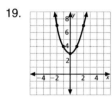
9. $x \geq 0$ **11.** all real numbers
13. Answers will vary.
15. $C(x) = 1.5 + 0.75(x - 1)$
17. If the domain is the set of natural numbers, there is no y-intercept. **19.** 23 hr
21. 665,280 **23.** 336
25. 36 **27.** ±7

Lesson 6-8, pages 276–279

1. 28 **3.** $91 **5.** 10 **7.** 50 **9.** 66 **11.** 7 **13.** 1760
15. 18.2 lb **17.** 78 m **19.** Answers will vary. **21.** direct
23. neither **25.** y is quadrupled; y is multiplied by 9
27. $\frac{y}{x} = \frac{a}{b}$; $by = ax$; $by + xy = ax + xy$; $y(b + x) =$
$x(a + y)$; $\frac{y}{x} = \frac{a + y}{b + x}$ **29.** increases **31.** decreases
33.

0	4 8
1	2 5 7 8 8 9 9
2	0 4 7 7 9
3	1

0 | 4 represents 4 calls.

35. −3b − 12 **37.** 8c − 2d
39. 4.5 **41.** $-\frac{3}{2}$

Review and Practice Your Skills, pages 280–281

1. c **3.** b **5.** (−4, −9), (−2, −7), (0, −5), (5, 0), (10, 5) **7.** (−6, −4), (−3, −1), (0, 2), (3, −1), (6, −4)
9. $f(x) = 3x$ **11.** $f(x) = |x|$ **13.** 24 **15.** 39 **17.** −4

19. $15.84 **21.** b **23.** d **25.** yes **27.** yes
29.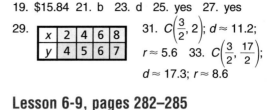

x	2	4	6	8
y	4	5	6	7

31. $C\left(\frac{3}{2}, 2\right)$; $d \approx 11.2$;
$r \approx 5.6$ **33.** $C\left(\frac{3}{2}, \frac{17}{2}\right)$;
$d \approx 17.3$; $r \approx 8.6$

Lesson 6-9, pages 282–285

1. 6 **3.** increase **5.** 5 yd/sec **7.** inverse variation
9. 320 **11.** 2.5 **13.** 1.6 min **15.** $24 **17.** inverse
19. inverse **21.** 1 newton **23.** 500 **25.** 0.8 **27.** y varies
directly as z^2 **29.** $h = \frac{132}{\pi r^2}$ **31.** 42° **33.** 117°
35. 70° **37.** 33° **39.** 19°

Chapter 6 Review, pages 286–288

1. i **3.** f **5.** l **7.** k **9.** h **11.** 1 **13.** 5 **15.** 2 **17.** 4.5
19. 2.2 **21.** $\left(-1\frac{1}{2}, 0\right)$ **23.** −4 **25.** 6 **27.** $\frac{5}{3}$ **29.** undefined
31. $y = -\frac{2}{5}x + 3$ **33.** $y = x + 7$ **35.** $y = x + 4$ **37.** $y = -2x + 11$ **39.** $y = -\frac{1}{2}x + 2$ **41.** $y = \frac{2}{3}x + 4$ **43.** yes
45. no **47.** $y \geq -2x + 2$ **49.**

51. No; vertical line test fails. **53.** Yes; vertical line test passes. **55.** 0 **57.** (−3, −7), (−2, −6), (0, −4), (1, −3), (2, −2) **59.** $y = 2x^2$ **61.** 15 **63.** 14 **65.** 7.5 **67.** $\frac{1}{4}$

Chapter 7: Coordinate Graphing and Transformations

Lesson 7-1, pages 296–299

1.  **3.**

5. $D'(-560, -154)$, $E'(-558, -155)$, $F'(-560, -159)$, $G'(-562, -158)$ **7.** $(x, y) \rightarrow (x + 6, y - 1)$
9. 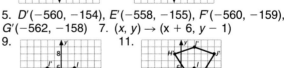 **11.**

13. 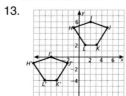 **15.** No; angles are not all congruent, sides are not all congruent. **17.** $(x, y) \rightarrow (x + 5, y - 2)$ **19.** The image should be translated right and up. **21.** The sliding door moves horizontally the appropriate number of units to open the door

wide enough for a person to access the deck. A move to the right or left depending on the position of the door is a translation.   23. Mandy translated the preimage 4 units up and 2 units right. The correct vertices are $A(1, 4) \rightarrow A'(1 + 4, 4 + 2) = A'(5, 6); B(4, -2) \rightarrow B'(4 + 4, -2 + 2) = B'(8, 0); C(7, 9) \rightarrow C'(7 + 4, 9 + 2) = C'(11, 11)$
25. No; the vertices did not all move the same number of units to the right.   27. Answers will vary.   29. $\frac{4}{13}$
31. 4   33. 12   35. 8   37. −3.5   39. 14   41. 6

Lesson 7-2, pages 300–303

1.   3.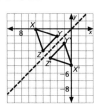

5. 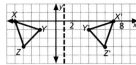   7. $y = -x + 1$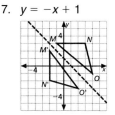

9. $(x, y) \rightarrow (x, -y)$; Quadrant I
11. 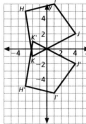   13.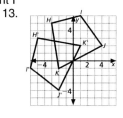

15. 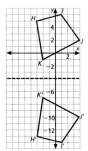   17. $(0, -4)$
19. $(8, -6)$
21. $(3, -3)$
23. $y = x + 2$

25. Draw a horizontal line at least twice the distance from A to line m from point A through the line m. Measure the distance from line m to point A. Locate point A' which is to the right of line m by measuring this distance from line m. Repeat for vertices B' and C'. Connect the points A', B', and C'.

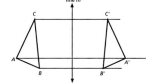

27.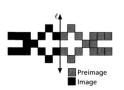

29. $(-10, 0) \rightarrow (8, 0)$
$(-5, 0) \rightarrow (3, 0)$
$(-10, 11) \rightarrow (8, 11)$
$(-5, 11) \rightarrow (3, 11)$
$(-7.5, 14) \rightarrow (5.5, 14)$

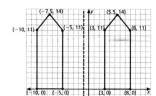

31. It is the same as $\triangle ABC$.   33. mean = 152; median = 152; mode = 145, 152; range = 118   35. mean = 32.1; median = 32; mode = 32; range = 32   37. 59°, 149°
39. 11°, 101°   41. 68°, 158°   43. 6°, 96°

Review and Practice Your Skills, pages 304–305

1.   3.

5. $(x, y) \rightarrow (x + 2, y - 4)$ 7. $(x, y) \rightarrow (x - 3, y + 2)$
9. $A'(-2, 3), B'(-1, 2), C'(-3, -1)$ 11. $A'(2, -8), B'(3, -9), C'(1, -12)$
13.   15.

17. 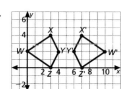   19. $(4, 4)$ 21. $(-6, -1)$
23. $(3, 0)$
25. $(x, y) \rightarrow (x + 3, y - 4)$
27. $(5, -4)$ 29. $(6, 0)$
31. $(0, -1)$

Lesson 7-3, pages 306–309

1. 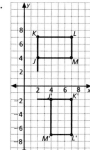   3.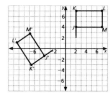

5. Draw a line segment connecting any two corresponding vertices between the image and the preimage. Construct the perpendicular bisector of the line segment. Repeat for two other corresponding vertices. The intersection of these perpendicular bisectors is the center of rotation. 7. 90° clockwise or 270° counterclockwise

9. 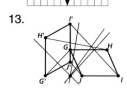 11.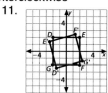

13. 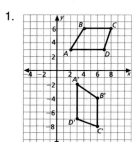 90° counterclockwise or 270° clockwise

15. $L(0, 500) \rightarrow L'(500, 0)$
$C(100, 200) \rightarrow C'(200, -100)$
$E(400, 200) \rightarrow E'(200, -400)$
$K(200, 0) \rightarrow K'(0, -200)$
$O(0, 0) \rightarrow O'(0, 0)$

17. *F* 19. *H*
21. counterclockwise
23. ASA 25. not congruent 27. 3.6

Lesson 7-4, pages 310–313

1. 1 line 3. none 5. 8 7. Yes. Answers will vary but Exercise 4 is one example. 9. yes 11. 3 13. 5
15. Any line through the center of the circle is a line of symmetry. Since a person can draw an infinite number of lines through a point, a circle would have an infinite number of lines of symmetry. 17. Answers will vary. Any palindrome which uses the digits 0, 1, 3 and 8 has line symmetry. A palindrome which uses the digits 0, 1, or 8 has rotational symmetry of order 2. 19. There are 6 lines of symmetry. This figure has order 6 rotational symmetry.
21. Each of the letters B, C, D, E, H, I, K, O and X has one horizontal line of symmetry. Each of the letters A, H, I, M, O, T, U, V, W, X and Y has one vertical line of symmetry. 23. Answers will vary. 25. Answers will vary. 27. Answers will vary. 29. Answers will vary.
31. *n* 33. 35 35. 495 37. 56 39. 108° 41. 60°

Review and Practice Your Skills, pages 314–315

1. 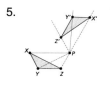 3.

5. 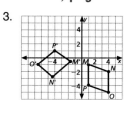 7. 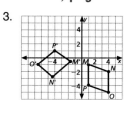 90° counterclockwise or 270° clockwise

9. 2 lines 11. 1 line 13. none 15. 4 17. 6 19. 4
21. 2 23. none

Lesson 7-5, pages 316–319

1. 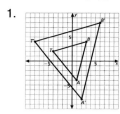 3. 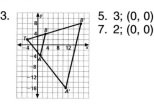 5. 3; (0, 0)
7. 2; (0, 0)

9. 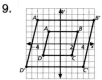 11.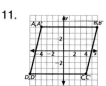

13. 13.8 15. ≈ 342.86 17. A dilation changes each figure into a similar figure not necessarily congruent to the original figure, while translations, reflections, and rotations change each figure into a congruent figure. All corresponding angles are equal in translations, reflections, rotations, and dilations. Dilations and translations have parallel corresponding sides between the preimage and the image. 19. $\frac{1}{3}$; (0, 0)
21. $5\frac{1}{3}$ in. 23. Answers will vary. Possible solution: Translation to the left 3 and up 2. Then use a dilation with center $(-1, 1)$ and a scale factor of 2. 25. 9 square units; 36 square units; 2.25 square units 27. tennis ball: 292.5 cm ≈ 3 m; baseball: 343 cm ≈ 3.43 m; basketball: 1080 cm ≈ 10.8 m 29. -5 31. 9
33. 22.5 35. -45 37. -16 39. 3; 5 41. 1; -4
43. $-\frac{5}{2}$; 3 45. $\frac{1}{2}$; 5

Lesson 7-6, pages 320–321

1. yes 3. yes 7. Sample answer:

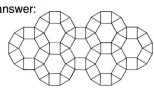

9. yes 11. yes 13. Answers will vary.
15.  17. tessellating rhombuses and four-pointed stars. 19. Answers will vary. 21. 28 23. $\frac{13}{5}$
25. 3 27. $\frac{25}{6}$ 29. $\frac{16}{37}$

Review and Practice Your Skills, pages 322–323

1. 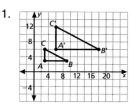 3.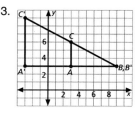

5. $G'(4,12)$, $H'(16,4)$, $I'(4,4)$ 7. $N'(-12,0)$, $O'(0,0)$, $P'(0,-6)$, $Q'(-12,-6)$ 9. yes 11. yes

13.   yes

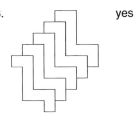

15.

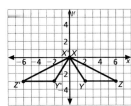

17. 2 lines
19. 1 line

Chapter 7 Review, pages 324–326

1. f 3. a 5. l 7. j 9. i 11.
13. $(x, y) \rightarrow (x - 5, y - 3)$

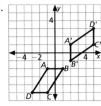
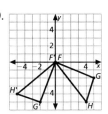

15.

17.
19.

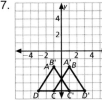
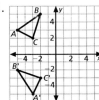

21.

23. 2 lines 25. 6 27. 2
29. $(x, y) \rightarrow (x - 5, y - 3)$
31. 93.75
33.

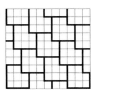

Chapter 8: Systems of Equations and Inequalities

Lesson 8-1, pages 334–337

1. $-\dfrac{5}{2}, \dfrac{2}{5}$ 3. 1, -1 5. neither 7. $y = 3x - 13$
9. $y = \dfrac{5}{7}x - \dfrac{17}{7}$ 11. $y = -\dfrac{4}{3}x + 11$ 13. $-\dfrac{1}{2}, 2$

15. $-\dfrac{7}{5}, \dfrac{5}{7}$ 17. undefined, 0 19. If the line is a horizontal line, then the y-coordinates will be the same in each point. If the line is a vertical line, then the x-coordinates will be the same in each point. 21. perpendicular 23. parallel
25. neither 27. $y = -\dfrac{2}{3}x - \dfrac{44}{3}$ 29. $y = 7x - 6$
31. $y = -\dfrac{5}{3}x - 3$ 33. $y = -\dfrac{1}{5}x - 1$ 35. $y = \dfrac{7}{2}x + 14$
37. $y = \dfrac{3}{2}x + \dfrac{3}{2}$
39.

No. The planes will not crash since their paths are parallel.
41. $y = -\dfrac{2}{3}x + 3$

43.

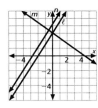

slope of $\overline{AD} = -\dfrac{5}{4}$, slope of $\overline{AC} =$ 0, slope of $\overline{BD}$ is undefined Opposite sides are parallel. Diagonals are perpendicular. The quadrilateral is a rhombus.
45. 14 47. 71°

Lesson 8-2, pages 338–341

1. no 3. yes
5.

$(3, -1)$
7. Julian: $d = 8t$;
Leticia: $d = 12t - 10$
9. no 11. yes 13. no

15.

$(4, 3)$ 17.

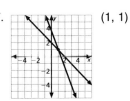

$(1, 1)$

19.

$(5, 3)$ 21.

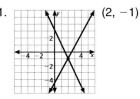

$(2, -1)$

23. no solution 25. (8.14, −0.29) 27. c, the police car
29. 1 solution 31. no solution
33. $y = -x + 7$ $m_1 = -1, b_1 = 7$
    $y = 2x - 5$ $m_2 = 2, b_2 = -5$
    $x = \dfrac{b_2 - b_1}{m_1 - m_2}$
    $x = \dfrac{-5 - 7}{-1 - 2} = \dfrac{-12}{-3} = 4$
35. 9.2 37. 11.2

Selected Answers

Review and Practice Your Skills, pages 342–343

1. $\frac{5}{4}, -\frac{4}{5}$ 3. $-\frac{2}{3}, \frac{3}{2}$ 5. $\frac{3}{7}, -\frac{7}{3}$ 7. neither   9. parallel
11. neither   13. $y = 3x - 11$ 15. $y = 2x + 5$
17. $y = -3x + 6$ 19. no   21. yes   23. no

25. 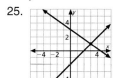   (3, 1)   27. 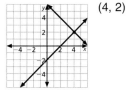   (4, 2)

29.   (0, 3)   31. (4, 3)
33. (1, 1)   35. no

37. 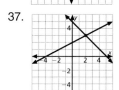   (2, 3)   39. 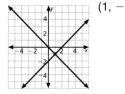   (1, −1)

41. $y = x + 1$ 43. $y = -\frac{2}{3}x + 2$ 45. $y = -3x + 5$

Lesson 8-3, pages 344–347

1. (7, 0)   3. (4, 3)   5. $\left(-\frac{1}{4}, \frac{5}{4}\right)$ 7. A hotdog cost $2
and a drink costs $1.   9. (0, 1)   11. (3, 1)   13. no
solution   15. (4, 1)   17. (−1, 2)   19. (2, 1)   21. (9, 15)
23. Answers will vary.   25. 1500 men   27. infinitely many
solutions   29. no solution   35. $x = 2, y = -1, z = 5$
37. $r = -3, s = 2, t = 1$ 39. Answers will vary.
41.   43.   45.

47. 924

Lesson 8-4, pages 348–351

1. (4, −3)   3. (−4, 6)   5. (1, 3)   7. length = 6 m;
width = 10 m   9. (0, −1)   11. (5, −3)   13. infinitely
many solutions   15. (1, 0)   17. (8, −1)   19. $\left(\frac{29}{10}, \frac{-2}{15}\right)$
21. (−9, −5)   23. Answers will vary.   25. 41 hits
27. 3 pepperoni and 2 extra cheese   29. infinitely many
solutions   31. One equation is a multiple of the other.
These are infinitely many solutions in each case.
33. Answers will vary.   35. 5   37. $\frac{5}{8}$ 39. $\frac{5}{3}$ 41. 138°
43. 42°   45. 42°

Review and Practice Your Skills, pages 352–353

1. (2, 4)   3. $\left(3, \frac{3}{2}\right)$ 5. (2, 3)   7. (2, −3)   9. $\left(\frac{1}{2}, -1\right)$
11. (0, 3)   13. (3, 2)   15. no solution   17. (−1, 1)

19. (4, 3)   21. $\left(0, \frac{1}{9}\right)$ 23. (−2, 4)   25. (2, 2)   27. (2, −2)
29. (10, 6)   31. $\left(4, \frac{18}{5}\right)$ 33. 55°, 125°   35. $y = 3x + 1$
37. (3, 2)   39. (2, 5)   41. (4, −2)   43. (2, 0)   45. (1, 5)

Lesson 8-5, pages 354–357

1. 8   3. −4.65   5. (−3, −2)   7. (−62, 39)   9. (−1, 4)
11. 35 female members   13. 32   15. −171   17. $\left(4, \frac{18}{5}\right)$
19. (−3.5, 2.5)   21. (5, −2)   23. (−4, −2)   25. (2, 3)
27. (−4, −1)   29. The congruent sides are each 10 m.
The third side is 15 m.   31. tennis: 50 min, swimming:
20 min 33. Answers will vary.   35.

37.  $c > -6$

39. $g \le 1$

41. $d > 6$

45. $a \ge \frac{8}{3}$

Lesson 8-6, pages 358–359

1.   3.   5. 34

7. 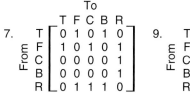   9.

T represents Towertop Restaurant
F represents Forest Theater
C represents Cascade Falls Water Ride
B represents Bootjack Camp Museum
R represents Redwood Rodeo

11. 75   13.   15.

17.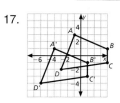

Review and Practice Your Skills, pages 360–361

1. 2 3. -31 5. -44 7. 47 9. $(3, -1)$ 11. $(-1, 3)$
13. $(7, -1)$ 15. $(1, 1)$ 17. $(1, -2)$ 19. $(-2, 2)$

21.

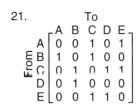

23.

25. Denver and Philadelphia 27. $-\dfrac{1}{2}, 2$ 29. $-\dfrac{3}{4}, \dfrac{4}{3}$

31.

$(-2, -4)$ 33. $(2, 3)$ 35. $(8, -7)$

Lesson 8-7, pages 362–365

1. $y \le -3x + 2$; $y > \dfrac{1}{3}x - 1$ 3. $y < 2x + 2$; $y > -\dfrac{1}{2}x$
$+ 1$ 7. Answers will vary, but a possible answer follows.
Choose a point above or below the line. Usually choosing
the origin is a good point. Substitute the x and y values
of the coordinate into the original inequality. If it is a true
statement, shade the area that includes the point. If it is a
false statement, shade on the other side of the line. To
check the solution, choose another point in the shaded area
to verify that it results in a true statement.

5.

9. $y \ge -4x + 2$; $y \le -x - 2$

11.

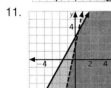

13.

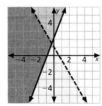

15.

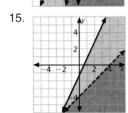

17.

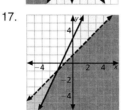

19.

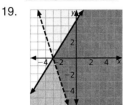

21.

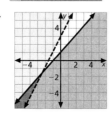

23. $x + y > 10$; $10x + 15y < 200$ 25. Answers will vary,
but a possible answer follows. A coordinate that satisfies
the inequalities is $(3, 9)$. A coordinate that does not satisfy
the inequality is $(5, 3)$. Jill can buy three \$10 CD's and
nine \$15 CD's. 27. $x + y \le 2000$; $0.05x + 0.10y > 150$
29. $(400, 1400)$; $(1000, 600)$ 31. $x > -2$; $x < 5$
33. $y \ge -3$; $y \le \dfrac{2}{3}x - 2$; $y > -\dfrac{1}{3}x + 3$

35.

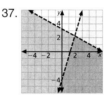

37.

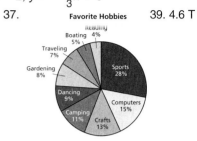

39. 4.6 T

Chapter 8 Review, pages 366–368

1. f 3. h 5. a 7. k 9. i 11. perpendicular
13. $y = 2x - 5$ 15.

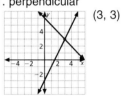

$(3, 3)$

17. $(-3, -1)$ 19. $(0, 5)$
21. $(2, 1)$ 23. $(7, 4)$
25. $(2, -1)$ 27. -6 29. -7
31. 30 CDs, 10 videos 33. 1
35. $y < x + 1$, $y \le -x + 1$

37.

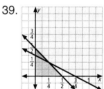

39.

Chapter 9: Polynomials

Lesson 9-1, pages 376–379

1. $3x^3 + 9x^2 - 2x + 1$ 3. x^2y^2 xy^3 5. $-k^3 + k^2 -$
k 7. $11h^2 + 3h$ 9. $-7jk^2 + 3jk$ 11. $8x + 5$ 13. $2x^4$
$+ 3x^3 + x^2 + x$ 15. $-5x^3 - 3x^2y + 2xy^2 + y^3$

17. $x^3y + x^2y^2 + xy^4$ 19. $\dfrac{19}{28}x^2$ 21. $6y^3 - 4y^2$

23. $2x^3 + 11x$ 25. $4n^2 + 8n$ 27. $4.6x - 2.2$
29. $6hk + 10k$ 31. $12x^2y + 10xy - 6$ 33. $-1.9m +$
$11.6n - 2.3$ 35. $x - 2.4$ 37. $15r + 46t$ 39. $(3x^2 -$
$22y - 71)$ ft 41. $16x - 10$ 43. $b^3 + 4a^2b^2 - 4$

45. Answers will vary. 47. 9 in. 49. $y = 3x - \dfrac{5}{2}$

51. $y = -3x + 9$ 53. $y = 6x + 21$

Lesson 9-2, pages 380–383

1. $27ef$ 3. $-8abc$ 5. $-\dfrac{4}{15}y^7$ 7. $16p^{20}$ 9. $64a^9b^6c^{15}$

11. $2x(-y - x)$ 13. Both monomials involve exponents.
The monomial $a^m \cdot a^n$ calls for the use of the product rule
for exponents where m and n are added to form a^{m+n}.
The monomial $(a^m)^n$ calls for the use of the power rule for
exponents where m and n are multiplied to form a^{mn}. The
difference to note is that in one case m and n are added
and in the other case they are multiplied. 15. $-8ab$

Selected Answers

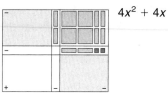

17. $5hkxy$ 19. $12tw$ 21. $10a^5b^6$ 23. $-24x^2y^2$
25. $2.58xy^6$ 27. $15x^7y^3$ 29. $-a^2b^2c^2$ 31. $-\dfrac{3}{8}p^3q^3$
33. $8h^9$ 35. $25x^4y^2$ 37. $a^6b^6c^{12}$ 39. $8a^{16}$ 41. $-40x^{11}$
43. $-a^{18}b^9$ 45. $128h^{13}$ 47. $6a^2b^3$ 49. $30x^2$ 51. $3x + x = 4x$ 53. $100x^4$ ft^2 55. $2a^3b$, $3ab$ 57. $(2x^2)(3y^2) - (2x)(2xy) = 6x^2y^2 - 4x^2y$ 59. $32a^4b^3$ 61. 3   63. $16a^4b^8$
65a. $1.5x$ mi   b. $2.5x$ mi   c. $2x$ mi   d. $3x$ mi   e. $2.5x$ mi
f. $2x$ mi   67. mean = 41.9, median = 40, mode = 35
69.

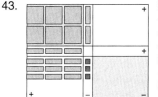

Review and Practice Your Skills, pages 384–385

1. $5x^3 + 2x^2 + 4$ 3. $-3x^3 + 7x^2y - xy^2 + y^3$ 5. $6x^3 - 4x^2 + 9x - 10$ 7. $5c$ 9. $4m$ 11. $5b + 4d$ 13. $-k^2 + 4k$ 15. $2a^2$ 17. $5.3x - 9$ 19. $2x - 7$ 21. $2ab^2 - 7ab + 6$ 23. $36ab$ 25. $-9cb$ 27. $8b^4c^8$ 29. $-2s^2t^2u^2$
31. $10g^7$ 33. $125g^6$ 35. x^{10} 37. $e^4f^8g^4$ 39. $675h^{14}$
41. $81p^{12}$ 43. $x^4y^8z^4$ 45. $-18a^6b^6$ 47. $72n^5$ ft^2
49. $\dfrac{27}{2}x^3$ m^2 51. $4t - 10$ 53. $a - 9b$ 55. $-27x^3y^9$
57. $31p^7$ 59. $-a^6$ 61. $6b^3c^5$

Lesson 9-3, pages 386–389

1. $-3a$ 3. $\dfrac{-3yz}{4}$ 5. $p - 2r$ 7. $2k^2m - 1$
9. $-3c^2d^2 + d - 2$ 11. $\dfrac{2xy - 6x}{2x}$ 13. $4x$ 15. $-\dfrac{10}{7}$
17. $\dfrac{b}{2c}$ 19. $15xy$ 21. $11abc^3$ 23. 3   25. $-6b^4$
27. $a + 5$ 29. $3a - 2$ 31. $2 - 3x$ 33. $1 - 3a^2 + 5a^4$
35. $5w^2 - 3w - 6$ 37. $-5ab^2 + 2a^2b^3 - b$ 39. $3x - 4y + 2z - 1$ 41. $6p$

43.

| 43. | | $2x - 3$ |

45. $4p$
47. $(4x^2 + 2x)$ mi
49. $18xy$ 51. 4   53. $\dfrac{a}{2}$
55. 2   57. DI, GI, FI, JI, OI, TI, NI, SI, MI, RI, HI
59. E

Lesson 9-4, pages 390–393

1. $6x^2 + 24x$ 3. $12k^5 + 12k^7 + 36k^3$
5. $-4p^3 + 8p^2 - 20p$ 7. $-3a^2b^2 - 4a^6b$
9. $21a^2 - 18a$ 11. $d = 5(56 - m)$
13. $6m^2 + 3mn$ 15. $6a^2 + 8a$ 17. $24p^3 - 88p^2$
19. $15x^2 + 10x^3$ 21. $-33n^{10} - 132n^8$
23. $27c^2 + 9c^3 + 54c^4$ 25. $7x^3 + 14x^2 + 7x$
27. $-28r^4 + 8r^2 - 40r$ 29. $-24x^3y^3 + 12xy^4$
31. $12k^7 - 3k^6 + k^5$ 33. $3.51c^7 + 1.17c^6 - 4.42c^5$
35. $-96t^3 - 40t^4 + 8t^9$ 37. $24m^3 + 12m^2n + 4mn^2$
39. $5h^2 + 4$ 41. $10x - 15$ 43. $5w^2 - 20$
45. $26x^2y + 6xy^2$ 47. $16(8.50 + d) = 136 + 16d$ dollars
49. $w(2w - 40) = (2w^2 - 40w)$ m^2 51. \$14

53. $48x^3 - 20x^2$ 55.

| 55. | | $4x^2 + 4x$ |

57.

| 57. | | $3xy + 3x$ |

59. $2.5x^5 - 23x^4 - 11.5x^3$ 61. $-5x^6y^6 - 10x^9y^4 + 5x^3y^6 - 15x^4y^3$ 63. $12(16 + n) = (192 + 12n)$ in.^{2}
65. $13.5x$ mi   67. 192   69. $\dfrac{1}{3}$ 71. 9   73. 36   75. 18

Review and Practice Your Skills, pages 394–395

1. $4b$ 3. $\dfrac{7}{4}$ 5. $12x^2 - 9x + 6$ 7. $12x^5yz^2$ 9. $-7a + 4$ 11. $9xy$ 13. $\dfrac{-e^2 + 6e + 4}{e}$ 15. w^2 17. $4q$
19. $18a^2 + 6ab$ 21. $-28b^2 + 35bd^2$ 23. $7v^5 + 4v^4$
25. $3xy^3 - 9x^2y + 3xy$ 27. $10m^3 + 2m^2n - 8mn^2$
29. $\dfrac{2}{3}d^4 + \dfrac{1}{10}d^3 + \dfrac{3}{2}d^2$ 31. $10b + 35$ 33. $13r^2s - 4rs^2$ 35. $-12x^4 + 24x^3 - 96x^2 + 84x$ 37. $63n + 36$
39. $21b^2 - 18b$ 41. $18x^2 + 54x$ 43. $12y^3 + 2y^2$
45. $2a^2b$ 47. $15a^3b + 3a^2b^2$ 49. $27x^3y^9$ 51. $-5b^2 + 6b$ 53. $3h - 7$

Lesson 9-5, pages 396–399

1. $x^2 + 7x + 10$ 3. $p^2 - 36$ 5. $y^2 + 23y + 132$
7. $p^2 + 11p - 42$ 9. $7y^2 + 35y + 36$
11. $(x + 3)(1 - 2x) = -2x^2 - 5x + 3$ 13. $a^2 + 5a + 6$
15. $p^2 - 2p - 3$ 17. $h^2 + 4h - 21$ 19. $x^2 - 8x + 15$
21. $c^2 - 25$ 23. $k^2 - 18k + 81$ 25. $-x^2 + 36$ 27. $t^2 + 19t + 84$ 29. $w^2 - 17w + 72$ 31. $10p^2 + 38p - 60$
33. $x^2 + 16x - 80$ 35. $16k^2 + 32k + 16$ 37. $49y^2 - 7y - 12$ 39. $30x^2 + 61x + 30$ 41. Yes, she is correct.
43. $x^3 - 2x^2 - 3x$ 45. $25n^3 - 42n^2 - 102n$ 47. $(5 - v)^2 = 25 - 10v + v^2$ 49. 12 in. by 18 in.   51. $5k^2 + 3k - 2$
53. $x^3 + 3x^2 + 3x + 1$ 55. \$1123.60
57. $\left(\dfrac{3}{2}, 5\right)$ 59. $(33, -42)$

Lesson 9-6, pages 400–401

1. 13   3. \$12.60   5. 1545 mi^2 7. 7:00 A.M.
9. Allegheny   11. 13.5 ft   13. x, 5   15. c^2, $3c$, 6
17. $4w^2y$, $5wk^2$, 1   19. $7x$ 21. 180   23. $-336x^3z^2$
25. $-4xz$

Review and Practice Your Skills, pages 402–403

1. $x^2 + 8x + 15$ 3. $m^2 - 4m - 21$ 5. $n^2 + 5n + 6$
7. $8x^2 - 9x - 14$ 9. $20x^2 - 22x + 6$ 11. $9x^2 + 18x + 9$
13. $25y^2 - 120y + 144$ 15. $30a^3 - 38a^2 + 12a$
17. $11x^2 - 47x + 1$ 19. $n^3 - 3n^2 - 10n$ 21. $b^4 + 7b^3 - 30b^2$ 23. $4y^2 + 32y - 596$ 25. 150   27. 333

29. b, a 31. $3c^2, 9c, 1$ 33. $4xy^2, 3xy, 1$ 35. $125a^6b^3$
37. $\dfrac{9x^2 - 1}{y}$ 39. $30m^2 - 18m$ 41. $y^2 - 16y + 64$
43. $3c^2 - 22c - 16$ 45. $4x^2 + 2x + \dfrac{1}{4}$

Lesson 9-7, pages 404–407

1. $7(w - 3)$ 3. $3a(3a - 2)$ 5. $9ab(5a - 3b)$
7. $A = P(1 + r)$ 9. $P = 2(l + w)$ 11. $3(4k + 5)$
13. $7e(e + 3)$ 15. $x(x - y)$ 17. $5y^3(y - 4)$ 19. $100(c - 2)$
21. $7x(y - 8z)$ 23. $5(n^3 - 6m^2 - 3)$ 25. $x(y + z + 2)$
27. $2xy(a - 2b + 3c)$ 29. $8(h^2 - 2h + 3)$
31. $5ab(3a^2 + 4a - 2)$ 33. $m^2n^2(6mn + 3m + 1)$
35. 90   37. Exercise 35   39. $D = \dfrac{1}{2}n(n - 3)$
41. $\dfrac{1}{2}(Z - N)$

43. 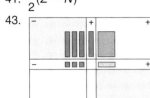 $y(x - 3)$ 45. $9m^2n(2mn + 5n^2 + 3m^2 - 6n)$
47. $4ab(12 - 10a^2b + 6ab^2 + 7ab)$ 49. $16n + 6$;
$2(8n + 3)$ 51. $16c^2d + 10cd^2$; $2cd(8c + 5d)$ 53. no
55. no   57. no   59. $x^2 - 25$
61. $p^2 - 16$ 63. $y^2 - 121$
65. $(v + 3)(v - 3)$ 67. $(m + 6)(m - 6)$ 69. $(2x + 5)(2x - 5)$ 71. 2   73. -1 75. 1   77. $x + (x + 2) + (x + 4) = -54; -20, -18, -16$

Lesson 9-8, pages 408–411

1. no   3. yes   5. $(m + 9)^2$ 7. not factorable
9. not factorable   11. $x^2 - 8x + 16 = (x - 4)^2$ 13. yes
15. no   17. no   19. $(k - 9)^2$ 21. $(x + 8)(x - 8)$
23. $(h + 15)(h - 15)$ 25. $(y + 1)(y - 1)$ 27. $(e - 3)^2$
29. $(p + 16)(p - 16)$ 31. no   33. $(h + 12)^2$
35. $(p + 12)(p - 12)$ 37. no   39. $(c - 4)^2$ 41. no
43. $(b + 13)^2$ 45. no, $(x + 3)^2$ 47. $3(m + 3)(m - 3)$
49. $2(m + 8)^2$ 51. $3a(k + 10)(k - 10)$
53. 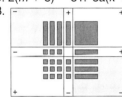 $(y - 3)^2$
55. Sumi is correct. It is a perfect square trinomial.
57. $\dfrac{(x - 6)5 + 30}{x}$
59. $(2n + 5)(2n - 5)$
61. $(3m + 5)^2$ 63. $(5x - 4)^2$
65. 7.5 mi   67. ≈ 6.2 mi
69. $36\dfrac{2}{3}$ yd^2 71. $-99n^3 + 27n^2$ 73. $12x^3y - 16xy^2$
75. $27a - 39b$

Chapter 9 Review, pages 412–414

1. g   3. f   5. a   7. e   9. d   11. $x^4 + 2x^2 - 6$
13. $5x^7 - 10x^6 + 3wx^2 + 6w^3x$ 15. $4k^2 + 3k$
17. $-6m^3n$ 19. $-6a^3b^7c$ 21. $-20x^4y^3$ 23. $\dfrac{3y^2}{4}$
25. $p^3 + 4p^2q + 5p$ 27. $5y^3 + 4xy^2 - x^2z$ 29. $3x^2y$
31. $2x^2 + 4$ 33. $3w^2 - 108$ 35. $-2d^2 + 19d$
37. $-6c^3 - 19c^2 - 8$ 39. $160 - 0.02x$ 41. $2t^2 - 6t + 4$
43. $e^2 + 10e + 21$ 45. $4y + 108$ 47. 16   49. 7:40 A.M.
51. $4(3x - 4)$ 53. $3m^2n(8 - m + 5m^2n^2)$ 55. $5x(x^2y^2 + 2xy + 5)$ 57. 42   59. 72   61. -50 63. $(k + 6)^2$
65. $(x - 8)(x + 8)$ 67. $(m - 5)^2$

Chapter 10: Three-Dimensional Geometry

Lesson 10-1, pages 422–425

1. Right rectangular prism; any face can be considered a base   3. Answers will vary.   5.   7.

9. right square pyramid; square $DFGH$ 11. right hexagonal prism; hexagons $ABCDEF$ and $GHIJKL$
13. Answers will vary.
15. 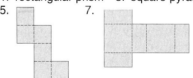   19.

21. Prisms and pyramids are both polyhedrons and are both named by the shape of their base. However, prisms have two parallel bases and pyramids have one base. The lateral faces of a prism are parallelograms and the faces of a pyramid are triangles.   23. cube or rectangular prism; 6 faces, 8 vertices, and 12 edges   25. right cone; no faces, vertices or edges   27. right square pyramid; 5 faces, 5 vertices, and 8 edges   29. cylinder
31. cone with cut-off tip   33. True. A cube is a closed three-dimensional figure made up of square faces.
35. False. The faces must be polygonal.   37. Yes. Every face is a polygon.   39. Yes. Every face is a polygon.
41. 6, 8, 12; 2   43. 6, 6, 10; 2   45. The number of edges in a polyhedron is two less than the sum of faces and vertices.   47. $-\dfrac{1}{2}$ 49. $\dfrac{5}{3}$ 51. vertical line; undefined   53. 0; horizontal line   55. 80°   57. 100°
59. 100°

Lesson 10-2, pages 426–429

1. rectangular prism   3. square pyramid
5. 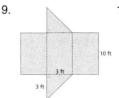   7.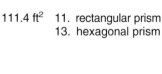

9. 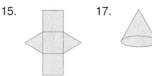   111.4 ft^2 11. rectangular prism
13. hexagonal prism

15.    17.

19. 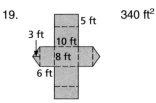 340 ft²

21. 164.9 ft²   23. Nets help you visualize the three-dimensional object in two dimensions. Nets also help you account for every part of a three-dimensional figure, especially figures with curved surfaces. Nets can transform some curved surfaces into flat surfaces.   25. 84 m²
27. 113 ft²   29. Answers will vary.   31. 61   33. 13
35. 8   37. 57   39. 350   41. $24r^5g^2$ 43. $10c^4d^5$
45. $-18s^4t^5q$

Review and Practice Your Skills, pages 430–431

1. d   3. b   5. f   7. square pyramid; 5, 5, 8
9. pentagonal pyramid; 6, 6, 10   11. cone
13.   15.   17. 156 cm²

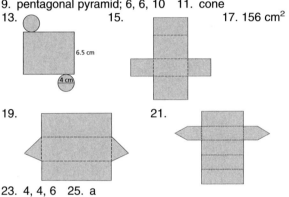

23. 4, 4, 6   25. a

Lesson 10-3, pages 432–435

1. 572 ft²   3. 1752.12 ft²   5. 1631.25 ft²   7. 10.88 cm²
9. 71.22 m²   11. 3.19 m²   13. 406.19 m²   15. 251.2 m²
17. 279 in.²   19. Wrapping a gift requires more paper than just the surface area of the box.   21. 282.6 cm²;
678.24 cm²;1808.64 cm²; 5425.92 cm²   23. 112 m²
25. 336 m²   27. 115 cm²   29. $y = 3x + 1$ 31. $y = 2x + 3$
33. $r^2 - r - 12$ 35. $b^2 - 8b + 16$ 37. $v^2 - 4v - 32$
39. $6k^2 + 10k - 4$

Lesson 10-4, pages 436–439

1.   3.   5. Answers will vary.

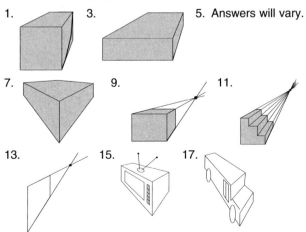

19.   21. anywhere that the box does not obstruct it   23.

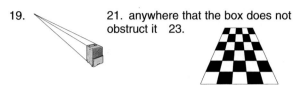

25. yes   27. no   29. yes   31. yes
33.

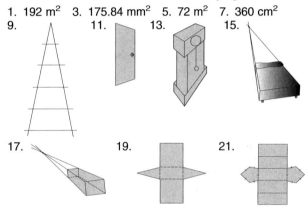

35. 1   37. 2
41. yes   43. yes
45. yes   47. yes
49. $3z(5z + 6)$
51. $4(2y^5 - 6y^4 + 7)$
53. $28b^2c(b - 3c^4)$
55. $8g^2(3g^2h - 5h^2 - 2g^3)$

Review and Practice Your Skills, pages 440–441

1. 192 m²   3. 175.84 mm²   5. 72 m²   7. 360 cm²
9.   11.   13.   15.

17.   19.   21.

23. 7

Lesson 10-5, pages 442–445

1.   3.   5.

7. 9; 2   9. 32 yd²
11.   13.   15.

17.   27.   29.

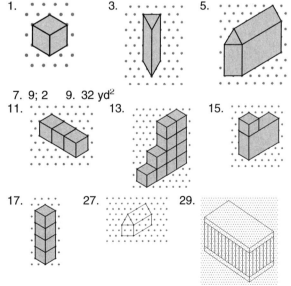

19. 34   21. 8; 24   23. 6; 22   25. 5; 22   31. 60
33. 72   35. (3, −1)   37. (−3, 5)   39. (2, 3)   41. HH, HT, TH, TT   43. 11, 12, 13, 14, 15, 21, 22, 23, 24, 25, 31, 32, 33, 34, 35, 41, 42, 43, 44, 45, 51, 52, 53, 54, 55, 61, 62, 63, 64, 65

Lesson 10-6, pages 446–449

1.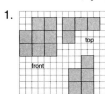

3.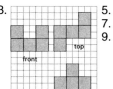

5. right-side
7. front
9.

3	2
1	1
1	

11.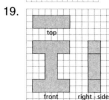

13.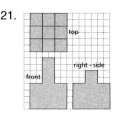

15.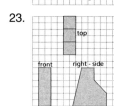

17. Answers will vary.

19.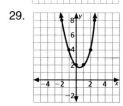

21.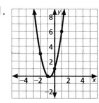

23.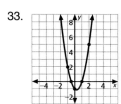

25. Answers will vary.
27. 1280 in.²

29.

31.

33.

35.

37. $\dfrac{6x}{3}$
39. $12n + 2$
41. $x - 9$

Review and Practice Your Skills, pages 450–451

1. 12   **3.** 313.6 in.²   **5.**

7.

9.

3	3	2
1	1	1
1	1	

11.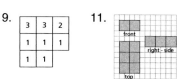

13.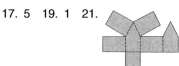

15.

17. 5   **19.** 1   **21.**

23.

Lesson 10-7, pages 452–455

1. 552 m³   **3.** 1078 m³   **5.** 6 in.   **7.** 4.5 cm
9. 658.8 m³   **11.** 29,791 cm³   **13.** 1122 ft³   **15.** 480 m³
17. Answers will vary.   **19.** 5625 cm³   **21.** $288
23. 99,944 m³   **25.** 1976 cm³   **27.** 330 in.³   **29.** 469 in.³
31. $8z^2 + 2zw$ **33.** $15r^2s - 20rs^2$ **35.** $43p^2 - 34pr$
37. $16y^2z - 4yz^2$

39.   **41.**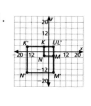

Lesson 10-8, pages 456–459

1. 3151 m³   **3.** 816 cm³   **5.** 5 cm   **7.** 117,750 lb
9. 7235 m³   **11.** 1417 m³   **13.** 1884 ft³   **15.** 2 in.
17. 359.0 cm³   **19.** 489.8 cm³   **21.** approximate volume
of Earth ≈ 1,086,230,341,000 ≈ 1.086 · 10¹² early
astronomers' estimate ≈ 588,678,827 ≈ 5.887 · 10⁸
The early astronomers were not even close.   **23.** 15 cm
25. 41.6 cm³   **27.** 332.9 cm³   **29.** a and b
31.

radii	volume
2 m	33.49 m³
4 m	267.95 m³
8 m	2143.57 m³
16 m	17,148.59 m³

When the radius is increased by a factor of 2, the
volume is increased by a factor of 8.
33. 21   **35.** 112   **37.** 6   **39.** 18   **41.** 3   **43.** 35°   **45.** 270°

Review and Practice Your Skills, pages 460–461

1. 90 m³   **3.** 390 m³   **5.** 208 m³   **7.** 216 cm³   **9.** 540 cm³
11. 2512 m³   **13.** 4924 in.³   **15.** 707 m³   **17.** 16 mm
19. 17 ft²
21.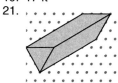

23. 1205.8 in.²; 2411.5 in.³
25. 456 m²; 408 m³
27. 74.4 ft²; 37.4 ft³
29. square pyramid; 5, 5, 8
31. triangular prism; 5, 6, 9

33.

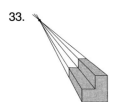

Lesson 10-9, pages 462–463

1. 168.0 ft 3. 98.9 in.² 5. 7 lb 7. 907.5 ft³ 9. Answers will vary. 11. 7*b* 13. 3*g*² 15. $\frac{7}{2}$ 17. 20 19. 151,200
21. 19,958,400 23. 720 25. 990

Chapter 10 Review, pages 464–466

1. g 3. h 5. k 7. j 9. d 11. right triangular prism
13. rectangles *ACFD*, *BCFE*, and *ABED* 15. 9 edges
17.

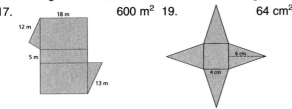

600 m² 19. 64 cm²

21. 703.38 m² 23. ≈ 270.43 cm²
25. 27.

29. 8; 28
31. 33. 35.

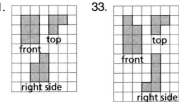 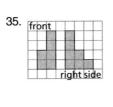

37. 1725 cm³ 39. the rectangular pan 41. 203.1 cm³
43. 392.5 in.³ 45. 2 gal 47. ≈ 91,394,008 ft³

Chapter 11: Right Triangle Trigonometry

Lesson 11-1, pages 474–477

1. yes 3. no 5. 7 in. 7. Answers will vary. 9. yes
11. yes 13. 24 m 15. 30° 17. 132° 19. 28 in.
21. *AB* : *DE*, *BC* : *EF*, *CA* : *FD* 23. Matt; △*ABC* ~ △*DBE*
25. always 27. sometimes 29. 675 cm² 31. 64
33. 42 35. SSS

Lesson 11-2, pages 478–481

1. 35 ft 3. 250 ft 5. 70.8 ft 7. 30 ft 9. 30 ft 11. 60 m
13. ≈ 6.7 m 15. 4.5 ft 17. 139.5 ft 19. 8 : 7; Yes. This ratio will not exist if the sun is directly over either or both objects or if the sun is not shining. 21. ≈ 17.3 m

23. (−4, 6) 25. (−3, −7) 27. $\frac{1}{2}$

Review and Practice Your Skills, pages 482–483

1. yes 3. no 5. 4 7. 70° 9. 15° 11. 8 cm 13. 25 m

Lesson 11-3, pages 484–487

1. 9.4 m 3. 4.9 m 5. 37 ft 7. 14.4 ft 9. 11.7 m
11. 26 ft 13. 22.4 cm 15. 41 ft 17. 52 ft 19. 127 ft
21. 4.2 in. 23. 12 ft 25. 6.3 ft 27. 17 in. 29. 555 ft
31. (4, 1) 33. (−1, 4) 35. (−3, 2) 37. (2, −4)
39. dashed, below 41. dashed, above 43. solid, below
45. solid, above

Lesson 11-4, pages 488–491

1. $\frac{3}{5}$ 3. $\frac{3}{5}$ 5. $\frac{4}{3}$ 7. $\frac{3}{5}$ 9. $\frac{3}{5}$ 11. 0.7431 13. $\frac{5}{13}$
15. $\frac{5}{13}$ 17. $\frac{12}{5}$ 19. 0.6157 21. 11.4301
23. 0.9877 25. 0.8387 27. ≈ 0.5396 29. $\overline{FG}$ 31. $\frac{4}{5}$
33. sin ≈ 0.2678, cos ≈ 0.9635, tan ≈ 0.2780
35. tan ≈ 0.1409, cos ≈ 0.9902 37. increases
39. increases 41. false 43. true 45. 18*r*³ + 24*r*
47. 45*p*³ − 53*p* 49. 24*g*²*h*² − 12*g*³*h* + 20*g*²*h*³
51. 56 53. 3 55. 15 57. −2

Review and Practice Your Skills, pages 492–493

1. 36.3 3. 10.2 5. 13.9 7. 7.1 in. 9. $\frac{8}{15}$ 11. $\frac{15}{17}$
13. $\frac{15}{8}$ 15. $\frac{3}{5}$ 17. 0.8660 19. 0.5 21. 0.9877
23. 0.7071 25. The sine of an acute angle is equal to the cosine of its complement. 27. $\frac{4}{5}$ 29. no 31. yes 33. 9

Lesson 11-5, pages 494–497

1. 14.3 3. 31.9 5. 28.8 m 7. ≈ 36.8 cm 9. 52.3
11. 13.6 13. 132.5 ft 15. 4.9 m, 7.7 m 17. 6.8 m
19. 443.2 in.² 21. 4.4 mi 23. 12 ft 25. 11 cm²
27. Triangles will vary. The tangents are 1.25. 29. 1 line
31. none 33. 4 35. −8 37. 81 39. 81

Lesson 11-6, pages 498–501

1. 20° 3. 21° 5. 25° 7. 84° 9. ≈ 5° 11. ≈ 53°
13. 42° 15. 77° 17. 33° 19. 22° 21. 30° 23. 69°
25. ≈ 18° 27. safe, ≈ 79° 29. safe, ≈ 77° 31. 45°
33. ≈ 26° 35. 27°, 63° 37. 44° 39. 37.7 cm²
41. 256.4 m² 43. 6*f*²(2*f*² + *f* + 3)

Review and Practice Your Skills, pages 502–503

1. 8.16, sine 3. 22.03, sine 5. 2.12 7. 7.37 9. 5.87
11. 77° 13. 10° 15. 30° 17. 54° 19. 24° 21. 35°
23. 9.8 25. 16.6 27. $\frac{24}{25}$ 29. ≈ 1545 ft

Lesson 11-7, pages 504–507

1. 10 3. 17.0 5. 727.5 ft 7. $10\sqrt{3}$ m, 20 m 9. 16
11. 19 13. $\frac{3\sqrt{2}}{2}$ 15. $\frac{14\sqrt{3}}{3}$ 17. 17,321 ft 19. 4.1 m
21. 30 m 23. Answers will vary. One possible explanation follows: The octagon is composed of four rectangles, four triangles and one square. Let *x* represent the length of each side of the octagon. This *x* also represents the length of the hypotenuse of each 45°-45°-90° right triangle, the

length of each side of the square, and the length of the two longer sides of the rectangles. The length of each leg of the triangles, which is also the length of the shorter sides of the rectangles is $\frac{x\sqrt{2}}{2}$. Now we have all the dimensions of the figure and can compute the area of the regions.

$$\underset{\text{triangles}}{\text{area of the}} \quad + \quad \underset{\text{rectangles}}{\text{area of}} \quad + \quad \underset{\text{square}}{\text{area of}}$$

$$4 \cdot \frac{1}{2}\left(\frac{x\sqrt{2}}{2}\right)\left(\frac{x\sqrt{2}}{2}\right) + 4 \cdot (x)\left(\frac{x\sqrt{2}}{2}\right) + x \cdot x$$

$$= \quad x^2 \quad + \quad 2\sqrt{2}x^2 \quad + \quad x^2$$

$$= 2x^2 + 2x^2\sqrt{2} = x^2(2 + 2\sqrt{2}) \approx 4.83x^2 \quad 25.\ 106.1\ \text{ft}$$

27.

	sine	cosine	tangent
30°	$\frac{1}{2}$	$\frac{\sqrt{3}}{2}$	$\frac{\sqrt{3}}{3}$
45°	$\frac{\sqrt{2}}{2}$	$\frac{\sqrt{2}}{2}$	1
60°	$\frac{\sqrt{3}}{2}$	$\frac{1}{2}$	$\sqrt{3}$

29. $\frac{\sqrt{3}}{2}x^2$

31. $\sqrt{3}\,x^2$

33. 169.6 cm³ 35. 463.0 mm³ 37.

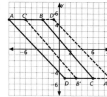

Lesson 11-8, pages 508–509

1. 10.4 3. no; $BC \approx 17.9$ and $AB \approx 22.1$
5. 15.2 7. yes 9. 3.4 km, 4.0 km 11. $(d - 6)^2$
13. $(p - 6)(p + 6)$ 15. 1080°

Chapter 11 Review, pages 510–512

1. b 3. k 5. i 7. h 9. l 11. similar 13. 224 ft
15. 15 m 17. 35.7 ft 19. 14.3 cm 21. $\frac{8}{17}$ 23. $\frac{15}{17}$
25. $\frac{8}{15}$ 27. 0.6018 29. 20.0 31. 22.0 33. ≈ 19.3 ft
35. $\approx 75.5°$ 37. 45.6° 39. $\approx 22.2°$ 41. $7\sqrt{2}$ 43. $9\sqrt{3}$
45. 6.5 in., $6.5\sqrt{2}$ or about 9.2 in. 47. 27.1 49. 13.9

Chapter 12: Logic and Sets

Lesson 12-1, pages 520–523

1. Roster notation: $L = \{0, 1, 2, 3, 4, 5, 6\}$; Set-builder notation: $L = \{x \mid x$ is a whole number less than 7$\}$ finite set
3. Roster notation: $P = \{1, 2, 3, 4, 5, \ldots, 21\}$;
Set-builder notation: $P = \{x \mid x$ is a positive integer less than 22$\}$ finite set 5. $3 \in \{3, 6, 9, 12, \ldots\}$ 7. $P, Q,$ and R are equivalent. Q and R are equal and equivalent.
9. Roster: $U = \{0, 1, 2, 3, \ldots, 14\}$; Set-builder: $U = \{x \mid x$ is a whole number less than 15$\}$; finite set 11. Roster: $M = \{\ \}$ Set-builder: $M = \{x \mid x$ is a month having 32 days$\}$ finite set 13. Roster: $K = \{-3, -2, -1, 0, 1, \ldots\}$ Set-builder: $K = \{x \mid x$ is an integer greater than $-4\}$ infinite set 15. $4 \in \{4, 6, 8, 10\}$ 17. $\varnothing \subseteq B$ 19. equal
21. Answers will vary but set must have four elements.
23. $\varnothing$, {cello}, {harp}, {flute}, {cello, harp}, {cello, flute}, {harp, flute}, {cello, harp, flute} 25. true 27. true
29. true 31. False; equivalent sets may not be equal.

33. Marcellus has put brackets around the 5. He should delete the brackets. 35. $\varnothing$, {1} 37. $\varnothing$, {m}, {a}, {t}, {m, a}, {m, t}, {a, t}, {m, a, t} 39. 2 41. 8 43. 32; 64; 1 45. Answers will vary. 47. 180 49. 5.1
51. 10 53. 8.5

Lesson 12-2, pages 524–527

1. {6, 8, 9} 3. {4, 6, 8, 9} 5. {1, 2, 4, 6, 9} 7. $\varnothing$ 9. {1, 2, 3, 4, 5, 6, 7} 11. {4, 6} 13. $\varnothing$ 15. Answers will vary.
17. {−1, 0, 4, 8} 19. {−3, −2, 0, 2, 8} 21. {8} 23. {−2, −1, 0, 2, 4, 8} 25. {chain saw, tractor} 27. $\varnothing$
29. {tractor, table saw, edger, backhoe, chain saw, hedge trimmer, forklift} 31. $\varnothing$ 33. $\varnothing$ 35. {1, 2, 3, 5, 7, 8, 9}
37. {1, 3, 5, 7, 9} 39. $\varnothing$ 41. {eggs, chocolate, vanilla}
43. {milk, water, oil, sugar, flour, eggs, chocolate}
45. $\varnothing$ 47. {6, 7, 8, 9, 10, 11, 12, 13, 14, 15, 16, 17, 18, 19, 20} 49. {1, 2, 3, 4, 5, 6, 7, 8, 9, 10, 11, 12, 13, 14, 15, 16, 17, 18, 19, 20} 51. {1, 2, 3, 4, 5} 53. {1, 2, 3, 4, 5, 7, 11, 13, 17, 19} 55. $A \cup B = \{x \mid x$ is a real number$\}$
57. $(A')' = A$ 59. {flute, guitar, piccolo, violin, tuba, keyboard} 61. {flute, guitar, tuba, violin, tamborine}
63. true 65. false; $R \cup S = U$ 67. yes 69. yes
71. no 73. yes 75. $x = 25°$, $y = 65°$

Review and Practice Your Skills, pages 528–529

1. $M = \{$April, June, September, November$\}$ $M = \{x \mid x$ is a month having 30 days$\}$; finite 3. $S = \{-9, -8, -7, \ldots\}$ $S = \{x \mid x$ is an integer greater than $-10\}$ infinite 5. $P = \{1, 2, 3, \ldots, 15, 16, 17\}$; $P = \{x \mid x$ is a positive integer less than 18$\}$ finite 7. $C \subseteq E$ 9. $M = \{\ \}$ 11. $\varnothing$, {e}, {f}, {g}, {e, f}, {e, g}, {f, g}, {e, f, g} 13. {0, 1, 3, 5} 15. {−5, −3, −1, 1, 3, 5} 17. {1} 19. {0, 1, 3, 5} 21. {2, 4, 5, 6, 7, 8, 10, 12} 23. $\varnothing$ 25. {3, 5, 6, 7, 9, 12, 15} 27. {0, 2, 3, 4, 6, 8, 9} 29. $D \subseteq E$ 31. {a, e, i, o, u, y} 33. {a, e, i, o, u, y, m, n, p, s, t}

Lesson 12-3, pages 530–531

1. If the product of two numbers is positive, then the two numbers are negative. (Converse); Statement: true; Converse: false; Example: $3 \cdot 4 = 12$ 3. If a triangle is isosceles, then it has two congruent sides. (Converse); Statement: true; Converse: true 5. If an object is thrown into the air, then it will fall back down to the ground.
7. If $b = 6$, then $3b + 7 = 25$. (Converse); Statement: true; Converse: true 9. If a figure is a square, then it is a rectangle. 11. If you intend to climb the wall, then you will need a ladder. 13. A conditional statement has a hypothesis and a conclusion. The converse of the original statement interchanges the conclusion and the hypothesis. If a statement is true, the converse is not necessarily true.
15. a human 17. 2002 19. (4, 3)

Lesson 12-4, pages 532–535

1. The trumpet is not a brass instrument. It is not the case that the trumpet is a brass instrument. 3. Cars do not all have 4 wheels. It is not the case that all cars have four wheels. 5. Converse: If a plant has leaves, then it is a tree. false; Counterexample: geranium; Inverse: If a plant is not a tree, then it does not have leaves. false; Counterexample: fern Contrapositive: If a plant does not have leaves, then it is not a tree. false; Counterexample: pine tree 7. Converse: If you play a stringed instrument, then you play the violin. false;

Counterexample: viola Inverse: If you do not play the violin, then you do not play a stringed instrument. false; Counterexample: cello Contrapositive: If you do not play a stringed instrument, then you do not play a violin. true   9. History is not Jasmine's favorite subject. It is not the case that history is Jasmine's favorite subject.   11. The factory does not recycle unused paper. It is not the case that the factory recycles unused paper.   13. Converse: If the object has wings, then it is an airplane. false; Counterexample: bird; Inverse: If the object is not an airplane, then it does not have wings. false; Counterexample: wasp Contrapositive: If an object does not have wings then it is not an airplane. true   15. Converse: If two lines are not parallel to each other, then they intersect. false; Counterexample: skew lines Inverse: If two lines do not intersect, then they are parallel to each other. false; Counterexample: skew lines Contrapositive: If two lines are parallel to each other, then they do not intersect. true   17. Answers will vary.   19. hypothesis: You wear Swift shoes. conclusion: You run fast.   21. Inverse: If you do not wear Swift shoes, then you do not run fast. This statement is not necessarily true, as you may wear another brand of shoes and run fast.   23. Answers will vary.   25. hypothesis: You play the french horn. conclusion: You play a brass instrument.   27. Inverse: If you do not play the french horn, then you do not play a brass instrument. false; Counterexample: trombone 29. My car is a convertible.   31. The bananas are ripe.   33. Inverse: If the bill is paid, then the phone service will not be turned off. Contrapositive: If the phone service is not turned off, then the bill is paid.   35. Inverse: If Akiko studies, then he will pass the test. Contrapositive: If Akiko will pass the test, then he will study.   37. inverse 39. inverse   41. Answers will vary.   43. SAS   45. $\{-18, -13, -8, -3, 2, 7, 12\}$ 47. $\{-16, -13, -10, -7, -4, -1, 2\}$

Review and Practice Your Skills, pages 536–537

1. If Mahala lives south of Canada, then he lives in Alabama; true; false; Counterexample: Kansas   3. If a triangle has 3 congruent sides, then it has 3 congruent angles.   true   5. If a quadrilateral is a square, then it is a rhombus.   7. If the music is composed by Mozart, then it is classical. 9. If Adam is to get an A for the semester, then he must get 95% on the final exam.   11. The house is not made of brick. 13. Soccer is not a popular sport in Brazil.   15. If you play the oboe, then you play a wind instrument.   17. You play a wind instrument.   19. If you do not play the oboe, then you do not play a wind instrument. false; Counterexample: saxophone 21. Answers will vary. example: $N = \{g, l, o, v, e\}$ 23. If a figure is a parallelogram, then the sum of the measures of the interior angles is 360°.   25. true; true

Lesson 12-5, pages 538–541

1. 1234321   3. 21   5. Ramon may apply for a driver's license.   7. inductive   9. 48   11. 0.001   13. Alita will join us for dinner.   15. Our appliances do not work. 17. inductive   19. Red sports cars are often stopped and given speeding tickets.   21. He could check police records. 23. Answers will vary.   25. no conclusion 27. No. Chloe could be a native of France and did not need to take the class to translate the passage.   29. ≈ 0.33 31. 0.14   33. $2y^4 - 2y^2 + 3y + 6$ 35. $x^5 + x^4 + 2x^2 + 4x$

Lesson 12-6, pages 542–545

1. valid; Law of Detachment   3. invalid   5. invalid 7. invalid   9. valid; Law of Detachment   11. valid: Law of the Contrapositive; sound   13. valid; Law of the Contrapositive; sound   15. invalid 17. invalid   For 19–22, the unstated key premise may be given in one of two ways depending on if the stated premise is taken to be the affirmation of p or the denial of q.   19. If you feel tired, then you should eat Pep Crackles for breakfast. If you eat Pep Crackles for breakfast, then you won't feel tired.   21. If you don't like rich coffee flavor, then don't buy Golden Bean coffee. If you buy Golden Bean coffee, you like rich coffee flavor.   23. Answers will vary.   25. 552 ft²
27. $\dfrac{3}{5}$ 29. $\dfrac{4}{5}$ 31. $\dfrac{4}{5}$

Review and Practice Your Skills, pages 546–547

1. 256   3. 750   5. Courtney is older than Al. 7. Sakima can babysit.   9. valid; Law of the Contrapositive 11. invalid   13. If angles are a linear pair, then they are supplementary. Converse: If angles are supplementary, then they are a linear pair. false 15. 15, 17   17. $\{f, o, t, b, a, l, s\}$

Lesson 12-7, pages 548–551

1. $\angle a$ and $\angle b$ are supplementary angles.
$m\angle a + m\angle b = 180°$ $a + b = 180°$
3. $2n + 1 + 2m + 1 = 2n + 2m + 2$
$\qquad\qquad\qquad\quad = 2(n + m + 1)$
This sum is even since it is a multiple of 2.
5. $\angle A \cong \angle B$ given
   $\angle C$ is supplement of $\angle A$;
   $\angle D$ is supplement of $\angle B$ given
   $m\angle A + m\angle C = 180°$;
   $m\angle B + m\angle D = 180°$ definition of supplementary angles
   $m\angle A + m\angle C = m\angle B + m\angle D$
   transitive property
   $m\angle C = m\angle D$ subtraction property
   $\angle C \cong \angle D$ definition of congruent angles
7. $m\angle 2 + 148° = 180° \rightarrow m\angle 2 = 32°$
$m\angle 1 = m\angle 2 = 32°$ 9. $2n - 2m = 2(n - m)$, and $2(n - m)$ is a multiple of 2.   11. 1st trip across river: take goat across, leaving wolf and cabbage on 1st bank; 1st trip back: go back across river alone; 2nd trip across: take cabbage across, leaving wolf on 1st bank; 2nd trip back: leave cabbage on 2nd bank, but bring goat back to 1st bank; 3rd trip across: take wolf across, leaving goat on 1st bank; 3rd trip back: go back across river alone; 4th trip across: take goat across; Now all three have been transported across without leaving goat and wolf alone or goat and cabbage alone.   13.

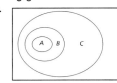

15. 100°, 95°, 115°, 130°, 125°, 155°   17. Answers will vary.   19. Inductive. The conclusion is based on a pattern of examples.   21. yes; $\dfrac{a}{b} \div \dfrac{c}{d} = \dfrac{a}{b} \cdot \dfrac{d}{c} = \dfrac{ad}{bc}$, assuming $c \neq 0$, and $\dfrac{ad}{bc}$ has the form of a rational number.

23. Answers will vary, but possible answers are listed.

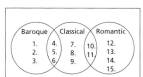

1. Henry Purcell
2. Monte Verdi
3. Johann Sebastian Bach
4. George Philipp Telemann
5. George Frideric Handel
6. Carl Philip Emmanuel Bach
7. Joseph Haydn
8. Wolfgang Amadeus
9. Christoph Willibald Gluck
10. Ludwig van Beethoven
11. Felix Mendelssohn
12. Peter Tchaikovsky
13. Johannes Brahms
14. Gustav Mahler
15. Richard Strauss

25. $x^2 + 4x + 3$ 27. $x^2 + 6x + 9$ 29. $x^2 - 25$
31. $6x^2 - 5x - 4$ 33. $16x^2 + 16x + 4$ 35. $9x^2 + 18x + 9$ 37. 264 ft^3

Chapter 12 Review, pages 552–554

1. c   3. b   5. k   7. j 9. l 11. $A = \{0, 1, 2, 3, 4\}$;
$A = \{x | x$ is a whole number less than 5$\}$; finite
13. $C = \{-7, -6, -5, ...\}$; $C = \{x | x$ is a whole number $>$
$-8\}$; infinite 15. $R = \{-99, -98, -97, ..., -1\}$;
$R = \{x | x$ is a negative integer $> -100\}$; infinite
17. $5 \in \{5, 10, 15, 20, 25, ...\}$ 19. $\{5, 6, 7, 8, 9, 10, 11, 12\}$
21. $\{3\}$ 23. $\{6, 9, 12\}$ 25. $\{1, 2, 3, 4, 5, 7, 9, 10\}$ 27. $\{9\}$
29. If a figure is a rectangle, then it is a quadrilateral.
31. The statement is true. The converse is false. A
counterexample would be any concave quadrilateral.
33. If people are using umbrellas, then it is raining. 35. If a
figure is a rhombus, then it has perpendicular diagonals.
37. The statement is true. The converse is false. A kite
has perpendicular diagonals and is not a rhombus. 39. If
you're not studying Russian, then you are not studying a
foreign language. 41. If you like broccoli then you like
vegetables. 43. If you don't like broccoli, then you don't
like vegetables. 45. Jacquine lives in Illinois. 47. valid;
Law of Contrapositive 49. invalid 51. Since the sum of
the interior angles of a triangle equal 180°, $a° + b° + c° =$
180°. Since the triangle is an equiangular triangle it is also
equilateral. Therefore, $a° = b° = c°$. By subsitituion, $3a° =$
60°. Therefore, $a° = b° = c° = 60°$.

Photo Credits

Cover (tl)MTPA Stock/Masterfile, (tr)Jim Cummings/ Getty Images, (bl)Stewart Cohen/Getty Images, (br)Mark Adams/Getty Images; Endsheet File photo; iv File photo; vii Aaron Haupt; viii CORBIS; ix Getty Images; x Mark Ransom; xi Getty Images; xii CORBIS; xiii through xviii Getty Images; 1 Mark Ransom; 2 Aaron Haupt; 3 (tl)Aaron Haupt, (tr)Mark Burnett, (c)Getty Images, (b)CORBIS; 6 7 Getty Images; 8 (t)Mark Ransom, (b)Getty Images; 9 12 15 Getty Images; 16 File photo; 18 20 Getty Images; 22 CORBIS; 31 Getty Images; 36 CORBIS; 37 File photo; 48 Getty Images; 49 (t)Getty Images, (b)CORBIS; 55 Getty Images; 56 Aaron Haupt; 59 MAK-1; 61 Getty Images; 63 Laura Sifferlin; 65 66 69 Getty Images; 73 Geoff Butler; 77 78 Getty Images; 81 Cessna Aircraft Co.; 85 through 101 Getty Images; 104 Ryan J. Hulvat; 111 113 Getty Images; 115 Masterfile; 125 127 129 Getty Images; 131 Doug Martin; 135 138 Getty Images; 146 (t)Getty Images, (b)Doug Martin; 147 through 157 Getty Images; 161 Mark Ransom; 164 Getty Images; 169 Light Source; 170 CORBIS; 172 Mark Ransom; 174 Stephen Webster; 177 through 189 Getty Images; 192 Ryan J. Hulvat; 194 197 Getty Images; 205 CORBIS; 209 215 Getty Images; 225 Skip Comer; 240 245 Getty Images; 247 Laura Sifferlin; 257 Getty Images; 268 Mark Ransom; 270 Getty Images; 273 CORBIS; 275 276 Getty Images;

277 CORBIS; 278 Courtesy of Tiara Observatory; 279 through 312 Getty Images; 317 Mark Steinmetz; 318 through 331 Getty Images; 334 Images 100; 340 343 345 Getty Images; 347 CORBIS; 351 356 Getty Images; 361 Tim Fuller; 362 Getty Images; 363 Mark Steinmetz; 365 372 379 Getty Images; 380 Ryan J. Hulvat; 385 389 Getty Images; 390 Ryan J. Hulvat; 392 Getty Images; 396 Mark Ransom; 406 411 418 Getty Images; 419 Library of Congress; 427 Getty Images; 429 Dynamic Graphics, Inc.; 431 CORBIS; 432 Matt Meadows; 433 through 453 Getty Images; 454 Elaine Shay; 456 457 459 462 Getty Images; 463 Tom Stack; 470 (l)Skip Comer, (r)Getty Images; 471 Getty Images; 477 CORBIS; 478 through 500 Getty Images; 501 CORBIS; 503 Getty Images; 504 File photo; 506 CORBIS; 508 516 517 520 Getty Images; 523 Aaron Haupt; 525 through 561 Getty Images; 562 (t)Getty Images (c)Aaron Haupt, (b)Geoff Butler; 563 (tr, tl, b)Getty Images, (tc)Jone Mason; 564 Getty Images; 565 (l)Tim Courlas, (r)Getty Images 566 567 568 Getty Images; 569 (t)Mullinix/KS Studio, (b)Getty Images; 570 (t)Tim Fuller (b)Getty Images; 571 (t)Getty Images, (b)Geoff Butler; 572 Getty Images; 573 (t)Masterfile, (b)Getty Images; 574 575 Getty Images

Index

Index

Index

Index

Index

Index

Index

Index

W

Weighted mean, 13
Whiskers, 29
Workplace Knowhow, *see* MathWorks
Writing
    linear equations, 254–257
    linear inequalities, 258–261
    variable expressions, 62–65
Writing Math, 9, 12, 18, 19, 23, 27, 31,
    37, 40, 41, 54, 58, 65, 69, 74, 78,
    84, 93, 106, 110, 115, 118, 124,
    125, 128, 134, 135, 138, 152, 153,
    155, 161, 164, 165, 170, 171, 175,
    180, 195, 199, 204, 208, 209, 215,
    218, 225, 228, 233, 247, 251, 256,
    261, 267, 269, 275, 278, 279, 284,
    298, 302, 308, 312, 318, 321, 336,
    341, 347, 351, 357, 359, 364, 378,
    382, 392, 399, 401, 406, 410, 424,
    429, 435, 438, 444, 448, 449, 454,
    458, 463, 476, 477, 480, 486, 490,
    496, 497, 500, 501, 507, 522, 523,
    526, 527, 531, 534, 541, 544, 545,
    551

X

x-axis, 244
x-coordinates, 244

Y

y-axis, 244
y-coordinates, 244
y-intercept, 254

Z

Zero, 52
    exponents of, 86–89
    factorial, 173
Zero pairs, 67, 396
Zero property of exponents, 86

Formulas

Coordinate Geometry

Slope	$m = \dfrac{y_2 - y_1}{x_2 - x_1}$
Distance	on a coordinate plane: $d = \sqrt{(x_2 - x_1)^2 + (y_2 - y_1)^2}$
Midpoint	on a number line: $M = \dfrac{a + b}{2}$ on a coordinate plane: $M = \left(\dfrac{x_1 + x_2}{2}, \dfrac{y_1 + y_2}{2} \right)$

Perimeter and Circumference

square	$P = 4s$
rectangle	$P = 2\ell + 2w$
circle	$C = 2\pi r$ or $C = \pi d$

Area

square	$A = s^2$
rectangle	$A = \ell w$ or $A = bh$
parallelogram	$A = bh$
trapezoid	$A = \dfrac{1}{2}h(b_1 + b_2)$
triangle	$A = \dfrac{1}{2}bh$
circle	$A = \pi r^2$

Pythagorean Theorem	$a^2 + b^2 = c^2$
Quadratic Formula	$x = \dfrac{-b \pm \sqrt{b^2 - 4ac}}{2a}$

Total Surface Area

rectangular prism	$SA = 2(\ell w + \ell h + wh)$
cylinder	$SA = 2\pi rh + 2\pi r^2$
pyramid	$SA = \left(\dfrac{1}{2}bh\right)n + B$
cone	$SA = \pi rs + \pi r^2$
sphere	$SA = 4\pi r^2$

Volume

cube	$V = s^3$
rectangular prism	$V = \ell wh$
prism	$V = Bh$
cylinder	$V = \pi r^2 h$
pyramid	$V = \dfrac{1}{3}Bh$
cone	$V = \dfrac{1}{3}\pi r^2 h$
sphere	$V = \dfrac{4}{3}\pi r^3$

Equations for Figures on a Coordinate Plane

slope-intercept form of a line	$y = mx + b$
point-slope form of a line	$y - y_1 = m(x - x_1)$
circle	$(x - h)^2 + (y - k)^2 = r^2$

Addition and Multiplication Properties

Additive Identity	For any number a, $a + 0 = 0 + a = a$.
Multiplicative Identity	For any number a, $a \cdot 1 = 0 \cdot a = a$.
Additive Inverse	For any number a, there is exactly one number $-a$ such that $a + (-a) = 0$.
Multiplicative Inverse	For any number $\dfrac{a}{b}$, where $a, b \neq 0$, there is exactly one number $\dfrac{b}{a}$ such that $\dfrac{a}{b} \cdot \dfrac{b}{a} = 1$.
Commutative (+)	For any numbers a and b, $a + b = b + a$.
Commutative (×)	For any numbers a and b, $a \cdot b = b \cdot a$.
Associative (+)	For any numbers a, b, and c, $(a + b) + c = a + (b + c)$.
Associative (×)	For any numbers a, b, and c, $(a \cdot b) \cdot c = a \cdot (b \cdot c)$.
Distributive	For any numbers a, b, and c, $a(b + c) = ab + ac$ and $a(b - c) = ab - ac$.